Bienvenue

ABOUT THE AUTHORS

Conrad J. Schmitt

Conrad J. Schmitt received his B.A. degree magna cum laude from Montclair State College, Upper Montclair, NJ. He received his M.A. from Middlebury College, Middlebury, VT. He did additional graduate work at Seton Hall University and New York University.

Mr. Schmitt has taught French and Spanish at the elementary, junior, and senior high school levels. He was Coordinator of Foreign Languages for the Hackensack, New Jersey, Public Schools. He also taught French at Upsala College, East Orange, NJ; Spanish at Montclair State College; and Methods of Teaching a Foreign Language at the Graduate School of Education, Rutgers University, New Brunswick, NJ. He was editor-in-chief of Foreign Languages and Bilingual Education for McGraw-Hill Book Company and Director of English Language Materials for McGraw-Hill International Book Company.

Mr. Schmitt has authored or co-authored more than eighty books, all published by Glencoe/McGraw-Hill or by McGraw-Hill. He has addressed teacher groups and given workshops in all states of the U.S. and has lectured and presented seminars throughout the Far East, Europe, Latin America, and Canada. In addition, Mr. Schmitt has travelled extensively throughout France, French-speaking Canada, North Africa, the French Antilles, and Haiti.

Katia Brillié Lutz

Ms. Lutz was Executive Editor of French at Macmillan Publishing Company. Prior to that, she taught French language and literature at Yale University and Southern Connecticut State College. Ms. Lutz also served as a senior editor at Harcourt Brace Jovanovich and Holt, Rinehart and Winston. She was a news translator and announcer for the BBC Overseas Language Services in London. Ms. Lutz has her *Baccalauréat* in Mathematics and Science from the Lycée Molière in Paris and her *Licence ès lettres* in Languages from the Sorbonne. She was a Fulbright Scholar at Mount Holyoke College. Ms. Lutz is the author of many foreign language textbooks at all levels of instruction. She presently devotes her time to teaching French at the United Nations and to writing.

Glencoe French 1

Bienvenue

Conrad J. Schmitt

Katia Brillié Lutz

 **Glencoe McGraw-Hill**

New York, New York   Columbus, Ohio   Mission Hills, California   Peoria, Illinois

Send all inquiries to:
Glencoe/McGraw-Hill
15319 Chatsworth Street
P.O. Box 9609
Mission Hills, CA 91346-9609

ISBN 0-02-636679-7
ISBN 0-02-636682-7
ISBN 0-02-636684-3

Printed in the United States of America.

1 2 3 4 5 6 7 8 9 QPK 03 02 01 00 99 98 97

CONTENTS

INTRODUCTION

Welcome to **Glencoe French**, the junior high and high school French series from Glencoe/McGraw-Hill. Every element in this series has been designed to help you create an atmosphere of challenge, variety, cooperation, and enjoyment for your students. From the moment you begin to use **Glencoe French**, you will notice that not only is it packed with exciting, practical materials and features designed to stimulate young people to work together towards language proficiency, but that it goes beyond by urging students to use their new skills in other areas of the curriculum.

Glencoe French uses an integrated approach to language learning: from the introduction of new material, through reinforcement, evaluation and review, its presentations, exercises and activities are designed to span all four language skills. Another characteristic of this series is that students use and reinforce these new skills while developing a realistic, up-to-date awareness of French culture. **Glencoe French** also incorporates a new feature in which French is used as the medium of instruction for a series of interdisciplinary presentations in the areas of natural sciences, social sciences, and the arts and humanities.

The Teacher's Wraparound Edition you are reading has been developed based on the advice of experienced foreign language educators throughout the United States in order to meet your needs as a teacher both in and out of the foreign language classroom. Here are some of the features and benefits which make **Glencoe French** a powerful set of teaching tools:

- flexible format
- student-centered instruction
- balance among all four language skills
- contextualized vocabulary
- thorough, contextual presentation of grammar
- an integrated approach to culture

FEATURES AND BENEFITS

Flexible Format While we have taken every opportunity to use the latest in pedagogical developments in order to create a learning atmosphere of variety, vitality, communication and challenge, we have also made every effort to make the **Glencoe French** series "teacher-friendly." This is where flexibility comes in.

The Student Textbook and the Teacher's Wraparound Edition provide an instructional method. However, every minute of every class period is not laid out. Plenty of room has been built in for you the teacher to be flexible: to draw on your own education, experience and personality in order to tailor a language program that is suitable and rewarding for the individual "chemistry" of each class.

A closer look at the most basic component, the Student Textbook, serves as an example of this flexibility. Each chapter opens with two sections of vocabulary (*Vocabulaire: Mots 1* and *Mots 2)* each with its own set of exercises. *Vocabulaire* is followed by the *Structure* consisting of a series of grammar points, each with accompanying exercises. But, there is nothing which says that the material must be presented in this order. The items of vocabulary and grammar are so well integrated that you will find it easy, and perhaps preferable, to move back and forth between them. You may also wish to select from the third and fourth sections of each chapter (the *Conversation* and *Lecture et culture* sections) at an earlier point than that at which they are presented, as a means of challenging students to identify or use the chapter vocabulary and grammar to which they have already been introduced.

These options are left to you. The only requirement for moving successfully through the Student Textbook is that the vocabulary and grammar of each chapter eventually be presented in their entirety, since each succeeding chapter builds on what has come before.

In the Student Textbook, there is a marked difference between learning exercises (*Exercices*) and communication-based activities (*Activités de communication*), both of which are provided in each chapter. The former serve as their name implies, as exercises for the acquisition and practice of new vocabulary and structures, while the latter are designed to get students communicating in open-ended contexts using the French they have learned. You can be selective among these, depending on the needs of your students.

We have been looking only at the Student Textbook. The abundance of suggestions for techniques, strategies, additional practice, chapter projects, independent (homework) assignments, informal assessment, and more, which are provided in this Teacher's Wraparound Edition—as well as the veritable banquet of resources available in the wide array of ancillary materials provided in the series—are what make **Glencoe French** truly flexible and "teacher-friendly." They guarantee you a great pool of ideas and teaching tools from which to pick and choose in order to create an outstanding course.

Student-Centered Instruction Teaching a foreign language requires coping with different learning styles and special student needs. It

requires the ability to capitalize on the great cultural and economic diversity present in many classrooms and to turn this diversity into an engine for learning by putting students together in goal-oriented groups. It often requires effective techniques for managing large classes.

Glencoe French anticipates these requirements by offering ideas for setting up a co-operative learning environment for students. Useful suggestions to this end accompany each chapter, under the heading Cooperative Learning, in the bottom margin of the Teacher's Wraparound Edition. Additional paired and group activities occur in the Student Textbook (*Activités de communication*), and in headings such as Additional Practice in the Teacher's Wraparound Edition.

Besides cooperative learning strategies, **Glencoe French** contains many other student-centered elements that allow students to expand their learning experiences. Here are a few examples: suggestions are offered in the Teacher's Wraparound Edition for out-of-class projects on topics related to the chapter theme. There is a topic called "For the Younger Student," with activities aimed primarily at stimulating the middle school/junior high student. In the Student Textbook, new grammatical material is divided into "bite-sized" lessons, so as not to be intimidating. The Writing Activities Workbook provides a self-test after every fourth chapter, so that students can prepare alone or in study groups for teacher-administered quizzes and tests. The Audio Cassette Program allows students to work at their own pace, stopping the tape whenever necessary to make directed changes in the language or to refer to their activity sheets in the Student Tape Manual. The Computer Software element consists of not only a Test Generator for the teacher, but also a Practice Generator for students, with which they can practice vocabulary and grammar items at their own pace.

These and other features discussed elsewhere in this Teacher's Manual have been designed with the student in mind. They assure that each individual, regardless of learning style, special need, background, or age, will have the necessary resources for becoming proficient in French.

Balance Among All Four Language Skills

Glencoe French provides a balanced focus on the listening, speaking, reading, and writing skills throughout all phases of instruction. And since it is "teacher-friendly," it gives you leeway if you wish to adjust the integration of these skills to the needs of a particular individual, group or class. Several features of the series lend themselves to this: the overall flexibility of format, the abundance of suggested optional and additional activities and the design of the individual activities themselves. Flexibility was discussed above. Let's look at some sections of a typical chapter as examples of the other two characteristics mentioned.

If the suggested presentation is followed, students are introduced to new words and phrases in *Vocabulaire* by the teacher, and/or by the audio cassette presentation. The focus is on listening and speaking through modeling and repetition. The *Exercices* which accompany the *Vocabulaire* section can be done with books either closed (accentuating listening and speaking) or open (accentuating reading, listening and speaking). However, these *Exercices* can just as well be assigned or reassigned as written work if the teacher wishes to have the whole class or individuals begin to concentrate on reading and writing. Throughout the *Vocabulaire* section, optional and additional reinforcement activities are suggested in the Teacher's Wraparound Edition.

These suggestions address all four language skills. Later in each chapter, students are asked to combine the material learned in *Vocabulaire* with material from the grammar section (*Structure*) using a combination of listening, reading, writing and speaking skills in the process.

Reading and writing activities are brought into play early in the **Glencoe French** series. The authors realize that communication in French includes the use of reading and writing skills and that these skills are indispensable for the assimilation and retention of new language and the organization of thought. Students are launched into writing, for example, as early as Chapter 1, through the use of brief assignments such as lists, labeled diagrams, notetaking or short answers. Longer writing activities are added in later chapters. These textbook activities are further reinforced in the Writing Activities Workbook.

Let's take a closer look at how each of the four skills is woven into the Student Textbook, the Teacher's Wraparound Edition and the ancillary materials.

Listening You the teacher are the primary source for listening, as you model new vocabulary, dialogues, structure and pronunciation, share your knowledge of French culture, history and geography, talk to students about their lives and your own, or engage in culturally oriented activities, and projects. As always, it is your ability to use French as much as possible with your students, both in and outside of the classroom, which determines how relevant and dynamic their learning experience will be.

Glencoe French offers numerous ways in which to develop the listening skill. There are teacher-focused activities, which provide the consistent modeling that students need. Teachers who use the Audio Cassette Program will find that these recordings help students become accustomed to a variety of voices, as well as rates of speech. There are also activities in which students interact with each other to develop listening spontaneity and acuity.

In the Student Textbook, new vocabulary will be modeled by the teacher. Students' attention to the sounds of the new words can be maximized by presenting this material with books closed and using the Vocabulary Transparencies to convey meaning. Following each *Mots* segment are several *Exercices* for practicing the new vocabulary. These can also be done with books closed. After the two *Mots* segments come *Activités de communication*, in which students may work in pairs or groups and must listen to each other in order to find information, take notes or report to others on what was said in their group. In *Structure*, students listen as the teacher models new grammatical material and then are given a chance to practice each structure in several *exercices*. Once again, closing the book will provide increased focus on the listening skill. The next section of each chapter is *Conversation*, in which a real-life dialogue is modeled either by the teacher or by playing the recorded version from the Audio Cassette Program. The dialogue is followed by several communication-based activities, where students must listen to and interact with their peers. In *Bienvenue*

(Level 1), *Conversation* also contains a *Prononciation* segment, covering an aspect of pronunciation related to the chapter material. Here again, students will be listening either to the teacher or recorded models. The last section of each chapter, *Culmination*, offers more listening-intensive *Activités de communication orale*, where students must be able to understand what their partners say in order to play their role.

In addition to the Student Textbook, the Teacher's Wraparound Edition offers several other listening-based activities correlated to the chapters, the most intensive of which occur under the heading "Total Physical Response." Here students must perform an action after listening to a spoken command. There are further listening-based activities suggested under the heading "Cooperative Learning" and often under "Additional Practice," both of which occur in the bottom margins in each Teacher's Wraparound Edition chapter.

The Audio Cassette Program has two main listening components. The first is practice-oriented, wherein students further reinforce vocabulary and grammar, following directions and making changes in utterances. They can self-check their work by listening to the correctly modeled utterances, which are supplied after a pause.

The second part of the program places more attention on the receptive listening skills. Students listen to language in the form of dialogues, announcements, or advertisements—language delivered at a faster pace and in greater volume—and then are asked to demonstrate their understanding of the main ideas and important details of what they have heard. The Student Tape Manual contains activity sheets for doing this work, and the Teacher Edition contains the complete transcript of all audio materials to assist you in laying out listening tasks for your class.

More listening practice is offered through the Videocassette Program. This material corresponds to and enriches that in the Student Textbook, and gives students a chance to hear variations of the language elements they have been practicing, as spoken by a variety of native speakers from different parts of France and other francophone countries. Students'

listening comprehension can be checked and augmented by using the corresponding print activities in the Video Activities Booklet.

Speaking Most of the areas of the Student Textbook and the Teacher's Wraparound Edition mentioned above which develop listening skills simultaneously develop the speaking skill. After hearing a model in the *Vocabulaire* or *Structure* sections, students will repeat it, either as a whole class, in small groups, or as individuals. From these modeled cues, they will progress to visual ones, supplied by the Vocabulary Transparencies or the photos and graphics in the textbook. The real thrust in the *Exercices* accompanying these two sections is to get students to produce this new material actively. Then, in the *Activités de communication*, students have the opportunity to adapt what they have learned by asking for and giving information to their classmates on a given topic. Here, and in the *Conversation* sections, students are engaged in meaningful, interesting sessions of sharing information, all designed to make them want to speak and experiment with the language. The Student Textbook regularly enriches this by offering expressions and mannerisms of speech currently popular in French culture, especially among teenagers, so that from the start your students will be accustomed to speaking in a way that is reflective of contemporary French. In Chapter 2, for example, popular adjectives of pleasure or displeasure are taught, such as *chouette, moche, extra,* and *terrible*. Previously presented material is constantly recycled in the communication-based activities, so that students' speaking vocabularies and knowledge of structure are always increasing. For this purpose, beginning with Chapter 3, there is a *Réintroduction et recombinaison* segment in the *Culmination* section. Another feature of the Student Textbook is that the length of utterances is increased over time, so that when students complete Level 1 (*Bienvenue*) they will have acquired an appreciation of the intonation and inflection of longer streams of language. To assist you in fine-tuning your students' speech patterns, the *Prononciation* section occurs in each chapter of the Level 1 Student Textbook.

The speaking skill is stressed in the first part of each recorded chapter of the Audio Cassette Program, where pauses are provided for the student to produce directed, spoken changes in the language. This is an excellent opportunity for those students who are self-conscious about speaking out in class. The Audio Cassette Program gives these students a chance to work in isolation. The format of making a change in the language, uttering the change and then listening for the correct model may improve the speaking skill. Sensitively administered, the Audio Cassette Program can serve as a confidence-builder for such students, allowing them to work their way gradually into more spontaneous speech with their classmates.

The packet of Situation Cards provides students with yet another opportunity to produce spoken French. They put the student into a contextualized, real-world situation. Students must ask and/or answer questions in order to perform successfully.

Reading Each chapter of the Student Textbook has a *Lecture et culture* section containing two readings based on the chapter theme. The first reading is accompanied by a comprehension check and an exercise called *Étude de mots*, which focuses on useful strategies for vocabulary-building and recognizing word relationships, which students can carry over into other readings. The second reading, *Découverte culturelle*, is optional and is to be read for more specific and detailed information about the theme of the chapter and as a stimulus for discussion on this theme. In the next section of each chapter, *Réalités*, students again use their reading skills albeit to a lesser degree. While the *Réalités* section is primarily visual in nature, students nevertheless are referred to numbered captions to learn more about the photographs shown in this two-page spread.

After every four chapters of the Student Textbook, **Glencoe French** provides a unique section called *Lettres et sciences*. This presentation is designed to use reading as a means of bridging the gap between French and other areas of the curriculum. Three separate readings are offered, one in each of the three areas of natural sciences, social sciences, arts and humanities. Here students have a chance to stretch their reading abilities in French by reading basic information they may have already learned in other academic subjects. Although the material has been carefully written to include themes (as well as words

and structures) which students have learned in previous chapters, it contains the most challenging readings. The *Lettres et sciences* sections are optional.

The Writing Activities Workbook offers additional readings under the heading *Un Peu Plus*. These selections and the accompanying exercises focus on reading strategies such as cognate recognition, related word forms and the use of context clues.

In addition to the reading development above, students are constantly presented with authentic French texts such as announcements from periodicals, telephone listings, transportation schedules, labeled diagrams, floor plans, travel brochures, school progress reports and many others, as sources of information. Sometimes these documents serve as the bases for language activities, and other times they appear in order to round out a cultural presentation, but, in varying degrees, they all require students to apply their reading skills.

Writing Written work is interwoven throughout the language learning process in **Glencoe French**. The exercises, which occur throughout the *Vocabulaire* and *Structure* sections of each chapter in the Student Textbook, are designed in such a way that they can be completed in written form as well as orally. Frequently, you may wish to reassign (as written homework) exercises which you have gone through orally in class. The Teacher's Wraparound Edition makes special note of this under the topic "Independent Practice." At the end of each chapter of the Student Textbook, direct focus is placed on writing in the *Culmination* section, under the heading *Activités de communication écrite*. Here there are one or more activities that encourage students to use the new vocabulary and structures they have learned in the chapter to create their own writing samples. These are short, and may be descriptive, narrative, argumentative, analytical or in the form of dialogues or interviews. Often a context is set up and then students are asked to develop an appropriate written response.

The Writing Activities Workbook is the component in which writing skills receive the most overt attention. All of the exercises in it are writing-based, and they vary in length from one-word answers to short compositions. They are designed to focus on the same vocabulary

and grammar presented in the corresponding chapter of the Student Textbook, but they are all new and all contextualized around fresh visual material or situational vignettes. Since they often have students making lists, adding to charts, and labeling, they provide an excellent means for them to organize the chapter material in their minds and make associations which will help them retain it. As students' knowledge of French increases, longer written pieces are required of them. One workbook section entitled *Mon Autobiographie* has students write installments of their own autobiographies. This is an effective way of stretching student writing skills. It also challenges students to personalize the French they have been studying.

Besides these major sources of writing, students are asked to make implicit use of writing almost everywhere in the series. They are constantly taking notes, listing, categorizing, labeling, summarizing, comparing or contrasting on paper. Even the Audio Cassette Program and the Videocassette Program involve students in writing through the use of activity sheets. By choosing among these options, you can be sure that your students will receive the practice they need to develop their writing skills successfully.

Contextualized Vocabulary

From the moment students see new words at the beginning of each chapter in **Glencoe French**, they see them within an identifiable context. So from the start, students learn to group words by association, thereby enhancing their ability to assimilate and store vocabulary for long-term retention. This contextualization remains consistent throughout the practice, testing and recycling phases of learning.

In the *Vocabulaire* section, each of the *Mots* segments contains a short exchange or a few lead-in sentences or phrases which, together with interesting, colorful visuals, establish the context. Other vocabulary items which occur naturally within this context are laid out among additional visuals, often as labels. The result is that students see at a glance the new language set into a real-life situation which provides "something to talk about"—a reason for using it. The accompanying exercises enrich this context. Each *exercice* practice item

is related to the others within the set, so that when taken together they form a meaningful vignette or story. In later sections of the chapter, i.e., *Structure, Conversation, Lecture et culture, Réalités* and *Culmination*, these words and phrases are reintroduced frequently.

Moreover, future chapters build on vocabulary and grammar from previous ones. Chapter themes introduced in Level 1 are reintroduced in Level 2 along with additional related vocabulary. Special attention has been given vocabulary in the reading sections of the series as well. For example, in *Lecture et culture*, students are encouraged to stretch their vocabularies in order to get as much meaning as possible from the selections. In addition to glossed words and frequent use of cognate recognition, the corresponding *Étude de mots* is there to help them with this. Another example is the *Lettres et sciences* section after every four chapters. The selections here include glossaries of the most important new vocabulary items, and the accompanying activities put implicit understanding of vocabulary to the test.

Thorough, Contextual Presentation of Grammar

A quick look through the chapters of *Bienvenue* will show the role grammar plays in the overall approach of the **Glencoe French** series. Although grammar is by no means the driving force behind the series, it is indeed an important aspect. In **Glencoe French**, grammar is presented as one of seven sections in each chapter. What makes this series particularly effective is that, as well as being thorough, the presentation of grammar runs concurrent with, and is embedded in, the chapter-long situational themes. Students are presented with French structure both directly, as grammar, and also as a set of useful functions that will aid them in communication, in expanding and improving their French across the four skills, and in learning about French culture as well as other areas of the school curriculum. Another important series characteristic is that the presentation of grammar has been divided into short, coherent "doses," which prevent grammar from becoming overwhelming to the student.

As you use this series, you will see as you teach the various grammar topics, student interest is kept high due to the presence of meaningful context and the diversity of the tasks that are given. As is the case with the vocabulary exercises, the individual practice items in the grammar section are related to each other contextually, in order to heighten student interest while assimilating and personalizing a new structure.

You will find that it is easy to move in and out of the teaching of grammar, dipping into the other sections of a chapter or other components as you see fit. This is true for several reasons: the grammar segments are short and intelligently divided, each one providing a good sense of closure; language elements (including grammar) taught in one section have been included as much as possible in the others; and again, there is a coherent contextual theme.

Aside from the Student Textbook and Teacher's Wraparound Edition, with their focus on grammar in the *Structure* section of each chapter and in the *Révision* after every four chapters, **Glencoe French** offers students opportunities to practice grammar in other components as well. Chapter by chapter, the Writing Activities Workbook provides ample tasks in which students must put into writing the new structures on which they have been working in class. The Audio Cassette Program includes recorded sections in every chapter of the Student Tape Manual which correspond directly to *Structure* in the Student Textbook. The Computer Software Program's Practice Generator contains additional grammar-based exercises. Of course students' knowledge of grammar is evaluated in the Chapter Quizzes and in the Testing Program, and each grammatical structure is practiced in other components, such as the Communication Activities Masters, Situation Cards and Videocassette Program.

An Integrated Approach to Culture

True competence in a foreign language cannot be attained without simultaneous development of the awareness of the culture in which the language is spoken. That is why **Glencoe French** places such great importance on culture. Accurate, up-to-date information on French culture is present either implicitly or explicitly throughout every phase of language

learning and in every component of the series.

The presentation of French in each chapter of the Student Textbook is embedded in running contextual themes, and these themes richly reflect the culture of France and areas of the world influenced by France. Even in chapter sections which focus primarily on vocabulary or grammar, the presence of culture comes through in the language used as examples or items in exercises, as well as in the content of the accompanying illustrations, photographs, charts, diagrams, maps or other reproductions of authentic, French documents. This constant, implicit inclusion of cultural information creates a format which not only aids in the learning of new words and structures, but piques student interest, invites questions, and stimulates discussion of the people behind the language.

Many culturally oriented questions raised by students may be answered in the two sections per chapter devoted to culture: *Lecture et culture* and *Réalités*. Through readings, captioned visuals and guided activities, these sections provide fundamental knowledge about such topics as French family life, school, restaurants, markets, sports, transportation, food, hotels, offices and hospitals, among many others. This information is presented with the idea that culture is a product of people—their attitudes, desires, preferences, differences, similarities, strengths and weaknesses—and that it is ever changing. Students are always encouraged to compare or contrast what they learn about French culture with their own, thereby learning to think critically and progress towards a more mature vision of the world. In addition to the presence of cultural material in each chapter of the Student Textbook, its importance is particularly apparent in the *Lettres et sciences* section which follows every four chapters. The readings here serve as valuable sources of information on the influence of the French people in the natural and social sciences and the arts and humanities. For more information on this unique feature, see the Teacher's Manual section immediately following, and also the section entitled ORGANIZATION OF THE STUDENT TEXTBOOK.

All of the cultural material described in the Student Textbook can be augmented by following a variety of suggestions in the Teacher's Wraparound Edition. There are guidelines for culturally rich instruction and activities, as well as useful, interesting facts for the teacher under headings such as Chapter Projects, Geography Connection, History Connection, Critical Thinking Activity, Did You Know? and others.

INTERDISCIPLINARY READINGS: LETTRES ET SCIENCES

This distinctive feature of **Glencoe French** allows students to use their French skills to expand their knowledge in other areas of the school curriculum. The interdisciplinary readings, called *Lettres et sciences*, occur in the Student Textbook after Chapters 4, 8, 12, and 16. They consist of three different readings on topics chosen from the Natural Sciences, the Social Sciences, the Arts and Humanities. Each reading topic is accompanied by pre- and post-reading activities. In the *Lettres et sciences* sections, students may read about important French explorers of the New World, for example, and begin to make associations between these men, their stories, and well-known names of cities or states in the United States. They may read and talk about the Impressionist movement in painting and a few of the great French artists who created it, as well as learn details which help to put the movement in perspective *vis à vis* other major events in world history. They may also learn about French scientists responsible for discoveries which are nowadays taken for granted.

Aside from providing basic information about the above topics—*[Pasteur] fonde une nouvelle science, la microbiologie*, for example—the readings have a French perspective. They include insights that students might not receive if they were reading about the same topic in an American textbook: *Au collège, [Pasteur] n'est pas très bon élève… il aime le dessin. On l'appelle «l'artiste»*. By using these interdisciplinary *Lettres et sciences* readings, you can open up two-way avenues of exchange between the French classroom and other subject areas in the school curriculum. These readings will also allow your students to exercise critical thinking skills, draw conclusions, and begin to interrelate in a mature way the knowledge coming to them from fields which they formerly considered unrelated to French. Perhaps the social studies, art, or science teachers in your school will have the pleasure of hearing from your students, "I learned in French class that …" or conversely, students will have outside knowledge about a topic to bring to discussions in your class.

It is hoped that these readings with interdisciplinary content will make this kind of cognitive connection more common in the overall learning process. Of course, while learning about the other subject areas, students are building their French language skills. The selections in *Lettres et sciences* recycle as much as possible the structures and vocabulary from previous chapters. Glossaries contribute to vocabulary building, and the accompanying activities are designed to encourage discussion in French around the topic.

SERIES COMPONENTS

In order to take full advantage of the student-centered, "teacher-friendly" curriculum offered by **Glencoe French**, you may want to refer to this section to familiarize yourself with the various resources the series has to offer. Both Levels 1 and 2 of **Glencoe French** contain the following components:

- Student Edition
- Teacher's Wraparound Edition
- Writing Activities Workbook & Student Tape Manual, Student Edition
- Writing Activities Workbook, Teacher's Annotated Edition
- Student Tape Manual, Teacher's Edition (tapescript)
- Audio Program (Cassette or Compact Disc)
- Overhead Transparencies
- Video Program (Videocassette or Videodisc)
- Video Activities Booklet
- Interactive Conversation Video
- Computer Software: Practice and Test Generator
- Communication Activities Masters
- Lesson Plans with Block Scheduling
- Internet Activities Booklet
- Bell Ringer Review Blackline Masters
- Situation Cards
- Chapter Quizzes with Answer Key
- Testing Program with Answer Key
- Performance Assessment
- CD-ROM Interactive Textbook

LEVEL 1 BIENVENUE IN TWO VOLUMES

At the junior high and intermediate school levels, where the material in *Bienvenue* is normally presented in two years, a two-volume edition is available, consisting of *Bienvenue* **Part A** and *Bienvenue* **Part B**. This two-volume edition may also be more suitable for other types of language programs where students are studying French for limited periods of time, or where student aptitude varies from the norm or for those programs where the teacher chooses to modify the pacing for other reasons. In addition to the *Bienvenue* Student Edition, the components of Level 1 (which are also available in two volumes) are the Teacher's Wraparound Edition, and the Writing Activities Workbook and Student Tape Manual, Student Edition. All other Level 1 components are completely compatible with this "split" edition of *Bienvenue*.

Bienvenue **Part A** consists of Chapters 1 through 8. *Bienvenue* **Part B** opens with 33 pages of *Révision*, a review section containing new activities designed to reenter the material in **Part A**. It then continues with Chapters 9 through 18.

ORGANIZATION OF THE STUDENT TEXTBOOK

Bienvenue preliminary lessons Chapter 1 of the Level 1 textbook (*Bienvenue*) is preceded by a group of eight preliminary lessons which bear the same title as the Level 1 textbook. These short lessons, A through H, will help orient your students to some of the routines of the foreign language classroom at the beginning of the term. They prime students with a few essential question words and get them speaking high-frequency French phrases used in greetings and leave-takings as well as in moving about the classroom in French and identifying basic classroom objects. Each preliminary lesson contains exercises and activities to help students retain this introductory material. If you guide them through all of the preliminary lessons in the *Bienvenue* section before beginning Chapter 1, your students will be able to make a smooth transition into the regular chapter material, and you will be able to conduct more of the classroom activities, including giving directions, in French.

Following the eight preliminary lessons, each chapter of *Bienvenue* and *À bord* (Level 2) is divided into the following sections:

- *Vocabulaire (Mots 1 & Mots 2)*
- *Structure*
- *Conversation*
- *Lecture et Culture*
- *Réalités*
- *Culmination*

After every fourth chapter in Level 1, the following special sections appear:

- *Le Monde francophone*
- *Révision*
- *Lettres et sciences* (interdisciplinary readings)

Vocabulaire The new vocabulary is laid out in two segments, *Mots 1* and *Mots 2*. Each of these presents new words in a cultural context in keeping with the theme of the chapter. Ample use is made of labeled illustrations to convey meaning and to provide an interesting introduction to the new vocabulary. The contextual vignettes into which the vocabulary items are embedded make use of the same grammatical structures which will be formally addressed later in the chapter, and recycle words and structures from previous chapters. Accompanying each *Mots* segment are a series of *Exercices* requiring students to use the new words in context. These *Exercices* employ techniques such as short answer, matching, multiple choice and labeling. They are always contextual, forming coherent vignettes, and they lend themselves well to any variations you might wish to apply to their delivery (books open, books closed, done as a class, in groups or pairs, written for homework). Wrapping up the *Vocabulaire* section are the *Activités de communication*, a segment consisting of communication-based activities which combine the new words from both *Mots* sections. These are more open-ended activities, requiring students to personalize the new language by performing such tasks as gathering information from classmates, interviewing, taking notes, making charts or reporting to the class.

Structure This is the grammar section of each chapter. It is conveniently and logically divided into four or five segments to aid in student assimilation of the material. Each segment provides a step-by-step description in English of how the new grammatical structure is used in

French, accompanied by examples, tables and other visuals. Each segment's presentation is followed by a series of flexible *Exercices*, designed along the same lines as those which accompany the *Vocabulaire* section, but focusing on the grammar point. As in *Vocabulaire*, the presentation of the new structures and the subsequent exercises is contextualized: examples as well as items in the exercises are never separate and unrelated, but always fit together in vignettes to enhance meaning. These vignettes are always directly related to the overall chapter theme. The *Structure* section also makes regular use of the new vocabulary from *Mots 1* and *Mots 2*, allowing for free interplay between these two sections of the chapter. This thorough yet manageable layout allows you to adapt the teaching of grammar to your students' needs and to your own teaching personality.

Conversation   Now that students have had a chance to see and practice the new items of vocabulary and grammar for the chapter, this section provides a recombined version of the new language in the form of an authentic, culturally rich dialogue under the heading *Scènes de la vie*. This can be handled in a variety of ways, depending on the teacher and the class and as suggested by accompanying notes in the Teacher's Wraparound Edition. Teacher modeling, modeling from the recorded version, class or individual repetitions, reading aloud by students, role-playing or adaptation through substitution are some of the strategies suggested. The dialogue is accompanied by one or more exercises which check comprehension and allow for some personalization of the material. Then students are invited once again to recombine and use all the new language in a variety of group and paired activities via the *Activités de communication*. New vocabulary and expressions are sometimes offered here, but only for the sake of richness and variation, and not for testing purposes. Every chapter in Level 1 also contains a *Prononciation* segment, which appears in the *Conversation* section. It provides a guide to the pronunciation of one or more French phonemes, a series of words and phrases containing the key sound(s), and an illustration which cues a key word containing the sound(s). These pronunciation illustrations are part of the Overhead Transparency package

accompanying the series. *Prononciation* can serve both as a tool for practice as students perform the chapter tasks, and as a handy speaking-skills reference to be used at any time.

Lecture et culture   This is a reading about people and places from France and the francophonic world, offering further cultural input to the theme of the chapter and providing yet another *recombinaison* of the chapter vocabulary and grammar. As is always the case with **Glencoe French**, material from previous chapters is recycled. Following the reading and based on it is *Étude de mots*—an exercise that gives students a chance to experiment with and expand their French vocabularies by using strategies such as searching for synonyms, identifying cognates, completing cloze exercises, matching and others. Next comes a series of comprehension exercises based on the reading (*Compréhension*), and finally the *Découverte culturelle*, where more cultural information is offered in the form of a shorter reading. The *Découverte culturelle* is optional in each chapter.

Réalités   These pages are intended as brief but enjoyable visual insights into the French-speaking world. The two pages of this section are filled with photographs that are pertinent to the chapter theme. Each photograph is identified with a caption, thereby providing some additional reading practice. Students are encouraged to formulate questions about what they see, and to compare and contrast elements of French culture with their own. The *Réalités* section is optional in each chapter.

Culmination   This wrap-up section requires students to consolidate material from the present as well as from previous chapters in order to complete the tasks successfully. The *Culmination* provides an opportunity for students to assess themselves on their own and to spend time on areas in which they are weak. You the teacher can pick and choose from them as you see fit. The first segment of *Culmination* consists of *Activités de communication orale*, where students must use the French they have learned to talk about various aspects of themselves: likes, dislikes, favorite activities, hobbies or areas of expertise, among others. This is followed by *Activités de communication écrite*, which encourage students to apply their knowledge

of French in written form. The *Réintroduction et recombinaison* segment recalls selected items of vocabulary and grammar from previous chapters. It is short and not meant as a comprehensive review, but rather as a quick reminder of important words, expressions and structures. Finally, the vocabulary words and expressions taught in the current chapter are listed categorically under the heading *Vocabulaire*, serving as a handy reference resource for both the student and the teacher.

Le Monde francophone This section occurs after Chapters 4, 8, 12 and 16 in the Student Textbook. It is designed to make students aware that, in addition to France, the French language is spoken in many other countries around the world. Because French is the official or second language in numerous and diverse countries, knowledge of the French language and culture is an important tool for meeting the career demands of the 21st century.

Each francophone section illustrates a topic presented in one of the four preceding chapters. Through photos and text, students are able to appreciate the cultural diversity of the francophone world. Each section incorporates the active vocabulary from the four preceding chapters. These pages are optional.

Révision This review section, designed to coincide with the more comprehensive Review Tests in the Testing Program, occurs after Chapters 4, 8, 12, and 16 in the Student Textbook. In each *Révision*, the main vocabulary and grammar points from the previous four chapters are recycled through a variety of new exercises, activities and dialogues. While in the individual chapters new grammar was divided into smaller, "bite-sized" portions to aid in the planning of daily lessons and help students assimilate it, now it is reviewed in a more consolidated format. This allows students to see different grammatical points side by side for the first time, to make new connections between the different points, and to progress toward a generative, "whole grammar." For example, in the first *Révision* following Chapter 4 (Level 1), indefinite articles, definite articles and possessive adjectives are reviewed

together on two pages, accompanied by explanations and various exercises. From these pages, students are able to conclude, among other things, that all of these structures have something in common—namely that they are all noun determiners, whose changes in number and gender depend upon the noun with which they are associated. Previously, these concepts were distributed over Chapters 1 through 4, and this point may have been missed by some students. Of course every possible combination of vocabulary and grammar does not reappear in the *Révision*. However, by carefully going through these exercises and activities and referring to the preceding chapters, students will be encouraged to make necessary connections and extrapolations themselves and therefore develop a true, working knowledge of the French they have studied. *Révision* is designed to be used by students studying alone, in unguided study groups or as a whole class with teacher guidance.

Lettres et sciences This is a unique, interdisciplinary feature of **Glencoe French** which allows students to use their French language skills to obtain, reinforce, and further their knowledge of other subject areas, namely the natural sciences, social studies, arts and humanities. This material is presented in the form of three readings, one from each of the above areas, accompanied by photos and illustrations. To stimulate discussion and aid in comprehension, there are pre-reading and post-reading activities. The reading selections are more vocabulary intensive than those in the regular chapters, and a French–English glossary is provided for each one. The focus here, however, is on the interdisciplinary content rather than the language itself. By engaging your students in some or all of these readings, you will encourage them to stretch their French reading skills in order to obtain useful, interesting information which will be of great service to them in their other academic courses. Also, you will be giving students the opportunity to judge for themselves the added insight that the study of French offers to their overall education.

SUGGESTIONS FOR TEACHING THE STUDENT TEXTBOOK

Teaching the Preliminary Lessons A through H in *Bienvenue* (Level 1)

The first day of class, teachers may wish to give students a pep-talk concerning the importance of the language they have chosen to study. Some suggested activities are:

- Show students a map (the maps located in the back of the Student Textbook can be used) to give them an idea of the extent of the French-speaking world.
- Have students discuss the areas within North America in which there is a high percentage of French speakers.
- Make a list of place names such as Baton Rouge, Terre Haute, Des Moines, Vermont, or names in your locality that are of French origin.
- Explain to students the possibility of using French in numerous careers such as: government, teaching, business, (banking, import/export), tourism, translating.
- The first day teachers will also want to give each student a French name. In the cases of students with names such as Kevin and Erica, teachers may want to give them a French nickname.

The short Preliminary Lessons A through H in *Bienvenue* are designed to give students useful, everyday expressions that they can use immediately. Each lesson is designed to take one day. The topics present students with easily learned expressions such as *Salut, Bonjour, Ça va?, Au revoir, etc.*, but do not confuse the students by expecting them to make structural changes such as the manipulation of verb endings. Formal grammar begins with Chapter 1. No grammar is taught in the *Bienvenue* Preliminary Lessons.

Teaching Various Sections of the Chapter

One of the major objectives of the **Glencoe French** series is to enable teachers to adapt the material to their own philosophy, teaching style, and students' needs. As a result, a variety of suggestions are offered here for teaching each section of the chapter.

Vocabulaire

The *Vocabulaire* section always contains some words in isolation, accompanied by an illustration that depicts the meaning of the new word. In addition, new words are used in contextualized sentences. These appear in the following formats: 1) one to three sentences accompanying an illustration, 2) a short conversation, 3) a short narrative or paragraph. In addition to teaching the new vocabulary, these contextualized sentences introduce, but do not teach, the new structure point of the chapter.

A vocabulary list appears at the end of each chapter in the Student Textbook.

General Techniques

- The Vocabulary Transparencies contain all illustrations necessary to teach the new words and phrases. With an overhead projector, they can easily be projected as large

visuals in the classroom for those teachers who prefer to introduce the vocabulary with books closed. The Vocabulary Transparencies contain no printed words.

■ All the vocabulary in each chapter (*Mots 1* and *Mots 2*) is recorded on the Audio Cassette Program. Students are asked to repeat the isolated words after the model.

Specific Techniques

Option 1 Option 1 for the presentation of vocabulary best meets the needs of those teachers who consider the development of oral skills a prime objective.

■ While students have their books closed, project the Vocabulary Transparencies. Point to the item being taught and have students repeat the word after you or the audio cassette several times. After presenting several words in this way, project the transparencies again and ask questions such as:

C'est une table libre?
Qu'est-ce que c'est?
C'est le serveur?
Qui est-ce? (Level 1, Chapter 5)

■ To teach the contextualized segments in the *Mots*, project the Vocabulary Transparency in the same way. Point to the part of the illustration that depicts the meaning of any new word in the sentence, be it an isolated sentence or a sentence from a conversation or narrative. Immediately ask questions about the sentence. For example, the following sentence appears in Level 1, Chapter 6:

Jean fait les courses.
Il fait les courses le matin.

Questions to ask are:
Jean fait les courses?
Qui fait les courses?
Où est-ce que Jean fait les courses?
Est-ce qu'il fait les courses le matin?
Quand est-ce qu'il fait les courses?

■ Dramatizations by the teacher, in addition to the illustrations, can also help convey the meaning of many words such as *chanter*, *danser*, etc.

■ After this basic presentation of the *Mots* vocabulary, have students open their books and read the *Mots* section for additional reinforcement.

■ Go over the exercises in the *Mots* section orally.

■ Assign the exercises in the *Mots* section for homework. Also assign the corresponding vocabulary exercises in the Writing Activities Workbook. If the *Mots* section should take more than one day, assign only those exercises that correspond to the material you have presented.

■ The following day, go over the exercises that were assigned for homework.

Option 2 Option 2 will meet the needs of those teachers who wish to teach the oral skills but consider reading and writing equally important.

■ Project the Vocabulary Transparencies and have students repeat each word once or twice after you or the audio cassette.

■ Have students repeat the contextualized sentences after you or the audio cassette as they look at the illustration.

■ Ask students to open their books. Have them read the *Mots* section. Correct pronunciation errors as they are made.

■ Go over the exercises in each *Mots* section.

■ Assign the exercises of the *Mots* section for homework. Also assign the vocabulary exercises in the Writing Activities Workbook.

■ The following day, go over the exercises that were assigned for homework.

Option 3 Option 3 will meet the needs of those teachers who consider the reading and writing skills of utmost importance.

■ Have students open their books and read the *Mots* items as they look at the illustrations.

■ Give students several minutes to look at the *Mots* words and vocabulary exercises. Then go over the exercises.

■ Go over the exercises the following day.

Expansion activities

Teachers may use any one of the following activities from time to time. These can be done in conjunction with any of the options previously outlined.

■ After the vocabulary has been presented, project the Vocabulary Transparencies or have students open their books and make up as many original sentences as they can, using

the new words. This can be done orally or in writing.

- Have students work in pairs or small groups. As they look at the illustrations in the textbook, have them make up as many questions as they can. They can direct their questions to their peers. It is often fun to make this a competitive activity. Individuals or teams can compete to make up the most questions in three minutes. This activity provides the students with an excellent opportunity to use interrogative words.
- Call on one student to read to the class one of the vocabulary exercises that tells a story. Then call on a more able student to retell the story in his/her own words.
- With slower groups you can have one student go to the front of the room. Have him or her think of one of the new words. Let classmates give the student the new words from the *Mots* until they guess the word the student in the front of the room has in mind. This is a very easy way to have the students recall the words they have just learned.

Structure

The *Structure* section of the chapter opens with a grammatical explanation in English. Each grammatical explanation is accompanied by many examples. With verbs, complete paradigms are given. In the case of other grammar concepts such as object pronouns, many examples are given with noun vs. pronoun objects. Irregular patterns are grouped together to make them appear more regular. For example, *sortir, partir, dormir,* and *servir* are taught together in Chapter 7, as are *pouvoir* and *vouloir* in Chapter 6. Whenever the contrast between English and French poses problems for students in the learning process, a contrastive analysis between the two languages is made. Two examples of this are the reflexive construction in Level 1 and the subjunctive in Level 2. Certain structure points are taught more effectively in their entirety and others are more easily acquired if they are taught in segments. An example of the latter is the direct and indirect object pronouns. In Chapter 15 *me, te, nous, vous* (as direct object pronouns) are presented, immediately followed by *le, la, l', les,* (Chapter 16), followed by *lui, leur* (Chapter 17).

Learning Exercises

The exercises that follow the grammatical explanation are plateaued or phased in to build from simple to more complex. In the case of verbs with an irregular form, for example, emphasis is placed on the irregular form, since it is the one students will most often confuse or forget. However, in all cases, students are given one or more exercises that force them to use all forms at random. The first few exercises that follow the grammatical explanation are considered **learning exercises** because they assist the students in grasping and internalizing the new grammar concept. These learning exercises are immediately followed by test exercises—exercises that make students use all aspects of the grammatical point they have just learned. This format greatly assists teachers in meeting the needs of the various ability levels of students in their classes. Every effort has been made to make the grammatical explanations as succinct and as complete as possible. We have purposely avoided extremely technical grammatical or linguistic terminology that most students would not understand. Nevertheless, it is necessary to use certain basic grammatical terms.

Certain grammar exercises from the Student Textbook are recorded on the Audio Cassette Program. Whenever an exercise is recorded, it is noted with an appropriate icon ( ) in the Teacher's Wraparound Edition.

The exercises in the Writing Activities Workbook also parallel the order of presentation in the Student Textbook. The Resource boxes and the Independent Practice topics in the Teacher's Wraparound Edition indicate when certain exercises from the Writing Activities Workbook can be assigned.

Specific Techniques for Presenting Grammar

Option 1 Some teachers prefer the deductive approach to the teaching of grammar. When this is the preferred method, teachers can begin the *Structure* section of the chapter by presenting the grammatical rule to students or by having them read the rule in their textbooks. After they have gone over the rule, have them read the examples in their textbooks or write the

examples on the chalkboard. Then proceed with the exercises that follow the grammatical explanation.

Option 2 Other teachers prefer the inductive approach to the teaching of grammar. If this is the case, begin the *Structure* section by writing the examples that accompany the rule on the chalkboard or by having students read them in their textbooks. Let us take, for example, the direct object pronouns *le, la, l', les*. The examples the students have in their books are:

Je sais le nom du film.	Je le sais.
Je vois le film.	Je le vois.
J'aime le film.	Je l'aime.
Je ne connais pas la vedette.	Je ne la connais pas.
Je lis les sous-titres.	Je les lis.
J'admire les costumes.	Je les admire.

In order to teach this concept inductively, teachers can ask students to do or answer the following:

- Have students find the object of each sentence in the first column. Say it or underline the object if it is written on the board.
- Have students notice that these words disappeared in the sentences in the second column. Have students give (or underline) the word that replaced each one.
- Ask students what word replaced *le nom du film, le film, la vedette, les sous-titres, les costumes*.
- Ask: What do we call a word that replaces a noun?
- Ask: What direct object pronoun replaces a masculine noun? A feminine noun, etc.?
- Have students look again. Ask: What word replaces *le film, la vedette?*
- Ask: Can *le* or *la* be used to replace a person or a thing?
- Ask: Where do the direct object pronouns *le, la, l', les* go, before or after the verb?

By answering these questions, students have induced, on their own, the rule from the examples. To further reinforce the rule, have students read the grammatical explanation and then continue with the grammar exercises that follow. Further suggestions for the inductive presentation of the grammatical points are given in the Teacher's Wraparound Edition.

Specific techniques for Teaching Grammar Exercises

In the development of the **Glencoe French** series, we have purposely provided a wide variety of exercises in the *Structure* section so that students can proceed from one exercise to another without becoming bored. The types of exercises they will encounter are: short conversations, answering questions, conducting or taking part in an interview, making up questions, describing an illustration, filling in the blanks, multiple choice, completing a conversation, completing a narrative, etc. In going over the exercises with students, teachers may want to conduct the exercises themselves or they may want students to work in pairs. The *Structure* exercises can be gone over in class before they are assigned for homework or they may be assigned before they are gone over. Many teachers may want to vary their approach.

All the *Exercices* and *Activités de communication* in the Student Textbook can be done with books open. Many of the exercises such as question-answer, interview, and transformation can also be done with books closed.

Types of Exercises

Question exercises The answers to many question exercises build to tell a complete story. Once you have gone over the exercise by calling on several students (Student 1 answers items numbered 1,2,3; Student 2 answers items numbered 4,5,6, etc.), you can call on one student to give the answers to the entire exercise. Now the entire class has heard an uninterrupted story. Students can ask one another questions about the story, give an oral synopsis of the story in their own words, or write a short paragraph about the story.

Personal questions or interview exercises Students can easily work in pairs or teachers can call a student moderator to the front of the room to ask questions of various class members. Two students can come to the front of the room and the exercise can be performed—one student takes the role of the interviewer and the other takes the role of the interviewee.

Completion of a conversation See Chapter 5, *Exercice D*, page 135, as an example. After students complete the exercise, they can be given time either in class or as an outside assignment to prepare a skit for the class based on the conversation.

Conversation

Specific Techniques Teachers may wish to vary the presentation of the *Conversation* from one chapter to another. In some chapters, the dialogue can be presented thoroughly and in other chapters it may be presented quickly as a reading exercise. Some possible options are:

- Have the class repeat the dialogue after you twice. Then have students work in pairs and present the dialogue to the class. The dialogue does not have to be memorized. If students change it a bit, all the better.
- Have students read the dialogue several times on their own. Then have them work in pairs and read the dialogue as a skit. Try to encourage them to be animated and to use proper intonation. This is a very important aspect of the *Conversation* section of the chapter.
- Rather than read the dialogue, students can work in pairs, having one make up as many questions as possible related to the topic of the dialogue. The other student can answer his/her questions.
- Once students can complete the exercise(s) that accompany the dialogue with relative ease, they know the dialogue sufficiently well without having to memorize it.
- Students can tell or write a synopsis of the dialogue.

Prononciation

Specific Techniques Have students read on their own or go over with them the short explanation in the book concerning the particular sound that is being presented. For the more difficult sounds such as *r, u, eu, en, in, un,* etc., teachers may wish to demonstrate the tongue and lip positions. Have students repeat the words after you or the model speaker on the audio cassette recording.

Activités de communication

Specific Techniques The *Activités de communication* presents activities that assist students in working with the language on their own. All the *Activités* are optional. In some cases, teachers may want the whole class to do them all. In other cases, teachers can decide which ones the whole class will do. Another possibility is to break the class into groups and have each one work on a different activity.

Lecture et culture

Specific Techniques: Option 1 Just as the presentation of the dialogue can vary from one chapter to the next, the same is true of the *Lecture*. In some chapters teachers may want students to go over the reading selection very thoroughly. In this case all or any combination of the following techniques can be used.

- Give students a brief synopsis of the story in French.
- Ask questions about the brief synopsis.
- Have students open their books and repeat several sentences after you or call on individuals to read.
- Ask questions about what was just read.
- Have students read the story at home and write the answers to the exercises that accompany the *Lecture*.
- Go over the *Étude de mots* and the *Compréhension* in class the next day.
- Call on a student to give a review of the story in his/her own words. If necessary, guide students to make up an oral review. Ask five or six questions, the answers to which review the salient points of the reading selection.
- After the oral review, the more able students can write a synopsis of the *Lecture* in their own words.

It should take less than one class period to present the *Lecture* in the early chapters. In later chapters, teachers may wish to spend two days on those reading selections they want students to know thoroughly.

Option 2 With those *Lectures* that teachers wish to present less thoroughly, the following techniques may be used:

- Call on an individual to read a paragraph.
- Ask questions about the paragraph read.

- Assign the *Lecture* to be read at home. Have students write the exercises that accompany the *Lecture*.
- Go over the *Etude de mots* and the *Compréhension* the following day.

Option 3 With some reading selections, teachers may wish merely to assign them to be read at home and then go over the exercises the following day. This is possible since the only new material in the *Lecture* consists of a few new vocabulary items that are always footnoted.

Découverte culturelle

The optional *Découverte culturelle* is a reading selection which is designed to give students an in-depth knowledge of many areas of the French-speaking world. You can omit any or all of this reading or they may choose certain selections that they would like the whole class to read. The same suggestions given for the *Lecture* of each chapter can be followed. Teachers may also assign the reading selections to different groups. Students can read the selection outside of class and prepare a report for those students who did not read that particular selection. This activity is very beneficial for slower students. Although they may not read the selection, they learn the material by listening to what their peers say about it. The *Découverte culturelle* can also be done by students on a voluntary basis for extra credit.

Réalités

Specific Techniques The purpose of the *Réalités* section is to permit students to look at photographs from the French-speaking world and to acquaint them with the many areas where French is spoken. The *Réalités* section contains no exercises. The purpose is for students to enjoy the material as if they were browsing through pages of a magazine. Items the students can think about are embedded in the commentary that accompanies the photographs. Teachers can either have students read the extended captions in class or students can read the captions on their own.

ORGANIZATION OF THE TEACHER'S WRAPAROUND EDITION

One important component, which is definitive of **Glencoe French** and adds to the series' flexible, "teacher-friendly" nature, is the Teacher's Wraparound Edition (TWE), of which this Teacher's Manual is a part. Each two-page spread of the TWE "wraps around" a slightly reduced reproduction of the corresponding pages of the Student Textbook and offers in the expanded margins a variety of specific, helpful suggestions for every phase in the learning process. A complete method for the presentation of all the material in the Student Textbook is provided—basically, a complete set of lesson plans—as well as techniques for background-building, additional reinforcement of new language skills, creative and communicative recycling of material from previous chapters and a host of other alternatives from which to choose. This banquet of ideas has been developed and conveniently laid out in order to save valuable teacher preparation time and to aid you in designing the richest, most varied language experience possible for you and your students. A closer look at the kinds of support in the TWE, and their locations, will help you decide which ones are right for your pace and style of teaching and the differing "chemistries" of your classes.

The notes in the Teacher's Wraparound Edition can be divided into two basic categories:

1. Core notes, appearing in the left- and right-hand margins, are those which most directly correspond to the material in the accompanying two-page spread of the Student Textbook.

2. Enrichment notes, in the bottom margin, are meant to be complementary to the material in the Student Textbook. They offer a wide range of options aimed at getting students to practice and use the French they are learning in diverse ways, individually and with their classmates, in the classroom and for homework. The enrichment notes also include tips to the teacher on clarifying and interconnecting elements in French language and culture—ideas that have proved useful to other teachers and which are offered for your consideration.

Description of Core Notes in the Teacher's Wraparound Edition

Chapter Overview At the beginning of each chapter a brief description is given of the language functions which students will be able to perform by chapter's end. Mention is made of any closely associated functions presented in other chapters. This allows for effective articulation between chapters and serves as a guide for more successful teaching.

Chapter Objectives This guide immediately follows the Chapter Overview and is closely related to it. Here the focus is on grammatical objectives for the chapter, which are stated in a concise list.

Chapter Resources The beginning of each chapter includes a reference list of all the ancillary components of the series that are applicable to what is being taught in the chapter, including the Writing Activities Workbook and

Student Tape Manual, Audio Cassette Program, Overhead Transparencies, Communication Activities Masters, Videocassette Program, Computer Software: Practice and Test Generator, Situation Cards, Chapter Quizzes and Test Booklets. A more precise version of this resource list will be repeated at the beginning of each section within the chapter, so that you always have a handy guide to the specific resources available to you for each and every point in the teaching process. Using these chapter and section resource references will make it easier for you to plan varied, stimulating lessons throughout the year.

Bell Ringer Reviews These short activities recycle vocabulary and grammar from previous chapters and sections. They serve as effective warm-ups, urging students to begin thinking in French, and helping them make the transition from their previous class to French. Minimal direction is required to get the Bell Ringer Review activity started, so students can begin meaningful, independent work in French as soon as the class hour begins, rather than wait for the teacher to finish administrative tasks, such as attendance, etc. Bell Ringer Reviews occur consistently throughout each chapter of Levels 1 and 2.

Presentation Step-by-step suggestions for the presentation of the material in all segments of the six main section headings in each chapter—*Vocabulaire, Structure, Conversation, Lecture et culture, Réalités,* and *Culmination* are presented in the left- and right-hand margins. They offer the teacher suggestions on what to say, whether to have books open or closed, whether to perform tasks individually, in pairs or in small groups, expand the material, reteach, and assign homework. These are indeed suggestions. You may wish to follow them as written or choose a more eclectic approach to suit time constraints, personal teaching style and class "chemistry." Please note however, that the central vocabulary and grammar included in each chapter's *Vocabulaire* and *Structure* sections is intended to be taught in its entirety, since this material is built into that which occurs in succeeding chapters. In addition, answers for all the *Exercices* in each segment are conveniently located near that exercise in the Student Textbook.

Because the answers will vary in the *Activités de communication*, they are usually not provided. However, the Presentation notes do offer other topics for enrichment, expansion and assessment. A brief discussion of these may help you incorporate them into your lesson plans.

Geography Connection These suggestions encourage students to use the maps provided in the Student Textbook as well as refer them to outside sources in order to familiarize them with the geography of France and the francophone world. These optional activities are another way in which **Glencoe French** crosses boundaries into other areas of the curriculum. Their use will instill in students the awareness that French Class is not just a study of language but an investigation into a powerful culture that has directly or indirectly affected the lives of millions of people all over the globe. Besides studying the geography within France itself, they will be urged to trace the presence of French culture throughout Europe, Africa, the Americas, Asia, and the Pacific. The notes also supply you the teacher with diverse bits of geographical and historical information which you might not have known, and you may decide to pass these on to your students.

Vocabulary Expansion These notes provide the teacher handy access to vocabulary items which are thematically related to those presented within the Student Textbook. They are offered to enrich classroom conversations, allowing students more varied and meaningful responses when talking about themselves, their classmates or the topic in question. Note that none of these items, or for that matter any information, in the TWE is included in the Chapter Quizzes, or in the Testing Program accompanying **Glencoe French**.

Cognate Recognition Since the lexical relationship between French and English is so rich, these notes have been provided to help you take full advantage of the vocabulary-building strategy of isolating them. The suggestions occur in the *Vocabulaire* section of each chapter and are particularly frequent in Level 1 in order to train students from the very beginning in the valuable strategy of recognizing cognates. Various methods of pointing out cognates are used, involving all four language skills, and the activities frequently encourage

students to personalize the new words by using them to talk about things and people they know. Pronunciation differences are stressed between the two languages. The teacher notes also call attention to false cognates when they occur in other chapter sections.

Informal Assessment Ideas are offered for making quick checks on how well students are assimilating new material. These checks are done in a variety of ways and provide a means whereby both teacher and students can monitor daily progress. By using the Informal Assessment topic, you will be able to ascertain (as you go along) the areas in which students are having trouble, adjust your pace accordingly, or provide extra help for individuals, either by making use of other activities offered in the TWE or devising your own. The assessment strategies are simple and designed to help you elicit from students the vocabulary word, grammatical structure, or other information you wish to check. Because they occur on the same page as the material to which they correspond, you may want to come back to them again when it is time to prepare students for tests or quizzes.

Reteaching These suggestions provided yet another approach to teaching a specific topic in the chapter. In the event some students were not successful in the initial presentation of the material the reteaching activity offers an alternate strategy. At the same time, they allow other students to further consolidate their learning.

History Connection Following these suggestions can be a very effective springboard from the French classroom into the history and social studies areas of the curriculum. Students are asked to focus their attention on the current world map, or historical ones, then they are invited to discuss the cultural, economic and political forces which shape the world with an eye on French influence. The notes will assist you in providing this type of information yourself or in creating projects in which students do their own research, perhaps with the aid of a history teacher. By making the history connection, students are encouraged to either import or export learning between the French classroom and the History or Social Studies realms.

Description of Enrichment Notes in the Teacher's Wraparound Edition

The notes in the bottom margin of the TWE enrich students' learning experiences by providing additional activities to those in the Student Textbook. These activities will be helpful in meeting each chapter's objectives, as well as in providing students with an atmosphere of variety, cooperation and enjoyment.

Chapter Projects Specific suggestions are given at the start of each chapter for launching individual students or groups into a research project in keeping with the chapter theme. Students are encouraged to gather information by using resources in school and public libraries, visiting local French institutions or interviewing French people or other persons knowledgeable in the area of French culture whom they may know. In Chapter 1, for example, they are asked to compare/contrast the French educational system with their own. These projects may serve as another excellent means for students to make connections between their learning in the French classroom and other areas of the curriculum.

Learning from Photos and Realia Each chapter of **Glencoe French** contains many colorful photographs and reproductions of authentic French documents, filled with valuable cultural information. In order to help you take advantage of this rich source of learning, notes of interesting information have been provided to assist you in highlighting the special features of these up-to-date realia. The questions that appear under this topic have been designed to enhance learners' reading, and critical thinking skills.

Total Physical Response (Level 1) At least one Total Physical Response (TPR) activity is provided with each *Mots* segment that makes up the *Vocabulaire* section of the chapter. The Total Physical Response approach to language instruction was developed by James J. Asher. Students must focus their attention on commands spoken by the teacher (or classmates) and demonstrate their comprehension by performing the physical task commanded. This strategy has proven highly successful for concentrating on the listening skill and assimilating new vocabulary. Students are relieved

momentarily of the need to speak—by which some may be intimidated—and yet challenged to show that they understand spoken French. The physical nature of these activities is another of their benefits, providing a favorable change of pace for students, who must move about the room and perhaps handle some props in order to perform the tasks. In addition, Total Physical Response is in keeping with cooperative learning principles, since many of the commands require students to interact and assist each other in accomplishing them.

Cooperative Learning At least one cooperative learning activity has been included in each chapter. These activities include guidelines both on the size of groups to be organized and on the tasks the groups will perform. They reflect two basic principles of cooperative learning: (a) that students work together, being responsible for their own learning, and (b) that they do so in an atmosphere of mutual respect and support, where the contributions of each peer are valued. For more information on this topic, please see the section in this Teacher's Manual entitled COOPERATIVE LEARNING.

Additional Practice There are a variety of Additional Practice activities to complement and follow up the presentation of material in the Student Textbook. Frequently the additional practice focuses on personalization of the new material and employs more than one language skill. Examples of Additional Practice activities include having students give oral or written descriptions of themselves or their classmates; asking students to conduct interviews around a topic and then report their findings to the class; using possessive forms to identify objects in the classroom and their owners. The additional practice will equip you with an ample, organized repertoire from which to pick and choose should you need extra practice beyond the Student Textbook.

Independent Practice Many of the exercises in each chapter lend themselves well to assignment or reassignment as homework. In addition to providing extra practice, reassigning on paper exercises that were performed orally in class makes use of additional language skills

and aids in informal assessment. The suggestions under the Independent Practice heading in the bottom margin of the TWE will call your attention to exercises that are particularly suited to this. In addition to reassigning exercises in the Student Textbook as independent practice, additional sources are suggested from the various ancillary components, specifically the Writing Activities Workbook and the Communication Activities Masters.

Critical Thinking Activities To broaden the scope of the foreign language classroom, suggestions are given that will encourage students to make inferences and organize their learning into a coherent "big picture" of today's world. These and other topics offered in the enrichment notes provide dynamic content areas to which students can apply their French language skills and their growing knowledge of French culture. The guided discussion suggestions derived from the chapter themes invite students to make connections between what they learn in the French program and other areas of the curriculum.

Did You Know? This is a teacher resource topic where you will find additional details relevant to the chapter theme. You might wish to add the information given under this topic to your own knowledge and share it with your students to spur their interest in research projects, enliven class discussions and round out their awareness of French culture, history or geography.

For the Younger Student Because Level 1 (*Bienvenue*) is designed for use at the junior high and intermediate level as well as the high school level, this topic pays special attention to the needs of younger students. Each chapter contains suggestions for meaningful language activities and tips to the teacher that cater to the physical and emotional needs of these youngsters. There are ideas for hands-on student projects, such as creating booklets or bringing and using their own props, as well as suggestions for devising games based on speed, using pantomime, show and tell, performing skits and more.

ADDITIONAL ANCILLARY COMPONENTS

All ancillary components are supplementary to the Student Textbook. Any or all parts of the following ancillaries can be used at the discretion of the teacher.

The Writing Activities Workbook and Student Tape Manual

The Writing Activities Workbook and Student Tape Manual is divided into two parts: all chapters of the Writing Activities Workbook appear in the first half of this ancillary component, followed by all chapters of the Student Tape Manual.

Writing Activities Workbook The consumable workbook offers additional writing practice to reinforce the vocabulary and grammatical structures in each chapter of the Student Textbook. Workbook exercises are presented in the same order as the material in the Student Textbook. The exercises are contextualized, often centering around line art illustrations. Workbook activities employ a variety of elicitation techniques, ranging from short answers, matching columns, and answering personalized questions, to writing paragraphs and brief compositions. To encourage personalized writing, there is a special section in each chapter entitled *Mon Autobiographie*. The workbook provides further reading skills development with the *Un Peu Plus* section, where students are introduced to a number of reading strategies such as scanning for information, distinguishing fact from opinion, drawing inferences and reaching conclusions, for the purpose of improving their reading comprehension and

expanding their vocabulary. The *Un Peu Plus* section also extends the cultural themes presented in the corresponding Student Textbook chapter. The Writing Activities Workbook includes a Self-Test after Chapters 4, 8, 12, 16, and 18. The Writing Activities Workbook, Teacher Annotated Edition provides the teacher with all the material in the student edition plus the answers—wherever possible—to the activities.

Student Tape Manual The Student Tape Manual contains the activity sheets which students will use when listening to the audio recordings. The Teacher Edition of the Student Tape Manual contains, in addition, the answers to the recorded activities, plus the complete tapescript of all recorded material.

The Audio Program (Cassette or CD)

The recorded material for each chapter of **Glencoe French**, Levels 1 and 2 is divided into two parts—*Première partie* and *Deuxième partie*. The *Première partie* consists of additional listening and speaking practice for the *Vocabulaire (Mots 1 & 2)* and the *Structure* sections of each chapter. There is also a dramatization of the *Conversation* dialogue from the Student Textbook, and a pronunciation section. The *Première partie* concludes with a *dictée*.

The *Deuxième partie* contains a series of activities designed to further stretch students' receptive listening skills in more open-ended, real-life situations. Students indicate their understanding of brief conversations, advertisements, announcements, et cetera, by

making the appropriate response on their activity sheets located in the Student Tape Manual.

Overhead Transparencies

There are five categories in the package of Overhead Transparencies accompanying **Glencoe French**, Level 1. Each category of transparencies has its special purpose. Following is a description:

Vocabulary Transparencies These are full-color transparencies reproduced from each of the *Mots* presentations in the Student Textbook. In converting the *Mots* vocabulary pages to transparency format, all accompanying words and phrases on the *Mots* pages have been deleted to allow for greater flexibility in their use. The Vocabulary Transparencies can be used for the initial presentation of new words and phrases in each chapter. They can also be reprojected to review or reteach vocabulary during the course of teaching the chapter, or as a tool for giving quick vocabulary quizzes.

With more able groups, teachers can show the Vocabulary Transparencies from previous chapters and have students make up original sentences using a particular word. These sentences can be given orally or in writing.

Pronunciation Transparencies In the *Prononciation* section of each chapter of *Bienvenue* (Level 1), an illustration has been included to visually cue the key word or phrase containing the sound(s) being taught, e.g., Chapter 3, page 79. Each of these illustrations has been converted to transparency format. These Pronunciation Transparencies may be used to present the key sound(s) for a given chapter, or for periodic pronunciation reviews where several transparencies can be shown to the class in rapid order. Some teachers may wish to convert these Pronunciation Transparencies to black and white paper visuals by making a photocopy of each one.

Communication Transparencies For each chapter in Levels 1 and 2 of the series there is one original composite illustration which visually summarizes and reviews the vocabulary and grammar presented in that chapter. These transparencies may be used as cues for addi-tional communicative practice in both oral and written formats. There are 18 Communication Transparencies for Level 1, and 16 for Level 2.

Map Transparencies The full-color maps located at the back of the Student Textbook have been converted to transparency format for the teacher's convenience. These can be used when there is a reference to them in the Student Textbook, or when there is a history or geography map reference in the Teacher's Wraparound Edition. The Map Transparencies can also be used for quiz purposes, or they may be photocopied in order to provide individual students with a black and white version for use with special projects.

Fine Art Transparencies These are full-color reproductions of works by well-known French artists including Matisse, Renoir, and others. Teachers may use these transparencies to reinforce specific culture topics in the *Réalités* sections as well as in the optional *Lettres et sciences* sections of the Student Textbook.

The Video Program (Cassette or Videodisc)

The video component for each level of **Glencoe French** consists of one hour-long video cassette and an accompanying Video Activities Booklet. Together, they are designed to reinforce the vocabulary, structures, and cultural themes presented in the corresponding Student Textbook. The **Glencoe French** Videocassette Program encourages students to be active listeners and viewers by asking them to respond to each video *Scène* through a variety of previewing, viewing and post-viewing activities. Students are asked to view the same video segment multiple times as they are led, via the activities in their Video Activities Booklet, to look and listen for more detailed information in the video segment they are viewing. The videocassette for each level of **Glencoe French** begins with an Introduction explaining why listening to natural, spoken French can be a difficult task and therefore why multiple viewings of each video *Scène* are required. The Introduction also points out the importance of using the print activities located in the Video Activities Booklet in order to use the Videocassette Program successfully.

Video Activities Booklet

The Video Activities Booklet is the vital companion piece to the hour-long video cassette for each level of **Glencoe French**. It consists of a series of pre-viewing, viewing, and post-viewing activities on Blackline Masters. These activities include specific instructions to students on what to watch and listen for as they view a given *Scène* on the videocassette. In addition to these student activities, the Video Activities Booklet also contains a Teacher's Manual, Culture Notes, and a complete transcript of the video soundtrack.

Computer Software: Practice and Test Generator

Available for Apple II, Macintosh and IBM-compatible machines, this software program provides materials for both students and teacher. The Practice Generator provides students with new, additional practice items for the vocabulary, grammar and culture topics in each chapter of the Student Textbook. All practice items are offered in a multiple choice format. The computer program includes a randomizer, so that each time a student calls up a set of exercises, the items are presented in a different order, thereby discouraging rote memorization of answers. Immediate feedback is given, along with the percent of correct answers, so that with repeated practice, students can track their performance. For vocabulary practice, illustrations from the *Vocabulaire* section of the Student Textbook have been scanned into the software to make practice more interesting and versatile.

The Test Generator allows the teacher to print out ready-made chapter tests, or customize a ready-made test by adding or deleting test items. The computer software comes with a Teacher's Manual as well as a printed transcript of all practice and test items.

Communication Activities Masters with Answer Key

This is a series of Blackline Masters, which provide further opportunities for students to practice their communication skills using the French they have learned. The contextualized, open-ended situations are designed to encourage students to communicate on a given topic, using specific vocabulary and grammatical structures from the corresponding chapter of the Student Textbook. The use of visual cues and interesting contexts will encourage students to ask questions and experiment with personalized responses. In the case of the paired communication activities, students actively work together as they share information provided on each partner's activity sheet. Answers to all activities are given in an Answer Key at the back of the Communication Activities Masters booklet.

Situation Cards

This is another component of **Glencoe French** aimed at developing listening and speaking skills through guided conversation. For each chapter of the Student Textbook, there is a corresponding set of guided conversational situations printed on hand-held cards. Working in pairs, students use appropriate vocabulary and grammar from the chapter to converse on the suggested topics. Although they are designed primarily for use in paired activities the Situation Cards may also be used in preparation for the speaking portion of the Testing Program or for informal assessment. Additional uses for the Situation Cards are described in the Situation Cards package, along with specific instructions and tips for their duplication and incorporation into your teaching plans. The cards are in Blackline Master form for easy duplication.

Bell Ringer Reviews on Blackline Masters

These are identical to the Bell Ringer Reviews found in each chapter of the Teacher's Wraparound Edition. For the teacher's convenience, they have been converted to this (optional) Blackline Master format. They may be either photocopied for distribution to students, or the teacher may convert them to overhead transparencies. The latter is accomplished by placing a blank acetate in the paper tray of your photocopy machine, then proceeding to make a copy of your Blackline Master (as though you were making a paper copy).

Interactive Conversation Video

An interactive video allows students to listen to and watch a real-life dramatization of each *Conversation* in the Student Textbook. Students may choose to participate in the video by taking the role of one of the characters.

Lesson Plans with Block Scheduling

Glencoe French offers flexible lesson plans for both 45- and 55-minute schedules. In addition, a separate set of lesson plans has been developed for those schools operating within a block scheduling arrangement.

The various **Glencoe French** support materials are incorporated into these lesson plans at their most logical point of use, depending on the nature of the presentation material on a given day. For example, the Vocabulary Transparencies and the Audio (Cassette or Compact Disc) Program can be used most effectively when presenting the chapter vocabulary. On the other hand, the corresponding Chapter Quiz is recommended for use one or two days after the initial presentation of vocabulary, or following a specific chapter structure point. Because student needs and teacher preferences vary, space has been provided on each lesson plan page for the teacher to write additional notes and comments, adjusting the day's activities as required.

Some Advantages of Block Scheduling

This type of scheduling differs from traditional scheduling in that fewer class sessions are scheduled for larger blocks of time over fewer days. For example, a course might meet for 90 minutes a day for 90 days, or half a school year.

For schools themselves, the greatest advantage of block scheduling is that there is a better use of resources. No additional teachers or classrooms may be needed, and more efficient use is made of those presently available in the school system. The need for summer school is greatly reduced because the students that do not pass a course one term can take it the next term. These advantages are accompanied by an increase in the quality of teacher instruction and student's time on-task.

There are many advantages for teachers who are in schools that use block scheduling. For example, teacher-student relationships are improved. With block scheduling, teachers have responsibility for a smaller number of students at a time, so students and teachers get to know each other better. With more time, teachers are better able to meet the individual needs of their students. Teachers can also be more focused on what they are teaching. Block scheduling may also result in changes in teaching approaches, classrooms that are more student-centered, improved teacher morale, increased teacher effectiveness, and decreased burn-out. Teachers feel free to venture away from discussion and lecture to use more productive models of teaching.

Block scheduling cuts in half the time needed for introducing and closing classes. It also eliminates half of the time needed for class changes, which results in fewer discipline problems. Flexibility is increased because less complex teaching schedules create more opportunities for cooperative teaching strategies such as team teaching and interdisciplinary studies.

Internet Activities Booklet

This booklet of blackline masters serves as a dynamic, real-world connection between cultural themes introduced in **Glencoe French**, and related topics available via the Internet. For example, **Bienvenue**, Chapter 2 is titled *Les copains et les cours*. The Internet activity for this chapter asks students to explore several WWW pages developed by French students, noting students' characteristics and interests. Using the vocabulary taught in Chapter 2, students are also asked to send a message to one of the student websites in France, describing himself/herself briefly, and mentioning a few personal interests.

In addition to serving as an innovative avenue for cultural reinforcement, the activities encourage both students and teachers to view the Internet as an engaging and valuable tool for learning the French language. Through this medium, students are able to further their knowledge of the French language, as well as increase their opportunities for participating in French-speaking communities around the world. The Internet activities encourage students to establish an ongoing keypal/pen pal relationship with French-speaking teenagers abroad.

The Internet Activities Booklet contains directions for the activities, student response sheets, and accompanying background teacher information, all on a chapter-by-chapter basis. Students will find the information required to complete each Internet activity by going to one or more of the websites whose addresses are provided on the Glencoe Foreign Language Home Page.

Chapter Quizzes with Answer Key

This component consists of short (5 to 10 minute) quizzes, designed to help both students and teachers evaluate quickly how well a specific vocabulary section or grammar topic has been mastered. For both Levels 1 and 2, there is a quiz for each *Mots* section (vocabulary) and one quiz for each grammar topic in the *Structure* section. The quizzes are on Blackline Masters. All answers are provided in an Answer Key at the end of the Chapter Quizzes booklet.

Testing Program with Answer Key

The Testing Program consists of three different types of Chapter Tests, two of which are bound into a testing booklet on Blackline Masters. The third type of test is available as part of the computer software component for **Glencoe French**.

1. The first type of test is discrete-point in nature, and uses evaluation techniques such as fill-in-the-blank, completion, short answers, true/false, matching, and multiple choice. Illustrations are frequently used as visual cues. The discrete-point tests measure vocabulary and grammar concepts via listening, speaking, reading, and writing formats. (As an option to the teacher, the listening section of each test has been recorded on cassette by native French speakers.) For testing cultural information, an optional section is included on each test corresponding to the *Lecture et Culture* section of the Student Textbook. For the teacher's convenience, the speaking portion of the tests has been physically separated from the listening, reading, and writing portions, and placed at the back of the testing booklet. These chapter tests can be administered upon the completion of each chapter. The Unit Tests can be administered upon the completion of each *Révision* (after every four chapters).

2. The Blackline Master testing booklet also contains a second type of test, namely the Chapter proficiency tests. These measure students' mastery of each chapter's vocabulary and grammar on a more global, whole-language level. For both types of tests above, there is an Answer Key at the back of the testing booklet.

3. In addition to the two types of tests described above, there is a third type which is part of the Computer Software: Practice and Test Generator Program (Macintosh; IBM; Apple versions). With this software, teachers have the option of simply printing out ready-made chapter tests, or customizing a ready-made test by selecting certain items, and/or adding original test items.

Performance Assessment

In addition to the tests described above, the Performance Assessment tasks provide an alternate approach to measuring student learning, compared to the more traditional paper and pencil tests. These tasks include individual student assignments, student interviews, and individual and small-group research projects with follow-up presentations. The Performance Assessment tasks can be administered following the completion of every fourth chapter in the Student Textbook.

GLENCOE FRENCH 1 CD-ROM INTERACTIVE TEXTBOOK

The **Glencoe French 1 CD-ROM Interactive Textbook** is a complete curriculum and instructional system for high school French students. The four-disc CD-ROM program contains all elements of the textbook plus photographs, videos, animations, a student portfolio feature, self-tests, and games, all designed to enhance and deepen students' understanding of the French language and culture. Although especially suited for individual or small-group use, it can be connected to a large monitor or LCD panel for whole-class instruction. With this flexible, interactive system, you can introduce, reinforce, or remediate any part of the French 1 curriculum at any time.

The CD-ROM program has four major sections: **Contents, Games, References**, and **Portfolio**. Of these four sections, the Games, Portfolio, and a special Contents feature—self-tests—are unique to the CD-ROM program.

Contents

The Contents section contains all the components of the French 1 *Bienvenue* textbook. The following selections can be found under Contents:

- **Vocabulaire** Vocabulary is introduced in thematic contexts. New words are introduced, and communication activities based on real-life situations are presented.

- **Structure** Students are given explanations of French structures. They then practice through contextualized exercises. One of the structure points in each chapter is enhanced with an electronic comic strip with which students can interact.

- **Conversation** Interactive videos, which were shot in France, enhance this feature comprised of real-life dialogues. Students may listen to and watch a conversation and then choose to participate as one of the two characters as they record their part of the dialogue.

- **Prononciation** Students are able to hear French pronunciation and then record the words and sentences themselves. They can then compare their pronunciation to that of native speakers.

- **Lecture et culture** Readings give students the opportunity to gain insight into French culture. They are also able to hear the readings in French. The similarities and differences between French and American culture are emphasized.

- **Réalités** In *Réalités*, students see glimpses of everyday life in France. The *Réalités* act as a starting point for discussions about similarities and differences that exist between life in France and in the United States.

- **Culmination** Chapter-end activities require students to integrate the concepts they have learned. There are oral and written activities as well as activities aimed at building skills, and a vocabulary review linked to the glossary.

- **Self-Test** The self-test, *Contrôle de révision*, provides a means for students to evaluate their own progress.

At the end of each four chapters are three features: *Le Monde francophone, Révision,* and *Lettres et sciences*. These selections may also be found in the Contents section.

- **Le Monde francophone** This section provides students with additional information about the various countries which comprise the francophone world. The text has been recorded to allow students to both read and hear authentic speech.

- **Révision** In the *Révision* section, students participate in a variety of review activities.

- **Lettres et sciences** This selection gives students the opportunity to practice their French reading skills through interdisciplinary readings that provide insights into French culture.

Games

The Games section gives users access to *Pour en savoir plus* (Discs 1 and 3) and to *Le Labyrinthe* (Discs 2 and 4). Each game reviews the vocabulary, structure, and culture topics that have been presented in the four chapters contained on that particular CD-ROM disc.

References

Maps, verb charts, and the French-English/English-French glossaries can be selected from this tab. The maps include France, Paris, and a world map with the French-speaking countries highlighted.

Portfolio

The electronic portfolio feature may be accessed by clicking on this tab. Students can access a photo library, and choose from a variety of "stationery" templates to create original written work which they can then save on their portfolio document.

For more information, see the User's Guide accompanying the **Glencoe French 1 CD-ROM Interactive Textbook.**

COOPERATIVE LEARNING

Cooperative learning provides a structured, natural environment for student communication that is both motivating and meaningful. The affective filter that prevents many students from daring to risk a wrong answer when called upon to speak in front of a whole class can be minimized when students develop friendly relationships in their cooperative groups and when they become accustomed to multiple opportunities to hear and rehearse new communicative tasks. The goal of cooperative learning is not to abandon traditional methods of foreign language teaching, but rather to provide opportunities for learning in an environment where students contribute freely and responsibly to the success of the group. The key is to strike a balance between group goals and individual accountability. Group (team) members plan how to divide the activity among themselves, then each member of the group carries out his or her part of the assignment. Cooperative learning provides each student with a "safe," low-risk environment rather than a whole-class atmosphere. As you implement cooperative learning in your classroom, we urge you to take time to explain to students what will be expected of every group member—listening, participating, and respecting other opinions.

In the Teacher's Wraparound Edition, cooperative learning activities have been written to accompany each chapter of the Student Textbook. These activities have been created to assist both the teacher who wants to include cooperative learning for the first time, and for the experienced practitioner of cooperative learning as well.

Classroom Management: implementing Cooperative Learning activities

Many of the suggested cooperative learning activities are based on a four-member team structure in the classroom. Teams of four are recommended because there is a wide variety of possible interactions. At the same time the group is small enough that students can take turns quickly within the group. Pairs of students as teams may be too limited in terms of possible interactions, and trios frequently work out to be a pair with the third student left out. Teams of five may be unwieldy in that students begin to feel that no one will notice if they don't really participate.

If students sit in rows on a daily basis, desks can be pushed together to form teams of four. Teams of students who work together need to be balanced according to as many variables as possible: academic achievement in the course, personality, ethnicity, gender, attitude, etc. Teams that are as heterogeneous as possible will ensure that the class progresses quickly through the curriculum.

Following are descriptions of some of the most important cooperative learning structures, adapted from Spencer Kagan's Structural Approach to Cooperative Learning, as they apply to the content of *Bienvenue*.

Round-robin Each team member answers in turn a question, or shares an idea with teammates. Responses should be brief so that students do not have to wait too long for their turn.

Example from *Bienvenue*, Chapter 2, Days of the week:

Teams recite the days of the week in a round-robin fashion. Different students begin additional rounds so that everyone ends up needing to know all the names of the days. Variations include starting the list with a different day or using a race format, i.e., teams recite the list three times in a row and raise their hands when they have finished.

Roundtable Each student in turn writes his or her contribution to the group activity on a piece of paper that is passed around the team. If the individual student responses are longer than one or two words, there can be four pieces of paper with each student contributing to each paper as it is passed around the team.

A to Z Roundtable Using vocabulary from *Bienvenue*, Chapters 7 and 8, students take turns adding one word at a time to a list of words associated with plane or train travel in A to Z order. Students may help each other with what to write, and correct spelling. Encourage creativity when it comes to the few letters of the alphabet that don't begin a specific travel word from their chapter lists. Teams can compete in several ways: first to finish all 26 letters; longest word; shortest word; most creative response.

Numbered Heads Together Numbered Heads Together is a structure for review and practice of high consensus information. There are four steps:

Step 1: Students number off in their teams from 1 to 4.

Step 2: The teacher asks a question and gives the teams some time to make sure that everyone on the team knows the answer.

Step 3: The teacher calls a number.

Step 4: The appropriate student from each team is responsible to report the group response.

Answers can be reported simultaneously, i.e., all students with the appropriate number can stand by their seats and recite the answer together, or go to the chalkboard and write the answer at the same time. Answers can also be reported sequentially. Call on the first student to raise his or her hand or have all the students with the appropriate number stand. Select one student to give the answer. If the other students agree, they sit down; if not, they remain standing and offer a different response.

Example from *Bienvenue*, Chapter 2, Telling time:

Step 1: Using a blank clock face on the overhead transparency or chalkboard, the teacher adjusts the hands on the clock.

Step 2: Students put their heads together and answer the question: *Quelle heure est-il?*

Step 3: The teacher calls a number.

Step 4: The appropriate student from each team is responsible to report the group response.

Pantomimes Give each team one card. Have each team decide together how to pantomime for the class the action identified on the card. Each team presents the pantomime for ten seconds while the rest of the teams watch without talking. Then each of the other teams tries to guess the phrase and writes down their choice on a piece of paper. (This is a good way to accommodate kinesthetic learning styles as well as vary classroom activities.)

Example from *Bienvenue*, Chapter 3 vocabulary: The teacher writes the following sentences on slips of paper and places them in an envelope:

1. *Ils parlent.*
2. *Ils parlent au téléphone.*
3. *Ils écoutent des cassettes.*
4. *Ils écoutent des disques compacts.*
5. *Ils écoutent un walkman.*
6. *Ils écoutent la radio.*
7. *Ils regardent la télé.*
8. *Ils dansent.*
9. *Ils chantent.*
10. *Ils rigolent.*

Each team will draw one slip of paper from the envelope and decide together how to pantomime the action for the class. As one team pantomimes their action for 30 seconds, the other teams are silent. Then the students within each team discuss among themselves

what sentence was acted out for them. When they have decided on the sentence, each team sends one person to write it on the chalkboard.

Inside/Outside Circle Students form two concentric circles of equal number by counting off 1-2, 1-2 in their teams. The "ones" form a circle shoulder to shoulder and facing out. The "twos" form a circle outside the "ones" to make pairs. With an odd number of students, there can be one threesome. Students take turns sharing information, quizzing each other, or taking parts of a dialogue. After students finish with their first partners, rotate the inside circle to the left so that the students repeat the process with new partners. For following rounds, alternate rotating the inside and outside circles so that students get to repeat the identified tasks, but with new partners. This is an excellent way to structure 100% student participation combined with extensive practice of communication tasks.

Other suggested activities are similarly easy to follow and to implement in the classroom. Student enthusiasm for cooperative learning activities will reward the enterprising teacher. Teachers who are new to these concepts may want to refer to Dr. Spencer Kagan's book, Cooperative Learning, published by Resources for Teachers, Inc., Paseo Espada, Suite 622, San Juan Capistrano, CA 92675.

SUGGESTIONS FOR CORRECTING HOMEWORK

Correcting homework, or any tasks students have done on an independent basis, should be a positive learning experience rather than mechanical "busywork." Following are some suggestions for correcting homework. These ideas may be adapted as the teacher sees fit.

1. Put the answers on an overhead transparency. Have students correct their own answers.

2. Ask one or more of your better students to write their homework answers on the chalkboard at the beginning of the class hour. While the answers are being put on the chalkboard, the teacher involves the rest of the class in a non-related activity. At some point in the class hour, take a few minutes to go over the homework answers that have been written on the board, asking students to check their own work. You may then wish to have students hand in their homework so that they know this independent work is important to you.

3. Go over the homework assignment quickly in class. Write the key word(s) for each answer on the chalkboard so students can see the correct answer.

4. When there is no correct answer, e.g., "Answers will vary," give one or two of the most likely answers. Don't allow students to inquire about all other possibilities, however.

5. Have all students hand in their homework. After class, correct every other (every third, fourth, fifth, etc.) homework paper. Over several days, you will have checked every student's homework at least once.

6. Compile a list of the most common student errors. Then create a worksheet that explains the underlying grammar points and practices on these topics.

STUDENT PORTFOLIOS

The use of student portfolios to represent long-term individual accomplishments in learning French offers several benefits. With portfolios, students can keep a written record of their best work and thereby document their own progress as learners. For teachers, portfolios enable us to include our students in our evaluation and measurement process. For example, the content of any student's portfolio may offer an alternative to the standardized test as a way of measuring student writing achievement. Assessing the contents of a student's portfolio can be an option to testing the writing skill via the traditional writing section of the chapter or unit test.

There are as many kinds of portfolios as there are teachers working with them. Perhaps the most convenient as well as permanent portfolio consists of a three-ring binder which each student will add to over the school year and in which the student will place his or her best written work. In the **Glencoe French** series, selections for the portfolio may come from the Writing Activities Workbook; Communication Activities Masters; the more open-ended activities in the Student Tape Manual and the Video Activities Booklet, as well as from written assignments in the Student Textbook, including the *Activités de communication écrite* sections. The teacher is encouraged to refer actively to students' portfolios so that they are regarded as more than just a storage device. For example, over the course of the

school year, the student may be asked to go back to earlier entries in his or her portfolio in order to revise certain assignments, or to develop an assignment further by writing in a new tense, e.g., the *passé composé*. In this way the student can appreciate the amount of learning that has occurred over several months' time.

Portfolios offer students a multidimensional look at themselves. A "best" paper might be the one with the least errors or one in which the student reached and synthesized a new idea, or went beyond the teacher's assignment. The Student Portfolio topic is included in each chapter of the Teacher's Wraparound Edition as a reminder that this is yet another approach the teacher may wish to use in the French classroom.

CD-ROM Electronic Portfolio

The Student Portfolio topic is a regular feature in each chapter of the Teacher's Wraparound Edition. In addition, the CD-ROM version of the student textbook (see page T37) includes an electronic portfolio feature. Students may access a photo library, and choose from a variety of stationery templates to create original written works which they can then store as a portfolio document. For more information, see the User's Guide accompanying the **Glencoe French 1 CD-ROM Interactive Textbook.**

PACING

Sample Lesson Plans

Level 1 (*Bienvenue*) has been developed so that it may be completed in one school year. However, it is up to the individual teacher to decide how many chapters will be covered. Although completion of the textbook by the end of the year is recommended, it is not necessary. Most of the important structures of Level 1 are reviewed in a different context in the early chapters of Level 2 (*À bord*). The establishment of lesson plans helps the teacher visualize how a chapter can be presented. However, by emphasizing certain aspects of the program and deemphasizing others, the teacher can change the focus and the approach of a chapter to meet students' needs and to suit his or her own teaching style and techniques. Sample lesson plans are provided below. They include some of the suggestions and techniques that have been described earlier in this Teacher's Manual.

STANDARD PACING

	Days	Total Days
(Preliminary Lessons A–H)	5 days	5
Chapitres 1–15	9 days per chapter	135
Testing	1 day per test	15
Révision (3)	3 days each	9
Lettres et sciences (3 [optional])	2 days each	6

	Class	Homework
Day 1	*Mots 1* (with transparencies) exercises (Student Textbook)	*Mots* exercises (written) Writing Activities Workbook: *Mots 1*
Day 2	*Mots 2* (with transparencies) exercises (Student Textbook)	*Mots* exercises (written) exercises from Student Textbook (written) prepare *Activités de communication* Writing Activities Workbook: *Mots 2*
Day 3	present *Activités de communication* one *Structure* topic exercises (Student Textbook)	exercises from Student Textbook (written) Writing Activities Workbook (written)

Day 4	two *Structure* topics exercises (Student Textbook)	exercises from Student Textbook (written) Student Tape Manual exercises
Day 5	one *Structure* topic exercises (Student Textbook)	exercises from Student Textbook (written) Writing Activities Workbook (written)
Day 6	*Conversation* (pronunciation) *Activités de communication* Audio Cassette Program	read *Lecture et culture* *Étude de mots*
Day 7	review *Lecture et culture* *Compréhension* questions Video Program	read *Découverte culturelle* and *Réalités*
Day 8	review homework *Activités de communication orale* Situation Cards	*Activités de communication écrite*
Day 9	Communication Activities Masters Communication transparency	review for test
Day 10	Test	after Chapters 4, 8, 12: *Révision* conversation

Révision (review) and *Lettres et sciences* (optional) sections

Day 1	grammar review	exercises in Student Textbook and Workbook
Day 2	correct homework *Activités de communication*	review for Test
Day 3	Unit Test	pre-read *Lettres et sciences*, first selection (optional)
Day 4	*Lettres et sciences*, first selection (optional)	*Lettres et sciences*, second selection (optional)
Day 5	*Lettres et sciences,* third selection (optional)	*Lettres et sciences*, second selection in-depth (optional)

ACCELERATED PACING

	Days	Total Days
(Preliminary Lessons A–H)	5 days	5
Chapters 1–8	8 days per chapter	64*
Chapters 9–18	7 days per chapter	70
Test	1 day per test	18
Révision (4)	2 days each	8
Lettres et Sciences (4 [optional])	2 days each	8

	Class	Homework
Day 1	*Mots 1* (with transparencies) exercises (Student Textbook)	*Mots* exercises (written) Writing Activities Workbook: *Mots 1*
Day 2	*Mots 2* (with transparencies) exercises (Student Textbook)	*Mots* exercises (written) exercises from Student Textbook (written) prepare *Activités de communication* Writing Activities Workbook: *Mots 2*
Day 3	present *Activités de communication* two *Structure* topics	exercises from Student Textbook (written) Writing Activities Workbook (written)
Day 4	two *Structure* topics	exercises from Student Textbook (written) Writing Activities Workbook (written)
Day 5	*Conversation* present *Activité de communication* Audio Cassette Program	read *Lecture et culture* *Étude de mots* and *Compréhension*
Day 6	*Découverte cultrelle* (optional) *Réalités* (optional) Videocassette Program Communication Activities Masters	*Culmination*
Day 7	review *Culmination* Situation Cards Communication transparency	review for test
Day 8	Test	After Chapters 4, 8, 12, 16: *Révision*

*Note: After Chapter 8, the teacher may choose among the *Culmination* activities on Days 6 and 7, thereby eliminating one day, or the teacher omits Chapter 18.

Révision (review) and *Lettres et sciences* (optional) sections

Day 1	*Révision* exercises	review for test
Day 2	Unit Test	
Day 3	*Lettres et sciences*, first selection (optional)	*Lettres et sciences*, second selection (optional)
Day 4	*Lettres et sciences*, third selection (optional)	

USEFUL CLASSROOM WORDS
AND EXPRESSIONS

Below is a list of the most frequently used words and expressions needed in conducting a French class.

Words

le papier	paper
la feuille de papier	sheet of paper
le cahier	notebook
le cahier d'exercices	workbook
le stylo	pen
le stylo-bille	ballpoint pen
le crayon	pencil
la gomme	(pencil) eraser
la craie	chalk
le tableau noir	blackboard
la brosse	blackboard eraser
la corbeille	waste basket
le pupitre	desk
le rang	row
la chaise	chair
l'écran (m.)	screen
le projecteur	projector
la cassette	cassette
le livre	book
la règle	ruler

Commands

Both the singular and the plural command forms are provided.

Viens.	Venez.	Come.
Va.	Allez.	Go.
Entre.	Entrez.	Enter.
Sors.	Sortez.	Leave.
Attends.	Attendez.	Wait.

Mets.	Mettez.	Put.
Donne-moi.	Donnez-moi.	Give me.
Dis-moi.	Dites-moi.	Tell me.
Apporte-moi.	Apportez-moi.	Bring me.
Répète.	Répétez.	Repeat.
Pratique.	Pratiquez.	Practice.
Étudie.	Étudiez.	Study.
Réponds.	Répondez.	Answer.
Apprends.	Apprenez.	Learn.
Choisis.	Choisissez.	Choose.
Prépare.	Préparez.	Prepare.
Regarde.	Regardez.	Look at.
Décris.	Décrivez.	Describe.
Commence.	Commencez.	Begin.
Prononce.	Prononcez.	Pronounce.
Écoute.	Écoutez.	Listen.
Parle.	Parlez.	Speak.
Lis.	Lisez.	Read.
Écris.	Écrivez.	Write.
Demande.	Demandez.	Ask.
Suis le modèle.	Suivez le modèle.	Follow the model.
Joue le rôle de…	Jouez le rôle de…	Take the part of…
Prends.	Prenez.	Take.
Ouvre.	Ouvrez.	Open.
Ferme.	Fermez.	Close.
Tourne la page.	Tournez la page.	Turn the page.
Efface.	Effacez.	Erase.
Continue.	Continuez.	Continue.
Assieds-toi.	Asseyez-vous.	Sit down.
Lève-toi.	Levez-vous.	Get up.
Lève la main.	Levez la main.	Raise your hand.
Tais-toi.	Taisez-vous.	Be quiet.
Fais attention.	Faites attention.	Pay attention.
Attention.		Attention.
Attention, s'il vous plaît.		Your attention, please.
Silence.		Quiet.
Fais attention.	Faites attention.	Careful.
Encore.		Again.
Encore une fois.		Once again.
Un à un.		One at a time.
Tous ensemble.		All together.
À haute voix.		Out loud.
Plus haut, s'il vous plaît.		Louder, please.
En français.		In French.
En anglais.		In English.

ADDITIONAL FRENCH RESOURCES

Pen pal sources Following is a list of French and American organizations that assist in finding French pen pals:

1. American Association of Teachers of French
   Bureau de Correspondance Scolaire
   57 East Armory Avenue
   Champaign, IL 61820
   tel: (217) 333-2842
2. Fédération Internationale des Organisations de Correspondance et d'Échanges Scolaires (FIOCES)
   29, rue d'Ulm
   75230 Paris CEDEX 05
   France
3. Contacts
   55, rue Nationale
   37000 Tours
   France
4. Office National de la Coopération à l'École
   101 bis, rue du Ranelagh
   75016 Paris
   France
5. Mairie
   Maison des Sociétés
   Square Weingarten
   69500 Bron
   France

French Embassy and Consulates in the United States

French Embassy
Press and Information Service
4101 Reservoir Road, N.W.
Washington, DC 20007
tel: (202) 944-6060

French Consulates
Atlanta: (404) 522-4226
Boston: (617) 542-7374
Chicago: (312) 787-5359
Honolulu: (808) 599-4458
Houston: (713) 528-2181
Los Angeles: (310) 235-3200
Miami: (305) 372-9799
New York: (212) 606-3688
New Orleans: (504) 523-5772
San Francisco: (415) 397-4330
San Juan, Puerto Rico: (809) 753-1700

The Embassy of France distributes neither French flags nor posters. To purchase a flag contact one of the following manufacturers:

U.N. Association
Capital Area Division
1319 18th Street N.W.
Washington, DC 20036-1802
(202) 785-2640

Abacrome
1-B Quaker Ridge Rd.
New Rochelle, NY 10804
(914) 235-8152

Glencoe French 1

Bienvenue

Conrad J. Schmitt

Katia Brillié Lutz

Glencoe
McGraw-Hill

New York, New York   Columbus, Ohio   Mission Hills, California   Peoria, Illinois

Photography
Front Cover: © Craddock, Erika/Tony Stone Images.
Allsport USA/Vandystadt: 252/2; Air France: 193, 194/1; Amantini, S./ANA: 111/3; Antman, M./Scribner: xM, xiT, 16T, 26B, 55, 66B, 67, 72, 76, 83/4, 98, 99, 101, 107/3, 117, 136, 138, 144/2, 161, 187, 205, 212B, 214B, 243, 276/3, 289, 294, 295, 322/1/2/3, 323/5, 324, 350, 359/3, 374T, 389, 397, 428, 439, 486, 494, 499; Archive Photos/Archive France: 233; Art Resource, NY: 122BL; Ascani M./Hoa-Qui: 221/5; Banahan, Lawrence/Allsport/Vandystadt: xB, 300/3; Bayer, Carol/La Photothèque SDP: 381/6; Billow, Nathan/Allsport USA: 380/3; Boehm, M./La Photothèque SDP: 111/4; Bohin, JL/Explorer: 31B; Brun, J./Explorer: 113/10; Bureau de Poids et Mésures: 229R; © California Newsreel, San Francisco: 437/9, 437/10; Canedi, Daniel/La Photothèque SDP: 359/5; Carle, Eric/Bruce Coleman: 253/4; Chadefaux, A./Top Agence: 420; Chardon, Phillipe/Option Photo: 119BR; Château d'Agneaux Hôtel, Eliophot, Aix en Provence: 472BR; Cogan, Michel/Top Agence: 491; Collection Lausat/Explorer: 446L; Collection Violet/Roger Viollet/Gamma Liaison: 446R; Comnet/Westlight: 30; Costa, S./Explorer: 250; Courlas, Tim/Horizons: 60, 172T, 183T, 218T, 297, 312T, 355, 393; Crallé, Gary/The Image Bank: 223/9; Christian, Erwin/Leo de Wys Inc.: 113/11; Cuny, C./Rapho/Gamma Liaison: 252/1; Damn, Fridmar/Leo de Wys Inc.: 463; Deschamps, Hervé/Gamma Liaison: 372; Ducasse, F./Rapho/Gamma Liaison: 404/1; Dung, Yo Trung/Gamma Liaison: 40; Duomo: 300-301, 357, 359/4, 438; Fischer, Zviki/La Photothèque: 380/1; Fischer, Curt: iv, v, viM, viB, vii, viiiT, ixT, xii, 2TR, 4R, 5, 6, 9T, 16B, 20B, 32-33, 32/1, 33/6, 34, 35R, 48T, 50/1, 50/3, 50/5, 59/1, 102B, 105B, 106-107, 109, 115, 116, 124-125, 135, 142L, 145/3, 160, 166B, 168B, 169, 170-171, 170/2, 170/3, 173, 179, 183, 186T, 188, 194-195, 195/3, 207, 212, 216-217, 216/1, 217/3, 218, 227, 256-257, 265, 275, 276/2, 277/4, 279, 293, 325, 352, 380, 406, 430-431/1, 470-471, 483, 502; Fleurent, C./Rapho/Gamma Liaison: 299; Florenz, David/Option Photo: 32/3; FPG: 249; Ford, Matthew/FSP/Gamma Liaison: 329/7; Foto World/The Image Bank: 436/7; Freed, Leonard/Magnum: 417; Gabriel/Explorer: 50/4; Gaveau, Alain: viT, ixM, ixB, xT, xiii, 90, 105T, 106/2, 280-281, 316, 320, 321, 354, 400, 418, 429, 456, 472TR, 474-475, 487, 497/4; Geiersperger, W./Explorer: 335B; Gely/Imapress: 434/1; Gerda, Paul/Leo de Wys: 110/2; Gerometta Soncin, Roberto/Photo 20-20: 50/6, 66T, 107/4, 134; Gibson, Mark/Photo 20-20: 327/3; Giraudon/Art Resource, NY: 58/2, 123B, 444, 445L, 445R; Gordon, Larry D./The Image Bank: 202; Gossler/Schuster/Explorer: 222/6, 327/4; Gritscher, Helmut/Peter Arnold: 339; Gschiedle, Gerhard/Scribner: 70, 238, 276-277; Guichaqua, Yann/Allsport/Vandystadt: viiiB; Gunn, F./Canapress: 348; Harlingue/Viollet/ Gamma-Liaison: 329/7; Hazat, V./Explorer: 20T; Heaton, Dallas & John/Westlight: 460; Hinous, Pascal/Top Agence: 217/4; Hoa-Qui: 220/1, 221/3; Holmes, Robert/Photo 20-20: 377, 381/5; Horowitz, Ted/The Stock Market:

121R; Hôtel de Paris, Cannes: 472L; Hôtel Idéal, Mont Blanc: 472M; Huet, M./Hoa-Qui: 165; Index Stock Photography: 120; Jalain, F./Explorer: 423; Jeffrey, David/The Image Bank: 3; Jenny, Andre/Photo 20-20: 112/8; Jenny, Andre/International Stock: 220/2; Joana M./La Photothèque SDP: 253, 358-359, 379; Jones, Spencer/Bruce Coleman: 214T; Kenny, Gill C./The Image Bank: 59/3; Kent, Keith/Peter Arnold: 339T; Kiki, Ozu/La Photothèque SDP: 29T; Kirtley, M&A/ANA: 33/4, 221/4, 329/8; Louvet, AM/Explorer: 31T; Lenfant, J.P./Vandystadt/Allsport USA: 251; Lessing, Eric/Art Resource, NY: 435/4; Lisl, Dennis/The Image Bank: 436/8; Losito, Brian/Air Canada: 185; Machatschek, Charles/La Photothèque SDP: 374, 382; Manceau, M./Rapho/Gamma Liaison: 145/4; Marché, Guy/FPG: 75; Marché, Guy/La Photothèque: 332; Martin, Richard/Allsport USA: 300/1; McCurry, Steve/Magnum: 334; Menzel, Peter/Peter Menzel: 27, 31M, 83/2, 431/4, 470/2; Merlin/La Photothèque SDP: 416; Messerschmidt, Joachim/FPG: 501; Meyer, Carl F.: 108; Moati/Kleinefenn/Opéra de Paris-Bastille: 123T; Naci, Jean Paul/Leo de Wys Inc: 366; N'Diaye, Jean Claude/Imapress: 314; Neumiller, Roberto/ANA: 113/9, 435/3; Nouvel, Daniel/Option Photo: 119BL; Parks, Claudia/The Stock Market: 326/1; Paireault, J.P./ANA: 328/6; Pelletier, M./Gamma Liaison: 434/2; Petit, Christian/Allsport USA/Vandystadt: 298; Photothèque de l'Institut Pasteur: 404/3L, 442, 443TL, 443TR; Pronin, Anatoly/Art Resource, NY: 419; Radford, Ben/Allsport USA: 351; Rega/Rapho/Gamma Liaison: 322-323; Renard, Éric/Agence Temp Sport: 301/4, 358/2; Renaudeau, M./Hoa-Qui: 97, 189, 326/2, 328/5, 436/5; Rey, Jean/ANA: 56; Romanelli, Marc/The Image Bank: 427; Rondeau, Pascal/Allsport USA: 380/4; Roux, Aimé/Explorer: 301/5; Rowe, Wayne: xiv-1, 2TL, 2B, 3, 4L, 9B, 10, 12-13, 19, 22, 26T, 28T, 32/2, 35L, 36-37, 45, 50/3, 54, 58-59, 62-63, 70, 78, 80, 82-83, 82/1, 83/4, 84, 86-87, 96, 102T, 103, 122BR, 140, 141, 142R, 144-145, 148-149, 152, 153, 158, 163, 166T, 168T, 170/1, 172, 174-175, 196, 198-199, 211, 234-235, 245, 248, 261, 266, 269, 270, 272T, 274, 288, 296T, 304-305, 312B, 313, 318, 319, 329/9, 340-341, 347, 353, 376, 384-385, 395, 396, 402, 403, 408-409, 416, 417, 425, 426, 430/2, 448-449, 466, 471/3, 478, 484, 487, 492, 496-497; Sanson, Nanette/Profolio: 362-363, 378; Scala/Art Resource, NY: 59/4, 421; Simmons, Ben/The Stock Market: 33/6; Sioen, Gérard/Rapho/Gamma Liaison: 48B; SNCF: 215; Streshinsky, Ted/Photo 20-20: 255; Stock, Dennis/Magnum: 197; Suaiton, Ken/The Stock Market: 436/6; SuperStock: 254, 278T, 371, 457; Talby, I/Rapho/Gamma Liaison: 83/3; Tauquer, Siegfried/Leo de Wys Inc.: 111/5; Tauqueuer, Siegried/Leo de Wys, Inc.: 380-381/2; Testelin, X/Rapho/Gamma Liaison: 404-405/2; Tetefolle, F./Explorer: 121L; Thomas, Marc: xiM, xiB, 114, 143, 147, 190B, 254B, 272B, 291, 296B, 344, 360, 370, 388, 394, 458, 489; Tovy, Adina/Photo 20-20: 29B, 119TR, 252-253; TPH/La Photothèque SDP: 278B; Travelpix/FPG: 110/1; Truchot, R./Explorer: 119TL; V&A/Art Resource, NY: 123M; Vaisse, C./Hoa-Qui: 111/6; Valentin, C./Gamma Liaison; 223/8; Vanni/Art Resource, NY: 122TR; Vidler, Steve/Leo de Wys Inc.: 224; Vielcanet, Patrick/Allsport-Vandystadt: 50T; Viollet Collection/Roger-Viollet/Gamma Liaison: 229TL; Watts, Ron/Westlight: 112/7; Weiss, S./Rapho/Gamma Liaison: 277/5; Wolf, A./Explorer: 192, 4043R, 443L; Wysocki, P./Explorer: 323/4, 335T, 405/4; Zuckerman, Jim/Westlight: 28.

Realia

Air Canada: 185; Air France: 191, 194, Air Inter: 179, 182; Air Orient, ©ARS, New York/ADAGP, Paris, illustration Paul Colin: 195; Allo Pizza: 146; A.N. Rafting, Le Grand Liou: 246; Banque Industrielle et Mobilière Privée: 497; Banque Nationale de Paris: 479L; Caisse d'Épargne Écurreuil: 496B; Cartotec, illustration Yannick Intesse: 303; Christian Dior: 277; Collections de la Comédie-Française: 413; Collège Eugène Delacroix: 85; Crédit Agricole: 496T; Éditions Gallimard: 233; Éditions Les Quatre Zéphires: 13; © Éditions S.A.E.P., 1993, Elle Magazine: 260; Espace Soleil: 244; France Télécom: 465; Galeries Lafayette, illustration Mats Gutafson: 267; Hachette-Gautier Languereau, illustration M. Boutet de Monvel: 14; Jazz Magazine: 139; Laboratoire Conseil Oberlin: 398, 399; Les ÉDITIONS ALBERT RENÉ/GOSCINNY-UDERZO, Carte postale éditée par ADMIRA: 91L, *Les Lauriers de César*, Dargaud Éditeur: 429T; Le Train Bleu Restaurant, Gare de Lyon: 219; Ligue Française pour les Auberges de la Jeunesse: 469; Locapark: 311; Michelin Red Guide France, 1992 Edition, Pneu Michelin, Services de Tourisme: 471; Ministère de l'Éducation Nationale: 81; Monoprix: 271; Okapi Magazine: 43; Pariscope Magazine, Backdraft, © by Universal City Studios, Inc. courtesy of MCA Publishing Rights, a Division of MCA Inc: 424; Pomme de Pain: 9; La Poste: 479R; La Redoute Catalogue: 371; Restaurant Marty: 133; Rev'Vacances: 440; SNCF: 206; Société IAG: 472; © Télérama: 60, 354; Vélo Sprint 2000 Magazine: 348.

Fabric designs by *Les Olivades*.

Maps

Eureka Cartography, Berkeley, CA.

In appreciation

Special thanks to the following people in France for their cordial assistance and participation in the photo illustration:

M. le Maire d'Ansouis; M. le Proviseur, les professeurs et les élèves du Lycée Henri IV; M. le Proviseur, les professeurs et les élèves du Lycée Val de Durance; Mme le Principal, les professeurs, en particulier Mlle Marie-Claude Éberlé, et les élèves du Collège Mignet; M. le Principal, les professeurs et les élèves du Collège du Pays d'Aigues; M. Jacques Lefèbvre et les élèves du Lycée du Parc Impérial; Groupe Scolaire Sainte-Anne

Dr Christian Amat, Marie-Françoise, Camille, Emmanuel et Alexandre Amat/Jean-Pierre Antoine et Bébé le caniche/Helena Appel/La Famille Baud/Sonia Benaïs/La Famille Bérard/Jérôme Bernard/Sylvain Casteleiro/Adelaïde Chanal/Amy Chang/Andréa Clément/Émilie Cusset/La Famille Dandré et Josué/Michèle Descalis/Denise Deschamps/Mme Duclos/Élisabeth Éberlé/Jeanne Grisoli/Hélène Guion/David Hadida/Amelle Hafafsa, Amar et Riad/Thomas Hardy/Simone Kayem/Marie-France Lamy/Olivier Lucas/Harry Magdaléon/Dr Francis Maguet/Barbara Marone et Jessie le collie/Katy Martin/Jean Martinez/Dr Jean Mori/Claudette Mori/Elarif M'Ze/Magali Parola/Daniel Pauchon/Olivier Perrière/Élodie Perrin/Nelly Pouani/Estelle et Hélène Puigt/Claude Rivière/Nadège Rivière/Elzéar, Foulques et Amic de Sabran-Pontevès/Maître Frédéric Sanchez, avocat/Kalasea Sanchez/Martine Serbin/Michel Skwarczewski/Florence Vareilles/Maître Marie-Christine Viard-Vassiliev, avocate/Jonathan Viretto/Bernard et Jacqueline Vittorio

Air France (M. Philippe Boulze)/L'Art Glacier/Banque Marseillaise de Crédit/Boutique Frenchy's/Cabinet du Dr Amat/Cabinet du Dr Maguet/Charcuterie Guers/Compact Club/Complexe Sportif du Val de l'Arc/Fromagerie Gérard Paul/les Gendarmes de Beaumont/Grand Café Thomas/Le Grand Véfour (M. Guy Martin)/Hôtel Le Moulin de Lourmarin/ Pâtisserie Chambost/Pharmacie de l'Europe/Restaurant La Récréation/Restaurant Le Viêt-Nam/Salon de Coiffure Sylvie

About the Cover

Notre-Dame Cathedral, a masterpiece of Gothic architecture, is located in the heart of Paris on the Île de la Cité in the Seine. Pope Alexander III laid the first stone in 1163 and construction was completed about 1330. The spectacular flying buttresses shown here are characteristic of Gothic architecture.

Acknowledgments

We wish to express our deep appreciation to the numerous individuals throughout the United States and France who have advised us in the development of these teaching materials. Special thanks are extended to the people whose names appear below.

Esther Bennett
Notre Dame High School
Sherman Oaks, California

Brillié Family
Paris, France

Kathryn Bryers
French teacher
Berlin, Connecticut

G. Gail Castaldo
The Pingry School
Martinsville, New Jersey

Veronica Dewey
Brother Rice High School
Birmingham, Massachusetts

Lyne Flaherty
Hingham High School
Hingham, Massachusetts

Marie-Jo Hoffmann
Poudre School District
Fort Collins, Colorado

Marcia Brown Karper
Fayetteville-Manlius Central Schools
Manlius, New York

Annette Lowry
Ft. Worth Independent School District
Ft. Worth, Texas

Fabienne Raab
Paris, France

Sally Schneider
Plano Independent School District
Plano, Texas

Faith Weldon
Schalmont Central School District
Schenectady, New York

TABLE DES MATIÈRES

BIENVENUE

CHAPITRE 1

UNE AMIE ET UN AMI

v

LA FAMILLE ET LA MAISON

AU CAFÉ ET AU RESTAURANT

CHAPITRE 6

ON FAIT LES COURSES

CHAPITRE 7

L'AÉROPORT ET L'AVION

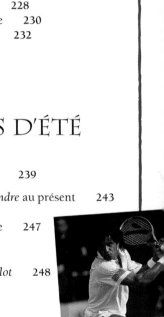

CHAPITRE 10

LES BOUTIQUES ET LES VÊTEMENTS

CHAPITRE 11

LA ROUTINE ET LA FORME PHYSIQUE

LA VOITURE ET LA ROUTE

LES SPORTS

CHAPITRE 14

L'HIVER ET LES SPORTS D'HIVER

CHAPITRE 15

LA SANTÉ ET LA MÉDECINE

xi

CHAPITRE 16

LES LOISIRS CULTURELS

CHAPITRE 17

L'HÔTEL

CHAPITRE 18

L'ARGENT ET LA BANQUE

APPENDICES

xiii

OVERVIEW

In these eight brief preliminary sections, students will learn some of the expressions needed for greeting and taking leave of people as well as some expressions of politeness. They will learn to form simple questions and identify people. They will also learn the names of common classroom objects and some items commonly found on café menus. The numbers from 1 to 60 are also presented. Note that no grammar is taught in these preliminary sections.

OBJECTIVES

By the end of these preliminary sections students will know:

1. how to greet and say good-bye to one another
2. how to ask who someone is and identify a person
3. the names of some common classroom objects
4. how to ask where something is
5. how to describe the location of an object
6. how to ask how much something costs
7. the numbers from 1 to 60
8. some common expressions of politeness
9. how to order food and drink at a café

BIENVENUE

DID YOU KNOW?

You may give your students the following information about the photos on this page.

Photo 1 (top right): This beautifully crafted, ornate gate graces one of the entrances to the Tuileries Gardens in Paris.

Photos 2 and 5 (left, p. xiv, and top left, p. 1): Two views of the market in Aix-en-Provence, with its colorful profusion of fruits and vegetables.

Photo 3 (middle): *Baguettes* like these are made on the premises of fewer and fewer French bakeries today. Those that still bake bread on-site often bear a sign in their window affirming their status as *artisans*.

Photo 4 (bottom right, p. xiv—bottom left, p. 1): These lovely lavender fields are typical of Provence. Harvested in July, lavender is used in the making of soap and perfume.

(continued on next page)

1

Pacing

Each preliminary section will take approximately half a class session, for a total of four class sessions for the entire preliminary unit. Pacing will depend on class length and the age and aptitude of the students.

Note The Lesson Plans offer guidelines for 45- and 55-minute classes and **Block Scheduling.**

Exercices vs. *Activités*

The exercises and activities are color coded. Exercises, which provide guided practice to prepare students for independent communication, are coded in blue. Communicative activities, which give students the opportunity for creative, open-ended expression, are coded in red.

GEOGRAPHY CONNECTION

In connection with the photos on these pages, you may wish to have students locate Paris, Lyon, and Aix-en-Provence on the map of France, page 504 (Part A, page 237), or use the Map Transparency.

(continued)

Photo 6 (top right): These teens are seated at the well-known Aix-en-Provence café *Aux Deux Garçons*. It is on the Cours Mirabeau, a beautiful, tree-lined street in the center of town.

Photo 7 (bottom right): The futuristic architecture of the Satolas train station outside of Lyon was designed by Calatrava, a Spanish architect. (See page 199, DID YOU KNOW? for further information.)

INTERNET ACTIVITIES

(optional)

These activities, student worksheets, and related teacher information are in the *Bienvenue* Internet Activities Booklet and on the Glencoe Foreign Language Home Page at **http://www.glencoe.com/secondary/fl**

A: Bonjour! 🎧

PRESENTATION *(page 2)*

A. Go around the room and give each student his or her French name. (See list below.) Have the class repeat each name after you. This provides an excellent introduction to French pronunciation, since so many sounds are used in names.

B. Have students repeat *Bonjour* or *Salut* with each name.

Culture note The following are some French boys' and girls' names that you may wish to give your students.

BOYS

Adam	Frédéric	Nicolas
Alain	Gaël	Olivier
Albert	Gauthier	Pascal
Alexandre	Georges	Patrice
Alfred	Gérard	Patrick
André	Gilbert	Paul
Antoine	Gilles	Philippe
Arnaud	Grégoire	Pierre
Arthur	Guillaume	Raoul
Benoît	Gustave	Raphaël
Bernard	Guy	Raymond
Bertrand	Hector	Régis
Bruno	Henri	Rémi
Charles	Hervé	René
Christian	Hugues	Richard
Christophe	Jacques	Robert
Claude	Jean	Roger
Daniel	Jérôme	Roland
David	Joseph	Sébastien
Denis	Julien	Serge
Didier	Laurent	Stéphane
Dominique	Léon	Sylvain
Édouard	Loïc	Thierry
Emmanuel	Louis	Thomas
Éric	Luc	Tristan
Étienne	Marc	Victor
Eugène	Marcel	Vincent
Fabrice	Martin	Xavier
François	Mathieu	Yann
Franck	Michel	Yves

GIRLS

Adèle	Andrée	Annick
Agnès	Angèle	Antoinette
Alice	Anne	Arlette

BONJOUR!

—Salut, Daniel!
—Salut, Stéphanie!

—Bonjour, Jean-Paul!
—Bonjour, Pierre!

When greeting a friend in French, you say *Salut* or *Bonjour.* *Salut* is a less formal way of saying hello.

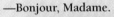

—Bonjour, Madame.

—Bonjour, Monsieur.

—Bonjour, Mademoiselle.

1. When greeting an adult in French, you say *Bonjour* with the person's title. You do not use the person's name with the title.

2. The following are abbreviations for these titles.

M.	Monsieur	Mme	Madame	Mlle	Mademoiselle

Activités de communication orale

A **Salut!** Choose a partner. Greet each other. Be sure to shake hands.

B **Bonjour.** Greet your French teacher.

C **Monsieur, Madame, Mademoiselle.** Choose a partner. Greet the following people. Your partner answers for the other person.

1. the principal of your school
2. your English teacher
3. a young saleswoman at the record store
4. your neighbor, Mr. Smith
5. your parents' friend, Mrs. Jones

B
ÇA VA?

—Salut, Marc.
—Salut, Valérie. Ça va?

—Ça va bien, merci. Et toi?
—Pas mal!

1. When you want to find out from a friend how things are going, you ask:

   Ça va?

2. Responses to *Ça va?* include:

   Ça va, merci.
   Bien, merci.
   Pas mal! Et toi?

Béatrice   Florence   Maryse
Bénédicte   Francine   Michèle
Bernadette   Françoise   Mireille
Brigitte   Gabrielle   Monique
Carole   Geneviève   Nadine
Caroline   Germaine   Nathalie
Catherine   Hélène   Nicole
Cécile   Irène   Odile
Célia   Isabelle   Pascale
Chantal   Jacqueline   Patricia
Christiane   Janine   Paule
Christine   Jeanne   Pauline
Claire   Josiane   Régine
Claude   Julie   Renée
Claudine   Laure   Sabine
Colette   Laurence   Sandrine
Corinne   Liliane   Sara
Danielle   Lise   Simone
Denise   Lisette   Solange
Diane   Louise   Sophie
Dominique   Madeleine   Stéphanie
Dorothée   Marguerite   Suzanne
Élisabeth   Marianne   Sylvie
Éliane   Marie   Thérèse
Élise   Mariel   Valérie
Émilie   Marlène   Véronique
Ève   Marthe   Virginie
Évelyne   Martine

Activités de communication orale

ANSWERS

Activité A
Salut (Bonjour), (first name)!

Activité B
Bonjour, Monsieur (Madame, Mademoiselle).

Activité C
1. Bonjour, Madame (Mademoiselle, Monsieur).
2. Bonjour, Madame (Mademoiselle, Monsieur).
3. Bonjour, Mademoiselle.
4. Bonjour, Monsieur.
5. Bonjour, Madame.

B: Ça va?

PRESENTATION *(page 3)*

A. Have the class repeat the conversation after you or Cassette 1B/CD-1 with as much expression as possible.
B. Have students stand and shake hands with one another using a brisk, French-style handshake.

PAIRED ACTIVITY

Have students work in pairs to make up their own exchanges greeting each other. Call on volunteers to present their skits to the class.

3

Activités de communication orale

ANSWERS

Activité A

Students will use the greetings in the text. Responses may include:

1. Salut, ___!
2. Ça va, merci. Et toi? (Ça va bien, merci. Bien, merci. Pas mal!)

Activité B

1. Salut (Bonjour), ___!/Salut (Bonjour), ___!
2. Ça va?/Ça va?

C: Au revoir

PRESENTATION *(pages 4–5)*

A. Have students repeat the expressions after you or Cassette 1B/CD-1.

B. Have a student near the front of the room say good-bye to a student seated nearby, using an appropriate expression. Continue around the room, having each student say good-bye to a neighbor using any appropriate expression.

Culture note You may wish to explain to students that *ciao* is actually an Italian word that has been adopted in many European countries as an informal way to say "good-bye" or "so long." It is used in Spain and Germany as well as in France.

4

Activités de communication orale

A **Salut!** Greet a classmate using the following expressions. Then reverse roles.

1. Salut!
2. Ça va?

B **Ça va?** You are walking down a street in Arles in southern France when you run into one of your French friends (your partner).

1. Greet each other.
2. Ask each other how things are going.

C
AU REVOIR

—Au revoir, Didier.
—Au revoir, Martine.

—Ciao, Gérard.
—Ciao. À tout à l'heure.

1. A common expression to use when saying good-bye is:

   Au revoir!

2. If you plan to see someone later in the day you say:

   À tout à l'heure!

3. An informal expression that you will hear frequently is:

   Ciao!

—Au revoir, tout le monde! À demain.
—Au revoir, Madame.

4. If you plan to see someone the next day, you say:

   À demain.

Conversation

—Salut, Christian.
—Salut, Francine. Ça va?

—Ça va bien, et toi?
—Pas mal, merci.

—Ciao, Christian.
—Ciao. À tout à l'heure!

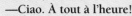ctivités de communication orale

A **Salut!**   Say the following to a classmate. He or she answers.

1. Salut!   3. Au revoir.
2. Ça va?   4. Ciao!

B **Au revoir!**

1. Say good-bye to your French teacher. Say you'll see him or her tomorrow.
2. Say good-bye to a friend. Say that you'll see him or her later in the day.

Conversation
PRESENTATION *(page 5)*

A. Have students repeat each line of the conversation after you or Cassette 1B/CD-1.
B. Call on pairs of students to read the conversation aloud with as much expression as possible.

Activités de communication orale
ANSWERS

Activité A
Answers will vary but may include the following.
1. É1: Salut!/É2: Salut!
2. É1: Ça va?/É2: Ça va bien, et toi?
3. É1: Au revoir./É2: Au revoir. (Ciao.)
4. É1: Ciao!/É2: Ciao! (À tout à l'heure./À demain.)

Activité B
1. Au revoir, Madame (Monsieur, Mademoiselle). À demain!
2. Ciao (Au revoir), ___. À tout à l'heure!

D: Qui est-ce?

PRESENTATION *(page 6)*

A. Go around the room. Point to a student as you ask: *Qui est-ce?* Look at the class and then give the person's name: *C'est ___.* After repeating the procedure with several students, ask *Qui est-ce?* and call on a student to respond with *C'est ___* and the person's name.

B. Point to each girl in the class and say *la fille.* Point to each boy and say *le garçon.* Have the class repeat. Point to a person in the distance as you say *la fille là-bas, le garçon là-bas.*

Conversation

PRESENTATION *(page 6)*

A. Ask two students to come to the front of the room. Introduce them to each other as in the second part of the conversation, and have them greet each other. Then have the entire class repeat the conversation after you or Cassette 1B/CD-1.

B. Call on three volunteers to dramatize the conversation in front of the class.

Note Students may watch a video of this conversation on the CD-ROM.

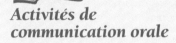

Activités de communication orale

ANSWERS

Activité A
Él: Qui est-ce?
É2: C'est ___.

Activité B
___, c'est ___.

Activité C
Answers will vary, but may include the following.
1. Salut, ___.
2. (Le garçon/La fille là-bas,) Qui est-ce?
3. Salut, ___.
4. Ça va?
5. Ciao, ___./Ciao.

D

QUI EST-CE?

Conversation

GARÇON 1: Qui est-ce?
GARÇON 2: Qui ça?
GARÇON 1: La fille là-bas.
GARÇON 2: C'est Mireille Claudel.

(She comes up to them.)
GARÇON 2: Mireille, c'est Guillaume.
FILLE: Salut, Guillaume.
GARÇON 1: Salut, Mireille.

1. When you want to know who someone is, you ask:

    Qui est-ce?

2. When you want to identify a person or introduce a person to someone else, you use *c'est* + the person's name.

    C'est Mireille Claudel.

Activités de communication orale

A **Qui est-ce?** Ask a classmate who someone else in the class is.

B **C'est…** Introduce someone you know to another person in the class.

C **Qui ça?** Prepare the following conversation with two classmates.

1. Greet your classmate.
2. Ask him or her who someone else in the class is.
3. Say hello to the new person.
4. Ask him or her how things are going.
5. Say good-bye to one another.

COOPERATIVE LEARNING

Have students work in groups of three to make up their own conversations using *la fille* or *le garçon* and real names.

E

QU'EST-CE QUE C'EST?

un cahier

un crayon

une chaise

un stylo

une table

un ordinateur

un livre

un autre livre

une calculatrice

un sac à dos

une feuille de papier

une autre feuille de papier

un bureau

un tableau

un morceau de craie

un devoir

1. When you want to know what something is, you ask:

   Qu'est-ce que c'est?

2. When you want to identify the object, you use *C'est* + the name of the object.

   C'est un cahier.

Activité de communication orale

A **C'est un (une)…** Work with a partner. He or she will hold up or point out five classroom objects and ask you what each one is.

E: Qu'est-ce que c'est?

PRESENTATION (*page 7*)

A. You may wish to use props when first presenting these nouns. Hold each item up or point to it and have students repeat the appropriate French word after you or Cassette 1B/CD-1.

B. Ask *Qu'est-ce que c'est?* At first give the response and then call on a student to give the response.

C. Have the students repeat *Qu'est-ce que c'est?* Then have them hold up an item and ask you what it is.

D. Show Vocabulary Transparency BV-E and have the students identify each item. Then have them open their books and read the words for reinforcement.

E. Have students pick up any item they have that they can identify in French, ask *Qu'est-ce que c'est?* and call on a classmate to respond.

Activité de communication orale

ANSWERS

É1 questions will be **Qu'est-ce que c'est?** É2 answers will all begin with **C'est un/une…**

F: Où est… ? 🎧

PRESENTATION (page 8)

A. Pretend you are looking for something and say *Où?*
B. Point to a desk top as you say *sur le bureau. Où? Sur le bureau.*
C. Open a drawer, point inside, and say *dans le bureau. Où? Dans le bureau.*
D. Call students to the front of the room and have them form a line. Point to the appropriate students as you teach *derrière* and *devant.*

Exercices

PRESENTATION (page 8)

Exercices A and B
You may wish to ask the questions in the exercises first and then have a student ask them.

ANSWERS

Exercice A

1. Dans le bureau.
2. Sur le bureau.
3. Sur le bureau.
4. Sur la table.
5. Dans le bureau.

Exercice B

1. Suzanne.
2. Marc.
3. Paul.
4. Marie.

Activités de communication orale

ANSWERS

Activités A and B
Answers will vary.

8

F
OÙ EST… ?

Où est le livre?

sur le bureau

dans le bureau

Où est Pierre?

devant Paul
derrière Monique

Paul   Pierre   Monique

Exercices

A **Où est… ?** Répondez d'après le dessin. *(Answer according to the illustration.)*

1. Où est le livre?
2. Où est le crayon?
3. Où est le cahier?
4. Où est l'ordinateur?
5. Où est la calculatrice?

B **Qui est devant ou derrière?** Répondez d'après le dessin. *(Answer according to the illustration.)*

1. Qui est devant Marie?
2. Qui est derrière Paul?
3. Qui est devant Marc?
4. Qui est derrière Suzanne?

A

B

Suzanne   Marie   Paul   Marc

Activités de communication orale

A **Où est… ?** Place a classroom object somewhere in the room. Have a classmate tell you where the item is using *sur, dans, devant,* or *derrière.*

B **Devant ou derrière?** Choose a row of students and tell where each person is seated in relation to another classmate in the row.

G

C'EST COMBIEN?

—C'est combien, Madame?
—Six francs, Mademoiselle.
—Merci, Madame.

1. When you want to find out how much something is, you ask:

   C'est combien?

2. In order to understand the answer, you must know some numbers. On the right are the numbers in French from zero to sixty.

LES NOMBRES DE ZÉRO À SOIXANTE		
0 zéro		
1 un	21	vingt et un
2 deux	22	vingt-deux
3 trois	23	vingt-trois
4 quatre	24	vingt-quatre
5 cinq	25	vingt-cinq
6 six	26	vingt-six
7 sept	27	vingt-sept
8 huit	28	vingt-huit
9 neuf	29	vingt-neuf
10 dix	30	trente
11 onze	31	trente et un
12 douze	40	quarante
13 treize	41	quarante et un
14 quatorze	50	cinquante
15 quinze	51	cinquante et un
16 seize	60	soixante
17 dix-sept		
18 dix-huit		
19 dix-neuf		
20 vingt		

Dix francs.

Activités de communication orale

A **C'est combien?** Tell how much French money is in each picture.

1.

2.

3.

4.

G: C'est combien?

PRESENTATION *(page 9)*

A. Have students repeat the numbers from 1 to 60 after you. Start with the numbers from 1 to 10, go on to 11 to 20, and then 20 to 60 by tens. Then have students repeat the numbers from 21 to 30, 31 to 40, and so on. Do not stay with the lesson until all students can count perfectly. The numbers will be reintroduced many times in future chapters.

B. Have students repeat *C'est combien?* To clarify the meaning, say *Combien? Deux? Non? Combien? Cinq? Oui.*

Activités de communication orale

PRESENTATION *(pages 9–10)*

Activité A

Have students be careful not to pronounce the *c* in *franc*.

ANSWERS

1. **Vingt francs.**
2. **Quinze francs.**
3. **Quarante francs.**
4. **Cinquante francs.**

PRESENTATION (continued)

Activité B

Go over the activity once with the class, giving the correct noun with *le* or *la*:

C'est combien, le stylo?
C'est combien, le cahier?
C'est combien, le livre?
C'est combien, la calculatrice?
C'est combien, le crayon?

ANSWERS

Él questions will begin with *C'est combien… ?* É2 answers will be:

Six francs.
Quatorze francs.
Cinquante-deux francs.
Soixante-quatre francs.
Deux francs.

H: Un café, s'il vous plaît

PRESENTATION *(pages 10–11)*

A. Have students repeat the expressions of politeness after you.

B. Read the conversation to the class once or have them listen to Cassette 1B/CD-1. Then call on students who have good pronunciation to read it.

B **À la papeterie.** You are spending the school year in France and are buying the following supplies at the stationery store. Ask the saleperson (your partner) how much each item is.

H
UN CAFÉ, S'IL VOUS PLAÎT

—Bonjour.
—Un café, s'il vous plaît.

(The waiter brings the coffee.)
—Merci.
—Je vous en prie.

(He's ready to leave.)
—C'est combien, le café, s'il vous plaît?
—Dix francs, Monsieur.

1. Expressions of politeness are always appreciated. Below are the French expressions for "Please," "Thank you," and "You're welcome."

FORMAL	INFORMAL
S'il vous plaît.	S'il te plaît.
Merci.	Merci.
Je vous en prie.	Je t'en prie.

PAIRED ACTIVITY

Have students work in pairs and make up their own conversations. One plays the part of a waiter/waitress and the other plays the part of a customer. Then call on volunteers to present their conversations to the class.

2. Other formal ways to say "You're welcome" are:

Ce n'est rien.   **Il n'y a pas de quoi.**

Other informal ways to say "You're welcome" are:

De rien.   **Pas de quoi.**

Activités de communication orale

A **Un coca, s'il vous plaît.** You are at a café in Dinard, a lovely resort in Brittany. Order the following items from the waiter or waitress (your partner).

1. un coca
2. un café
3. un sandwich
4. une limonade
5. un thé
6. une soupe à l'oignon
7. une salade
8. une omelette
9. une tarte aux fruits

B **C'est combien, la limonade?** You are now ready to leave the café. Ask the waiter or waitress how much you owe for the following items. (He or she can check the prices on the menu on the right.)

1. le café
2. le sandwich
3. la limonade
4. le dessert
5. la soupe
6. la salade

Café de Dinard

Sandwich	18,00
Salade	20,00
Omelette	24,00
Soupe à l'oignon	25,00
Tarte aux fruits	15,00
Coca	10,00
Café	6,00
Limonade	11,00
Thé	6,00

Vocabulaire

NOMS
un tableau
un morceau de craie
un bureau
un ordinateur
une calculatrice
une table
une chaise
un crayon
un stylo
une feuille de papier
un devoir
un cahier
un livre
un sac à dos

une fille
un garçon
Madame (Mme)
Mademoiselle (Mlle)
Monsieur (M.)

PRÉPOSITIONS
derrière
devant
dans
sur

AUTRES MOTS
ET EXPRESSIONS
Bonjour.
Salut.

Ça va.
bien
Pas mal.
Au revoir.
ciao
À tout à l'heure.
À demain.

s'il vous plaît
s'il te plaît
Merci.
Je vous en prie.
Je t'en prie.
Ce n'est rien.
De rien.
(Il n'y a) pas de quoi.

autre
c'est
là-bas
oui
tout le monde
combien
où
Qu'est-ce que c'est?
Qui est-ce?

NOMBRES
zéro–soixante (0–60)

PRESENTATION *(page 11)*

Activité A

Before doing Activity A, have the students repeat the food items once after you or Cassette 1B/CD-1. Since they are all cognates, it is possible that students will tend to anglicize the pronunciation if they try to read them without hearing them first.

ANSWERS

Activité A

Answers will vary.

Activité B

É1 will ask **C'est combien, le/la ___?** for each of the items listed. É2 answers will be:

1. Six francs.
2. Dix-huit francs.
3. Onze francs.
4. Quinze francs.
5. Vingt-cinq francs.
6. Vingt francs.

Vocabulaire

The words and phrases in the *Vocabulaire* have been taught for productive use. They are summarized here as a convenient resource for both students and teacher.

CHAPTER OVERVIEW

Students will learn to describe themselves and a friend using *être* and high frequency adjectives. The plural forms will be presented in Chapter 2.

The cultural focus is on the educational system in France.

CHAPTER OBJECTIVES

By the end of this chapter, students will know:

1. singular forms of *être*
2. singular subject pronouns
3. adjective agreement
4. singular definite and indefinite articles
5. negation

CHAPTER 1 RESOURCES

1. Workbook
2. Student Tape Manual
3. Audio Cassette 2A/CD-1
4. Bell Ringer Review Blackline Masters
5. Vocabulary Transparencies
6. Pronunciation Transparency P-1
7. Grammar Transparency G-1
8. Communication Transparency C-1
9. Communication Activities Masters
10. Map Transparencies
11. Situation Cards
12. Conversation Video
13. Videocassette/Videodisc, Unit 1
14. Video Activities Booklet, Unit 1
15. Lesson Plans
16. Computer Software: Practice/Test Generator
17. Chapter Quizzes
18. Testing Program
19. Internet Activities Booklet
20. CD-ROM Interactive Textbook

CHAPITRE

1

UNE AMIE ET UN AMI

OBJECTIFS

In this chapter you will learn to do the following:

1. ask or tell where someone is from
2. ask what someone is like
3. describe yourself or someone else
4. name people and things
5. tell some differences between French and American schools

CHAPTER PROJECTS

(optional)

Have one or more students research in as much detail as possible the French educational system. Have students find out what some of the major differences are between schools in the U.S. and those in France.

COMMUNITIES

To do the Chapter Project, have students use any of these resources:

1. a local library
2. a native French person whom they could interview
3. a local French Consulate (when available)
4. French pen pals on the Internet (For information, see COMMUNITIES, page 36.)

13

Pacing

This chapter will require eight to ten class sessions. Pacing will depend on the length of the class and the age and aptitude of the students.

For more information on planning your class, see the Lesson Plans, which offer guidelines for 45- and 55-minute classes and **Block Scheduling**.

NOTE ON INTERROGATIVES

1. In conversational French, the most common way to form a question is to raise one's voice at the end of the sentence. This rising intonation pattern can be used with yes/no questions and many question words: *Richard est américain? Il est de quelle nationalité?* Note that in spoken French the question word is often placed at the end of the sentence.

2. Another common way to ask a question is to begin a statement with *est-ce que* or with a question word + *est-ce que*: *Est-ce que Marie est française? Où est-ce qu'elle habite?*

3. A question can also be formed using inversion: *D'où est Jean?*

In the early chapters, questions are formed in the above three ways. Most frequently the rising intonation pattern is used.

The only form not used in the early chapters is the formal inversion: *Où Robert va-t-il?* We have not used this inversion since it is not often used in conversation.

Tell your students that they will hear questions in the three ways outlined above. When they ask questions, they can use any option they please. A complete summary of question formation is presented in Chapter 12, page 316.

Exercices vs. *Activités*

All exercises (which provide guided practice) are coded in blue. All communicative activities are coded in red.

INTERNET ACTIVITIES

(optional)

These activities, student worksheets, and related teacher information are in the *Bienvenue* Internet Activities Booklet and on the Glencoe Foreign Language Home Page at http://www.glencoe.com/secondary/fl

LEARNING FROM PHOTOS

Ask students if they think there is much difference between their own appearance and that of French students, based on the photo.

Bell Ringer Review

Write the following on the board or use BRR Blackline Master 1-1: How would you greet the following people?

1. your best friend
2. your French teacher
3. the cashier at the store
4. the principal of your school
5. another player on your team

PRESENTATION (*pages 14–15*)

A. Have students close their books. Present the vocabulary by using the Vocabulary Transparencies 1.1 (A & B), pictures of famous people from magazines, or appropriate student models.

B. Present one word or phrase at a time, then build to a complete sentence as in the following example: using Transparency 1.1-A, point to Yvonne Delacroix as you say *française*. Point to the *Arc de Triomphe* as you say *Yvonne Delacroix est française*. Point to each

---

VOCABULAIRE

MOTS 1

Comment est la fille?

petite   grande   brune

contente   amusante

Voici Yvonne Delacroix.
Yvonne Delacroix est française.
Salut, Yvonne!

D'où est Yvonne?
Elle est de Paris.

14   CHAPITRE 1

TOTAL PHYSICAL RESPONSE

(*following the Vocabulary presentation*)

Getting Ready

Before doing this activity, make sure students understand each of the following commands by acting them out: *levez-vous, promenez-vous, arrêtez-vous, montrez,* and *asseyez-vous.*

TPR

___ , levez-vous s'il vous plaît.
Promenez-vous dans la salle de classe.
Arrêtez-vous.
Montrez-moi un garçon.
Montrez-moi un garçon blond.
Montrez-moi un garçon brun.
Montrez-moi un garçon content.
Montrez-moi une fille.
Montrez-moi une fille blonde.
Merci. Asseyez-vous, s'il vous plaît.

Comment est le garçon?

petit   grand

brun

LYON

content   amusant

Voici Jean-Luc Charpentier.
Jean-Luc Charpentier est français
   aussi.
Salut, Jean-Luc!

D'où est Jean-Luc?
Il est de Lyon.

CHAPITRE 1   **15**

individual on pages 14–15 and
model the accompanying word
or phrase. Have students repeat
each word or phrase in unison.

C. Have students open their books
to page 14 and look at the new
vocabulary as they repeat after
you or Cassette 2A/CD-1.

D. Ask the following questions as
you point to Yvonne Delacroix:
*C'est Yvonne Delacroix? C'est
Yvonne Delacroix ou Jean-Luc
Charpentier? Yvonne Delacroix
est française ou américaine? Qui
est française?* (as you point to
Yvonne). *Qui est français?* (as
you point to Jean-Luc). *De
quelle nationalité est Yvonne?*
(Repeat *de quelle nationalité* in
case students have not understood. Say *française, américaine,
italienne?*) Ask similar questions about Jean-Luc.

E. Check for comprehension by
asking *Comment est Yvonne/
Jean-Luc/la fille/le garçon?*, etc.
and have students describe
characteristics of that person.
Students may answer in unison
or individually.

Vocabulary Expansion

When students ask for
additional related vocabulary
in the early chapters, it is
strongly recommended that
they not be provided with
material that will complicate
or confuse the concept being
presented. For example, give
students only the oral forms
of any of the following adjectives that they may request.
roux (rousse)

mexicain(e)	**philippin(e)**
libanais(e)	**iranien(ne)**
arménien(ne)	**portoricain(e)**

COOPERATIVE LEARNING

Work in teams of four. Each team member
has one piece of blank paper. After each
round, each team member passes his/her
paper to the next team member.

Round 1: Draw the face and head of a
French or American boy or girl and write a sentence describing his or her hair color. *Il est blond.*

Round 2: Draw a stick body and describe
the person's height. *Il est grand.*

Round 3: Draw an item of ethnic clothing
(beret, baseball cap, etc.). *Il est français.*

Round 4: Draw something in the hand
of the person, maybe a book, and describe a
personality trait. *Il est intelligent.*

Students read the descriptions of the completed drawings to their other team members.

Note For additional information on
cooperative learning, see the Teacher's
Manual at the front of this Teacher's
Wraparound Edition.

Note: You have already seen that many words in French and English look alike even though they are pronounced quite differently. Such words are called cognates. The following are some cognates used to describe people.

américaine	américain
blonde	blond
impatiente	impatient
intelligente	intelligent
intéressante	intéressant
patiente	patient
confiante	confiant

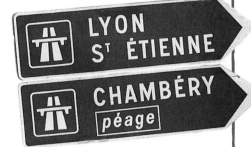

Exercices

A **Une Française, Yvonne Delacroix.** Répondez. *(Answer.)*

1. Yvonne Delacroix est française?
2. Elle est grande ou petite?
3. Elle est amusante?
4. Elle est contente?
5. Yvonne est brune?
6. Elle est de Paris?

B **Salut, Jean-Luc!** Répondez. *(Answer.)*

1. Jean-Luc Charpentier est français ou américain?
2. Il est brun ou blond?
3. Il est intelligent?
4. Il est amusant aussi?
5. Il est content?
6. Jean-Luc est intéressant?
7. Il est de Lyon?

C **Un Français et un Américain.** Répondez d'après les photos. *(Answer according to the photos.)*

1. Qui est américain, Marc ou Paul?
2. Qui est français?
3. Qui est de Paris?
4. Qui est de New York?
5. Qui est brun?
6. Qui est blond?
7. Qui est content?
8. Qui est impatient?

Marc Hugot

Paul Green

16 CHAPITRE 1

VOCABULAIRE

MOTS 2

une amie

un ami

une école américaine

PACIFIC
HIGH SCHO

une élève

LYCÉE HENRI

un élève

un lycée français

Vocabulary Teaching Resources

1. Vocabulary Transparencies 1.2 (A & B)
2. Audio Cassette 2A/CD-1
3. Student Tape Manual, Teacher's Edition, *Mots 2: E–G*, pages 9–10
4. Workbook, *Mots 2: C–F*, pages 1–2
5. Communication Activities Masters, *Mots 2: B*, page 2
6. Chapter Quizzes, *Mots 2: Quiz 2*, page 2
7. Computer Software, *Vocabulaire*
8. CD-ROM, Disc 1, *Mots 2*: pages 17–20

Bell Ringer Review

Write the following on the board or use BRR Blackline Master 1-2: How would you write the following adjectives so that they refer to a girl: *blond, impatient, petit, content, amusant?*

PRESENTATION *(pages 17–18)*

A. Have students close their books. Present the new vocabulary on pages 17–18 by using one of the following:
   1. Vocabulary Transparencies 1.2 (A & B)
   2. pictures of famous people from magazines
   3. student models that are appropriate
B. Model each new word or phrase. Have students repeat each word or phrase in unison after you or Cassette 2A/CD-1. Emphasize the difference between *une/un*.

TOTAL PHYSICAL RESPONSE

(following the Vocabulary presentation)

Getting Ready

Before doing this activity, make sure students understand the following command by acting it out: *Retournez à votre place.*

TPR

___, levez-vous, s'il vous plaît.
Promenez-vous un peu dans la classe.
Arrêtez-vous et montrez-moi un ami (un élève).
Qui est-ce?
Promenez-vous encore dans la classe.
Et maintenant, arrêtez-vous.
Montrez-moi une amie (une élève).
Qui est-ce? C'est ___.
Très bien, ___. Et merci.
Retournez à votre place.
Asseyez-vous, s'il vous plaît.

C. Now ask students to open their books to pages 17–18. Point to each person and ask individual students to read the accompanying word or phrase. On page 18, have individual students read each sentence beginning with *Yvonne Delacroix est française.*

D. Vocabulary presentation for page 18 (bottom): If you have a student whose pronunciation is quite good, call him or her to the front of the room. Say to the class, for example, *C'est Richard Williams.* Then tell students, in English, that Richard is going to tell you something about himself. Now have the student read the sentences to the class.

E. Have students give the opposite of the following nouns: *une sœur, une amie, une élève,* etc.

INFORMAL ASSESSMENT
(Mots 2)

Check for comprehension by asking yes/no and either/or questions, calling on the entire class as well as individual students to respond. For example: *Yvonne est la sœur ou le frère de Paul? Paul est la sœur ou le frère d'Yvonne? Yvonne est élève? Paul est élève aussi? Yvonne est élève dans un lycée français? Yvonne est la sœur de Claude Gautier ou une amie de Claude Gautier?*

Yvonne Delacroix
la sœur

Paul Delacroix
le frère

Claude Gautier
un ami

Yvonne Delacroix est française.
Yvonne est élève dans un lycée.
Yvonne est la sœur de Paul Delacroix.
Yvonne est une amie de Claude Gautier.
Paul est un ami de Claude aussi.

Bonjour, tout le monde.
Je suis Richard, Richard Williams.
Moi, je suis américain.
Je ne suis pas français.
Je suis de Miami.
Je suis élève dans une école secondaire américaine.
Je suis très populaire, n'est-ce pas?

18   CHAPITRE 1

Note: The following are other cognates used to describe people.

aimable	fantastique
désagréable	énergique
timide	populaire
comique	célèbre
sincère	

There are many French words for which there is no exact English equivalent. Such a word is *sympathique*. It has the meanings "nice," "pleasant," and "friendly." In informal French *sympathique* is often shortened to *sympa*. Its opposite is *antipathique*.

Exercices

A **Une élève française.** Choisissez. (*Choose the best answer.*)

1. ___ est française.
   a. Yvonne Delacroix
   b. Claude Gautier

2. Yvonne est élève dans ___.
   a. une école américaine
   b. un lycée français

3. Yvonne est ___.
   a. de Paris   b. de Miami

4. Yvonne est ___ de Paul Delacroix.
   a. une amie   b. la sœur

5. Yvonne est ___ de Claude Gautier.
   a. une amie   b. la sœur

6. Paul Delacroix est ___ d'Yvonne.
   a. un ami   b. le frère

7. Et Claude Gautier est ___ d'Yvonne.
   a. un ami   b. le frère

B **Comment est Richard Williams?** Répondez. (*Answer.*)

1. Richard est français ou américain?
2. D'où est Richard?
3. Il est élève dans une école secondaire américaine ou dans un lycée français?
4. Comment est Richard? Il est brun ou blond?
5. Il est petit ou grand?
6. Il est aimable ou désagréable?
7. Richard est sympathique ou antipathique?

COGNATE RECOGNITION

🎧 Have students close their books. Read aloud the list of cognates on page 19. Have individual students give the English equivalent. Then have students open their books, follow along, and repeat after you or Cassette 2A/CD-1 as you pronounce the list again.

Exercices

PRESENTATION (*pages 19–20*)

Exercice A

Go over the exercises in class before you assign them as homework. Exercise A can be done with books open. This exercise checks comprehension and makes students focus on several indicators such as the adjective form in number 1 and the relationship in number 5.

Exercice B

You may wish to do Exercise B with books closed the first time.

Extension of *Exercice B*

After completing Exercise B, have one student retell the story in his/her own words, describing a boy in class.

ANSWERS

Exercice A

1. Yvonne Delacroix
2. un lycée français
3. de Paris
4. la sœur
5. une amie
6. le frère
7. un ami

Exercice B

1. Richard est américain.
2. Il est de Miami.
3. Il est élève dans une école secondaire américaine.
4. Richard est blond.
5. Il est grand.
6. Il est aimable.
7. Il est sympathique.

PAIRED ACTIVITY

Have students work in pairs and write a description of their partner that includes information modeled in the presentations of Yvonne and Richard, page 18. Student 1 will make statements beginning with *Je suis...* Student 2 will write *Il/Elle est...* Partners will then reverse roles. Finally, they will take turns introducing each other to the rest of the class.

INDEPENDENT PRACTICE

Have students write a short self-description. Ask several students to present their self-description to the class. Vary the activity by having the students describe famous individuals orally or in writing.

Exercice C

If you do Exercise C with books open, first call on individual students to do one item each.

Extension of *Exercice C*

You can do Exercise C again by having one student do the first half and another student do the second half.

ANSWERS

Exercice C

1. française, américaine
2. lycée français
3. brune
4. désagréable
5. sympathique

Activités de communication orale
Mots 1 et 2

PRESENTATION *(page 20)*

The *Activités de communication orale* allow students to use the chapter vocabulary in open-ended situations. It is not necessary that you do all of the *Activités*. You may select those you consider most appropriate.

ANSWERS

Activités A, B, and C

Answers will vary.

RETEACHING *(Mots 1 and 2)*

1. Ask students the following questions about their classmates: *Marc est américain ou français? Il est élève? Il est intelligent? Paul est élève dans une école secondaire française ou américaine? Quelle école?* (You may wish to give the names of two schools to help convey the meaning of *quelle*.) *Anne est de ___?* (Your town or city) *Anne est une amie de ___?*

2. Bring to class a magazine picture of a well-known personality all students will recognize. Have them describe the personality using vocabulary they know.

C **Élisabeth Gautier.** Complétez. *(Complete.)*

1. Élisabeth Gautier est la sœur de Claude Gautier. Elle est de quelle ville? Elle est de Paris. Elle est ___. Elle n'est pas ___.
2. Élisabeth est élève dans un ___. Elle n'est pas élève dans une école secondaire américaine.
3. Élisabeth est ___. Elle n'est pas blonde.
4. Elle est aimable. Elle n'est pas ___.
5. Élisabeth est une amie ___. Elle n'est pas antipathique.

Activités de communication orale
Mots 1 et 2

A **Gilles Baud.** Here's a photo of Gilles Baud. He's a student from Strasbourg. Say as much as you can about Gilles.

B **Caroline Baud.** The blond girl in the photo below is Gilles Baud's sister, Caroline. She's a student in Strasbourg, too. Say a few things about her.

C **Qui est-ce?** Describe a classmate but don't mention his or her name. Someone in the class has to guess who it is.

20 CHAPITRE 1

ADDITIONAL PRACTICE	**INDEPENDENT PRACTICE**
Student Tape Manual, Teacher's Edition, *Activités F–G,* pages 9–10.	Assign any of the following: 1. Exercises and activities, pages 19–20 2. Workbook, *Mots 2: C–F,* pages 1–2 3. Communication Activities Masters, *Mots 2: B,* page 2 4. Computer Software, *Vocabulaire* 5. CD-ROM, Disc 1, pages 17–20

Les articles indéfinis et définis au singulier

Talking about One Person or Thing

LES ARTICLES INDÉFINIS

1. The name of a person, place, or thing is a noun. In French, every noun has a gender, either masculine or feminine. Many words that accompany nouns can indicate their gender in French. They are called gender markers. An article is such a word.

2. The French words *une* and *un* are indefinite articles. They correspond to *a (an)* in English. You use an indefinite article when speaking about a non-specific person or thing: *a girl, an exam.* Study the following examples with the indefinite article.

FÉMININ	MASCULIN
une fille	un garçon
une sœur	un frère
une école	un lycée
une calculatrice	un ordinateur

3. You use the indefinite article *une* before all feminine nouns. You use the indefinite article *un* before all masculine nouns.

Exercices

A Alain et Charles. Complétez avec «un» ou «une». (*Complete with* un *or* une.)

1. Alain est ___ garçon très sympathique.
2. Alain est ___ ami de Charles.
3. Charles est ___ élève très intelligent.
4. Il est élève dans ___ école secondaire à New York.
5. Annette est la petite sœur d'Alain. Elle est élève dans ___ école primaire.
6. Suzanne est ___ amie d'Alain, pas d'Annette.

ADDITIONAL PRACTICE

After completing Exercise A, reinforce the lesson with the following:

Give each student two index cards or papers with *un* or *une* written on them. Call out a familiar noun, e.g., *le stylo*, and have students raise the card showing the correct indefinite article. Have the class repeat in unison the noun and article together, e.g., *un stylo*.

INDEPENDENT PRACTICE

After doing Exercise A in class, assign it as homework and go over it quickly the next class session. (See suggestions for homework correction in the Teacher's Manual.)

Structure Teaching Resources

1. Workbook, *Structure: A–H,* pages 3–5
2. Student Tape Manual, Teacher's Edition, *Structure: A–G,* pages 10–13
3. Audio Cassette 2A/CD-1
4. Grammar Transparency G-1
5. Communication Activities Masters, *Structure: A–E,* pages 3–6
6. Computer Software, *Structure*
7. Chapter Quizzes, *Structure: Quizzes 3–6,* pages 3–6
8. CD-ROM, Disc 1, pages 21–27

Bell Ringer Review

Write the following on the board or use BRR Blackline Master 1-3: List five or six adjectives that could be given in response to the following question: Comment est le professeur?

Les articles indéfinis et définis au singulier

PRESENTATION (*page 21*)

Ask students to listen to the differences in the pronunciation of masculine and feminine nouns accompanied by the indefinite article. Pronounce words in pairs. For example: *un ami/une amie, un élève/une élève.* Ask students: How can one tell if a word is masculine or feminine in French?

Exercices

ANSWERS

Exercice A

1. un	3. un	5. une
2. un	4. une	6. une

Extension of *Exercice B*

1. Point to classroom objects and ask: *Qu'est-ce que c'est?* to see if students can identify the objects and give the correct indefinite article.

2. Emphasize the fact that inanimate objects are also masculine or feminine and remind students that they must memorize the gender of all nouns.

ANSWERS

Exercice B

1. stylo
2. cahier
3. ordinateur
4. livre
5. tableau (noir)
6. chaise
7. sac à dos

Bell Ringer Review

Write the following on the board or use BRR Blackline Master 1-4: On your paper make two columns. Write *un* at the top of the first one and *une* at the top of the second one. Then write the following words under the appropriate column: *bureau, lycée, école, sœur, ami, fille, cahier, frère, amie, tableau, livre, table, feuille, chaise, calculatrice.*

Les articles définis au singulier

PRESENTATION (page 22)

A. Contrast the usage of a definite article to refer to a specific person or thing with the indefinite article to refer to any one person or thing. Act out the differences with TPR commands such as *Donnez-moi un stylo* and *Donnez-moi le stylo de Marie.*

B. **Listening Activity.** Ask students to listen as you say the following: *le lycée, la fille, l'école, le livre, le stylo, la chaise, l'élève.* Ask for student observations of what they have heard.

B **Qu'est-ce que c'est?** Répondez d'après les photos. (*Answer according to the photos.*)

1.
2.
3.
5.
4.
6.
7.

1. C'est un stylo ou un crayon?
2. C'est un cahier ou une feuille de papier?
3. C'est une calculatrice ou un ordinateur?
4. C'est un livre ou un cahier?
5. C'est un tableau ou un bureau?
6. C'est une table ou une chaise?
7. C'est une feuille de papier ou un sac à dos?

LES ARTICLES DÉFINIS AU SINGULIER

1. You use the definite article when referring to a definite or specific person or thing: *the boy, the desk.* Study the following examples of definite articles.

FÉMININ	MASCULIN
la fille	le garçon
la sœur	le frère
la chaise	le bureau

2. You use the definite article *la* before a feminine noun. You use the definite article *le* before a masculine noun.

3. You use the definite article *l'* before a masculine or feminine noun that begins with a vowel or silent *h*. The vowels are *a, e, i, o, u.*

l'élève   l'école
l'ami   l'hôtel
l'amie

ADDITIONAL PRACTICE

Have students volunteer six nouns they know, giving the indefinite articles. Write them on the board and ask students to replace the indefinite articles with the correct definite articles. Then have the class repeat them.

INDEPENDENT PRACTICE

Assign any of the following:
1. Exercises, pages 21–23.
2. Workbook, *Structure: A–C,* page 3
3. Communication Activities Masters, *Structure: A–B,* page 3
4. CD-ROM, Disc 1, pages 21–23

Exercice

A Richard Williams et Claudine Simonet. Complétez avec *le, la* ou *l'*. *(Complete with* le, la, *or* l'.*)*

___ garçon, Richard Williams, est américain mais ___ fille, Claudine Simonet,
  1   2
n'est pas américaine. Elle est française. Claudine est ___ amie de Gilbert
                                                                                  3
Duhamel et ___ sœur de Christian Simonet. Richard n'est pas ___ ami de
                       4   5
Claudine: il est de Miami et Claudine est de Lille. Richard est ___ ami de
                                                                                              6
Suzanne Jackson et ___ frère de Cassandra Williams. Richard est élève et
                                  7
Claudine est élève aussi. ___ école de Richard est à Miami et ___ lycée de
                                          8   9
Claudine est à Lille.

L'accord des adjectifs au singulier

Describing a Person or Thing

---

1. A word that describes a noun is an adjective. The italicized words in the following sentences are adjectives.

> **La fille est *française*.** **Le garçon est *français*.**
> **Yvette est *intelligente*.** **Robert est *intelligent*.**

2. In French, an adjective must agree with the noun it describes or modifies. If the noun is masculine, then the adjective must be in the masculine form. If the noun is feminine, the adjective must be in the feminine form. An adjective is therefore a gender marker. Most adjectives follow the noun.

> **une fille blonde** **un garçon blond**

3. Many feminine adjectives end in *e*. When the *e* follows a consonant, you pronounce the consonant.

> **peti*te* gran*de* intelligen*te***

4. Many masculine adjectives end in a consonant. Since the consonant is not followed by an *e*, you do not pronounce the final consonant.

> **peti*t̸* gran*d̸* intelligen*t̸***

5. Certain feminine adjectives, such as *brune*, end in *ne*. You pronounce the *n* in these words. The masculine form is written without the *e*. The vowel that goes before the *n* is nasal.

> **brune brun**

FOR THE YOUNGER STUDENT

Using large cards labeled with feminine adjectives, call girls to the front of the class to hold the cards. Pronounce the words in unison with the class.

Give the boys cards with the letter *e* crossed out. Ask them to stand next to the girls to form the masculine adjective forms. Now pronounce the masculine forms of the words.

Bell Ringer Review

Write the following on the board or use BRR Blackline Master 1-5: Draw a circle on your paper. Write *le, la, l'* in this circle. Now make a list of nouns outside the circle. Connect each noun to the appropriate article in the circle.

L'accord des adjectifs au singulier

PRESENTATION *(pages 23–24)*

A. Draw two stick figures on the board. Name them Marie and Paul. Point to Marie as you say *française, intelligente, intéressante,* etc. Then point to Paul as you say *français, intelligent, intéressant,* etc. Ask students if they hear a difference in the sound and why there is a difference. Remind students that words used to describe males and females are pronounced differently.

B. Model the examples given in the structure explanation on page 23 and have students repeat the examples after you.

C. To demonstrate that adjectives ending in *-e* can refer to feminine or masculine words, have female and male students come to the front of the class. Each pair will hold his or her own adjective card with the same word written on each one. Make statements about each student and then ask questions. For example: *André est sympathique. Yvonne est sympathique aussi. Qui est sympathique? (André.) Qui est sympathique aussi? (Yvonne.)*

Note In the CD-ROM version, this structure point is presented via an interactive electronic comic strip.

PRESENTATION *(page 24)*

Exercice A

You may do Exercise A twice—once with books closed and a second time with books open as students look at the written forms of the words.

Extension of *Exercice B*

1. Have one student read all of Exercise B. Then have another student retell the story in the exercise in his/her own words.

2. Have students substitute the names of students in the class for the names in the exercise and have them ask the questions about people in their class.

ANSWERS

Exercice A

1. Emphasize the pronunciation differences between the feminine and masculine forms.

2. Ask students why the pronunciation is different and what the rule is.

Exercice B

1. française
2. brune
3. petite
4. oui
5. oui
6. blond
7. grand
8. oui
9. dans un lycée français
10. dans un lycée français aussi

Exercice C

1. amusante, sincère
2. amusant, aimable, désagréable
3. français, célèbre
4. Answers will vary. Adjectives will end in -e for females.
5. secondaire américaine
6. français

INFORMAL ASSESSMENT

Give students a masculine or feminine adjective and have them describe someone in the class using that adjective. For example, you say: *blonde.* Students must say: *Marie est blonde,* etc.

---

6. Many adjectives that end in an *e* have only one singular form. You use this form with both masculine and feminine nouns. Study the following examples.

> **Charles est un garçon sincère. Il est sympathique.**
> **Carole est une fille sincère. Elle est très sympathique.**

Exercices

A **Prononciation.** Répétez après votre professeur. *(Repeat after your teacher.)*

FÉMININ	MASCULIN	FÉMININ	MASCULIN
1. américaine	américain	5. petite	petit
2. blonde	blond	6. intelligente	intelligent
3. brune	brun	7. intéressante	intéressant
4. grande	grand	8. française	français

B **Marie-Thérèse et François.** Répondez d'après le dessin. *(Answer according to the illustration.)*

1. Marie-Thérèse est française ou américaine?
2. Elle est blonde ou brune?
3. Elle est grande ou petite?
4. Elle est amusante?
5. François est le frère de Marie-Thérèse?
6. François est blond ou brun?
7. Il est grand ou petit?
8. Il est amusant?
9. Marie-Thérèse est élève dans un lycée français ou dans une école américaine?
10. Et le frère de Marie-Thérèse est élève dans un lycée français ou dans une école américaine?

François et Marie-Thérèse Leroux

C **Carole et André.** Complétez. *(Complete.)*

1. Carole Colbert est une amie ___ et ___. (amusant, sincère)
2. André est le frère de Carole. André est ___ aussi. Il est ___. Il n'est pas ___. (amusant, aimable, désagréable)
3. Carole est élève dans un lycée ___ ___. (français, célèbre)
4. Et moi, je suis ___ *(your name)*. Je suis ___. Je ne suis pas ___. (américain, français)
5. Je suis élève dans une école ___ ___. (secondaire, américain)
6. Je ne suis pas élève dans un lycée ___. (français)

INDEPENDENT PRACTICE

Assign any of the following:

1. Exercises, page 24 (See suggestions for homework correction in the Teacher's Manual.)
2. Workbook, *Structure:* D–E, page 4
3. Communication Activities Masters, *Structure: C,* page 4
4. CD-ROM, Disc 1, pages 23–24

Le verbe *être* au singulier

Identifying People and Things

1. The verb "to be" in French is *être*. Note that the form of the verb changes with each person. Study the following.

ÊTRE			
je suis	tu es	il est	elle est

2.

> Je suis française.

> Tu es américain?

> Il est intelligent.

> Elle est intelligente.

You use *je* to talk about yourself.

You use *tu* to address a friend.

You use *il* or the person's name to talk about a male.

You use *elle* or the person's name to talk about a female.

3. You also use *il* and *elle* when referring to things.

> **Le stylo est sur la table. Il est sur la table.**
> **La calculatrice est dans le bureau. Elle est dans le bureau.**

Exercices

A **En France.** Répétez la conversation. *(Practice the conversation.)*

> Salut!

> Salut! Tu es Chantal Binand, n'est-ce pas?

> Oui, je suis Chantal. Et toi, tu es David, non?

> Oui, je suis David Butler.

> Tu es américain, David?

> Oui, je suis de Saint-Louis.

ADDITIONAL PRACTICE

1. Tell where a student is seated in the classroom using his or her name and *devant* or *derrière. Marie est devant Mark.* First have students replace the noun subject with the correct subject pronoun: *Elle est devant Mark.* Then increase the difficulty by asking: *Où est Marie?* and having students respond with the correct subject pronoun: *Elle est devant Mark.*

2. Use the same method as in #1 using inanimate objects and *dans, sur.*
   Le livre est sur la table.
   Il est sur la table.
   Où est le livre? Il est sur la table.

3. Student Tape Manual, Teacher's Edition, *Activités A–E, G,* pages 10–13.

Le verbe *être au singulier*

PRESENTATION *(page 25)*

Note Before presenting the verb *être*, go over the meaning of the personal pronouns *je, tu, il, elle* by having students do the following:

1. point to themselves as they say *je*
2. look at a neighbor as they say *tu*
3. point to a boy as they say *il*
4. point to a girl as they say *elle*

A. Write the following examples on the board: *Je suis Mireille. Je suis de Nice. Je suis intelligente et énergique.*
   Lead students through these examples by explaining that the verb *être* is used to tell who you are, where you are from, and to describe yourself and others.
B. Model the singular forms of *être* on page 25. Ask students to repeat after you in unison.
C. Use Grammar Transparency G-1 and have students create a dialogue.

Exercices

PRESENTATION *(pages 25–26)*

Exercice A

The purpose of this mini-conversation is to let students hear, see, and use the singular forms of the verb *être* in context before they use them on their own.

1. Read the conversation to the class, or use Cassette 2A/CD-1, having students repeat after you or the recorded speaker.
2. Ask students to take the roles.

Exercice B

This exercise allows students to hear *je suis* prior to using this phrase actively in Exercises C and D.

Exercice E

This exercise combines both the *je* and the *tu* forms of the verb.

ANSWERS

Exercice B

1.–4. *Pardon, tu es d'où?*

As pairs of students do this exercise orally, emphasize the pattern *je suis... , tu es...*

Exercice C

Answers will vary but may include the following:

1. name of student
2. name of city
3. student's nationality
4. Je suis élève.

Exercice D

Answers will vary but may include the following:

1. Je suis américain(e).
2. Je suis de (name of city).
3. Oui, je suis élève dans une école secondaire.
4. Je suis impatient(e) (patient[e]).

Exercice E

1. Jean-Paul, tu es élève? Oui, je suis élève.
2. ... de Nîmes? Oui, je suis de Nîmes.
3. ... élève dans un lycée à Nîmes? Oui, je suis élève dans un lycée à Nîmes.
4. ... content? Oui, je suis content.
5. ... intelligent? Oui, je suis intelligent.

Exercice F

1. Germaine est canadienne.
2. Elle est blonde.
3. Elle est sympathique.
4. Elle est de Québec.
5. Elle est étudiante universitaire à Québec.

B **Pardon!** Répondez d'après le modèle. (*Answer according to the model.*)

> Élève 1: Je suis de Paris.
> Élève 2: Pardon, tu es d'où?

1. Je suis de Nice.
2. Je suis d'Antibes.
3. Je suis de Lille.
4. Je suis de Strasbourg.

C **Je suis…** Donnez des réponses personnelles. (*Give your own answers.*)

1. Je suis ___ (*name*).
2. Je suis de ___ (*place*).
3. Je suis ___ (*nationality*).
4. Je suis ___ (*occupation*).

D **Une interview.** Posez des questions à un(e) ami(e). (*Ask a friend the following questions.*)

1. Tu es français(e) ou américain(e)?
2. Tu es de quelle ville? De New York? De Chicago?
3. Tu es élève dans une école secondaire?
4. Tu es impatient(e) ou patient(e)?

E **Jean-Paul.** Voici une photo de Jean-Paul Tonone. Il est de Nîmes. Posez des questions à Jean-Paul d'après le modèle. (*Ask Jean-Paul questions following the model. Your partner will answer for him.*)

> français
> Élève 1: Jean-Paul, tu es français?
> Élève 2: Oui, je suis français.

1. élève
2. de Nîmes
3. élève dans un lycée à Nîmes
4. content
5. intelligent

F **Germaine.** Voici une photo de Germaine LeBlanc. Décrivez-la d'après les indications. (*Here's a photo of Germaine LeBlanc. Describe her using the following cues.*)

1. canadienne
2. blonde
3. sympathique
4. de Québec
5. étudiante universitaire à Québec

COOPERATIVE LEARNING

On a 3x5 card have students write three questions they will use to interview other students in the class. Have students form two concentric circles of equal number by counting off 1-2-1-2 in their teams. The ones form a circle shoulder to shoulder and facing out. The twos form a circle outside the ones to make pairs. With an odd number of students, there can be one threesome.

Students take turns interviewing each other by asking the questions on their cards. When both circles have been interviewed, rotate the inside circle to the left. Then, have students exchange cards and rotate the outside circle to the right.

G **Luc Delacourt.** Complétez. *(Complete.)*

Voici Luc Delacourt. Il ___ français. Il ___ de Lyon. Moi aussi, je ___ de Lyon.
Lyon ___ une ville importante en France. Luc ___ élève dans un lycée à Lyon.
Le lycée ___ assez grand. Et toi? Tu ___ français(e) ou américain(e)? Tu ___
de quelle ville? Tu ___ élève dans une école secondaire? L'école ___ petite?

La négation   *Making a Sentence Negative*

1. The sentences in the first column are affirmative and the sentences in the
   second column are negative.

AFFIRMATIF	NÉGATIF
Je suis américain.	Je *ne* suis *pas* français.
Tu es sympathique.	Tu *n'es pas* antipathique.
Il est aimable.	Il *n'est pas* désagréable.
Elle est de Lyon.	Elle *n'est pas* de Paris.

> *ne* + verb + *pas*

2. You place *ne* before the verb and *pas* after the verb.
   Ne becomes *n'* before a vowel. This is called elision.

   Il est américain?   **Non, il *n'est pas* américain.**

Exercices

A **Non, Marie-France n'est pas américaine.** Mettez à la forme négative.
(Change to the negative.)

1. Marie-France est américaine.
2. Elle est de San Francisco.
3. Elle est élève dans une école
   secondaire à San Francisco.
4. Et moi, je suis français(e).
5. Je suis de Paris.
6. Je suis élève dans un lycée
   à Paris.

B **Tu es français(e)?** Donnez des réponses personnelles. *(Give your own
answers.)*

1. Tu es français(e)?
2. Tu es de Lyon?
3. Tu es élève dans un lycée célèbre à Lyon?
4. Tu es très, très désagréable?
5. Tu es l'ami(e) de Claudine Simonet?

MEMORY AID

Dramatize the negative concept by comparing the *ne... pas* construction with a sandwich. The *ne... pas* is the bread, the verb is the filling for the sandwich.

INDEPENDENT PRACTICE

Assign any of the following:
1. Exercises, pages 26–27
2. Workbook, *Structure: F–H*, page 5
3. Communication Activities Masters,
   Structure: D–E, pages 5–6
4. Computer Software, *Structure*
5. CD-ROM, Disc 1, pages 25–27

Exercice G
1. est	5. est	9. es
2. est	6. est	10. est
3. suis	7. es	
4. est	8. es	

Bell Ringer Review

*Write the following on the
board or use BRR Blackline
Master 1-7:* Match each subject
with the correct form of the
verb *être*. Then finish each sentence. Keep your paper.

tu	
le garçon	est
je	suis
elle	es
l'école	

La négation

PRESENTATION *(page 27)*

A. Explain how to make sentences
   negative by choosing a student
   to read the affirmative statements on page 27. After each
   one, model the corresponding
   negative statement.
B. Repeat the procedure. This
   time model the affirmative
   statements and have the class
   say the negative version in
   unison.

Exercices

ANSWERS

Exercice A
1. Marie-France n'est pas
   américaine.
2. ... n'est pas...
3. ... n'est pas...
4. ... ne suis pas...
5. ... ne suis pas...
6. ... ne suis pas...

Exercice B
1. Je ne suis pas français(e).
2. Je ne suis pas de Lyon.
3. Je ne suis pas élève dans un
   lycée célèbre à Lyon.
4. Je ne suis pas très, très
   désagréable.
5. Je ne suis pas l'ami(e) de
   Claudine Simonet.

RETEACHING

Use the sentences that students
wrote for the Bell Ringer Review.
Students read them aloud; others
make the sentences negative.

Bell Ringer Review

Write the following on the board, or use BRR Blackline Master 1-8: On a sheet of paper answer the following questions in the negative:

1. Tu es français(e)?
2. Tu es désagréable?
3. Paris est une ville américaine?
4. Un lycée est une école primaire?

PRESENTATION (*page 28*)

A. Tell students they are going to hear a conversation between Christian, a French boy, and Carole, an American girl, who are meeting each other for the first time.

B. Have them close their books and watch the Conversation Video or have them listen as you read the conversation or play the recorded audio version.

C. Ask students to open their books and look at the dialogue as you lead them through what is said.

D. Have students work in pairs to practice the conversation. Then have several pairs present the conversation to the class.

Note In the CD-ROM version, students can play the role of either one of the characters and record the conversation.

ANSWERS

Exercice A

1. Carole est américaine.
2. Christian est français.
3. Stéphanie est l'amie de Christian.
4. Christian est de Nice.
5. Nice est sur la Côte d'Azur.
6. C'est fantastique.

Scènes de la vie   *Tu es d'où?*

CHRISTIAN: Bonjour.
CAROLE: Bonjour.
CHRISTIAN: Tu es Carole, n'est-ce pas?
CAROLE: Oui, je suis Carole Winters. Et tu es l'ami français de Stéphanie, n'est-ce pas?
CHRISTIAN: Oui, je suis Christian.
CAROLE: Tu es de Paris, Christian?
CHRISTIAN: Non, je ne suis pas de Paris. Je suis de Nice, sur la Côte d'Azur.
CAROLE: La Côte d'Azur? Oh, là, là! C'est fantastique ça!

A **Christian est niçois.** Répondez d'après la conversation. (*Answer according to the conversation.*)

1. Qui est américain?
2. Qui est français?
3. Qui est l'amie de Christian?
4. Christian est de quelle ville?
5. Où est Nice?
6. Comment est la Côte d'Azur?

La Côte d'Azur

Prononciation   *L'accent tonique*

1. In English, you stress certain syllables more than others. In French you pronounce each syllable evenly. Compare the following.

fantastic	**fantastique**	popular	**populaire**
timid	**timide**	impatient	**impatient**

2. Repeat the following sentence. Notice how each word is linked to the next so that the sentence sounds like one long word.

   Élisabeth est l'amie de Nathalie.

DID YOU KNOW?

Because of its temperate climate and beautiful setting on the Mediterranean, the French Riviera is a popular vacation destination of the French and other Europeans particularly during the month of August. The English and Russians were the first to discover the French Riviera as a winter resort in the late 1800's. The main street of Nice is called *la Promenade des Anglais*.

INDEPENDENT PRACTICE

Assign any of the following:
1. Exercise and activities, pages 28–29
2. CD-ROM, Disc 1, pages 28–29

Activités de communication orale

A **Un élève français.** You've just met Laurent Dumas (your partner), an exchange student from Toulouse. You strike up a conversation with him.

1. Greet him and tell him who you are.
2. You think you know who he is, but ask him anyway.
3. Ask him where he's from.

B **Le café Rive Gauche.** You are at a café near Notre-Dame Cathedral in Paris. A student at the next table strikes up a conversation with you. Answer her questions.

1. Bonjour.
2. Tu es des États-Unis, n'est-ce pas?
3. Tu es d'où?

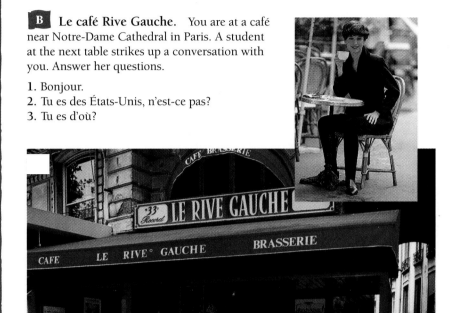

C **Il/Elle est intéressant(e) ou pas?** Describe someone to a classmate. He or she will say whether the person is interesting or not.

Prononciation

PRESENTATION (*page 28*)

You may wish to use the following resources to present this section:

1. Cassette 2A/CD-1, *Prononciation*
2. Student Tape Manual, Teacher's Edition, *Activité J*, page 14.
3. Pronunciation Transparency P-1.

Bell Ringer Review

Write the following on the board or use BRR Blackline Master 1-9: Sketch the following:

1. **Le garçon est de Paris.**
2. **La fille est grande.**
3. **Paul est l'ami de Jacques.**
4. **L'élève est dans le Lycée Jeanne d'Arc.**

Activités de communication orale

PRESENTATION (*page 29*)

Activité B

In the CD-ROM version of this activity, students can interact with an on-screen native speaker.

ANSWERS

Activité A

Answers will vary; however, students should use the appropriate forms of *être*.

Activités B and C

Answers will vary.

GEOGRAPHY CONNECTION

On the map of France, page 504 (Part A, page 237) point out to students the location of Paris and Nice, or have a student find them. Then indicate to students the area referred to as *la Côte d'Azur*. Ask them what *Côte d'Azur* means in English. Have them look at the photo of *la Côte d'Azur* on page 28. See if they can figure out why it is called this.

LEARNING FROM PHOTOS

Ask students if they know the meaning of the name of the café, *Le Rive Gauche*. Explain that the river Seine flows through Paris, dividing it into two parts: *la Rive Gauche* and *la Rive Droite* (Left and Right Banks). Explain that the Left Bank has always been a student quarter. Have the students turn to the map of Paris on page 505 (Part A, page 238) or use the Map Transparency. Have them find the Right and Left Banks as well as Île de la Cité and Notre-Dame Cathedral.

Note to teacher Although the word *rive* is feminine, the café is called *Le Rive Gauche* since the word *café* is understood.

LECTURE ET CULTURE

Bell Ringer Review

Write the following on the board or use BRR Blackline Master 1-10: Working in teams, list at least ten items you would expect to find in your French class. Which team can make the longest list?

READING STRATEGIES

(page 30)

Pre-reading

Using the Map Transparency or the map on page 504 (Part A, page 237), review the locations of Paris and *la Côte d'Azur.*

Reading

A. Lead students through the *Lecture* by reading it aloud. Have students repeat each sentence in the first paragraph after you.

B. After every two sentences, ask questions such as the following: *Qui est français? Jacques est de Nice? D'où est Jacques? Il est de la capitale? Quelle est la capitale de la France?*

C. Have students read the selection on page 30 aloud, or have them work with a partner.

Post-Reading

A. Have students complete the *Étude de mots* and the *Compréhension* exercises, pages 30–31, in writing.

B. Assign individual students to report briefly on the geography references in the reading: Paris; Èze; *La Côte d'Azur;* Nice.

Note Students may listen to a recorded version of the *Lecture* on the CD-ROM.

Étude de mots

ANSWERS

Exercice A

Answers will vary but should include three of the following: *capitale, intelligent, excellent, pittoresque, célèbre, petit, village.*

30

UN PARISIEN ET UNE PROVENÇALE

Jacques Poulain est français. Il est de Paris, la capitale de la France. Jacques est un garçon intelligent. Il est très sympathique aussi. Il est élève dans un lycée à Paris, le Lycée Henri IV. Le Lycée Henri IV est très célèbre. C'est un lycée excellent.

Chantal Lévêque est française aussi. Mais Chantal n'est pas de Paris. Elle est d'Èze, un petit village pittoresque sur la Côte d'Azur. Elle est élève dans un lycée à Nice. Chantal est une amie de Jacques. Maintenant[1] Jacques est en vacances à Èze.

[1] Maintenant *Now*

Èze Village

Étude de mots

A **Le français, c'est facile.** Trouvez au moins trois mots apparentés dans la lecture. *(Find at least three cognates in the reading.)*

B **C'est quel mot?** Trouvez la définition. *(Find the definition.)*

la capitale   la Côte d'Azur   un lycée   un village

1. une école secondaire française
2. une région de la France sur la mer Méditerranée
3. la ville principale d'une nation, où le gouvernement est situé
4. une très petite ville dans une zone rurale

Compréhension

C **C'est Jacques ou Chantal?** Décidez. *(Decide if it's Jacques or Chantal.)*

1. Elle/Il est de Paris.
2. Elle/Il est de la capitale.
3. Elle/Il est de la Côte d'Azur.
4. Elle/Il est d'un petit village pittoresque.
5. Elle/Il est d'Èze.

30　CHAPITRE 1

CRITICAL THINKING ACTIVITY

(Thinking skills: drawing conclusions; making inferences)

After reading the *Lecture*, write the following on the board or on an overhead transparency:

1. Jacques Poulain est de Paris.
   a. Il est américain.
   b. Il est français.

2. Èze est un village important?
   a. Oui, pour l'éducation.
   b. Oui, pour le tourisme. Pour les vacances.

3. Paris est une ville importante.
   a. Paris est un lycée célèbre.
   b. Paris est la capitale.

D À Paris et à Èze. *Répondez. (Answer.)*

1. Jacques est élève dans quel lycée?
2. Comment est le Lycée Henri IV?
3. Qui est d'Èze?
4. Où est Èze?
5. Èze est grand ou petit?
6. Qui est un ami de Chantal?
7. Où est Jacques maintenant?
8. Il est en vacances à Èze?

E Des faits. *Trouvez les renseignements suivants dans la lecture. (Find the following information in the reading.)*

1. the capital of France
2. a famous *lycée* in Paris
3. a small town on the French Riviera

F Un peu de géographie. *Trouvez les lieux suivants. (Locate the following places on the map of France on page 504.)*

1. Paris
2. la Seine
3. la Côte d'Azur
4. Nice
5. la mer Méditerranée

DÉCOUVERTE CULTURELLE

AUX ÉTATS-UNIS	EN FRANCE
L'éducation est obligatoire.	L'éducation est obligatoire.
l'école primaire ou «élémentaire»	l'école primaire
l'école «intermédiaire»	le collège
l'école secondaire	le lycée
l'université	l'université

Point essentiel! En France le collège et le lycée sont des écoles secondaires. Un collège en France n'est pas une université.

Compréhension
ANSWERS

Exercice C

1. Il...   4. Elle...
2. Il...   5. Elle...
3. Elle...

Exercice D

1. ... dans le Lycée Henri IV.
2. Il est vraiment célèbre.
3. Chantal est d'Èze.
4. Èze est sur la Côte d'Azur.
5. Èze est petit.
6. Jacques est un ami de Chantal.
7. Jacques est à Èze.
8. Oui, il est en vacances à Èze.

Exercice E

1. Paris
2. le Lycée Henri IV
3. Èze

Exercice F

Answers require students to indicate places on a map.

Bell Ringer Review

Write the following on the board or use BRR Blackline Master 1-11: Draw a simple map of France. Label the capital, the neighboring countries, and all cities mentioned in this chapter.

OPTIONAL MATERIAL

Découverte culturelle

PRESENTATION *(page 31)*

Read the information in the chart on page 31 aloud.

A. Have students compare the educational systems in the U.S. and France. Ask individual students to point out similarities and differences. For example: *L'éducation est obligatoire en France et aux États-Unis.*

B. Point out that *collège* is the equivalent of junior high or middle school in the U.S.

Note Students may listen to a recorded version of the *Découverte culturelle* on the CD-ROM.

DID YOU KNOW?

1. France has a national system of education. Everything related to education, i.e., schedules, courses, graduation requirements, etc., is determined by the Ministry of Education in Paris.
2. Children in France attend *l'école primaire* between the ages of five and eleven; then *le collège*, which is the equivalent of junior high, followed by *le lycée* (high school).
3. The 9th grade is called *troisième*; the 10th grade is *deuxième*; the 11th grade is *première*; and the 12th grade is *terminale*.
4. Students are graded on a scale of 1–20 with 20 being the highest. Grades of 20 are rarely given. An excellent grade is 16–18.

Bell Ringer Review

Write the following on the board or use BRR Blackline Master 1-12: Copy the following sentences, supplying the missing words.

1. ___ est le professeur.
2. ___ est-ce? C'est Marie!
3. Chantal Dubois est ___ .
4. Minnie Mouse est ___ .
5. Donald Duck est ___ .

PRESENTATION (*pages 32–33*)

Before doing the reading, ask students to name various countries where French is spoken. Write these on the board. Have students categorize the countries by their respective continents and locate them on the Map Transparency of the *Monde francophone* or on the map on page 506 (Part A, page 239). Now lead students through the captions on page 32.

Note In the CD-ROM version, students can listen to the recorded captions and discover a hidden video behind one of the photos.

RÉALITÉS

Voici Gilbert Bertrand **1**. Il est de Strasbourg, une ville française importante. Gilbert est élève dans un lycée. Gilbert est aimable ou désagréable?

Salut. Je suis Anne André **2**. Je suis de Schœlcher, un petit village martiniquais. La Martinique est une île dans la mer des Caraïbes. Je suis élève au Lycée Bellevue à Fort-de-France, la capitale. La Martinique est un département français d'outre-mer.

Voici Karim Ashour **3**. Il est de Tunis, la capitale de la Tunisie. La Tunisie est un pays en Afrique du Nord.

Bonjour. Moi, je suis Yvonne Senghor **4**. Je suis d'Abidjan. Abidjan est la capitale de la Côte-d'Ivoire. La Côte-d'Ivoire est un pays africain francophone.

Bonjour. Je suis Raymond LeClerc **5**. Je suis canadien. Je suis étudiant à l'Université Laval à Québec.

Tiare Teuira est de Bora Bora, une île volcanique en Polynésie française **6**. La Polynésie française est dans le Pacifique Sud. Tiare Teuira est sympathique?

32

INDEPENDENT PRACTICE

Assign any of the following:
1. *Étude de mots* and *Compréhension* exercises, pages 30–31
2. Workbook, *Un Peu Plus,* pages 6–8
3. Situation Cards, Chapter 1
4. CD-ROM, Disc 1, pages 30–33

PAIRED ACTIVITY

Have students work in pairs to create questions for each photo in *Réalités*. Then have several pairs of students read their questions aloud for other students to answer.

HISTORY CONNECTION

Ask students why French is spoken in so many different countries. (France colonized the New World during the 17th and 18th centuries, and much of Africa during the 19th century.)

THE FRANCOPHONE WORLD

For photos of the countries mentioned here and more information on other French-speaking countries, see *Le Monde francophone*, pages 110–113; pages 220–223; pages 326–329; and pages 434–437.

DID YOU KNOW?

France is divided into 95 administrative metropolitan *départements* and has nine overseas possessions, called either *départements d'outre-mer* (D.O.M.) or *territoires d'outre-mer* (T.O.M.). *La Martinique, la Guadeloupe, la Réunion* and *la Guyane* are *départements* and *la Nouvelle-Calédonie, la Polynésie française, Wallis et Futuna, Mayotte,* and *Saint-Pierre-et-Miquelon* are *territoires*. These possessions are considered part of France. Their citizens vote in French presidential elections and send representatives to both houses of the French parliament.

RECYCLING

The *Activités de communication orale* and *écrite* allow students to use the vocabulary and grammar from this chapter in open-ended, real-life situations. In subsequent chapters, they also give students the opportunity to recycle the vocabulary and structure from earlier chapters.

INFORMAL ASSESSMENT

The oral activities on page 34 may be used to assess students' speaking skills. You may wish to emphasize the message and downplay grammatical accuracy. The following are sugggestions to help you determine how to grade each student's response:

5(A):	Complete message conveyed, precise structural and vocabulary control.
4-3(B-C):	Complete message conveyed, some structural or vocabulary errors.
2-1(D):	Message partially conveyed, frequent errors.
0(F):	No message conveyed.

Activités de communication orale

PRESENTATION *(page 34)*

A. Have students do Activities A, B, and C with a partner. These activities are also appropriate as review before testing.

B. Call on pairs of students to present Activities A and C to the class.

ANSWERS

Activité A

Answers will vary but may include the following:

1. **Je suis américain(e).**
2. **Je suis élève dans une école secondaire.**
3. **Je suis de** (name of city).

Activités de communication orale

A Roissy-Charles-de-Gaulle. You are going through Immigration at the Roissy-Charles-de-Gaulle Airport on the outskirts of Paris. Give the immigration officer the following information.

1. your nationality
2. your occupation
3. where you are from in the U.S.

B Mireille Gaudin. Here's a photo of Mireille Gaudin. She's a French student from Cannes, which is near Nice. Describe Mireille and say as much about her as you can.

C Un(e) élève français(e). You've just met a French student (your partner) who's visiting the U.S. Ask him or her some questions using the following words.

d'où
grande ville ou petite ville
élève
lycée

Nice, Côte d'Azur

34   CHAPITRE 1

Activités de communication écrite

A **Qui est-ce?** Write down four things about yourself on a piece of paper. Your teacher will collect everyone's descriptions and have students read them to the class. You'll all try to guess who's being described.

> **Je suis blond. Je ne suis pas brun.**
> **Je suis très amusant et très populaire.**
>
> **Qui est-ce? C'est ___.**

B **Une lettre.** You have a new pen pal in France. She just sent you this photo and you want to answer her immediately—in French, of course! Write and tell her who you are, your nationality, where you're from, and where you're a student. Give her a brief description of yourself and be sure to include your photo.

Le - septembre, 199-

Chère Sophie,

Je suis...

Bien amicalement,

Vocabulaire

NOMS
le frère
la sœur
l'ami (m.)
l'amie (f.)
l'élève (m. et f.)
l'école (f.)
le lycée

ADJECTIFS
aimable
amusant(e)
comique
célèbre

confiant(e)
content(e)
désagréable
énergique
fantastique
patient(e)
impatient(e)
intelligent(e)
intéressant(e)
populaire
sincère
sympathique
antipathique
timide

grand(e)
petit(e)
brun(e)
blond(e)
français(e)
américain(e)

AUTRES MOTS ET EXPRESSIONS
aussi
moi
n'est-ce pas
ou
voici

Activité B
Answers will vary but may include the following: *Mireille Gaudin est française. Elle est de Cannes. Elle est élève. Elle est brune. Elle est intelligente et sympathique.*

Activité C
Answers will vary.

Activités de communication écrite
ANSWERS

Activités A and B
Answers will vary.

ASSESSMENT RESOURCES

1. Chapter Quizzes
2. Testing Program
3. Situation Cards
4. Communication Transparency C-1
5. Computer Software: Practice/Test Generator

VIDEO PROGRAM

INTRODUCTION À LA VIDÉO   (00:42)

INTRODUCTION   (06:36)

EMMANUEL   (07:11)

INTRODUCTION   (08:07)

OLIVIER   (08:23)

INTRODUCTION   (11:05)

SALIMA   (11:31)

FOR THE YOUNGER STUDENT

The following may be especially appropriate for beginning French students at the junior-high/middle school level. Have students draw a series of faces that indicate the meaning of the adjectives presented in this lesson. Have them label each drawing with the appropriate French word. Select the most attractive ones and put them on a bulletin board entitled *Caractéristiques*.

STUDENT PORTFOLIO

Have students keep a notebook with their best written work from the textbook and ancillaries. In the Workbook, students will develop an organized autobiography *(Mon Autobiographie)* which may also become part of their portfolio.

Note Students may create and save both oral and written work using the Electronic Portfolio feature on the CD-ROM.

CHAPTER OVERVIEW

In this chapter students will learn to describe people and things using the plural forms of articles, adjectives, and the verb *être*. (The singular forms were taught in Chapter 1.) Active vocabulary from Chapter 1 is recycled in this chapter as new descriptive adjectives and school-related terms are presented.

The cultural focus of Chapter 2 is on the Paris suburbs and two small French towns in different regions of France—Antibes and Giverny. The cultural reading talks about time zones and the French use of the 24-hour clock.

CHAPTER OBJECTIVES

By the end of this chapter, students will know:

1. plural forms of *être*
2. the difference between *vous* and *tu*
3. agreement of adjectives in the plural
4. the names of academic subjects common in most schools
5. the days of the week
6. how to ask yes/no questions using intonation or *est-ce que*
7. how to tell time

Pacing

This chapter will require eight to ten class sessions. Pacing will depend on the length of the class, the age of the students, and student aptitude.

For more information on planning your class, see the Lesson Plans, which offer guidelines for 45- and 55-minute classes and **Block Scheduling**.

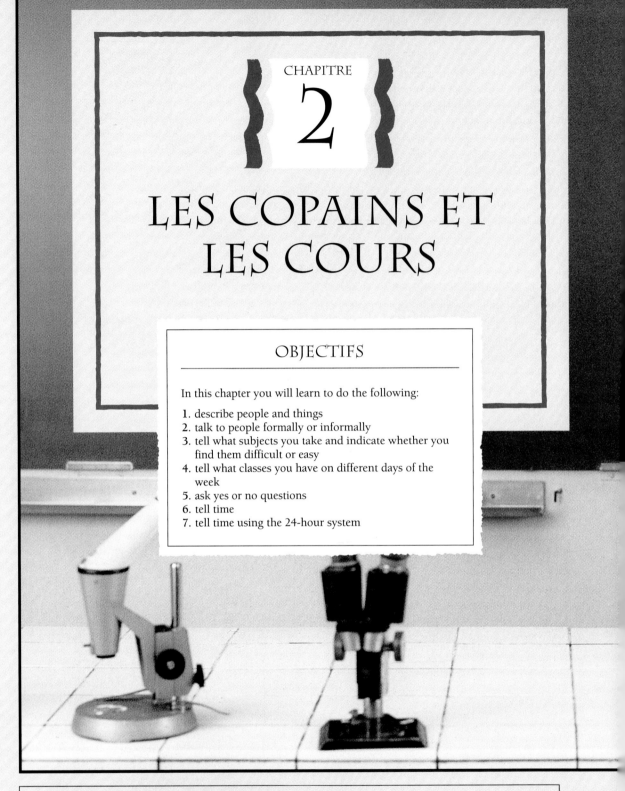

CHAPITRE

2

LES COPAINS ET LES COURS

OBJECTIFS

In this chapter you will learn to do the following:

1. describe people and things
2. talk to people formally or informally
3. tell what subjects you take and indicate whether you find them difficult or easy
4. tell what classes you have on different days of the week
5. ask yes or no questions
6. tell time
7. tell time using the 24-hour system

COMMUNITIES

Have students begin a correspondence with a French pen pal. (For information on obtaining pen pals, see the Resource List in the Teacher's Manual.) By the end of this chapter they will be able to say something about themselves and about the courses they are taking in school.

Note If your school has an Internet connection, please see the *Bienvenue* Internet Activities Booklet, Chapter 2, or the Glencoe Foreign Language Home Page (address, page 37) for information on electronic keypals.
.

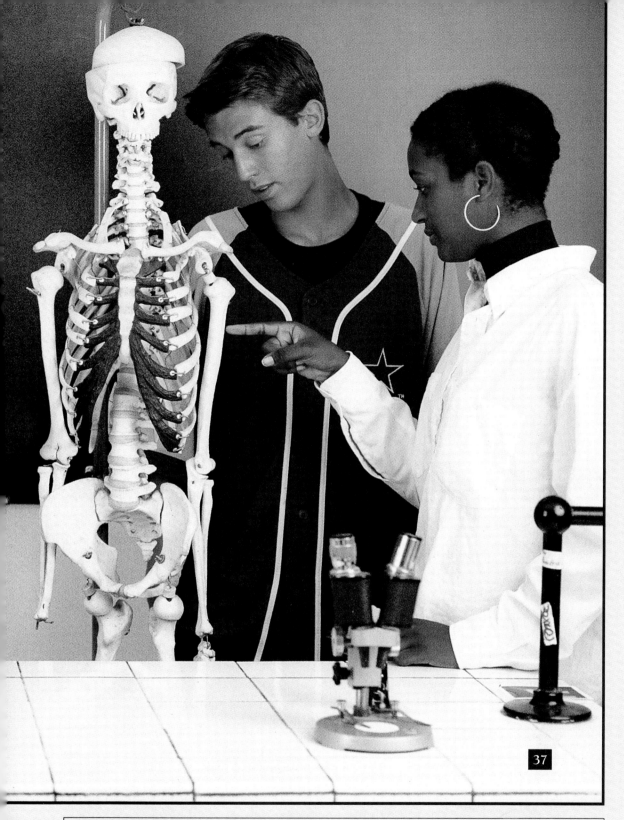

NOTE ON COGNATES

The large number of cognates in this chapter will help students learn the new words quickly, but, because they are so similar to English, students will try to anglicize their pronunciation. Model the pronunciation of cognates just as carefully as you do that of other new vocabulary. Have students repeat the words several times after you or the cassette/CD recording.

INTERNET ACTIVITIES

(optional)

These activities, student worksheets, and related teacher information are in the *Bienvenue* Internet Activities Booklet and on the Glencoe Foreign Language Home Page at **http://www.glencoe.com/secondary/fl**

LEARNING FROM PHOTOS

1. After presenting *Mots 1* and *2*, ask students the following questions about the photo: *C'est quel cours? À ton avis, le cours de biologie (d'anatomie) est intéressant? Le professeur est un homme ou une femme? Elle est sympathique ou antipathique? Comment est l'élève?*
2. You may wish to give students the words: *le squelette, les os,* and *le microscope.*

Exercices vs. *Activités*

All exercises (which provide guided practice) are coded in blue. All communicative activities are coded in red.

VOCABULAIRE

MOTS 1

les professeurs (les profs)

un homme

une femme

les élèves

brunes

françaises

Anne   Lise

Guy   Alain

bruns

français

les amies = les copines

les amis = les copains

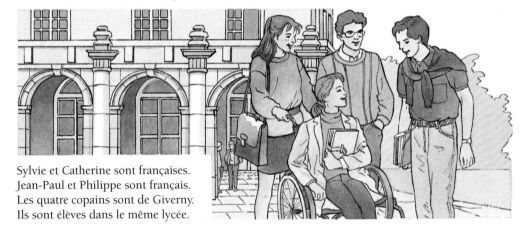

Sylvie et Catherine sont françaises.
Jean-Paul et Philippe sont français.
Les quatre copains sont de Giverny.
Ils sont élèves dans le même lycée.

Vocabulary Teaching Resources

1. Vocabulary Transparencies 2.1 (A & B)
2. Audio Cassette 2B/CD-2
3. Student Tape Manual, Teacher's Edition, *Mots 1: A–C*, pages 17–19
4. Workbook, *Mots 1: A–C*, page 9
5. Communication Activities Masters, *Mots 1: A*, page 7
6. Chapter Quizzes, *Mots 1: Quiz 1*, page 7
7. CD-ROM, Disc 1, *Mots 1:* pages 38–40

Bell Ringer Review

Write the names Christine and Paul on the board or use BRR Blackline Master 2-1: Write two sentences about Christine and two sentences about Paul using the following words: élève, ami, lycée, français.

PRESENTATION (pages 38–39)

A. Have students close their books. Using Vocabulary Transparencies 2.1 (A & B), have students repeat the words and sentences in *Mots 1* after you, or play the recording on Cassette 2B/CD-2.

B. Prepare large cards on which you have written plural adjectives and nouns from *Mots 1*. Have pairs of students hold up a card together. Now ask yes/no and either/or questions referring to the pairs of students: *Guy et Alain sont élèves? Anne et Lise sont françaises ou américaines?*

°TOTAL PHYSICAL RESPONSE

(following the Vocabulary presentation)

Getting Ready

Before doing this activity, make sure students understand each of the following words by acting them out: *marchez, lentement,* and *vite.*

TPR

(To one or more students)

Levez-vous.
Promenez-vous dans la salle de classe.
Marchez vite. Allez.
Et maintenant, marchez lentement.
Arrêtez-vous.
Montrez-moi un garçon.
Montrez-moi deux garçons.
Montrez-moi deux garçons bruns.
Montrez-moi une fille.
Et maintenant, montrez-moi deux filles.

(continued on next page)

la salle de classe

le cours

la classe

Bonjour! Nous sommes élèves dans la classe de Monsieur Bétancourt.
M. Bétancourt est le prof de français.
Maintenant, nous sommes dans la salle de classe 21.

facile

difficile

Le cours de français est très facile.
Mais le cours d'anglais est vraiment difficile.
Tu es d'accord ou pas?

CHAPITRE 2 **39**

C. Intersperse the following questions as you lead students through the *Mots 1* presentation: *Jean-Paul et Philippe sont français? Ils sont élèves? Ils sont amis? Sylvie et Catherine sont françaises? Elles sont élèves? Elles sont amies? Les quatre amis sont élèves dans un lycée? Ils sont élèves dans le même lycée?*

D. Using Vocabulary Transparency 2.1 (B), have two students read the text in their book under the top illustration on page 39. This helps to convey the meaning of *nous*.

E. Check students' comprehension of the *Mots 1* vocabulary by asking the following questions: *Qui est français(e)? Qui est brun(e)? Qui sont copains (copines)? Qui est de Giverny? Qui est le professeur? Le cours de français est dans quelle salle de classe? Comment est le cours de français? Comment est le cours d'anglais?*

COGNATE RECOGNITION

Ask volunteers to identify words in *Mots 1* that are cognates: *le professeur, le cours, la classe, difficile, facile.*

Teaching Tip Use a French beret or a small American flag to signal the difference in pronunciation between French and English words.

Vocabulary Expansion

You may give students the following classroom words. It is recommended, however, that you limit the amount of additional vocabulary.

la gomme	**la brosse** (*black-*
le pupitre	*board eraser*)
la porte	**la fenêtre**
le rang	**le drapeau**
le mur	

TPR (*continued*)
Promenez-vous dans la salle de classe et montrez-moi tous les garçons blonds.
Et maintenant, continuez. Promenez-vous dans la salle de classe et montrez-moi toutes les filles blondes.
Merci, ___ , retournez à votre place et asseyez-vous.

LEARNING FROM ILLUSTRATIONS

Ask students if they know who Victor Hugo was (a 19th-century French writer) and if they can name another novel he wrote (*The Hunchback of Notre-Dame*). You may want to ask them if they know the names of any other French writers, novels, or plays.

Exercices

PRESENTATION (*page 40*)

Exercices A and B

Have students close their books and do Exercises A and B orally. Then read the exercises with them for additional reinforcement.

Extension of *Exercice A*

Call on one student to answer all the questions. Then have another student retell the story of the exercise in his or her own words.

Extension of *Exercice B*

Read the answers to the class. Then have the class ask you questions about the story without looking at it.

Extension of *Exercice C*

After doing Exercise C orally, you may wish to have students close their books and write a short paragraph about their French class.

ANSWERS

Exercice A

1. Sylvie et Catherine sont les deux amies.
2. Elles sont françaises.
3. Non, elles ne sont pas de Paris. (Non, elles sont de Giverny.)
4. Elles sont élèves dans un lycée.
5. Oui, elles sont élèves dans le même lycée à Giverny.

Exercice B

1. Oui, Jean-Paul et Philippe sont copains.
2. Oui, les deux copains sont contents.
3. Oui, Jean-Paul et Philippe sont lycéens.
4. Oui, Jean-Paul et Philippe sont élèves dans le même lycée.
5. Non, le lycée n'est pas à Paris.
6. Oui, le lycée est à Giverny.

Exercice C

Answers will vary.

Exercice D

1. c	3. b	5. d
2. a	4. f	6. e

Exercices

A **Sylvie et Catherine.** Répondez. (*Answer.*)

1. Qui sont les deux amies?
2. Elles sont françaises ou américaines?
3. Elles sont de Paris?
4. Elles sont élèves dans un lycée ou étudiantes à l'université?
5. Elles sont élèves dans le même lycée à Giverny?

B **Jean-Paul et Philippe.** Répondez. (*Answer.*)

1. Jean-Paul et Philippe sont copains?
2. Les deux copains sont contents?
3. Jean-Paul et Philippe sont lycéens (élèves dans un lycée)?
4. Ils sont élèves dans le même lycée?
5. Le lycée est à Paris?
6. Le lycée est à Giverny?

C **Le cours de français.** Donnez des réponses personnelles. (*Give your own answers.*)

1. Qui est le prof ou la prof de français?
2. Le professeur est un homme ou une femme?
3. Il/Elle est sympa?
4. Le cours de français est difficile ou facile?
5. Les élèves sont en classe maintenant?
6. Le cours de français est intéressant?

D **Des mots.** Trouvez les mots qui correspondent. (*Find the corresponding word or phrase.*)

1. français	a. la copine
2. l'amie	b. l'étudiant, l'écolier
3. l'élève	c. de France
4. brun	d. le copain
5. l'ami	e. le professeur
6. le prof	f. le contraire de blond

Le professeur explique comment utiliser l'ordinateur.

ADDITIONAL PRACTICE

1. After completing Exercises A–D, have students work in pairs. One makes incorrect statements about the French class and the teacher. The other corrects them. For example, É1: *La classe de français est grande.* É2: *La classe de français n'est pas grande. Au contraire, elle est petite.*
2. Student Tape Manual, Teacher's Edition, *Activités B–C,* page 18

INDEPENDENT PRACTICE

Assign any of the following:

1. Exercises, page 40
2. Workbook, *Mots 1: A–C*, page 9
3. Communication Activities Masters, *Mots 1: A*, page 7
4. CD-ROM, Disc 1, pages 38–40
(See suggestions for homework correction in the Teacher's Manual.)

VOCABULAIRE

MOTS 2

LES MATIÈRES (f.)

Les sciences (f.)

Les maths (f.)

la géométrie

la chimie

l'algèbre (f.)

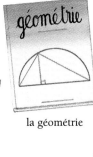

la trigonométrie

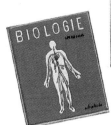
la biologie

la physique

Les langues (f.)

D'autres cours (m.)

l'art (m.)

l'anglais (m.)

l'espagnol (m.)

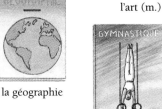
la géographie

la gymnastique

la musique

le latin

le français

la littérature

l'histoire (f.)

l'informatique (f.)

TOTAL PHYSICAL RESPONSE

(following the Vocabulary presentation)

Getting Ready
   Dramatize *Levez la main.*

TPR 1
Les élèves d'espagnol, levez la main.
Les élèves de maths, levez la main.
Les élèves de biologie, levez la main.
Les élèves de latin, levez la main.

Les élèves d'histoire, levez la main.
Merci, tout le monde.

TPR 2
Tout le monde, levez-vous.
Si vous êtes élèves d'italien, asseyez-vous.
Si vous êtes élèves d'art, asseyez-vous.
Si vous êtes élèves de musique, asseyez-vous.
Si vous êtes élèves de français, asseyez-vous.
C'est tout le monde, n'est-ce-pas?

Vocabulary Teaching Resources

1. Vocabulary Transparencies 2.2 (A & B)
2. Audio Cassette 2B/CD-2
3. Student Tape Manual, Teacher's Edition, *Mots 2: D–G*, pages 19–21
4. Workbook, *Mots 2: D–F*, page 10
5. Communication Activities Masters, *Mots 2: B*, page 8
6. Chapter Quizzes, *Mots 2: Quiz 2*, page 8
7. Computer Software, *Vocabulaire*
8. CD-ROM, Disc 1, *Mots 2*: pages 41–44

Bell Ringer Review

Write the following on the board, or use BRR Blackline Master 2-2: Complete with *C'est un* or *C'est une.*
____ cahier.
____ bureau.
____ salle de classe.
____ feuille de papier.
____ livre.
____ autre salle de classe.

PRESENTATION *(pages 41–42)*

A. Most of the words on page 41 are cognates. Because students will tend to anglicize their pronunciation, have them repeat each word twice after you, or use Cassette 2B/CD-2 as the model.

B. Using pictures or props (an equation for algebra, a triangle for geometry, paint brushes for art, etc.), ask students either/or questions. For example: *C'est un cours d'art ou un cours de physique? C'est un cours de géométrie ou un cours d'algèbre?*

PRESENTATION (*page 42*)

Have students look at the datebook. Read the text below it, pointing to the appropriate days on the datebook as you read. Practice the days by saying: *Aujourd'hui, c'est _____ . Et demain?*

Writing Tip Alert students to the fact that in French, days of the week are not capitalized.

CROSS-CULTURAL COMPARISON

1. Using the schedule on page 43, point out that in France classes do not meet every day and that they meet at different times on different days.
2. Point out that on a French calendar the week begins with Monday, not Sunday. Tell students that in France people say *huit jours*, not *sept jours*, when they mean a week. (*Lundi* to *lundi* = *huit jours*.)

Vocabulary Expansion

You may want to give students the following additional vocabulary to allow them to talk about other school subjects.

> le calcul
> l'allemand
> l'éducation physique
> l'économie domestique
> les arts manuels
> le dessin
> le cours de conduite
> la science économique
> le cours de dactylo

UN AGENDA

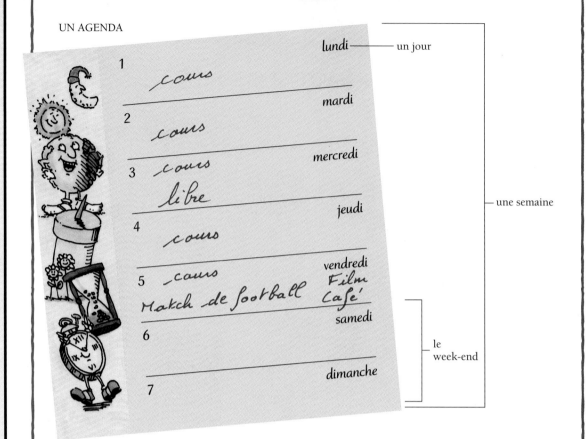

C'est quel jour, aujourd'hui? C'est lundi.
Et demain? Demain, c'est mardi.

Voilà l'agenda de Jean-Paul.
Il est très occupé vendredi, n'est-ce pas?
Mais samedi et dimanche, il n'est pas occupé. Il est libre.

Note: *Vendredi* means "on Friday." *Le vendredi* means "every Friday" or "on Fridays."

On the right are some informal words in French which you may use to describe people and things.

Note that *terrible* can have either a positive or a negative meaning, depending on the tone of voice or intonation.

POSITIF	NÉGATIF
chouette	moche
super-chouette	
terrible	terrible
extra	
super	

42   CHAPITRE 2

COOPERATIVE LEARNING

Form teams of four. Have each student write the name of a teacher on a piece of paper. (They should check to make sure they have each named different teachers.) Students then pass their paper to the teammate to their left.

Pass 1: Add the classroom number of the teacher named on the paper.

Pass 2: Add the subject he/she teaches.
Pass 3: Add a physical characteristic.
Pass 4: Add a description of the course.

After the last pass, students take turns reading the composite descriptions to their teammates or to the whole class.

Exercices

A **C'est quel cours?** Identifiez le cours. *(Identify the course.)*

Un problème, une solution, une équation—c'est quel cours?
C'est le cours d'algèbre.

1. la littérature, la composition, la grammaire
2. la conversation, la culture française
3. un poème, une pièce de théâtre, une fable
4. un microbe, un animal, une plante, un microscope
5. un cercle, un rectangle, un triangle, un parallèlogramme
6. un piano, un violon, un concert, un opéra
7. les montagnes, les villes, les villages, les capitales, les océans, les produits agricoles
8. le gouvernement, les partis politiques, l'État, la communauté
9. la peinture, la statue, la sculpture, les artistes célèbres
10. une disquette, un moniteur, un bit, un microprocesseur

B **C'est quel jour?** Répondez. *(Answer.)*

Aujourd'hui, c'est lundi. Et demain?
Demain, c'est mardi.

1. Aujourd'hui, c'est mercredi. Et demain?
2. Aujourd'hui, c'est vendredi. Et demain?
3. Aujourd'hui, c'est samedi. Et demain?
4. Aujourd'hui, c'est mardi. Et demain?
5. Aujourd'hui, c'est dimanche. Et demain?
6. Aujourd'hui, c'est jeudi. Et demain?
7. Aujourd'hui, c'est lundi. Et demain?

C **L'emploi du temps de David.** Répondez d'après l'emploi du temps de David. *(Answer according to David's schedule.)*

le cours de maths
Le cours de maths est le lundi, le mercredi et le vendredi.

1. le cours d'anglais
2. le cours de physique
3. le cours de latin
4. le cours de musique
5. le cours de français
6. le cours d'éducation civique

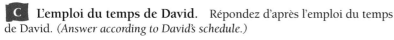

	Lundi	Mardi	Mercredi	Jeudi	Vendredi
8h–8h30		Sc. Nat	MATH	PHYSIQUE	MATH
8h30–9h30	HIST/GÉO				
9h30–10h30	MATH	ALLEMAND	ED. CIVIQUE	ANGLAIS	HIST/GÉO
10h30–11h30	INFOR-MATIQUE	ANGLAIS	1er SEMESTRE / 2e SEMESTRE	MUSIQUE	E.P.S.
11h30–12h30		LATIN	E.P.S. / DESSIN	ALLEMAND	
12h30–13h					
13h30–14h	FRANÇAIS	HIST/GÉO		FRANÇAIS	LATIN
14h–15h				LATIN	
15h–16h	ALLEMAND	FRANÇAIS		FRANÇAIS	
16h–17h	ANGLAIS				
17h–18h					

CRITICAL THINKING

Ask the following questions about David's schedule. (You may have to write some of the cognates on the board to get the meaning across.) What can students conclude about the French school system? *David a combien de cours? Les cours durent combien de minutes? Ils commencent à quelle heure? Ils terminent à quelle heure? Les élèves sont à l'école pendant combien d'heures par jour?*

INDEPENDENT PRACTICE

Assign any of the following:
1. Exercises, page 43
2. Workbook, *Mots 2: D–F*, page 10
3. Communication Activities Masters, *Mots 2: B*, page 8
4. Computer Software, *Vocabulaire*
5. CD-ROM, Disc 1, pages 41–43

Exercices

PRESENTATION *(page 43)*

Exercice A: Cognate Recognition

The vocabulary here is not to be learned. The aim is to show students how many new words they can recognize. You may want to have them say the words once.

Exercice B

Cue the answers by referring to the days of the week on page 42, using Vocabulary Transparency 2.2 (B), or playing Cassette 2B/CD-2.

ANSWERS

Exercice A

1. C'est le cours d'anglais. (... de langues.)
2. ... de français.
3. ... de littérature.
4. ... de biologie.
5. ... de géométrie.
6. ... de musique.
7. ... de géographie.
8. ... d'éducation civique. (... d'histoire.)
9. ... d'art.
10. ... d'informatique.

Exercice B

1. Demain, c'est jeudi.
2. ... samedi.
3. ... dimanche.
4. ... mercredi.
5. ... lundi.
6. ... vendredi.
7. ... mardi.

Exercice C

1. Le cours d'anglais est le lundi, le mardi et le jeudi.
2. Le cours de physique est le jeudi.
3. Le cours de latin est le mardi, le jeudi et le vendredi.
4. Le cours de musique est le jeudi.
5. Le cours de français est le lundi, le mardi et le jeudi.
6. Le cours d'éducation civique est le mercredi.

RETEACHING *(Mots 2)*

Have students list their courses and indicate if they consider each one *facile* or *difficile*.

RECYCLING

This material recycles the singular of nouns, adjectives, and the verb *être* from Chapter 1.

PRESENTATION *(page 44)*

You may have students do any number of the activities you wish.

Extension of *Activité A*

Have students do the same activity basing their statements on their French class.

Extension of *Activité B*

Have each student write down the description given by his or her partner and read it to the class as a follow-up activity.

ANSWERS

Activité A

Answers will vary. However, true statements may include the following: *Aujourd'hui c'est le jeudi 15 octobre. Les élèves sont dans la salle de classe 15, etc.*

Activités B and C

Answers will vary.

INFORMAL ASSESSMENT

(Mots 1 and 2)

Ask individual students the following questions: *Qui est le prof de français? Qui sont deux élèves dans le cours de français? Comment est le cours de français, facile ou difficile? Le cours de français, c'est le lundi etc.?* Repeat with other courses.

44

Activités de communication orale
Mots 1 et 2

A **La classe de Mme Martin.** Make up a few true or false statements about the illustration. Your partner will either agree with your statement or correct it.

> Élève 1: Les élèves sont dans la classe de Monsieur Laurent.
> Élève 2: Non, ils sont dans la classe de Madame Martin.
> (Oui, je suis d'accord.)

B **Mon prof favori.** Describe your favorite teacher to a classmate.

> M. Jones est le prof de biologie. Il est…

C **À ton avis.** Make a chart like the one below. List all your classes and rate them—*pas difficile, assez difficile, très difficile*. Compare your chart with a classmate's and see whether the two of you agree or not.

> Élève 1: Pour moi, le français est facile. Tu es d'accord?
> Élève 2: Oui, je suis d'accord. (Non, je ne suis pas d'accord. Pour moi, le français est très difficile.)

COURS	Pas difficile	Assez difficile	Très difficile
l'anglais	X	X	X
le français	X	X	

COOPERATIVE LEARNING

After completing Activities A–C, reinforce the lesson with the following:
Have students rate their classes using additional criteria: *très difficile, très facile, amusant, intéressant.* Have them compare the results.

PAIRED ACTIVITY

Have students work in pairs and make out a weekly *agenda* of their own activities similar to the one on page 42. The students find out what days their partner is free and what days their partner is busy.

Le pluriel: Articles et noms

Talking About More than One Person or Thing

1. Plural means more than one. To make most nouns plural in French, you add *s*, as you do in English. You do not pronounce this final *s*. If the noun ends in *s* in the singular, you do not add another *s* in the plural.

2. The plural form of the definite articles *le, la, l'* is *les*. You do not pronounce the *s* of *les* when it is followed by a consonant. When *les* is followed by a vowel or silent *h*, you pronounce the *s* like a *z*, connecting the sound to the next word. This is called "liaison."

SINGULIER	PLURIEL
le garçon	les garçons
le cours	les cours
la fille	les filles
la classe	les classes
l'amie	les amies
l'élève	les élèves

Exercice

A **Tous les deux.** Mettez au pluriel. *(Give the plural.)*

> **Le garçon est blond.**
> **Les garçons sont blonds.**

1. La fille est blonde.
   ___ sont blondes.

2. Le garçon est brun.
   ___ sont bruns.

3. Le professeur est intelligent.
   ___ sont intelligents.

4. Le cours est difficile.
   ___ sont difficiles.

5. Le livre est intéressant.
   ___ sont intéressants.

6. La classe de M. Dupont est petite.
   ___ de M. Dupont sont petites.

7. L'ami de Paul est sympathique?
   ___ de Paul sont sympathiques?

8. La copine de Marie est très amusante.
   ___ de Marie sont très amusantes.

9. L'élève de M. Bétancourt est vraiment intelligent.
   ___ de M. Bétancourt sont vraiment intelligents.

10. L' amie de Sophie est très populaire.
    ___ de Sophie sont très populaires.

CHAPITRE 2   45

ADDITIONAL PRACTICE

Before class, prepare two sets of index cards: one set with singular and plural forms of nouns students already know, the other set with the definite articles. (Duplicate sets will be needed.) Students work in pairs: Student 1 receives noun cards, and Student 2 the articles. Partners match each noun with the correct article. They then read their article/noun combinations to the class.

INDEPENDENT PRACTICE

Assign any of the following:
1. Exercise A, page 45
2. Workbook, *Structure: A*, page 11
3. Communication Activities Masters, *Structure: A*, pages 9–10
4. CD-ROM, Disc 1, page 45

Structure Teaching Resources

1. Workbook, *Structure: A–H*, pages 11–15
2. Student Tape Manual, Teacher's Edition, *Structure: A–H*, pages 21–24
3. Audio Cassette 2B/CD-2
4. Grammar Transparencies G-2 (A–C)
5. Communication Activities Masters, *Structure: A–E*, pages 9–13
6. Computer Software, *Structure*
7. Chapter Quizzes, *Structure*: Quizzes 3–7, pages 9–13
8. CD-ROM, Disc 1, pages 45–53

Le pluriel: Articles et noms
RECYCLING

This material reintroduces the singular of nouns, adjectives, and the verb *être*, as well as vocabulary from Chapter 1.

PRESENTATION *(page 45)*

To present the grammar point deductively, have students close their books. Write the singular nouns from the chart on page 45 on the board. Ask students for the plural forms. (They know plural forms from *Mots 1*, pages 38–39.) Then have students open their books and read paragraph 2.

Note In the CD-ROM version, this structure point is presented via an interactive electronic comic strip.

ANSWERS

Exercice A

1. Les filles	6. Les classes
2. Les garçons	7. Les amis
3. Les professeurs	8. Les copines
4. Les cours	9. Les élèves
5. Les livres	10. Les amies

Le verbe être *au pluriel*

PRESENTATION (*pages 46–47*)

A. Have students keep their books closed as you write *je, tu, il/elle* on the board with the appropriate forms of *être*. Use the standard conjugation format shown in the chart on page 46. Remind students that they learned the singular forms of the verb *être* in Chapter 1, page 25.

B. Now write in *ils/elles* and ask students if they remember the corresponding verb form (from *Mots 1*, page 38). Do the same with *nous*, and then introduce *vous êtes*, the only form they have not yet seen. (The difference between *tu* and *vous* for informal and formal terms of address will be explained in the next structure section, page 49. For now, use *vous* in its plural meaning only.)

C. Using groups of students in front of the class, convey the meaning of *nous* (include yourself in the meaning), then *vous*, *ils*, and *elles*. If there is enough space, have the groups stand in front of the chalkboard, and write the appropriate subject pronoun above each group in large letters.

D. Have students open their books and lead them through each of the six steps on pages 46–47.

Le verbe *être* au pluriel

Talking About More than One Person or Thing; Asking Yes-or-No Questions

You have already learned the singular forms of the verb *être*, "to be." Now study the plural forms of *être*.

SINGULIER	PLURIEL
je suis	nous sommes
tu es	vous êtes
il/elle est	ils/elles sont

1. You use *nous* when referring to yourself and other people.

2. You use *vous* when talking to two or more people.

3. You use *elles* when referring to two or more females.

---

4. You use *ils* when referring to two or more males or when referring to a group of males and females.

> Ils sont assez intéressants.

> Ils sont de Nice.

5. You also use *ils* and *elles* when referring to things.

Les stylos sont sur la table.	**Ils sont sur la table.**
Les chaises sont dans la salle 21.	**Elles sont dans la salle 21.**

6. Note that in order to form a yes-or-no question, you can raise the tone of your voice at the end of the statement or put *est-ce que* in front of the statement. *Est-ce que* becomes *est-ce qu'* in front of a vowel.

Vous êtes français?	**Est-ce que vous êtes français?**
Il est américain?	**Est-ce qu'il est américain?**

Exercices

A **Le cours d'histoire.** Répondez d'après le modèle en utilisant «il(s)» ou «elle(s)». (*Answer with* il[s] *or* elle[s] *according to the model.*)

> **Est-ce que les garçons sont derrière les filles?**
> *Oui, ils sont derrière les filles.*

1. Est-ce que la prof est devant la classe?
2. Est-ce que Paul est devant Monique?
3. Les élèves sont intelligents?
4. Les filles sont sympathiques?
5. Est-ce que Paul et Pierre sont copains?
6. Est-ce que Monique et Paul sont amis?
7. Monique et Marie sont brunes?
8. Est-ce que les quatre copains sont dans le même cours?

Exercices
PRESENTATION (*page 47*)

Exercice A
The purpose of Exercise A is to have students respond to a noun subject with a subject pronoun. Do the exercise with books open.

ANSWERS

Exercice A
1. Oui, elle est devant la classe.
2. Oui, il est...
3. Oui, ils sont...
4. Oui, elles sont...
5. Oui, ils sont...
6. Oui, ils sont...
7. Oui, elles sont...
8. Oui, ils sont...

ADDITIONAL PRACTICE

Have individuals and various groups made up of two, three, or four students stand in different locations in the room. Some groups should be all one sex, some mixed. Choose one base sentence, such as *Il est amusant.* Have different students apply this sentence to various groups or individuals which you point to, changing subject pronoun, verb, and adjective as necessary. For example: *Elles sont amusantes. Vous êtes amusantes. Je suis amusant. Nous sommes amusants.* The student speaking should point to the person or persons he or she is referring to.

Exercice B

In the mini-conversation, students hear a natural exchange with *vous* and *nous*. The exercise items practice *sont*. Do the exercise with books open. You may play the recording of this conversation on Cassette 2B/CD-2.

Extension of *Exercice B*

Students do the conversation again, using information about themselves.

Exercice C

Here students practice responding with *nous sommes* when they hear *vous êtes*. Practice is important because students frequently hear *vous êtes* and want to respond with the same form.

Exercices D and E

These are more challenging. Students use all forms of *être*. Exercise D can be done first orally.

Extension of *Exercice E*

Students work in groups to answer the questions and compose a dialogue. One student then retells the story of the dialogue to the class.

ANSWERS

Exercice B

1. sont	5. sont
2. sont	6. sont
3. sont	7. sont
4. est	8. est

Exercice C

Answers will vary but should begin with *Nous sommes...*

Exercice D

1. suis	8. sommes
2. est	9. est
3. sommes	10. est
4. sommes	11. est
5. est	12. est
6. sommes	13. sont
7. est	

Exercice E

1. êtes	4. est
2. êtes	5. sont
3. êtes	6. est

B **Vous êtes d'où?** Répétez la conversation. *(Practice the conversation.)*

LES FILLES: Vous êtes d'où?
LES GARÇONS: Nous? Nous sommes de New York.
LES FILLES: Ah, alors vous êtes américains?
LES GARÇONS: Oui, nous sommes américains. Et vous?
LES FILLES: Nous sommes françaises. Nous sommes de Grenoble.

Complétez d'après la conversation. *(Complete according to the conversation.)*

1. Les deux garçons ___ américains.
2. Ils ne ___ pas de Chicago.
3. Ils ___ de New York.
4. New York ___ une très grande ville américaine.
5. Les deux filles ne ___ pas américaines.
6. Elles ___ françaises.
7. Elles ___ de Grenoble.
8. Grenoble ___ une grande ville française.

C **À votre tour.** Répondez en utilisant «nous». *(Answer with nous.)*

1. Vous êtes américains?
2. Vous êtes de quelle ville?
3. Vous êtes élèves?
4. Vous êtes élèves dans une école secondaire?
5. Vous êtes très intelligents?
6. Vous êtes maintenant dans la classe de quel professeur?

D **L'ami de Christophe.** Complétez avec «être». *(Complete with être.)*

Je ___$_1$ un ami de Christophe. Christophe ___$_2$ très sympa et très amusant. Nous ___$_3$ français, Christophe et moi. Nous ___$_4$ de Cancale, un petit village breton (en Bretagne). Cancale ___$_5$ vraiment très pittoresque.

Nous ___$_6$ élèves dans un lycée. Où ___$_7$ le lycée? À Dinard. Nous ___$_8$ élèves d'anglais. Mademoiselle Fielding ___$_9$ la prof d'anglais. Elle ___$_{10}$ anglaise. Elle ___$_{11}$ de Liverpool. Le cours d'anglais ___$_{12}$ assez difficile. Mais les élèves dans la classe de Mademoiselle Fielding ___$_{13}$ très intelligents.

E **Et vous?** Complétez avec «être». *(Complete with être.)*

1. Et vous? Vous ___ américains, n'est-ce pas?
2. Vous ___ élèves dans une école secondaire?
3. Vous ___ maintenant dans quel cours?
4. Qui ___ le professeur?
5. Les élèves ___ intelligents?
6. Pour vous, le cours ___ facile ou difficile?

Cancale, en Bretagne

48   CHAPITRE 2

INDEPENDENT PRACTICE	LEARNING FROM PHOTOS

Assign any of the following:

1. Exercises, pages 47–48
2. Workbook, *Structure: B–C*, pages 11–12
3. Communication Activities Masters, *Structure: B*, page 11
4. Have students write a brief description of their classmates or friends using as many different subject pronouns as possible.
5. CD-ROM, Disc 1, pages 47–48

Ask students to describe the boy in the photo on page 48.

Vous et tu

Talking to People Formally or Informally

As you already know, in French there are two ways to say "you:" *tu* and *vous*.

1. You use *tu* when talking to one friend, one person your own age, or to a family member.

> Paul, tu es libre maintenant?

> Maman, tu es contente?

2. You use *vous* when talking to two or more people.

> Paul et Marie, vous êtes occupés maintenant?

3. You also use *vous* when talking to an older person, a person whom you do not know well, or to anyone to whom you wish to show respect.

> Monsieur Moreau, vous êtes occupé?

3y - 2y = 1
3x = 10

Vous *et* tu

PRESENTATION (*pages 49–50*)

A. Have students open their books to page 49. Explain how the two forms of "you" are used, leading students through the examples on page 49. Explain that *tu* is also used when talking to a pet.

B. You may wish to present *vous* and *tu* using puppets or stuffed animals. Show a puppet of a child when using *tu* and a puppet of an adult when using *vous*. Show both puppets when teaching the plural usage of *vous*. Hold up the puppets as students ask *tu* and *vous* questions.

C. Use magazine pictures of pets, young children, and adults labeled with names and ask students to respond in unison with either *vous* or *tu*.

D. Give the pictures to various students. Each student will assume the role of the person whose picture he/she is holding. Now ask the person questions, for example: *Mme Béjart, vous êtes intelligente? Brigitte, tu es libre vendredi?*

E. Now give one-word cues to other students and have them ask the same kind of questions of the students holding the pictures, who then respond. For example: **Prof** (designating student holding picture): *Intelligente.* É1: *Mme. Béjart, vous êtes intelligente?* É2 (holding picture): *Oui, je suis intelligente.*

F. Have students close their books. Use Grammar Transparency G-2B. Have individual students make up original dialogues using *vous* and *tu* appropriately.

ADDITIONAL PRACTICE

1. Check to see if the students have understood the difference between *vous* and *tu* by giving them the following cues. Have them respond with either *vous* or *tu*.
   Grand-mère; deux amis, Paul et Georges; un chien (barking noise to clarify the meaning); *M. le Président* ___ (give his name to clarify the meaning); *deux femmes, Mme X et Mme Y; deux petites filles, Anne et Sophie; le chat de Marc* (clarify meaning with a meow).

2. Student Tape Manual, Teacher's Edition, *Activités C–F*, pages 22–23.

Exercices

Exercices

PRESENTATION (*page 50*)

Exercice A: Speaking

This exercise can be done in pairs. One student calls out the number of a picture and the other student asks the question. The first student may answer for the person or people in the picture. Then the partners reverse roles.

Exercice B

This exercise works best as a teacher-led, whole-class activity.

ANSWERS

Exercice A

1. Tu es française?
2. Vous êtes français?
3. Vous êtes française?
4. Vous êtes française?
5. Vous êtes français?
6. Tu es française?

Exercice B

1. Vous êtes français(e)?
2. Tu es français(e)?
3. Tu es français(e)?
4. Tu es française?
5. Vous êtes français?

L'accord des adjectifs au pluriel

RECYCLING

This grammar section reviews vocabulary taught in Chapter 1.

PRESENTATION (*pages 50–51*)

A. This concept should be easy for students to understand. Students merely have to learn that all the definite articles change to *les* in the plural, and all these nouns add an *-s* to form their plural.

B. Lead students through steps 1–3 on pages 50–51.

Exercices

A **Ils sont français?** Regardez les photos et posez la question en utilisant «tu» ou «vous». (*Ask the people in each of the pictures if they are French.*)

Tu es français?

1.

2.

3.

4.

5.

6.

B ***Tu ou vous?*** Posez la même question. (*Ask the following people in your class if they are French.*)

un élève
Tu es français?

1. le professeur
2. la personne devant vous
3. la personne derrière vous
4. une fille
5. deux garçons

L'accord des adjectifs au pluriel *Describing More Than One Person or Thing*

---

1. When a noun is in the plural, any adjective that describes or modifies the noun must also be in the plural. Study the following sentences.

Les deux filles sont américaine**s**.
Les deux filles sont sympathique**s**.
Les classes sont petite**s**.

Les garçons aussi sont américain**s**.
Les garçons aussi sont très sympathique**s**.
Les livres sont intéressant**s**.

INDEPENDENT PRACTICE

Assign any of the following:
1. Exercises, page 50
2. Workbook, *Structure: D*, page 12
3. Communication Activities Masters, *Structure: C*, page 12
4. CD-ROM, Disc 1, page 50

LEARNING FROM PHOTOS

Ask students to describe the people in the photos on page 50 using the French they already know. For example, for Photo 1: *C'est une jeune fille. Elle est sympa.*

2. To form the plural of most French adjectives, you add *s* to the singular masculine or feminine form of the adjective. This *s* is not pronounced.

3. If a singular adjective ends in *s*, you do not add another *s* to the plural form.

> Le garçon est français.
> Les garçons sont français.

Exercices

A **Érica et Brigitte.** Décrivez les deux filles. *(Describe the two girls.)*

> populaire
> *Érica et Brigitte sont populaires.*

1. français
2. timide
3. brun
4. énergique
5. américain

B **Jean-François et Yann.** Décrivez les deux garçons. *(Describe the two boys.)*

> intéressant
> *Jean-François et Yann sont intéressants.*

1. français
2. américain
3. brun
4. musclé
5. content

C **Luc et Anne.** Récrivez le paragraphe d'après le modèle. *(Rewrite the paragraph according to the model.)*

> Sophie et Marie sont françaises.
> *Luc et Anne sont français.*

Sophie et Marie sont élèves dans un lycée à Paris. Les deux amies sont très amusantes. Elles sont aussi très énergiques. Maintenant elles sont en vacances. Elles sont à Nice. C'est chouette ça, des vacances à Nice. Vous n'êtes pas d'accord?

INDEPENDENT PRACTICE

Assign any of the following:
1. Exercises, page 51
2. Workbook, *Structure: E–F,* page 13
3. Communication Activities Masters, *Structure: D,* page 13
4. CD-ROM, Disc 1, pages 50–51

C. Write several singular sentences on the board or on an overhead transparency. Call on students to make them plural. For example: *Le professeur est américain. Les professeurs sont américains.* Be sure to emphasize that the *s* is silent.

Exercices

PRESENTATION *(page 51)*

Since the formation of plurals (when there is no liaison) is a written rather than an oral problem, all exercises can be done with books open. Go over the exercises in class and have the students write them for homework.

Exercices A and B: **Writing**

Exercises A and B can be done as whole-class oral and written activities. After going over the exercises orally, you may wish to have one of your better students write the answers on the board, or use an overhead transparency. Students can then check their own written responses. It is important that all students spell the nouns and adjectives correctly.

ANSWERS

Exercice A

1. Érica et Brigitte sont françaises.
2. Elles ne sont pas timides.
3. Elles ne sont pas brunes. Elles sont blondes.
4. Elles sont énergiques.
5. Elles ne sont pas américaines. Elles sont françaises.

Exercice B

1. Jean-François et Yann sont français.
2. Ils ne sont pas américains.
3. Ils sont bruns.
4. Ils sont musclés.
5. Ils ne sont pas contents.

Exercice C

<u>Luc et Anne</u> sont élèves dans un lycée à Paris. Les deux <u>amis</u> sont très <u>amusants</u>. Ils sont aussi très <u>énergiques</u>. Maintenant ils sont en vacances. <u>Ils</u> sont à Nice. C'est chouette ça, des vacances à Nice. Vous n'êtes pas d'accord?

1. Observe the following examples of how to tell time.

Il est une heure.   Il est deux heures.   Il est trois heures.

Il est sept heures dix.   Il est huit heures vingt-cinq.

Il est neuf heures moins dix.   Il est dix heures moins cinq.

Il est quatre heures et quart.   Il est cinq heures moins le quart.   Il est six heures et demie.

Il est midi.   Il est minuit.

2. To indicate a.m. and p.m. in French, you use the following expressions.

Il est cinq heures du matin.   Il est trois heures de l'après-midi.   Il est onze heures du soir.

3. Note the way times are abbreviated in French.

9h30	**neuf heures et demie**
11h15	**onze heures et quart**
3h45	**quatre heures moins le quart**

4. To ask what time it is, you say: **Il est quelle heure?**
A more formal way to ask the time is: **Quelle heure est-il?**

5. Note how to ask and tell what time something (such as French class) takes place.

Le cours de français est _à_ quelle heure?
Le cours de français est _à_ neuf heures.

6. Note how to give the duration of an event (to indicate from when until when).

Le cours de français est _de_
neuf heures _à_ dix heures.

Exercices

A **Il est quelle heure?** Répondez d'après le modèle. (*Answer according to the model.*)

> **2h**
> Élève 1: **Il est quelle heure?**
> Élève 2: **Il est deux heures.**

1. 9h	3. 5h10	5. 7h55	7. 10h25	9. 6h40	11. 1h30
2. 3h35	4. 8h15	6. 12h ☀	8. 9h45	10. 2h05	12. 12h30 ☾

B **Quand?** Posez les questions suivantes à un copain ou une copine. (*Ask a classmate the following questions. Then reverse roles.*)

1. Il est quelle heure maintenant?
2. Le cours de français est à quelle heure?
3. Le cours de maths est à quelle heure?
4. Le cours d'anglais est le matin ou l'après-midi?
5. Le cours d'histoire est le matin ou l'après-midi?

C **À quelle heure sont les cours?** Répondez. (*Tell when four of your classes begin and end.*)

> **Le cours d'anglais est de dix heures et quart à onze heures.**

Teaching Tip Continue to recycle the topic of telling time throughout the year. It is reintroduced frequently in the text through illustrations and schedules. It is not necessary to wait until all students can tell time perfectly before going on to other material.

Exercices

PRESENTATION (*page 53*)

Exercices A, B, and C

Exercises A, B, and C can be done in pairs to maximize opportunities for oral practice.

Extension of *Exercice C*

Prepare a sample school schedule on an overhead transparency and have students make statements of time duration. Make incorrect statements about the schedule and have students correct them.

ANSWERS

Exercice A

1. É1: Il est quelle heure?
   É2: Il est neuf heures.
2. ... quatre heures moins vingt-cinq.
3. ... cinq heures dix.
4. ... huit heures et quart.
5. ... huit heures moins cinq.
6. ... midi.
7. ... dix heures vingt-cinq.
8. ... dix heures moins le quart.
9. ... sept heures moins vingt.
10. ... deux heures cinq.
11. ... une heure et demie.
12. ... minuit et demi.

Exercices B and C

Answers will vary. However, students should use the phrases *il est...* , *à... heures*, or *de... à...* correctly.

INDEPENDENT PRACTICE

Assign any of the following:
1. Exercises, page 53
2. Workbook, *Structure: G–H*, pages 14–15
3. Communication Activities Masters, *Structure: E*, page 13
4. Computer Software, *Structure*
5. CD-ROM, Disc 1, pages 52–53

CONVERSATION

PRESENTATION (*page 54*)

A. To prepare students, have them answer the following questions with one word: *Vous êtes américains? Vous êtes de quelle ville? ___ est une grande ville ou une petite ville?*

B. Tell students they will hear a conversation between Sylvie, Catherine, Mark, and David. Have them watch the Conversation Video or have them repeat after you or the recording on Cassette 2B/CD-2.

C. Call on pairs of students to read the conversation.

Note In the CD-ROM version, students can play the role of either Sylvie or David and record the conversation.

ANSWERS

Exercice A

1. De Los Angeles.
2. De Nice.
3. Sur la mer Méditerranée. (Sur la Côte d'Azur.)
4. Une assez grande ville.
5. Une grande ville.

Prononciation

PRESENTATION (*page 54*)

A. Using Pronunciation Transparency P-2, model the key word *l'art*. Have students say it in unison and individually. Write the word on the board and ask students what they notice about its pronunciation. (The last letter is silent.) Explain that this is common in French.

B. Now lead students through steps 1–2 on page 54, modeling the examples.

C. For additional pronunciation practice, use Cassette 2B/CD-2: *Prononciation* and the Student Tape Manual, Teacher's Edition, *Activités K–L*, page 26.

54

Scènes de la vie *Vous êtes de quelle nationalité?*

> SYLVIE: Vous êtes américains?
> MARK: Oui, nous sommes américains. Et vous, vous êtes françaises, n'est-ce pas?
> CATHERINE: Oui, nous sommes de Nice.
> DAVID: Nice? C'est où ça?
> SYLVIE: Sur la mer Méditerranée.
> MARK: Nice est une grande ville ou une petite ville?
> CATHERINE: C'est une assez grande ville sur la Côte d'Azur.
> SYLVIE: Et vous, vous êtes d'où?
> DAVID: Nous sommes de Los Angeles.
> CATHERINE: Los Angeles! C'est chouette, ça!

A **Nice et Los Angeles.** Répondez d'après la conversation. (*Answer according to the conversation.*)

1. D'où sont les Américains?
2. Et les Françaises?
3. Où est Nice?
4. Nice est une grande ville ou une petite ville?
5. Et Los Angeles?

Prononciation *Les consonnes finales*

1. In French, you do not usually pronounce the final consonant you see at the end of words. Repeat the following.

   salu~~t~~ devan~~t~~ maintenan~~t~~ un restauran~~t~~ l'anglai~~s~~

2. In the same way, you do not pronounce the final s you add to a word to make it plural. This is why a singular noun and its plural sound alike. Repeat the following.

   l'ar~~t~~

   le copain les copain~~s~~ le livre les livre~~s~~ la fille les fille~~s~~

   Les garçons et les filles sont devant le restaurant.
   Ils sont impatients.

54 CHAPITRE 2

<table>
<tr><td>

LEARNING FROM PHOTOS

Ask students to look at the *Conversation* photos and describe each of the four characters, using the French they know.

</td><td>

MEMORY AID

Tell students that the final consonants that are usually pronounced in French are those in the English word <u>CaReFuL</u>, for example: *Luc, jour, prof, espagnol.*

</td></tr>
</table>

Activités de communication orale

A **Aux États-Unis.** You and your partner are French students visiting the U.S. Ask two other students for the following information, then reverse roles.

1. their nationality
2. where they're from
3. if it's a large city or a small town
4. if they're high school students
5. if classes are easy or difficult
6. what the teachers are like
7. what the students are like

B **«Mieux vaut tard que jamais».** Anne's philosophy is "Better late than never," but she's trying hard to be more punctual. With a partner, compare her arrival times with her schedule and tell if she's on time (*à l'heure*), late (*en retard*), or early (*en avance*).

> (7. le dentiste)
> Anne arrive à: 4h
>
> Élève 1: Il est quelle heure?
> Élève 2: Il est quatre heures.
> Élève 1: Anne est en avance.

Anne arrive à:

1. 8h05   6. 3h20
2. 9h13   7. 4h30
3. 10h10   8. 6h30
4. 12h   9. 9h
5. 12h47

1. le cours de français	8h
2. le cours de maths	9h15
3. la récréation	10h10
4. la cantine/cafétéria	12h
5. le cours de biologie	12h45
6. le tennis	3h15
7. le dentiste	4h15
8. le dîner	6h30
9. un programme à la télé	8h55

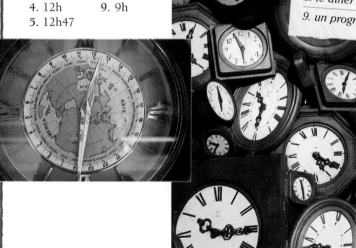

INDEPENDENT PRACTICE

1. Exercise and activities, pages 54–55
2. Workbook, *Un Peu Plus*, page 16
3. CD-ROM, Disc 1, pages 54–55

LEARNING FROM PHOTOS

1. The right-hand photo is a close-up of the clock sculpture *L'heure de tous* by Arman in front of the Gare Saint-Lazare in Paris.
2. Have students write down the times shown on each clock in the photos on page 55. See who can come up with the most correct times.

Activités de communication orale

PRESENTATION (*page 55*)

Extension of *Activité A*

Have one of the "French students" follow up by explaining to the class orally (or in writing) the information given. The narration should be in the third person.

ANSWERS

Activité A

Answers will vary.

Activité B

1. É1: Il est quelle heure?
   É2: Il est huit heures cinq.
   É1: Elle est en retard.
2. É1: Il est quelle heure?
   É2: Il est neuf heures treize.
   É1: Elle est en avance.
3. ... É2: Il est dix heures dix.
   É1: Elle est à l'heure.
4. ... É2: Il est midi.
   É1: Elle est à l'heure.
5. ... É2: Il est une heure moins treize.
   É1: Elle est en retard.
6. ... É2: Il est trois heures vingt.
   É1: Elle est en retard.
7. ... É2: Il est quatre heures et demie.
   É1: Elle est en retard.
8. ... É2: Il est six heures et demie.
   É1: Elle est à l'heure.
9. ... É2: Il est neuf heures.
   É1: Elle est en retard.

READING STRATEGIES
(page 56)

Pre-reading

A. Have students locate Paris and the Côte d'Azur on the map on page 504 (page 237, Part A) or use the Map Transparency.

B. Have students scan the reading and list at least five cognates.

Reading

A. Have students close books and listen to the letter on Cassette 2B/CD-2. Then have them listen again as they follow along in their books.

B. Call on students to read several sentences. Ask questions after every three sentences or so.

C. Write statements about the reading on the board or on a transparency. If students disagree, they must state why. For example: *Christian est d'Antibes. Non, je ne suis pas d'accord. Il est de Saint-Germain-en-Laye.*

Post-reading

A. Assign Exercise A on page 56.

B. Have students write a letter to Christian about themselves.

Note Students may listen to a recorded version of the *Lecture* on the CD-ROM.

Étude de mots

ANSWERS

Exercice A

1. f	3. d	5. e	7. b
2. c	4. a	6. g	

LECTURE ET CULTURE

UNE LETTRE

Antibes, le 15 juillet

Chers amis,

Salut! Je suis Christian Capet. Je suis de Saint-Germain-en-Laye. Saint-Germain-en-Laye est une petite ville près de Paris, dans la banlieue. Je suis élève dans un lycée. Mais maintenant je ne suis pas à Saint-Germain. Je suis à Antibes avec la famille de Gilbert Berthollet. Gilbert et moi, nous sommes copains. Nous sommes élèves dans le même lycée. Et nous sommes dans le même cours d'anglais. Le prof d'anglais est très sympa, mais l'anglais, ce n'est pas très facile. Mais maintenant, pas de prof, pas de classe! Nous sommes libres! Nous sommes en vacances à Antibes. Antibes est une petite ville très pittoresque sur la Côte d'Azur. Pour moi les vacances, c'est toujours extra. Vous êtes d'accord?

Affectueusement,
Christian

¹ près de   *near*   ² dans la banlieue   *in the suburbs*

Étude de mots

A **Quel est le mot?** Trouvez les mots qui correspondent. *(Find the corresponding word or phrase.)*

1. Bonjour!
2. super
3. pas difficile
4. une langue
5. pas différent
6. période de temps libre
7. une petite ville près d'une grande ville

a. l'anglais
b. la banlieue
c. extra
d. facile
e. le même
f. Salut!
g. les vacances

Antibes: La vieille ville

56   CHAPITRE 2

Compréhension

 Vous avez compris? Répondez d'après la lecture. *(Answer according to the reading.)*

1. D'où est Christian?
2. Où est Saint-Germain-en-Laye?
3. Où est Christian maintenant?
4. Il est à Antibes avec qui?
5. Les deux garçons sont copains?
6. Ils sont dans le même cours d'anglais?
7. Comment est le prof d'anglais?
8. Le cours d'anglais est facile ou difficile?
9. Les deux copains sont en vacances? Où?

C **Un peu de géographie.** Oui ou non? *(Answer "yes" or "no.")*

1. Saint-Germain-en-Laye est dans la banlieue parisienne.
2. Antibes est aussi dans la banlieue parisienne.
3. Les villages et les villes de la Côte d'Azur sont très agréables pour les vacances.

DÉCOUVERTE CULTURELLE

LES 24 HEURES ET LE DÉCALAGE HORAIRE

DANS LA CONVERSATION	SUR LES HORAIRES
huit heures du matin	8h (huit heures)
deux heures de l'après-midi	14h (quatorze heures)
quatre heures et demie de l'après-midi	16h30 (seize heures trente)
dix heures et quart du soir	22h15 (vingt-deux heures quinze)

L'heure n'est pas la même partout. À New York il est midi. À Paris il est dix-huit heures. La différence entre l'heure de New York et l'heure de Paris (le décalage horaire) est de six heures. Il est midi à Paris. Quelle heure est-il à New York? Il est neuf heures à San Francisco. Quelle heure est-il à New York?

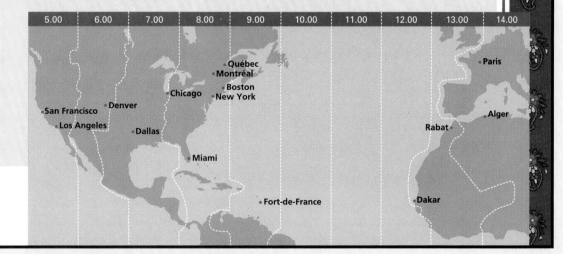

CRITICAL THINKING ACTIVITY

(Thinking skills: making inferences; drawing conclusions)

Put the following on the board or on a transparency:

1. C'est un cours très difficile. Pourquoi? Pour quelle raison?
   a. C'est une matière très compliquée.
   b. Le professeur n'est pas intéressant.
   c. C'est un cours de maths.

2. Robert n'est pas sérieux. Il n'aime pas étudier et il n'étudie pas. Quelles en sont les conséquences?

Compréhension

ANSWERS

Exercice B

1. De Saint-Germain-en-Laye.
2. Près de Paris.
3. Il est à Antibes.
4. Avec la famille de Gilbert Berthollet.
5. Oui, ils sont copains.
6. Oui, ils sont dans le même cours d'anglais.
7. Il est très sympa.
8. Il est difficile.
9. Oui. À Antibes.

Exercice C

1. Oui.   2. Non.   3. Oui.

Bell Ringer Review

Write the following on the board or use BRR Blackline Master 2-8: Make sentences from the following words:

1. filles Paris sont ne les de deux pas
2. dans Grenoble est Alpes françaises les ville une

OPTIONAL MATERIAL

Découverte culturelle

PRESENTATION *(page 57)*

A. Explain the 24-hour clock using the examples in the chart. The French use a 24-hour clock in official schedules (planes, trains, buses, radio or television programs) rather than A.M. and P.M. For example, 4:00 P.M. would be 16:00 (sixteen hours after midnight).

B. Answers to the questions in the *Découverte: Il est midi à Paris; il est 6 h du matin à New York. Il est 9 h à San Francisco; il est minuit (midi) à New York.*

Note Students may listen to a recorded version of the *Découverte culturelle* on the CD-ROM.

58

Bell Ringer Review

Write the following on the board or use BRR Blackline Master 2-9: Write the French name of a school course, with its article, that you associate with the categories below.

1. Chopin, Mozart
2. Einstein, Pasteur
3. Paris, Washington D.C., the Mediterranean
4. Shakespeare, Twain, Hugo
5. Franklin Roosevelt, Louis XIV, World War I
6. Renoir, Picasso

OPTIONAL MATERIAL

PRESENTATION (*pages 58–59*)

The purpose of this section is to allow students to enjoy the photographs in a relaxed manner in order to gain an appreciation of French culture. However, if you would like to do more with the section, you may do any of the following activities.

Before reading the captions, have students do the following:

Photo 1: Have students locate Rouen and Paris on the map of France on page 504 (Part A, page 237) or use the Map Transparency. Tell them that the village of Giverny is located halfway between these two cities in a region of France called *la Normandie*. The house and gardens of French Impressionist painter Monet are located in this village.

Note In the CD-ROM version, students can listen to the recorded captions and discover a hidden video behind one of the photos.

Bonjour **1**. Nous sommes Béatrice, Nathalie, Gilles et Christian. Nous sommes tous français. Nous sommes de Giverny. Giverny n'est pas très loin de Paris, mais ce n'est pas dans la banlieue parisienne comme Saint-Germain-en-Laye. Giverny est un village très pittoresque.

Voici le jardin de Claude Monet, un peintre français célèbre **2**. Le jardin de Monet est à Giverny. Monet peint beaucoup de tableaux de son jardin.

Bonjour **3**. Je suis Maeva. Je suis de Papeete, une ville sur l'île de Tahiti. Tahiti est une île de la Polynésie française, un territoire français d'outre-mer.

Voici un tableau de Paul Gauguin **4**. Gauguin, comme Monet, est un peintre français célèbre. Le tableau représente une scène de la vie en Polynésie française. Gauguin est considéré comme un des initiateurs de la peinture moderne.

DID YOU KNOW?

Claude Monet and Eugène Henri Paul Gauguin were French Impressionist painters. Monet was the leader of the Impressionist movement. He was fascinated with the effects of sunlight on a subject. In 1883 he settled in his country house in Giverny. Many of his paintings depict his house and gardens. The large photo on this page is an actual photo of the lily pond in Monet's garden. It became the subject of a famous series of large mural-size paintings called *Les Nymphéas,* which hang in the Orangerie Museum in Paris.

Gauguin lived in Tahiti for many years. He portrayed Tahiti's lush beauty and its peaceful atmosphere in many of his paintings. In his work Gauguin idealized the people of the South Sea Islands. His tropical settings were painted with brilliant colors.

Photos 2 and 4: Find out if any of your students recognize the names of the famous French painters Monet and Gauguin. If so, what can they say about them? (Photos of paintings by other Impressionist artists can be found in Chapter 16, pages 420–421, and in *Lettres et Sciences*, page 444.)

Note For art activities related to the paintings and artists on pages 58–59, see Fine Art Transparencies F-1 and F-2.

Photo 3: Have students locate Tahiti on the map on page 506 (Part A, page 239) or use the Map Transparency.

THE FRANCOPHONE WORLD

For more information on French Polynesia and Gauguin in Tahiti, see *Le Monde francophone*, pages 110–113 and pages 434–437.

59

INTERDISCIPLINARY TOPICS

Read the following paragraph to students.

Qu'est-ce qu'une biographie? Une biographie est l'histoire d'une personne célèbre. Voici quelques détails biographiques à propos d'une personne célèbre, Claude Debussy. Claude Debussy est un compositeur français. Il est de Saint-Germain-en-Laye. Il compose des préludes et des études pour le piano. Il compose aussi le poème symphonique, *La Mer.*

After reading the paragraph, have a student answer the following question.

What was or is Debussy's field?

a. *les sciences*

b. *la littérature, surtout la biographie*

c. *la musique*

You may wish to obtain a recording of *La Mer* and play it for the class.

RECYCLING

The *Activités de communication orale* and the *Activités de communication écrite* allow students to use the vocabulary and grammar from this chapter in open-ended, real-life situations. They also give students the opportunity to reuse, as much as possible, the vocabulary and structure from Chapter 1.

You may have students do as many of the *Activités* as you wish.

INFORMAL ASSESSMENT

Oral activities A and B may be used to evaluate the speaking skill informally. For suggestions on how to grade each student's response, see page 34 of this Teacher's Wraparound Edition.

Activités de communication orale

PRESENTATION (page 60)

Activité A

Have students work in groups of three. Then have several groups present their dialogue to the class.

Note In the CD-ROM version of this activity, students can interact with an on-screen native speaker.

Activité B

Have students work on Activity B in pairs.

Extension of *Activité B*

Ask pairs of students to write, practice, and present short dialogues about their TV preferences.

ANSWERS

Activités A and B

Answers will vary.

Activités de communication orale

A **Roissy-Charles-de-Gaulle.** You and your friend Suzanne have just arrived at Roissy-Charles-de-Gaulle Airport on the outskirts of Paris. Since she doesn't speak French, answer the immigration officer's questions for both of you.

1. Vous êtes de quelle nationalité?
2. Vous êtes d'où?
3. Vous êtes en vacances?

B **Mes programmes favoris.**

1. Name a few TV shows you watch.
2. Give the time and day each program is on.
3. Give your opinion of each show.
4. Ask a classmate if he or she agrees with you.

Samedi 16 Novembre

9.50 La5 10.20
Les animaux du soleil
Documentaire français. Rediffusion. Rives de Cunene.

10.00

10.00 M6 10.05 **Infoprix**
10.05 M6 10.30
M6 Boutique
Présentation : Pierre Dhostel et Julie.
10.20 La5 10.55
Chevaux et casaques
Magazine de Patrice Dominguez et Jean-Louis Burgat. Présentation : Caroline Avon. Sauts d'obstacles à Auteuil.
10.30 FR3 12.00
Espace 3 entreprises
11.50 L'homme du jour.
10.30 M6 12.00 **Multitop**
Présentation : Laurent Petitguillaume.
10.35 C+ 10.40
Journal du cinéma
10.40 C+ 12.30
T Susie et les Baker Boys
Film américain de Steve Kloves (1989). 110 mn. Voir Tra 2182 page 111. La «vie d'artiste», calamiteuse, remarquablement décrite par un cinéaste doué et chaleureux. Les acteurs sont parfaits, l'actrice, une révélation : Michelle Pfeiffer, éclatante. Rediffusions : mardi 19 à 13.35, samedi 23 en v.o à 0.35.
10.50 TF1 11.15

à la recherche d'un mode de garde pour son enfant. La crèche modèle de Lille. L'adaptation et les problèmes d'infection dans les crèches. Les livres et les objets transitionnels des enfants. Question aux enfants : «Est-ce que tu es content que tes parents travaillent ?»

10.55 La5 11.50
Mille et une pattes
Magazine animalier. Présentation : Pierre Rousselet-Blanc et Pétra. Réalisation : P. Lumbroso. Invités : André Pittion-Rossillon, de la Société Centrale Canine, et Claude Fargeon, spécialiste du comportement animal. Gros plan : le dogue argentin. Reportages.

11.00

11.15 TF1 11.50
Auto moto
Magazine de Jacques Bonnecarrère. Supercross à Bercy. L'essai de la Mazda MX3. Salon de Tokyo.
11.20 A2 11.45
Motus
Jeu. Présentation : Patrice Laffont.
11.45 A2 11.55
Flash infos
11.50 TF1 12.25
Tournez manège
Jeu de Noël Coutisson et Claude Savarit. Présentation : Evelyne Leclercq, Simone Garnier et Charly Oleg.
11.50 La5 11.55
TT Ecrire contre l'oubli

12.25 A2 12.50
T Le français tel qu'on le parle
Documentaire français de Pierre Nivollet. A Antananarivo, capitale de l'île de Madagascar, on parle un français très coloré, un peu créole et souvent remarquable. Jean et sa femme Laura nous font visiter la ville, de l'école française au «zoma», le plus grand marché de «Tana». Ou quand la langue française facilite les rapports entre les gens et permet à certains de s'ouvrir sur le monde.
12.30 C++ 12.35
Flash infos
12.30 M6 13.00
Cosby show
Série américaine. Redif.
12.35 C++ 13.30
T 24 heures
Magazine d'Hervé Chabalier, Erik Gilbert, Claude Chelli. Programme non communiqué.
12.45 La5 13.20
Journal
12.50 A2 13.00
1, 2, 3, théâtre
Reprise.

13.00

13.00 TF1 13.15
Journal
13.00 A2 13.25
Journal
13.00 M6 13.55
O'Hara
Série américaine.
13.15 TF1 13.50
T Reportages

13.30 C++ 13.35
Journal du cinéma
13.35 C+ 15.10
Désastre à la centrale 7
Téléfilm américain de Larry Elikann (1988). Michael O'Keefe : Le sergent Fitzgerald. Perry King : Le commandant Hicks. Peter Boyle : Le général Sanger. Patricia Charbonneau : Kathy Fitzgerald. Deux soldats maladroits endommagent un missile dans une base militaire du Texas. Adulé par les siens, le sergent Fitzgerald arrive sans se presser pour constater les dégâts. Stupeur et terreur : le missile fuit et vrombit. Il risque d'exploser...
13.45 A2 14.15
T Objectif jeunes
Magazine de Raymond Tortora et du service éducation de la rédaction. Présentation : Dominique Laury et Philippe Lefait. Réalisation : Roger Gomez.
Etudier en Europe (Patrick Redslob). Grâce au programme «Erasmus», soixante mille étudiants, dont dix mille Français, fréquentent les universités d'Europe. Reportage à Grenoble II, qui accueille Anglais, Allemands, Hollandais... Louvain-la-Neuve (Marc Maisonneuve). L'université de Louvain, en Belgique, qui reçoit des étudiants d'Europe entière, est réputée pour ses filières Sciences agronomiques, Génie civil

---

STUDENT PORTFOLIO

Have each student keep a notebook containing his/her best written work from each chapter of *Bienvenue*. These selected writings can be based on assignments from the Student Textbook, the Writing Activities Workbook, and the Communication Activities Masters.

In the Workbook, students will develop an ongoing organized autobiography (*Mon*

Autobiographie). These Workbook pages may also become a part of each student's portfolio. (See the Teacher's Manual for more information on the Student Portfolio.)

Note Students may create and save both oral and written work using the Electronic Portfolio feature on the CD-ROM.

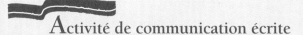

Aᴄtivité de communication écrite

A **Mon emploi du temps.** Make a chart like the one below and fill it out in French based on your weekly schedule. For each class give the time, the teacher, and your opinion of the teacher and the class itself.

Cours	Jours	Heure	Prof	Opinion: Prof	Opinion: Cours
anglais	le lundi le mardi le mercredi le vendredi	de 9h à 9h45	Mlle Shaw	assez intéressante	difficile

Vocabulaire

NOMS
le copain
la copine
le prof
la prof
le professeur
l'homme (m.)
la femme

la classe
la salle de classe
le cours
l'agenda (m.)
la matière
les maths (f.)
l'algèbre (f.)
la géométrie
la trigonométrie
l'informatique (f.)
les sciences (f.)
la biologie

la chimie
la physique
la littérature
la langue
le français
l'anglais (m.)
l'espagnol (m.)
le latin
l'histoire (f.)
la géographie
la musique
l'art (m.)
la gymnastique

le jour
lundi
mardi
mercredi
jeudi
vendredi
samedi

dimanche
aujourd'hui
demain
le week-end
la semaine
l'heure (f.)
midi
minuit
le matin
l'après-midi (m.)
le soir

ADJECTIFS
difficile
facile
chouette
extra
superx
terrible
moche
libre

occupé(e)
même

VERBE
être

AUTRES MOTS
ET EXPRESSIONS
être d'accord
maintenant
vraiment

PRESENTATION (*page 61*)

Extension of *Activité A*

Have students hand in their charts without putting their names on them. Redistribute the charts so that everyone will read someone else's chart aloud. The class then tries to guess to whom each chart belongs. As a student reads the chart aloud, encourage the others to comment on the person's opinions. Try to elicit such statements as: *Je suis dans la même classe! Non! Le cours d'anglais n'est pas difficile! Le prof est sympa!*

ANSWERS
Activité A
Answers will vary.

ASSESSMENT RESOURCES

1. Chapter Quizzes
2. Testing Program
3. Situation Cards
4. Communication Transparency C-2
5. Computer Software: Practice/Test Generator

VIDEO PROGRAM

INTRODUCTION   (11:51)

LES COURS   (12:33)

INDEPENDENT PRACTICE

Assign any of the following:
1. Activities, pages 60–61
2. Workbook, *Mon Autobiographie,* page 17
3. Situation Cards, Chapter 2
4. Communication Activities Masters, pages 7–13
5. CD-ROM, Disc 1, pages 60–61

FOR THE YOUNGER STUDENT

Play an oral game that is fun but at the same time checks auditory discrimination. Give any adjective the students have learned. Have them raise their right hand (if it describes something feminine) or their left hand (if it describes something masculine). Use only adjectives that have a sound difference.

CHAPTER OVERVIEW

In this chapter students will learn to talk about their school life and some after-school leisure activities. In order to do this, they will learn several *-er* verbs, the use of the pronoun *on*, the plural and negative forms of indefinite articles, and the use of the infinitive after verbs such as *aimer, détester,* and *adorer.* The forms of the verb *être* and vocabulary having to do with school life and telling time are recycled from Chapters 1 and 2. The cultural context of the chapter is the life of French *lycéens.*

CHAPTER OBJECTIVES

By the end of this chapter, students will know:

1. the pronoun *on*
2. verbs ending in *-er*
3. indefinite articles in the plural
4. negation with indefinite articles
5. conjugated verbs followed by the infinitive

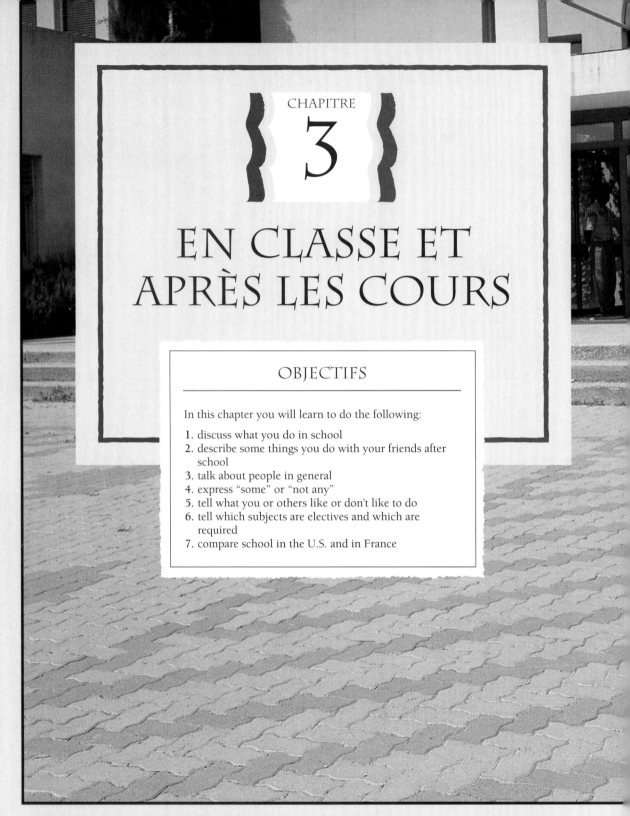

CHAPITRE

3

EN CLASSE ET APRÈS LES COURS

OBJECTIFS

In this chapter you will learn to do the following:

1. discuss what you do in school
2. describe some things you do with your friends after school
3. talk about people in general
4. express "some" or "not any"
5. tell what you or others like or don't like to do
6. tell which subjects are electives and which are required
7. compare school in the U.S. and in France

CHAPTER PROJECTS

(optional)

Have students in your class survey the student body for information about after-school activities. If the survey is restricted to the activities taught in the *Mots* presentation, pages 68–69, students can report back in French. Each person interviewed should initial the interview sheet to avoid repetition.

INTERNET ACTIVITIES

(optional)

These activities, student worksheets, and related teacher information are in the *Bienvenue* Internet Activities Booklet and on the Glencoe Foreign Language Home Page at **http://www.glencoe.com/secondary/fl**

Pacing

This chapter requires eight to ten class sessions. Pacing will depend on class length and the age and aptitude of the students.

Note The Lesson Plans offer guidelines for 45- and 55-minute classes and **Block Scheduling**.

Exercices vs. *Activités*

All exercises (which provide guided practice) are coded in blue. All communicative activities are coded in red.

DID YOU KNOW?

The girl in the photo is riding *un scooter*. French teenagers can't drive cars until they are 18 (or 16 if they have a special permit and are accompanied by an adult). However, they can drive smaller vehicles, such as mopeds and motorbikes (*vélomoteurs* and *cyclomoteurs*) at 16. Since the driving age in France is set by the government, it's the same all over the country, unlike the situation in the U.S., where each state has the right to establish its own driving age. For further information about driving various types of vehicles in France, see the *Découverte culturelle* on page 321 and the DID YOU KNOW? box at the bottom of the same page in this Teacher's Wraparound Edition.

Vocabulary Teaching Resources

1. Vocabulary Transparencies 3.1 (A & B)
2. Audio Cassette 3A/CD-3
3. Student Tape Manual, Teacher's Edition, *Mots 1: A-C*, pages 29–31
4. Workbook, *Mots 1: A–C*, page 18
5. Communication Activities Masters, *Mots 1: A*, page 14
6. Chapter Quizzes, *Mots 1: Quiz 1*, page 14
7. CD-ROM, Disc 1, *Mots 1:* pages 64–67

Bell Ringer Review

Draw five clocks (with hands) showing the times listed below, then write the following on the board, or use BRR Blackline Master 3-1: Give the time shown on each clock.

1. 2:20
2. 6:45
3. 8:30
4. 11:50 (A.M.)
5. 1:15

PRESENTATION (*pages 64–65*)

A. Have students close their books. Model the new vocabulary on pages 64–65 using Vocabulary Transparencies 3.1 (A & B). Have students repeat each word two or three times after you or Cassette 3A/CD-3.

B. Act out the *-er* verbs on page 64 using a sequence of actions and statements. For example: *quitter la salle de classe* (leave the room); *entrer dans la salle de classe* (enter the room); *parler* (say something); *écouter* (point to your ears); *travailler* (imitate physical work); *étudier* (pore over some books). Write

64

VOCABULAIRE

MOTS 1

habiter à Paris

une maison

une rue

quitter la maison

écouter

parler

arriver

entrer

étudier

travailler

regarder le tableau noir

poser une question

passer un examen

TOTAL PHYSICAL RESPONSE

(*following the Vocabulary presentation*)

Getting Ready

Before doing this activity, make sure students understand each of the following verbs by acting them out: *étudier, quitter, arriver, travailler, regarder, poser une question, écouter, parler, entrer.*

TPR

(Student[s]), **levez-vous, s'il vous plaît.**
Écoutez, s'il vous plaît.
Quittez la salle de classe.
Entrez dans la salle de classe.
Étudiez.
Regardez un garçon.
Regardez une fille.
Parlez français.
Maintenant, parlez anglais.

Voici Paul Lafontaine.
Paul habite à Paris.
Il habite rue Saint-Dominique.

Paul quitte la maison à sept heures et demie.

Il arrive à l'école à huit heures.

À huit heures et quart, il entre dans la salle de classe.

Il quitte l'école à cinq heures et il rentre à la maison.

Note: The expression *passer un examen* is a false cognate. A false cognate is a word that looks like an English word but means something different. *Passer un examen* means "to take an exam," not "to pass an exam."

Quand est-ce que Paul étudie?
Paul étudie beaucoup le soir.

CHAPITRE 3   **65**

---

the infinitive of each verb after you demonstrate it.
C. Repeat the actions, this time having students act out each verb using the Total Physical Response approach.
D. Now model the phrases on page 65 that put the words just taught into meaningful sentences. As you present these sentences, intersperse the presentation with comprehension questions, building from simple to more complex. The natural progression is yes/no, choice, question word. For example: *Paul habite à Paris? Il habite à Paris ou à Lyon? Qui habite à Paris? Où est-ce que Paul habite? Il habite rue Saint-Dominique? Il habite quelle rue? Paul quitte la maison? Il quitte la maison ou l'école?*

Teaching Tip The type of questioning described above allows students to hear and use the words so they become an active part of their vocabulary in a natural way. It also lets you take into account individual differences when presenting new material. Ask the easy yes/no questions of the less able students and the more difficult questions with interrogative words of the more able students.

RECYCLING

Review time by pointing to each clock on page 65 and asking students: *Quelle heure est-il?*

Exercices

Exercice A

Do Exercise A with books closed. Then have students open their books and do the exercise again as a reading activity. Finally, have students retell the story in their own words.

Extension of *Exercice A*

1. To practice both affirmative and negative constructions, cue *oui* or *non* answers by nodding or shaking your head.
2. After students have completed the exercise, have them write their own daily routines.

Exercice B

Do this exercise with books open. It is recommended that you do all the exercises in class before assigning them as independent practice.

ANSWERS

Exercice A

1. Oui, Paul habite à Paris.
2. Oui, il habite rue Saint-Dominique.
3. Il quitte la maison à sept heures et demie.
4. Il arrive à l'école à huit heures.
5. Oui, il entre dans la salle de classe (à huit heures et quart).
6. Oui, le professeur est là.
7. Oui, le professeur parle.
8. Oui, Paul écoute le professeur.
9. Oui, Paul regarde le tableau noir.
10. Oui, il pose une question.
11. Oui, il étudie le français.
12. Oui, il passe un examen.
13. Il quitte l'école à cinq heures.
14. Oui, il travaille beaucoup le soir.

Exercice B

1. c, e	6. d, f, h
2. b, g, m	7. l
3. f	8. j
4. a, g	9. k
5. d, i, k	10. g, m

Exercices

A **À l'école le matin.** Répondez. *(Answer.)*

1. Est-ce que Paul habite à Paris?
2. Il habite rue Saint-Dominique?
3. Le matin, il quitte la maison à quelle heure?
4. Il arrive à l'école à quelle heure?
5. Il entre dans la salle de classe?
6. Le professeur est là?
7. Le professeur parle?
8. Paul écoute le professeur?
9. Paul regarde le tableau noir?
10. Il pose une question?
11. Il étudie le français?
12. Il passe un examen?
13. L'après-midi, il quitte l'école à quelle heure?
14. Il travaille beaucoup le soir?

L'Arc de Triomphe

B **Des expressions.** Trouvez les mots qui correspondent aux verbes. *(Find the words or phrases that correspond to the verbs.)*

1. parler	a. avenue des Champs-Élysées
2. arriver	b. à l'école
3. quitter	c. français
4. habiter	d. le prof
5. écouter	e. anglais
6. regarder	f. la maison
7. passer	g. à Paris
8. entrer	h. le tableau noir
9. poser	i. quand le prof parle
10. rentrer	j. dans la salle de classe
	k. une question
	l. un examen
	m. à la maison

LEARNING FROM PHOTOS

Ask the following questions about the photos on this page: *C'est quelle rue? L'avenue des Champs-Élysées est à Paris? L'Arc de Triomphe est un monument? Il est à Paris? L'Arc de Triomphe est un monument célèbre à Paris?*

DID YOU KNOW?

The Avenue des Champs-Élysées was named after the Elysian Fields, or Elysium, of Greek mythology, where heroes favored by the gods went after death. The broad, elegant avenue on the Right Bank was refurbished at considerable expense in the mid-1990's to rid it of the tawdry commercialization to which it had fallen prey.

C **Le lycéen, Paul.** Choisissez la bonne réponse. (*Choose the correct answer.*)

1. Où habite Paul?
   a. À Paris.   b. Le matin.   c. À l'école.

2. Quand est-ce que Paul quitte la maison?
   a. Rue Saint-Dominique.   b. Le matin.   c. Avec un copain.

3. Où est-ce qu'il arrive?
   a. À huit heures.   b. À l'école.   c. Le matin.

4. Qui parle?
   a. Le prof.   b. La salle de classe.   c. Français.

5. Qui écoute quand le prof parle?
   a. Le prof.   b. La salle de classe.   c. La classe.

6. Quand est-ce que Paul arrive à l'école?
   a. Le matin.   b. L'après-midi.   c. Le soir.

7. Quand est-ce qu'il quitte l'école?
   a. Le matin.   b. L'après-midi.   c. Le soir.

D **Qu'est-ce que... ?** Répondez d'après les indications. (*Answer according to the cues.*)

1. Qu'est-ce que Paul regarde? (le livre)
2. Qu'est-ce qu'il étudie? (le vocabulaire)
3. Qu'est-ce qu'il passe? (un examen)
4. Qu'est-ce qu'il parle? (français)
5. Qu'est-ce qu'il pose? (une question)

Exercice C: Speaking

Do Exercise C with books open. After doing Exercise C, have students retell the story in their own words.

ANSWERS

Exercice C

1. a
2. b
3. b
4. a
5. c
6. a
7. b

Exercice D

1. Paul regarde le livre.
2. Il étudie le vocabulaire.
3. Il passe un examen.
4. Il parle français.
5. Il pose une question.

INFORMAL ASSESSMENT (*Mots 1*)

Have students open their books. Check for comprehension by asking questions that require either a correction or a negative answer. For example: *C'est Paul Leclerc? Paul habite à Lyon? Paul habite avenue des Champs-Élysées? Il quitte la maison à sept heures et demie du soir?*

RETEACHING (*Mots 1*)

With books opened to page 64, have several students act out each of the verbs presented in *Mots 1*. Other students will guess the verb or verb phrase that is being acted out.

LEARNING FROM PHOTOS

Say: *C'est Serge. Il est parisien.*
Then ask: *Serge est français ou américain? Il est blond? Il habite à Paris? Serge est élève dans un lycée? Il étudie? Il est intelligent?*

INDEPENDENT PRACTICE

Assign any of the following:
1. Exercises, pages 66–67
2. Workbook, *Mots 1: A–C*, page 18
3. Communication Activities Masters, *Mots 1: A*, page 14
4. CD-ROM, Disc 1, pages 64–67

Vocabulary Teaching Resources

1. Vocabulary Transparencies 3.2 (A & B)
2. Audio Cassette 3A/CD-3
3. Student Tape Manual, Teacher's Edition, *Mots 2: D–F*, pages 31–32
4. Workbook, *Mots 2: D–G*, pages 19–20
5. Communication Activities Masters, *Mots 2: B*, page 14
6. Chapter Quizzes, *Mots 2: Quiz 2*, page 15
7. Computer Software, *Vocabulaire*
8. CD-ROM, Disc 1, *Mots 2: pages 68–71*

Bell Ringer Review

Write the following on the board or use BRR Blackline Master 3-2: On a piece of paper, make a sketch that depicts the following words or phrases.

1. arriver à la maison
2. étudier l'histoire
3. regarder la télévision
4. une rue
5. parler français

PRESENTATION *(pages 68–69)*

A. Have students close books. Model the new vocabulary on pages 68–69 using Vocabulary Transparencies 3.2 (A & B). Have students repeat each word two or three times after you or Cassette 3A/CD-3.

B. On the board, write: *À plein temps = 40 heures. À mi-temps = 10 heures, 15 heures, etc., mais pas 40 heures.*

C. Act out the following *-er* verbs using gestures or dramatization: *parler au téléphone, écouter, regarder, danser, chanter, rigoler.*

VOCABULAIRE

MOTS 2

APRÈS LES COURS

un magasin de disques

des cassettes

une cassette

des magazines

un magazine

un walkman

une vidéo(cassette)

un compact disc

l'argent

Voici Pauline.
Pauline travaille après les cours.
Elle travaille dans un magasin de disques à Montréal.

Pauline
lundi
mardi   de 16h à 20h
mercredi
jeudi   de 16h à 20h
vendredi
samedi   de 10h à 17h
dimanche

Elle travaille quinze heures par semaine.
Elle travaille à mi-temps.
Elle ne travaille pas à plein temps.

Elle gagne cinquante dollars par semaine.

68 CHAPITRE 3

TOTAL PHYSICAL RESPONSE

(following the Vocabulary presentation)

TPR 1
(Female student), **levez-vous.**
Venez ici un moment.
Regardez ___ (a male student).
Parlez à ___ .
___ (name of male student), **levez-vous.**
Invitez ___ **à danser.**

Dansez avec ___ .
Dansez lentement. (Gesture for *lentement.*)
Et maintenant, arrêtez-vous.
___ , **retournez à votre place.**
Merci, vous dansez très bien. Et maintenant, asseyez-vous, s'il vous plaît.

la télé

Les copains parlent.
Ils parlent au téléphone.

Ils regardent la télé.
Ils n'écoutent pas la radio.

la radio

une fête

Vendredi soir Caroline donne une fête.
Caroline aime (adore) les fêtes.
Elle invite des amis.

Pendant la fête les amis dansent.
Ils écoutent des cassettes.

Ils rigolent.

Note: The verb *rigoler* is an informal word which means "to joke around," "to have a good time."

Ils chantent.

CHAPITRE 3   **69**

RECYCLING

To recycle words presented in *Mots 1*, ask plural questions using verbs from *Mots 1: Les élèves quittent la maison à quelle heure? Ils arrivent à l'école à quelle heure? Ils entrent dans la salle de classe? Ils regardent le tableau noir? Ils posent des questions?*

Culture Note Since the word *rigoler* ("to fool around," "to laugh") is an informal word, it is fine to use it with friends. Similarly, students may prefer using the informal word for a party, *une boum,* instead of *une fête.*

COGNATE RECOGNITION

Ask students to identify the following cognates on pages 68–69. Have them repeat the cognates carefully, as these are words they are prone to anglicize.

la cassette	**le magazine**
le walkman	**la vidéocassette**
le compact disc	**adorer**
la radio	**danser**
le téléphone	**inviter**

Vocabulary Expansion

Solely for recognition purposes and rapid expansion of vocabulary, you may give the students some noun forms of the verbs they have learned in this chapter. They do not have to learn to produce them.

arriver	**l'arrivée**
entrer	**l'entrée**
travailler	**le travail**
étudier	**l'étude**
rentrer	**la rentrée**
inviter	**une invitation**
danser	**une danse**
chanter	**une chanson**

TPR 2

___ , levez-vous, s'il vous plaît.
Vous êtes Marcel Marceau, le célèbre mime français.
Prenez le téléphone.
Téléphonez à un(e) ami(e).
Parlez à votre ami(e).
Chantez pour votre ami(e).
Vous adorez votre petit(e) ami(e). Ah, mais quelque chose se passe.

Vous n'êtes pas content(e). Vous n'aimez pas votre ami(e).
Parlez sévèrement à votre ami(e).
Très bien. Ça change. Vous aimez encore votre ami(e).
Parlez doucement à votre ami(e).
Merci, Marcel. Vous êtes un très bon mime.
Et maintenant, retournez à votre place et asseyez-vous.

CROSS-CULTURAL COMPARISON

The term *surprise-partie* is still heard, but it is no longer common. It was borrowed from the English and is used for any party. No surprise is involved.

Exercices

PRESENTATION (*page 70*)

Extension of *Exercice A*

Review vocabulary from this and previous chapters by having one or more students walk around the room pointing to objects whose names they have already learned. The student asks: *Qu'est-ce que c'est?* Other students respond.

ANSWERS

Exercice A

1. un walkman
2. un magazine
3. une radio
4. une vidéo(cassette)
5. une cassette
6. un compact disc

Exercice B

1. Les copains écoutent des cassettes.
2. Ils aiment la musique populaire.
3. Oui, les copains regardent la télé.
4. Caroline donne une fête.
5. Oui, elle invite des amis.
6. Elle donne la fête vendredi soir.

Exercice C

1. Pauline est canadienne.
2. Non, après les cours elle n'est pas libre. (Elle travaille.)
3. Oui, elle travaille.
4. Elle travaille dans un magasin de disques (à Montréal).
5. Elle travaille à mi-temps.
6. Elle gagne cinquante dollars par semaine.

INFORMAL ASSESSMENT

(*Mots 2*)

Hold up appropriate objects, or show visuals that represent the vocabulary on pages 68–69. Ask students to give the word each object or visual represents.

RETEACHING (*Mots 1* and 2)

Using Vocabulary Transparencies 3.1 and 3.2, quickly have students say anything they can about the images without your help. This gives students a chance to use many of the words they have just learned in Chapter 3.

70

Exercices

A **Qu'est-ce que c'est?** Identifiez. (*Identify each item.*)

1. 2. 3.

4. 5. 6.

B **Après les cours.** Répondez. (*Answer.*)

1. Après les cours les copains écoutent des compact discs ou des cassettes?
2. Ils aiment la musique classique ou populaire?
3. Les copains regardent la télé?
4. Qui donne une fête?
5. Elle invite des amis?
6. Quand est-ce qu'elle donne la fête?

C **Pauline travaille!**
Répondez. (*Answer.*)

1. Pauline est française ou canadienne?
2. Après les cours elle est libre?
3. Elle travaille?
4. Où est-ce qu'elle travaille?
5. Elle travaille à mi-temps ou à plein temps?
6. Elle gagne combien d'argent par semaine?

Armand et Justine travaillent au Chalet Suisse.

COOPERATIVE LEARNING

Write the following sentences on slips of paper and place them in an envelope:

Ils parlent.
Ils parlent au téléphone.
Ils écoutent des cassettes.
Ils écoutent des compact discs.
Ils écoutent un walkman.
Ils écoutent la radio.
Ils regardent la télé.
Ils dansent.
Ils chantent.
Ils rigolent.

Each team draws a slip of paper. Teams take turns pantomiming the action in front of the class. The other teams discuss what has been acted out and then send one person to write the sentence on the board.

Activités de communication orale
Mots 1 et 2

A **Au magasin de disques.** You're in a record store in Quebec. Ask the salesperson (your partner) the price of each of the following items.

> un disque ($5)
>
> Élève 1: S'il vous plaît, Mademoiselle (Monsieur).
>   C'est combien, le disque?
> Élève 2: C'est cinq dollars.

1. 2. 3.
4. 5. 6.

B **En classe ou après les cours?** Tell a few things a friend of yours does almost every day. Your partner will decide whether your friend does these things in class or after school.

> Élève 1: Il/Elle regarde la télé.
> Élève 2: Il/Elle regarde la télé après les cours.
>
> Élève 1: Il/Elle passe un examen.
> Élève 2: Il/Elle passe un examen en classe.

INDEPENDENT PRACTICE

Assign any of the following:
1. Exercises and activities, pages 70–71
2. Workbook, *Mots 2: D–G,* pages 19–20
3. Communication Activities Masters, *Mots 2: B,* page 14
4. Computer Software, *Vocabulaire*
5. CD-ROM, Disc 1, pages 68–71

ADDITIONAL PRACTICE

Student Tape Manual, Teacher's Edition, *Activités E–F,* page 32.

Activités de communication orale
Mots 1 et 2

These activities allow students to use the chapter vocabulary and grammar in open-ended situations. It is not necessary that you do them all. You may select those you consider most appropriate.

RECYCLING

These activities recycle vocabulary from previous chapters, including numbers, and practice *-er* verbs in the third person.

PRESENTATION *(page 71)*
Extension of *Activité B*

After completing Activity B, reinforce by writing one or more situations on the board, e.g., *En classe, À la fête, Après les cours.* Students in pairs compose sentences to say what people do NOT do in that situation, e.g., *À la fête, ils ne regardent pas le prof. Ils ne travaillent pas.* Compile a list for each situation on the board.

ANSWERS
Activité A
1. É1: S'il vous plaît, Mademoiselle (Madame, Monsieur), c'est combien, le compact disc? É2: C'est quinze dollars.
2. ... la cassette? ... huit dollars.
3. ... la vidéo(cassette)? ... trente dollars.
4. ... la radio? ... quarante-cinq dollars.
5. ... le walkman? ... soixante dollars.
6. ... le magazine? ... trois dollars.

Activité B
Answers will vary.

STRUCTURE

Structure Teaching Resources

1. Workbook, *Structure: A–I*, pages 21–24
2. Student Tape Manual, Teacher's Edition, *Structure: A–H*, pages 33–36
3. Audio Cassette 3A/CD-3
4. Grammar Transparency G-3
5. Communication Activities Masters, *Structure: A–E*, pages 15–18
6. Computer Software: *Structure*
7. Chapter Quizzes, *Structure: Quizzes 3–6*, pages 16–19
8. CD-ROM, Disc 1, pages 72–77

Bell Ringer Review

Write the following on the board or use BRR Blackline Master 3-4: Write words or phrases associated with:

la musique   **la fête**
la salle de classe   **un garçon**

Le pronom on

PRESENTATION (*page 72*)

Lead students through steps 1–3 on page 72. You may wish to put additional examples on the board.

ANSWERS

Exercice A

Answers will vary, but each response should use *on*.

RETEACHING

Have students write sentences using *on* and each of the following verbs: *entrer, écouter, étudier, gagner, parler, habiter.*

72

Le pronom *on*   *Talking About People in General: "We," "People," "They"*

1. You will use the word *on* a great deal in French. It can have many different meanings. One of its most common meanings is "we." Its other equivalents in English are words such as "people" and "they."

   On parle français en France.   *They (People) speak French in France.*

2. You can also use *on* to make suggestions about doing something.

   On regarde la télé?   *Let's watch TV. (Are we going to watch TV?)*

   On écoute la cassette?   *Shall we listen to the cassette?*

3. With *on* you use the same form of the verb as you do with *il* and *elle.*

   Il parle français en classe.
   On parle français en Belgique.

Exercice

A   Aux États-Unis.   Un(e) élève français(e) pose des questions à un(e) élève américain(e). *(You are a French student. Ask a classmate about life in the U.S.)*

   On arrive à l'école à quelle heure?
   On arrive à l'école à huit heures.

1. On entre dans la salle de classe à quelle heure?
2. On quitte l'école à quelle heure?
3. On déteste les examens?
4. On travaille beaucoup à l'école?
5. On aime le cinéma?
6. On travaille après les cours?
7. On écoute des disques rock?
8. On regarde la télé?
9. On parle au téléphone?

72   CHAPITRE 3

ADDITIONAL PRACTICE

After completing Exercise A on page 72, tell students you are going to give them some action words. Let them suggest that you all do it, using the pronoun *on.*

danser: On danse?   parler
regarder le   donner une fête
   tableau noir   écouter (un disque)
chanter   étudier

LEARNING FROM PHOTOS

Have students look at the photo on page 72. How many original statements can they make about it?

Les verbes réguliers en -er au présent

Describing People's Activities

1. A verb is a word that expresses an action or a state of being. Words such as *parler, travailler,* and *aimer* are verbs. These are called regular verbs because they all follow the same pattern and have the same endings.

2. The infinitive form of these verbs ends in *-er.* The infinitive is the basic form of the verb that you find in the dictionary.

parler	*to speak, to talk*
travailler	*to work*
aimer	*to like*

3. You drop the *-er* of the infinitive to form the stem.

parler	parl-
aimer	aim-

4. You add the endings for each subject to this stem. Study the following chart.

INFINITIVE	PARLER	AIMER	
STEM	parl-	aim-	ENDINGS
	je parle	j'aime	-e
	tu parles	tu aimes	-es
	il parle	il aime	
	elle parle	elle aime	-e
	on parle	on aime	
	nous parlons	nous aimons	-ons
	vous parlez	vous aimez	-ez
	ils parlent	ils aiment	
	elles parlent	elles aiment	-ent

5. You pronounce the *je, tu, il, elle, on, ils,* and *elles* forms of the verb the same even though they are spelled differently.

6. When a verb begins with a vowel or a silent *h, je* is shortened to *j'.*

   J'aime Paris.
   J'habite à Lyon.

7. In the negative, you shorten the *ne* to *n'* before a vowel or a silent *h.*

   Je n'aime pas les maths.
   Je n'habite pas à Paris.

MEMORY AID

To help students remember the pronunciation of the *-er* verb endings, write out the conjugation of the verb *parler* on the board, listing the singular and plural forms in two columns. Draw a large circle around the three singular forms and the third person plural form. The resulting boot-shaped outline will help students remember which forms sound the same. They're all in the same "boot!"

Les verbes réguliers en -er au présent

PRESENTATION *(pages 73–74)*

A. Draw two stick figures on the board and label one Mireille and the other Didier. Write the following verbs on the board and have students make up sentences about either Mireille or Didier: *travailler, étudier, passer un examen, quitter la maison, arriver à l'école, donner une fête, inviter des amis, danser, chanter, rigoler.* This reviews the *-er* verbs from the *Mots* section in Chapter 3.

B. Lead students through steps 1–8 on pages 73–74. Write the present tense forms of *parler* and *aimer* on the board. Point to the pronouns *je, tu, il, elle, on, ils, elles* as you pronounce the forms. Students should come to realize that although the forms are spelled differently, they are pronounced the same.

C. Point out that the oral forms of the *-er* verbs are quite easy. They do, however, present a spelling problem. You may wish to emphasize this by writing the endings in a different color chalk. Then ask students to note which endings remain silent and which ones are pronounced.

D. Point out elision and liaison in the pronunciation of the forms of *aimer.*

Note In the CD-ROM version, this structure point is presented via an interactive electronic comic strip.

RETEACHING

Review the subject pronouns by having students point to themselves as they say *je;* look at a friend as they say *tu;* point to a boy as they say *il;* and point to a girl as they say *elle.* Have students do the same with the plural pronouns.

Exercices

PRESENTATION (pages 74–75)

Note Exercises A–G on pages 74–76 can be done with books closed the first time and open the second time. When calling on individuals to respond, you may have each student do one item or several at a time.

Exercice A

Exercise A reviews verb forms students already know from the *Mots* sections of Chapter 3.

Extension of *Exercice A*

After completing Exercise A, have several students combine the sentences into a short narration. For example, *Thérèse est française. Elle habite à Paris. Elle habite avenue Saint-Pierre,* etc.

Exercice B

Exercise B reviews *qui*. This exercise also points out that *qui* with a singular verb form is used to elicit plural responses.

ANSWERS

Exercice A

1. Thérèse est française.
2. Elle habite à Paris.
3. Elle habite avenue Saint-Pierre.
4. Elle parle français.
5. Elle quitte la maison le matin.
6. Elle arrive à l'école à huit heures dix.

Exercice B

1. Les profs et les élèves entrent dans la salle de classe.
2. Les profs et les élèves parlent en classe.
3. Les profs et les élèves écoutent en classe.
4. Les élèves regardent le tableau noir.
5. Les profs donnent les examens.
6. Les élèves passent les examens.
7. Les profs corrigent les examens.
8. Les élèves étudient beaucoup.
9. Les élèves et les profs posent des questions.

---

8. With all verbs beginning with a vowel or a silent *h* there is a liaison between the subject and the verb with the plural forms *nous, vous, ils,* and *elles.* The *s* is pronounced like a *z.*

> nous étudions   vous aimez   ils habitent

Exercices

A **Thérèse parle français.** Répondez d'après les dessins. (*Answer according to the illustrations.*)

1. Thérèse est américaine ou française?
2. Elle habite à Chicago ou à Paris?
3. Elle habite avenue Gambetta ou avenue Saint-Pierre?
4. Elle parle anglais ou français?
5. Elle quitte la maison le matin ou l'après-midi?
6. Elle arrive à l'école à quelle heure?

B **Les élèves ou les profs?** Dites si ce sont les professeurs, les élèves ou les deux. (*Tell who is doing the following activities—the students, the teachers, or both.*)

> Qui arrive à l'école le matin?
> *Les profs et les élèves arrivent à l'école le matin.*

1. Qui entre dans la salle de classe?
2. Qui parle en classe?
3. Qui écoute en classe?
4. Qui regarde le tableau noir?
5. Qui donne les examens?
6. Qui passe les examens?
7. Qui corrige les examens?
8. Qui étudie beaucoup?
9. Qui pose des questions?

LEARNING FROM ILLUSTRATIONS

Ask students to volunteer additional questions based on the illustrations on page 74. For example, *Thérèse a un sac à dos ou un morceau de craie? Thérèse parle avec une copine ou un copain?*

ADDITIONAL PRACTICE

Ask students questions such as the following about your own class:
Qui arrive à l'école à sept heures?
Qui donne les examens d'anglais?
Qui corrige les examens de maths?
Qui étudie beaucoup?
Qui écoute en classe?
Qui parle beaucoup en classe?
Qui passe des examens?

C Tu parles français? Répétez la conversation. *(Practice the conversation.)*

BARBARA: René, tu n'es pas français, n'est-ce pas?
RENÉ: Non, je ne suis pas français.
BARBARA: Mais tu parles français.
RENÉ: Bien sûr, je parle français.
BARBARA: Mais comment ça, si tu n'es pas français?
RENÉ: Mais je suis belge.
BARBARA: Ah, c'est vrai. On parle français en Belgique.

D À votre tour. Donnez des réponses personnelles. *(Give your own answers.)*

1. Tu habites dans quelle ville?
2. Tu quittes la maison à quelle heure le matin?
3. Tu arrives à l'école à quelle heure?
4. Est-ce que tu parles français avec les copains?
5. Tu parles quelle langue dans la classe de maths?
6. Tu aimes quels cours? quels profs?
7. Tu détestes quels cours?
8. Est-ce que tu travailles après les cours?
9. Est-ce que tu chantes quand tu écoutes la radio?
10. Quand est-ce que tu regardes la télé?

La Grand-Place à Bruxelles, en Belgique

E Pardon? Posez des questions d'après le modèle. *(Ask questions according to the model.)*

> Nous écoutons des compact discs.
> *Pardon? Qu'est-ce que vous écoutez?*

1. Nous détestons la musique classique.
2. Nous regardons la télé.
3. Nous regardons les magazines.
4. Nous écoutons la radio.
5. Nous aimons les fêtes.
6. Nous donnons une fête.

F Vous donnez une fête? Donnez des réponses personnelles avec «nous». *(Give your own answers with* nous.*)*

1. Vous donnez une fête?
2. Pendant la fête, vous dansez?
3. Vous chantez?
4. Vous écoutez des disques?
5. Vous regardez la télé?

INDEPENDENT PRACTICE

Assign any of the following:
1. Exercises, pages 72, 74–76
2. Workbook, *Structure: A–F,* pages 21–23
3. Communication Activities Masters, *Structure: A & B,* pages 15–16
4. CD-ROM, Disc 1, pages 72–76

DID YOU KNOW?

Tell students that La Grand-Place, the main square, is in the heart of Brussels, the capital of Belgium. It offers a variety of sights, including flower markets, museums, concerts and sound-and-light shows. Especially entertaining is the people-watching that can be engaged in from any of the cafés located on the square.

PRESENTATION *(page 75)*

Exercice C

This mini-conversation lets students hear, see, and say the first- and second-person singular of *-er* verbs before they use them actively in Exercise D.

1. Read the conversation and have students repeat after you.
2. Ask for volunteers to take the parts.

Exercice E

This exercise gives students practice using the interrogative *Qu'est-ce que* and the *nous* form of the verb. Have students look at one another to make this exercise more realistic.

Note Go over all exercises in class before assigning them as independent practice.

ANSWERS

Exercice D

Answers will vary. However, students should respond with *je* and the correct form of each verb.

Exercice E

1. Pardon? Qu'est-ce que vous détestez?
2. ... vous regardez?
3. ... vous regardez?
4. ... vous écoutez?
5. ... vous aimez?
6. ... vous donnez?

Exercice F

Answers will vary, but students should respond with *nous* and the correct form of the verb.

RETEACHING

Place groups and individuals corresponding to the subject pronouns around the room. You, the teacher, will be *vous*. Draw a dog or cat on the board for *tu*. Give a student a base sentence, such as *J'écoute la radio.* The student walks around the room designating different groups or individuals and changing the sentence accordingly. For example: Student (pointing to him/herself): *J'écoute la radio.* (pointing to a girl): *Elle écoute la radio.* (pointing to a mixed group): *Ils écoutent la radio.* You may wish to use props.

Exercice G

1. donnons	7. danse,
2. donnons	chante
3. invitons	8. aimez
4. arrive	9. aimez
5. arrivent	10. rentrez
6. rigolent	

L'article indéfini au pluriel; La négation des articles indéfinis

PRESENTATION *(page 76)*

A. Review the singular articles *un* and *une* on page 21 before introducing the plural *des*.

B. Lead students through step 1 on page 76. Have them repeat the examples after you.

C. Model singular and plural forms using familiar objects: *un élève, des élèves, un livre, des livres.* Then have students supply the plural forms of: *une feuille de papier, un garçon, une fille, un examen, un magazine, une cassette.*

D. Now lead students through step 2 on page 76. Point out that *un*, *une*, and *des* are not used in negative constructions. Instead, *de* is used. Be sure to point out the fourth example of step 2 on page 76, which demonstrates that *de* elides to *d'* before a vowel or a silent *h*.

Exercices

ANSWERS

Exercice A

Answers will vary. However, students should use *de* if they answer in the negative.

G   **Notre fête.**   Complétez. *(Complete.)*

1. Nous ___ une fête. (donner)
2. Nous ___ la fête pour célébrer l'anniversaire *(birthday)* de Claude. (donner)
3. Nous ___ les amis de Claude. (inviter)
4. Claude ___ à l'heure. (arriver)
5. Les amis ___ à la fête. (arriver)
6. Pendant la fête les amis ___. (rigoler)
7. On ___ et on ___. (danser, chanter)
8. Et vous, vous ___ les fêtes? (aimer)
9. Vous ___ danser? (aimer)
10. Vous ___ à quelle heure? (rentrer)

L'article indéfini au pluriel; La négation des articles indéfinis

Expressing "Some" and "Not Any"

1. You have already learned the singular indefinite articles *une* and *un*. The plural of *une* and *un* is *des*, which means "some" or "any" in English.

Il regarde un magazine.	**Il regarde *des* magazines.**
Elle écoute une cassette.	**Elle écoute *des* cassettes.**
Il invite un(e) ami(e).	**Il invite *des* ami(e)s.**

2. In the negative all the indefinite articles change to *de*. Note that *de* is shortened to *d'* before a vowel or silent *h*.

J'écoute un disque.	**Je *n'*écoute *pas de* disque.**
Tu regardes une vidéo.	**Tu *ne* regardes *pas de* vidéo.**
Nous invitons des copains.	**Nous *n'*invitons *pas de* copains.**
Les élèves passent des examens.	**Mais ils *ne* passent *pas d'*examens aujourd'hui.**

Exercices

A   **Le temps libre.**   Donnez des réponses personnelles. *(Give your own answers.)*

1. Quand tu es libre, est-ce que tu regardes des livres scolaires ou des magazines?
2. Tu écoutes des cassettes ou des disques après les cours?
3. Pendant le week-end, tu regardes des livres ou des vidéos?
4. Quand tu donnes une fête, tu invites des amis?
5. Pendant une fête tu regardes des vidéos?

76   CHAPITRE 3

B **En classe.** Répondez négativement. *(Answer in the negative.)*

1. Tu écoutes des compact discs?
2. Tu regardes une vidéo?
3. Tu chantes une chanson populaire?
4. Tu regardes un magazine?
5. Le professeur passe des examens?
6. Il donne des devoirs amusants?

Le verbe + l'infinitif

Talking About What You Like or Don't Like to Do

1. In French when the verbs *aimer, adorer,* and *détester* are followed by another verb, the second verb is in the infinitive form.

   J'aime chanter.   **J'adore danser.**   **Je déteste étudier.**

2. In a negative sentence the *ne… pas* goes around the first verb.

   Il *n'*aime *pas* chanter.

Exercices

A **Tu aimes danser?** Posez les questions suivantes à un copain ou à une copine. *(Ask a classmate the following questions.)*

   Élève 1: **Tu aimes danser?**
   Élève 2: **Bien sûr. J'aime beaucoup danser. (Mais non! Pas du tout.**
     Je déteste danser.)

1. Tu aimes écouter la radio?
2. Tu aimes regarder la télé?
3. Tu aimes étudier?
4. Tu aimes parler au téléphone?
5. Tu aimes rigoler?
6. Tu aimes chanter?

B **Ils aiment danser?** Décidez si ces personnes aiment ces activités. *(Decide if these people like the following activities.)*

   Elle aime (adore) chanter.

1.   2.   3.   4.

Exercice B
1. Je n'écoute pas de compact discs.
2. Je ne regarde pas de vidéo.
3. Je ne chante pas de chanson populaire.
4. Je ne regarde pas de magazine.
5. Le professeur ne passe pas d'examens.
6. Il ne donne pas de devoirs amusants.

Le verbe + l'infinitif

PRESENTATION *(page 77)*

A. Lead students through steps 1–2 on page 77. Have them repeat the examples after you. Ask them to volunteer other examples stating what they like and dislike doing.
B. Ask questions using familiar verbs, for example: *Tu aimes danser? Tu détestes écouter la musique d'Elvis?*

Exercices

PRESENTATION *(page 77)*

Exercice A
You may wish to use the recorded version of this exercise.

ANSWERS

Exercice A
Answers will vary.

Exercice B
1. Il aime danser.
2. Elle aime parler.
3. Ils aiment danser.
4. Il n'aime pas étudier.

CONVERSATION

Bell Ringer Review

Write the following on the board or use BRR Blackline Master 3-6: Write three things you like to do in school and three things you don't like to do.

PRESENTATION (*page 78*)

A. Tell students they will hear a conversation between Jeanne and Charles about their French class.

B. Have them close their books and watch the Conversation Video or listen as you read the conversation or play the recorded audio version.

C. Call on two students to read the conversation with as much expression as possible.

D. Have students practice reading the conversation in pairs, then call on two students to present it to the class.

Note In the CD-ROM version, students can play the role of either one of the characters and record the conversation.

Exercices

ANSWERS

Exercice A

1. Charles aime le français. C'est extra.
2. Le prof est très intéressant.
3. Oui, Charles aime beaucoup parler français.
4. Oui, Charles parle bien.
5. Oui, Jeanne est française.
6. Non, Charles n'est pas français.

Exercice B

Answers will vary.

CONVERSATION

Scènes de la vie   *Après le cours de français*

JEANNE: Charles, tu aimes le français?
CHARLES: Beaucoup. C'est extra, vraiment.
JEANNE: Pourquoi ça?
CHARLES: Le prof est très intéressant.
JEANNE: Et tu aimes parler?
CHARLES: Beaucoup.
JEANNE: Et tu parles très, très bien le français, Charles.
CHARLES: Merci, Jeanne.

A **Charles et Jeanne.**   Répondez d'après la conversation. (*Answer according to the conversation.*)

1. Charles aime quel cours? Pourquoi?
2. Comment est le prof?
3. Charles aime parler français?
4. Charles parle bien ou pas?
5. Jeanne est française ou pas?
6. Et Charles, il est français ou pas?

B **À votre tour.**   Donnez des réponses personnelles. (*Give your own answers.*)

1. Tu aimes le français? Pourquoi?
2. Comment est le professeur?
3. Tu aimes parler français?
4. Tu parles bien ou pas?

78   CHAPITRE 3

LEARNING FROM PHOTOS

Have students describe Jeanne and Charles and where they are in the photo on page 78.

Prononciation   *Les sons /é/ et /è/*

There is an important difference in the way French and English vowels are pronounced. When you say the French word *des*, your mouth is tense, in one position. You can actually repeat the vowel sound /é/ as many times as you want without moving your mouth at all. But when you pronounce the English word "day," your mouth is relaxed and you actually say two vowel sounds.

Now listen to the word *élève*. There are two distinct vowel sounds. The sound /é/ is a "closed" sound and /è/ an "open" sound. This describes the positions of the mouth for each of these sounds. Repeat the following.

Le son /é/   la télé   le café   l'école   écoutez
Le son /è/   après   la fête   vous êtes   la cassette

Après l'école, les élèves aiment écouter des cassettes.

élève

Activités de communication orale

A   Des préférences.   Ask a classmate which courses he or she likes or dislikes and why. Then tell the class what he or she said.

B   Un copain français ou une copine française.   You're spending the summer in France and you've just met a French student (your partner) who'd like to know more about you. Tell him or her:

1. where you're from
2. where you're a student
3. what time you leave home in the morning
4. what your French class is like
5. what your French teacher is like
6. when you leave school
7. some things you do after school

C   Vous aimez ou vous n'aimez pas... ?   Divide into small groups and choose a leader. Using the list below, the leader finds out what each person in the group likes and doesn't like to do and then reports to the class.

> étudier
>
> Élève 1: Tu aimes étudier?
> Élève 2: Moi, j'aime étudier. (Je déteste étudier.), etc.
> Élève 1 (*à la classe*): Martin et Anne aiment étudier...

1. danser
2. chanter
3. passer des examens
4. regarder la télé
5. donner des fêtes
6. parler au téléphone

Prononciation

PRESENTATION (*page 79*)

A. Model the key word *élève* and have students repeat in unison and individually.
B. Now model the words and phrases in similar fashion.
C. You may wish to give students the following *dictée*:
   La télé est dans le café. Vous êtes dans une fête. Écoutez la cassette. Écoutez la cassette après la fête.
D. For additional practice, you may use the Pronunciation section on Cassette 3A/CD-3 and *Activités K–M* in the Student Tape Manual, Teacher's Edition, pages 37–38.
E. Use Pronunciation Transparency P-3 for practice.

Bell Ringer Review

Write the following on the board or use BRR Blackline Master 3-7: Make sentences from the following word scrambles.

1. une prof habite maison le
2. beaucoup la rigolent les fête pendant amis
3. au vous aimez téléphone parler
4. la nous ne télé regardons pas

Activités de communication orale

PRESENTATION (*page 79*)

Extension of Activité B
   How would you ask this French student these same questions?

ANSWERS

Activités A, B and C
   Answers will vary.

LECTURE ET CULTURE

Bell Ringer Review

Write the following on the board or use BRR Blackline Master 3-8: Write the question word or phrase associated with each of the following. For example: *le matin = quand?*

1. le prof
2. à la maison
3. à huit heures
4. l'après-midi
5. à Paris

READING STRATEGIES
(page 80)

Pre-reading

Have students locate the Sorbonne (5ᵉ *arrondissement*) on the map of Paris on page 505 (page 238, Part A). You may want to have students look at the photos on pages 82–83 now.

Reading

A. Have students close their books. Relate the story on page 80 in your own words. Follow up by asking a few questions about what you said.
B. With books open, have the class repeat two or three sentences after you. Ask comprehension questions.
C. Go over the story a second time, calling on individuals to read.

Post-reading

Have students write answers to the *Étude de Mots*, page 80, and the *Compréhension*, page 81.

Note Students may listen to a recorded version of the *Lecture* on the CD-ROM.

Étude de mots

ANSWERS

Exercice A

1. arrive
2. commencent
3. cours
4. obligatoire

Exercice B

1. c
2. d
3. b
4. e
5. a
6. f

UNE ÉLÈVE PARISIENNE

Geneviève habite rue Saint-Julien-le-Pauvre à Paris. La rue Saint-Julien-le-Pauvre est près de la Sorbonne. La Sorbonne est une université célèbre à Paris. Geneviève quitte la maison à huit heures moins le quart. Elle est élève au Lycée Saint-Louis. Les cours commencent à huit heures et demie. Geneviève arrive au lycée à huit heures. Elle aime arriver de bonne heure[1]! Avant[2] les cours elle parle avec les copains dans la cour[3]. Elle aime ça. Elle quitte le lycée à cinq heures.

Les lycéens français passent à peu près[4] trente heures par semaine à l'école. En France la plupart[5] des matières sont obligatoires et très peu de[6] matières sont facultatives. On est libre le mercredi après-midi.

En France les élèves passent un examen difficile, le baccalauréat (le bachot ou le bac) avant d'être diplômés.

[1] de bonne heure *early*
[2] Avant *Before*
[3] cour *courtyard*
[4] à peu près *about*
[5] la plupart *most*
[6] peu de *few*

Étude de mots

A **Des mots apparentés.** Choisissez le bon mot. (*Choose the correct word.*)

arrive   commencent
cours   obligatoire

1. J'___ à l'école à sept heures et demie du matin.
2. Les cours ___ à huit heures.
3. Le ___ de Madame Benoît est très intéressant.
4. L'anglais est un cours ___.

B **En France.** Trouvez les mots qui correspondent. (*Find the corresponding word or phrase.*)

1. célèbre
2. le lycée
3. la matière
4. une matière facultative
5. le baccalauréat
6. les vacances

a. le bachot, le bac
b. la discipline, le cours
c. fameux, illustre
d. une école secondaire française
e. le contraire d'une matière obligatoire
f. la période de temps où on est libre

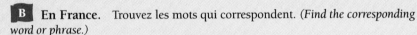

5ᵐᵉ ARR.
RUE SAINT JULIEN LE PAUVRE

FINDING DIFFERENCES

After reading the *Lecture* and going over the *Compréhension* exercises, have students make a short list in French of those things that are different in French schools.

CRITICAL THINKING ACTIVITY

(*Thinking skills: making inferences*)
After reading the *Lecture*, write the following on the board or on an overhead transparency:
Pourquoi est-il important d'arriver à l'école à l'heure?
a. Le prof est content.
b. C'est une des responsabilités des élèves.

Compréhension

C Vous avez compris? Répondez. (*Answer.*)

1. Où est-ce que Geneviève habite?
2. Elle habite dans quelle rue?
3. Où est la rue?
4. Geneviève quitte la maison à quelle heure?
5. Les cours commencent à quelle heure?

D Les écoles en France. Trouvez les renseignements suivants dans la lecture. (*Find the following information in the reading.*)

1. the name of a university in Paris
2. a test taken by French students
3. the number of hours spent weekly by French students in school
4. when French students are free
5. the time school begins in France

E Aux États-Unis. Répondez. (*Answer.*)

1. Les cours commencent à quelle heure?
2. Les élèves américains passent combien d'heures par semaine à l'école?
3. On est libre quels jours aux États-Unis?

DÉCOUVERTE CULTURELLE

Aux États-Unis beaucoup d'élèves travaillent après les cours. Ils travaillent à mi-temps. Ils travaillent, par exemple, dans un magasin, dans un supermarché ou dans un restaurant. Ils gagnent de l'argent—quarante ou cinquante dollars par semaine. Ils dépensent[1] l'argent pour aller au cinéma, pour acheter[2] des cassettes ou des jeans.

En France, au contraire, relativement peu de lycéens travaillent après les cours. Les élèves ne travaillent pas à mi-temps. C'est assez rare.

[1] dépensent *spend*
[2] acheter *to buy*

RÉPUBLIQUE FRANÇAISE

MINISTÈRE DE L'ÉDUCATION NATIONALE

ACADÉMIE DE PARIS

DÉPARTEMENT DE PARIS

LE DIPLÔME NATIONAL DU BREVET

SÉRIE : COLLÈGE

VU les textes en vigueur   VU le procès verbal du jury

EST DELIVRÉ

à   **MADEMOISELLE BRILLIÉ**   **MARINA DIANE AVIVA**

né(e)   le 01 AVRIL 1982   à 075 PARIS 14

fait à   **ARCUEIL**   le 22 OCTOBRE 1997

Signature du Titulaire,

Le Directeur du Service Interacadémique des examens et concours

J. KOOIJMAN

No. 7506094

Vous êtes priés de faire des photocopies certifiées conformes a l'original: il ne sera pas délivré de duplicata

---

DID YOU KNOW?

French students have the longest school day (six hours) of students in any European country. Their school day usually ends at four or four-thirty. French schools are closed on Wednesday afternoon; however, students go to school on Saturday morning. French students spend many hours doing homework. Schools use a grading scale from 0 to 20. Teachers in general are strict graders; therefore, students consider a 10 to be satisfactory and a 14 and above very good. Since French teenagers rarely have part-time jobs, they must depend on a weekly allowance for their spending money.

Bell Ringer Review

Write the following on the board or use BRR Blackline Master 3-10: Are the following subjects required or elective in our school? Use *obligatoire/pas obligatoire*.

1. l'anglais
2. la géographie
3. la biologie
4. l'histoire
5. la gymnastique
6. l'art
7. la musique
8. l'algèbre
9. l'informatique

OPTIONAL MATERIAL

PRESENTATION (*pages 82–83*)

A. Have students look at the photographs for enjoyment. If they would like to talk about them, let them say anything they can.

B. Before doing the reading, have students look at the map of Paris on page 505 (page 238, Part A). Have them locate the *Quartier Latin*, the *5^e arrondissement*, the *Sorbonne*, and the *Jardin du Luxembourg*. Then lead the students through the captions on page 82.

Note In the CD-ROM version, students can listen to the recorded captions and discover a hidden video behind one of the photos.

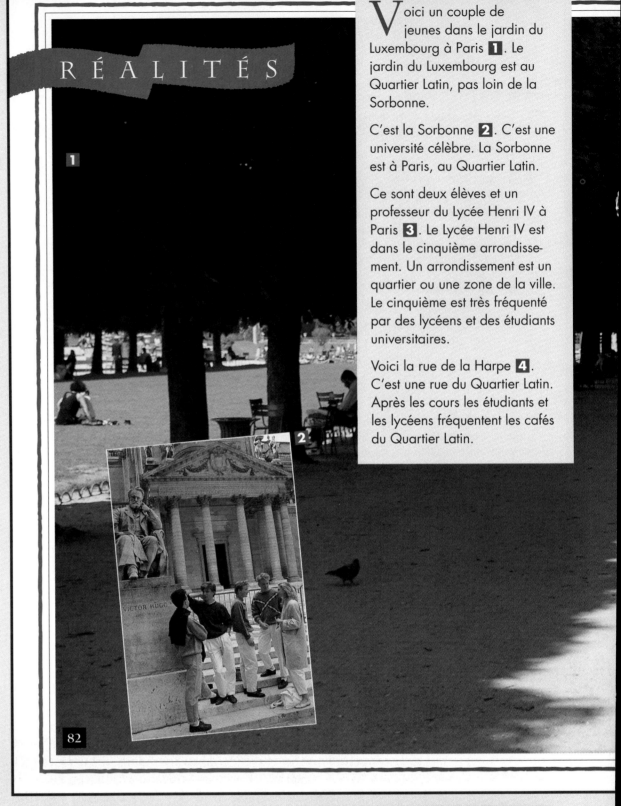

Voici un couple de jeunes dans le jardin du Luxembourg à Paris **1**. Le jardin du Luxembourg est au Quartier Latin, pas loin de la Sorbonne.

C'est la Sorbonne **2**. C'est une université célèbre. La Sorbonne est à Paris, au Quartier Latin.

Ce sont deux élèves et un professeur du Lycée Henri IV à Paris **3**. Le Lycée Henri IV est dans le cinquième arrondissement. Un arrondissement est un quartier ou une zone de la ville. Le cinquième est très fréquenté par des lycéens et des étudiants universitaires.

Voici la rue de la Harpe **4**. C'est une rue du Quartier Latin. Après les cours les étudiants et les lycéens fréquentent les cafés du Quartier Latin.

82

LEARNING FROM PHOTOS

1. Point out the statue of Victor Hugo on page 82 (photo #2). Are statues of literary figures common in the U.S.? Ask students to name any of Hugo's works. The musical *Les Misérables* is based on his novel.
2. The Jardin du Luxembourg, with its fountains, marionette theater, tennis courts, etc., is a wonderful place to stroll, read, or people-watch in the heart of the Latin Quarter.

ADDITIONAL PRACTICE

Student Tape Manual, Teacher's Edition, *Deuxième Partie*, pages 39–40

HISTORY CONNECTION

You may wish to explain to students that the name of the girl in the *Lecture* on page 80, Geneviève, is a famous name in the history of Paris. Geneviève is the patron saint of Paris. Why? You may wish to read the following story to your class and help them get whatever information they can from it.

Un peu d'histoire

En 52 avant Jésus-Christ l'empereur romain Jules César attaque une petite ville gauloise, Lutèce, et Lutèce devient une ville romaine. Au troisième siècle on change le nom de Lutèce en Paris. En 450 les Huns attaquent Paris. Une femme, Geneviève, galvanise la résistance des Parisiens contre les Huns. Elle assure la victoire des Parisiens sur les Huns. Geneviève est une sainte—la sainte patronne de la ville de Paris.

DID YOU KNOW?

Paris is divided into twenty districts called *arrondissements*. The affairs of each *arrondissement* are handled by a commission. The 16th *arrondissement*, on the right bank of the Seine, is probably the most elegant of all the districts. The Sorbonne and the Rue de la Harpe are in the 5th *arrondissement*, which, like the 6th, has many restaurants and shops frequented by students.

INDEPENDENT PRACTICE

1. Workbook: *Un Peu Plus*, page 25
2. Choose a few students to research Paris. Have them research the place names shown on the map of Paris, page 505 (page 238, Part A). They can use the Paris Map Transparency to present their findings to the class.
3. Situation Cards, Chapter 3
4. CD-ROM, Disc 1, page 80–83

RECYCLING

The *Activités de communication orale* and *écrite* allow students to use the vocabulary and grammar in this chapter in open-ended, real-life situations. They also give students a chance to reuse the vocabulary and structure from earlier chapters.

You may have students work on as many of the Activities as you wish.

INFORMAL ASSESSMENT

Oral activities A and B may be used to evaluate the speaking skill. For suggestions on how to grade each student's response, see page 34 of this Teacher's Wraparound Edition.

Activités de communication orale

PRESENTATION (page 84)

Divide students into pairs or groups. Assign one activity to each pair or group. Encourage students to elaborate on the basic theme and to be creative. They may use props, pictures, or posters, if they wish.

Activité A

In the CD-ROM version of this activity, students can interact with an on-screen native speaker.

ANSWERS

Activités A and B

Answers will vary.

Activités de communication orale

A **Au café.** You're seated at a café in Aix-en-Provence. You're chatting with a French student, Paul, who wants to know about life in the U.S. Answer his questions.

1. Tu passes combien d'heures par semaine à l'école?
2. Les cours commencent à quelle heure?
3. Tu es libre quel jour?
4. Les jeunes Américains travaillent après les cours?
5. Tu travailles à mi-temps?

Paul

B **La fête.** You're showing your French friend Alain Dumont a photo of a party at your house. Tell him what an American party is like.

Activité de communication écrite

A **Et toi?** Write a paragraph about yourself by answering the following questions.

1. Où est-ce que tu habites?
2. Où est-ce que tu es élève?
3. Tu arrives à l'école à quelle heure?
4. Les cours commencent à quelle heure?
5. Tu aimes le cours de français?
6. Qui est le prof?
7. Tu aimes quelle autre matière?
8. Tu quittes l'école à quelle heure?
9. Tu travailles à mi-temps après les cours?
10. Tu aimes donner des fêtes pendant le week-end?

84 CHAPITRE 3

Ask the following questions about the bottom photo: *À ton avis, est-ce que nous sommes aux États-Unis ou en France? Qui est sur la photo? Ce sont des lycéens ou des étudiants?*

Assign any of the following:
1. Activities and exercises, pages 84–85
2. CD-ROM, Disc 1, pages 84–85
3. Communication Activities Masters, pages 14–22

Réintroduction et recombinaison

A **À l'école.** Donnez des réponses personnelles. *(Give your own answers.)*

1. Tu es de quelle ville?
2. Où est-ce que tu es élève?
3. Tu étudies quelles matières?
4. Tu aimes quels cours?
5. Tu n'aimes pas quels cours?
6. Tu passes des examens?
7. Les examens sont faciles ou difficiles? Et les devoirs?
8. Qui donne les examens?
9. Qui est le prof de français?

B **Un autoportrait.** Complétez. *(Complete.)*

1. Bonjour! Je suis ___.
2. Je ___ de ___. (ville)
3. Je parle ___ et ___.
4. J'habite ___.
5. J'arrive à l'école ___.
6. À l'école j'étudie ___, ___ et ___.
7. Je quitte l'école ___.
8. Avec les copains j'aime ___ et ___.

Vocabulaire

NOMS	VERBES	
la maison	aimer	rigoler
la rue	adorer	travailler
la fête	détester	à mi-temps
la télé	arriver	à plein temps
la radio	chanter	
le walkman	danser	AUTRES MOTS ET EXPRESSIONS
le magazine	donner	parler au téléphone
le disque	écouter	poser une question
le compact disc	entrer	passer un examen
la cassette	étudier	par jour
la vidéo(cassette)	gagner	par semaine
le magasin	habiter	pendant
l'argent (m.)	inviter	après
l'examen (m.)	parler	beaucoup
	quitter	quand
	regarder	
	rentrer	

CHAPITRE 3 **85**

Activité de communication écrite

PRESENTATION *(page 84)*

Extension of *Activité A*

Students work in pairs, editing each other's paragraphs before handing them in. They should check their partners' verb forms, word order, and spelling, including accents. Writers then write a second draft.

ANSWERS

Answers will vary.

OPTIONAL MATERIAL

Réintroduction et recombinaison

PRESENTATION *(page 85)*

The tasks in this section are designed to blend and recycle vocabulary and structures from previous chapters as well as from Chapter 3.

ANSWERS

***Exercices A* and B**

Answers will vary.

ASSESSMENT RESOURCES

1. Chapter Quizzes
2. Testing Program
3. Situation Cards
4. Communication Transparency C-3
5. Computer Software: Practice/Test Generator

VIDEO PROGRAM

INTRODUCTION (13:23)

APRÈS LES COURS (13:54)

FOR THE YOUNGER STUDENT

1. Students love to talk about their teachers. If you are willing, have them use adjectives they have learned that they feel describe you. They can either use your name or say *le/la prof de français*.
2. Have students dramatize or pantomime the following words: *regarder, écouter, parler, chanter, danser*.

STUDENT PORTFOLIO

Have students keep a notebook with their best written work from the textbook and ancillaries. Written assignments that may be included in students' portfolios are Exercises A and B on page 85 and the *Mon Autobiographie* section of the Workbook on page 26.

Note Students may create and save both oral and written work using the Electronic Portfolio feature on the CD-ROM.

CHAPTER OVERVIEW

Students will learn to talk about family and housing. They will also recycle earlier chapters' vocabulary and learn *avoir; mon, ton, son;* and some irregular adjectives. The cultural focus is French family living situations.

CHAPTER OBJECTIVES

By the end of this chapter, students will know:

1. family relationships
2. months of the year and how to express today's date
3. rooms in a house or apartment
4. *avoir* in the present tense
5. possessive adjectives
6. common irregular adjectives
7. types of housing in France

CHAPTER 4 RESOURCES

1. Workbook
2. Student Tape Manual
3. Audio Cassette 3B/CD-3
4. Bell Ringer Review Blackline Masters
5. Vocabulary Transparencies
6. Pronunciation Transparency P-4
7. Communication Transparency C-4
8. Communication Activities Masters
9. Map Transparencies
10. Situation Cards
11. Conversation Video
12. Videocassette/Videodisc, Unit 1
13. Video Activities Booklet, Unit 1
14. Lesson Plans
15. Computer Software: Practice/Test Generator
16. Chapter Quizzes
17. Testing Program
18. Internet Activities Booklet
19. CD-ROM Interactive Textbook
20. Performance Assessment

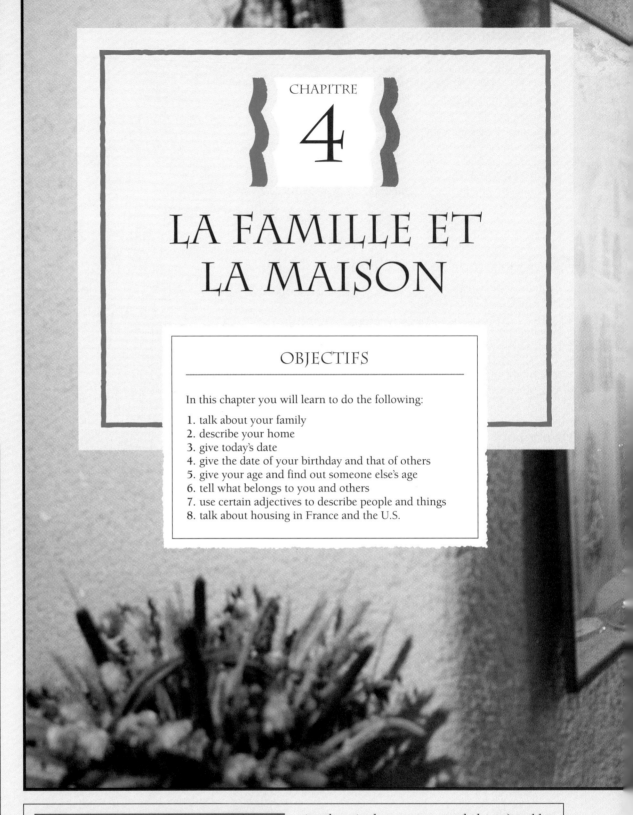

CHAPITRE

4

LA FAMILLE ET LA MAISON

OBJECTIFS

In this chapter you will learn to do the following:

1. talk about your family
2. describe your home
3. give today's date
4. give the date of your birthday and that of others
5. give your age and find out someone else's age
6. tell what belongs to you and others
7. use certain adjectives to describe people and things
8. talk about housing in France and the U.S.

CHAPTER PROJECTS

(optional)

Students use family photos to "introduce" their family to the class, or they create a family out of magazine pictures mounted on poster board to be presented to the class and later displayed. They can also make floor plans of their house/apartment and give a "tour" to other classmates. For cooperative activities, put students in a "family" (tradi- tional or single-parent, extended, etc.) and let them decide the role for each member.

87

Pacing

This chapter requires eight to ten class sessions. Pacing will depend on class length and the age and aptitude of the students.

Note Lesson Plans offer guidelines for 45- and 55-minute classes and **Block Scheduling**.

NOTE ON FAMILY RELATIONSHIPS

In French "half-brother" and "half-sister" are *demi-frère* and *demi-sœur*. (Note that there is no *e* on *demi* in *demi-sœur*.) Many people would say *mon frère* or *ma sœur* for "stepbrother" or "stepsister" and let it go at that. As for stepparents, *ma belle-mère* and *mon beau-père* are sometimes heard, but these terms are more commonly used for "mother-in-law" and "father-in-law." (*Ma belle-mère* or *mon beau-père* might be used for a stepparent if the natural parent were still alive, but if the latter were deceased, the stepparent might be referred to as *ma mère* or *mon père*, depending on the personal situation.) To be precise about a step relationship, a French person would say: *le mari de ma mère, la femme de mon père, le fils (la fille) du mari de ma mère, le fils (la fille) de la femme de mon père.*

As in the U.S., the divorce rate in France has risen over the years. New terms for a "blended family" may be coined soon. Students may want to know the following expressions dealing with:

1. **divorce and death** *Ma mère est divorcée. Mon père est divorcé. Mes parents sont divorcés. Mon père est mort. Ma mère est veuve. Ma mère est morte. Mon père est veuf.*
2. **adoption** *Je suis adopté(e). Ma mère adoptive, mon père adoptif, mes parents adoptifs.*

LEARNING FROM PHOTOS

Ask students the following questions about the photo. *C'est la famille Boivin. M. et Mme Boivin ont combien d'enfants? Ils ont combien de filles? Et combien de fils? Les fils sont blonds ou bruns? Ils sont français?*

Exercices **vs.** *Activités*

All exercises (which provide guided practice) are coded in blue. All communicative activities are coded in red.

VOCABULAIRE

MOTS 1

Vocabulary Teaching Resources

1. Vocabulary Transparencies 4.1 (A & B)
2. Audio Cassette 3B/CD-3
3. Student Tape Manual, Teacher's Edition, *Mots 1: A–C*, pages 41–43
4. Workbook, *Mots 1: A–D*, pages 27–28
5. Communication Activities Masters, *Mots 1: A*, page 19
6. Chapter Quizzes, *Mots 1*: Quiz 1, page 20
7. CD-ROM, Disc 1, *Mots 1*: pages 88–91

Bell Ringer Review

Write the following on the board or use BRR Blackline Master 4-1: Make sentences of the following word scrambles.

1. **sont filles Paris ne les de deux pas**
2. **dans Grenoble est française les ville une Alpes**
3. **les maison huit quittent élèves heures la à**

PRESENTATION *(pages 88–89)*

A. Have students close their books. Using Vocabulary Transparencies 4.1 (A & B), have them repeat the names of the members of the Debussy family after you or Cassette 3B/CD-3. Be sure that they pronounce the many cognates correctly.

B. Ask the following questions as students look at the transparencies or pages 88–89: *C'est la famille Debussy? C'est la famille Debussy ou la famille Gaudin? C'est quelle famille? M. et Mme Debussy ont deux enfants? Ils ont un fils? Ils ont une fille? Ils ont deux enfants ou trois enfants?*

les grands-parents

M. Girard — le grand-père
Mme Girard — la grand-mère

les parents

M. Revel — l'oncle
Mme Revel — la tante
Mme Debussy — la mère / la femme
M. Debussy — le père / le mari

les enfants

Guy — le neveu / le cousin
Anne — la nièce / la cousine
Philippe — le fils / le petit-fils
Monique — la fille / la petite-fille

Minou — le chat
Médor — le chien

88   CHAPITRE 4

TOTAL PHYSICAL RESPONSE

(following the Vocabulary presentation)

Getting Ready

For **TPR 1** draw a family tree on the board, writing in the name of each family member.

For **TPR 2** dramatize very quickly the meaning of *écrivez*.

TPR 1

___, levez-vous, s'il vous plaît.

Allez au tableau noir.
Regardez l'arbre généalogique.
Prenez la règle.
Montrez-moi Philippe.
Montrez-moi le père de Philippe.
Montrez-moi la tante de Philippe.
Montrez-moi sa sœur.
Et son oncle.
Montrez-moi le cousin de Philippe.
Montrez-moi ses grands-parents.
Retournez à votre place, s'il vous plaît.

Voici la famille Debussy.

M. et Mme Debussy ont deux enfants, un fils et une fille.

La famille Debussy a un appartement à Paris.

Les Debussy ont un chien, Médor.

Ils n'ont pas de chat.

Philippe Debussy est le fils de M. et Mme Debussy.

Il a quel âge?

Il a seize ans.

Monique est la fille de M. et Mme Debussy.

Elle a quatorze ans.

C'est quand, l'anniversaire de Monique?

L'anniversaire de Monique est le 4 novembre.

C'est aujourd'hui!

Monique et Philippe sont jeunes.

Ils ne sont pas vieux.

Les mois de l'année sont:

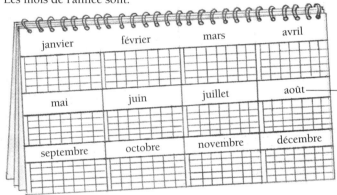

janvier	février	mars	avril
mai	juin	juillet	août — un mois
septembre	octobre	novembre	décembre

Quelle est la date aujourd'hui?
C'est le 4 novembre.

Note: For the first day of the month, you say *le premier*.

le 1er avril → le premier avril

Combien d'enfants ont les Debussy? Ils ont un chien? Ils ont un chat? Qu'est-ce qu'ils ont?

C. When teaching the sentences on page 89, break them into parts. For example, point to the dog and have students repeat *un chien*. Then have them repeat *pas de chat*. Now have students repeat the complete sentences: *Les Debussy ont un chien. Ils n'ont pas de chat.*

D. Have students read the *Mots* for additional reinforcement.

E. Introduce the months of the year by using labeled calendar pictures or flash cards. At this time you may also want to practice dates and recycle the days of the week.

CROSS-CULTURAL COMPARISON

In French *les* is used with the family name to refer to the entire family. No "s" is added to the last name as is done in English.

Teaching Tip During the presentation, ask questions that build from easy to more complex. The natural sequence is yes/no, choice questions, then interrogative word.

Writing Tip Remind students that in French, names of days of the week and months are not capitalized.

Vocabulary Expansion

You may want to give students the following additional vocabulary in order to talk about the family and pets.

fille unique
fils unique
un hamster
des poissons rouges
un oiseau

TPR 2

Si vous avez un frère, levez-vous.

Et maintenant, asseyez-vous.

Si vous avez une sœur, levez la main.

___, vous avez une sœur? Levez-vous, s'il vous plaît.

Allez au tableau noir.

Prenez la craie.

Vous avez une sœur, n'est-ce pas? Écrivez son nom au tableau noir.

Elle a quel âge? Écrivez son âge.

Elle habite où? Écrivez son adresse.

Elle va à quelle école? Écrivez le nom de son école.

Merci, ___. Mettez la craie ici, s'il vous plaît.

Et maintenant, retournez à votre place et asseyez-vous.

Exercices

PRESENTATION (page 90)

Exercice A: Listening

Focus on the listening skill by asking students the questions in Exercise A with books closed. Ask a question and call on one student to respond. Ask the next question of another student, and so on.

Exercice A: Reading

Have students open their books. Let one student read the question and call on another to respond.

Exercice B

Exercise B can be done immediately with books open.

Note Exercise B introduces students to the forms *mon, ma, mes* but doesn't require them to use these forms.

ANSWERS

Exercice A

1. Oui, la famille Debussy a un appartement à Paris.
2. Oui, M. et Mme Debussy ont deux enfants.
3. La famille Debussy est petite.
4. Le fils a seize ans.
5. La fille a quatorze ans.
6. Les Debussy ont un chien.
7. Oui, les enfants de M. et Mme Debussy ont des cousins.
8. Oui, ils ont un oncle et une tante.
9. Oui, ils ont des grands-parents.

Exercice B

1. oncle
2. tante
3. oncle
4. tante
5. cousin
6. cousine
7. cousins
8. la nièce (le neveu)
9. grand-père
10. grand-mère
11. la petite-fille (le petit-fils)... la fille (le fils)

Exercices

A **La famille Debussy.** Répondez. (*Answer.*)

1. La famille Debussy a un appartement à Paris?
2. M. et Mme Debussy ont deux enfants?
3. La famille Debussy est grande ou petite?
4. Le fils a quel âge?
5. La fille a quel âge?
6. Les Debussy ont un chien ou un chat?
7. Les enfants de M. et Mme Debussy ont des cousins?
8. Ils ont des oncles et des tantes?
9. Ils ont des grands-parents?

B **Ma famille et moi.** Complétez. (*Complete.*)

1. Le frère de mon père est mon ___.
2. La sœur de mon père est ma ___.
3. Le frère de ma mère est mon ___.
4. La sœur de ma mère est ma ___.
5. Le fils de mon oncle et de ma tante est mon ___.
6. Et la fille de mon oncle et de ma tante est ma ___.
7. Les enfants de mes oncles et de mes tantes sont mes ___.
8. Et moi, je suis ___ de mon oncle et de ma tante.
9. Le père de ma mère est mon ___.
10. La mère de mon père est ma ___.
11. Je suis ___ de mes grands-parents et ___ de mes parents.

COOPERATIVE LEARNING

On a piece of paper the reporter in each team writes down a month of the year. Students pass the paper around the team, with each team member adding a month, until all months of the year have been written down. After putting the months in proper order, a student from each team reads its list to the class.

LEARNING FROM PHOTOS

Have the students say as much about the family in the photograph as they can.

C Les anniversaires. Indiquez l'anniversaire de chaque personne d'après le carnet d'anniversaires. *(Give the date of each person's birthday according to the birthday book.)*

Maman
L'anniversaire de Maman est le 4 mars.

1. Papa
2. Philippe
3. Oncle Pierre
4. Tante Marie
5. Céline
6. Grégoire
7. Marie-France
8. Grand-mère

mars
4 Maman

janvier
10 Papa

octobre
20 Philippe

avril
12 Céline

mai
17 Grégoire

juin
22 Marie-France

juillet
31 Grand-mère

décembre
15 Tante Marie

février
1 Oncle Pierre

JOYEUX ANNIVERSAIRE!

ADDITIONAL PRACTICE

1. After completing the exercises, reinforce the lesson by having students write profiles of their family members. For example: *Billy Larsen est mon frère. Il est jeune. Il a dix ans. L'anniversaire de Billy est le 15 décembre. Il est grand, brun et timide.*
2. Student Tape Manual, Teacher's Edition, *Activités B–C*, pages 42–43.

INDEPENDENT PRACTICE

Assign any of the following:
1. Exercises, pages 90–91
2. Workbook, *Mots 1: A–D*, pages 27–28
3. Communication Activities Masters, *Mots 1: A*, page 19
4. CD-ROM, Disc 1, pages 88–91

PRESENTATION *(page 91)*
Extension of *Exercice C*: Speaking

Have students write their birthday in the following way in large letters on a piece of paper: *le 3/7* (July 3), for example. As each student holds up his or her paper, call on volunteers to say the birthday of their classmate.

CROSS-CULTURAL COMPARISON

In French, the number of the day is written first and the number of the month second. For example, *10-8-94* is August 10, 1994.

ANSWERS
Exercice C

1. L'anniversaire de Papa est le 10 janvier.
2. …Philippe est le 20 octobre.
3. …l'Oncle Pierre est le 1er février.
4. … la Tante Marie est le 15 décembre.
5. … Céline est le 12 avril.
6. … Grégoire est le 17 mai.
7. … Marie-France est le 22 juin.
8. … Grand-mère est le 31 juillet.

RETEACHING *(Mots 1)*

Have students create their family tree and label it to show how the people in it are related to them. For example: *George Miller—mon grand-père.* Students who don't wish to share this information about their own families can create an imaginary family tree using pictures of people cut out from magazines.

INFORMAL ASSESSMENT *(Mots 1)*

Using Vocabulary Transparency 4.1(A), have students identify as many family members as they can.

PRESENTATION *(pages 92–93)*

A. Using Vocabulary Transparencies 4.2 (A & B), have students repeat each word after you or Cassette 3B/CD-3 two or three times.

B. Ask the question *Qu'est-ce que c'est?* as you point to various objects on the transparencies.

C. When teaching *près de* and *loin de* on page 92, draw arrows on the board—a short one for *près de*; a long arrow for *loin de*.

D. Have students open their books to pages 92–93 and ask individuals to read the words and sentences aloud.

E. Ask yes/no, either/or, and then interrogative word questions about the visuals on pages 92–93. For example: *La maison a un balcon? L'immeuble a trois étages ou quatre étages? La maison a combien de pièces? Où est l'ascenseur? Avec qui est-ce que M. Debussy bavarde?*

F. Ask personal questions based on the *Mots 2* vocabulary. For example: *Qui a un jardin à la maison? Est-ce que tu aimes les voisins?*

92

une vieille maison

un garage

une voiture

un jardin

une terrasse

un immeuble

un quartier

un appartement

le troisième étage

le deuxième étage

un balcon

le premier étage

le rez-de-chaussée

une entrée

une station de métro

Les Debussy ont un joli appartement près de la station de métro.
L'appartement n'est pas loin de la station de métro.
Il y a dix appartements dans l'immeuble.

92 CHAPITRE 4

TOTAL PHYSICAL RESPONSE

(following the Vocabulary presentation)

Getting Ready

Use desks to represent different rooms and label them. Pictures of a stove, table, television set, and telephone may be placed on the desks. Dramatize the meaning of *venez ici.*

TPR

___, levez-vous, s'il vous plaît.

Venez ici.
Allez dans la cuisine.
Préparez le dîner.
Mettez le dîner sur la table.
Mettez-vous à table.
Et maintenant, levez-vous.
Venez ici.
Entrez dans la salle de séjour.
Mettez la télévision.

(continued on next page)

un ascenseur

bavarder

une cour

un voisin

une voisine

Il y a un ascenseur dans l'immeuble.
Il n'y a pas de garage dans l'immeuble.
M. Debussy bavarde avec les voisins.

Il y a six pièces dans l'appartement.

les toilettes (f.)

la salle de bains

la chambre à coucher

la salle à manger

la cuisine

la salle de séjour

dîner

préparer le dîner

Ta maison a une salle de bains ou deux salles de bains?

G. Have students describe magazine pictures of various types of homes.

Teaching Tip Repetition is important for introducing new material, but it can quickly become tiresome. To maintain student interest, alternate between whole-class and individual repetition. Walk around the room during repetition, and switch to another activity if you sense that students are becoming bored. Come back to repetition later.

CROSS-CULTURAL COMPARISON

1. In France the ground floor of a building is called *le rez-de-chaussée*. What the French call the "first floor," *le premier étage*, is our second floor.

2. Have students look at the illustration at the top of page 93. Point out to them that, in comparison to U.S. cities, French cities are very old. In many older French buildings, elevators have been installed in the stairwell since these buildings were constructed long before the invention of elevators.

TPR (*continued*)
Asseyez-vous.
Regardez la télévision.
Allez au téléphone.
Téléphonez à un ami.
Parlez. Dites «bonjour».
Merci, ___. Levez-vous maintenant.
Retournez à votre place et asseyez-vous, s'il vous plaît.

PAIRED ACTIVITY

Have students work in pairs. Each student draws and labels a floor plan of his or her house or apartment. Then, without showing the drawing to their partner, students describe their house or apartment. Each student draws a floor plan according to the description provided by his or her partner. When finished, they compare the two plans and discuss differences.

Exercice B

Exercise B teaches the use of the expression *il y a* followed by either a singular or plural noun.

ANSWERS

Exercice A

1. C'est le rez-de-chaussée.
2. … la cour.
3. … l'entrée.
4. … le premier étage.
5. … le balcon.

Exercice B

1. La maison a cinq pièces.
2. Non, les pièces ne sont pas grandes.
3. Il y a un étage.
4. Non, il y a une petite cuisine.
5. Il y a deux chambres à coucher.
6. Il y a une salle de bains.
7. Il y a un cabinet de toilette.
8. Non, il n'y a pas de balcon.

Exercices

A **Un très bel immeuble.**
Identifiez. *(Identify.)*

1. C'est le premier étage ou le rez-de-chaussée?
2. C'est le balcon ou la cour?
3. C'est l'entrée ou le rez-de-chaussée?
4. C'est le premier étage ou le deuxième étage?
5. C'est le balcon ou l'ascenseur?

B **La vieille maison.** Répondez d'après le dessin. *(Answer according to the illustration.)*

1. La maison a combien de pièces?
2. Les pièces sont grandes?
3. Il y a combien d'étages?
4. Il y a une grande cuisine?
5. Il y a combien de chambres à coucher?
6. Il y a combien de salles de bains?
7. Il y a combien de toilettes?
8. Il y a un balcon?

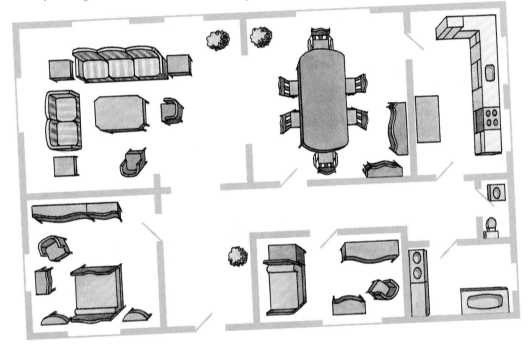

C Quelle pièce? Choisissez la bonne réponse. *(Choose the correct answer.)*

1. On regarde la télé dans ___.
   a. la salle à manger   b. la salle de bains   c. la salle de séjour

2. On prépare le dîner dans ___.
   a. la salle à manger   b. la cuisine   c. la chambre à coucher

3. On bavarde avec les voisins dans ___.
   a. la chambre à coucher   b. la cour   c. la salle de bains

4. On dîne dans ___.
   a. la salle de séjour   b. la cuisine ou la salle à manger
   c. la chambre à coucher

5. On dîne sur ___ de la maison en juillet et en août.
   a. la terrasse   b. l'étage   c. la pièce

D Ma maison. Donnez des réponses personnelles. *(Give your own answers.)*

1. Où habites-tu?
2. Tu habites dans quelle rue?
3. Tu habites dans un appartement ou dans une maison?
4. Il y a combien de pièces dans l'appartement ou la maison?
5. Il y a combien de chambres à coucher?
6. Il y a une terrasse ou un balcon?
7. Il y a un ascenseur dans l'immeuble?
8. La maison ou l'immeuble a un garage?
9. La voiture est dans le garage le soir?
10. Tu bavardes avec les voisins?

Activités de communication orale
Mots 1 et 2

A La famille Lapeyre. Here's a picture of the Lapeyre family. Say as much as you can about them.

B Ma maison. During a visit to Nîmes in Southern France you meet a French student. He or she wants to know:

1. whether you live in a house or an apartment
2. what your house or apartment is like
3. whether there's a yard and what it's like

ADDITIONAL PRACTICE

1. After doing the activities, reinforce as follows. Students work in pairs. One thinks of a place in his/her home where the family pet is hiding. His/her partner tries to guess where the pet is.
   Élève 1: (Le chat) est dans le jardin?
   Élève 2: Non, il n'est pas dans le jardin.
   (Oui, il est dans le jardin.)

INDEPENDENT PRACTICE

Assign any of the following:
1. Exercises and activities, pages 94–95
2. Workbook, *Mots 2: E–G,* pages 29–30
3. Communication Activities Masters, *Mots 2: B,* page 20
4. Computer Software, *Vocabulaire*
5. CD-ROM, Disc 1, pages 92–95

PRESENTATION *(continued)*

Exercice C

Exercise C must be done with books open. Note the use of the cognates *dîner* and *préparer*. Students should have no difficulty determining their meaning.

Extension of *Exercice D:* Speaking

After completing Exercise D with the whole class, have students work in pairs. Partners use the questions to interview each other, taking notes. Then they report to the class on their partner's home, using the third person.

ANSWERS

Exercice C

1. c
2. b
3. b
4. b
5. a

Exercice D

Answers will vary.

RETEACHING *(Mots 2)*

Have students draw an apartment building, labeling the floors and other features. Allow them to refer to pages 92–94.

Activités de communication orale
Mots 1 et 2

PRESENTATION *(page 95)*

The *Activités de communication orale* allow students to use the chapter vocabulary and grammar in open-ended situations.

ANSWERS

Activité A

Answers will vary but may include the following:

Il y a six personnes. Il y a une mère, un père, trois enfants et une grand-mère. Ils ont un chien.

Activité B

Answers will vary.

 STRUCTURE

Structure Teaching Resources

1. Workbook, *Structure: A–I*, pages 31–33
2. Student Tape Manual, Teacher's Edition, *Structure: A–D*, pages 46–48
3. Audio Cassette 3B/CD-3
4. Communication Activities Masters, *Structure: A–C*, pages 21–22
5. Computer Software, *Structure*
6. Chapter Quizzes, *Structure:* Quizzes 3–5, pages 22–24
7. CD-ROM, Disc 1, pages 96–101

Le verbe avoir *au présent*

PRESENTATION (*page 96*)

A. Review *il/elle a* and *ils/elles ont* from the *Mots*.
B. Draw a stick figure labeled *Robert* on the board and write *une voiture, un chat, deux frères*. Now draw *Corinne et Luc*, with *une voiture, un chien, deux sœurs*. Have students say what Robert has and what Corinne and Luc have.
C. Write the forms of *avoir* on the board. Have students repeat them, noting the liaisons.
D. Now lead students through steps 1–2 on page 96.

Note In the CD-ROM version, this structure point is presented via an interactive electronic comic strip.

Exercices

ANSWERS

Exercice A

1. Oui, elle a un frère.
2. Oui, Jacques a une sœur.
3. Oui, ils ont deux enfants.
4. Non, ils n'ont pas d'appartement à Paris.
5. Oui, ils ont un jardin.
6. Oui, les Lefèvre ont un chat.

Le verbe *avoir* **au présent** · *Telling What You and Others Have; Telling People's Ages*

1. The verb *avoir*, "to have," is an irregular verb. Study the present tense forms of this verb. Note that there is a liaison in the plural. The *s* is pronounced like a *z*.

AVOIR	
j'ai	nous avons
tu as	vous avez
il a	ils ont
elle a	elles ont

2. You also use the verb *avoir* to express age in French.

> **Tu as quel âge?**
> **Moi, j'ai seize ans.**

Exercices

A **Les Lefèvre.** Répondez d'après la photo. (*Answer according to the photo.*)

1. Catherine Lefèvre a un frère?
2. Jacques a une sœur?
3. Monsieur et Madame Lefèvre ont deux enfants?
4. Ils ont un appartement à Paris?
5. Ils ont un jardin?
6. Les Lefèvre ont un chat?

B **Qu'est-ce qu'il a, Robert?** Répondez d'après le modèle. (*Answer according to the model.*)

> **Robert a une cassette?**
> *Non, il n'a pas de cassette.*

1. Il a un stylo?
2. Il a une voiture?
3. Il a un chat?
4. Il a un chien?
5. Il a un éléphant?
6. Il a des livres?

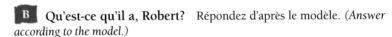

ADDITIONAL PRACTICE

Using Vocabulary Transparency 4.1(A), label the family picture with years appropriate for the ages of various members. Ask students to tell you how old everyone is in the picture. For example: *Le fils de M. Debussy a quel âge?*

LEARNING FROM PHOTOS

Have students say as much as they can about the photo of the Lefèvre family.

C **Tu as un frère?** Répétez la conversation. (*Practice the conversation.*)

THÉRÈSE: René, tu as un frère?
RENÉ: Non, je n'ai pas de frère, mais j'ai une sœur.
THÉRÈSE: Tu as une sœur? Elle a quel âge?
RENÉ: Elle a quatorze ans.
THÉRÈSE: Et toi, tu as quel âge?
RENÉ: Moi, j'ai seize ans.
THÉRÈSE: Ta sœur et toi, vous avez un chien?
RENÉ: Non, nous n'avons pas de chien. Mais nous avons un petit chat.

Complétez d'après la conversation. (*Complete according to the conversation.*)

1. René n'___ pas ___ frère.
2. Mais il ___ une sœur.
3. Sa sœur ___ quatorze ans.
4. René ___ seize ans.
5. René et sa sœur n'___ pas ___ chien.
6. Mais ils ___ un petit chat.

D **J'ai.** Donnez des réponses personnelles. (*Give your own answers.*)

1. Tu as des frères? Tu as combien de frères?
2. Tu as des sœurs? Tu as combien de sœurs?
3. Tu as un chien?
4. Tu as un chat?
5. Tu as des amis?
6. Tu as des cousins?
7. Tu as combien de cousins?
8. Tu as combien d'oncles?
9. Tu as combien de tantes?
10. Tu as une petite ou une grande famille?
11. Tu as quel âge?

E **Dans ton sac à dos.** Posez des questions à un copain ou à une copine d'après le modèle. (*Ask a classmate questions according to the model.*)

un crayon
Élève 1: Tu as un crayon dans ton sac à dos?
Élève 2: Oui, j'ai un crayon. (Non, je n'ai pas de crayon.)

1. un stylo
2. une calculatrice
3. un livre d'espagnol
4. des cassettes
5. un chien
6. un cahier
7. des devoirs
8. un ordinateur

F **Qu'est-ce que vous avez?** Posez des questions d'après le modèle. (*Ask questions according to the model.*)

une maison ou un appartement
Maurice et Pauline, vous avez une maison ou un appartement?

1. un chien ou un chat
2. un frère ou une sœur
3. un neveu ou une nièce
4. des disques ou des cassettes
5. une voiture ou une bicyclette

Exercice B

1. Non, il n'a pas de stylo.
2. ... de voiture.
3. ... de chat.
4. ... de chien.
5. ... d'éléphant.
6. ... de livres.

PRESENTATION (*continued*)

Exercice C

Exercise C lets students hear, see, and use *avoir* in a natural context. Have students repeat the conversation, then have several pairs read it aloud. Now call on individuals to complete the sentences that follow it.

Extension of Exercice C

Have two students present the the mini-conversation as a skit.

Exercice D

Exercise D is very important since it helps students practice answering with *j'ai* when they hear *tu as*. They often repeat what they hear (*tu as*) rather than respond correctly with the *je* form of the verb.

It is recommended that you do this exercise first with books closed for ear training. Call on individuals to answer one item. Since the exercise relates to the student's personal situation, you may do it again with one student answering several questions.

Exercice E

You may wish to use the recorded version of this exercise.

ANSWERS

Exercice C

1. a... de
2. a
3. a
4. a
5. ont... de
6. ont

Exercice D

Answers will vary; however, students should use *j'ai* in each response.

Exercice E

Questions and answers should follow the model and cues.

Exercice F

Questions should follow the model and cues.

PAIRED ACTIVITY

Have each student ask the age of his or her neighbor.

LEARNING FROM PHOTOS

Ask students the following questions about the photo on page 97: *C'est une photo d'un père et d'un fils? Ils sont sympathiques? À ton avis, le fils a quel âge? Et le père?*

97

Exercise G helps students practice responding with *nous avons* when they hear *vous avez*. Students often repeat the form they hear (*vous avez*), rather than respond correctly with the *nous* form of the verb.

Exercice H

Exercise H requires students to use all forms of the verb *avoir*.

Extension of *Exercice H*

Have students make up as many questions as they can about the Duhamels.

ANSWERS

Exercice G

Answers will vary; however, students should use *nous avons* in each response.

Exercice H

1. a	8. ont
2. a	9. as
3. ont	10. as
4. a	11. as
5. a	12. a
6. a	13. avez
7. a	

RETEACHING

Have students work in groups of three. They tell one another what members of their family have and then compare. For example, someone says he/she has a dog. The others can say: *Ah, vous avez un chien. Nous avons un chien aussi.* Or: *Ah, vous avez un chien. Nous n'avons pas de chien. Nous avons un chat.*

Les adjectifs possessifs

PRESENTATION (pages 98–99)

A. Quickly go over Exercise B, page 90, to review the vocabulary used in the following exercises and to introduce the possessive adjectives *mon, ma, mes.*

G **Ma famille et moi.** Donnez des réponses personnelles en utilisant «nous». (*Give your own answers about you and your family using* nous.)

1. Vous avez une maison?
2. Vous avez un appartement?
3. Vous avez un chien?
4. Vous avez un chat?
5. Vous avez une voiture?
6. Vous avez un jardin?

H **La famille Duhamel.** Complétez avec «avoir». (*Complete with* avoir.)

Voici la famille Duhamel. La famille Duhamel ___ dans un très joli appartement à Paris dans le cinquième arrondissement. L'appartement ___ six pièces. Les Duhamel ___ aussi une maison à Juan-les-Pins. La maison à Juan-les-Pins est une petite villa ou un bungalow où la famille Duhamel passe les vacances. La villa ___ cinq pièces.

Il y a quatre personnes dans la famille Duhamel. Olivier est le fils. Olivier ___ une sœur, Gabrielle. Gabrielle ___ dix-sept ans et son frère ___ quinze ans. Olivier et Gabrielle ___ un petit chien, Milou. Ils adorent Milou.

Tu ___ un chien? Si tu n'___ pas de chien, tu ___ un chat? Ta famille ___ un appartement ou une maison? Ta famille et toi, vous ___ une petite villa ou un bungalow où vous passez les vacances?

Les adjectifs possessifs

Telling What Belongs to You and Others

1. You use possessive adjectives to show possession or ownership. Like other adjectives, the possessive adjectives must agree with the nouns they modify. For example, if the noun is feminine, the adjective is feminine. If the noun is plural, the adjective is plural.

2. Study the following forms of the possessive adjectives: *mon, ma, mes* (my); *ton, ta, tes* (your); *son, sa, ses* (his *or* her).

MASCULIN SINGULIER	FÉMININ SINGULIER	PLURIEL
mon père ton père son père	ma mère ta mère sa mère	mes parents tes parents ses parents

---

ADDITIONAL PRACTICE

Give each student a card with a noun. List the nouns on the board. Students find out who has what by asking questions, e.g., *Jean et Luc, vous avez une radio? Non, nous n'avons pas de radio. Marie, tu as une radio? Oui, j'ai une radio.* Write the name of the possessor(s) next to the word on the board. When all have been identified, practice *qui.* For example: *Qui a une radio? Marie a une radio.*

3. You use *mon, ton, son* before a masculine singular noun.
You use *ma, ta, sa* before a feminine singular noun.
You use *mes, tes, ses* before a plural noun.

4. Note that *son, sa, ses* can mean either "his" or "her." The agreement is with the item owned, not the owner.

> **le chien de Charles → son chien**
> **la maison de Charles → sa maison**

5. Before a masculine or feminine singular noun that begins with a vowel or silent *h*, you use *mon, ton,* or *son.*

MASCULIN	FÉMININ
mon ami	*mon* amie
ton ami	*ton* amie
son ami	*son* amie

Exercices

A À votre tour. Donnez des réponses personnelles. *(Give your own answers.)*

1. Où est ta maison ou ton appartement?
2. Ta maison (Ton appartement) a combien de pièces?
3. Ta maison est grande ou petite? (Ton appartement est grand ou petit?)
4. C'est quand, ton anniversaire? Tu as quel âge?
5. Quel âge a ton frère, si tu as un frère?
6. Quel âge a ta sœur, si tu as une sœur?
7. Il y a combien de personnes dans ta famille?
8. Tes oncles et tes tantes habitent près ou loin de ta ville (ton village)?

B J'ai une question pour toi. Complétez avec «ton», «ta» ou «tes» et posez les questions à un copain ou à une copine d'après le modèle. *(Complete with* ton, ta, *or* tes *and then ask a classmate the questions according to the model.)*

> Où est ___ maison?
>
> Élève 1: Où est ta maison?
> Élève 2: Ma maison est près de l'école.

1. Qui est ___ amie?
2. Qui est ___ ami?
3. Où habitent ___ grands-parents?
4. ___ frère a quel âge?
5. ___ sœur a quel âge?
6. Où est ___ maison ou ___ appartement?
7. Tu aimes ___ cours de français?
8. ___ prof de français est sympa?

CHAPITRE 4   **99**

B. Write the examples with the possessive adjectives on the board. Call on individual students to read aloud steps 1–5 on pages 98 and 99.

C. Make ownership statements by using items in the classroom. For example: *mon livre, ma radio, mes papiers, ton livre, ta chaise, tes cahiers.*

D. To demonstrate *son, sa, ses,* use students' items. For example: *le livre de Marcel, son livre; la chaise de Patrick, sa chaise.*

Note The possessive adjectives *notre, votre,* and *leur* will be taught in Chapter 5.

Exercices

PRESENTATION *(page 99)*

Exercice A: **Listening**
  Do Exercise A first with books closed. Call on individual students to answer one item each. You may want to do the exercise a second time, having one student respond to several consecutive items before calling on the next student.

Extension of *Exercice A:* Writing
  You may wish to have students write a paragraph about their family, based on the responses to Exercise A.

Exercice B
  Exercise B may be done as a paired activity.

ANSWERS
Exercice A
  Answers will vary.

Exercice B
  É2 answers will vary. É1 questions are as follows:

1. ton
2. ton
3. tes
4. Ton
5. Ta
6. ta ... ton
7. ton
8. Ton (Ta)

PRESENTATION (continued)

Exercice C

This exercise reinforces the concept that the possessive adjective agrees with the item, not the possessor.

ANSWERS

Exercice C

1. son père
2. sa sœur
3. sa sœur
4. sa maison
5. son appartement
6. ses cousins
7. ses grands-parents
8. ses oncles
9. son amie

RETEACHING

The following activity may be done orally, or the words may be written on the board. Have students volunteer a series of nouns. Then have them use these nouns with *mon, ma, mes*. Next, have them ask a question using the same noun with *ton, ta,* or *tes*.

Adjectifs qui précèdent le nom: Adjectifs réguliers

PRESENTATION (page 100)

Lead students through the explanation on page 100. Emphasize that *joli, jeune, petit,* and *grand* are exceptions to the usual rule of placing adjectives after the noun.

ANSWERS

Exercice A

1. Marie-France est une jeune fille.
2. Elle a une petite famille.
3. Elle a un grand chat.
4. Oui, elle a un joli appartement à Paris.
5. Oui, il y a un petit restaurant près de l'appartement.

RETEACHING

Show students magazine pictures that illustrate *joli, jeune, petit,* and *grand*. Have students write two sentences about each picture.

C **Le frère de Suzanne ou de Jacques.** Changez d'après le modèle. *(Change according to the model.)*

> le frère de Suzanne
> *son frère*

1. le père de Suzanne
2. la sœur de Suzanne
3. la sœur de Jacques
4. la maison de Jacques
5. l'appartement de Suzanne
6. les cousins de Jacques
7. les grands-parents de Jacques
8. les oncles de Jacques
9. l'amie de Suzanne

Adjectifs qui précèdent le nom   *Describing People and Things*

ADJECTIFS RÉGULIERS

In French most adjectives follow the noun they modify. However, some frequently used adjectives come before the noun. You already know a few of them: *joli, jeune, petit, grand.*

> **Ils ont un petit appartement à Paris.**
> **L'appartement est près d'une grande station de métro.**
> **Marlène est une jeune fille.**
> **Elle a un joli petit chien.**

Exercice

A **Marie-France.** Répondez d'après le dessin. *(Answer according to the illustration.)*

1. Marie-France est une jeune fille ou une jeune femme?
2. Elle a une grande famille ou une petite famille?
3. Elle a un petit chien adorable ou un grand chat?
4. Marie-France a un joli appartement à Paris?
5. Il y a un petit restaurant près de l'appartement?

ADJECTIFS IRRÉGULIERS

1. The adjectives *beau* (beautiful), *nouveau* (new), and *vieux* (old) also come before the noun. These adjectives have several forms.

FÉMININ SINGULIER	MASCULIN SINGULIER + VOYELLE	MASCULIN SINGULIER + CONSONNE
une belle maison	un bel appartement	un beau quartier
une nouvelle maison	un nouvel appartement	un nouveau quartier
une vieille maison	un vieil appartement	un vieux quartier

FÉMININ PLURIEL	MASCULIN PLURIEL	
de belles maisons	de beaux appartements	de beaux quartiers
de nouvelles maisons	de nouveaux appartements	de nouveaux quartiers
de vieilles maisons	de vieux appartements	de vieux quartiers

2. Note the special singular forms *bel, nouvel,* and *vieil* that come before masculine singular nouns beginning with a vowel or silent *h*.

3. In the masculine plural form, you pronounce the *x* like a *z* when it is followed by a vowel or silent *h*.

4. When an adjective comes before a plural noun, *des* becomes *de*.

> **Il y a** *de* petites et *de* grandes stations de métro dans la ville.
> **Il y a** *de* nouveaux et *de* vieux immeubles dans la ville.

Exercice

A **Le bel appartement des Dubois.** Complétez. *(Complete.)*

1. Les Dubois ont un ___ appartement dans un ___ immeuble dans un ___ quartier. (beau, vieux, beau)
2. Il y a de ___ et de ___ quartiers à Paris. (nouveau, vieux)
3. L'appartement des Dubois est près d'une ___ ou d'une ___ station de métro? (vieux, nouveau)
4. L'appartement des Dubois a de ___ pièces. (beau)
5. Il a de ___ pièces et un très ___ balcon. (grand, beau)
6. De l'appartement il y a une ___ vue sur la ville. (beau)
7. Les Dubois ont une ___ voiture. (nouveau)
8. La ___ voiture est ___. (nouveau, beau)

LEARNING FROM PHOTOS

Ask students: *Votre opinion, s'il vous plaît: C'est une vieille ou une nouvelle station de métro?*

Adjectifs irréguliers

PRESENTATION *(page 101)*

Note This presentation begins with the feminine forms since students tend to have an easier time dropping the final sound in the oral form and the final letters in the written form, e.g., *belles, belle, bel, beau.*

A. Lead students through steps 1–4 on page 101, studying the chart and providing additional examples.
B. In more able groups, you may wish to explain the difference between *nouvelle/nouveau* and *neuve/neuf* ("new" and "brand-new").

Exercice A

PRESENTATION *(page 101)*

Extension of *Exercice A*
After completing the exercise, have students retell the story of the Dubois family in their own words.

ANSWERS

1. bel... vieil... beau
2. nouveaux... vieux
3. vieille... nouvelle
4. belles
5. grandes... beau
6. belle
7. nouvelle
8. nouvelle... belle

INFORMAL ASSESSMENT

Write the following adjective/noun combinations on the board and have individuals make sentences with *C'est* or *Ce sont.*
joli/maisons
joli/maison
chien/sympathique
élève/timide
jeune/garçon
petit/voiture
grand/immeubles
petit/ascenseurs

CONVERSATION

PRESENTATION *(page 102)*

A. Tell students they'll hear a conversation between Michel and his neighbor. Have them watch the Conversation Video or listen to Cassette 3B/CD-3 with their books closed.

B. Have them open their books and repeat the conversation after you.

C. Have two students read the conversation aloud with as much expression as possible.

D. Do the comprehension exercise on page 102. You may also use the multiple-choice questions following the *Conversation* on Cassette 3B/CD-3.

Note In the CD-ROM version, students can play the role of either one of the characters and record the conversation.

ANSWERS

Exercice A

1. Michel habite en France.
2. La vieille voisine n'est pas la mère de Danielle.
3. Les nouveaux voisins sont américains.
4. Michel a quinze ans.
5. Le frère de Danielle est adorable.
6. Les Smith sont de Dallas.
7. Ils ont un vieux chien.
8. La vieille voisine n'aime pas les chiens.

Scènes de la vie *Danielle, la nouvelle voisine*

MICHEL: Bonjour, Madame. Il y a une nouvelle famille dans le quartier?

LA VOISINE: Oui, les Smith. Ils sont américains.

MICHEL: Ils sont d'où?

LA VOISINE: De Dallas.

MICHEL: Il y a des enfants?

LA VOISINE: Oui, il y a deux enfants. David, le fils, a sept ans. Danielle, la fille, a quinze ans.

MICHEL: Ah, juste comme moi! Et... comment est Danielle?

LA VOISINE: Elle est très jolie et très sympathique. Le petit David est adorable. Mais ils ont un vieux chien et moi je n'aime pas beaucoup les chiens, surtout les vieux chiens.

MICHEL: Mais Danielle, elle aime les chiens?

LA VOISINE: Oui, elle est comme toi, mon petit Michel! Vous les jeunes, vous aimez beaucoup les chiens et les chats.

A **Les voisins.** Corrigez les phrases. *(Correct the sentences.)*

1. Michel habite à Dallas.
2. La vieille voisine est la mère de Danielle.
3. Les nouveaux voisins sont français.
4. Michel a sept ans.
5. Le frère de Danielle est désagréable.
6. Les Smith sont de New York.
7. Ils ont un vieux chat.
8. La vieille voisine adore les chiens.

102   CHAPITRE 4

Prononciation *Le son /ã/*

There are three nasal vowel sounds in French: /ã/ as in *cent*, /õ/ as in *sont* et /ĕ/ as in *cinq*. They are called "nasal" because some air passes through the nose when they are pronounced. In this chapter, you will practice only the sound /ã/ as in *cent*.

Repeat the following. Notice that there is no /n/ sound after the nasal vowel.

Jean	cent	grand	amusant
français	parent	fantastique	

Voilà les grands-parents, les parents et les enfants.
Jean-François est fantastique. Il est français, grand, amusant.

grand

Activités de communication orale

A **Les nouveaux voisins.** Imagine that your family is living for a while in Paris. A new family, the Lamberts, has just moved into your apartment building. Make up a few questions that you'd want to ask one of your neighbors to find out about the Lambert family.

B **Une rencontre.** As you're walking along the Seine in Paris, a friendly French woman strikes up a conversation with you. Answer her questions.

1. Tu habites où?
2. Tu habites dans une maison ou dans un appartement?
3. La plupart *(majority)* des Américains habitent dans une maison ou dans un appartement?
4. La plupart des familles américaines sont petites ou grandes?

C **Ton quartier.** Ask a classmate if his or her neighborhood has the following. Your partner will then ask you about your neighborhood.

> Élève 1: Il y a une station de métro?
> Élève 2: Non, il n'y a pas de station de métro.

1. Un cinéma?
2. Un parc?
3. Des immeubles?
4. Une discothèque?
5. Un lycée?
6. Des restaurants?
7. Des cafés?
8. Une banque?

Bell Ringer Review
Write the following on the board or use BRR Blackline Master 4-4: Give the opposite of the following adjectives. Do not change the gender or number.

1. jeune	4. moches
2. grande	5. faciles
3. nouvelles	6. occupées

Prononciation

PRESENTATION *(page 103)*

A. Using Pronunciation Transparency P-4 or textbook page 103, model the key word *grand*. Have students repeat in unison and individually.
B. Model the words and phrases on page 103 in similar fashion.
C. Give students following *dictée*: **Jean est français. Les parents de Jean sont amusants. Les grands-parents ont cent ans.**
D. For additional pronunciation practice, use the *Prononciation* section on Cassette 3B/CD-3 and *Activités G–I,* pages 50–51 in the Student Tape Manual, Teacher's Edition.

Activités de communication orale

PRESENTATION *(page 103)*

Activité B

In the CD-ROM version of this activity, students can interact with an on-screen native speaker.

ANSWERS

Activités A and B
Answers will vary.

Activité C
É2 answers will vary. É1 questions are as follows:

1. É1: (Est-ce qu') il y a un cinéma dans ton quartier?
2. É1: … un parc dans ton quartier?
3. É1: … des immeubles… ?
4. É1: … une discothèque… ?
5. É1: … un lycée… ?
6. É1: … des restaurants… ?
7. É1: … des cafés… ?
8. É1: … une banque… ?

LECTURE ET CULTURE

RECYCLING

This reading selection recycles vocabulary and structures students have learned actively in this and previous chapters.

READING STRATEGIES
(*page 104*)

Pre-reading

A. Tell students they are going to read about a typical Parisian family.

B. Have students locate the 7ᵉ *arrondissement* on the map on page 104.

Reading

A. Break the first paragraph of the reading into two parts. Have the class repeat the first three sentences after you. Then ask comprehension questions. For example: *Où habitent les Debussy? Beaucoup de familles à Paris habitent dans un appartement? Les Debussy ont un bel appartement?* Follow the same procedure with the next two sentences.

B. Call on two individuals to read the second paragraph. Ask questions after every two or three sentences.

Post-reading

A. Assign the reading, including the exercises that follow, as homework.

B. Based on the description in the reading, have students draw a floor plan of the Debussy apartment.

Note Students may listen to a recorded version of the *Lecture* on the CD-ROM.

Étude de Mots

ANSWERS

Exercice A

1. c	4. a
2. b	5. b
3. a	

104

UNE FAMILLE FRANÇAISE

*L*es Debussy habitent à Paris. Comme[1] beaucoup de familles à Paris, ils habitent dans un appartement. Ils ont un très bel appartement dans un vieil immeuble dans le septième. Le septième arrondissement[2] à Paris est un très beau quartier. Le septième est un quartier assez résidentiel.

Dans l'immeuble il y a six étages. Les Debussy habitent au quatrième. Il y a six pièces dans l'appartement. Le salon donne sur[3] la rue mais les chambres à coucher donnent sur la cour. L'appartement a un balcon. Du balcon il y a une très belle vue sur la tour Eiffel.

[1] comme   *like*
[2] arrondissement   *district in Paris*
[3] donne sur   *faces*

Étude de mots

A **Quel est le mot?** Choisissez la bonne réponse. (*Choose the correct answer.*)

1. Une partie (une zone) d'une ville est ___.
   a. un village   b. une banlieue   c. un quartier

2. Un bâtiment (un building) qui a des appartements est ___.
   a. une école   b. un immeuble   c. un arrondissement

3. Il y a des maisons ou des appartements privés dans un quartier ___.
   a. résidentiel   b. industriel   c. universitaire

4. Le père, la mère et les enfants sont ___.
   a. une famille   b. une classe   c. un groupe de copains

5. La ville de Paris est divisée en vingt ___.
   a. étages   b. arrondissements   c. pièces

Les arrondissements de Paris

CRITICAL THINKING ACTIVITY

(*Thinking skills: drawing conclusions, making inferences*)

Write the following on the board or on an overhead transparency:

1. **La maison n'a pas de garage. C'est un problème pour la famille Boivin. Mais ce n'est pas un problème pour la famille Berthollet. Pourquoi?**

2. **Quelles sont les caractéristiques des personnes qui aiment les chats et les chiens? Discutez.**

3. **Pourquoi les chiens et les chats sont-ils des amis excellents? Discutez.**

Compréhension

B **La famille Debussy.** Corrigez les phrases. (*Correct the statements.*)

1. Les Debussy habitent dans la banlieue de Paris.
2. Ils ont une grande maison.
3. L'appartement est dans un nouvel immeuble.
4. Ils habitent dans le sixième arrondissement.
5. Le septième arrondissement est commercial et industriel.
6. Il y a huit pièces dans l'appartement de la famille Debussy.
7. Les chambres à coucher donnent sur la rue.
8. L'appartement est au rez-de-chaussée.

DÉCOUVERTE CULTURELLE

*B*eaucoup de Français habitent dans un appartement, surtout[1] les habitants des grandes villes. Il y a des appartements de grand standing pour les gens[2] riches et il y a des H.L.M. (Habitations à Loyer Modéré[3]) pour les gens qui n'ont pas beaucoup d'argent. Les H.L.M. sont généralement à l'extérieur des villes, à la périphérie ou en banlieue.

Il y a aussi des catégories de maisons privées. Pour les très riches il y a de grands châteaux à la campagne et pour les gens plus modestes il y a des pavillons, de petites maisons confortables en banlieue.

En France, comme aux États-Unis, dans beaucoup de familles la mère et le père travaillent. Il y a des crèches municipales[4] où les petits enfants passent la journée[5] quand les deux parents travaillent. Comme aux États-Unis, le taux de divorces[6] augmente en France. Beaucoup d'enfants habitent avec un seul parent, la mère ou le père. Il y a beaucoup de familles à parent unique.

[1] surtout *especially*	[4] crèches municipales *day-care centers*
[2] gens *people*	[5] journée *day*
[3] H.L.M. *low-income housing*	[6] taux de divorces *divorce rate*

Compréhension

ANSWERS

Exercice B

1. **Les Debussy habitent à Paris.**
2. **Ils ont un appartement.**
3. **L'appartement est dans un vieil immeuble.**
4. **Ils habitent dans le septième arrondissement.**
5. **Le septième arrondissement est résidentiel.**
6. **Il y a six pièces dans l'appartement de la famille Debussy.**
7. **Les chambres à coucher donnent sur la cour (intérieure).**
8. **L'appartement est au quatrième étage.**

OPTIONAL MATERIAL

Découverte culturelle

PRESENTATION (*page 105*)

Have students read the selection silently. Ask them to prepare a question or two about something they don't understand in the content. Have them share their questions with the class and let other students try to answer.

Note Students may listen to a recorded version of the *Découverte culturelle* on the CD-ROM.

LEARNING FROM PHOTOS

Have the students answer the following questions about the photo of the house. *La maison est grande ou petite? La maison a un jardin? Le jardin est devant la maison ou derrière la maison? Il y a des chaises dans le jardin? Il y a une table? Il y a une bicyclette dans le jardin? Il y a un garage? Il y a une voiture?*

INDEPENDENT PRACTICE

Assign any of the following:
1. *Étude de mots* and *Compréhension* exercises, pages 104–105
2. Workbook, *Un Peu Plus*, pages 34–35
3. CD-ROM, Disc 1, pages 104–105

OPTIONAL MATERIAL

PRESENTATION *(pages 106–107)*

Have students look at the photographs for enjoyment. If you wish to do something more thorough with this section, you may do some of the following activities.

A. Before reading the captions on page 106, have students look for the *7ᵉ arrondissement* on the map of Paris on page 505 (page 238, Part A).

B. Ask students to volunteer cognates as they scan the captions on page 106.

C. Now have students look at the photos and read the captions.

Note In the CD-ROM version, students can listen to the recorded captions and discover a hidden video behind one of the photos.

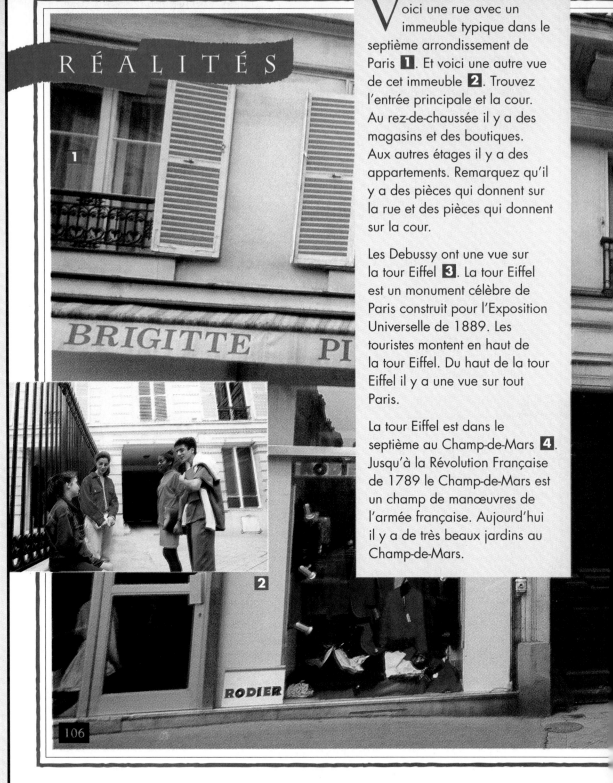

RÉALITÉS

Voici une rue avec un immeuble typique dans le septième arrondissement de Paris **1**. Et voici une autre vue de cet immeuble **2**. Trouvez l'entrée principale et la cour. Au rez-de-chaussée il y a des magasins et des boutiques. Aux autres étages il y a des appartements. Remarquez qu'il y a des pièces qui donnent sur la rue et des pièces qui donnent sur la cour.

Les Debussy ont une vue sur la tour Eiffel **3**. La tour Eiffel est un monument célèbre de Paris construit pour l'Exposition Universelle de 1889. Les touristes montent en haut de la tour Eiffel. Du haut de la tour Eiffel il y a une vue sur tout Paris.

La tour Eiffel est dans le septième au Champ-de-Mars **4**. Jusqu'à la Révolution Française de 1789 le Champ-de-Mars est un champ de manœuvres de l'armée française. Aujourd'hui il y a de très beaux jardins au Champ-de-Mars.

106

HOMME

3

4

107

COGNATE RECOGNITION

Have three volunteers go to the board and write down all the cognates they can find in captions 1 and 2, 3, and 4, respectively. They should be able to tell the class what these words mean in English.

Captions 1 and 2: *typique/principale/la boutique/l'appartement*

Caption 3: *le monument/le (la) touriste/l'exposition/universelle*

Caption 4: *la révolution/ la manœuvre/l'armée*

Point out the difference between *le tour* (the tour) and *la tour* (the tower).

THE FRANCOPHONE WORLD

For photos of houses and apartment buildings in other French-speaking countries, see the *Les Pays* section of *Le Monde francophone,* pages 110–113.

DID YOU KNOW?

The Eiffel Tower was built for the *Exposition universelle* in 1889 by Gustave Eiffel, a structural and aeronautical engineer. The Tower is 984 feet high and 330 feet square at the base. It cost more than one million dollars to build. The construction costs were paid for from admission receipts to the *Exposition.*

ADDITIONAL PRACTICE

1. Student Tape Manual, Teacher's Edition, *Deuxième Partie,* pages 52–54
2. Situation Cards, Chapter 4

RECYCLING

The *Activités de communication orale* and *écrite* allow students to use the vocabulary and grammar in this chapter in open-ended, real-life situations. They also give students a chance to reuse the vocabulary and structure from earlier chapters.

Have students work on as many of the activities as you wish.

INFORMAL ASSESSMENT

The situations in oral Activities A and B may be used to assess speaking ability. Use the evaluation criteria given on page 34 of this Teacher's Wraparound Edition.

Activités de communication orale

PRESENTATION *(page 108)*

Divide the class into groups or pairs to role-play the situations in the *Activités de communication orale*.

Activité A

Students will have to decide on the make-up of their "families" first. Remind them to do their presentation using *nous*. Encourage them to develop names and personalities for the family members.

Activité B

This activity can be done individually or in groups. If the group format is chosen, all members should have something to say. Encourage students to assume a role, and not just give a report.

ANSWERS

Activités A and B

Answers will vary.

Activités de communication écrite

ANSWERS

Activités A and B

Answers will vary.

Activités de communication orale

A **Quelle maison?** You and your family are planning to spend a month in France. Which of the following houses, as described in the newspaper ads below, would suit your family best? Explain why, using the model as a guide.

> J'ai une grande famille. Nous sommes six. Nous aimons la jolie villa avec quatre chambres. Nous aimons aussi les chats et les chiens.

Appartement

dans bel immeuble, cinq pièces (deux chambres à coucher), avec grande cuisine moderne, bien situé au centre de la ville, près d'une banque et d'un cinéma.

Petit bungalow

dans un vieux quartier, beaucoup de charme. Trois pièces (une chambre à coucher), salle à manger avec belle table et chaises anciennes. Vingt minutes de la ville.

Jolie villa

avec jardin et balcon avec vue sur la mer. Huit pièces (quatre chambres à coucher), garage pour deux voitures, chien et chat inclus. Située dans une rue très calme, assez loin de la ville.

B **Une nouvelle identité.** Imagine you're someone else. Describe your new family members and their personalities, your house or apartment, and yourself.

Une maison de campagne dans la Creuse

Activités de communication écrite

A **Mon arbre généalogique.** Draw your own family tree. Give the names of all your relatives and their relationship to you.

B **Mon parent favori.** On your family tree, circle the name of your favorite relative and write a short paragraph about him or her. Be sure to include the following information.

1. name
2. relationship to you
3. age
4. physical description
5. personality
6. what he or she likes to do and doesn't like to do

ADDITIONAL PRACTICE

After completing the oral activities, reinforce as follows: Have students choose one of the homes described on page 108 and imagine what kind of family lives there. Students should give a description of the family members. They can also describe pets, neighbors, etc.

STUDENT PORTFOLIO

A possible written assignment for inclusion in students' portfolio is the *Mon Autobiographie* section of the Workbook on page 36.

Note   Students may create and save both oral and written work using the Electronic Portfolio feature on the CD-ROM.

Réintroduction et recombinaison

A **À votre tour.** Donnez des réponses personnelles. (*Give your own answers.*)

1. Tu es élève dans une école primaire?
2. Les élèves dans ton cours de français sont intelligents?
3. Tu as des cours le samedi?
4. Tes copains et toi, vous étudiez quelles matières?
5. Tes parents adorent écouter de la musique rock? Et toi?
6. Où est-ce que tu regardes la télé?
7. Tu invites des copains pendant le week-end?

B **Ma famille et ma maison.** Complétez. (*Complete with your own answers.*)

1. Il y a ___ personnes dans ma famille.
2. Je ressemble à ___.
3. J'ai ___ ans.
4. Nous habitons à ___.
5. Notre maison (appartement) a ___ pièces.
6. Ma pièce favorite est ___.
7. Dans ma chambre, il y a ___ et ___.

Vocabulaire

NOMS	la maison	le métro	ADJECTIFS
la famille	l'appartement (m.)	la station de métro	beau (bel), belle
le père	l'immeuble (m.)	le quartier	nouveau (nouvel),
la mère	l'ascenseur (m.)		nouvelle
les parents (m.)	le balcon	l'âge (m.)	vieux (vieil), vieille
la femme	la cour	l'année (f.)	joli(e)
le mari	l'entrée (f.)	la date	jeune
l'enfant (m.)	l'étage (m.)	l'anniversaire (m.)	premier, première
le fils	le rez-de-chaussée		deuxième
la fille	la pièce	le mois	troisième
la grand-mère	les toilettes (f.)	janvier	
le grand-père	la salle de bains	février	VERBES
les grands-parents	la chambre à coucher	mars	avoir
le petit-fils	la cuisine	avril	bavarder
la petite-fille	le dîner	mai	dîner
l'oncle (m.)	la salle à manger	juin	préparer
la tante	la salle de séjour	juillet	
le cousin	le garage	août	AUTRES MOTS
la cousine	la voiture	septembre	ET EXPRESSIONS
le neveu	le jardin	octobre	avoir... ans
la nièce	la terrasse	novembre	il y a
le chat	le voisin	décembre	loin de
le chien	la voisine		près de

CHAPITRE 4   **109**

Réintroduction et recombinaison

PRESENTATION (*page 109*)

These exercises incorporate previously learned vocabulary and structures.

ANSWERS

Exercices A and B

Answers will vary.

ASSESSMENT RESOURCES

1. Chapter Quizzes
2. Testing Program
3. Situation Cards
4. Communication Transparency C-4
5. Performance Assessment
6. Computer Software: Practice/Generator

VIDEO PROGRAM

INTRODUCTION (15:11)

À LA MAISON (16:44)

FOR THE YOUNGER STUDENT

1. Ask students to bring in pictures of their family and pets, if they have any. Have them identify each person or animal in the picture and write a sentence about each.
2. Have students make a birthday card (*Joyeux anniversaire!*) out of construction paper and prepare a list of things they would like to receive as birthday gifts.

INDEPENDENT PRACTICE

1. Activities and exercises, pages 108–109
2. CD-ROM, Disc 1, pages 108–109
3. Communication Activities Masters, pages 19–22

LE MONDE FRANCOPHONE

OPTIONAL MATERIAL

Les Pays

PRESENTATION *(pages 110–113)*

This section is designed to give students information about regions and countries in the francophone world. The captions are simple and incorporate the active vocabulary from Chapters 1–4. If you wish to present the material in class, here are some suggestions.

A. Have students scan the captions and give the names of five regions where French is spoken. Then have them locate these regions on the map of the Francophone World on page 506 (Part A, page 239), or you may prefer to point them out to students on the Francophone World Map Transparency. Write the names of the regions on the board.

B. Read the captions with students and write the name of each country under the appropriate region. Have students add the names of other French-speaking countries in the same region. (You may refer them to the map of the Francophone World for this information.)

C. Now have them look at the photos of students from these countries in Chapter 1, *Réalités*, pages 32–33.

MORE ABOUT THE FRANCOPHONE WORLD

Paris is the largest French-speaking city in the world and Montreal, Canada, the next-largest. The chart on page 111 shows the countries where French is spoken. It is sometimes the official language of a country—even though the citizens speak languages other than French, as in Senegal and the Ivory Coast in West Africa.

In other countries, such as those in North Africa, French is not the official language, but it is spoken as

LES PAYS

Le français est une langue importante. Plus de 120.000.000 (cent vingt millions) de personnes parlent français dans le monde: en Europe, en Amérique du Nord, en Amérique du Sud, en Afrique et en Asie. Incroyable!

Regardez la carte du monde à la page 506. Identifiez les pays francophones.

On parle français dans beaucoup de pays. Pourquoi? Parce que, pendant 300 ans, la France explore et colonise une grande partie du monde. Aujourd'hui on continue à parler français dans un grand nombre d'ex-colonies françaises et dans certains pays européens proches de la France.

L'EUROPE

1 De ces immeubles à Dinant, en Belgique, il y a une très belle vue sur la Meuse. En Belgique on parle deux langues. Au sud les Wallons parlent français et au nord les Flamands parlent néerlandais (flamand).

2 En Suisse il y a trois langues officielles: l'allemand, le français et l'italien. À Lausanne, sur le Lac Léman, on parle français.

a second language by a large percentage of the population.

MORE ABOUT THE PHOTOS

Photo 1 Belgium's population is split into two groups. The Flemish, in the north, comprise about 60% of the population and speak Flemish. The Walloons, in the south, make up the other 40% and speak French. Brussels, the capital, is predominantly French-speaking (although located in the north and officially bilingual). It is the headquarters of both the European Union and NATO.

Photo 2 German, French, and Italian are the three official federal languages of Switzerland. A fourth language, Romansh, spoken mainly in the canton of Craubünden, is recognized as an official national—not federal—language. Derived from Latin, Romansh has survived in the isolation of mountain valleys.

L'AFRIQUE

3 Abidjan est la ville principale de la Côte-d'Ivoire, en Afrique occidentale. Abidjan est un port très actif. C'est aussi une ville moderne de plus d'un million d'habitants. Beaucoup d'Abidjanais habitent dans des immeubles modernes. Ces grands immeubles ont plus de 20 étages.

4 Voilà des Haoussas devant une maison typique du Niger, un autre pays de l'Afrique occidentale. Les Haoussas sont des cultivateurs, des artisans et des commerçants. Ils habitent la frontière Niger-Nigeria.

5 Il y a de la *neige* en Afrique? Mais oui! Voilà une vue superbe de la vallée d'Ourika au pied de l'Atlas, une chaîne de montagnes en Afrique du Nord. Ce petit village isolé est situé au Maroc. Remarquez le minaret de la mosquée. Les Marocains, qui parlent arabe et français, sont des musulmans.

6 L'île Maurice est dans l'océan Indien à l'est de Madagascar. Ces petites Mauriciennes sont dans la cour d'une école primaire à Port-Louis, la capitale de l'île Maurice. Ici on parle deux langues: le français et l'anglais.

LE MONDE FRANCOPHONE

LES PAYS FRANCOPHONES

L'EUROPE
la France, la Belgique, la Suisse, le Luxembourg, Monaco

L'AFRIQUE

L'AFRIQUE DU NORD
le Maroc, l'Algérie, la Tunisie

L'AFRIQUE OCCIDENTALE
la Mauritanie, le Mali, le Niger, le Sénégal, le Burkina-Faso, la Guinée, la Côte d'Ivoire, le Togo, le Bénin, le Cameroun

L'AFRIQUE ÉQUATORIALE
le Gabon, le Congo

L'AFRIQUE CENTRALE
le Tchad, la République centrafricaine, le Zaïre, le Rouanda, le Burundi

L'AFRIQUE ORIENTALE
Djibouti

L'OCÉAN INDIEN
les Seychelles, les Comores, l'île Maurice, la Réunion, Mayotte, Madagascar

L'AMÉRIQUE

L'AMÉRIQUE DU NORD
le Canada, la Louisiane, Saint-Pierre-et-Miquelon

LES ANTILLES
Haïti, la Martinique, la Guadeloupe

L'AMÉRIQUE DU SUD
la Guyane

L'OCÉANIE
la Nouvelle-Calédonie, la Polynésie française, les îles Wallis-et-Futuna, Vanuatu

L'ASIE
le Viêt-Nam, le Cambodge, le Laos, Pondichéry

Photo 3 Some 60 ethnic groups make up the population (13.8 million) of the Ivory Coast, which is still mostly a rural country. Many African languages are spoken here in addition to French, the official language. The market language spoken everywhere is Dioula.

Photo 4 A black Islamic people, the Hausa (*Haoussa*, in French) make up over half the population of Niger. The Hausa language is spoken by over 20 million people in Africa.

Commerce is one of the Hausa's most important activities, especially during the dry season. They are also skilled farmers who devote themselves to agriculture during the wet season.

Photo 5 Morocco was a French protectorate until 1956, when it got its independence. Though Morocco's official language is Arabic, in school all students learn French, which is spoken as a second language by most people.

Tourism is an important industry of Morocco owing to its varied attractions—beaches, cities, mountains, and oases. From mid-December until April the snow on the Rif, Middle, and High Atlas mountains is at its best, and one can enjoy good skiing.

Photo 6 A beautiful island in the Indian Ocean, Mauritius is a very popular tourist destination thanks to its wonderful climate and pretty beaches.

Photo 7 Canada is a bilingual country. English and French are its official languages. French-speaking communities exist throughout Canada, and French is the only official language of the province of Quebec. There is a very strong independence movement in Quebec. French is also widely spoken in New Brunswick (*le Nouveau-Brunswick*), which is officially a bilingual province. New Brunswick is one of the Atlantic provinces, along with Nova Scotia (*la Nouvelle-Écosse*), Prince Edward Island (*l'île du Prince Édouard*), and Newfoundland (*Terre-Neuve*).

Photo 8 When the English took possession of *la Nouvelle-France* (Canada) from the French in the 18th century, they expelled the *Acadiens*, as the French settlers were called, from the maritime provinces. Many of these *Acadiens* went to Louisiana, which was still a French territory at that time. The term "Cajun" is a deformation of the French word *Acadien*.

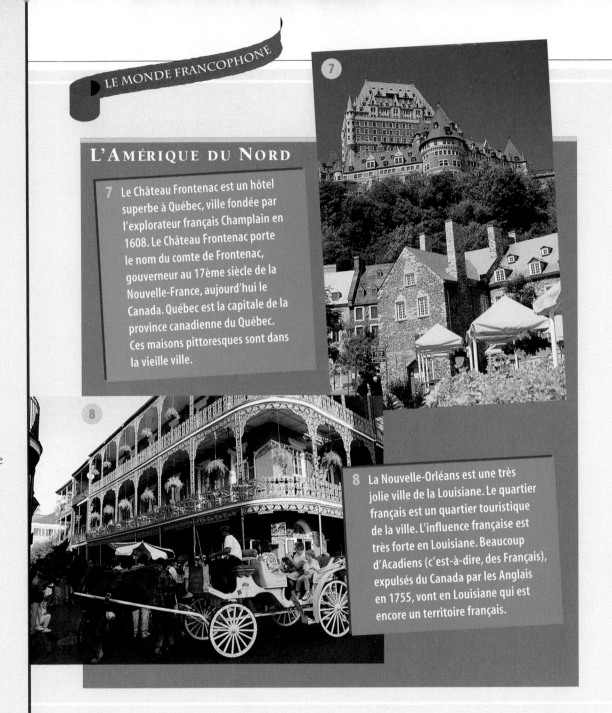

LE MONDE FRANCOPHONE

L'AMÉRIQUE DU NORD

7 Le Château Frontenac est un hôtel superbe à Québec, ville fondée par l'explorateur français Champlain en 1608. Le Château Frontenac porte le nom du comte de Frontenac, gouverneur au 17ème siècle de la Nouvelle-France, aujourd'hui le Canada. Québec est la capitale de la province canadienne du Québec. Ces maisons pittoresques sont dans la vieille ville.

8 La Nouvelle-Orléans est une très jolie ville de la Louisiane. Le quartier français est un quartier touristique de la ville. L'influence française est très forte en Louisiane. Beaucoup d'Acadiens (c'est-à-dire, des Français), expulsés du Canada par les Anglais en 1755, vont en Louisiane qui est encore un territoire français.

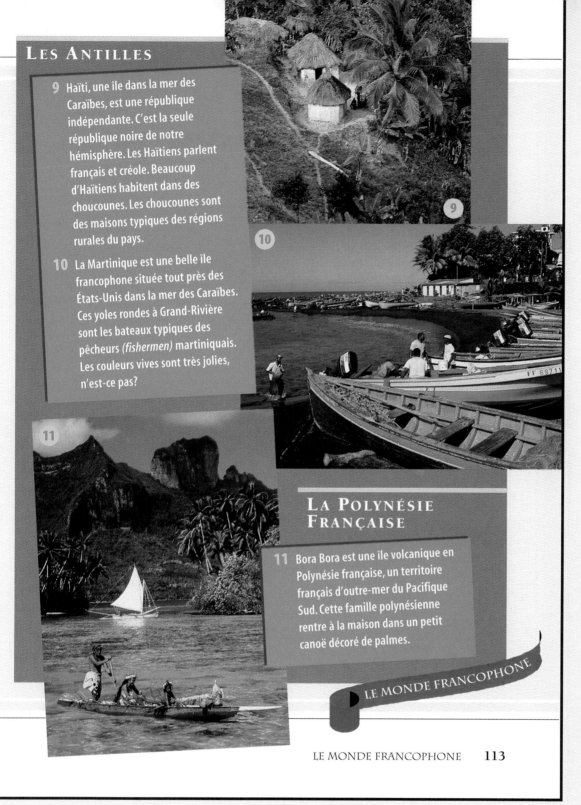

LES ANTILLES

9 Haïti, une île dans la mer des Caraïbes, est une république indépendante. C'est la seule république noire de notre hémisphère. Les Haïtiens parlent français et créole. Beaucoup d'Haïtiens habitent dans des choucounes. Les choucounes sont des maisons typiques des régions rurales du pays.

10 La Martinique est une belle île francophone située tout près des États-Unis dans la mer des Caraïbes. Ces yoles rondes à Grand-Rivière sont les bateaux typiques des pêcheurs *(fishermen)* martiniquais. Les couleurs vives sont très jolies, n'est-ce pas?

LA POLYNÉSIE FRANÇAISE

11 Bora Bora est une île volcanique en Polynésie française, un territoire français d'outre-mer du Pacifique Sud. Cette famille polynésienne rentre à la maison dans un petit canoë décoré de palmes.

LE MONDE FRANCOPHONE

Photo 9 Martinique and Guadeloupe are French *départements d'outre-mer (D.O.M.).* The inhabitants of Martinique and Guadeloupe are French citizens. These two Caribbean islands are popular tourist destinations for French, Canadian, and American tourists.

Photo 10 Haiti is an independent nation. It shares the island of Hispaniola with the Dominican Republic. Haiti declared its independence from France in 1804 when Jacques Dessaline proclaimed himself emperor. Haiti, the only black republic in this hemisphere, has suffered from severe political problems and economic hardships due to a continuous series of dictatorships. Democracy was recently restored when, with the help of the U.S., Haiti's first free elections were held.

The official language of Haiti is French, but the general population speaks creole. Creole is a language derived from French, with strong African influences. To a lesser degree it has been influenced by Spanish and English.

Photo 11 Polynesia is a name given to a vast expanse of islands of Oceania in the South Pacific to the east of Australia and Micronesia. It includes French Polynesia, New Zealand, Samoa, and Hawaii. French Polynesia is a *territoire d'outre-mer (T.O.M.).*

RÉVISION

CHAPITRES 1–4

OVERVIEW

This section reviews key grammatical structures and vocabulary from Chapters 1–4. The structure topics were first presented on the following pages: *-er* verbs, page 73; *avoir*, page 96; *être*, pages 25 and 46; indefinite articles, pages 21 and 76; definite articles, pages 21 and 45; agreement of adjectives, pages 23 and 50; possessive adjectives, page 98.

REVIEW RESOURCES

1. Workbook, Self-Test 1, pages 37–39
2. Videocassette/Videodisc, Unit 1
3. Video Activities Booklet, Unit 1: Chapters 1–4, pages 1–17
4. Computer Software, Chapters 1–4
5. Testing Program, Unit Test: Chapters 1–4, pages 22–27
6. Performance Assessment
7. CD-ROM, Disc 1, *Révision:* Chapters 1–4, pages 114–117
8. CD-ROM, Disc 1, Self-Tests 1–4
9. CD-ROM, Disc 1, Game: *Pour en savoir plus*
10. Lesson Plans

Conversation

PRESENTATION *(page 114)*

Read the conversation aloud as the class listens with books closed. Ask questions to check for comprehension. Repeat the procedure with books open. After students have practiced the conversation in pairs, call on several pairs to dramatize it for the class.

Conversation *Paul est français.*

ANNICK: Paul, tu es canadien ou français?
PAUL: Moi, je suis français.
ANNICK: Tu habites à Paris?
PAUL: Non, je n'habite pas à Paris. J'habite à Toulouse.
ANNICK: Tu as une grande famille?
PAUL: Oui, ma famille est grande. Nous sommes six.
ANNICK: Ta famille habite dans une maison ou dans un appartement?
PAUL: Nous avons une maison.

A **Paul et sa famille.**   Complétez d'après la conversation. (*Complete according to the conversation.*)

1. Paul ___ français.
2. Il n'est pas ___.
3. Il habite à ___.
4. Il ___ à Paris.
5. Il n'a pas une petite famille. Il a une ___ famille.
6. Il y a six personnes dans ___ famille.
7. Paul et sa famille n'ont pas ___. Ils ___ une maison.

Structure

Les verbes en *-er*

Review the following forms of regular *-er* verbs.

ÉTUDIER	
j' étudie	nous étudions
tu étudies	vous étudiez
il/elle/on étudie	ils/elles étudient

LEARNING FROM PHOTOS

Ask students the following questions about the *Famille Duchamp* in the photo. *Il y a combien de personnes dans cette famille? Il y a combien d'enfants? Combien de filles? Combien de garçons? Qui est blond? Qui est brun? Comment est la mère? Le père?*

PAIRED ACTIVITY

Have students work together in pairs to make up their own conversations about their nationality, where they live, and their family. Have several pairs present their conversation to the class.

A A la fête.   Choisissez un verbe pour compléter les phrases. *(Choose a verb to complete the sentences.)*

aimer	étudier	parler
chanter	gagner	regarder
danser	inviter	travailler

1. Alain et Catherine ___ vraiment bien ensemble.
2. Du courage! J'___ Marie-Claire à danser.
3. J'aime beaucoup cette cassette. Qui ___?
4. Nous ___ beaucoup la musique classique.
5. Vous ___ la télé?
6. Où est Véronique? Elle ___ au téléphone?
7. Olivier et Philippe ne sont pas là. Olivier a un examen, alors il ___. Philippe ___ au magasin de disques.
8. Il ___ beaucoup d'argent.
9. Tu ___ à mi-temps? Tu ___ beaucoup d'argent?

Les verbes *avoir* et *être*

Review the following forms of the irregular verbs *avoir* and *être*.

AVOIR	
j' ai	nous avons
tu as	vous avez
il/elle/on a	ils/elles ont

ÊTRE	
je suis	nous sommes
tu es	vous êtes
il/elle/on est	ils/elles sont

B Ma famille.   Complétez avec *avoir* ou *être*. *(Complete with* avoir *or* être.)

Dans ma famille nous ___ cinq. Il y ___ mon père, ma mère, mon frère
                         1   2
Christophe, ma sœur Stéphanie et moi. Moi, j'___ quatorze ans, mon frère ___
                                              3   4
dix-sept ans et ma sœur ___ dix-huit ans. Mon frère ___ sympa. Ma sœur
                        5   6
aussi, et elle ___ beaucoup d'amis, alors elle n'___ pas souvent à la maison.
               7   8
Nous ___ des parents sympathiques. Je ___ content. J'___ une famille très
     9   10   11
chouette. Tu ___ content(e) aussi?
             12

ANSWERS

Exercice A

1. est		5. grande	
2. canadien		6. sa	
3. Toulouse		7. d'apparte-	
4. n'habite pas		ment, ont	

Structure
Les verbes en -er
PRESENTATION *(page 114)*

A. Quickly write the model verb *étudier* on the board and underline the endings. Have students pronounce each form after you. Repeat the *je, tu, il/elle/on, ils/elles* forms, emphasizing that they are all pronounced the same way in spite of their spelling differences.

B. Ask a student to give you another -*er* verb. Write its forms on the board alongside *étudier* and have the class quickly repeat the verb. Then do Exercise A.

ANSWERS

Exercice A

1. dansent	7. étudie,
2. invite	travaille
3. chante	8. gagne
4. aimons	9. travailles,
5. regardez	gagnes
6. parle	

Les verbes avoir et être
PRESENTATION *(page 115)*

Have students repeat the forms of *avoir* and *être* after you as they read along in their books.

ANSWERS

Exercice B

1. sommes	7. a
2. a	8. est
3. ai	9. avons
4. a	10. suis
5. a	11. ai
6. est	12. es

ADDITIONAL PRACTICE

You may wish to ask students personalized questions about French class or their family. *Où es-tu élève? Qui est ton professeur de français? Comment est le cours de français? Quel âge as-tu? Il y a combien de personnes dans ta famille? Quel âge a ton frère? Et ta sœur? Ta famille et toi, vous avez un chien ou un chat?*, etc.

INDEPENDENT PRACTICE

Have students write a paragraph for homework (modeled after Exercise B) in which they substitute information about their own family for the information in the text.

Les articles et les adjectifs

PRESENTATION (page 116)

Have students open their books to page 116. Read the explanation aloud with them. Have them repeat the model words and sentences after you. Then do Exercises C and D.

Exercices

ANSWERS

Exercice C

1. un, une
2. Des, des
3. une
4. un, un
5. une
6. un, une
7. un
8. un

Exercice D

1. Sa sœur est contente aussi!
2. Sa sœur est amusante aussi!
3. Sa sœur est sympathique aussi!
4. Sa sœur est énergique aussi!
5. Sa sœur est intéressante aussi!
6. Sa sœur est brune aussi!

Les articles et les adjectifs

1. Review the following forms of the indefinite and definite articles.

un garçon	une fille	un(e) ami(e)	des enfants
le garçon	la fille	l'ami(e)	les enfants

2. Adjectives that end in a consonant have four forms.

Le garçon est blond.	La fille est blonde.
Les garçons sont blonds.	Les filles sont blondes.

3. Adjectives that end in *e* have only two forms, singular and plural.

un ami sympathique	une amie sympathique
des amis sympathiques	des amies sympathiques

C **La famille de Christian.** Complétez avec *un, une* ou *des*. (*Complete with* un, une, *or* des.)

1. Christian a une grande famille. Il a ___ père et ___ mère.
2. ___ frères et ___ sœurs? Oui, il a trois frères et quatre sœurs.
3. Il a aussi sept cousins, mais ___ seule cousine.
4. Il a ___ chien, Médor, et ___ chat, Minouche.
5. Christian et sa famille habitent dans ___ petite maison à Pontchartrain.
6. Pontchartrain est ___ village, ou ___ petite ville, près de Paris.
7. Christian est élève dans ___ lycée de la région.
8. C'est ___ élève excellent.

D **Sa sœur aussi.** Répondez d'après le modèle. (*Answer according to the model.*)

> Il est très intelligent.
> *Sa sœur est très intelligente aussi!*

1. Il est content.
2. Il est amusant.
3. Il est sympathique.
4. Il est énergique.
5. Il est très intéressant.
6. Il est brun.

ADDITIONAL PRACTICE

Have students give you adjectives that they know. Write them on the board. Then call on students to give you original sentences using these adjectives. To conserve time, ask for volunteers.

LEARNING FROM PHOTOS

Point to the armchair in the photo and say: *C'est un fauteuil.* Then ask the following questions: *Est-ce que le chat est dans le fauteuil? Est-ce que le chat est beau? Il est adorable? Il est grand ou petit, le chat? Il est content? Il est énergique? Tu aimes les chats? Tu as un chat?*

Les adjectifs possessifs

1. Review the following forms of the possessive adjectives.

mon livre	mes livres	ma cassette	mes cassettes
ton cousin	tes cousins	ta cousine	tes cousines
son appartement	ses appartements	sa maison	ses maisons

2. Remember that you use *mon, ton,* and *son* before a masculine or feminine noun beginning with a vowel or a silent *h: mon ami, mon amie.*

E **La famille de Marc.** Complétez. *(Complete.)*

ANNE: Marc, qui est ___ sœur?

MARC: ___ sœur? Je n'ai pas de sœur.

ANNE: Qui est ___ frère alors?

MARC: ___ frère? Je n'ai pas de frère. Je suis enfant unique. ___ parents n'ont pas d'autres enfants.

Marc n'a pas de sœur et il n'a pas de frère. ___ famille est très petite. Ils sont trois. ___ parents ont un seul fils, c'est Marc. ___ mère et ___ père adorent Marc.

Activités de communication orale et écrite

A **Un(e) jeune Français(e).** Imagine a French teenager. Describe him or her as well as his or her family and house or apartment.

B **Un(e) ami(e).** Describe one of your friends and his or her family and house or apartment.

C **Une conversation.** Imagine that the friend you described in *Activité B* and the French teenager in *Activité A* meet. Write the conversation they might have.

LETTRES ET SCIENCES

Les Sciences humaines

OVERVIEW

The three readings in this *Lettres et sciences* section are related topically to material in Chapters 2 and 3. You may wish to allow students to choose which of the three selections they want to read according to their own interests, or you may wish to have the entire group read a particular selection.

Each reading may be presented at different levels of intensity:

1. The least intensive treatment would be to assign the selection as independent reading and the post-reading exercises as homework. This treatment requires no class time and minimal teacher involvement.
2. For a more intensive treatment, the reading and post-reading exercises can be assigned for homework, which will be gone over orally in class the next day.
3. The most intensive treatment includes a pre-reading presentation of the text by the teacher, an in-class reading and discussion of the passage, the assignment of the exercises for homework, and a discussion of the assignment in class the following day.

Avant la lecture

PRESENTATION *(page 118)*

A. Ask students to list some of the social sciences. Most of the terms are cognates: *les sciences naturelles, l'histoire, la géographie, la sociologie, l'économie, la psychologie, les sciences politiques, l'anthropologie.*
B. Give students several minutes to read the *Avant la lecture* section in their book.

You have seen that French teenagers, like you, study many subjects. In this part of the textbook, we will introduce you to topics related to the subjects you are now studying or may study in the future. Who knows, you may soon have the opportunity to discuss them with some new French-speaking friends.

LES SCIENCES HUMAINES

Avant la lecture

The social sciences are fields that deal with history, human behavior, and social customs and interactions. One important social science is geography, which is the study of the surface of the earth. For a moment, think about the geography of your own state—its rivers, mountains, size, etc.

Lecture

Les sciences humaines étudient l'homme, son histoire, ses institutions et son comportement[1]. La sociologie étudie l'homme et ses rapports avec les autres membres de la société: la famille, le mariage, le divorce. L'anthropologie étudie l'homme, ses coutumes, son travail, ses cérémonies. L'histoire étudie le passé[2]. La géographie étudie la surface de la terre[3], des États-Unis ou de la France, par exemple.

Quand on parle de la France on utilise le mot «hexagone». Un hexagone est une forme géométrique qui a six côtés. La France est très bien située, en pleine zone tempérée (latitude entre[4] le 42e et le 51e parallèle Nord, longitude entre le 5e méridien Ouest et le 8e méridien Est).

La France n'est pas un grand pays; elle a une superficie de 551 695 km², mais elle a des paysages[5] très variés. Au sud-est et au sud il y a de très hautes montagnes, les Alpes et les Pyrénées. À l'ouest et au nord il y a des plaines. Au centre on trouve des plateaux et des montagnes pas très hautes, le Massif Central.

La France a cinq fleuves[6]. Le Rhin est la frontière entre l'Allemagne et la France. La Seine est un fleuve calme qui passe par Paris; la Loire est un fleuve très long; la Garonne est un fleuve «violent» et le Rhône est une grande source d'énergie électrique. Trouvez ces fleuves sur la carte. La France a des mers[7] sur trois des six côtés de l'hexagone. Trouvez les mers sur la carte— la Manche, l'océan Atlantique et la mer Méditerranée.

Map labels: nord, ouest, est, sud, LA MANCHE, la Seine, le Rhin, la Loire, L'OCÉAN ATLANTIQUE, le Rhône, la Garonne, Le Massif Central, Les Pyrénées, LA MER MÉDITERRANÉE

ADDITIONAL PRACTICE

1. Survey the students as to their likes and dislikes regarding the social sciences they have already identified. They can rate their preferences: *(1) J'adore; (2) J'aime bien; (3) J'aime assez; (4) Je n'aime pas; (5) Je déteste.*
2. Give each student a map of France and have him/her label its important cities, rivers, mountains, and borders. (See the *"Bienvenue"* preliminary chapter in the *Bienvenue* Internet Activities Booklet for an outline map of France.)

La France est un vieux pays, mais c'est aussi un pays très moderne qui occupe une place importante dans le monde.

¹ comportement *behavior*
² passé *past*
³ terre *earth*
⁴ entre *between*
⁵ paysages *landscapes*
⁶ fleuves *rivers*
⁷ mers *seas*

Strasbourg, en Alsace

Un paysage d'hiver en Haute-Savoie

Un port de pêche en Bretagne

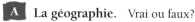

Des vignobles en Bourgogne

Après la lecture

A **La géographie.** Vrai ou faux?

1. La France est un pays très grand.
2. Il y a cinq fleuves en France.
3. La France a des paysages très variés.
4. La France est un vieux pays.
5. La France n'est pas un pays moderne.

B **En Amérique du Nord.** Répondez.

1. Nommez deux ou trois fleuves américains.
2. Quelles sont les montagnes qui séparent l'est de l'ouest?
3. En quoi sont divisés les États-Unis?
4. Nommez des grandes villes.
5. Quels sont les océans?

C **Votre état.** Vous décrivez votre état à des amis français. Dites où sont les montagnes, les plaines, les grandes villes, les fleuves, les lacs, etc.

LETTRES ET SCIENCES **119**

C. You may wish to make a list of the cognates in the reading (there are approximately 30) and share them with your students.
D. Have students look at the map. Say: *La France est un hexagone. Pourquoi? Regardez. La France, comme un hexagone, a six côtés.* Point out the six sides on the map.

Lecture

PRESENTATION *(pages 118–119)*

A. Have students read the selection silently.
B. Ask them to come up with an alternate title for the reading selection.

Après la lecture

PRESENTATION *(page 119)*

You might like to ask some simple questions in French about geography and other social sciences. For example: *Tu aimes l'histoire? La géographie? Quelle est la capitale de… ? Quelles sont les montagnes… ?*, etc.

Exercices

ANSWERS

Exercice A

1. faux
2. vrai
3. vrai
4. vrai
5. faux

Exercice B

1. Answers will vary.
2. **Les montagnes Rocheuses séparent l'est de l'ouest.**
3. **Les États-Unis sont divisés en états.**
4. Answers will vary.
5. **L'océan Pacifique et l'océan Atlantique.**

Exercice C

Answers will vary.

LETTRES ET SCIENCES

OPTIONAL MATERIAL

Les Sciences naturelles
Avant la lecture

PRESENTATION *(page 120)*

A. Have students take a few minutes to do the *Avant la lecture* activity.

B. Ask students to repeat the cognates in the reading: *les sciences naturelles, la biologie, la physique, une catégorie, importante, l'anatomie, la zoologie, la botanique, humain, un animal, une plante, l'énergie, un laboratoire, un instrument, un microscope.*

Lecture

PRESENTATION *(page 120)*

Developing reading recognition skills

Explain to students that one of the tactics they will use to guess the meaning of unknown words is guessing from context. They can often guess the meaning from the entire sentence. Go through the following word-attack procedures.

1. **la chimie:** This word is only a near-cognate. Students should be able to guess the meaning from the context: *Les sciences naturelles sont la biologie, la physique et la chimie.* Ask them what the natural sciences are: biology, physics, and what?

2. **le corps:** This word is unknown to students, but the context makes it clear: *L'anatomie étudie le corps humain.* In the sentence, *anatomie* and *humain* are cognates, so students should be able to guess the meaning of *corps* from the context. Ask them what the field of anatomy studies.

3. **le savant:** The context is: *Les savants travaillent dans un laboratoire avec un microscope.* Ask students what type of people work in laboratories.

LES SCIENCES NATURELLES

Avant la lecture

The natural sciences are divided into three major categories—physics, chemistry, and biology. Each of these can be divided into subcategories. List as many subcategories and subspecialties as you can.

Lecture

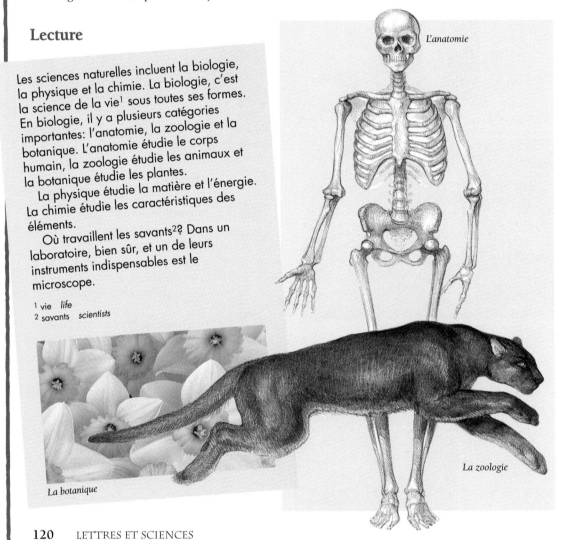

Les sciences naturelles incluent la biologie, la physique et la chimie. La biologie, c'est la science de la vie[1] sous toutes ses formes. En biologie, il y a plusieurs catégories importantes: l'anatomie, la zoologie et la botanique. L'anatomie étudie le corps humain, la zoologie étudie les animaux et la botanique étudie les plantes.

La physique étudie la matière et l'énergie. La chimie étudie les caractéristiques des éléments.

Où travaillent les savants[2]? Dans un laboratoire, bien sûr, et un de leurs instruments indispensables est le microscope.

[1] vie *life*
[2] savants *scientists*

L'anatomie

La zoologie

La botanique

LEARNING FROM PHOTOS

1. You may wish to teach students the following words: *un squelette, un os, une fleur, une jonquille, une panthère.*

2. Using these words, ask questions such as: *Est-ce qu'il y a un squelette dans la classe de biologie? Vous aimez les fleurs? Les jonquilles sont de belles fleurs? La panthère est un animal domestique ou sauvage?*

SCIENCES

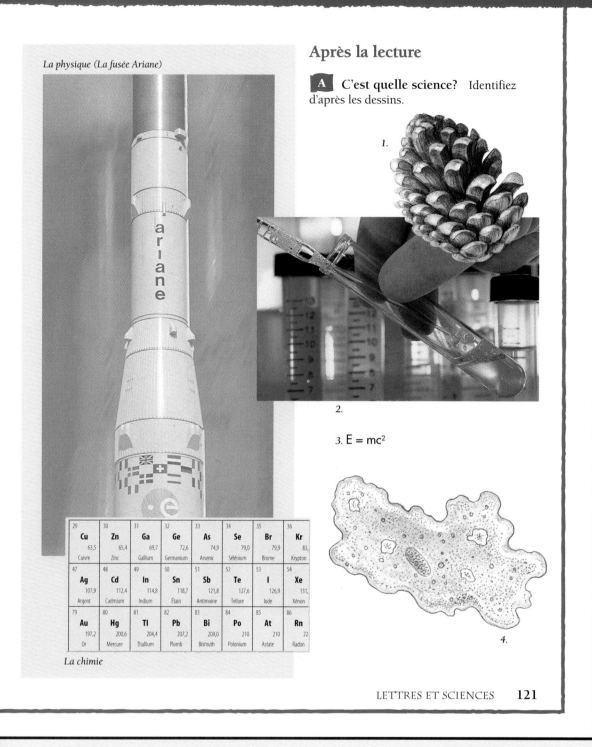

La physique (La fusée Ariane)

La chimie

29 **Cu** 63,5 Cuivre	30 **Zn** 65,4 Zinc	31 **Ga** 69,7 Gallium	32 **Ge** 72,6 Germanium	33 **As** 74,9 Arsenic	34 **Se** 79,0 Sélénium	35 **Br** 79,9 Brome	36 **Kr** 83, Krypton
47 **Ag** 107,9 Argent	48 **Cd** 112,4 Cadmium	49 **In** 114,8 Indium	50 **Sn** 118,7 Étain	51 **Sb** 121,8 Antimoine	52 **Te** 127,6 Tellure	53 **I** 126,9 Iode	54 **Xe** 131, Xénon
79 **Au** 197,2 Or	80 **Hg** 200,6 Mercure	81 **Tl** 204,4 Thallium	82 **Pb** 207,2 Plomb	83 **Bi** 209,0 Bismuth	84 **Po** 210 Polonium	85 **At** 210 Astate	86 **Rn** 22 Radon

Après la lecture

A **C'est quelle science?** Identifiez d'après les dessins.

1.

2.

3. $E = mc^2$

4.

PRESENTATION *(page 121)*

You might ask your students some simple questions about science in French. For example: *Tu aimes quelles sciences? Tu détestes quelles sciences? Tu aimes étudier les animaux? Le corps humain? Les fleurs? Les plantes? Tu aimes travailler dans le laboratoire? Tu aimes utiliser un microscope? Tu utilises un microscope dans le cours de biologie? Quelle science étudie les animaux? Quelle science étudie le corps humain? Quelle science étudie les plantes et les fleurs? Quelle science étudie la matière et l'énergie?*

ANSWERS

Exercice A

1. la botanique
2. la chimie
3. la physique
4. la biologie

LEARNING FROM PHOTOS

1. You may wish to teach the students the following words: *une pomme de pin, une éprouvette, la Classification périodique des éléments, une amibe.*
2. Have students look at the Periodic Table of Elements. The chemical symbols (Cu, Zn, Ga, etc.) are the same in all languages. Tell them that if they know what the symbol means, they can tell what the element name means in French. Have students identify as many elements as they can.

LETTRES ET SCIENCES

Les Beaux-Arts
Avant la lecture

PRESENTATION *(page 122)*

A. Discuss the *Avant la lecture* activities with the students.

B. You may wish to go over the cognates that appear in this selection: *l'art, la sculpture, l'architecture, la musique, la danse, le théâtre, les activités culturelles, célèbre, le domaine artistique, un exemple.*

Lecture

PRESENTATION *(page 122)*

This reading selection is very easy and can be read quickly. Ask students to read it silently, or assign it to be read at home.

LES BEAUX-ARTS

Avant la lecture

1. In your opinion, who are the best American painters and writers?
2. Do you know any French artists or writers? Which ones?

Lecture

Les Beaux-Arts, c'est le nom donné aux arts plastiques, c'est-à-dire, la peinture, la sculpture et l'architecture. Mais on inclut aussi souvent la musique, la danse et le théâtre. Les Beaux-Arts et les activités culturelles intéressent beaucoup les Français. Et il y a beaucoup de Français célèbres dans tous les domaines artistiques. En voici quelques exemples.

LA SCULPTURE
Auguste Rodin: «Le Penseur»

L'ARCHITECTURE
Pierre Lescot: Le Louvre

LA PEINTURE
Marc Chagall: «La Promenade»

LEARNING FROM PHOTOS

You may wish to give students the following information about the photos in this section.

La Promenade est une peinture de Marc Chagall. Marc Chagall est un peintre. *Le Penseur* est une sculpture d'Auguste Rodin. Auguste Rodin est un sculpteur. Victor Hugo compose de la poésie, des poèmes. Victor Hugo est un poète.

SCIENCES

LA MUSIQUE
Jacques Offenbach: «Les Contes d'Hoffmann»

LE THÉÂTRE
Molière: «Le Bourgeois gentilhomme»
(tableau de William Powell Frith)

LA POÉSIE
«Victor Hugo» par Bonnat

Après la lecture

A **D'autres Américains et Français célèbres.** Faites des recherches.

1. Trouvez un Américain ou une Américaine célèbre pour chacune des catégories ci-dessus (*above*).
2. Trouvez un autre Français ou une autre Française pour ces mêmes catégories.

LETTRES ET SCIENCES **123**

CHAPTER OVERVIEW

In this chapter students will learn how to order food in a café or restaurant. In order to be able to function in this communicative situation, they will learn vocabulary related to foods and beverages. They will also learn the verb *aller,* the contractions of *à* and *de* with definite articles, the possessive adjectives *notre, votre,* and *leur,* and the future with *aller.*

The cultural focus of Chapter 5 is on customs and traditions associated with dining in France.

CHAPTER OBJECTIVES

By the end of this chapter students will know:

1. the present indicative forms of *aller*
2. contractions of the prepositions *à* and *de* with the definite articles
3. the *futur proche*
4. all forms of the possessive adjectives *notre, votre,* and *leur*

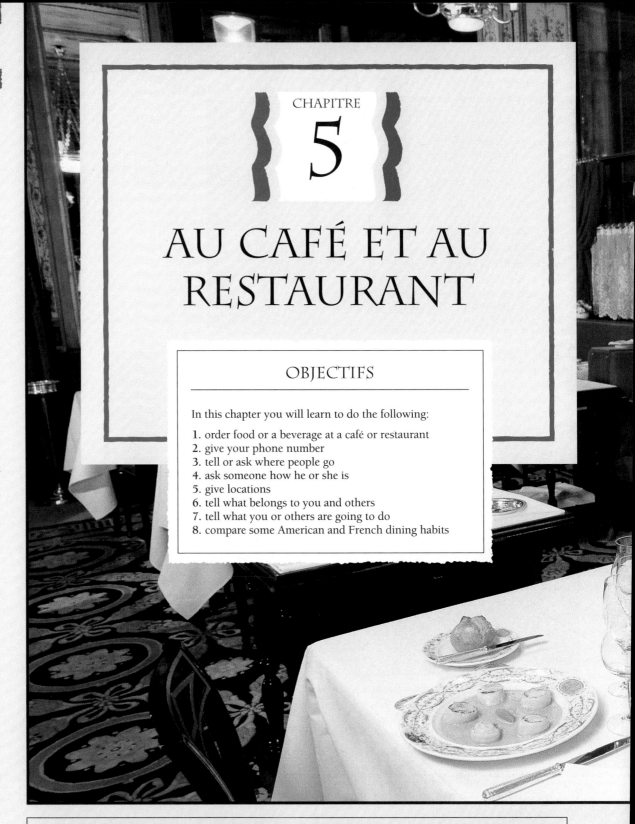

CHAPITRE

5

AU CAFÉ ET AU RESTAURANT

OBJECTIFS

In this chapter you will learn to do the following:

1. order food or a beverage at a café or restaurant
2. give your phone number
3. tell or ask where people go
4. ask someone how he or she is
5. give locations
6. tell what belongs to you and others
7. tell what you or others are going to do
8. compare some American and French dining habits

CHAPTER PROJECTS

(optional)

Prepare a French dish and share it, or have students prepare French foods to bring to class on a given day.

INTERDISCIPLINARY CONNECTIONS

Have your students ask the Home Economics class to assist them in preparing a French dish. (Consult the list of regional dishes in the CROSS-CULTURAL COMPARISON on page 145 for suggestions.)

125

CHAPTER 5 RESOURCES

1. Workbook
2. Student Tape Manual
3. Audio Cassette 4A/CD-4
4. Bell Ringer Review Blackline Masters
5. Vocabulary Transparencies
6. Pronunciation Transparency P-5
7. Communication Transparency C-5
8. Communication Activities Masters
9. Map Transparencies
10. Situation Cards
11. Conversation Video
12. Videocassette/Videodisc, Unit 2
13. Video Activities Booklet, Unit 2
14. Lesson Plans
15. Computer Software: Practice/Test Generator
16. Chapter Quizzes
17. Testing Program
18. Internet Activities Booklet
19. CD-ROM Interactive Textbook

Pacing

This chapter requires eight to ten class sessions. Pacing will depend on class length and the age and aptitude of the students.

Note The Lesson Plans offer guidelines for 45- and 55-minute classes and **Block Scheduling.**

Exercices vs. *Activités*

All exercises (which provide guided practice) are coded in blue. All communicative activities are coded in red.

INTERNET ACTIVITIES

(optional)

These activities, student worksheets, and related teacher information are in the *Bienvenue* Internet Activities Booklet and on the Glencoe Foreign Language Home Page at **http://www.glencoe.com/secondary/fl**

LEARNING FROM PHOTOS

Ask students: *C'est un restaurant français? Il est élégant? Il est beau, le restaurant? Le serveur est jeune? Il est content?*

This is a photo of *Le Grand Véfour*, a well-known Paris restaurant that is over 200 years old. Founded under the reign of Louis XV, it has been patronized by many famous people (Napoleon, Victor Hugo, Colette, Cocteau, et al.).

125

MOTS 1

Vocabulary Teaching Resources

1. Vocabulary Transparencies 5.1 (A & B)
2. Audio Cassette 4A/CD-4
3. Student Tape Manual, Teacher's Edition, *Mots 1: A–C*, pages 55–57
4. Workbook, *Mots 1: A–C*, pages 40–41
5. Communication Activities Masters, *Mots 1: A*, page 23
6. Chapter Quizzes, *Mots 1: Quiz 1*, page 25
7. CD-ROM, Disc 2, *Mots 1: pages 126–128*

Bell Ringer Review

Write the following on the board or use BRR Blackline Master 5-1: The newspaper of your sister school in France is doing a story about you. Give the following information:

1. your name and age
2. how many brothers and sisters you have
3. brief description of yourself
4. a brief description of your home
5. one important personal trait

COGNATE RECOGNITION

Students already know the French words for many foods. Have them concentrate on their pronunciation as they repeat these *Mots 1* items after you or Cassette 4A/CD-4: *un sandwich, une salade, un steak, une omelette, une soupe, un coca, un café, un thé.*

PRESENTATION *(pages 126–127)*

A. Use Vocabulary Transparencies 5.1 (A & B), point to each item, and model the individual words. Build to complete sentences.

126

VOCABULAIRE

MOTS 1

À LA TERRASSE D'UN CAFÉ

une table prise

une table libre

trouver une table

chercher une table

Guillaume va au café.
Il va au café avec Marie-France.
Les deux copains vont au café ensemble.

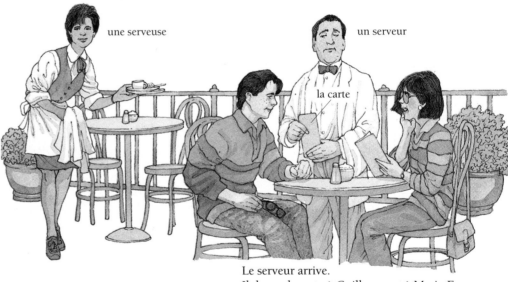

une serveuse

un serveur

la carte

Le serveur arrive.
Il donne la carte à Guillaume et à Marie-France.

126 CHAPITRE 5

TOTAL PHYSICAL RESPONSE

(following the Vocabulary presentation)

TPR
(Student 1), **levez-vous, s'il vous plaît.**
Venez ici. Vous allez mimer les actions suivantes:
Vous êtes à la terrasse d'un café.
Maintenant vous êtes assis(e).
(Student 2), **levez-vous et allez à la terrasse.**

Vous êtes la serveuse/le serveur.
(Student 1), **vous commandez un sandwich au fromage et un citron pressé.**
Maintenant, la serveuse/le serveur arrive avec le sandwich et le citron pressé.
Miam-miam! C'est délicieux.
Retournez à vos places, s'il vous plaît.

Guillaume commande une boisson.

Marie-France regarde la carte.

J'ai soif. Je voudrais un coca.

un café (un express)   un crème   un Orangina

un citron pressé   un thé citron

J'ai faim. Je voudrais quelque chose à manger.

un sandwich au jambon   un sandwich au fromage   un sandwich au pâté   un croque-monsieur

une soupe à l'oignon   une omelette nature   une omelette aux fines herbes

une salade   une saucisse de Francfort   une glace au chocolat   une glace à la vanille

des frites   une crêpe au chocolat

CHAPITRE 5   127

PRESENTATION *(page 128)*

Exercices A, B, and C

These exercises can be done with books either open or closed.

Extension of *Exercice B*

After completing Exercise B, call on two students. Each one will do half of Exercise B. Then call on a third student to retell the story in his or her own words.

ANSWERS

Exercice A

1. J'ai soif.	7. J'ai soif.
2. J'ai soif.	8. J'ai faim.
3. J'ai faim.	9. J'ai faim.
4. J'ai faim.	10. J'ai faim.
5. J'ai faim.	11. J'ai faim.
6. J'ai soif.	

Exercice B

1. Oui, Guillaume et Marie-France (ils) sont copains.
2. Oui, après les cours, Guillaume (il) va au café.
3. Oui, Marie-France (elle) va au café aussi.
4. Oui, ils vont au café ensemble.
5. Oui, ils cherchent une table.
6. Oui, ils trouvent une table libre.
7. Oui, le serveur (il) arrive.
8. Oui, il a la carte.
9. Oui, Marie-France (elle) regarde la carte.
10. Guillaume (il) commande un coca (une boisson).

Exercice C

1. Les tables sont à la terrasse du café.
2. La table est prise.
3. Les clients commandent.
4. La jeune fille commande un express.
5. Oui, elle commande une boisson et quelque chose à manger.
6. Son copain commande un sandwich au jambon.
7. Il a faim.
8. Il commande quelque chose à manger.
9. Elle commande une glace au chocolat.

Exercices

A **Tu as faim ou soif?** Choisissez d'après le modèle. *(Choose according to the model.)*

> une salade
> *J'ai faim.*
> un coca
> *J'ai soif.*

1. un citron pressé
2. un thé citron
3. un sandwich au jambon
4. une soupe à l'oignon
5. un croque-monsieur
6. un Orangina
7. un crème
8. une saucisse de Francfort
9. une omelette nature
10. une glace à la vanille
11. une crêpe au chocolat

B **Au café.** Répondez. *(Answer.)*

1. Guillaume et Marie-France sont copains?
2. Après les cours Guillaume va au café?
3. Marie-France va au café aussi?
4. Ils vont au café ensemble?
5. Ils cherchent une table?
6. Ils trouvent une table libre?
7. Le serveur arrive?
8. Il a la carte?
9. Marie-France regarde la carte?
10. Guillaume, qu'est-ce qu'il commande?

C **Un café typique.** Répondez d'après le dessin. *(Answer according to the illustration.)*

1. Les tables sont à la terrasse ou à l'intérieur du café?
2. La table est prise ou libre?
3. Qui commande, le serveur ou les clients?
4. La jeune fille commande un citron pressé ou un express?
5. Elle commande une boisson et quelque chose à manger?
6. Son copain commande une omelette ou un sandwich au jambon?
7. Il a faim ou soif?
8. Il commande une boisson ou quelque chose à manger?
9. Elle préfère la glace au chocolat, pas la glace à la vanille. Elle commande quel parfum?

LEARNING FROM ILLUSTRATIONS

Have students look at the illustration and say as much about it as they can.

INDEPENDENT PRACTICE

Assign any of the following:
1. Exercises, page 128
2. Workbook, *Mots 1: A–C*, pages 40–41
3. Communication Activities Masters, *Mots 1: A*, page 23
4. CD-ROM, Disc 2, pages 126–128

VOCABULAIRE

MOTS 2

AU RESTAURANT

Charles va au restaurant.
Il ne va pas au restaurant tout seul.
Il y va avec ses copains.
Ils y vont à pied.

Ils arrivent au restaurant.
Charles parle au maître
d'hôtel.

Vous avez notre table?

Ah oui, Monsieur.
J'ai votre table.

LE COUVERT

un verre — une tasse
une assiette
une nappe — une serviette
une fourchette   un couteau   une cuillère

CHAPITRE 5   **129**

Vocabulary Teaching Resources

1. Vocabulary Transparencies 5.2 (A & B)
2. Audio Cassette 4A/CD-4
3. Student Tape Manual, Teacher's Edition, *Mots 2: D–G,* pages 57–59
4. Workbook, *Mots 2: D–G,* page 42
5. Communication Activities Masters, *Mots 2: B,* page 24
6. Chapter Quizzes, *Mots 2:* Quiz 2, page 26
7. Computer Software, *Vocabulaire*
8. CD-ROM, Disc 2, *Mots 2:* pages 129–132

Bell Ringer Review

Write the following on the board or use BRR Blackline Master 5-2: Under each category write appropriate drinks.
Hot drinks   Cold drinks

PRESENTATION *(pages 129–130)*

A. Review *Mots 1* by asking: *Guillaume va au restaurant? Il va au restaurant avec Marie-France? Ils vont au restaurant ensemble? Ils vont au restaurant à pied? Ils trouvent une table? Le serveur arrive?*

B. Use a restaurant set-up in the classroom. Model the new words in *Mots 2* as you set a table with real objects. Demonstrate the meaning of *à gauche de, à droite de, à côté de.*

C. With books closed, students repeat the rest of the vocabulary (except the numbers) after you or Cassette 4A/CD-4. Ask several students: *Tu vas commander un steak? Tu voudrais le steak comment?*

TOTAL PHYSICAL RESPONSE

(following the Vocabulary presentation)

Getting Ready

Set up a restaurant table with a place setting. Dramatize the meaning of *coupez.*

TPR
(Student 1), **levez-vous.**
Allez au restaurant.
Cherchez une table.

Vous trouvez une table.
Asseyez-vous.
Et maintenant, (Student 2), **levez-vous.**
Allez dans le restaurant.
Vous êtes le serveur/la serveuse.
(Student 1), **demandez la carte au serveur/à la serveuse.**
Regardez la carte.
Commandez un steak.
(Student 1), **servez le steak.**

(continued on page 130)

129

D. Go over the numbers on page 130 quickly. Point out the rationale behind 70, 80, 90 (60 + 10; 4 × 20; 4 × 20 + 10).

E. Now introduce the numbers within each group of ten (71, 72, etc.) and have students repeat them. After a few examples, let volunteers follow the logic and come up with the new numbers themselves. Do not insist upon mastery. Review the numbers frequently.

Note A colloquial way of saying *le restaurant* is *le restau.*

CROSS-CULTURAL COMPARISON

1. In France, the tip is included in the check, although many people leave a little extra.
2. The French tend to eat meat rare. There is a category even rarer than *saignant*, called *bleu*. The meat is almost raw.
3. When people pass others in a restaurant who are eating, they often say *«Bon appétit».*

RECYCLING

The verbs *arriver, parler, dîner, déjeuner, aimer, avoir,* and *être* are recycled.

Vocabulary Expansion

You may want to give students the following additional food-related vocabulary.
le rond de serviette
la petite cuillère
le bol
la soucoupe
l'eau
le vin
les hors d'œuvre
le pain
le dessert
végétarien(ne)

INFORMAL ASSESSMENT

Ask students to give as many words as they can related to a restaurant or café.

Vous aimez le steak comment?

saignant   à point

bien cuit

Le service est compris.

Charles déjeune à midi.

un steak frites

C'est le soir. On dîne.

L'addition, s'il vous plaît.

laisser un pourboire

LES NOMBRES DE SOIXANTE ET UN À CENT (61–100)

61	soixante et un	80	quatre-vingts	90	quatre-vingt-dix
62	soixante-deux…	81	quatre-vingt-un	91	quatre-vingt-onze
69	soixante-neuf	82	quatre-vingt-deux	92	quatre-vingt-douze
70	soixante-dix	83	quatre-vingt-trois	93	quatre-vingt-treize
71	soixante et onze	84	quatre-vingt-quatre	94	quatre-vingt-quatorze
72	soixante-douze	85	quatre-vingt-cinq	95	quatre-vingt-quinze
73	soixante-treize	86	quatre-vingt-six	96	quatre-vingt-seize
74	soixante-quatorze	87	quatre-vingt-sept	97	quatre-vingt-dix-sept
75	soixante-quinze	88	quatre-vingt-huit	98	quatre-vingt-dix-huit
76	soixante-seize	89	quatre-vingt-neuf	99	quatre-vingt-dix-neuf
77	soixante-dix-sept			100	cent
78	soixante-dix-huit				
79	soixante-dix-neuf				

130 CHAPITRE 5

TPR (continued from page 129)
(Student 1), **prenez la fourchette.**
Prenez le couteau.
Coupez le steak.
Mangez le steak.
Demandez l'addition au serveur/à la serveuse.
Regardez l'addition.
Payez l'addition. Donnez de l'argent au serveur/à la serveuse.

Laissez un pourboire pour le serveur/ la serveuse.
Levez-vous.
Et maintenant, retournez à vos places.
Merci.

Exercices

A **Qu'est-ce que c'est?** Identifiez. *(Identify.)*

B **On arrive au restaurant.** Choisissez la bonne réponse. *(Choose the correct answer.)*

1. Charles ne va pas au restaurant tout seul. Il y va ___.
   a. avec ses copains   **b.** avec le serveur   **c.** avec son prof
2. Ils arrivent au restaurant. Charles parle ___.
   a. au serveur   **b.** au chef de cuisine   **c.** au maître d'hôtel
3. Charles et ses copains vont ___.
   a. à une table   **b.** à la salle à manger   **c.** à la cuisine
4. Les copains de Charles regardent ___.
   a. le pourboire   **b.** le café   **c.** la carte
5. Le service est compris. Mais Charles laisse ___ pour le serveur.
   a. une addition   **b.** un pourboire   **c.** un verre

C **Personnellement.** Donnez des réponses personnelles. *(Give your own answers.)*

1. Tu as faim maintenant?
2. Tu aimes manger?
3. Tu aimes aller au restaurant?
4. En général, tu déjeunes à quelle heure?
5. Tu regardes la carte au restaurant?
6. Qu'est-ce que tu commandes?
7. Tu aimes le steak comment?
8. Tu demandes l'addition?
9. Le service est compris aux États-Unis?
10. Tu laisses un pourboire?

D **En bus ou à pied?** Dites comment les élèves vont à l'école. *(Tell how the students go to school.)*

1. en bus   2. en voiture   3. à pied   4. en métro

CRITICAL THINKING ACTIVITY

(Thinking skills: Comparing & Contrasting)
   Compare and contrast an American coffee shop with a French café.

Exercices

PRESENTATION *(pages 131–132)*

Exercice A: Paired Activity
   With books open, partners can quiz each other on Exercise A, one randomly stating the numbers and asking the question, the other responding. Then they switch roles. When finished, they both write out the items, checking each other's spelling by referring to page 129.

Exercice B: Reading
   Call on individuals to read each item aloud and give the correct answer.

Exercice C: Listening
   Have students close their books. Focus on the listening skill by asking the questions in Exercise C of individuals. It can be done again with books open, if you wish.

Extension of *Exercice D*
   Have students make up their own sentences using these expressions.

ANSWERS

Exercice A
1. une nappe
2. une tasse
3. une fourchette
4. une cuillère
5. une assiette
6. une serviette
7. un verre
8. un couteau

Exercice B
1. a
2. c
3. a
4. c
5. b

Exercice C
   Answers will vary.

Exercice D
1. Elle va à l'école en bus.
2. Ils vont à l'école en voiture.
3. Ils vont à l'école à pied.
4. Il va à l'école en métro.

Exercice E: Speaking

Have students say the phone numbers the French way, as a series of two-digit numbers (*soixante-dix-huit, quatre-vingt-quatre,* etc.).

ANSWERS

Exercice E

1. ... de «l'Éléphant»... ?
   ... 42.76.08.06.
2. ... du «Lion»... ?
   ... 45.51.41.77.
3. ... du «Loft»... ?
   ... 46.34.29.95.
4. ... du «Liberté»... ?
   ... 43.44.80.79.
5. ... du «Longchamp»... ?
   ... 43.43.49.39.

INFORMAL ASSESSMENT
(*Mots 2*)

Check comprehension by making false statements about the illustrations in *Mots 2* and calling on students to correct the statements. For example: *On dîne le matin. (Non, on ne dîne pas le matin. On dîne le soir.)*

Activités de communication orale

Mots 1 et 2

PRESENTATION (page 132)

The *Activités de communication orale* give students the opportunity to use the chapter vocabulary in open-ended situations and to recycle words and structures from previous chapters.

ANSWERS

Activité A

Answers will vary.

E **Renseignements, bonjour.** Demandez le numéro de téléphone du restaurant d'après le modèle. (*Ask for the phone number of each restaurant according to the model.*)

«Chez Pauline»
Élève 1: Quel est le numéro de téléphone de «Chez Pauline», s'il vous plaît?
Élève 2: C'est le 78.84.65.91.

1. L'Éléphant   4. Le Liberté
2. Le Lion   5. Le Longchamp
3. Le Loft

2504 restaurants

Restaurants (suite)

LE LAUMIÈRE
voir annonce même page
4 r Petit
75019 Paris - - - - - - - (1) 42 02 46 71

LE LAZARE 68 r Quincampoix 3ᵉ (1) 48 87 99 34
L'ÉLEPHANT 10 r Trésor 4ᵉ - - - (1) 42 76 08 06
LE LIBAN A LA MOUFFETARD
16 r Mouffetard 5ᵉ - - - - - - (1) 47 07 30 72
LE LIBERTÉ 35 r Sibuet 12ᵉ - - (1) 43 44 80 79
LE LIMOURS
RESTAURANT-LEFÈVRE
7 pl Denfert Rochereau 14ᵉ - * (1) 43 27 20 66
LE LION (Sté Le Barbecue de la Tour)
23 r Duvivier 7ᵉ - - - - - - (1) 45 51 41 77
LE LITEAU 14 r Washington 8ᵉ (1) 42 89 90 43
LE LOFT 95 bd St Michel 5ᵉ - - (1) 46 34 29 95
L'ÉLOGE DE LA FOLIE
37 bis r Montpensier 1ᵉʳ - - (1) 42 96 08 42
L'ÉLOGE DE LA FOLIE
37 B r Montpensier 1ᵉʳ - - (1) 42 96 25 49
LE LONGCHAMP
5 r Serg Bauchat 12ᵉ - - - (1) 43 43 49 39

LE MANDARIN DE RAMBUTEAU
11 r Rambuteau 4ᵉ - - - - (1) 42 72 87 22
LE MANDARIN DE LA TOUR MAUBOURG
SPECIALITES CHINOISES
CUISINE RAFFINEE SALLE CLIMATISEE
23 bd Latour Maubourg
75007 Paris - - - - - - (1) 45 51 25 71
LE MANDARIN DE LA TOUR MAUBOURG
23 bd Latour Maubourg 7ᵉ - (1) 45 51 25 71
LE MANGE TARD
17 r Jouffroy 17ᵉ - - - - (1) 46 22 12 38
LE MANGE TOUT
24 bd Bastille 12ᵉ - - - - (1) 43 43 95 15
LE MANGUIER
67 av Parmentier 11ᵉ - - (1) 48 07 03 27
LE MANOIR DE PARIS
6 r Pierre Demours 17ᵉ - (1) 45 72 25 25
— 6 r Pierre Demours 17ᵉ
Télécopieur - - - - - - (1) 45 74 80 98
LE MARAICHER
5 r Beautreillis 4ᵉ - - - - (1) 42 71 42 49
LE MARAICHER

Activités de communication orale

Mots 1 et 2

A **A mon avis…** Make a chart like the one below. Put an *x* under the heading that best describes your opinion of each of the foods listed.

	J'adore	J'aime assez	Je déteste
1. le pâté			x
2. la pizza	x		
3. la glace au chocolat			
4. la soupe à l'oignon			
5. le café			
6. l'omelette nature			
7. les frites			
8. le fromage			
9. les saucisses de Francfort			

Now compare your chart with a classmate's and see if they're similar. Follow the model below.

Élève 1: Moi, j'adore le pâté. Et toi?
Élève 2: Moi, je déteste le pâté.

B **Au restaurant.** You and your classmates, accompanied by your teacher, go to a local French restaurant and order your meal in French.

132 CHAPITRE 5

ADDITIONAL PRACTICE

1. Students role-play a restaurant scene in Paris between a waiter/waitress and some customers. (Students will need a simple menu.) The customers should:
   • order something to eat and drink.
   • ask for the check and take turns paying.
2. Student Tape Manual, Teacher's Edition, *Activités F–G*, page 59.

INDEPENDENT PRACTICE

Assign any of the following:
1. Exercises and activities, pages 131–132
2. Workbook, *Mots 2: D–G*, page 42
3. Communication Activities Masters, *Mots 2: B*, page 24
4. Computer Software, *Vocabulaire*
5. CD-ROM, Disc 2, pages 129–132

STRUCTURE

Le verbe *aller* au présent

Telling and Asking Where People Go; Asking How Someone Is

1. All verbs that end in *-er* are regular verbs, with one exception. That exception is the verb *aller*, "to go."

ALLER	
je vais	nous allons
tu vas	vous allez
il elle }va on	ils elles }vont

> Je vais au café et mon petit frère va à l'école.
> Tu vas à la fête de ta copine?
> Nous n'allons pas à Paris pendant les vacances.
> Vous allez au café après les cours mais elles
>   vont à la maison.

2. You also use *aller* to ask how someone is. You have already learned *Ça va*. Here are other ways to ask how a person is and some possible responses.

> Comment vas-tu?
> Pas mal, merci. Et toi?
>
> Comment allez-vous?
> Je vais très bien, merci. Et vous?

3. You will often use the word *y* (referring to a place already mentioned) with the verb *aller*. If you use the verb *aller* without mentioning the place you are going to, you must put *y* in front of the verb. *Aller* cannot stand alone.

> Tu vas au restaurant?
> Oui, j'y vais.
> Et Robert y va aussi.
> Mais il n'y va pas avec ses copains. Il y va tout seul.

4. *On y va* is a very useful expression. It can mean "Let's get going," "Let's go," or, as a question, "Do you want to go?"

E. Marty
20, avenue des Gobelins, 75005 PARIS
Tél. 43.31.39.51

CHAPITRE 5   **133**

LEARNING FROM REALIA

1. Ask the students the following questions about the restaurant bill:
   Quel est le nom du restaurant? Quelle est son adresse? Quel est son numéro de téléphone? On commande combien de cafés? Un café coûte combien dans ce restaurant?
2. Explain to the students that by looking at the postal code (75005), you can tell that the Restaurant E. Marty is in the 5th *arrondissement*. See if they can find this *arrondissement* and the avenue des Gobelins on the Paris map on page 505 (Part A, page 238) or use the Map Transparency.

Structure Teaching Resources

1. Workbook, *Structure: A–K,* pages 43–46
2. Student Tape Manual, Teacher's Edition, *Structure: A–F,* pages 60–62
3. Audio Cassette 4A/CD-4
4. Communication Activities Masters, *Structure: A–D,* pages 25–27
5. Computer Software, *Structure*
6. Chapter Quizzes, *Structure: Quizzes 3–6,* pages 27–30
7. CD-ROM, Disc 2, pages 133–139

Le verbe aller *au présent*

PRESENTATION *(pages 133–134)*

A. Write the forms of *aller* on the board and have students repeat them. Pay particular attention to the liaison.
B. Write the forms of *avoir* on the board next to those of *aller,* point out the similarity between them, and have students repeat them. Draw lines through the *v* in the *je, tu, il(s), elle(s)* forms of *aller* and write a *v* over the *ll* in the *nous* and *vous* forms to further emphasize the similarity between the two verbs.
C. Lead students through steps 1 & 2 with individuals reading the examples.
D. Have each student ask a neighbor how he or she is. That student responds and asks another student, and so on.
E. Lead students through step 3. The word *y* is introduced here simply as a completion to *aller,* since it is difficult to avoid. (See Chapter 18 for the formal presentation of *y,* along with other, less frequent uses of it.)

F. Now lead students through steps 4–6 on pages 133–134.

Note In the CD-ROM version, this structure point is presented via an interactive electronic comic strip.

Exercices

PRESENTATION *(page 134)*

Exercice A

You may wish to have students listen to the mini-conversation on Cassette 4A/CD-4.

Extension of *Exercice A*

Have pairs of students present the mini-conversation to the class. After going over the comprehension exercise that follows, have students make up questions about Simone and Paul.

Exercices A, B, C, and D

The exercises on pages 134–135 can be done with books either open or closed.

ANSWERS

Exercice A

1. va
2. va
3. va
4. va
5. va
6. vont
7. vont

Exercice B

Answers will vary.

5. Question words such as *où, quand, comment, avec qui* can be used with *est-ce que* or with the subject and verb inverted.

Où *est-ce que* tu vas?	Je vais au café.
Où vas-tu?	Je vais au café.
Quand *est-ce que* tu vas au café?	J'y vais demain.
Quand vas-tu au café?	J'y vais demain.

6. The words *toujours* (always), *souvent* (often), *quelquefois* (sometimes), and *maintenant* (now) are frequently used with the verb *aller.*

Je vais toujours au café le mardi.
Ton copain y va souvent aussi?
Non, pas souvent. Mais il y va quelquefois.
Et nous y allons maintenant.

Exercices

A **Au restaurant!** Répétez la conversation avec un copain ou une copine. *(Practice the conversation with a classmate.)*

SIMONE: Salut, Paul. Comment vas-tu?
PAUL: Pas mal, et toi?
SIMONE: Très bien, merci. Où vas-tu maintenant?
PAUL: Je vais au Café de Flore.
SIMONE: Tu y vas tout seul?
PAUL: Oui. On y va ensemble?
SIMONE: Pourquoi pas?

Complétez d'après la conversation. *(Complete according to the conversation.)*

1. Simone ___ très bien.
2. Où ___ Paul?
3. Il ___ au Café de Flore.
4. Il n'y ___ pas tout seul.
5. Son amie Simone y ___ aussi.
6. Les deux copains y ___ ensemble.
7. Ils y ___ à pied, pas en métro.

B **Tu vas au restaurant?** Donnez des réponses personnelles. *(Give your own answers.)*

1. Tu vas souvent au restaurant?
2. Avec qui est-ce que tu vas au restaurant? Avec ta famille?
3. Tu vas quelquefois dans un restaurant français, italien ou chinois?
4. Tu vas toujours au même restaurant?
5. Quand est-ce que tu vas au restaurant?

134   CHAPITRE 5

LEARNING FROM PHOTOS

Ask questions about the photo on this page: *C'est la terrasse d'un café ou c'est l'intérieur? C'est quel café? Est-ce qu'il y a des tables libres? C'est une photo relativement récente ou une photo prise juste après la Deuxième Guerre mondiale* (World War II)*?*

PAIRED ACTIVITY

Have students work in pairs. One student asks the other how he/she is, where he/she goes after school, and with whom. Then they reverse roles.

C **Tes copains et toi.** Donnez des réponses personnelles. *(Give your own answers.)*

1. Tes copains et toi, vous allez à l'école?
2. Vous allez à quelle école?
3. Vous allez à l'école à quelle heure?
4. Vous allez à l'école comment? À pied, en bus, en voiture ou en métro?
5. Après les cours vous allez au café?

D **On va dîner au restaurant.** Complétez la conversation. *(Complete the conversation.)*

ANNE: Ce soir je ___ dîner au restaurant «La Bonne Fourchette». J'y ___ toute seule.

PATRICK: Tu ___ à «La Bonne Fourchette»? C'est une excellente idée. On y ___ ensemble?

ANNE: Pourquoi pas? Mais on y ___ à pied ou en bus?

PATRICK: En bus? Tu rigoles! On y ___ en voiture! J'ai une nouvelle voiture.

ANNE: Elle est super, ta nouvelle voiture. Mais tu ne ___ pas trouver de place libre dans le parking.

Les contractions avec *à* et *de*

Giving Locations; Telling What Belongs to Others

1. The preposition *à* can mean "to," "in," or "at." *À* is contracted with *le* and *les* to form one word. *À + le* becomes *au*. *À + les* becomes *aux*. The preposition *à* does not change when used with the articles *la* and *l'*.

à + la = à la	Je vais *à la* salle à manger.
à + l' = à l'	J'étudie le français *à l'*école.
à + le = au	Je suis *au* café.
à + les = aux	Je parle *aux* élèves.

You make a liaison with *aux* and any word beginning with a vowel or silent *h*. The *x* is pronounced *z*.

2. You also use the preposition *à* with many food expressions.

   une glace à la vanille et une glace au chocolat
   une soupe à l'oignon
   un sandwich au jambon et au fromage
   une omelette aux fines herbes

CHAPITRE 5   **135**

COOPERATIVE LEARNING

Have students work in groups. Each person in the group writes down the name of a local restaurant, school, or café. Then students discuss who goes to which place. For example, *Mes copains et moi, nous allons au Burger King. Mais mes parents vont au restaurant Genghis Khan.*

ADDITIONAL PRACTICE

Student Tape Manual, Teacher's Edition, *Activités A–D*, pages 60–61

Exercice C
Answers will vary. However, all answers should include *nous allons.*

Exercice D
vais... vais... vas... va... va... va... vas...

RETEACHING

Write the words *toujours, souvent, quelquefois* on the board. Students use the number in parentheses as a guide to tell how often during a month the person goes to the place. For example: *Michel (30)/le café → Michel va toujours au café.*

1. M. Clemenceau (3)/l'aéroport
2. Tu (30)/le lycée
3. Jean-Pierre (2)/le café
4. Nous (10)/le restaurant

Bell Ringer Review

Write the following on the board or use BRR Blackline Master 5-3: You have just moved into your new apartment. List at least six items of tableware you will need for your kitchen.

Les contractions avec *à* et *de*

PRESENTATION *(pages 135–136)*

A. To help students understand the basic difference in meaning between *à* and *de*, draw a simple building on the board. Then draw an arrow going to the building and write *à* on the arrow. Draw an arrow coming from the building and write *de* on this arrow.

B. Now lead students through steps 1–6 on pages 135–136. For each example given, ask students why a particular form of *à* or *de* is used.

C. To practice *à*, use flash cards with locations and their definite articles written out (*le parc, l'école*). Give students a model sentence and have them change it according to the card you flash. For example: *Nous allons au parc.* (*Nous allons à l'école.*)

136

3. The following expressions denote place but do not take the preposition *à*.

> **Je vais chez René. (à la maison de René)**
> **Nous allons en ville.**
> **Les élèves vont en classe.**

4. In French the word *de* can mean "of" or "from." Like *à*, the preposition *de* is contracted with *le* and *les* to form one word. *De + le* becomes *du*. *De + les* becomes *des*. The preposition *de* does not change when used with the articles *la* and *l'*.

de + la = de la	*De la* terrasse on a une belle vue.
de + l' = de l'	On va *de l'*école à la maison en bus.
de + le = du	Quelle est votre opinion *du* film?
de + les = des	Ils rentrent *des* magasins à midi.

5. The following expressions of location with *de* contract in the same way: *près de, loin de, à côté de* (next to), *à gauche de* (to the left of), *à droite de* (to the right of).

> **Le café est près *du* cinéma.**
> **L'immeuble est loin *des* magasins.**

6. You also use the preposition *de* to indicate possession.

> **C'est la moto *de* Marc.**
> **Voici la voiture *du* professeur.**
> **Minou est le chat *des* voisins.**

Exercices

A **Où vas-tu?** Donnez des réponses personnelles. (*Give your own answers.*)

1. Tu vas au collège, au lycée ou à l'université?
2. Tu vas au cours de français le matin ou l'après-midi?
3. Tu vas à l'école à quelle heure?
4. Tu vas au cours d'anglais à quelle heure?
5. Après les cours tu vas chez un copain ou une copine?
6. Tu aimes aller au restaurant?

B **Je ne vais pas à la fête.** Complétez avec «à». (*Complete with à.*)

Ce soir je ne vais pas ___ (le concert). Je ne vais pas ___ (le parc), je ne vais pas ___ (le lycée), je ne vais pas ___ (le restaurant). Je ne vais pas parler ___ (les copains). Je ne vais pas ___ (la fête) de Suzanne. Je vais aller où alors? Je vais rentrer ___ (la maison). Pourquoi? Je suis fatigué.

136   CHAPITRE 5

C **Qu'est-ce que tu préfères?** Donnez des réponses personnelles. *(Give your own answers.)*

1. Tu préfères les sandwichs au jambon ou les sandwichs au fromage?
2. Tu préfères les omelettes au fromage ou les omelettes aux fines herbes?
3. Tu préfères la soupe à la tomate ou la soupe à l'oignon?
4. Tu préfères le café ou le thé?
5. Tu préfères la glace au chocolat ou la glace à la vanille?
6. Tu préfères les crêpes au chocolat ou les crêpes nature?

D **Où est… ?** Regardez le plan du quartier. Posez des questions à un copain ou à une copine d'après le modèle. *(Look at the map and ask a friend questions according to the model.)*

> Élève 1: Où est le théâtre?
> Élève 2: Le théâtre est à gauche du café.

1. le parc
2. l'école
3. la banque
4. le café
5. le restaurant
6. la discothèque

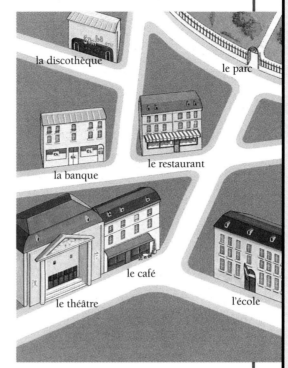

E **Le dîner des élèves.** Combinez d'après le modèle. *(Combine according to the model.)*

> c'est la voiture / les parents de Vincent
> *C'est la voiture des parents de Vincent.*

1. je vais à la table / les amis de Marc
2. ils sont à la terrasse / le café
3. nous regardons la carte / le restaurant
4. le sac à dos / l'élève est sur la chaise
5. c'est le pourboire / la serveuse

Le futur proche

Telling What You or Others Are Going to Do

1. You use the verb *aller* followed by an infinitive to tell what you or others are going to do in the near future.

> Demain Claude va donner une fête.
> Samedi soir il va inviter ses amis à la maison.
> Pendant le week-end je vais aller au cinéma.
> En décembre on va avoir des vacances.

LEARNING FROM ILLUSTRATIONS

Have students say as much about the illustration as they can. They may also mention some school, café, restaurant, park, and discotheque activities they can already describe in French.

INDEPENDENT PRACTICE

Assign any of the following:
1. Exercises, pages 134–137
2. Workbook, *Structure: A–H*, pages 43–45
3. Communication Activities Masters, *Structure: A–B*, pages 25–26
4. CD-ROM, Disc 2, pages 133–137

PRESENTATION *(page 137)*

Exercice C

This exercise can be done orally with books closed then read for reinforcement.

Extension of *Exercice D*

After completing Exercise D, practice expressions of location by asking the location of places known to the students. For example: *Où est la salle de classe de M. Jones? (Elle est à droite de la cafétéria.)*

ANSWERS

Exercice C
Answers will vary.

Exercice D
É1 questions follow the model and cues. Possible É2 answers are:

1. Le parc est près de la discothèque.
2. L'école est près du café.
3. La banque est à gauche du restaurant.
4. Le café est à côté du théâtre.
5. Le restaurant est à droite de la banque.
6. La discothèque est près du parc.

Exercice E

1. Je vais à la table des amis de Marc.
2. Ils sont à la terrasse du café.
3. Nous regardons la carte du restaurant.
4. Le sac à dos de l'élève est sur la chaise.
5. C'est le pourboire de la serveuse.

INFORMAL ASSESSMENT

Say where each student is going: *Pierre/le café; Anne/la fête; Bruno/les Champs-Élysées; Claire/l'école; Luc/le lycée.*

Le futur proche

PRESENTATION *(pages 137–138)*

A. Using fortune cookies or a crystal ball, pretend to predict the fortunes of several students. For example: *Delphine, tu vas être riche. Marc, tu vas aimer une jeune fille blonde.* Ask students to figure out the meanings.

B. Write several sentences on the board using the *futur proche*, but leaving out the infinitive. Ask for volunteers to fill in an appropriate infinitive. For example: *Tu vas ___ Jean et Marie? (inviter)* Call on other volunteers to make the sentences negative.

Exercices

ANSWERS

Exercice A

Answers will vary but should follow the model below.

1. Oui, je vais regarder la télé. (Non, je ne vais pas regarder la télé.)

Exercice B

1. **Nous n'allons pas au cours de français pendant le week-end.**
2. **Les chiens et les chats ne vont pas au cinéma.**
3. **Demain le/la prof de maths ne va pas chanter en français.**
4. **Vous n'allez pas manger pendant le cours d'algèbre.**
5. **Ce soir je ne vais pas parler au téléphone avec Elvis Presley.**

Les adjectifs possessifs
notre, votre, leur

PRESENTATION (page 138)

A. Briefly review the singular forms of the possessive adjectives from Chapter 4, pages 98–100, before presenting the plural forms.

B. Lead students through steps 1–3 on page 138.

138

---

2. Note that in negative sentences *ne… pas* goes around the verb *aller.*

> **Je *ne* vais *pas* travailler après les cours.**
> **Ce soir tu *ne* vas *pas* regarder la télé.**
> **Nous *n'*allons *pas* danser ensemble à la fête.**

Exercices

A **Ce soir!** Donnez des réponses personnelles. *(Give your own answers.)*

1. Ce soir tu vas regarder la télé?
2. Tu vas téléphoner à un copain ou à une copine?
3. Tu vas préparer le dîner?
4. Tu vas aller en classe?
5. Tu vas inviter tes professeurs au restaurant?

B **Absurdités.** Mettez à la forme négative. *(Change to the negative.)*

1. Nous allons au cours de français pendant le week-end.
2. Les chiens et les chats vont au cinéma.
3. Demain le/la prof de maths va chanter en français.
4. Vous allez manger pendant le cours d'algèbre.
5. Ce soir je vais parler au téléphone avec Elvis Presley.

Les adjectifs possessifs
notre, votre, leur

Telling What Belongs to You and Others

1. You have already learned the possessive adjectives *mon, ton,* and *son.* Study the following forms of the possessive adjectives *notre* (our), *votre* (your), and *leur* (their).

MASCULIN SINGULIER	FÉMININ SINGULIER	PLURIEL
notre ami	notre amie	nos ami(e)s
votre ami	votre amie	vos ami(e)s
leur ami	leur amie	leurs ami(e)s

2. The adjectives *notre, votre,* and *leur* are used with both masculine and feminine singular nouns. With plural nouns you use *nos, vos,* and *leurs.*

3. With the plural forms, you make a liaison before a vowel or silent *h.*

ADDITIONAL PRACTICE

1. After completing Exercises A and B, reinforce with the following: Have students write four fortunes, two in the affirmative and two in the negative, using the *futur proche.* They should fold each fortune and write *garçon, fille,* or *garçon ou fille* on the outside. Place the fortunes in a hat or box. Call on students to draw one and have them read their fortune to the class.

2. Divide the class into several groups and give each group different familiar objects. Groups should have objects that each student possesses (their own pencils, books, notebooks) as well as singular objects that the whole group "possesses" (one dictionary, one photo, etc.) Then question the groups about the objects to elicit the possessive adjectives. For example: *C'est votre photo? (Oui, c'est notre photo.) Ce sont vos livres? (Non, ce sont leurs livres.)*

Exercices

A **Notre maison.** Donnez des réponses personnelles d'après le modèle. *(Give your own answers according to the model.)*

> Votre voiture est nouvelle ou vieille?
> *Notre voiture est vieille. (Notre voiture est nouvelle.)*

1. Votre maison ou appartement est grand(e) ou petit(e)?
2. Votre maison ou appartement a combien de pièces?
3. Votre maison ou appartement est en ville?
4. Votre maison ou appartement est près de l'école?
5. Vous avez un chien ou un chat? Votre chien ou chat est adorable?

B **Nos cours.** Donnez des réponses personnelles avec «nos». *(Give your own answers with* nos.*)*

1. Vos profs sont sympa?
2. Vos amis sont sincères?
3. Vos cours sont intéressants?
4. Vos cassettes de musique rock sont fantastiques?
5. Vos devoirs sont longs?
6. Vos examens sont difficiles?

C **Leur maison.** Complétez avec «leur» ou «leurs». *(Complete with* leur *or* leurs.*)*

Georges et Paul sont frères. Ils sont dans ____ $_1$ chambre. Ils écoutent ____ $_2$ cassettes. ____ $_3$ collection de cassettes est surtout de jazz. ____ $_4$ amies Catherine et Véronique aiment aussi le jazz. Mais elles préfèrent la musique classique. Elles ont ____ $_5$ musiciens favoris. ____ $_6$ copains n'écoutent pas de musique classique. Samedi soir Georges et Paul vont aller au concert de jazz avec ____ $_7$ parents parce que ____ $_8$ mère et ____ $_9$ père adorent le jazz aussi.

Chaque mois dans Jazz Magazine

des interviews en profondeur (Pat Metheny, Charlie Haden, Herbie Hancock, Miles Davis, etc., etc.)

des signatures prestigieuses (Ben Sidran, Jacques Réda, Francis Marmande, Giuseppe Pino, Aldo Romano...)

des études sur l'histoire et l'actualité du jazz et des musiques périphériques une encyclopédie permanente en fiches à découper

et le bilan-panorama des événements et des productions phonographiques dans le monde

jazz magazine
Pour ceux qui aiment le jazz vraiment

139

PRESENTATION *(page 139)*

Exercices A and B

Exercises A and B can be done with books either open or closed.

Extension of Exercice A

Have students tell about their own family's house or apartment using *notre.*

Exercice C

Exercise C must be done with books open. After going over Exercise C, have students make up questions about *Georges et Paul.* You may also call on a student to retell the story in his or her own words.

ANSWERS

Exercices A and B

Answers will vary, but Exercise A answers should use *notre,* and Exercise B answers *nos.*

Exercice C

1. leur
2. leurs
3. Leur
4. Leurs
5. leurs
6. Leurs
7. leurs
8. leur
9. leur

INFORMAL ASSESSMENT

Check for comprehension of all possessive adjectives by walking around the room, picking up various objects, and asking if the object belongs to a particular person or persons. For example: *C'est le crayon de John? (Oui, c'est son crayon.) C'est le bureau des professeurs? (Oui, c'est leur bureau.) C'est ton cahier? (Non, ce n'est pas mon cahier.)*

PAIRED ACTIVITY

Students work in pairs. Student 1 tells the other something that a brother, sister, or friend is going to do. Student 2 asks if Student 1 is going to do the same thing. For example:

É1: Mon frère va regarder la télé.
É2: Et toi, tu vas regarder la télé aussi?
É1: Oui (Non), je (ne) vais (pas) regarder la télé.

INDEPENDENT PRACTICE

Assign any of the following:

1. Exercises, pages 138–139
2. Workbook, *Structure: I–K,* pages 45–46
3. Communication Activities Masters, *Structure: C–D,* page 27
4. Computer Software, *Structure*
5. CD-ROM, Disc 2, pages 137–139
6. Have students write sentences about their families/friends, using the verbs they know.

CONVERSATION

Bell Ringer Review

Write the following on the board or use BRR Blackline Master 5-4: Write five things you are going to do tomorrow.

PRESENTATION *(page 140)*

A. Tell students they will hear a conversation among Marie-Claire, Didier, and a waiter. Have them close their books and watch the Conversation Video or listen as you read or play Cassette 4A/CD-4.

B. After introducing the conversation (see suggestions in previous chapters), set up a café in the classroom and have groups of students act out the conversation for the class.

C. Have students retell the conversation in their own words.

Note In the CD-ROM version, students can play the role of either one of the characters and record the conversation.

ANSWERS

Exercice A

1. **Ils sont au restaurant.**
2. **Non, elle ne va pas commander immédiatement.**
3. **Elle va regarder la carte.**
4. **Il commande un steak frites et une petite salade.**
5. **Il aime le steak entre saignant et à point.**
6. **Elle commande le menu touristique.**
7. **Le serveur arrive avec l'addition.**
8. **La maison n'accepte pas les cartes de crédit.**

Prononciation

PRESENTATION *(pages 140–141)*

A. Model the key word *verre* and have students repeat chorally. Then do the same for the other words and phrases.

CONVERSATION

Scènes de la vie *Au restaurant*

DIDIER: Tu es prête à commander, Marie-Claire?
MARIE-CLAIRE: Non, je vais regarder la carte encore un moment.
SERVEUR: Vous désirez?
DIDIER: Pour moi, le steak frites et une petite salade.
SERVEUR: Et vous aimez le steak comment?
DIDIER: Entre saignant et à point.
SERVEUR: Et pour Madame?
MARIE-CLAIRE: Le menu touristique, s'il vous plaît.
(Après le dîner)
DIDIER: L'addition, s'il vous plaît.
SERVEUR: Oui, Monsieur. J'arrive.
MARIE-CLAIRE: Tu as ta Carte Bleue, Didier?
SERVEUR: Ah Madame, je regrette. La maison n'accepte pas les cartes de crédit.

A **Marie-Claire et Didier.** Répondez d'après la conversation. *(Answer according to the conversation.)*

1. Où sont Marie-Claire et Didier?
2. Marie-Claire va commander immédiatement?
3. Qu'est-ce qu'elle va regarder?
4. Qu'est-ce que Didier commande?
5. Il aime le steak comment?
6. Qu'est-ce que Marie-Claire commande?
7. Qui arrive avec l'addition?
8. La maison accepte les cartes de crédit?

Prononciation *Le son /r/*

The French /r/ sound is very different from the American /r/. When you say /r/, the back of your tongue should almost completely block the air going through the back of your throat. Repeat the following words and sentences.

le verre	la nature	la cour	la mère	l'art
la terrasse	le garage	adorer	arriver	terrible

J'adore la littérature, l'art, l'histoire et l'informatique.
Le serveur arrive avec un verre d'Orangina.

verre

DID YOU KNOW?

The French love to eat. To them, eating is necessary to both physical and mental health. Meals are a time to appreciate good food, socialize, and spend time with family and friends. In France, meals are leisurely affairs. Contrary to the prevailing American custom of putting all food for a meal (with the exception of dessert) on the table at the same time, in France, foods are served in courses starting with an hors d'œuvre, followed by a main dish, then a green salad, cheese, and dessert.

Activités de communication orale

A **Le menu touristique.** You've just ordered a *steak frites* in a small restaurant off the Boulevard Saint-Michel in Paris. Answer the waiter's questions.

1. Bon. Un steak frites et une petite salade. Et vous aimez le steak comment?
2. Et comme boisson?
3. Et comme dessert, la glace à la vanille ou les crêpes au chocolat?

B **Pendant le week-end.** Tell a classmate a couple of things you like to do when you're free and find out if he or she likes to do them too. Then say whether you're going to do them over the weekend. Reverse roles.

> Élève 1: Quand je suis libre, j'aime regarder la télé et aller au cinéma. Et toi?
> Élève 2: Moi aussi, j'aime regarder la télé et aller au cinéma. (Non, je n'aime pas regarder la télé, mais j'adore aller au cinéma.)
> Élève 1: Je vais regarder la télé et aller au cinéma pendant le week-end.

C **Les projets.** You and a classmate are making plans for this evening. Choose a place from the list below and see what your friend thinks. Use the model as a guide.

> à la fête de Patrick

> Élève 1: On va à la fête de Patrick ce soir?
> Élève 2: D'accord, on y va. J'adore danser. (Non, merci. Je déteste danser.)

> le café
> le cinéma
> le concert
> la discothèque
> chez (le nom d'un[e] ami[e])
> le restaurant français

D **Où est ton restaurant favori?** Name your favorite restaurant. Tell where it's located using some of the following expressions.

à côté de	derrière
à droite de	loin de
à gauche de	près de
devant	

B. You may wish to give students the following *dictée*:
J'adore ma mère. La voiture est dans le garage. Carole arrive à la terrasse avec un verre.

C. For additional practice of pronunciation, you may wish to use Pronunciation Transparency P-5; Cassette 4A/CD-4: *Prononciation;* and the Student Tape Manual, Teacher's Edition, *Activités I–K*, pages 63–64.

Cultural Note *Le menu touristique* features a complete meal for a fixed price (*un prix fixe*).

Bell Ringer Review

Write the following on the board or use BRR Blackline Master 5-5: State where in the house you usually find the following things:

le sofa	la voiture
le réfrigérateur	la table
le bureau	la télé

Activités de communication orale

PRESENTATION *(page 141)*

Activité A

 In the CD-ROM version of this activity, students can interact with an on-screen native speaker.

ANSWERS

Activités A, B, C, and D
Answers will vary.

INDEPENDENT PRACTICE

Assign any of the following:
1. Exercise and activities, pages 140–141
2. CD-ROM, Disc 2, pages 140–141

LECTURE ET CULTURE

LECTURE ET CULTURE

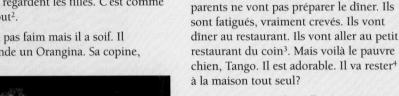

READING STRATEGIES
(*page 142*)

Note The *Lectures* will sometimes contain new expressions whose meaning is explained by the use of expressions that mean the same thing. The new expressions are not footnoted. Students should learn that, by making educated guesses, they will acquire additional receptive vocabulary. This technique is used twice in this *Lecture: Françoise a très faim:* <u>*Elle a une faim de loup.*</u> *Ils sont fatigués:* <u>*Ils sont vraiment crevés.*</u>

Pre-reading

A. Ask students if they have an after-school "hang-out." What is served there?

B. Introduce the *Lecture* by giving a brief oral overview.

Reading

A. Call on volunteers to read two or three sentences aloud. Ask questions after each turn. Continue in this way until the selection has been completed.

Post-reading

A. Ask students how French restaurants are different from those in the U.S.

B. Assign the reading selection and the exercises that follow for homework. Go over the exercises the next day.

Note Students may listen to a recorded version of the *Lecture* on the CD-ROM.

Étude de mots

ANSWERS

Exercice A

1. ... ses copains.
2. ... très faim.
3. Les parents...
4. ... très fatigués.
5. ... est bien élevé.
6. ... au petit restaurant du coin.

142

ON A SOIF ET ON A FAIM

*A*près les cours Paul va au café. Il y va avec ses copains. Au café ils aiment bien regarder les gens[1] qui passent. Les filles regardent les garçons et les garçons regardent les filles. C'est comme ça partout[2].

Paul n'a pas faim mais il a soif. Il commande un Orangina. Sa copine,

Françoise, a très, très faim. Elle a une faim de loup. Elle commande une omelette au fromage avec des frites.

Paul arrive à la maison. Ce soir ses parents ne vont pas préparer le dîner. Ils sont fatigués, vraiment crevés. Ils vont dîner au restaurant. Ils vont aller au petit restaurant du coin[3]. Mais voilà le pauvre chien, Tango. Il est adorable. Il va rester[4] à la maison tout seul?

Absolument pas! Il va aller au restaurant avec la famille. Il n'y a pas de problème. Il est très bien élevé[5], Tango.

[1] gens *people*
[2] partout *everywhere*
[3] petit restaurant du coin *neighborhood restaurant*
[4] rester *to stay*
[5] bien élevé *well-mannered*

Étude de mots

A **Synonymes.** Récrivez les phrases avec des synonymes. (*Rewrite the sentences using synonyms.*)

1. Il va au café avec *ses amis*.
2. Françoise a *une faim de loup*.
3. *Maman et Papa* ne vont pas préparer le dîner ce soir.
4. Ils sont *crevés*.
5. Le chien *a de bonnes manières*.
6. Ils vont *dans un bon petit restaurant modeste*.

142 CHAPITRE 5

(*Thinking skills: making inferences*)
Write the following on the board or on a transparency:
1. Madame Delacroix décide qu'elle ne va pas préparer le dîner ce soir. Quels sont les choix de la famille?
2. Les Berthollet vont aller au restaurant. Mais ils n'ont pas beaucoup de temps. Ils vont aller dans quel type de restaurant?

Photo 1 Ask students: *Il y a combien de copains au café? Ils sont contents? Ils commandent quelque chose à manger, ou ils commandent une boisson? Les copains parlent? Ils vont au même lycée?*

Photo 2 Ask students: *Le restaurant est en France ou aux États-Unis? Le chien est à la maison ou au restaurant? Il est à côté de la table ou sous la table? Il est bien élevé?*

Compréhension

B Paul et Françoise. Répondez. *(Answer.)*

1. Quand est-ce que Paul va au café?
2. Il y va avec qui?
3. Qui regarde les garçons?
4. Qui regarde les filles?
5. Ça arrive *(happens)* aux États-Unis ou uniquement en France?
6. Paul a faim ou soif?
7. Qu'est-ce qu'il commande?
8. Françoise a soif ou faim?
9. Qu'est-ce qu'elle commande?

C Pas vrai. Corrigez les phrases. *(Correct the statements.)*

1. Un Orangina est quelque chose à manger.
2. Une omelette est une boisson.
3. Ce soir Papa va préparer le dîner.
4. Paul et ses parents vont dîner à la maison.
5. Ils vont dîner dans un grand restaurant.
6. Tango est un chat.
7. Tango va rester à la maison tout seul.
8. Le chien n'est pas bien élevé.

D Au restaurant en France. Trouvez le renseignement suivant. *(Find the following information.)*

In this reading selection, you learned a cultural difference between the United States and France. What is that difference?

DÉCOUVERTE CULTURELLE

*E*n France on dîne vers[1] sept heures et demie ou huit heures. Si on va dîner au restaurant, on arrive au restaurant entre huit heures et dix heures.

En France, le lait c'est pour les enfants, pas pour les adultes. On sert le café après le dessert, pas avec le repas. On sert du vin avec le repas—du vin rouge ou du vin blanc[2]. On place le pain sur la nappe à côté de l'assiette, pas sur une assiette spéciale.

En général au déjeuner ou au dîner on ne mange pas de beurre[3] avec le pain.

[1] vers   *around*
[2] du vin rouge ou du vin blanc   *red or white wine*
[3] beurre   *butter*

CHAPITRE 5   **143**

Exercice B

1. Paul va au café après les cours.
2. Il y va avec ses copains.
3. Les filles regardent les garçons.
4. Les garçons regardent les filles.
5. Ça arrive aux États-Unis aussi.
6. Paul a soif.
7. Il commande un Orangina.
8. Françoise a faim.
9. Elle commande une omelette au fromage avec des frites.

Exercice C

1. Un Orangina est une boisson.
2. Une omelette est quelque chose à manger.
3. Ce soir Papa ne va pas préparer le dîner.
4. Paul et ses parents vont dîner au restaurant.
5. Ils vont dîner au petit restaurant du coin.
6. Tango est un chien.
7. Tango va aller au restaurant avec la famille.
8. Le chien est très bien élevé.

Exercice D

En France les chiens vont au restaurant.

OPTIONAL MATERIAL

Découverte culturelle

PRESENTATION *(page 143)*

A. Before they read the selection ask students about eating habits in their own household. What times do they eat different meals? Is bread important at all meals? How is it eaten?
B. Have students read the selection silently. Write two column headings on the board: *En France* and *Aux États-Unis*. Ask students to contribute items for each column, based on the reading and their knowledge of American customs.

Note Students may listen to a recorded version of the *Découverte culturelle* on the CD-ROM.

INDEPENDENT PRACTICE

Assign any of the following:
1. *Étude de mots* and *Compréhension* exercises, pages 142–143
2. Workbook, *Un Peu Plus*, pages 47–48
3. CD-ROM, Disc 2, pages 142–143
4. Pretend you own a restaurant that serves both American and French food. Draw up a menu for your restaurant.

LEARNING FROM PHOTOS

Have students say as much about the photo as they can.

RÉALITÉS

OPTIONAL MATERIAL

PRESENTATION *(pages 144–145)*

The purpose of this section is student enjoyment of the photos. You may wish to do the following activities.

A. Ask students if they have been to a French restaurant. Ask them to describe the ambience.

B. Using your own resources, show students French photos, cookbooks, and sample menus.

C. Compare restaurants mentioned to American restaurants.

D. If possible, plan a trip to an inexpensive French restaurant.

Note In the CD-ROM version, students can listen to the recorded captions and discover a hidden video behind one of the photos.

GEOGRAPHY CONNECTION

Geography influences local cuisines. Brittany is famous for its shellfish. Normandy has many dairy farms, so many Norman dishes have cream sauce. The red wine of Burgundy has inspired dishes prepared with it. Provençal cooking is flavored by olive oil and the herbs that are grown in the area.
Photo 2: Explain that *crêpes* are from Brittany *(les crêpes bretonnes)*.

RÉALITÉS

La France est vraiment un pays gastronomique. Il y a beaucoup de genres différents de restaurants en France.

Ces gens sont à la terrasse d'un petit restaurant **1**.

Il y a des crêperies partout en France **2**.

On mange bien dans les grands restaurants gastronomiques comme Le Train Bleu dans la Gare de Lyon **3**.

Une brasserie est un excellent choix pour un repas simple et rapide et pas très cher **4**.

Il y a aussi des restaurants fast-food comme McDonald's en France **5**.

144

ADDITIONAL PRACTICE

1. After reading and discussing the *Réalités* captions, reinforce the lesson with the following. Have students work in pairs and look at the list below. One student says what he/she would order. The other has to guess the type of restaurant his/her partner is in: *un restaurant fast-food* or *un petit restaurant du coin*.

un express
une mousse au chocolat
une omelette au fromage
une pizza
une saucisse de Francfort
une soupe à l'oignon
un steak frites
un hamburger

2. Student Tape Manual, Teacher's Edition, *Deuxième Partie*, pages 64–65

3. Situation Cards, Chapter 5

Flambées

145

145

CROSS-CULTURAL COMPARISON

French cuisine is famous throughout the world. As a result, there are many French restaurants in the U.S. and in many other countries. The following is a list of dishes often found on the menu in French restaurants. Find out if the students know what any of these items are. If not, give them a brief description.

Le coq au vin: chicken cooked in a red wine sauce.

Le bœuf bourguignon: pieces of beef cooked in red wine with mushrooms, white onions, and peas.

Le pot-au-feu: a typical, home-cooked French meal with chunks of beef, bone with marrow, cooked in a broth with many vegetables.

Le cassoulet: a dish from the Southwest with duck, sausage, some other meats, and white beans.

La bouillabaisse: popular in Provence, a wonderful type of fish chowder.

Le canard à l'orange: duck glazed with an orange sauce. Although duck is popular in France, the preparation with orange sauce is more common in French restaurants outside of France.

Le gigot d'agneau: leg of lamb. Many people eat it rare or medium in France, contrary to the U.S. custom of eating it well done.

La salade niçoise: a salad from Provence with lettuce, tomato, onion, black olives, tuna or anchovy, hard-boiled eggs, etc. (Ask students what city *niçoise* refers to.)

La choucroute: a specialty of Alsace-Lorraine made with sauerkraut and sausages.

Le couscous: A North African specialty made with semolina wheat, lamb, and vegetables served with a spicy sauce. This dish has become very popular due to the large number of North African immigrants in France.

DID YOU KNOW?

Fast-food restaurants have become very popular in France. One difference between American and French fast-food restaurants is that in France drinks other than soft drinks are served. McDonald's, which is now in many locations in France, is popularly referred to as *MacDo* and a Big Mac as *un beeg*.

CRITICAL THINKING ACTIVITY

(Thinking Skill: making inferences)

Ask students what kind of food they think would be served in the different types of restaurants mentioned in the *Réalités* section. Have them make a list of foods for each restaurant.

Note Activity F in the *Un Peu Plus* section of the Workbook uses many of these terms.

RECYCLING

The *Activités de communication* let students apply this chapter's vocabulary and grammar to open-ended, real-life situations and re-use, as much as possible, the vocabulary and structure from earlier chapters.

INFORMAL ASSESSMENT

The oral activities may be used to evaluate students' speaking ability. Activity A may be done individually with the teacher by those students who are struggling. Activity B may suit more able students. Use the evaluation criteria on page 34 of this Teacher's Wraparound Edition.

Activités de communication orale

PRESENTATION *(page 146)*

Activité B

Let groups of students act out their conversations for Activity B. Encourage them to use props and to vary the language as much as possible.

ANSWERS

Activité A

Answers should include the *futur proche* and the correct contraction of *à.*

Activité B

Answers will vary.

Activités de communication écrite

ANSWERS

Activités A and B

Answers will vary.

Activités de communication orale

A **Tu vas où?** Work with a classmate. Look at the following list of places, then take turns telling each other when you're going to each place, how you're going to get there, and who you're going with.

le concert de rock	le parc	le restaurant fast-food
le cinéma	le restaurant	le café

B **Au café.** Work in groups of three. You and another classmate are having a leisurely conversation in a café. The waiter or waitress (the third person) has to interrupt you once in a while to wait on you.

Activités de communication écrite

A **R.S.V.P.** The French club, *le Cercle français,* is giving a party after school. Design an invitation to send to your classmates. On your invitation include the following information.

1. the date of your party   3. the place
2. the time of your party   4. the French menu

B **Test: La nourriture et toi.** Take the following test to see what it reveals about your interest in food. Compare results with a classmate.

1. En général je préfère manger dans ___.
   a. les restaurants fast-food   b. les restaurants gastronomiques
2. Quand j'ai faim, l'essentiel c'est ___.
   a. la quantité   b. la qualité
3. Je préfère ___.
   a. les saucisses de Francfort   b. le pâté
4. Je préfère manger mon steak ___.
   a. sur une assiette en plastique   b. sur une belle assiette
5. Je préfère dîner ___.
   a. dans la cuisine   b. dans la salle à manger

If you answered *b* most of the time, you are a *gourmet,* a person who appreciates good food in a nice setting. If you answered *a* most of the time, you are a *gourmand,* someone who just likes to eat a lot.

POUR COMMANDER, RIEN DE PLUS SIMPLE!

LIVRAISON A DOMICILE
ALLO PIZZA EXPRESS
45 26 94 94

Allo Pizza!

ESSAYEZ-LA !

146   CHAPITRE 5

INDEPENDENT PRACTICE

1. Activities and exercises, pages 146–147
2. CD-ROM, Disc 2, pages 146–147
3. Communication Activities Masters, pages 23–27

Réintroduction et recombinaison

A **Faim ou soif?** Complétez. (*Complete.*)

1. J'___ soif. Je ___ commander un Orangina.
2. J'___ faim. Je ___ commander quelque chose à manger.
3. Si tu ___ soif, je propose un citron pressé.
4. Si tu ___ faim, je propose un sandwich au
   jambon ou une omelette au fromage.
5. On ___ faim. On ___ dîner.

Vocabulaire

NOMS

le restaurant
le café (*café*)
le maître d'hôtel
le serveur
la serveuse
la carte
l'addition (f.)
le service compris
le pourboire
la terrasse

le couvert
l'assiette (f.)
le couteau
la cuillère
la fourchette
la serviette
la nappe
la tasse
le verre

la boisson
le café (*coffee*)
le crème
l'express (m.)

le citron pressé
le thé citron
le coca
l'Orangina (m.)

la crêpe
le croque-monsieur
les frites (f.)
le fromage
le jambon
le steak frites
  saignant
  à point
  bien cuit
l'omelette (f.)
  aux fines herbes
  nature
le pâté
la salade
le sandwich
la saucisse de Francfort
la soupe à l'oignon
la glace
  à la vanille
  au chocolat

ADJECTIFS

pris(e)

ADVERBES

ensemble
maintenant
quelquefois
souvent
toujours

VERBES

aller
chercher
commander
déjeuner
laisser
manger
trouver

**AUTRES MOTS
ET EXPRESSIONS**

avoir faim
avoir soif
je voudrais
quelque chose
à côté de
à droite de
à gauche de
chez
tout(e) seul(e)
à pied
en bus
en métro
en voiture

NOMBRES

soixante et un à cent
(61–100)

Réintroduction et recombinaison (page 147)

Exercice A
PRESENTATION

Exercise A recycles the forms of *avoir* in conjunction with vocabulary from Chapter 5.

You may wish to refer students to appropriate pages in their textbook to look for sample answers to items with which they have trouble.

ANSWERS

1. **ai, vais**
2. **ai, vais**
3. **as**
4. **as**
5. **a, va**

ASSESSMENT RESOURCES

1. Chapter Quizzes
2. Testing Program
3. Situation Cards
4. Communication
   Transparency C-5
5. Computer Software:
   Practice/Test Generator

VIDEO PROGRAM

INTRODUCTION (17:59)

ON VA FÊTER LE (18:37)
RETOUR DE NATALIE

FOR THE YOUNGER STUDENT

1. Have students make lists in their notebook of the things they would and would not order in a restaurant. Encourage them to add to these lists as they learn more foods. Tell students to learn the words for the items they order.
2. Tell students they are going to open a restaurant. Have them name the restaurant and create a menu for it.

STUDENT PORTFOLIO

Written assignments for the students' portfolios may include a menu they have created and the *Mon Autobiographie* section of the Workbook, page 49.

Note Students may create and save both oral and written work using the Electronic Portfolio feature on the CD-ROM.

CHAPTER OVERVIEW

In this chapter students will learn vocabulary and structures associated with various foods and grocery shopping, including looking for items, expressing quantities, and talking about prices. They will also learn the forms of the irregular verb *faire* and the partitive. Students will learn to talk about what they and others can do and want to do using the verbs *pouvoir* and *vouloir*.

The cultural focus of Chapter 6 is a comparison and contrast of French and American food-shopping customs.

CHAPTER OBJECTIVES

By the end of this chapter students will know:

1. basic food vocabulary and the names of some important food categories
2. vocabulary associated with various types of shops common in France
3. the numbers 101–1,000
4. the partitive in the affirmative and the negative
5. present indicative forms of *faire*
6. present indicative forms of *vouloir* and *pouvoir*

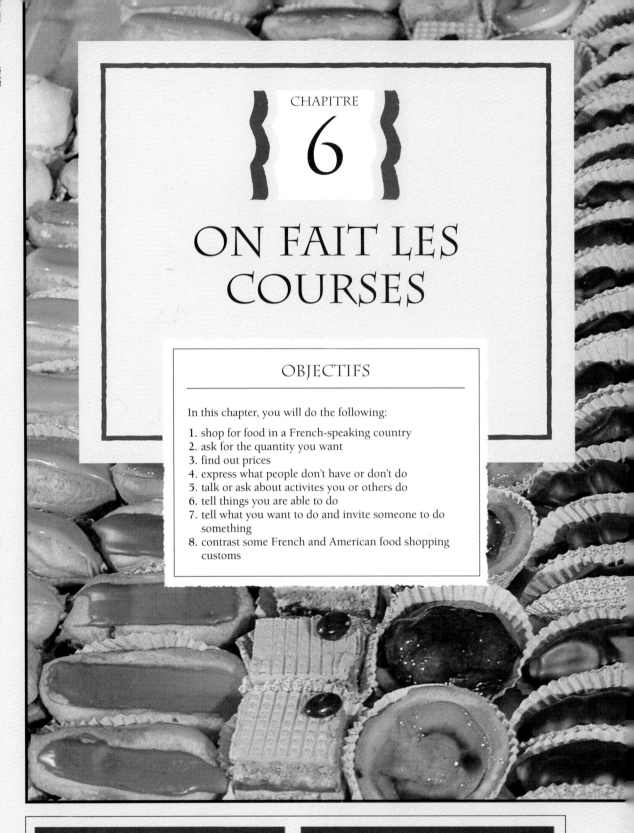

ON FAIT LES COURSES

OBJECTIFS

In this chapter, you will do the following:

1. shop for food in a French-speaking country
2. ask for the quantity you want
3. find out prices
4. express what people don't have or don't do
5. talk or ask about activites you or others do
6. tell things you are able to do
7. tell what you want to do and invite someone to do something
8. contrast some French and American food shopping customs

CHAPTER PROJECTS

(optional)

Tell students that as they progress through the chapter, they are to keep in mind the foods they learn to identify that they happen to like. These are the most important ones for them since they are the very items they will want to order. At the end of the chapter they are each going to make up a grocery shopping list and several menus.

DID YOU KNOW?

Besides the éclairs in the photo, the names of the other French pastries are (starting from the left, after the éclairs): *pavés croquants; tartelettes aux fruits; barquettes chocolat orange; babas à la crème* (or *savarins*); *barquettes aux noix; tartelettes aux framboises; millefeuilles* ("Napoleons," in English); *barquettes à la crème d'amandes* (last row, extreme right).

149

Pacing

This chapter requires eight to ten class sessions. Pacing will vary according to class length and the age and aptitude of the students.

Note The Lesson Plans offer guidelines for 45- and 55-minute classes and **Block Scheduling.**

Exercices vs. *Activités*

All exercises (which provide guided practice) are coded in blue. All communicative activities are coded in red.

INTERNET ACTIVITIES

(*optional*)

These activities, student worksheets, and related teacher information are in the *Bienvenue* Internet Activities Booklet and on the Glencoe Foreign Language Home Page at **http://www.glencoe.com/secondary/fl**

LEARNING FROM PHOTOS

After you have presented the vocabulary in the *Mots* section, ask students the following questions about this photo: *Ce sont des baguettes ou des pâtisseries? Où est-ce qu'on va pour acheter des pâtisseries? Il y a des éclairs au chocolat sur cette photo? Aimez-vous les éclairs au chocolat? Il y a aussi des tartelettes (de petites tartes)?*, etc.

VOCABULAIRE

MOTS 1

À LA BOULANGERIE-PÂTISSERIE

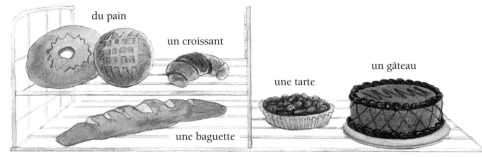

du pain

un croissant

un gâteau

une tarte

une baguette

À LA CRÉMERIE

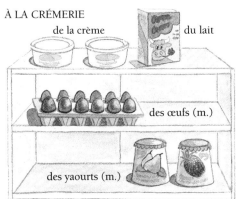

de la crème

du lait

des œufs (m.)

des yaourts (m.)

À LA BOUCHERIE

un poulet

de la viande

du bœuf

À LA CHARCUTERIE

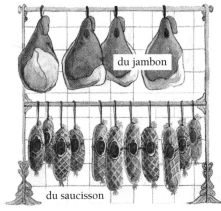

du jambon

du saucisson

À LA POISSONNERIE

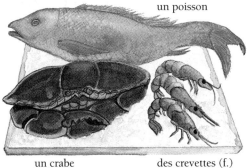

un poisson

un crabe

des crevettes (f.)

Vocabulary Teaching Resources

1. Vocabulary Transparencies 6.1 (A & B)
2. Audio Cassette 4B/CD-4
3. Student Tape Manual, *Mots 1: A–C*, pages 66–67
4. Workbook, *Mots 1: A–D*, pages 50–51
5. Communication Activities Masters, *Mots 1: A*, page 28
6. Chapter Quizzes, *Mots 1: Quiz 1*, page 31
7. CD-ROM, Disc 2, *Mots 1:* pages 150–153

Bell Ringer Review

Write the following on the board or use BRR Blackline Master 6-1: Answer the following questions:

1. **Comment vas-tu à l'école?**
2. **Avec qui vas-tu au cinéma?**
3. **Qu'est-ce que tu vas commander à McDonald's?**
4. **Où vas-tu en vacances?**

PRESENTATION *(pages 150–151)*

A. Tell students they are about to learn some new food items. Review the Chapter 5 food vocabulary using the Vocabulary Transparencies from that chapter. You may wish to have students use the words in a sentence.

B. Model the new words using Vocabulary Transparencies 6.1 (A & B), plastic replicas, or large pictures of food items. Have students repeat the words and phrases after you or Cassette 4B/CD-4. Review by pointing to the items as you ask *Qu'est-ce que c'est?*

TOTAL PHYSICAL RESPONSE

(following the Vocabulary presentation)

Getting Ready

Using posters or pictures of stores, set up areas of the classroom as *la pâtisserie, la boulangerie, la poissonnerie, la boucherie,* and *la crémerie.* Have pictures of food items from *Mots 1* and from Chapter 5 in each store. Dramatize the meaning of *montrez-moi.*

TPR

Allez à la boulangerie-pâtisserie, (Student 1).
Prenez une baguette.
Donnez la baguette à (Student 2).
Allez à la boucherie.
Achetez un poulet.
Montrez le poulet à (Student 3).
Allez à la poissonnerie.
Indiquez les crevettes.

(continued on next page)

Jean fait les courses.
Il fait les courses le matin.
Il ne fait pas ses courses au supermarché.
Il va à la boucherie, à la crémerie et à la
   boulangerie-pâtisserie.

un sac

la caisse

un filet

payer

Jean est à la boulangerie.
Il veut du pain.
Il achète une baguette et des croissants.
Il paie à la caisse.

CHAPITRE 6   **151**

C. Now have students open their books and read the new words.

D. Ask: *Jean fait les courses? Il fait les courses le matin? Il fait ses courses au supermarché? Il va à la crémerie? Il va à la boucherie?*

E. Now ask either/or questions such as: *C'est un croissant ou une baguette?* For the sentences on page 151, try using question words. For example: *Qui fait les courses? Quand est-ce que Jean fait les courses? Il va au supermarché? Non? Il va où alors? Où est-ce qu'il paie?*

F. Write the conjugation of *acheter* on the board, pointing out the forms with an *accent grave* and modeling the resulting change in pronunciation. Write out the forms of *payer*. Point out that the *je, tu, il(s), elle(s)* forms can be spelled with a *y* or an *i*.

COGNATE RECOGNITION

Point out the cognates in *Mots 1: une tarte, de la crème, un yaourt, du bœuf, un crabe, payer.*

RECYCLING

To further reinforce *-er* verbs, *acheter* and *payer* are presented in *Mots 1*. *Être* and *aller* are also reintroduced.

CROSS-CULTURAL COMPARISON

There are many types of French bread. *Une baguette* is used for sandwiches. *Un pain de mie*, similar to American bread, would be used to make an American-type sandwich. *Une ficelle* is a thin *baguette.* The large round loaf on page 150 is *un pain de campagne.*

TPR (*continued*)
Allez à la charcuterie.
Montrez-moi le saucisson.
Allez à la crémerie.
Achetez des œufs.
Allez à la caisse.
Payez.

COOPERATIVE LEARNING

(*For the younger student*)
   In teams of four, students sit in a circle. One student names a food item and simultaneously tosses a sponge ball to another student, who must name the place where that item can be purchased. That student then names a new item and tosses the ball to another student.

PRESENTATION (pages 152–153)

Exercice A: Listening

Focus on the listening skill by doing Exercise A with books closed. Then do the exercise again with books open.

Exercice B

After doing Exercise B, call on a student to retell the story of the exercise in his or her own words.

Extension of Exercice B: Listening

After completing Exercise B, focus on the listening skill by making false statements about the photos, which students correct. For example: *Mme Lenôtre achète six yaourts. Elle a un filet.*

ANSWERS

Exercice A

1. Oui, Jean fait les courses.
2. Oui, il fait les courses le matin.
3. Non, il ne va pas au supermarché.
4. Oui, il a un filet.
5. Oui, il va à la boulangerie-pâtisserie.
6. Oui, il veut du pain.
7. Oui, il achète une baguette.
8. Oui, il paie à la caisse.
9. Oui, il paie avec de l'argent.

Exercice B

1. C'est la crémerie.
2. On achète du fromage à la crémerie.
3. Elle achète du lait.
4. Elle achète des yaourts.
5. Elle va payer à la caisse.

RETEACHING (Mots 1)

Have students say whether they like or don't like each of the following items: *le pain, le lait, la crème, les œufs, le yaourt, la viande, le bœuf, le poulet.*

152

Exercices

A **Jean fait les courses.** Répondez. (*Answer.*)

1. Jean fait les courses?
2. Il fait les courses le matin?
3. Il va au supermarché?
4. Il a un filet?
5. Il va à la boulangerie-pâtisserie?
6. Il veut du pain?
7. Il achète une baguette?
8. Il paie à la caisse?
9. Il paie avec de l'argent?

B **À la crémerie.** Répondez d'après les photos. (*Answer according to the photos.*)

1. C'est la crémerie ou la boucherie?
2. On achète du fromage ou du pain à la crémerie?
3. Madame Lenôtre achète du lait ou du thé?
4. Elle achète des yaourts ou du fromage?
5. Elle va payer à la caisse ou au café?

ADDITIONAL PRACTICE

Put the following on the board. Have pairs practice it, using other stores and foods.

É1: Où vas-tu?
É2: Je vais à la boulangerie.
É1: À la boulangerie? Qu'est-ce que tu vas acheter à la boulangerie?
É2: De la crème.
É1: Mais on achète de la crème à la crémerie!

Now have groups of three practice the same dialogue using different verb forms. For example:

É1: Où allez-vous?
É2 & É3: Nous allons à la crémerie, etc. *or*
É1: Elle va où? (Où est-ce qu'elle va?)
É2 (*referring to É3*): Elle va au supermarché, etc.

Erase the board and let students do the dialogue from memory. Cue them if necessary.

C **Au supermarché.** Complétez d'après la photo. *(Answer according to the photo.)*

1. On achète du ___.
2. On achète du ___.
3. On achète du ___.
4. On achète du ___.
5. On achète des ___.

D **On va où pour acheter ça?** Complétez. *(Complete.)*

1. On veut du bœuf. On va ___.
2. On veut du lait. On va ___.
3. On veut des croissants et un gâteau. On va ___.
4. On veut de la viande. On va ___.
5. On veut du saucisson et du jambon. On va ___.
6. On veut de la crème et des œufs. On va ___.
7. On veut du poisson et du crabe. On va ___.
8. On veut une baguette. On va ___.

CHAPITRE 6   **153**

INDEPENDENT PRACTICE

Assign any of the following:
1. Exercises, pages 152–153
2. Workbook, *Mots 1: A–D,* pages 50–51
3. Communication Activities Masters, *Mots 1: A,* page 28
4. CD-ROM, Disc 2, pages 150–153

ADDITIONAL PRACTICE

Student Tape Manual, Teacher's Edition, *Activité B,* page 67

PRESENTATION *(continued)*

Exercice C

Exercise C should be done with books open.

Extension of *Exercice C:* Speaking

After completing Exercise C, mix up the items, cueing the partitive and the noun and having individuals say the whole sentence. For example: *des œufs. (On achète des œufs.)* Continue with other food items students have learned.

Exercice D

Exercise D should be done with books open, or you may wish to use the recorded version of this exercise.

ANSWERS

Exercice C

1. **lait** (Answers 1–4 can be in any order.)
2. **jambon**
3. **saucisson**
4. **bœuf (haché)**
5. **œufs**

Exercice D

1. **à la boucherie**
2. **à la crémerie**
3. **à la boulangerie-pâtisserie**
4. **à la boucherie**
5. **à la charcuterie**
6. **à la crémerie**
7. **à la poissonnerie**
8. **à la boulangerie-pâtisserie**

INFORMAL ASSESSMENT *(Mots 1)*

Check for comprehension by naming a store and having students mention as many items as possible that can be purchased there.

RETEACHING *(Mots 1)*

Have students open their books to pages 150–151 and name any stores on these pages that are found in their own local shopping area. Have them use *Il y a* and any location information they can give in French. For example: *Il y a une boucherie à côté du cinéma. Il y a une boulangerie rue Main.*

VOCABULAIRE

MOTS 2

Vocabulary Teaching Resources

1. Vocabulary Transparencies 6.2 (A & B)
2. Audio Cassette 4B/CD-4
3. Student Tape Manual, Teacher's Edition, *Mots 2: D–F*, pages 68–69
4. Workbook, *Mots 2: E–G*, pages 51–52
5. Communication Activities Masters, *Mots 2: B*, page 28
6. Chapter Quizzes, *Mots 2: Quiz 2*, page 32
7. Computer Software, *Vocabulaire*
8. CD-ROM, Disc 2, *Mots 2:* pages 154–157

PRESENTATION *(pages 154–155)*

A. Model the new words using Vocabulary Transparencies 6.2 (A & B), plastic replicas, food containers, or pictures of food. Follow the procedures outlined in *Mots 1*, page 150.
B. With the exception of the numbers on page 155, have students repeat the words and phrases after you or Cassette 4B/CD-4.

CROSS-CULTURAL COMPARISON

Explain to students that the metric system is used in France. Additional information about the metric system appears in the *Découverte culturelle* on page 169 of this chapter, and also in *Lettres et sciences* following Chapter 8.

VOCABULAIRE

MOTS 2

AU MARCHÉ

la marchande

le marchand de fruits et légumes

les fruits (m.)

des bananes (f.)

des pommes de terre (f.)

8ᶠ

6ᶠ 85

10ᶠ

des pommes (f.)

des carottes (f.)

5ᶠ

des oranges (f.)

une laitue

des oignons (m.)

7ᶠ 50   les légumes (m.)

des haricots verts (m.)   des tomates (f.)

Carole veut des légumes. Elle est au marché.
Elle va chez le marchand de fruits et légumes.
Elle achète un kilo de carottes et une livre de tomates.
Elle paie la marchande.

un kilo = 1.000 (mille) grammes
une livre = 500 (cinq cents) grammes

154 CHAPITRE 6

TOTAL PHYSICAL RESPONSE

(following the Vocabulary presentation)
Getting Ready
Tell students a supermarket cart is called either *un chariot* or *un caddie*. Now point to your pocket as you say *poche*.
TPR
____, levez-vous, s'il vous plaît.
Venez ici.
Vous allez imaginer que vous êtes au

supermarché.
Voilà un chariot. Prenez le chariot.
Poussez le chariot. *(gesture)*
Promenez-vous dans le supermarché.
Marchez lentement.
Arrêtez-vous.
Regardez les bouteilles d'eau minérale.
Prenez une bouteille.
Mettez la bouteille dans le chariot.
Maintenant regardez la boîte de conserve.
(continued on next page)

154

À L'ÉPICERIE

Richard est à l'épicerie.
Il veut de l'eau minérale et du lait.
Il achète une bouteille d'eau minérale
    et un litre de lait.

un paquet de
légumes surgelés

un pot de moutarde

500 grammes de beurre

un litre de lait

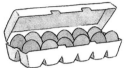
une douzaine d'œufs

une boîte de conserve

une bouteille
d'eau minérale

LES NOMBRES DE CENT UN À MILLE (101–1.000)	
101	cent un
102	cent deux
103	cent trois
200	deux cents
201	deux cent un
300	trois cents
301	trois cent un
400	quatre cents
500	cinq cents
600	six cents
700	sept cents
800	huit cents
900	neuf cents
1.000	mille

Note:

1. To ask the price (*le prix*), you use the following expressions.

C'est combien le bœuf?	**Vingt francs le kilo.**
C'est combien le beurre?	**Huit francs la livre.**

2. To find out how much you owe when you have purchased several items, you ask:

Ça fait combien?

3. The vendor often asks if you want something else. If you don't, you may use one of the expressions below.

MARCHAND(E)	CLIENT(E)
Autre chose?	**Rien d'autre, merci.**
Et avec ça?	**C'est tout.**
C'est tout?	

PRESENTATION *(continued)*

C. Briefly review the numbers 1–100 with flashcards or by writing them on the board. Model the new numbers on page 155 and have students repeat in unison.

D. Write three- and four-digit numbers on the board and challenge individuals to read them.

COGNATE RECOGNITION

Point out the following cognates that appear in this section: *des fruits, une banane, une orange, une carotte, un oignon, un gramme, un litre, une bouteille, une douzaine, minérale, un paquet.*

Vocabulary Expansion

When teaching *de l'eau minérale,* you may wish to add the expressions *gazeuse* and *plate* since they are frequently needed when ordering mineral water.

RETEACHING *(Mots 2)*

Have students open their books to pages 154–155 and ask them to say whether they like each item shown or not.

THE FRANCOPHONE WORLD

If you wish to show students photos of food and markets in several different francophone countries, see *Le Monde francophone* section on *La Cuisine,* pages 220–223.

TPR *(continued from page 154)*

Non, vous ne voulez pas la boîte de
    conserve.
Poussez le chariot jusqu'à la caisse.
Donnez la bouteille d'eau à la fille qui
    travaille à la caisse.
Cherchez de l'argent dans votre poche.
Donnez l'argent à la fille. Payez.
Prenez le sac.
Merci. Retournez à votre place et asseyez-
    vous.

PAIRED ACTIVITY

Have students work in pairs. One gives a quantity and the other says an item that can be bought in that quantity. For example:
É1: Une bouteille.
É2: Une bouteille d'eau minérale.

PRESENTATION *(page 156)*

Exercice A

The main objective of Exercise A is to give students practice with the singular gender marker.

Extension of *Exercices B* and *C*

After doing Exercises B and C, have a student retell the story of each exercise in his or her own words.

ANSWERS

Exercice A

1. C'est une banane. C'est un fruit.
2. C'est un oignon. C'est un légume.
3. C'est une laitue. C'est un légume.
4. C'est une orange. C'est un fruit.
5. C'est une pomme de terre. C'est un légume.
6. C'est une pomme. C'est un fruit.

Exercice B

Answers will vary but may include the following:

... la marchande de fruits et légumes... laitue... tomates ... carottes... c'est tout... marchande

Exercice C

1. bouteilles   5. caisse
2. paquet   6. fait
3. boîtes   7. boîtes
4. pot   8. pot

Exercice D

1. Élève 1: C'est combien, la boîte de conserve? (de haricots verts?)
   Élève 2: Quatre francs la boîte.
2. É1: C'est combien, les œufs?
   É2: Huit francs la douzaine.
3. É1: C'est combien, l'eau minérale?
   É2: Trois francs la bouteille.
4. É1: C'est combien, le lait?
   É2: Sept francs le litre.
5. É1: C'est combien, la moutarde?
   É2: Six francs le pot.

Exercices

A   **Un fruit ou un légume?**   Identifiez d'après le modèle. *(Identify according to the model.)*

C'est une carotte. C'est un légume.

1.   2.   3.   4.   5.   6.

B   **Nicole va au marché.**   Complétez. *(Complete.)*

Nicole veut préparer une grande salade. Elle va au marché. Elle va chez la ___. Elle achète une ___, des ___ et des ___. La marchande demande, «Pas d'oignons aujourd'hui?» Nicole répond: «Non merci, ___ ___ .» Elle donne de l'argent à la ___.

C   **Louis va à l'épicerie.**   Complétez. *(Complete.)*

Louis veut de la moutarde, de l'eau minérale, des boîtes de conserve et un paquet de légumes surgelés. Pour acheter tout ça il va à une épicerie. À l'épicerie Louis achète deux ___ d'eau minérale, un ___ de carottes surgelées et trois ___ de sardines. Et quelque chose d'autre—un ___ de moutarde. Louis va à la ___ où il paie. Ça ___ combien, les bouteilles d'eau minérale, le paquet de carottes, les ___ de sardines et le ___ de moutarde? Ça fait trente francs.

D   **C'est combien, s'il vous plaît?**   Demandez le prix à un copain ou à une copine. *(Ask a classmate how much the following items are.)*

1. la boîte de conserve
2. la douzaine d'œufs
3. la bouteille d'eau minérale
4. le litre de lait
5. le pot de moutarde

PAIRED ACTIVITY

After completing Exercises A–D reinforce the lesson with the following activity. Have students work in pairs and find out what their partner's favorite and least favorite foods are from each of these categories: *la viande, les légumes, les fruits, les desserts, les boissons.* Then have them report their findings to the class.

For example:
É1: J'aime le poulet, mais je déteste le bœuf.
É2: *(À la classe):* Elle aime le poulet, mais elle déteste le bœuf.

Activités de communication orale
Mots 1 et 2

A **À la boulangerie-pâtisserie.** Visit a French bakery in your community with your classmates. Ask the baker the French names of the pastries that appeal to you. Choose a few items and find out how much you owe. Be sure to speak French. (If there isn't a French bakery in your community, set one up in your classroom by bringing in baked goods or magazine photos of French pastries and breads. Take turns playing the roles of baker and customers.)

B **Au marché.** You're at a vegetable stand at the open-air market in Nice. Make a list of items you want to buy. Use the list of expressions below to talk to the *marchand(e)* (your partner).

Bonjour.	Et avec ça?
Vous désirez, (Monsieur, Mademoiselle)?	C'est tout?
Je voudrais…	Ça fait ___ francs.

C **Je fais les courses.** You've offered to do the shopping for the French family you're living with in Tours. You've got to buy the items on the grocery list below. Ask your French host (your partner) where you have to go to get each item.

des crevettes

Élève 1: Je vais où pour acheter des crevettes?
Élève 2: Tu vas à la poissonnerie.

D **À l'épicerie.** You're in a French *épicerie*. Ask the clerk (your partner) for some of the items on the list below. Be sure to tell him or her the quantity you want. Then reverse roles.

jambon

Élève 1: Je voudrais 500 grammes de jambon, s'il vous plaît.
Élève 2: Voilà. Et avec ça?

1. eau minérale	7. oranges
2. œufs	8. Orangina
3. pommes	9. fromage
4. frites surgelées	10. crème
5. pommes de terre	11. lait
6. beurre	12. bananes

crevettes
saucisson
tarte aux fruits
poulet
fromage
baguette
haricots verts
pommes de terre

INDEPENDENT PRACTICE

Assign any of the following:
1. Exercises and activities, pages 156–157
2. Workbook, *Mots 2: E–G*, pages 51–52
3. Communication Activities Masters, *Mots 2: B*, page 28
4. Computer Software, *Vocabulaire*
5. CD-ROM, Disc 2, pages 154–157

ADDITIONAL PRACTICE

Student Tape Manual, Teacher's Edition, *Activités E–F*, page 69.

Activités de communication orale
Mots 1 et 2

Bell Ringer Review
Write the following on the board or use BRR Blackline Master 6-2: You are planning a lunch for a friend. Make a shopping list and note where you are going to buy each item.

PRESENTATION *(page 157)*
Activité C
Students should refer to the handwritten note on page 157 to do this activity.

ANSWERS
Activité B
Answers will vary.

Activité C
1. É1: Je vais où pour acheter du saucisson? É2: Tu vas à la charcuterie.
2. É1: … une tarte aux fruits? É2: … à la boulangerie-pâtisserie.
3. É1: … un poulet? É2: … à la boucherie.
4. É1: … du fromage? É2: … à la crémerie.
5. É1: … une baguette? É2: … à la boulangerie-pâtisserie.
6. É1: … des haricots verts? É2: … chez le marchand de fruits et légumes.
7. É1: … des pommes de terre? É2: … chez le marchand de fruits et légumes.

Activité D
Answers will vary.

INFORMAL ASSESSMENT
(Mots 2)

Introduce the expression *C'est vrai* and its negative. Then make true and false statements about different grocery items and where they can be found and have students elaborate. For example: *On trouve du lait à la crémerie. (C'est vrai.) On trouve de l'eau minérale à la boucherie. (Ce n'est pas vrai. On trouve de l'eau minérale à l'épicerie.)*

STRUCTURE

Talking about an Indefinite Quantity or Things in General

1. You use the definite article (*le, la, l', les*) to refer to a specific item or items.

Le poisson est au réfrigérateur dans la cuisine.	*The fish is in the refrigerator in the kitchen.*
Voilà le dessert.	*Here's the dessert.*

2. You also use the definite article when talking about something in a general sense.

Le thé est délicieux.	*Tea is delicious.*
Les enfants aiment le lait.	*Children like milk.*
Je déteste les haricots verts.	*I hate green beans.*
Ils n'aiment pas la viande.	*They don't like meat.*

Note that the definite article is often used with verbs that express likes and dislikes—*aimer, détester, préférer, adorer.*

3. You use the partitive construction to express an unspecified amount or part of the whole. In English we often say "some" or "any" to express the partitive. We may omit those words in English, but in French the partitive construction must be used to express indefinite quantity. Study the following examples.

Vous avez du thé?	*Do you have (any) tea?*
Tu voudrais du lait?	*Would you like (some) milk?*
Il achète des haricots verts.	*He's buying (some) green beans.*
Je commande de la viande.	*I'm ordering (some) meat.*

4. You express the partitive in French by using *de* + the definite article. *De* combines with *le* to form *du*. *De* + *les* becomes *des*. *De la* and *de l'* remain unchanged. Study the following chart.

de + la = de la	J'ai *de la* crème.
de + l' = de l'	Je voudrais *de l'*eau.
de + le = du	Tu manges *du* pain?
de + les = des	Il achète *des* fruits et *des* légumes.

STRUCTURE

Structure Teaching Resources

1. Workbook, *Structure: A–H,* pages 53–55
2. Student Tape Manual, Teacher's Edition, *Structure: A–C,* pages 70–71
3. Audio Cassette 4B/CD-4
4. Communication Activities Masters, *Structure: A–C,* pages 29–31
5. Computer Software, *Structure*
6. Chapter Quizzes, *Structure: Quizzes 3–6,* pages 33–36
7. CD-ROM, Disc 2, pages 158–165

Le partitif et l'article défini

PRESENTATION *(page 158)*

A. Say each of the foods below and ask students to answer with "*Oui, je voudrais (+ item)*" if they'd like some: *des fruits, du yaourt, du poulet, des œufs, de l'eau, du beurre, du lait, de la moutarde, de la mayonnaise, de la glace.*

B. Lead students through steps 1 and 2 and discuss examples. Ask for additional examples.

C. Briefly review the contractions with *de* + the definite articles (page 135).

D. Lead students through steps 3 and 4 and discuss the examples. Write out other examples contrasting the definite article and the partitive. For example: *J'aime les bananes. Les bananes ici sont bonnes. Je vais acheter des bananes.*

E. Ask questions (such as *Qui aime le pain? Qui a du pain?*) that contrast the definite article and the partitive.

Note In the CD-ROM version, this structure point is presented via an interactive electronic comic strip.

In teams of four, have students write as many nouns as they can think of on small bits of paper. They should write only the noun, with no article. Students place all the papers in a hat or bag and pass it around the team. Each member draws one noun and makes a sentence that uses the noun in a partitive construction. Answers may begin with: *J'ai... , Il y a... , J'achète... , Je gagne... , Je mange...* , etc.

Tell students that these are spices—*des épices*—on display in a Provençal market. Ask them what word they have learned in this chapter that sounds like *épice.* (*une épicerie*) Originally, *épiceries* sold spices, but they gradually evolved into what we know as the grocery store, selling a variety of foods.

Exercices

A **Qu'est-ce que je vais acheter?** Répondez d'après le modèle. (*Answer according to the model.*)

> **Tu vas acheter des fruits?**
> *Oui, je vais acheter des fruits. J'aime les fruits.*

1. Tu vas acheter du pain?
2. Tu vas acheter du fromage?
3. Tu vas acheter des bananes?
4. Tu vas acheter de la glace?

B **Au marché.** Complétez. (*Complete.*)

Je vais acheter ___ légumes et ___ fruits chez le marchand de fruits et légumes.
Ensuite je vais aller à la boucherie où je vais acheter ___ bœuf et ___ poulet.
Et comme la famille aime bien manger ___ fromage après le dîner, je vais aller à
la crémerie pour acheter ___ fromage.

C **Des provisions.** Complétez. (*Complete.*)

Au marché Robert achète ___ pain, ___ jambon, ___ fromage, ___ bananes
et ___ crème. Il va préparer ___ sandwichs au jambon et au fromage. Pour le
dessert il va préparer ___ bananes avec ___ crème.

D **Des différences.** Complétez. (*Complete.*)

Janine Dupont a une sœur, Colette. Quand les deux sœurs vont au restaurant,
Colette commande toujours ___ poisson. Elle aime bien ___ poisson. Mais
Janine n'aime pas du tout ___ poisson. Elle aime ___ viande et elle commande
toujours ___ viande. Elle commande toujours ___ bœuf.

Le partitif à la forme négative

Expressing What People Don't Have or Don't Do

1. You have already seen that *un*, *une*, and *des* change to *de* (*d'*) in the negative.

AFFIRMATIF	NÉGATIF
J'ai un livre.	Je *n'ai pas de* livre.
Nous avons une voiture.	Nous *n'avons pas de* voiture.
Ils ont des frères.	Ils *n'ont pas de* frères.

CHAPITRE 6 **159**

INDEPENDENT PRACTICE

Assign any of the following:
1. Exercises, page 159
2. Workbook, *Structure: A*, page 53
3. CD-ROM, Disc 2, pages 158–159

Exercices

PRESENTATION (*page 159*)

Exercice A
You may wish to use the recorded version of this exercise. It can be done with books either closed or open.

Exercices B, C, and D
Exercises B, C, and D should be done with books open.

Note The partitive will be reintroduced often.

ANSWERS

Exercice A
1. Oui, je vais acheter du pain. J'aime le pain.
2. ... du fromage. ... le fromage.
3. ... des bananes. ... les bananes.
4. ... de la glace. ... la glace.

Exercice B

1. des	3. du	5. du
2. des	4. du (un)	6. du

Exercice C

1. du	5. de la
2. du	6. des
3. du	7. des
4. des	8. de la

Exercice D

1. du	3. le	5. de la
2. le	4. la	6. du

Le partitif à la forme négative

PRESENTATION (*pages 159–160*)

A. Lead students through step 1 and the chart on page 159. Using objects in the room, review the change from *un*, *une*, or *des* to *de* (*d'*) in the negative. For example, (holding a pencil): *J'ai un crayon.* (Hiding the pencil): *Je n'ai pas de crayon.*

B. From the chart on page 160, read the affirmative examples to the class and have students read the negative forms after you in unison. Continue with other affirmative sentences, having first the class, then individuals, give the negative.

Exercices

ANSWERS

Exercice A

1. É1: Tu as un(e) ami(e)? É2: Non, je n'ai pas d'ami(e). (Oui, j'ai un[e] ami[e].)
2. É1: ... de l'argent? É2: ... pas d'argent. (Oui, ...)
3. É1: ... des cassettes? É2: ... pas de cassettes. (Oui, ...)
4. É1: ... un chat? É2:... pas de chat. (Oui, ...)
5. É1: ... un chien? É2: ... pas de chien. (Oui, ...)
6. É1: ... des cousins? É2: ... pas de cousins. (Oui, ...)
7. É1: ... des cousines? É2: ... pas de cousines. (Oui, ...)
8. É1: ... des disques? É2: ... pas de disques. (Oui, ...)
9. É1: ... des frères? É2: ... pas de frères. (Oui, ...)
10. É1: ... des grands-parents? É2: ... pas de grands-parents. (Oui, ...)
11. É1: ... des livres? É2: ... pas de livres. (Oui, ...)
12. É1: ... des magazines? É2: ... pas de magazines. (Oui, ...)

Exercice B

1. Non, elle n'achète pas de pain à la boucherie. Elle achète du pain à la boulangerie-pâtisserie.
2. ... pas de fromage à la boulangerie-pâtisserie. ... du fromage à la crémerie.
3. ... pas de légumes à la charcuterie. ... des légumes chez le marchand de fruits et légumes.
4. ... pas de viande à la crémerie. ... de la viande à la boucherie.
5. ... pas d'œufs chez le marchand de fruits et légumes. ... des œufs à la crémerie.

2. Note that in the negative, all forms of the partitive (*du, de la, de l'*, and *des*) also change to *de* or *d'*.

AFFIRMATIF	NÉGATIF
J'achète du pain.	Je n'achète *pas de* pain.
J'ai de la crème.	Je n'ai *pas de* crème.
Je prépare des carottes.	Je ne prépare *pas de* carottes.
Il a des amis.	Il n'a *pas d'*amis.

Exercices

A **Qu'est-ce que tu as?** Posez une question d'après le modèle. (*Ask a question according to the model.*)

> des crayons
>
> Élève 1: Tu as des crayons?
> Élève 2: Non, je n'ai pas de crayons. (Oui, j'ai des crayons.)

1. un(e) ami(e)
2. de l'argent
3. des cassettes
4. un chat
5. un chien
6. des cousins
7. des cousines
8. des disques
9. des frères
10. des grands-parents
11. des livres
12. des magazines

B **Juliette fait ses courses.** Répondez d'après le modèle. (*Answer according to the model.*)

> Elle achète du poisson à la boucherie?
> *Non, elle n'achète pas de poisson à la boucherie. Elle achète du poisson à la poissonnerie.*

1. Elle achète du pain à la boucherie?
2. Elle achète du fromage à la boulangerie-pâtisserie?
3. Elle achète des légumes à la charcuterie?
4. Elle achète de la viande à la crémerie?
5. Elle achète des œufs chez le marchand de fruits et légumes?

Une poissonnerie dans la rue Mouffetard à Paris

ADDITIONAL PRACTICE

After completing Exercises A–C, reinforce the lesson by having students change the following sentences to the negative.

1. Le prof d'anglais mange du pâté en classe.
2. Au marché nous passons des examens d'histoire.
3. Tes parents achètent du pain à la poissonnerie.
4. Ma grand-mère donne une pomme à la nouvelle voisine.
5. Au restaurant les chiens commandent de la viande.
6. Il y a un œuf dans ma poche.
7. Elle mange de la glace avec ses légumes.

C Au supermarché. Complétez. *(Complete.)*

Quand Jacqueline va au supermarché elle n'achète pas ___₁ fruits. Elle n'aime pas ___₂ fruits au supermarché. Elle achète ___₃ fruits au marché, chez le marchand de fruits et légumes. Elle n'achète pas ___₄ café au supermarché. Elle n'achète pas ___₅ viande. Elle n'achète pas ___₆ haricots verts. Elle n'achète pas ___₇ oignons. Qu'est-ce qu'elle achète au supermarché alors? Elle achète seulement ___₈ boîtes de conserve et ___₉ bouteilles d'eau.

Le verbe *faire* au présent

Telling and Asking What You or Others Do

1. The verb *faire*, "to do" or "to make," is irregular. Study the following forms.

FAIRE			
je	fais	nous	faisons
tu	fais	vous	faites
il		ils	
elle }	fait	elles }	font
on			

2. *Qu'est-ce que tu fais?* or *Qu'est-ce que vous faites?* means "What are you doing?". Note that you can use verbs other than *faire* in your answer.

   Qu'est-ce que tu fais?   **Je regarde la télé.**
   Qu'est-ce que vous faites?   **Nous préparons le dîner.**

3. You also use the verb *faire* in many idiomatic expressions. An idiomatic expression is one that does not translate directly from one language to another. *Faire les courses* is an example of such an expression. In English we say "to go grocery shopping," while in French the verb *faire* is used.

4. You also use *faire* to tell what subjects you are taking.

   Je fais du français et mon frère fait de l'espagnol.
   Mon frère et moi faisons des maths.

CHAPITRE 6   **161**

C. Explain to students that the verb *faire* is an extremely useful verb because it is used in many expressions. Students will encounter it frequently.

Exercices

PRESENTATION (pages 162–163)

Exercice A

A. Exercise A uses several forms of the verb *faire* in a natural context.

B. If you call on students to read Exercise A, have them use as much expression as possible. You may also wish to go over this exercise once or twice orally. Then have several students retell the story in their own words.

Extension of *Exercice B*

After completing Exercise B, have students conduct a poll of which classes their classmates take. Then ask questions about the results. For example: *Qui fait de la biologie? (Marc, Luc et Carole font de la biologie.)*

ANSWERS

Exercice A

1. fait	4. font
2. fait	5. font
3. fait	

Exercice B

1. É1: Tu fais du français? É2: Oui, je fais du français. (Non, je ne fais pas de français.)
2. É1: ... de la géométrie? É2: ... de la géométrie. (... pas de géométrie.)
3. É1: ... de l'anglais? É2: ... de l'anglais. (... pas d'anglais.)
4. É1: ... des sciences naturelles? É2: ... des sciences naturelles. (... pas de sciences naturelles.)
5. É1: ... de l'histoire? É2: ... de l'histoire. (... pas d'histoire.)
6. É1: ... de la géographie? É2: ... de la géographie. (... pas de géographie.)

---

5. Here are some other expressions using *faire*. You can probably guess their meaning.

> Il fait ses études secondaires au lycée du Parc Impérial.
> Tu ne fais pas attention en classe.
> Nous n'aimons pas faire nos devoirs devant la télé.
> Maman prépare un bon dîner. Elle aime faire la cuisine.
> Nous aimons faire un pique-nique au parc.

6. Note that *de la, du, de l'*, and *des* following *faire* change to *de (d')* in the negative.

> Elle fait du français mais elle ne fait pas de maths.
> Nous ne faisons pas d'anglais.

Exercices

A On fait les courses. Répétez la conversation. (*Practice the conversation.*)

LUC: Salut, Robert. Qu'est-ce que tu fais?
ROBERT: Moi, je fais les courses.
LUC: Tiens! Quelle surprise! Moi aussi. Je vais au marché de la rue Cler. Tu veux y aller avec moi?
ROBERT: Pourquoi pas? Mais Annette va aussi faire les courses avec moi aujourd'hui.
LUC: Pas de problème! On fait les courses ensemble.

Complétez d'après la conversation. (*Complete according to the conversation.*)

1. Luc ___ ses courses.
2. Robert ___ ses courses aussi.
3. Et Annette ___ ses courses.
4. Luc, Robert et Annette ___ leurs courses ensemble.
5. Ils ___ leurs courses au marché de la rue Cler.

B Quels cours? Posez des questions à un copain ou à une copine d'après le modèle. (*Ask a classmate questions according to the model.*)

> de la gymnastique
>
> Élève 1: Tu fais de la gymnastique?
> Élève 2: Oui, je fais de la gymnastique. (Non, je ne fais pas de gymnastique.)

1. du français	4. des sciences naturelles
2. de la géométrie	5. de l'histoire
3. de l'anglais	6. de la géographie

DID YOU KNOW?

Exercise A on page 162 mentions the market in the rue Cler. Explain to students that there are many street markets in Paris. Tell them that the rue Cler is in the 7e *arrondissement*. Have them review what they know about this area of Paris. For example, *C'est un arrondissement résidentiel ou commercial? Comment s'appelle le monument célèbre du 7e arrondissement?*

PAIRED ACTIVITY

Have students work in pairs. One names a course he/she takes. They make up a conversation as per the model.
É1: **Moi, je fais du latin.**
É2: **Tu fais du latin.** (Name of another student) **aussi fait du latin. Vous faites du latin ensemble?**
É2: **Oui (Non), nous (ne) sommes (pas) dans le même cours.**

C **Tes copains et toi.** Donnez des réponses personnelles. *(Give your own answers.)*

1. Vous faites des études au lycée ou au collège?
2. Vous faites vos devoirs devant la télé?
3. Vous faites attention en classe?
4. Vous faites la cuisine française en classe?

D **Qu'est-ce que vous faites, Monsieur?** Posez des questions d'après le modèle. *(Ask questions according to the model.)*

> **Madame fait les courses au marché.**
> *Et vous, Monsieur? Vous faites aussi les courses au marché?*

1. Madame fait la cuisine le soir.
2. Madame fait un gâteau d'anniversaire.
3. Madame fait un sandwich à midi.
4. Madame fait les courses au supermarché.

E **Mon copain Yves.** Complétez. *(Complete.)*

Voilà Yves, mon copain du lycée. Il est très intelligent. Nous sommes dans le même cours d'anglais. Yves ___ toujours attention en classe. Moi, je ne ___ pas très attention. Yves et moi ___ nos devoirs ensemble après les cours. Yves ne ___ pas de fautes (erreurs). Mais moi, je ___ beaucoup de fautes.

$\underset{1}{\quad}$ $\underset{2}{\quad}$ $\underset{3}{\quad}$ $\underset{4}{\quad}$ $\underset{5}{\quad}$

Yves et son amie Monique ___ du français avec Madame Delacourt. Ils aiment beaucoup le cours de français. Qu'est-ce qu'ils ___ au cours de français? Ils parlent beaucoup et ils chantent des chansons françaises.

$\underset{6}{\quad}$ $\underset{7}{\quad}$

Vous ___ du français aussi, n'est-ce pas? Vous ___ du français avec qui? Qui est votre prof? Qu'est-ce que vous ___ au cours de français?

$\underset{8}{\quad}$ $\underset{9}{\quad}$ $\underset{10}{\quad}$

LEARNING FROM PHOTOS

Tell students that the type of sandwich on page 163 is a *pan bagnat*. It is a specialty of Provence. Its ingredients are basically the same as those in a *salade niçoise*. Ask students *Qu'est-ce qu'il y a dans un pan bagnat?* Give them the words they don't know: *du thon, des olives.* Then ask them questions such as: *Tu aimes le thon? Les olives? Tu voudrais manger un pan bagnat?*

INDEPENDENT PRACTICE

Assign any of the following:
1. Exercises, pages 162–163
2. Workbook, *Structure: D–F,* pages 54–55
3. Communication Activities Masters, *Structure: B,* page 30
4. CD-ROM, Disc 2, pages 161–163

PRESENTATION *(continued)*

Exercice C

Exercise C gives receptive practice with *vous faites* as students respond with *nous faisons.* This prepares them to actively use *vous faites* in Exercise D.

Exercice C: **Listening**

Exercise C is best done as a whole-class activity. To focus on the listening skill, have students close their books before doing the exercise orally.

Extension of *Exercice D:* **Listening**

After completing Exercise D, make comparisons between French and American young people using *faire.* For example: *En France, ils font du russe. Et vous?* (*Nous ne faisons pas de russe.* or *Oui,...*)

Exercice E

This exercise recombines all forms of *faire.* It should be done with books open. Then students retell the story in their own words.

ANSWERS

Exercice C

Answers will vary, but all answers should begin with *nous (ne) faisons (pas).*

Exercice D

1. Et vous, Monsieur? Vous faites aussi la cuisine le soir?
2. ... Vous faites aussi... ?
3. ... Vous faites aussi... ?
4. ... Vous faites aussi... ?
5. ... Vous faites aussi... ?

Exercice E

1. fait   6. font
2. fais   7. font
3. faisons   8. faites
4. fait   9. faites
5. fais   10. faites

RETEACHING

Have students write three questions using the verb *faire.* Then have them ask their questions of a partner.

Les verbes pouvoir et vouloir

PRESENTATION *(page 164)*

A. First have students repeat the *ils/elles* forms. Then have them repeat all the singular forms. The latter all sound alike, but they are spelled differently and they drop the consonant sound heard in the *ils/elles* form.

B. Now have students repeat the *nous* and *vous* forms, which again pick up the consonant sound of the infinitive. Many verbs follow this pattern in French.

C. Lead students through steps 2–4 on page 164. Now do the exercises that follow.

Exercices

PRESENTATION *(page 164)*

Exercice A

Exercise A can be done with books either open or closed.

ANSWERS

Exercice A

Answers 1–4 follow the same pattern: **Je veux bien, mais je ne peux pas.** Answers 5–8 follow the same pattern: **Il veut bien, mais il ne peut pas.**

Les verbes *pouvoir et vouloir*   *Describing What You or Others Can Do or Want to Do*

1. Study the following forms of the verb *pouvoir*, "to be able to," and *vouloir*, "to want."

POUVOIR			
je	peux	nous	pouvons
tu	peux	vous	pouvez
il elle on	peut	ils elles	peuvent

VOULOIR			
je	veux	nous	voulons
tu	veux	vous	voulez
il elle on	veut	ils elles	veulent

2. You use *pouvoir* and *vouloir* with an infinitive of another verb to express what one can do or wants to do.

> Michelle peut dîner au restaurant ce soir.
> Je veux inviter mes copains à la fête.
> Vous pouvez commander un steak frites pour moi?

3. As with other verbs that come before an infinitive, *ne... pas* goes around the verbs *pouvoir* and *vouloir* to form the negative.

> Je ne veux pas manger de frites avec mon steak.
> Nous ne pouvons pas aller à la discothèque.

4. To ask for something politely, you use *je voudrais*, "I'd like," instead of *je veux*, "I want."

> Je voudrais une livre de haricots verts, s'il vous plaît.

Exercices

A **Je veux bien, mais je ne peux pas.** Répondez d'après le modèle. *(Answer according to the model.)*

> **Tu veux aller au restaurant?**
> *Je veux bien, mais je ne peux pas.*

1. Tu veux aller au café?
2. Tu veux dîner avec Claude?
3. Tu veux travailler à plein temps?
4. Tu veux gagner de l'argent?
5. Ton frère veut faire les courses?
6. Il veut aller au marché?
7. Il veut préparer le dîner?
8. Il veut inviter ses amis?

ADDITIONAL PRACTICE

After completing Exercises A–D, reinforce with the following. Have students answer according to the model:

Moi, je ne peux pas aller au restaurant. →
Mais tes amis peuvent aller au restaurant, n'est-ce pas?

1. Moi, je ne peux pas aller au restaurant.
2. Moi, je ne veux pas arriver avant huit heures.
3. Moi, je ne veux pas préparer le dîner.
4. Moi, je ne peux pas faire les courses.

B **Si vous voulez, vous pouvez.** Répondez d'après le modèle. (*Answer according to the model.*)

> **Nous voulons travailler.**
> *Si vous voulez, vous pouvez travailler.*

1. Nous voulons travailler à mi-temps.
2. Nous voulons gagner de l'argent.
3. Nous voulons avoir des succès.
4. Nous voulons être riches.

C **Les garçons n'ont pas beaucoup d'argent.** Complétez avec «pouvoir» ou «vouloir». (*Complete with* pouvoir *or* vouloir.)

Pierre et son frère Jacques ont faim. Ils ___ aller dans un restaurant où ils ___
                                            1   2
dîner rapidement. Ils ___ commander deux hamburgers chacun (*each*) mais
                       3
ils ne ___ pas. Pierre insiste, mais Jacques crie: «Pas question! On n'a pas
        4
beaucoup d'argent! Tu ___ commander seulement un hamburger aujourd'hui.»
                       5

D **Qui peut préparer le dîner?** Complétez. (*Complete.*)

ANNE: Je ___ (vouloir) préparer le dîner ce soir, mais franchement je ne ___
      (pouvoir) pas.

JEAN: Tu ne ___ (pouvoir) pas? Pourquoi?

ANNE: Parce que je ___ (être) très fatiguée. Je ___ (être) vraiment crevée.

JEAN: On ___ (pouvoir) aller dîner au restaurant alors.

ANNE: Je ne ___ (vouloir) pas y aller ce soir.

JEAN: Si tu ne ___ (vouloir) pas, je ne ___ (vouloir) pas.

ANNE: J'___ (avoir) une idée. Tu ___ (pouvoir) aller faire les courses et tu ___
      (pouvoir) faire la cuisine. C'est une bonne idée, n'est-ce pas?

JEAN: Euh… D'accord. Je ___ (vouloir) bien. Qu'est-ce que tu ___ (vouloir)
      manger alors?

Le supermarché Printania à Dakar, au Sénégal

Exercice B
    All answers begin with **Si vous voulez, vous pouvez…**
1. … travailler à mi-temps.
2. … gagner de l'argent.
3. … avoir des succès.
4. … être riches.

Exercice C
1. veulent
2. peuvent
3. veulent
4. peuvent
5. peux

Exercice D
veux, peux
peux
suis, suis
peut
veux
veux, veux
ai, peux, peux
veux, veux

RETEACHING
    Have students write five things they want to do and five things they can do.

INDEPENDENT PRACTICE

    Assign any of the following:
1. Exercises, pages 164–165
2. Workbook, *Structure: G–H,* page 55
3. Communication Activities Masters, *Structure: C,* page 31
4. Computer Software, *Structure*
5. CD-ROM, Disc 2, pages 164–165

DID YOU KNOW?

    The name of this supermarket in Dakar is Printania. The word for "spring" in French is *le printemps,* and Printania comes from the same root. Ask why a supermarket might have a name evoking the springtime. (Spring is the season when fruits and vegetables ripen and start coming to market.)

CONVERSATION

🎧 ▶️◀️

Bell Ringer Review

Write the following on the board or use BRR Blackline Master 6-6: Write five things that might be said between a waiter/ waitress and his/her customers in a café.

PRESENTATION *(page 166)*

A. Tell students they will hear a conversation between a fruit vendor and Mme Garnier. Have them close their books and watch the Conversation Video or listen as you read the conversation to them or play it on Cassette 4B/CD-4.

B. Have them open their books and follow along as you read it or play Cassette 4B/CD-4 again.

C. Set up a small fruit stand in front of the classroom and have students act out the conversation.

D. Have them make up their own conversation based on the dialogue. They can change *des oranges* to anything they wish to buy.

Note In the CD-ROM version, 💿 students can play the role of either one of the characters and record the conversation.

ANSWERS

Exercice A

1. c	4. c
2. a	5. c
3. b	

CONVERSATION

Scènes de la vie *Chez la marchande de fruits*

LA MARCHANDE: Bonjour, Madame. Comment allez-vous aujourd'hui?
MADAME GARNIER: Très bien, merci. Et vous?
LA MARCHANDE: Très bien. Et qu'est-ce que Madame désire aujourd'hui?
MADAME GARNIER: Je voudrais des oranges. C'est combien, les oranges?
LA MARCHANDE: Les oranges d'Espagne? Elles sont exquises. Dix francs le kilo.
MADAME GARNIER: Une livre, s'il vous plaît.
LA MARCHANDE: Et avec ça?
MADAME GARNIER: Rien d'autre, merci.
LA MARCHANDE: Bien, Madame.
MADAME GARNIER: Ça fait combien?
LA MARCHANDE: Cinq francs. Merci, Madame. Et au revoir!

A **Les courses.** Choisissez la bonne réponse. *(Choose the correct answer.)*

1. Madame Garnier est ___.
   a. à la boucherie   b. à la crémerie   c. chez la marchande de fruits

2. Madame Garnier veut ___.
   a. des oranges   b. des légumes   c. une baguette

3. Les oranges sont ___.
   a. des légumes   b. d'Espagne
   c. dix francs la boîte

4. Madame Garnier achète ___ d'oranges.
   a. un kilo   b. un paquet
   c. cinq cents grammes

5. Les oranges sont ___.
   a. de France   b. la boîte
   c. dix francs le kilo

INDEPENDENT PRACTICE

Assign any of the following:
1. Exercise and activities, pages 166–167
2. CD-ROM, Disc 2, pages 166–167

Prononciation *Les sons /œ̸/ et /œ/*

Listen to the difference in the vowel sounds in *peut* and *peuvent*. The sound /œ̸/ in *peut* is a closed vowel sound and the sound /œ/ in *peuvent* is an open vowel sound. Repeat the following words with the sound /œ̸/.

> il peut   il veut   des œufs   deux

Repeat the following words with the sound /œ/.

> ils peuvent   ils veulent   un œuf   leur sœur   du beurre

Now repeat the following pairs of words. Be sure to distinguish between the two vowel sounds.

> il peut/ils peuvent   il veut/ils veulent

Now repeat the following sentences.

> Elle veut faire les courses, mais ils ne veulent pas.
> Elle veut du beurre et des œufs.
> Leur sœur est sérieuse.

un œuf

des œufs

Activités de communication orale

A **Je veux…**   Tell a classmate a few things you want to do and find out if he or she wants to do them too.

B **Je veux… mais je ne peux pas.**   Tell a classmate several things you want to do but can't do. He or she wants to know why you can't do these activities. Reverse roles.

> Élève 1: Je veux aller au cinéma ce soir, mais je ne peux pas.
> Élève 2: Pourquoi pas?
> Élève 1: J'ai un examen demain.

C **On est moderne chez toi?**   Divide into small groups and choose a leader. The leader asks the others the following questions and reports to the class.

1. Qui fait les courses dans ta famille?
2. Qui fait la cuisine généralement?
3. Qui fait la cuisine quand il y a des invités (*guests*)?

D **Moi, je fais la cuisine.**   You've invited a friend over for a birthday dinner. Make up the menu. Tell your friend what you're going to prepare. Find out whether he or she likes your menu.

> Élève 1: Je vais préparer une salade et du poulet.
> Élève 2: Très bien! J'adore la salade et le poulet.
>   (Euh… je déteste la salade et le poulet.)

CHAPITRE 6   **167**

DID YOU KNOW?

Students have already learned about the Parisian market in the rue Cler. There is another lively market in the rue Mouffetard, behind the Panthéon in the 5th *arrondissement* (see Photo 4, p. 170).

Most small towns have a market several days a week in the main square. Many foods are sold from stands, but there are also small trucks that are transformed into food shops.

CROSS-CULTURAL COMPARISON

1. It is considered common courtesy to chat a little before getting down to business at a French market. Politeness is the rule.
2. Each product's place of origin is often indicated. In this conversation the oranges are from Spain.

Prononciation

PRESENTATION (*page 167*)

A. Model the key words *un œuf/des œufs* and have students repeat chorally.
B. Now model the other words and phrases in similar fashion.
C. You may wish to give students the following *dictée:*
   Leur sœur veut du beurre et des œufs. Il faut aussi du bœuf et un peu de pain.
D. For additional pronunciation practice you may wish to use the following resources:
   1. Cassette 4B/CD-4: *Prononciation*
   2. Student Tape Manual, Teacher's Edition, *Activités F–H*, pages 72–73.
   3. Pronunciation Transparency P-6

Activités de communication orale

ANSWERS

Activités A, B, C, and D
Answers will vary.

HISTORY CONNECTION

For 800 years Les Halles was a huge wholesale food market that supplied all of Paris. From the 19th century on, it was housed in a glass and cast-iron building with a zinc roof, built by Baltard. In 1969 the building was torn down and Les Halles was moved to Rungis, in the suburbs. In 1979 the Forum des Halles, a modern pedestrian mall, opened where Les Halles once stood.

READING STRATEGIES
(page 168)

Pre-reading

Ask students who does the food shopping in their family. How many times a week? Where? Do they always go to the same store? Why? Do they chat with people who work in the stores?

Reading

A. Read the *Lecture* aloud, using as much expression as possible. Have students repeat each sentence after you.

B. Call on volunteers to read two to three sentences aloud at a time. Ask the other students content questions about the sentences read. For the first three sentences, some possible questions are: *Où sont les supermarchés en France, normalement? Les Français vont toujours au supermarché pour faire leurs courses?*

Post-reading

Assign the *Lecture* and the exercises on pages 168–169 as homework.

Note Students may listen to a recorded version of the *Lecture* on the CD-ROM.

Étude de mots

ANSWERS

Exercice A

1. d	3. f	5. c
2. a	4. b	6. e

LES COURSES EN FRANCE

Il y a des supermarchés en France? Bien sûr qu'il y a des supermarchés, surtout en dehors des[1] villes. Mais les Français ne vont pas toujours au supermarché pour faire leurs courses. Beaucoup de Français font leurs courses tous les jours—le lundi, le mardi, le mercredi, etc.—dans de petits magasins. En France, on n'achète pas tout[2] dans le même magasin. On achète de la viande à la boucherie et du pain à la boulangerie. On veut des boîtes de conserve, du détergent ou de l'eau minérale? On peut aller à l'épicerie du coin.

Les Français préfèrent aller d'un petit magasin à l'autre. Pourquoi? Premièrement, parce que la qualité est presque[3] toujours excellente dans les petits magasins. Et deuxièmement, les Français aiment bavarder un peu[4] avec le marchand ou la marchande. Ils trouvent ça sympathique.

[1] surtout en dehors des   *especially outside of*
[2] tout   *everything*
[3] presque   *almost*
[4] un peu   *a little*

Étude de mots

A   **Le contraire.**   Trouvez le contraire. *(Find the opposite.)*

1. en dehors de la ville	a. un jour par semaine
2. tous les jours	b. différent
3. petit	c. le supermarché
4. même	d. en ville
5. l'épicerie du coin	e. beaucoup
6. un peu	f. grand

CRITICAL THINKING ACTIVITY

(Thinking skill: drawing conclusions)
Write on the board or a transparency:

1. **La mère de Luc est très fatiguée. Il y a quelque chose qu'elle ne peut pas faire ce soir. C'est quoi? Luc veut aider sa mère. Qu'est-ce qu'il peut faire?**

2. **Expliquez pourquoi aux États-Unis les supermarchés sont plus populaires que les petits magasins spécialisés.**

LEARNING FROM PHOTOS

1. (top photo) Ask the following questions: *Dans quel magasin est Mme Bernier? Elle achète du jambon ou du saucisson? Qu'est-ce qu'on achète à la boucherie?*

2. (bottom photo) Ask the following questions: *C'est du fromage français? Quelle sorte de fromage?* (le Bleu d'Auvergne— which is not the same as le Roquefort, our "blue cheese.") *C'est combien le kilo?*

Compréhension

B **En France.** Répondez. *(Answer.)*

1. Où sont la plupart des supermarchés en France?
2. Les Français vont toujours au supermarché pour faire leurs courses?
3. Où est-ce qu'on achète du pain?
4. Où est-ce qu'on achète de la viande?
5. Qu'est-ce qu'on achète à l'épicerie?
6. Comment est la qualité des produits dans les petits magasins?
7. On peut bavarder avec qui?

C **Un peu de culture.** En France ou aux États-Unis? *(In France or in the U.S.?)*

1. On fait presque toujours les courses au supermarché.
2. On fait les courses tous les jours.
3. On fait les courses une ou deux fois par semaine, pas tous les jours.
4. Il y a des supermarchés surtout en dehors des villes.

DÉCOUVERTE CULTURELLE

*A*ujourd'hui les marchands français donnent des sacs à leurs clients. Les sacs sont en plastique ou en papier. Mais beaucoup de gens ont leur propre[1] sac ou filet pour leurs achats[2]. Tu as un filet? Tu vas au supermarché avec ton propre sac?

Quand les Français vont au marché, ils ne parlent pas de *pounds*. En France on utilise le système métrique. On parle de «kilos» dans le système métrique. Un kilo (un kilogramme) est l'équivalent de 2,2 *pounds*. Dans un kilo il y a mille grammes. Un demi-kilo (1/2 kg) est une livre. Une livre fait cinq cents grammes.

Le pain français est très célèbre. Tout le monde adore une baguette bien croustillante[3] avec son odeur délicieuse. Les Français mangent du pain à tous les repas[4]. Dans chaque[5] quartier il y a une

ou deux boulangeries où on achète du pain tous les matins. En France, il y a beaucoup de variétés de pain. Dans certaines boulangeries, faire du pain, c'est un art.

[1] propre *own*
[2] achats *purchases*
[3] croustillante *crusty*
[4] tous les repas *every meal*
[5] chaque *each*

DID YOU KNOW?

The French buy bread every day at the bakery. Bread is eaten many different ways. For breakfast, slices spread with butter are dunked in coffee, tea, or hot chocolate. At lunch and dinner, it is not buttered. A piece of bread is used to push food onto the fork or to mop up gravy. The French don't cut bread—they tear a piece off and place it by the plate on the tablecloth.

INDEPENDENT PRACTICE

Assign any of the following:
1. *Étude de mots* and *Compréhension* exercises, pages 168–169
2. Workbook, *Un Peu Plus*, pages 56–58
3. CD-ROM, Disc 2, pages 168–169

Compréhension

Exercice B

1. La plupart des supermarchés en France sont en dehors des villes.
2. Non, les Français ne vont pas toujours au supermarché pour faire leurs courses.
3. On achète du pain à la boulangerie-pâtisserie.
4. On achète de la viande à la boucherie.
5. On achète des boîtes de conserve, du détergent et de l'eau minérale à l'épicerie.
6. La qualité des produits est presque toujours excellente dans les petits magasins.
7. On peut bavarder avec le marchand ou la marchande.

Exercice C

1. aux États-Unis
2. en France
3. aux États-Unis
4. en France

OPTIONAL MATERIAL

Découverte culturelle

PRESENTATION *(page 169)*

A. Before reading the selection, have students discuss the following questions in groups:
   1. Americans use a lot of paper and plastic bags. What are some of the advantages and disadvantages of this practice?
   2. Does your family ever buy bread at a bakery, rather than in a package? Which kind do you prefer? Why?
B. Have students read the selection silently. Which topic does it contain that they had not previously discussed?
C. Ask questions in French about the selection, having students answer in French. If necessary, they can read phrases or sentences directly from the text to answer.

Note Students may listen to a recorded version of the *Découverte culturelle* on the CD-ROM.

RÉALITÉS

Bell Ringer Review

Write the following on the board or use BRR Blackline Master 6-8: Write down some food items that are served frequently at your home.

OPTIONAL MATERIAL

PRESENTATION *(pages 170–171)*

The main purpose of this section is to allow students to enjoy the photographs and develop an appreciation of French culture.

Note In the CD-ROM version, students can listen to the recorded captions and discover a hidden video behind one of the photos.

Pre-reading

Remind students where Nice and the *Côte d'Azur* are located by referring them to the map of France on page 504 (Part A, page 237), or you may use the France Map Transparency.

Reading

A. Have some of your better students read the captions on page 170.
B. Have pairs of students select one of the places in the photographs on pages 170–171 and write a short dialogue that would typically take place between a customer and a merchant there.

Post-reading

Share with students any photographs or slides of French shops and shopkeepers you may have in your own collection, from library books or from magazines.

RÉALITÉS

Voici le marché d'Aix-en-Provence **1**. C'est un marché typique avec beaucoup d'étals.

C'est une boulangerie-pâtisserie à Paris **2**. Les petits magasins en France sont très jolis. Ils ressemblent souvent à un tableau.

Voici un supermarché en France **3**. Il est en dehors de Nice sur la Côte d'Azur. Il y a beaucoup de supermarchés dans les centres commerciaux en France. Au Carrefour, on peut acheter des provisions et aussi toutes sortes d'autres produits pour la maison et le jardin.

Voici une poissonnerie dans la rue Mouffetard à Paris **4**. C'est une rue très animée où il y a un marché et beaucoup de petits magasins.

DID YOU KNOW?

Rue Mouffetard is one of Paris's original streets. It is a winding road lined with old houses. At the lower end of the street there are many small shops. The stores have picturesque, painted signs that date from long ago. Most of the shops open onto the street. There is always a great deal of activity on the street except on Mondays, when the shops are closed. At the upper end of the street there is an open-air market. Ernest Hemingway lived for a while above a cabaret in the neighborhood.

4

171

GEOGRAPHY CONNECTION

The following are foods that are produced in various areas of France.

la Bretagne:
les primeurs
  (early vegetables)
les artichauts
les petits pois
les carottes

la Normandie:
les pommes
la crème
le beurre
le fromage
  (le camembert)

le Périgord:
le pâté
les truffes

la Provence:
les fruits
les melons
les olives
les herbes

Bordeaux, la Bourgogne
l'Alsace:
le vin

THE FRANCOPHONE WORLD

The section on food and markets in the francophone world (see pages 220–223) contains information on various products found in these countries. Ask students to scan the material, then have them list several French-speaking countries and the products associated with them.

PAIRED ACTIVITY

Have students work in pairs to make up questions for each photo in the *Réalités* section. Then have several pairs of students read their questions aloud for other students to answer.

ADDITIONAL PRACTICE

1. Student Tape Manual, Teacher's Edition, *Deuxième Partie,* pages 74–76
2. Situation Cards, Chapter 6

RECYCLING

These *Activités de communication orale* and *écrite* allow students to apply the vocabulary and grammar of the chapter to open-ended, real-life situations and to re-use the vocabulary and structures from earlier chapters.

INFORMAL ASSESSMENT

Oral Activity A may be used to evaluate speaking. See page 34 of this Teacher's Wraparound Edition for evaluation criteria.

Activité de communication orale

PRESENTATION *(page 172)*

Activité A

In the CD-ROM version of this activity, students can interact with an on-screen native speaker. Activity A can also be done in groups in order to practice the *nous* and *vous* forms. Two students act as hosts and answer the questions of two others who act as the French exchange student guests.

ANSWERS

Answers will vary, but may include: C'est vendredi. C'est à huit heures. J'habite (au) 9, rue du Moulin. Je vais préparer un poulet avec des pommes de terre et des haricots verts. C'est très gentil à vous, mais ce n'est pas nécessaire. Ma mère adore faire le dessert.

Activité de communication écrite

ANSWERS

Activité A

Answers will vary.

172

Activité de communication orale

A **Une invitation.** You've invited the French exchange student from your school to dinner. He has a few questions for you about the plans.

1. C'est quel jour le dîner?
2. C'est à quelle heure?
3. Quelle est ton adresse?
4. Qu'est-ce que tu vas préparer?
5. Je peux faire un dessert si tu veux.

Activité de communication écrite

A **Un déjeuner français.** You're living in France and want to get some food for lunch but you don't have time.

1. Leave a note for your roommate and ask if he or she can do the shopping for you.
2. Make a list of at least six items.
3. Tell your roommate where each item can be bought.
4. Invite your roommate to have lunch with you.
5. Tell him or her what time you're going to eat.

De belles fraises au marché en plein air

172   CHAPITRE 6

ADDITIONAL PRACTICE

As was suggested in the Chapter Projects, have students each make up a shopping list of all the foods they like that they might buy at a market in France. Then have them make up menus for one or two days using these foods.

LEARNING FROM PHOTOS

Have students note the beautiful ripe strawberries in the bottom photo. Ask them how much the strawberries are. (They're sold by weight in France, not volume, as they often are in the U.S.) These berries come from Carpentras, a town in the south of France, not far from Avignon. Much of France's produce comes from the South owing to the climate.

Réintroduction et recombinaison

A La rue Cler. Complétez avec la forme convenable de à. *(Complete with the correct form of à.)*

1. Madame Dion va ___ marché de la rue Cler.
2. Elle ne va pas ___ supermarché.
3. Madame Dion va ___ boulangerie pour acheter du pain et elle va ___ épicerie pour acheter des boîtes de conserve.
4. ___ marché Madame Dion aime parler ___ marchands.
5. Aujourd'hui les marchands donnent des sacs ___ clients.

B Des différences. Complétez. *(Complete.)*

Éric a $\underset{1}{\underline{\quad}}$ sœurs mais il n'a pas $\underset{2}{\underline{\quad}}$ frères. Les sœurs d'Éric font $\underset{3}{\underline{\quad}}$ études universitaires à l'Université de Grenoble. Catherine fait $\underset{4}{\underline{\quad}}$ anglais mais Michèle ne fait pas $\underset{5}{\underline{\quad}}$ anglais. Elle fait $\underset{6}{\underline{\quad}}$ espagnol. Catherine fait toujours $\underset{7}{\underline{\quad}}$ gymnastique mais Michèle ne fait pas $\underset{8}{\underline{\quad}}$ gymnastique. Elle n'aime pas du tout $\underset{9}{\underline{\quad}}$ gymnastique.

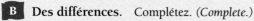

Vocabulaire

NOMS

le marché
le supermarché
l'épicerie
le/la marchand(e)
chez le marchand de
   fruits et légumes
la boucherie
la charcuterie
la boulangerie-pâtisserie
la crémerie
la poissonnerie
la caisse

le fruit
la banane
l'orange (f.)
la pomme
la tomate

le légume
la carotte
la laitue
l'oignon (m.)

la pomme de terre
les haricots (m.) verts

la viande
le bœuf
le poulet
le saucisson

le poisson
le crabe
la crevette

le lait
la crème
le beurre
le yaourt
le pain
la baguette
le croissant
le gâteau
la tarte

les légumes (m.) surgelés
l'eau (f.) minérale
la moutarde
l'œuf (m.)

le paquet
le filet
le sac
la boîte
la bouteille
la douzaine
le pot
le gramme
le kilo
le litre
la livre

VERBES

acheter
faire
payer
pouvoir
vouloir

**AUTRES MOTS
ET EXPRESSIONS**

faire attention
faire des études
faire la cuisine
faire les courses
faire un pique-nique
C'est tout.
C'est combien?
Rien d'autre.
Avec ça?
Ça fait combien?

NOMBRES

cent un à mille
   (101–1.000)

CHAPITRE 6   173

Réintroduction et recombinaison

PRESENTATION *(page 173)*

Exercices A and B

Exercise A focuses on contractions of *à* with the definite articles, while Exercise B focuses on the partitive.

ANSWERS

Exercice A

1. au
2. au
3. à la, à l'
4. Au, aux
5. aux

Exercice B

1. des
2. de
3. des
4. de l'
5. d'
6. de l'
7. de la
8. de
9. la

ASSESSMENT RESOURCES

1. Chapter Quizzes
2. Testing Program
3. Situation Cards
4. Communication
   Transparency C-6
5. Computer Software:
   Practice/Test Generator

VIDEO PROGRAM

INTRODUCTION   (20:44)

ON FAIT LES COURSES (21:12)

FOR THE YOUNGER STUDENT

Students form teams of three. Each team pretends it is opening one of the stores mentioned in *Mots 1* and 2. They should name their stores and then create window posters to advertise the things for sale in their store. Groups exchange "visits" to each other's stores, buying and selling items, asking about prices, explaining what is available, and summing up purchase totals.

STUDENT PORTFOLIO

Written assignments for the students' portfolio may include their personal grocery lists, menus they have created, and the *Mon Autobiographie* section of the Workbook, page 59.

Note Students may create and save both oral and written work using the Electronic Portfolio feature on the CD-ROM.

CHAPTER OVERVIEW

In this chapter students will learn vocabulary and structures necessary for situations that arise when traveling by air. Some basic terms are introduced for use at the point of departure, on board a flight, and at the point of arrival. (The topic of air travel will be expanded upon in *À bord*, Chapter 7.)

Students will also learn to use several important *-ir* verbs, to ask and answer questions using the forms of *quel* and *tout*, to use all forms of nouns and adjectives ending in *-al*, and to use irregular verbs such as *partir*.

The cultural focus of Chapter 7 is on airports and airlines in France.

CHAPTER OBJECTIVES

By the end of this chapter students will know:

1. vocabulary associated with airport check-in and flight boarding
2. vocabulary associated with services and procedures on board a flight and at the destination
3. present indicative forms of regular *-ir* verbs
4. present indicative forms of the verbs *partir, sortir,* and *dormir*
5. all forms of the adjectives *quel* and *tout*
6. all forms of adjectives and nouns ending in *-al*

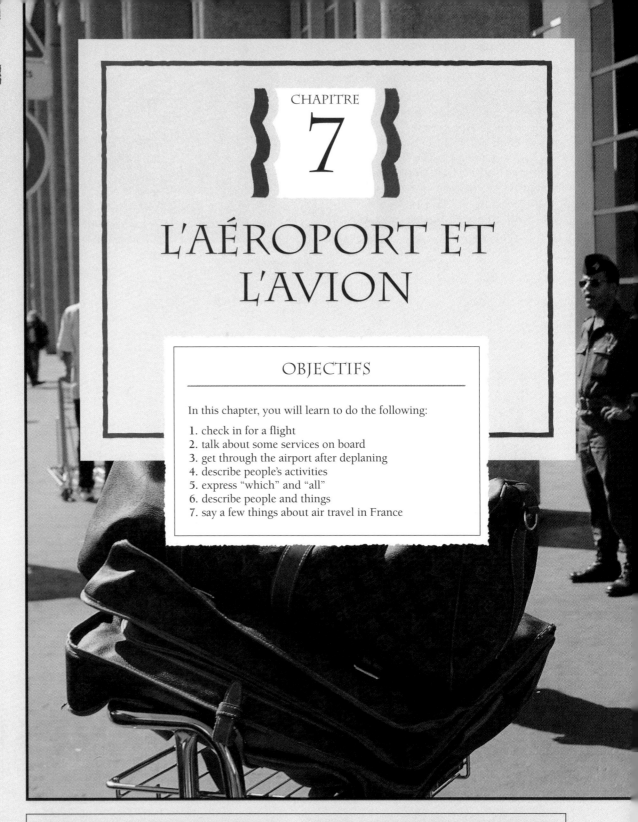

CHAPITRE

7

L'AÉROPORT ET L'AVION

OBJECTIFS

In this chapter, you will learn to do the following:

1. check in for a flight
2. talk about some services on board
3. get through the airport after deplaning
4. describe people's activities
5. express "which" and "all"
6. describe people and things
7. say a few things about air travel in France

CHAPTER PROJECTS

(optional)

Have students create their own French passports. Have them fill out forms in French with their name, address, nationality, birth date, metric weight, etc. Students could use real photos of themselves or draw a picture. Covers can be made of construction paper. The passports can be used for activities in this chapter as well as in Chapters 8 and 17.

COMMUNITIES

If your students have had little opportunity to travel, plan a field trip to a nearby airport. This is a wonderful enrichment experience for some students. As you tour the airport, have them use as much of their French vocabulary as possible.

175

Pacing

This chapter requires eight to ten class sessions. Pacing will vary according to class length and the age and aptitude of the students.

Note The Lesson Plans offer guidelines for 45- and 55-minute classes and **Block Scheduling.**

Exercices vs. *Activités*

All exercises (which provide guided practice) are coded in blue. All communicative activities are coded in red.

INTERNET ACTIVITIES

(optional)

These activities, student worksheets, and related teacher information are in the *Bienvenue* Internet Activities Booklet and on the Glencoe Foreign Language Home Page at **http://www.glencoe.com/secondary/fl**

LEARNING FROM PHOTOS

After presenting *Mots 1*, you may wish to ask the following questions about the photo: *Cet homme est à l'aéroport? Il a des bagages? Ses bagages sont derrière lui ou devant lui? Il fait enregistrer ses bagages ou il cherche une compagnie aérienne? Quelle compagnie aérienne est-ce qu'il cherche?*

Vocabulary Teaching Resources

1. Vocabulary Transparencies 7.1 (A & B)
2. Audio Cassette 5A/CD-5
3. Student Tape Manual, Teacher's Edition, *Mots 1: A–C*, pages 77–79
4. Workbook, *Mots 1: A–C*, page 60
5. Communication Activities Masters, *Mots 1: A*, pages 32–33
6. Chapter Quizzes, *Mots 1: Quiz 1*, page 37
7. CD-ROM, Disc 2, *Mots 1:* pages 176–179

Bell Ringer Review

Write the following on the board or use BRR Blackline Master 7-1: Complete the following paragraph with words that fit logically.

**Aujourd'hui je ___ un voyage.
Mes amis ___ le voyage aussi.
Nous ___ le voyage en avion.
Mais notre prof ___ un voyage en train.**

PRESENTATION *(pages 176–177)*

A. Write the words *près* and *loin* on the board. Working in groups, students tell one another where they'd like to go and categorize the destinations under *près* and *loin*. They then decide whether trips to these destinations would be taken by train or by plane.

B. Have students close their books. Using Vocabulary Transparencies 7.1 (A & B), introduce the vocabulary. Have students repeat after you or Cassette 5A/CD-5 as you point to the illustrations.

VOCABULAIRE

MOTS 1

Marc fait un voyage à Pointe-à-Pitre. Avant le voyage il fait ses valises.

un agent

un écran

À L'AÉROPORT

le comptoir de la compagnie aérienne

un passeport

des bagages (m.)

des valises (f.)

des bagages à main

une carte d'embarquement

un billet

vérifier le billet

faire enregistrer les bagages

Marc choisit sa place.
Il choisit une place côté couloir.
Il choisit le siège 16C.

TOTAL PHYSICAL RESPONSE

(following the Vocabulary presentation)

TPR 1
___, levez-vous, s'il vous plaît.
Prenez votre livre de français.
Venez ici avec votre livre.
Ouvrez le livre à la page 194.
Regardez la carte d'embarquement.
Allez au tableau noir. Prenez votre livre.

Regardez la carte d'embarquement et sur le tableau noir écrivez le nom de la compagnie aérienne.
Écrivez le numéro du vol.
Écrivez le numéro du siège du passager.
Écrivez la lettre du siège du passager.
Écrivez le numéro de la porte d'embarquement.
Merci, ___. Vous avez très bien fait.
Retournez à votre place et asseyez-vous.

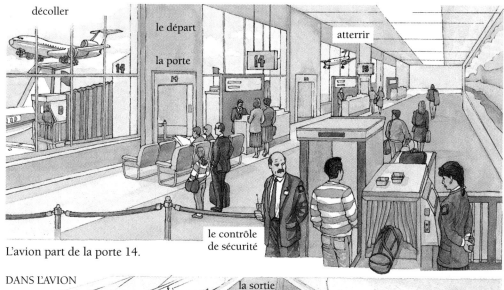

décoller

le départ

la porte

atterrir

le contrôle
de sécurité

L'avion part de la porte 14.

DANS L'AVION
la cabine

la sortie

(une zone) non-fumeurs

côté fenêtre

un passager

une passagère

un siège

côté couloir

un vol à destination de Paris	*un vol qui va à Paris*
un vol en provenance de Lyon	*un vol qui arrive de Lyon*
un vol intérieur	*un vol dans le même pays*
un vol international	*un vol d'un pays à un autre*

CHAPITRE 7   **177**

Teaching Tip Once students are familiar with the vocabulary, have a student play the teacher and ask questions of classmates.

Vocabulary Expansion

You may wish to give students the following additional information regarding the *Mots 1* vocabulary.

1. *Une carte d'embarquement* On a boarding pass, the words *carte d'accès à bord* often appear. (See p. 194, Photo 2, left-hand side of boarding pass.) *Une carte d'embarquement* is the term most frequently used in conversation, however.

2. The technical term for the window of a plane is *le hublot*. When you request a window seat, however, you ask for *une place côté fenêtre*.

3. The large arrival-and-departure board is *un tableau*. The TV monitor in many airports today is *un écran*.

THE FRANCOPHONE WORLD

Pointe-à-Pitre est la capitale de la Guadeloupe. Située dans la mer des Caraïbes, la Guadeloupe est une des Antilles françaises. La Guadeloupe est formée de deux îles: Basse-Terre et Grande-Terre.

TPR 2

___, vous allez à l'aéroport. On va imaginer un peu. C'est l'aéroport, d'accord?

Bon! Allez à l'aéroport.

Vous avez des valises. Portez les valises.

Allez au comptoir de la compagnie aérienne.

Mettez vos bagages au comptoir.

Et maintenant, cherchez votre billet.

Zut! Vous ne trouvez pas votre billet.

Cherchez encore.

Ah, bon. Voilà le billet.

Donnez le billet à l'agent au comptoir.

L'agent veut voir votre passeport. Cherchez le passeport.

Donnez le passeport à l'agent.

L'agent vous donne votre carte d'embarquement.

Prenez la carte.

Vérifiez la carte d'embarquement.

Ah, voilà le contrôle de sécurité.

Merci, ___. Et bon voyage!

Et maintenant, retournez à votre place.

PRESENTATION (pages 178–179)

The exercises on pages 178–179 use the new verbs only in the third-person singular so students do not have to change the forms. Students will learn the other forms in the *Structure* section.

Exercice A: Listening

To focus on listening, have students keep their books closed as you ask the questions. Then have students open their books and retell the story in their own words.

Teaching Tip Have students answer according to their abilities. For example, in Exercise B, first item, less able students may merely respond with *de grandes valises,* while more able students may say or write a complete answer.

ANSWERS

Exercice A

1. Oui, le passager est au comptoir de la compagnie aérienne.
2. Oui, le comptoir est à l'aéroport.
3. C'est (le comptoir d') Air Inter.
4. Oui, le passager a son billet et son passeport.
5. Oui, l'agent de la compagnie aérienne vérifie le billet.
6. Non, il ne vérifie pas le passeport.
7. Oui, le passager choisit sa place dans l'avion.
8. Il choisit le siège 16C dans la zone non-fumeurs.
9. Il veut une place côté couloir.
10. Oui, l'agent donne une carte d'embarquement au passager.
11. Oui, le passager fait enregistrer ses bagages.
12. Oui, il passe par le contrôle de sécurité.

Exercice B

1. Le passager a de grandes valises.
2. Il regarde son passeport.
3. L'agent vérifie son billet.

Exercices

A **Un voyage à Pointe-à-Pitre.** Répondez. (*Answer.*)

1. Le passager est au comptoir de la compagnie aérienne?
2. Le comptoir est à l'aéroport?
3. C'est le comptoir de quelle compagnie aérienne?
4. Le passager a son billet et son passeport?
5. L'agent de la compagnie aérienne vérifie le billet?
6. Il vérifie le passeport aussi?
7. Le passager choisit sa place dans l'avion?
8. Il choisit quel siège? Son siège est dans la zone non-fumeurs?
9. Il veut une place côté couloir ou côté fenêtre?
10. L'agent donne une carte d'embarquement au passager?
11. Le passager fait enregistrer ses bagages?
12. Il passe par le contrôle de sécurité?

B **À l'aéroport.** Répondez d'après les dessins. (*Answer according to the illustrations.*)

1. 2. 3.

4. 5. 6.

1. Le passager a de grandes valises ou des bagages à main?
2. Il regarde son billet ou son passeport?
3. L'agent vérifie son billet ou sa carte d'embarquement?
4. Le passager est au comptoir de la compagnie aérienne ou il passe par le contrôle de sécurité?
5. Le passager va au contrôle de sécurité ou à la porte d'embarquement?
6. L'avion décolle ou atterrit?

178 CHAPITRE 7

PAIRED ACTIVITY

Have students work in pairs. One student tells the other a flight number that he/she has made up. The second student asks if the flight is leaving or arriving, then tries to guess where the flight is leaving from or going to.

ADDITIONAL PRACTICE

Student Tape Manual, Teacher's Edition, *Activités B–C,* pages 78–79

C **Les arrivées et les départs.** Répondez d'après les écrans. (*Answer according to the arrival and departure screens.*)

ARRIVEES			AEROGARE TERMINAL **2**	
INFORMATIONS GENERALES				
HORS	**PROVENANCES**	**VOL**	**OBSERVATIONS**	**GARE**
0920	NAIROBI	MD 052	ARRIVE 1009	2A
0920	GENEVE	SR 722	ARRIVE 0952	2B
0925	ZURICH	AF 987	ARRIVE 0942	2B
0930	LON-HEATHROW	AF 807	ARRIVE 0939	2D
0940	LUGANO	LXAF 750	PREVU 1050	2B
0949	BERNE	LXAF 772	PREVU 1106	2C
0950	ROME	AF 639	ARRIVE 0945	2D
1000	MANCHESTER	AF 909	ARRIVE 0954	2D
1005	BRUXELLES	AF 1221	ARRIVE 1004	2B

DEPARTS			AEROGARE TERMINAL **2**	
INFORMATIONS GENERALES				
HORS	**DESTINATION**	**VOL**	**OBSERVATIONS**	**GARE**
0940	VENISE	AZ 297	TERMINE B33	2B
0955	OSLO	AF 1132	TERMINE	2B
1005	MILAN	AZ 345	TERMINE B33	2B
1010	ROME	AZ 283	TERMINE B33	2B
1015	VENISE	AF 670	TERMINE	2D
1020	PRAGUE	AF 2968	EMBARQT	2C
1020	BERNE	AFLX 972	B30	2D
1030	LON-HEATHROW	AF 810	EMBARQT D63	2D
1035	BRISTOL	BC 602	EMBARQT D69	2D

1. Le vol 987, c'est un vol de quelle compagnie aérienne?
2. Où va le vol 297?
3. Le vol 345 est à destination de quelle ville?
4. Le vol 810 part à quelle heure?
5. Le vol 772 est en provenance de quelle ville?
6. Il va arriver à quelle heure?
7. Quel vol est en provenance de Rome?

4. Le passager passe par le contrôle de sécurité.
5. Le passager va à la porte d'embarquement.
6. L'avion décolle.

Exercice C
1. C'est un vol d'Air France.
2. Le vol 297 va à Venise.
3. Le vol 345 est à destination de Milan.
4. Le vol 810 part à 10h30.
5. Le vol 772 est en provenance de Berne.
6. Il va arriver à 11h06.
7. Le vol 639 est en provenance de Rome.

INFORMAL ASSESSMENT
(*Mots 1*)

Check for understanding by making false statements about the items on Vocabulary Transparencies 7.1. Have students correct your statements. For example: *Avant le voyage, Marc fait ses devoirs.* (Non, il ne fait pas ses devoirs. Il fait ses valises.) *C'est un tableau noir.* (Ce n'est pas un tableau noir. C'est un écran.)

COGNATE RECOGNITION

Have students scan the *Mots 1* words again and then identify and pronounce each cognate.

COOPERATIVE LEARNING

Divide the class into three or four teams, depending on class size. Every member of each team contributes two airport activities. Each team writes down all its activities and then mimes them in front of the class. The other teams guess what they're doing.

INDEPENDENT PRACTICE

Assign any of the following:
1. Exercises, pages 178–179
2. Workbook, *Mots 1: A–C,* page 60
3. Communication Activities Masters, *Mots 1: A,* pages 32–33
4. CD-ROM, Disc 2, pages 176–179

VOCABULAIRE

MOTS 2

PENDANT LE VOL OU À BORD DE L'AVION

le personnel de bord

un steward   une hôtesse de l'air

On sert le dîner.

Un passager sort ses bagages du compartiment.

Un passager dort.

Une passagère remplit sa carte de débarquement.

180   CHAPITRE 7

Vocabulary Teaching Resources

1. Vocabulary Transparencies 7.2 (A & B)
2. Audio Cassette 5A/CD-5
3. Student Tape Manual, Teacher's Edition, *Mots 2: D–F,* pages 79–80
4. Workbook, *Mots 2: D–F,* page 61
5. Communication Activities Masters, *Mots 2: B,* page 33
6. Computer Software, *Vocabulaire*
7. Chapter Quizzes, *Mots 2:* Quiz 2, page 38
8. CD-ROM, Disc 2, *Mots 2:* pages 180–183

Bell Ringer Review

Write the following on the board or use BRR Blackline Master 7-2: Sketch the inside of an airplane as if the top had been cut off. Label as many parts of the plane as you can. Include people you would expect to see on a plane.

PRESENTATION (pages 180–181)

A. Briefly review the vocabulary from *Mots 1* by having volunteers point to illustrations on the *Mots 1* Vocabulary Transparencies as they ask other students *Qu'est-ce que c'est?* or *Qu'est-ce qu'il/elle fait?*
B. Have students look at their books as you introduce the *Mots 2* vocabulary. Model the words and phrases or play Cassette 5A/CD-5. Have students repeat first chorally and then individually.

180

TOTAL PHYSICAL RESPONSE

(following the Vocabulary presentation)

TPR
____, levez-vous, s'il vous plaît.
Vous êtes dans la petite salle d'attente de la porte d'embarquement numéro 6.
Trouvez une place et asseyez-vous dans la salle d'attente.

Vous entendez l'annonce du départ de votre avion. Levez-vous.
Cherchez votre carte d'embarquement.
Faites la queue devant la porte.
Maintenant, c'est votre tour. Voilà le steward. Montrez votre carte d'embarquement au steward.
Continuez, ____. Embarquez.
Voilà, vous êtes dans l'avion. Bienvenu(e) à bord.

(continued on next page)

L'ARRIVÉE À PARIS

On passe à l'immigration.

DOUANE

On récupère ses bagages.

On passe à la douane.

une aérogare en ville

un taxi

un autocar

C. As you present the new words, ask questions such as the following: *On sert le déjeuner ou le dîner à bord? Qu'est-ce qu'on sert? Le steward ou l'hôtesse de l'air sert le dîner? Qui sert le dîner? Qu'est-ce qu'il/elle sert?* These types of questions reinforce the meaning of the interrogative words. You may have students answer two ways: with isolated expressions or in complete sentences.

INFORMAL ASSESSMENT
(*Mots 2*)

Check for comprehension by miming activities and having students try to guess who you are and what you are doing by asking questions. For example: Teacher mimes the action of looking at a paper. Students ask: *Vous regardez quelque chose? Vous regardez un billet? un passeport? une carte d'embarquement? Vous êtes l'hôtesse de l'air?*

TPR (*continued*)

Montrez votre carte d'embarquement à l'hôtesse de l'air qui est à la porte d'entrée de l'avion.
Cherchez votre place. Vous avez le siège 5A.
Ah, vous avez une place côté fenêtre. Et la place côté couloir est prise. Le passager est assis. Pas de problème!
Mettez vos bagages à main dans le compartiment au-dessus de votre tête.
Indiquez au passager qui est assis dans le

siège 5C côté couloir que vous avez le siège 5A. Soyez poli(e).
Bon, et maintenant, prenez votre place dans l'avion. Asseyez-vous.
Merci, ——. Vous avez très bien fait. Et maintenant, retournez à votre place.

Exercices

Exercices

Exercices A 🎧 **and B**

The purpose of Exercises A and B is to increase the students' vocabulary and "word power" as much and as easily as possible, even if they only retain some of the words for recognition.

ANSWERS

Exercice A

1. c	4. f
2. a	5. b
3. e	6. d

Exercice B

1. c	5. f
2. a	6. e
3. h	7. b
4. g	8. d

Exercice C

1. Le vol de New York à Paris est un vol international.
2. Oui, on sert le dîner à bord.
3. Non, le steward ne sert pas le dîner.
4. Oui, l'hôtesse de l'air sert le dîner.
5. Oui, un passager dort pendant le vol.
6. Oui, le passager remplit une carte de débarquement.
7. Oui, à Paris, on passe à l'immigration.
8. Oui, on passe à la douane.

RETEACHING (Mots 2)

Show Vocabulary Transparencies 7.2 (A & B). Have students make up true/false statements about what they see. They may make up wildly illogical things that will make everyone laugh. (This is an excellent technique to check comprehension. If students laugh, you know they have understood.) Expand this activity and have the students correct the false statements.

Exercices

A **Un lexique aérien.** Trouvez le contraire. (*Find the opposite.*)

1. le steward	a. l'aéroport
2. l'aérogare en ville	b. international
3. l'embarquement	c. l'hôtesse de l'air
4. décoller	d. à destination de
5. intérieur	e. le débarquement
6. en provenance de	f. atterrir

B **Un autre lexique aérien.** Trouvez le nom qui correspond au verbe. (*Find the noun that corresponds to the verb.*)

1. arriver	a. le départ
2. partir	b. l'embarquement
3. atterrir	c. l'arrivée
4. décoller	d. le débarquement
5. servir	e. la sortie
6. sortir	f. le service
7. embarquer	g. le décollage
8. débarquer	h. l'atterrissage

C **À bord.** Répondez. (*Answer.*)

1. Le vol de New York à Paris est un vol intérieur ou un vol international?
2. On sert le dîner à bord?
3. Le steward sert le dîner?
4. L'hôtesse de l'air sert le dîner?
5. Un passager dort pendant le vol?
6. Avant l'arrivée ou l'atterrissage à Paris, le passager remplit une carte de débarquement?
7. À Paris, on passe à l'immigration?
8. On passe à la douane?

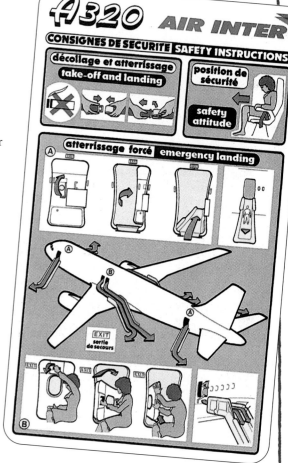

PAIRED ACTIVITY

Have students work in pairs to make a list of airport activities. They then decide under which of the following headings each activity belongs: *avant le départ, à bord de l'avion,* or *après l'arrivée.*

ADDITIONAL PRACTICE

Student Tape Manual, Teacher's Edition, *Activité F,* page 80

Activités de communication orale
Mots 1 et 2

A **Au comptoir d'Air Inter.** You're checking in for your flight at the Air Inter counter at Orly Airport. Answer the agent's questions.

1. Où allez-vous?
2. Vous avez combien de valises à enregistrer?
3. Vous avez combien de bagages à main?
4. Vous voulez une place fumeurs ou non-fumeurs?
5. Côté couloir ou côté fenêtre?

B **Des renseignements.** You're on a flight to Paris. Get some information from the flight attendant (your partner) about the following:

1. service on board, meals, etc.
2. arrival time in Paris
3. customs
4. getting from the airport to the city

C **Des arrivées et des départs.** Work with a classmate. Look at this arrival board at Charles-de-Gaulle Airport. Give as much information about the flights as you can, then ask each other questions about them.

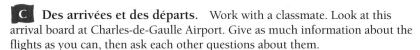

PRESENTATION *(page 183)*

Activité A

In the CD-ROM version of this activity, students can interact with an on-screen native speaker.

ANSWERS

Activité A

Answers will vary.

Activité B

Answers will vary, but may include the following:

1. **On sert le dîner à bord?**
2. **À quelle heure est-ce que nous arrivons à Paris?**
3. **On passe à la douane à Paris?**
4. **À l'aéroport il y a des taxis pour aller en ville? Il y a aussi un autocar pour aller en ville?**

Activité C

Answers will vary.

INDEPENDENT PRACTICE

Assign any of the following:
1. Exercises and activities, pages 182–183
2. Workbook, *Mots 2: D–F*, page 61
3. Communication Activities Masters, *Mots 2: B*, page 33
4. CD-ROM, Disc 2, pages 180–183

STRUCTURE

Structure Teaching Resources

1. Workbook, *Structure: A–I,* pages 62–64
2. Student Tape Manual, Teacher's Edition, *Structure: A–F,* pages 81–83
3. Audio Cassette 5A/CD-5
4. Communication Activities Masters, *Structure: A–E,* pages 34–37
5. Computer Software, *Structure*
6. Chapter Quizzes, *Structure:* Quizzes 3–6, pages 39–42
7. CD-ROM, Disc 2, pages 184–189

Les verbes en -ir au présent

PRESENTATION *(page 184)*

Note Most regular *-ir* verbs are not high-frequency words. Students will use them far less often than the *-er* verbs and the common irregular verbs. For this reason, do not spend a great deal of time on *-ir* verbs.

A. Write the forms of *choisir* and *finir* on the board. Now pronounce the plural forms.

B. Draw a line through the plural endings beginning with *ss* and pronounce what remains, i.e., the pronunciation for all the singular forms.

C. Point out the spelling of the singular endings: *-is, -is, -it.*

Exercices

ANSWERS

Exercice A

1. Madame Lauzier choisit Air France.
2. Elle choisit la classe économique.
3. Elle choisit une place dans la zone non-fumeurs.
4. Elle choisit un siège côté couloir.

184

STRUCTURE

Les verbes en -ir au présent   *Describing People's Activities*

1. Another group of regular verbs in French end in *-ir.* The two most commonly used verbs in this group are *choisir,* "to choose," and *finir,* "to finish." Study the following forms.

INFINITIVE	CHOISIR	FINIR	
STEM	chois-	fin-	ENDINGS
	je choisis	je finis	-is
	tu choisis	tu finis	-is
	il elle ⎱ choisit on ⎰	il elle ⎱ finit on ⎰	-it
	nous choisissons	nous finissons	-issons
	vous choisissez	vous finissez	-issez
	ils elles ⎱ choisissent	ils elles ⎱ finissent	-issent

Note that the final consonant sound of all singular forms is silent.

2. The following are some other common *-ir* verbs.

atterrir	to land	**réussir à**	to succeed, to pass (a test)
punir	to punish	**remplir**	to fill, to fill out (a form)
obéir à	to obey		

Exercices

A   Un vol à Paris.   Répondez d'après les indications. *(Answer according to the cues.)*

1. Madame Lauzier choisit quelle compagnie? (Air France)
2. Elle choisit quelle classe de service? (classe économique)
3. Elle choisit une place dans la zone fumeurs ou non-fumeurs? (non-fumeurs)
4. Elle choisit un siège côté couloir ou côté fenêtre? (côté couloir)
5. Son avion atterrit à quelle heure? (à huit heures du matin)
6. Il atterrit à quel aéroport? (Charles-de-Gaulle)
7. Qu'est-ce qu'elle remplit avant l'arrivée? (une carte de débarquement)

ADDITIONAL PRACTICE

Student Tape Manual, Teacher's Edition, *Activité A,* page 81

INDEPENDENT PRACTICE

Assign any of the following:
1. Exercises, pages 184–185
2. Workbook, *Structure: A–C,* pages 62–63
3. Communication Activities Masters, *Structure: A,* page 34
4. CD-ROM, Disc 2, pages 184–185

B **Au restaurant.** Donnez des réponses personnelles. (*Give your own answers.*)

1. Tu choisis un restaurant bon marché ou élégant?
2. Tu choisis le menu touristique ou le menu à la carte?
3. Tu choisis la viande ou le poisson?
4. Tu finis le dîner par un dessert ou un fromage?
5. Tu choisis un gâteau ou une glace?
6. Quand tu finis le dîner, tu laisses un pourboire pour le serveur?

C **Un bon dîner.** Mettez au pluriel d'après le modèle. (*Change to the plural according to the model.*)

> Je choisis un express et tu choisis un thé citron.
> *Nous choisissons un express et vous choisissez un thé citron.*

1. Je choisis un restaurant bon marché et tu choisis un restaurant gastronomique.
2. Je choisis le menu à prix fixe et tu choisis un dîner à la carte.
3. Je choisis un coca et tu choisis une bouteille d'eau minérale.
4. Je finis mon dîner par une tarte et tu finis ton dîner par des crêpes Suzette flambées.
5. Je finis mon dîner par un crème et tu finis ton dîner par un express.

D **Un autre vol.** Complétez avec «choisir» ou «remplir». (*Complete with* choisir *or* remplir.)

1. Les passagers ___ un vol direct?
2. Ils ___ un siège côté couloir ou côté fenêtre?
3. Ils ___ un siège dans la zone fumeurs ou non-fumeurs?
4. Ils ___ leur carte de débarquement à l'aéroport ou pendant le vol?

E **Qui obéit?** Complétez. (*Complete.*)

J'___ (obéir) toujours à mes parents et j'___ (obéir) toujours à mes profs. Les profs ___ (punir) les élèves qui n'___ (obéir) pas. Et vous, vous ___ (obéir) à vos parents? Vous ___ (obéir) à vos profs? Vous ___ (finir) toujours vos examens? Vous ___ (réussir) à tous les examens que vous passez?

EXECUTIVE FIRST | SUPER AFFAIRES

---

5. Son avion atterrit à huit heures du matin.
6. Il atterrit à l'aéroport Charles-de-Gaulle.
7. Elle remplit une carte de débarquement avant l'arrivée.

PRESENTATION *(page 185)*
Exercice C: **Paired Activity**

Have students do Exercise C in pairs. One student gives answers using the *nous* and *vous* forms. The other student supplies classmates' names instead and says: *Paul et Simone choisissent un restaurant bon marché et Christian et Virginie choisissent un restaurant gastronomique.*

ANSWERS
Exercice B

Answers will vary. You may wish to have students give longer answers. For example: **Je ne choisis pas un restaurant bon marché. Moi, je choisis toujours un restaurant élégant.**

Exercice C

1. **Nous choisissons un restaurant bon marché et vous choisissez un restaurant gastronomique.**
2. **Nous choisissons... vous choisissez...**
3. **Nous choisissons... vous choisissez...**
4. **Nous finissons... vous finissez...**
5. **Nous finissons... vous finissez...**

Exercice D

1. choisissent
2. choisissent
3. choisissent
4. remplissent

Exercice E

1. obéis
2. obéis
3. punissent
4. obéissent
5. obéissez
6. obéissez
7. finissez
8. réussissez

LEARNING FROM PHOTOS

You may wish to ask students questions such as these about the photo on page 185: *L'hôtesse de l'air sert un repas? Elle donne le repas à un passager ou à une passagère? La passagère est contente?*

Bell Ringer Review

Write the following on the board or use BRR Blackline Master 7-3: Complete the sentences with a logical word that agrees with the verb.

1. ___ atterrit à l'aéroport de Nice.
2. ___ obéissons à Maman.
3. ___ choisissent leurs places dans l'avion.
4. ___ finis la glace.

Les adjectifs quel et tout

PRESENTATION *(page 186)*

Lead students through steps 1–4 on page 186. Explain the spelling, pronunciation changes, and meanings of the adjectives *quel* and *tout*.

Exercices

ANSWERS

Exercice A

1. Moi? J'aime toutes les matières.
2. ... toutes les langues.
3. ... toutes les sciences.
4. ... tous les livres.
5. ... tous les disques.
6. ... tous (toutes) les profs.

Les adjectifs *quel* et *tout*   Expressing "Which" and "All"

1. You use the interrogative adjective *quel* + a noun when you want to ask "what?" or "which?". All forms of *quel* sound the same even though they are spelled differently.

	SINGULIER	PLURIEL
FÉMININ	Quelle compagnie?	Quelles compagnies?
MASCULIN	Quel vol?	Quels vols?

2. You make a liaison with the plural forms when they are followed by a vowel or silent *h*.

> Quelles‿amies?
> Quels‿hôtels?

3. You use the adjective *tout* with the definite article (*le, la, l'*) to express "the entire" or "the whole."

	SINGULIER
FÉMININ	Toute la classe regarde le tableau noir.
MASCULIN	Tout le livre est comique.

4. You use *tous* and *toutes* to express "all" or "every."

	PLURIEL
FÉMININ	Toutes les classes de M. Lapeyre sont amusantes.
MASCULIN	Tous les livres sont intéressants.

Exercices

A **Quel cours?** Répondez d'après le modèle. *(Answer according to the model.)*

> **Tu aimes quels cours?**
> *Moi? J'aime tous les cours.*

1. Tu aimes quelles matières?
2. Tu aimes quelles langues?
3. Tu aimes quelles sciences?
4. Tu aimes quels livres?
5. Tu aimes quels disques?
6. Tu aimes quels profs?

PAIRED ACTIVITY

Have students work in pairs and write the following nouns on a sheet of paper: *les classes, les matières, les cours, les professeurs, les voyages, les magazines, les sports, les croissants, les légumes.* One student makes up a question using *quel* with the noun.(For example, *Quels cours est-ce que tu préfères?*) The other student provides a response with *tout*.

B **Quel vol?** Complétez avec «quel». (*Complete with* quel.)

1. Tu fais un voyage? Ton vol est ___ jour?
2. Ton avion part à ___ heure?
3. Ton avion part de ___ porte?
4. Pendant le vol tu vas regarder ___ film?
5. Tu vas écouter ___ cassettes?
6. Tu aimes ___ magazines?

C **Toute la classe.** Complétez avec la forme convenable des adjectifs. (*Complete with the correct form of the adjectives.*)

1. ___ la classe passe ___ examen? (tout, quel)
2. ___ les élèves réussissent à l'examen. (tout)
3. ___ les élèves de ___ classe réussissent à ___ examen? (tout, quel, quel)
4. ___ cours sont difficiles? (quel)
5. ___ les cours de ___ professeur sont difficiles? (tout, quel)

D **Tous les vols pour quelle ville?** Complétez avec «tout». (*Complete with* tout.)

1. ___ les places sont occupées.
2. ___ l'avion est classe économique. Il n'y a pas de première classe.
3. ___ les cabines sont non-fumeurs.
4. Ce n'est pas vrai ça. ___ les vols internationaux ont une zone fumeurs.

Les noms et les adjectifs en -al   *Describing People and Things*

1. To form the plural of all feminine words ending in -ale you add -s to the singular.

une île tropicale	des îles tropicales
la ville principale	les villes principales
la capitale	les capitales

2. Note, however, that the plural form of most masculine words ending in -al is -aux.

un vol international	des vols internationaux
un parc national	des parcs nationaux
un animal	des animaux

3. Many adjectives that end in -al are cognates.

général	local	national	principal	spécial
international	municipal	original	social	tropical

4. The following are some common nouns that end in -al.

le terminal	le général	le journal (*newspaper*)

CHAPITRE 7   **187**

ADDITIONAL PRACTICE

Student Tape Manual, Teacher's Edition, *Activités C–E*, pages 82–83

INDEPENDENT PRACTICE

Assign any of the following:
1. Exercises, pages 186–187
2. Workbook, *Structure: D–E*, page 63
3. Communication Activities Masters, *Structure: B–C*, page 35
4. CD-ROM, Disc 2, pages 186–187

PRESENTATION (*page 187*)

Extension of *Exercice B*

After students complete Exercise B, have them answer the questions they formulated.

ANSWERS

Exercice B

1. quel	4. quel
2. quelle	5. quelles
3. quelle	6. quels

Exercice C

1. Toute, quel
2. Tous
3. Tous, quelle, quel
4. Quels
5. Tous, quel

Exercice D

1. Toutes
2. Tout
3. Toutes
4. Tous

LISTENING ACTIVITY

Have students practice writing the various forms by giving the following *dictée*:
Quelles voitures sont nouvelles?
Quel livre est vieux?
Où sont toutes les filles?
Quels élèves aiment tous les profs?
Toute la classe fait tous les devoirs.

Les noms et les adjectifs en -al

PRESENTATION (*page 187*)

A. Lead students through steps 1–4 on page 187.
B. In general students need a great deal of ear training before they remember to make the correct plural sound of these masculine nouns and adjectives. Have students repeat after you the masculine plural forms in step 2. Now, supply masculine plural nouns, and have students combine them with the appropriate forms of the adjectives in step 3.
C. Have students supply the plural forms of the nouns in step 4.

I. EXPRESS

TROIS MAGAZINES TOUTES LES SEMAINES

Exercices

ANSWERS

Exercice A

1. tropicale
2. principale
3. municipaux
4. international
5. internationaux
6. municipal
7. internationaux

Exercice B

1. journaux
2. journaux
3. animaux, animaux
4. terminaux
5. généraux, généraux

INFORMAL ASSESSMENT

Have students quickly give the plural of the following expressions:
le problème local
le problème médical
le problème régional
le problème social
une organisation locale
une organisation médicale
une organisation régionale
une organisation sociale

Les verbes sortir, partir, dormir et servir au présent

PRESENTATION (pages 188–189)

A. Begin your explanation with the plural forms of these verbs. Draw a line through the last four letters and show students that what is left is the sound for all the singular forms.
B. Now go over the singular forms and their spellings. Have students repeat the forms after you.
C. Now lead students through steps 2–3 on page 189.

Note In the CD-ROM version, this structure point is presented via an interactive electronic comic strip.

Exercices

A **La Martinique, une île tropicale.** Complétez. *(Complete.)*

La Martinique est une île ___ (tropical) dans la Mer des Caraïbes. Sa ville ___ (principal) est Fort-de-France. Fort-de-France est la capitale. Dans la capitale il y a plusieurs petits parcs ___ (municipal). L'aéroport ___ (international) est près de la ville. Tous les jours il y a des vols ___ (international) qui arrivent et partent de l'aéroport ___ (municipal). Il y a des vols ___ (international) à destination de Paris et de beaucoup de villes des États-Unis comme Miami et New York.

B **Quel est le mot que je veux?** Complétez avec un mot en -al. *(Complete with a word ending in -al.)*

1. Le *New York Times* et le *Washington Post* sont des ___ américains.
2. *France-Soir*, *Paris Presse* et *Le Figaro* sont des ___ français.
3. Il y a des ___ au Bronx Zoo à New York et il y a des ___ au jardin zoologique à Paris.
4. Il y a deux grands ___ d'autocars dans la ville.
5. Les ___ sont dans l'armée. Les ___ sont des militaires.

Les verbes *sortir, partir, dormir* et *servir* au présent

Describing People's Activities

1. The verbs *sortir*, "to go out," *partir*, "to leave," *dormir*, "to sleep," and *servir*, "to serve," are irregular. Study the forms below.

SORTIR	PARTIR	DORMIR	SERVIR
je sors	je pars	je dors	je sers
tu sors	tu pars	tu dors	tu sers
il / elle / on sort	il / elle / on part	il / elle / on dort	il / elle / on sert
nous sortons	nous partons	nous dormons	nous servons
vous sortez	vous partez	vous dormez	vous servez
ils / elles sortent	ils / elles partent	ils / elles dorment	ils / elles servent

---

LEARNING FROM PHOTOS

Ask students: *Il y a combien de personnes sur la photo? Il y a combien de filles et combien de garçons? Ils sont blonds ou bruns? Qu'est-ce qu'ils regardent? C'est quel journal? Est-ce qu'ils parlent français?*

INDEPENDENT PRACTICE

Assign any of the following:
1. Exercises, pages 188–189
2. Workbook, *Structure: F–I*, page 64
3. Communication Activities Masters, *Structure: D–E*, pages 36–37
4. CD-ROM, Disc 2, pages 187–189

2. The verb *sortir* has more than one meaning. Used alone, it means "to go out." *Sortir de* means "to leave" in the sense of "to go out of a place, to exit." When followed by a noun, *sortir* means "to take out."

> Après les cours j'aime sortir avec mes copains.
> Il sort de l'école.
> Le passager sort ses bagages du compartiment.

3. The verb *partir* means "to leave." *Partir de* means "to leave from a place." "To leave for a place" is *partir pour*.

> L'avion part ce soir.
> Il part de la porte trois.
> L'avion part pour Paris.

Exercices

A **Un vol Abidjan–Paris.** Répondez par «oui». *(Answer "yes.")*

1. L'avion pour Paris part de la porte 10?
2. Il part à midi?
3. On sert le déjeuner à bord?
4. Le passager sort ses bagages à main du compartiment?
5. Il dort pendant le vol?

La Côte-d'Ivoire: Des hommes d'affaires descendent de l'avion.

B **Qui part?** Donnez des réponses personnelles. *(Give your own answers.)*

1. Tu pars pour l'école à quelle heure?
2. Tu sors de la maison à quelle heure le matin?
3. Quand tu arrives à l'école, qu'est-ce que tu sors de ton sac à dos?
4. Tu dors en classe?
5. Pendant le week-end, tu sors avec tes copains? Où allez-vous?

C **On part demain.** Répétez la conversation. *(Practice the conversation.)*

JACQUES: Vous partez à quelle heure demain?
CHANTAL: Solange et moi, nous partons à onze heures.
JACQUES: L'avion part de quel aéroport?
CHANTAL: Il part du Bourget.
JACQUES: Vous partez pour Tunis, n'est-ce pas?
CHANTAL: Oui, et nous allons immédiatement après à Monastir.

Complétez d'après la conversation. *(Answer according to the conversation.)*

1. Chantal et sa copine ___ pour Tunis.
2. Elles ___ en avion.
3. Leur vol ___ à onze heures.
4. Il ___ du Bourget.

CHAPITRE 7 **189**

Exercices

PRESENTATION *(page 189)*

Exercices A and B

It is recommended that you go over the exercises in class before assigning them for homework. Students can retell the stories of Exercises A and B in their own words

Exercice C

You may wish to use the recorded version of this exercise. Have students act out the conversation.

ANSWERS

Exercice A

Answers begin with *Oui* and repeat the wording of the question.

Exercice B

Answers will vary.

Exercice C

1. partent   3. part
2. partent   4. part

RETEACHING

1. Write the verbs *sortir, partir, servir,* and *dormir* on the board. Have students make up original sentences using these verbs.
2. Students may enjoy discussing *Qui sort avec qui?* They can talk about celebrities if they feel uncomfortable talking about themselves.

THE FRANCOPHONE WORLD

Have students locate Tunisia either on the map of *Le Monde francophone,* page 506 (Part A, page 239) or on the Map Transparency. Then have them find the capital, Tunis, and the lovely coastal town of Monastir, a beach resort where many Europeans vacation. Monastir also has a university. (Students can learn more about the Maghreb in *Le Monde francophone,* pages 222, 327, 434, and 436 as well as in *À bord,* Chapter 14.)

COOPERATIVE LEARNING

Have students jot down some actions that can be mimed using *sortir, servir, partir,* and *dormir.* In teams of four, each one mimes an action while the others guess which it is by asking questions. To practice *il/elle* forms, have them ask you questions about the student doing the miming. Then, each group chooses one action to mime for the class, which guesses the action using *vous.*

LEARNING FROM PHOTOS

You may wish to ask students questions such as the following about the photo above: *Les passagers embarquent ou débarquent? Ils débarquent dans quel pays? La Côte d'Ivoire est un pays africain ou européen? L'avion est un avion de quelle compagnie aérienne?*

CONVERSATION

Bell Ringer Review

Write the following on the board or use BRR Blackline Master 7-4: Choose the correct form of either *sortir, servir, dormir,* or *partir.*

1. Tu ne ___ pas à l'école.
2. Ils ___ du coca-cola.
3. Les passagers ___ leurs passeports.
4. Papa ___ pour Paris.

PRESENTATION *(page 190)*

A. Tell students they will hear a conversation between an airline ticket agent and Alice, who is about to take a trip.

B. Have them open their books to page 190 as you either play the Conversation Video, read the dialogue aloud, or play Cassette 5A/CD-5.

C. Have several pairs of students role-play the conversation with their books open.

Note In the CD-ROM version, students can play the role of either one of the characters and record the conversation.

ANSWERS

Exercice A

1. Alice est au comptoir de la compagnie aérienne.
2. Elle parle à l'agent.
3. Elle va à Paris.
4. L'agent vérifie son passeport.
5. Alice a deux valises.
6. Elles sont petites.
7. Elle veut une place non-fumeurs.
8. Elle a le siège 20C.
9. L'avion part à 20h10.
10. Il part de la porte 15.

190

Scènes de la vie *Au comptoir de la compagnie aérienne*

L'AGENT: Votre billet, s'il vous plaît.
ALICE: Oui, Madame.
L'AGENT: Vous partez pour Paris ce soir? Votre passeport, s'il vous plaît.
ALICE: Voilà mon passeport.
L'AGENT: Merci.
ALICE: Je vous en prie.
L'AGENT: Vous avez combien de valises?
ALICE: Deux petites valises.
L'AGENT: Bien. Vous préférez une place fumeurs ou non-fumeurs?
ALICE: Non-fumeurs, s'il vous plaît. Je ne fume pas.
L'AGENT: J'ai une place côté couloir non-fumeurs.
ALICE: Très bien.
L'AGENT: Voilà votre carte d'embarquement. Vous avez le siège 20C. Embarquement 20 heures 10, porte 15.

A **Le départ.** Répondez d'après la conversation. *(Answer according to the conversation.)*

1. Où est Alice?
2. Elle parle à qui?
3. Où est-ce qu'elle va?
4. Qu'est-ce que l'agent vérifie?
5. Alice a combien de valises?
6. Elles sont grandes ou petites?
7. Elle veut une place fumeurs ou non-fumeurs?
8. Elle a quel siège?
9. L'avion part à quelle heure?
10. Il part de quelle porte?

190   CHAPITRE 7

DID YOU KNOW?

In France a passport is a common form of identification. The French passport pictured above has the words *Communauté Européenne, République Française* on the cover. Eventually, French passports will bear the words *Union Européenne,* reflecting the formation of the European Union. In the future, the European Union countries will have one passport among them and a common currency, the euro.

Prononciation   *Le son /l/ final*

The names Michelle and Nicole were originally French names, but today many American girls have these names. When you hear French people say the names Nicole and Michelle, the final /l/ sound is much softer than in English. Say "Michelle" and "Nicole" in French. Repeat the following words.

île	vol	général	elle
ville	décolle	journal	quel

Now repeat the following sentences.

> C'est un vol international spécial.
> Quelle est la ville principale de l'île?
> C'est Mademoiselle Michelle. Elle est belle.

l'île

Activités de communication orale

A **Tu vas où?**   You're at the airport waiting for your flight. The person sitting next to you (your partner) asks you where you're going, what your flight number is, what time your plane leaves, why you're going on this trip, and who's going with you. Answer his or her questions, then reverse roles.

B **Une carte d'accès à bord.**   You just got your boarding pass for your flight to Bordeaux. Tell your classmate everything you can about your flight.

CARTE D'ACCES A BORD/boarding pass
AIR FRANCE ////

NOM DU PASSAGER / name of passenger

DE / from
PARIS/C GAULLE 2 B

A / to
BORDEAUX

VOL / flight   CLASSE   DATE   DEPART / time
IT6117   Y   01OCT   08H55

EMBARQUEMENT / boarding
24   08H30

SIEGE / seat
X   NO

PORTE / gate   HEURE / time
Nß   POIDS / weight

007

HUMMEL II/160/01 PC

PRESENTATION *(page 191)*

A. Model the key word *l'île* and have students repeat in unison. Then do the same for the other words and phrases.

B. You may wish to give students the following *dictée:*
   Cette île n'a pas de ville.
   Le général a quel journal pendant le vol?
   Michelle, elle est belle.

C. For additional pronunciation practice you may wish to use Cassette 5A/CD-5: *Prononciation; Activités I–K* in the Student Tape Manual, Teacher's Edition, page 85; and Pronunciation Transparency P-7.

Bell Ringer Review

Write the following on the board or use BRR Blackline Master 7-5: Write questions using a form of quel. For example: Le garçon? Quel garçon?

1. **Les journaux?**
2. **Les filles?**
3. **L'avion?**
4. **Ton amie?**

Activités de communication orale

ANSWERS

Activité A
  Answers will vary.

Activité B
  Answers will vary, but may include the following:
Je vais à Bordeaux. Je pars de l'aéroport Charles-de-Gaulle le premier octobre à 8 heures 55. C'est le vol IT6117. L'embarquement est à 8 heures 30, porte 24. J'ai une place dans la zone non-fumeurs.

LECTURE ET CULTURE

READING STRATEGIES
(page 192)

Pre-reading

Ask students who among them has traveled by air. Did they take a domestic or an international flight? What are some of the differences between the two? What were the good and bad points about their flights?

Reading

Present the reading in two or three segments. Have one student read three sentences, then ask comprehension questions of the others. For example, questions for the second paragraph might be: *Par où passent les élèves? Ils vont où? Leur avion part de quelle porte? À quelle heure? Qu'est-ce qu'ils entendent?*

Post-reading

A. Have students retell the story, changing the French class to their own and adding personalized information.

B. Have pairs of students write and act out conversations that might take place at the airport or on the plane, either between themselves or between them and the airline personnel.

C. Have students talk about the airport nearest your city or town.

Note Students may listen to a recorded version of the *Lecture* on the CD-ROM.

TOUTE LA CLASSE VA À PARIS

*T*ous les élèves de la classe de français de Madame Bardot vont à Paris. Ils sont au comptoir d'Air France à l'aéroport JFK. L'agent vérifie tous leurs billets et tous leurs passeports. Il enregistre tous leurs bagages. Il donne toutes les cartes d'embarquement à Madame Bardot.

Les élèves passent par le contrôle de sécurité. Leur avion à destination de Paris part de la porte cinquante-deux à vingt heures dix. À vingt heures moins le quart on fait l'annonce du départ. Les élèves embarquent et trouvent leurs places à bord de l'avion. Ils placent leurs bagages à main dans le compartiment au-dessus de leur tête[1] ou sous[2] leur siège.

Quelle chance[3]! Leur avion décolle à l'heure[4]. Il n'a pas de retard. Pendant le vol les copains bavardent et regardent un film. Les hôtesses de l'air et les stewards passent dans la cabine et servent des boissons et un dîner. Avant l'arrivée à Paris les élèves remplissent une carte de débarquement.

On arrive à Paris à huit heures du matin. L'avion atterrit à l'aéroport Charles-de-Gaulle à Roissy. Charles-de-Gaulle est un des trois aéroports internationaux de Paris. La plupart des vols internationaux arrivent à Roissy ou partent de Roissy.

Les élèves de Madame Bardot débarquent et récupèrent leurs bagages. Ils passent à l'immigration et à la douane. Les formalités de douane sont très simples à Charles-de-Gaulle.

Quarante minutes après l'atterrissage les élèves sont dans l'autocar qui fait la navette[5] entre l'aéroport et le Terminal des Invalides, au centre de la ville de Paris. Tout le monde est crevé après le long voyage en avion et un décalage horaire de six heures. On va dormir, n'est-ce pas? Absolument pas! Le premier jour à Paris on ne dort pas. On flâne[6] dans les rues de Paris. On flâne le long de la Seine.

---

[1] compartiment au-dessus de leur tête *overhead compartment*
[2] sous *under*
[3] chance *luck*
[4] à l'heure *on time*
[5] fait la navette *goes back and forth*
[6] flâne *strolls*

La Pyramide du Louvre

CRITICAL THINKING ACTIVITY

(Thinking skill: drawing conclusions)

Put the following on the board or on an overhead transparency:
Julie fait un voyage en avion. Elle a combien de valises? Elle a une valise, c'est tout! Incroyable mais vrai! Quels sont les avantages de voyager avec une seule valise?

DID YOU KNOW?

As jets became larger, more space was needed to maneuver them and to accommodate increased passenger traffic in the terminal. At Charles-de-Gaulle Airport a large main terminal, surrounded by satellite terminals with their own arrival and departure gates, was built. Buses, automated trains, and Plexiglas-enclosed "people-movers" take passengers between terminals.

Étude de mots

A À l'aéroport. Trouvez le contraire. *(Find the opposite.)*

1. international
2. l'atterrissage
3. au-dessus de
4. faire enregistrer les bagages
5. à destination de
6. le départ
7. embarquer
8. décoller
9. au centre
10. simple

a. en provenance de
b. intérieur
c. atterrir
d. débarquer
e. l'arrivée
f. le décollage
g. sous
h. compliqué
i. récupérer les bagages
j. à la périphérie

Compréhension

B Un voyage. Répondez d'après la lecture. *(Answer according to the reading.)*

1. Qui va à Paris?
2. Où sont-ils maintenant?
3. Qu'est-ce que l'agent d'Air France fait?
4. À qui est-ce qu'il donne les cartes d'embarquement?
5. Par où passent les élèves?
6. Leur avion part de quelle porte?
7. Il part à quelle heure?
8. Leur avion décolle à l'heure ou avec un retard d'une heure?
9. Qui travaille à bord de l'avion?
10. Qu'est-ce qu'on sert à bord?
11. On regarde un film pendant le vol?

DÉCOUVERTE CULTURELLE

Air Inter est une des principales compagnies aériennes françaises. Air Inter dessert[1] à peu près cinquante villes françaises et quelques villes étrangères[2].

Les tarifs aériens[3] en France et dans les autres pays d'Europe sont très chers[4].

Il y a une grande industrie aérospatiale en France. Toulouse est le centre de l'industrie aérospatiale française. À Toulouse on assemble les Airbus.

Le Concorde est un avion français et anglais. Le Concorde est un avion supersonique. Il fait New York–Paris en trois heures et demie.

[1] dessert *serves*
[2] étrangères *foreign*
[3] tarifs aériens *airfares*
[4] chers *expensive*

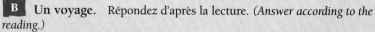

RÉALITÉS

PRESENTATION (*pages 194–195*)

The objective of this section is for students to enjoy the photographs in a relaxed fashion and gain an appreciation for France. You may, however, also wish to do the following activities.

A. Ask whether any students have flown on Air France. What can they remember about it? Ask what students remember about food they have had during a flight. Can they compare it to the menu on page 195?

B. Have students read aloud and answer the questions in Caption 2 about the realia on page 194 or have them write the answers with a partner. Then verify comprehension by calling on various pairs of students to answer the questions.

Note In the CD-ROM version, students can listen to the recorded captions and discover a hidden video behind one of the photos.

Voici le Concorde **1**. En combien d'heures fait-il New York–Paris? Tu veux prendre cet avion?

Voici une carte d'embarquement **2**. Quel est le numéro du vol? Quel est le numéro du siège? Quel est le nom du passager? L'avion part de quel aéroport?

Voici l'écran des arrivées de l'aérogare des Invalides à Paris **3**. Ce sont des vols intérieurs ou des vols internationaux? Tu veux visiter quelles villes?

Voici une carte d'une compagnie aérienne **4**. On sert combien de repas pendant le vol? Pourquoi?

194

DID YOU KNOW?

In 1962, the French and British governments signed an agreement to build the supersonic Concorde. The first test model flew in 1971, but the first passenger service wasn't until Jan. 21, 1976. Service to the U.S. began with flights from London and Paris to Washington, D.C. The Port Authority of New York originally refused to allow the Concorde to land at Kennedy International Airport until a court battle forced them to grant permission.

The supersonic plane is capable of flying round-trip across the North Atlantic in less time than an ordinary jet takes to fly one-way. The Concorde has been criticized for being uneconomical and extremely noisy. Operating losses in excess of $1 billion ended production of the plane in 1979.

1000 DUBLIN	FU	111 ARRIVE 1000	2 D
1005 BRISTOL	BC	601 ARRIVE 0957	2 D
1005 BUDAPEST	MAAF	558 ARRIVE 1002	2 D
1010 AMSTERDAM	AF	1251 ARRIVE 1021	2 D
1010 REUNION			
MAURICE			
SEYCHELLES			
DUBAI	AF	464 ARRIVE 1009	2 A
1015 COLOGNE	AF	785 ARRIVE 1020	2 D
1015 DUSSELDORF	AF	769 PREVU 1020	2 D
1020 MIAMI	AF	042 PREVU 1050	2 A
1020 GENEVE	AF	961 PREVU 1011	2 D
1025 STUTTGART	AF	723 PREVU 1016	2 D
1030 NEW YORK	AF	074 PREVU 1100	2 A
1030 BIRMINGHAM	AF	901 PREVU 1028	2 D
1035 AMSTERDAM	FU	021 PREVU 1028	

AIR ORIENT

PAUL COLIN

EUROPE. ORIENT
EXTRÊME-ORIENT
Service Hebdomadaire

MENU
*
ASSORTIMENT DE CHARCUTERIE
*
TOURNEDOS MAÎTRE D'HÔTEL
*
BOUQUETIÈRE DE PETITS LÉGUMES
*
SALADE COMPOSÉE
*
FROMAGE
*
DESSERT
*
CAFÉ DE COLOMBIE
*
PETIT DÉJEUNER
*

*Pour les boissons alcoolisées dont le service n'est pas gratuit
en classe Economique et pour les autres produits mis en vente
à bord, nous vous prions de vous reporter au tarif mis
à votre disposition.*

ADDITIONAL PRACTICE

1. Student Tape Manual, Teacher's Edition, *Deuxième Partie,* pages 86–87
2. Situation Cards, Chapter 7
3. Communication Transparency C-7

PAIRED ACTIVITY

Have students make a boarding pass and exchange it with another student. Partners then take turns asking questions such as: *Tu fais un voyage en Europe? Où vas-tu?*

RECYCLING

The *Activités de communication orale* and *écrite* allow students to use the vocabulary and grammar from this chapter in open-ended, real-life situations. They also give students the opportunity to reuse, as much as possible, vocabulary and structures from earlier chapters.

INFORMAL ASSESSMENT

Oral Activities A and B may serve to evaluate students' speaking skills. Use the evaluation criteria on page 34 of this Teacher's Wraparound Edition.

Activités de communication orale

ANSWERS

Activité A
Answers will vary.

Activité B
Answers will vary but É1's questions may include the following: *Tu sors souvent? (Tous les soirs? Deux ou trois fois par semaine? Le week-end seulement?, etc.) Tu vas où? Avec qui est-ce que tu sors? Qui paie?*

Activités de communication écrite

ANSWERS

Activités A and B
Answers will vary.

CULMINATION

Activités de communication orale

A **Avant, pendant ou après le vol?**   Play this game in groups of three. Each of you will think of several things a flight attendant, a passenger, and a ticket agent do. The first person says an activity and one of the other two players has to guess *who* does it. The third person then has to say *when* the activity is done—before, during, or after the flight.

> Élève 1: Elle sert le dîner.
> Élève 2: C'est l'hôtesse de l'air.
> Élève 3: Elle sert le dîner pendant le vol.

B **Tu aimes sortir?**   You're chatting with a classmate. Find out how often each of you goes out, where you go, with whom, and who pays.

Activités de communication écrite

A **Vive les vacances!**   You've just won two free plane tickets to an exciting foreign city. Write to a friend inviting him or her to join you.

B **Un horrible vol!**   Imagine you're taking a trip and everything goes wrong before, during, and after the flight. Write a brief paragraph describing your experience. For some possible topics to include, look at the list below.

> l'agent
> le déjeuner (le dîner)
> le film
> la personne à côté de vous
> le personnel de bord
> la place

FOR THE YOUNGER STUDENT

Have students prepare a large arrival-and-departure board that includes exotic French-speaking destinations. (Have them use *Le Monde francophone* map on page 506 [Part A, page 239] for ideas, or suggest they look at the various *Le Monde francophone* sections in this book on pages 110–113; 220–223; 326–329; and 434–437.) Use the board they create as a bulletin board display while doing this chapter. Have students refer to their "board" and make up questions, sentences, and stories about their arrivals and departures.

Réintroduction et recombinaison

A **À Paris.** Complétez. (*Complete.*)

1. Je ___ à Paris. (aller)
2. J'y ___ avec ma classe de français. (aller)
3. Notre prof de français ___ Madame Bardot. (être)
4. Nous ___ très contents. (être)
5. Nous ___ à Paris en avion. (aller)
6. Madame Bardot ___ le voyage avec nous. (faire)

B **À vous.** Écrivez des phrases originales. (*Write original sentences.*)

1. américain, français
2. joli
3. nouveau
4. beau
5. tout
6. quel

Vocabulaire

NOMS
l'aéroport (m.)
l'agent (m.)
l'arrivée (f.)
le départ
le comptoir
la compagnie aérienne
le billet
la carte d'embarquement
l'écran (m.)
le vol
le pays
le passager
la passagère
les bagages (m.)
les bagages à main
la valise
la porte
le contrôle de sécurité
l'immigration (f.)
le passeport
la douane
l'aérogare (f.)
le taxi
l'autocar (m.)

l'avion (m.)
la cabine
la sortie
le siège
la place
le côté couloir
le côté fenêtre
le compartiment
le personnel de bord
l'hôtesse (f.) de l'air
le steward
la zone non-fumeurs
   (fumeurs)
la carte de débarquement

ADJECTIFS
intérieur(e)
international(e)
quel(le)
tout(e), tous, toutes

VERBES
débarquer
embarquer
décoller
passer

récupérer
vérifier

atterrir
choisir
finir
remplir
réussir (à)
obéir (à)
punir

sortir
partir
dormir
servir

**AUTRES MOTS
ET EXPRESSIONS**
faire enregistrer
faire les valises
faire un voyage
à bord de
à destination de
en provenance de
avant

Réintroduction et recombinaison

RECYCLING

Exercise A recycles the forms of the irregular verbs *aller, être,* and *faire* in conjunction with vocabulary from Chapter 7.

Exercise B reviews the forms of different types of adjectives.

ANSWERS

Exercice A

1. vais
2. vais
3. est
4. sommes
5. allons
6. fait

Exercice B

Answers will vary.

ASSESSMENT RESOURCES

1. Chapter Quizzes
2. Testing Program
3. Situation Cards
4. Communication Transparency C-7
5. Computer Software: Practice/Test Generator

VIDEO PROGRAM

INTRODUCTION   (24:02)

NATALIE ARRIVE À   (25:27)
L'AÉROPORT

DID YOU KNOW?

Tell students: *Ce monument parisien, c'est l'Arc de Triomphe. Il commémore les victoires des armées de Napoléon. L'Arc de Triomphe est aux Champs-Élysées, sur la place Charles-de-Gaulle. Les Parisiens préfèrent l'ancien nom de cette place: la place de l'Étoile. Pourquoi «Étoile»? Parce qu'une douzaine d'avenues partent de cette place, qui ressemble à une étoile. Sur la photo il y a aussi le drapeau français.*

STUDENT PORTFOLIO

Written assignments that may be included in students' portfolios are the *Activités de communication écrite,* page 196, and the *Mon Autobiographie* section of the Workbook, page 69.

Note Students may create and save both oral and written work using the Electronic Portfolio feature on the CD-ROM.

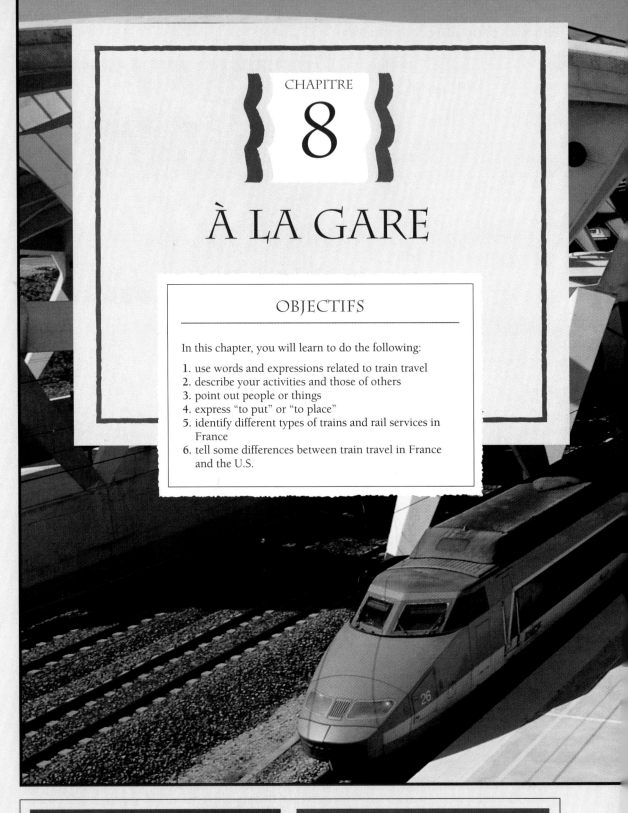

CHAPITRE 8

CHAPITRE
8

À LA GARE

OBJECTIFS

In this chapter, you will learn to do the following:

1. use words and expressions related to train travel
2. describe your activities and those of others
3. point out people or things
4. express "to put" or "to place"
5. identify different types of trains and rail services in France
6. tell some differences between train travel in France and the U.S.

CHAPTER OVERVIEW

In this chapter students will learn to communicate when traveling by train in a French-speaking country. They will learn the present tense of *-re* verbs, the demonstrative adjectives, and the verb *mettre*.

The cultural focus of this chapter is on trains and train travel in France.

CHAPTER OBJECTIVES

By the end of this chapter students will know:

1. vocabulary associated with facilities, personnel, and procedures at a railway station and on board a train
2. the present indicative forms of *-re* verbs
3. all forms of the demonstrative adjectives
4. the present indicative of the verb *mettre*

CHAPTER PROJECTS

(optional)

1. Groups plan a rail trip through France (arrival/departure times, length of each stop, etc.), with at least one night trip. They report to the class in French on their trip.
2. They research a city they've "visited" and report briefly on it to the class.

COMMUNITIES

A field trip to a local train station can be most instructive after students have learned the vocabulary in this chapter. As you tour the station, have students use as much of their French vocabulary as possible. Have them compare an American train station (facilities, procedures, etc.) with a French one.

199

Pacing

This chapter requires eight to ten class sessions. Pacing will vary according to class length and the age and aptitude of the students.

Note The Lesson Plans offer guidelines for 45- and 55-minute classes and **Block Scheduling.**

Exercices vs. *Activités*

All exercises (which provide guided practice) are coded in blue. All communicative activities are coded in red.

INTERNET ACTIVITIES

(optional)

These activities, student worksheets, and related teacher information are in the *Bienvenue* Internet Activities Booklet and on the Glencoe Foreign Language Home Page at **http://www.glencoe.com/secondary/fl**

DID YOU KNOW?

The photo shows a TGV (*Train à Grande Vitesse*) in Lyon's new station at Satolas, designed by Spanish architect Calatrava. In 1994 the Satolas airport became the first to hook up with the TGV network. The number of passengers has risen to about 500 a day, but the station will only become really relevant when the Mediterranean, Lyon-Turin, and Rhine-Rhone lines are finished.

VOCABULAIRE

MOTS 1

Vocabulary Teaching Resources

1. Vocabulary Transparencies 8.1 (A & B)
2. Audio Cassette 5B/CD-5
3. Student Tape Manual, Teacher's Edition, *Mots 1: A–C,* pages 88–90
4. Workbook, *Mots 1: A–C,* pages 70–71
5. Communication Activities Masters, *Mots 1: A,* page 38
6. Chapter Quizzes, *Mots 1: Quiz 1,* page 43
7. CD-ROM, Disc 2, *Mots 1:* pages 200–202

Bell Ringer Review

Write the following on the board or use BRR Blackline Master 8-1: Write as many sentences as you can about a plane trip to France.

Note The Bell Ringer Review activity above serves as a nice introduction to train travel.

PRESENTATION *(pages 200–201)*

A. Have students close books. Using Vocabulary Transparencies 8.1 (A & B), introduce the vocabulary. Have students repeat after you or Cassette 5B/CD-5 as you point to the illustrations.

B. Point to the items on the transparency at random and call on an individual to identify the item you are pointing to.

C. When presenting the sentences on page 201, ask: *Qu'est-ce qu'on vend au guichet? Où est-ce qu'on vend les billets? Qu'est-ce qu'on vend au kiosque? Où est-ce que Jean met ses bagages? Qui attend le train?*

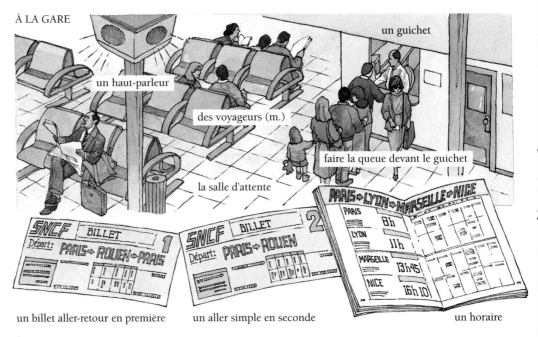

À LA GARE

un guichet

un haut-parleur

des voyageurs (m.)

faire la queue devant le guichet

la salle d'attente

un billet aller-retour en première

un aller simple en seconde

un horaire

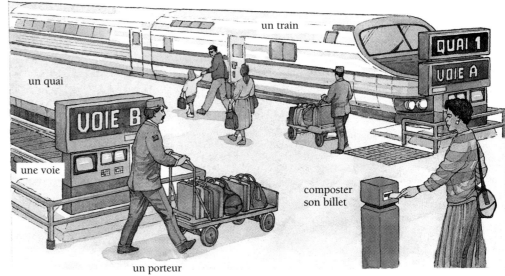

un train

un quai

QUAI 1 VOIE A

VOIE B

une voie

composter son billet

un porteur

TOTAL PHYSICAL RESPONSE

(following the Vocabulary presentation)

Getting Ready

Have your desk be *le guichet.* One student can be *l'employé(e).* Numbers on the board can represent the *quais.* Demonstrate *monter.*

TPR 1

____, venez ici, s'il vous plaît.
On va imaginer que c'est la gare.

Allez au guichet.
Faites la queue devant le guichet.
Sortez de l'argent de votre poche.
Achetez un billet. Prenez le billet.
Mettez le billet dans votre poche.
Cherchez le quai numéro deux.
Allez au quai numéro deux.
Attendez le train.
Le train arrive. Montez dans le train.
Au revoir, ____, et bon voyage!

On vend les billets.
On vend les billets au guichet.
On vend des journaux au kiosque.

Jean met ses bagages à la consigne automatique.
Marie-Claire laisse ses bagages à la consigne.

Les voyageurs attendent.
Ils sont en avance.
Ils attendent le train dans la salle d'attente.

On annonce le départ du train.
On entend l'annonce au haut-parleur.

Le train part à l'heure.

CROSS-CULTURAL COMPARISON

Students will be unfamiliar with the action of *composter un billet.* You may wish to explain that in a French railroad station, at the entrance to the platform area, there is a machine into which one puts one's ticket in order to validate it. Doing so marks the ticket with the station at which the passenger boarded the train. If a ticket is not *composté,* the conductor can charge the passenger the full fare from the first station on the line.

Note Note that the new structure points, the -*re* verbs and *mettre,* are introduced in these sentences in the third person. Students can immediately answer questions using the new verbs. In this way the meaning is reinforced without the students' having to manipulate endings. They will learn to manipulate the endings in the *Structure* section of the chapter.

RECYCLING

Note the reentry of (*journal*), *journaux* from the previous chapter. The verb *partir* is also reintroduced.

Vocabulary Expansion

The following are some additional words you may wish to give the students. It is recommended, however, that you keep the additional vocabulary to a minimum. Additional train-related vocabulary at a higher level of proficiency is presented in the Level 2 textbook, *À bord,* Chapter 4.
une horloge
une montre
les toilettes
le buffet de la gare
prendre la correspondance
   (changer de train)

TPR 2

Getting Ready

A piece of paper with the word *valise* written on it can represent a suitcase.
___, venez ici, s'il vous plaît.
Prenez cette valise. Ouvrez la valise.
Mettez un tee-shirt dans la valise.
Mettez un livre dans la valise.
Fermez la valise.
Allez au téléphone.

Téléphonez à un taxi.
Descendez au rez-de-chaussée.
Attendez le taxi. Le taxi arrive.
Mettez la valise dans le taxi.
Montez dans le taxi.
Descendez du taxi. Prenez la valise.
Sortez de l'argent de votre poche. Payez le chauffeur. Donnez un pourboire au chauffeur.

Exercices

PRESENTATION *(page 202)*

After doing the exercises orally, have students read them for reinforcement. They can then be assigned for homework.

Exercice A

After completing this exercise, call on a student to retell the information in his/her own words.

Exercice B

After completing this exercise, call on a student to talk about a real train trip he/she has taken or an imaginary one.

Exercice C

You may wish to use the recorded version of this exercise.

ANSWERS

Exercice A

1. **Les voyageurs voyagent.**
2. **On vend les billets au guichet.**
3. **On fait la queue devant le guichet.**
4. **Les voyageurs achètent les billets au guichet.**
5. **On met ses bagages à la consigne automatique.**
6. **Oui, on entend l'annonce du départ du train dans la salle d'attente.**
7. **Oui, on fait l'annonce au haut-parleur.**
8. **Oui, quand on entend l'annonce du départ de son train, on sort ses bagages de la consigne automatique.**
9. **Les voyageurs vont sur le quai.**

Exercice B

Answers will vary.

Exercice C

1. **C'est un aller simple.**
2. **C'est un billet aller-retour.**
3. **On vend les billets au guichet.**
4. **On vend des journaux au kiosque.**
5. **On met ses bagages à la consigne (automatique).**
6. **Le(s) porteur(s) aide(nt) les voyageurs avec leurs bagages.**
7. **Les voyageurs attendent le train dans la salle d'attente.**
8. **On consulte un horaire.**
9. **Le train part de la voie (du quai).**

202

Exercices

A **Un voyage en train.** Répondez. *(Answer.)*

1. Les porteurs ou les voyageurs voyagent?
2. On vend les billets au guichet ou à la consigne?
3. On fait la queue devant le guichet ou sur le quai?
4. Les voyageurs achètent ou compostent les billets au guichet?
5. On met ses bagages à la consigne automatique ou au kiosque?
6. On entend l'annonce du départ du train dans la salle d'attente?
7. On fait l'annonce au haut-parleur?
8. Quand on entend l'annonce du départ de son train, on sort ses bagages de la consigne automatique?
9. Les voyageurs vont sur le quai ou au guichet?

B **À la gare.** Donnez des réponses personnelles. *(Give your own answers.)*

1. Tu fais un voyage en train. Où vas-tu?
2. Tu arrives à la gare en avance?
3. Tu achètes ton billet?
4. Tu veux un billet aller-retour ou un aller simple?
5. Tu vas voyager en première ou en seconde?
6. Ton train part à quelle heure?
7. Tu regardes l'horaire?
8. Ton train part de quel quai? De quelle voie?
9. Tu achètes un journal? Où?
10. Ton train part à l'heure ou en avance?

C **Je cherche quel mot?** Répondez. *(Answer.)*

1. Qu'est-ce que c'est un billet Paris–Avignon?
2. Qu'est-ce que c'est un billet Paris–Avignon–Paris?
3. Où est-ce qu'on vend les billets?
4. Où est-ce qu'on vend des journaux?
5. Où est-ce qu'on met ses bagages?
6. Qui aide les voyageurs avec leurs bagages?
7. Où est-ce que les voyageurs attendent le train?
8. Qu'est-ce qu'on consulte pour vérifier l'heure du départ du train?
9. D'où part le train?

La Suisse: Un train rouge dans les Alpes

INDEPENDENT PRACTICE

Assign any of the following:
1. Exercises, page 202
2. Workbook, *Mots 1: A–C*, pages 70–71
3. Communication Activities Masters, *Mots 1: A*, page 38
4. CD-ROM, Disc 2, pages 200–202

VOCABULAIRE

MOTS 2

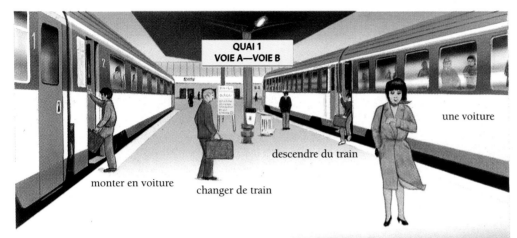

QUAI 1
VOIE A—VOIE B

une voiture

descendre du train

monter en voiture

changer de train

un contrôleur

assis

vérifier le billet

debout

SNCF

Les voyageurs attendent le prochain train.

La plupart des voyageurs sont assis.
Quelques voyageurs sont debout dans le couloir.

Vocabulary Teaching Resources

1. Vocabulary Transparencies 8.2 (A & B)
2. Audio Cassette 5B/CD-5
3. Student Tape Manual, Teacher's Edition, *Mots 2: D–F,* pages 90–92
4. Workbook, *Mots 2: D–F,* page 72
5. Communication Activities Masters, *Mots 2: B,* page 38
6. Chapter Quizzes, *Mots 2: Quiz 2,* page 44
7. Computer Software, *Vocabulaire*
8. CD-ROM, Disc 2, *Mots 2:* pages 203–206

Bell Ringer Review

Write the following on the board or use BRR Blackline Master 8-2: Draw a picture for each word or phrase, or use each one in a sentence.

1. **faire la queue**
2. **attendre le train sur le quai**
3. **le guichet**
4. **un porteur**
5. **la consigne**

PRESENTATION *(pages 203–204)*

A. Model the new words using Vocabulary Transparencies 8.2 (A & B). Have students repeat after you or Cassette 5B/CD-5.
B. After presenting *monter en voiture, descendre du train, changer de train,* and *attendre le prochain train,* have students make up their own sentences using these expressions. Call on the more able students first.
C. Now have students open their books and read the new vocabulary.

TOTAL PHYSICAL RESPONSE

(following the Vocabulary presentation)

Getting Ready

Set up an area of the classroom as *le compartiment* with four chairs facing four other chairs. Call on pairs of students, one to play the part of the passenger, the other the part of the conductor. Give the passenger a bag with the word *valise* written on it.

TPR

___, levez-vous et venez ici, s'il vous plaît.
Allez à la voie A.
Mettez votre valise dans le train.
Montez en voiture.
Dites «au revoir» à vos amis.
Allez à votre compartiment. Cherchez votre place.
Asseyez-vous.

(continued on next page)

COGNATE RECOGNITION

Have students scan the vocabu-lary on pages 203 and 204 and pick out the cognates: *changer, descendre, un voyageur, le train, (la) patience, (les) minutes.*

GAME

You may wish to play a guess-ing game with opposites. Have students give you the opposite of the following words:
monter (descendre)
assis (debout)
à l'heure (en retard)
l'impatience (la patience)
un billet aller-retour (un aller simple)

INFORMAL ASSESSMENT
(*Mots 1 and 2*)

Show Vocabulary Transparen-cies 8.1 and 8.2 (A & B). Have the students give you as many words as they can.

DANS LA VOITURE DU TRAIN

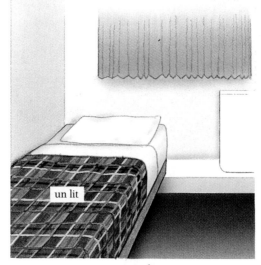

une voiture-lit   une couchette

Devant la gare Jean attend son ami.
Il attend son ami depuis quarante minutes.
Son ami est en retard.
Jean perd patience!

204   CHAPITRE 8

TPR (*continued from page 203*)
Le contrôleur arrive. Sortez votre billet de votre poche.
Donnez le billet au contrôleur.
Reprenez votre billet et mettez-le dans votre poche.
Prenez votre valise. Ouvrez la valise.
Sortez un livre de la valise.
Lisez votre livre.
Merci, ____. Retournez à votre place.

PAIRED ACTIVITY

Have pairs of students make up as many sentences as they can about travel. They then read their sentences to each other and decide if they deal with train travel or air travel. They should categorize their sentences under the headings *Voyages en avion* or *Voyages en train*. Finally, they reorganize their sentences to tell two stories, one about train travel and another about air travel.

Exercices

A **De Paris à La Baule.** Répondez d'après les indications. (*Answer according to the cues.*)

1. Pour aller de Paris à La Baule, on change de train? (oui)
2. Où est-ce qu'on change? (à Nantes)
3. Qui crie: «En voiture! En voiture!»? (le contrôleur)
4. Qui aide les voyageurs à descendre leurs bagages sur le quai? (le porteur)
5. Quand le contrôleur crie: «En voiture!», qui monte en voiture? (les voyageurs)
6. Où est-ce que les voyageurs qui vont à La Baule descendent? (à Nantes)
7. Toutes les places sont occupées? (oui)
8. Il y a quelques voyageurs debout? (oui)
9. Où sont-ils debout? (dans le couloir)
10. La plupart des voyageurs sont assis? (oui)
11. Où peut-on dormir dans le train? (dans une voiture-lit ou dans une couchette)

B **Pour aller à La Baule, s'il vous plaît?** Choisissez la bonne réponse. (*Choose the correct answer.*)

1. On change de train pour aller où?
   a. À Nantes.   b. À La Baule.   c. À la gare.
2. Qui aide les voyageurs avec leurs bagages à la gare?
   a. L'agent.   b. Le porteur.   c. Le contrôleur.
3. Qui travaille dans le train?
   a. L'agent.   b. Le porteur.
   c. Le contrôleur.
4. Qui crie: «En voiture!» avant le départ du train?
   a. L'agent.   b. Le porteur.
   c. Le contrôleur.
5. Qu'est-ce que les voyageurs font?
   a. Ils vendent leurs billets.
   b. Ils crient.
   c. Ils montent en voiture.
6. Qu'est-ce que le contrôleur fait?
   a. Il vend les billets.
   b. Il vérifie les billets.
   c. Il fait les valises.

Exercices

PRESENTATION (*page 205*)

Exercice A

This exercise can be done with books closed or open, or once each way. You may call on one student to do one or two items. If the student makes an error, call on another student to correct it.

Extension of *Exercice A*

After completing Exercise A, call on a student to retell the story in his/her own words.

Exercice B

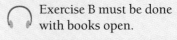 Exercise B must be done with books open.

ANSWERS

Exercice A

1. Oui, pour aller de Paris à La Baule, on change de train.
2. On change à Nantes.
3. Le contrôleur crie: «En voiture! En voiture!»
4. Le porteur aide les voyageurs à descendre leurs bagages sur le quai.
5. Les voyageurs montent en voiture quand le contrôleur crie: «En voiture!»
6. Les voyageurs qui vont à La Baule descendent à Nantes.
7. Oui, toutes les places sont occupées.
8. Oui, il y a quelques voyageurs debout.
9. Ils sont debout dans le couloir.
10. Oui, la plupart des voyageurs sont assis.
11. On peut dormir dans une voiture-lit ou dans une couchette.

Exercice B

1. b	3. c	5. c
2. b	4. c	6. b

INDEPENDENT PRACTICE

Assign any of the following:
1. Exercises, page 205
2. Workbook, *Mots 2: D–F*, page 72
3. Communication Activities Masters, *Mots 2: B*, page 38
4. Computer Software, *Vocabulaire*
5. CD-ROM, Disc 2, pages 203–205

DID YOU KNOW?

Have students locate Nantes on the map of France, page 504 (Part A, page 237) or use the Map Transparency. Tell them La Baule is just north of Nantes, on the Atlantic coast. La Baule and Dinard are the most popular beach resorts in Brittany. La Baule has a beautiful beach with very fine sand. It also has a harbor for pleasure craft. Pine trees protect the dunes from the north winds.

Allow the students to choose the activity they would like to take part in. Different groups can do different activities.

ANSWERS

Activités A and B

Answers will vary.

Activités de communication orale

Mots 1 et 2

A **Dans le train ou à la gare?** Look at the list of places below. Choose one, but don't tell your partner which one. Just tell him or her what you're doing there. He/She guesses where you are. Take turns until all the places have been used.

> Élève 1: J'achète mon billet.
> Élève 2: Tu es au guichet.

la consigne automatique	la salle d'attente
le guichet	la voiture-lit
le kiosque	la voiture-restaurant
le quai	

B **L'horaire.** Look at the information below. Take turns with a classmate asking and answering questions about it.

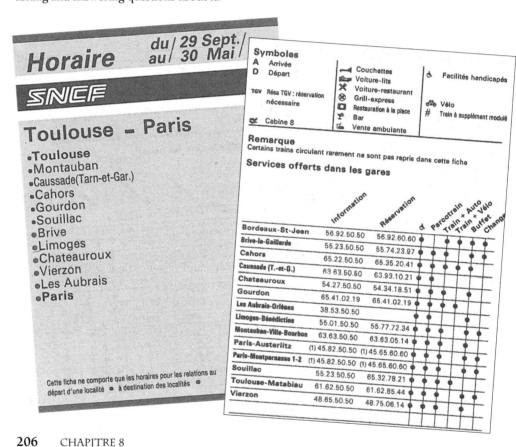

Have students note the symbols used for the different train facilities. *Vente ambulante* means that someone comes through the train car selling food and drinks. (They should recognize *vente*, since they know *vendre*. *Ambulante*, like ambulance, is derived from the Latin word meaning "to walk.")

Student Tape Manual, Teacher's Edition, *Activité F,* page 92

STRUCTURE

STRUCTURE

Les verbes en -re au présent   *Describing People's Activities*

1. Another group of regular verbs in French ends in *-re*. Study the following forms.

INFINITIVE	ATTENDRE	VENDRE	
STEM	**attend-**	**vend-**	ENDINGS
	j' **attends**	je **vends**	-s
	tu **attends**	tu **vends**	-s
	il elle } **attend** on	il elle } **vend** on	—
	nous **attendons**	nous **vendons**	-ons
	vous **attendez**	vous **vendez**	-ez
	ils **attendent**	ils **vendent**	
	elles **attendent**	elles **vendent**	-ent

2. Other regular verbs that end in *-re* are *entendre*, "to hear," *répondre*, "to answer," *perdre*, "to lose," and *descendre*, "to go down" or "to get off." Note that the verb *répondre* takes the preposition *à* when followed by a noun.

   Les voyageurs répondent à la question du contrôleur.

3. To increase your vocabulary, study the noun forms of these verbs.

attendre	l'attente
descendre	la descente
perdre	la perte
répondre	la réponse
vendre	la vente

Structure Teaching Resources

1. Workbook, *Structure: A–G,* pages 73–75
2. Student Tape Manual, Teacher's Edition, *Structure: A–C,* pages 92–94
3. Audio Cassette 5B/CD-5
4. Communication Activities Masters, *Structure: A–C,* pages 39–42
5. Computer Software, *Structure*
6. Chapter Quizzes, *Structure:* Quizzes 3–5, pages 45–47
7. CD-ROM, Disc 2, pages 207–211

Bell Ringer Review

Write the following on the board or use BRR Blackline Master 8-3: Put each of these verbs in a sentence.

sortir	servir
partir	dormir

Les verbes en -re au présent

PRESENTATION *(page 207)*

A. Begin with the third-person plural: Write *ils attendent* and *ils vendent* on the board. Draw a line through the ending: students now have the sound for all the singular forms. Point out that the *je* and *tu* forms take an *-s*. The third-person singular has no ending.

B. Have students pronounce the infinitives of the other *-re* verbs in step 2 of the grammar explanation.

C. Have them repeat the verbs and their noun forms in step 3.

Note In the CD-ROM version, this structure point is presented via an interactive electronic comic strip.

LEARNING FROM PHOTOS

1. Ask students: *Qu'est-ce qu'on vend dans ce magasin?*
2. Have them find the noun forms of *vendre* and *acheter* on the awning.
3. Ask them if they know what "K7" on the awning means. (*cassette*)
4. *Compact disc* is, of course, the anglicized version of *disque compact,* which is also used.

ADDITIONAL PRACTICE

Student Tape Manual, Teacher's Edition, *Activité A,* pages 92–93

PRESENTATION *(pages 208–209)*

Exercice A

This exercise can be done with books open, closed, or once each way.

Extension of *Exercice A*

After doing Exercise A, have students make up their own story based on the illustrations and the exercise.

Exercice B

Exercise B should be done with books open.

Extension of *Exercice B*

Have pairs of students present the mini-conversations. You can call on individuals to retell the situation of each mini-conversation in their own words.

Exercice C

Exercise C can be done with books open or closed, or once each way.

ANSWERS

Exercice A

1. Oui, les voyageurs attendent le train sur le quai.
2. Non, ils n'attendent pas le train dans la salle d'attente.
3. Oui, ils perdent leurs billets.
4. Oui, ils entendent l'annonce du départ de leur train.
5. Non, ils ne descendent pas du train. (Non, ils montent en voiture.)

Exercice B

1. attendez
   attendons
2. perdons
   perdez
3. descend
   descendons

Exercice C

1. Oui, j'attends le train pour Nice.
2. Oui, j'attends depuis quarante minutes.
3. Oui, je perds patience.
4. Oui, j'entends l'annonce du départ.
5. Oui, je réponds à la question du contrôleur.
6. Oui, quand je descends à Nice, je suis fatigué(e).

Exercices

A Les voyageurs. Répondez d'après les dessins. *(Answer according to the illustrations.)*

1. Les voyageurs attendent le train sur le quai?
2. Ils attendent le train dans la salle d'attente?

3. Ils perdent leurs billets?

4. Ils entendent l'annonce du départ de leur train?
5. Ils descendent du train?

B De petites conversations dans la gare. Complétez. *(Complete.)*

1. attendre
   - MARTIN: Vous ___ depuis combien de temps?
   - PIERRE: Nous ___ depuis cinq minutes. C'est tout.

2. perdre
   - CLAUDE: Le train pour Washington est en retard et nous ___ patience.
   - MARIE: Vous ___ patience? Pourquoi?
   - CLAUDE: Mais il a un retard de deux heures!

3. descendre
   - GEORGES: Le porteur ___ vos bagages du train?
   - ANNE: Absolument pas! Nous ___ nos bagages nous-mêmes.
   - GEORGES: Vous ne voulez pas d'aide?

C Je vais à Nice en train. Répondez par «oui». *(Answer "yes.")*

1. Tu attends le train pour Nice?
2. Tu attends depuis quarante minutes?
3. Tu perds patience?

4. Tu entends l'annonce du départ?
5. Tu réponds à la question du contrôleur?
6. Quand tu descends à Nice, tu es fatigué(e)?

208   CHAPITRE 8

COOPERATIVE LEARNING

Have students work in groups of three. Two students question each other to find out what each is in the habit of losing. The third student listens in and reports.

É1: ___, qu'est-ce que tu perds?
É2: Moi, je perds toujours mes ___. Et toi, qu'est-ce que tu perds?
É1: Moi, je perds mes ___.

É3: <u>(1)</u> perd ses ___ et <u>(2)</u> perd ses ___. Ils (ne) perdent (pas) la même chose.

D **Dans la salle d'attente.** Complétez. *(Complete.)*

Les voyageurs ___ (attendre) le train dans la salle d'attente. Marc ___
$\overline{1}$ $\overline{2}$
(attendre) le train pour Saint-Malo. Ah, voilà son ami, Luc.

MARC: Bonjour, Luc. Quelle surprise! Tu ___ (attendre) quel train?
$\overline{3}$

LUC: J'___ (attendre) le train pour Saint-Malo.
$\overline{4}$

MARC: Sans blague! Tu vas à Saint-Malo? Pas vrai. Moi aussi, j'y vais.

Les deux garçons ___ (entendre) l'annonce du départ de leur train. Leur train
$\overline{5}$
part du quai cinq. Ils vont au quai. Les voyageurs qui arrivent ___ (descendre)
$\overline{6}$
leurs bagages du train. Ils ___ (descendre) leurs bagages sur le quai.
$\overline{7}$

Le contrôleur crie: «En voiture! En voiture!» Tout le monde monte dans le train.

Le contrôleur demande aux garçons où ils vont. Luc ___ (répondre) à la
$\overline{8}$
question du contrôleur. Il ___ (répondre): «À Saint-Malo.»
$\overline{9}$

Les adjectifs démonstratifs   *Pointing Out People or Things*

1. You use the demonstrative adjectives to point out people or things. In English
   the demonstrative adjectives are "this," "that," "these," and "those." Study the
   following forms of the demonstrative adjectives in French.

	SINGULIER	PLURIEL
FÉMININ	cette voiture cette amie	ces voitures ces amies
MASCULIN	cet ordinateur cet horaire ce train ce billet	ces ordinateurs ces horaires ces trains ces billets

2. Note that you use *cet* before a masculine noun beginning with a vowel or
   silent *h.*

   cet élève   cet horaire   cet hôtel

3. There is only one plural form, *ces.* Note the liaison with words that begin
   with a vowel or silent *h.*

   ces élèves   ces horaires   ces hôtels

CHAPITRE 8   **209**

INDEPENDENT PRACTICE

Assign any of the following:
1. Exercises, pages 208–209
2. Workbook, *Structure: A–C,* pages 73–74
3. Communication Activities Masters,
   Structure: A, page 39
4. CD-ROM, Disc 2, pages 207–209

PRESENTATION *(continued)*
Exercice D
   Exercise D should be done with
books open.
Extension of *Exercice D*
   After completing the exercice,
have students make up questions
about the story in the exercise.
Then have them answer their own
questions. Finally, have them
retell the story in their own words.

ANSWERS
Exercice D
1. attendent
2. attend
3. attends
4. attends
5. entendent
6. descendent
7. descendent
8. répond
9. répond

Les adjectifs démonstratifs

PRESENTATION *(page 209)*
A. Go over the grammatical
   explanation with the class.
B. Have students repeat the following sequence as you write
   the forms of the demonstrative
   adjective on the board: *Cette
   voiture, cet ordinateur, ce train.*
   Have them take note how the
   sound gets softer.
C. Explain that there is only one
   plural form: *ces voitures, ces
   ordinateurs, ces trains.*
D. Collect as many objects as you
   can that students can identify
   in French and put them on a
   table. Call some students to the
   table and ask them to give you
   specific items: *Donne-moi ce
   cahier. Donne-moi cette cassette.*
   This gives students listening
   practice with demonstrative
   adjectives. After completing the
   exercises that follow, redo this
   activity as a paired activity and
   have the students produce all
   the forms of the demonstrative
   adjective on their own. (See
   PAIRED ACTIVITY, bottom
   of page 210.)
E. Point out the liaison in steps 2
   and 3.

209

Exercices

Exercices A and B

Exercises A and B can be done with books closed, open, or once each way. You may wish to use the recorded version of Exercise B.

ANSWERS

Exercice A

1. Oui, cette fille est intelligente.
2. Non, cette amie n'est pas sympa.
3. Non, cet élève n'est pas sérieux.
4. Non, cet ami n'est pas amusant.
5. Oui, ce copain est aimable.
6. Oui, ce prof est intéressant.
7. Oui, ces filles sont françaises.
8. Oui, ces garçons sont américains.
9. Oui (Non), ces copains (ne) vont (pas) au même lycée.

Exercice B

1. Je parle de ce garçon.
2. ... de cette amie.
3. ... de cet ami.
4. ... de ces élèves.
5. ... de ces profs.
6. ... de cette maison.
7. ... de ces voitures.
8. ... de ce livre.
9. ... de ces cassettes.
10. ... de ces journaux.

Exercices

A **Cette personne ou cet individu.** Répondez d'après les dessins. (*Answer according to the illustrations.*)

1. Cette fille est intelligente?   2. Cette amie est sympa?   3. Cet élève est sérieux?

4. Cet ami est amusant?   5. Ce copain est aimable?   6. Ce prof est intéressant?

7. Ces filles sont françaises?   8. Ces garçons sont américains?   9. Ces copains vont au même lycée?

B **Tu parles de qui?** Répondez d'après le modèle. (*Answer according to the model.*)

> **Tu parles de quelle fille?**
> *Je parle de cette fille.*

1. Tu parles de quel garçon?
2. Tu parles de quelle amie?
3. Tu parles de quel ami?
4. Tu parles de quels élèves?
5. Tu parles de quels profs?
6. Tu parles de quelle maison?
7. Tu parles de quelles voitures?
8. Tu parles de quel livre?
9. Tu parles de quelles cassettes?
10. Tu parles de quels journaux?

PAIRED ACTIVITY

Collect as many objects as possible that students can identify in French and put them on a table. Have students work in pairs. Pointing to the object, one says: *Donne-moi ce (cette)* ___. The second one picks it up and says, *Voilà le (la)* ___. They then reverse roles.

Le verbe *mettre* au présent *Describing People's Activities*

1. The verb *mettre,* "to put" or "to place," is irregular. Study the following forms.

METTRE			
je	mets	nous	mettons
tu	mets	vous	mettez
il		ils	
elle	met	elles	mettent
on			

Je mets les billets dans mon sac à dos.

2. The verb *mettre* has several additional meanings.

    a. **Il met le couvert.**   *He sets the table.*
    b. **Il met la télé.**   *He turns on the TV.*

On peut laisser ses affaires à la consigne automatique.

Exercices

A **On met le couvert.** Répondez. *(Answer.)*

1. Tu mets le couvert pour le dîner?
2. Tu mets le couteau à gauche ou à droite de l'assiette?
3. Tu mets la cuillère à côté du couteau ou à côté de la fourchette?
4. Tu mets une nappe et des serviettes?

B **La consigne automatique.** Complétez avec «mettre». *(Complete with mettre.)*

Le garçon est à la gare. Il ___ son billet dans son sac à dos. Il veut laisser son
                1
sac à dos à la consigne automatique. Il ___ son sac à la consigne. Il ___ une
                                2   3
pièce de cinq francs dans la consigne automatique.

C **Pas le garçon—les garçons!** Dans l'Exercice B, remplacez *le garçon* par *les garçons* et faites les changements nécessaires. *(Change* le garçon *to* les garçons *in Exercise B and make the necessary changes.)*

D **Vous mettez…** Répondez en utilisant «nous». *(Answer with* nous.)

1. Vous mettez la radio le soir?
2. Vous mettez la télé après les cours?
3. Vous mettez la chaîne 2 à la télé?
4. Vous mettez les magazines dans le sac à dos?

INDEPENDENT PRACTICE

Assign any of the following:
1. Exercises, pages 210–211
2. Workbook, *Structure: D–G,* pages 74–75
3. Communication Activities Masters, *Structure: B–C,* pages 40–42
4. Computer Software, *Structure*
5. CD-ROM, Disc 2, pages 210–211

Le verbe mettre *au présent*

PRESENTATION *(page 211)*

Go over the forms of the verb *mettre.* Write them on the board. Underline the double consonant in the plural forms.

Extension Students are given some practice with the verbs *permettre* and *promettre* in the Workbook. You may wish to teach these verbs very quickly and indicate to students that they are conjugated the same way as *mettre.*

Note The use of *mettre* meaning "to put on clothes" will be taught in Chapter 10.

RETEACHING

To reintroduce previously learned vocabulary, have students give items one can turn on, using the verb *mettre* (la télé, la radio, l'ordinateur).

Exercices

PRESENTATION *(page 211)*

Exercices A and D

Exercises A and D can be done with books closed, open, or once each way.

Exercices B and C

Exercises B and C are to be done with books open.

ANSWERS

Exercice A

Answers should employ **Je (ne) mets (pas)…**

Exercice B

Answers 1.–3. will be **met.**

Exercice C

Les garçons sont à la gare. Ils mettent leurs billets dans leurs sacs à dos. Ils veulent laisser leurs sacs à dos à la consigne automatique. Ils mettent leurs sacs à la consigne. Ils mettent une pièce de cinq francs dans la consigne automatique.

Exercice D

Answers will vary but should employ either **Oui, nous mettons…** or **Non, nous ne mettons pas…**

Bell Ringer Review

Write the following on the board or use BRR Blackline Master 8-4: Use each phrase in a sentence: un billet aller-retour, à quelle heure, de quel quai.

PRESENTATION *(page 212)*

A. Have students close their books and watch the Conversation Video or listen as you either read the conversation to them or play Cassette 5B/CD-5.

B. Have them repeat each line once.

C. Have two students read the conversation with as much expression as possible.

Note In the CD-ROM version, students can play the role of either one of the characters and record the conversation.

Exercices

PRESENTATION *(page 212)*

Extension of *Exercice A*

After completing the exercise, have the students retell the story in their own words.

ANSWERS

Exercice A

1. Marie va à Avignon.
2. Elle veut un aller-retour.
3. Elle voyage en seconde.
4. C'est cent vingt francs.
5. Le prochain train part à quatorze heures huit.
6. Il part du quai numéro sept.

Exercice B

1. Marie va à Avignon.
2. Elle veut un aller-retour.
3. Elle voyage en seconde classe.
4. Le billet coûte cent vingt francs.
5. Le prochain train part à deux heures huit de l'après-midi.
6. Il part du quai numéro sept.

212

Scènes de la vie *Au guichet*

MARIE: Un billet pour Avignon, s'il vous plaît.
L'EMPLOYÉE: Un aller simple ou un aller-retour?
MARIE: Un aller-retour en seconde, s'il vous plaît.
L'EMPLOYÉE: Bien, Mademoiselle.
MARIE: C'est combien, le billet?
L'EMPLOYÉE: Cent vingt francs, s'il vous plaît.
MARIE: Voilà. Le prochain train part à quelle heure?
L'EMPLOYÉE: À quatorze heures huit, quai numéro sept.
MARIE: Merci, Madame.

A **Un billet pour Avignon.** Répondez d'après la conversation. *(Answer according to the conversation.)*

1. Où va Marie?
2. Elle veut un aller simple ou un aller-retour?
3. Elle voyage en quelle classe?
4. C'est combien, le billet?
5. Le prochain train part à quelle heure?
6. Il part de quel quai?

B **À la gare.** Corrigez les phrases. *(Correct the sentences.)*

1. Marie va à Perpignan.
2. Elle veut un aller simple.
3. Elle voyage en première classe.
4. Le billet coûte vingt dollars.
5. Le prochain train part à deux heures du matin.
6. Il part du quai numéro huit.

212   CHAPITRE 8

DID YOU KNOW?

Have students locate Avignon on the map, page 504 (Part A, page 237) or use the Map Transparency. Tell them Avignon is a beautiful city in the south of France. It is surrounded by 14th-century ramparts. In that same century there was a schism in the Catholic Church: one pope lived in Rome and another (the so-called "anti-pope") lived in Avignon. Standing high on a hill, the Palais des Papes (the papal palace) is today one of the city's most-visited sites. Every summer there is a famous theater festival held outdoors in its courtyard. The bridge made famous by the children's song *Sur le pont d'Avignon* once spanned the Rhône but is now only a fragment of its former self.

Prononciation *Les sons /õ/ et /ẽ/*

Listen to the difference between the nasal sound /ã/ as in *cent* and the two other nasal sounds, /õ/ as in *sont*, and /ẽ/ as in *cinq: cent/sont/cinq*. Repeat the following words with the sounds /õ/ and /ẽ/.

annonce	cinq
consigne	copain
non	train

Now repeat the following sentences that combine all three nasal sounds.

On annonce le train dans combien de temps?
Nous attendons des copains.

son train

Activités de communication orale

A **Renseignements.** You're at the Information desk at one of the Paris train stations and need some information. Have a conversation with the SNCF agent (your partner) using the following words or expressions.

à quelle heure
le prochain train pour…
quelle voie
quel quai
voyager en première (en seconde)
c'est combien
changer de train

B **À la gare.** Work with a classmate. The two of you are traveling in France on a railpass. Using the following expressions, tell how you spent an hour at the train station.

aller au kiosque
regarder et acheter
vendre les billets au guichet
attendre le train
entendre l'annonce
composter le billet
partir de la voie 5
monter dans le train

Prononciation
PRESENTATION *(page 213)*

A. Model the key words *son train*. Have the students repeat in unison after you. Model the other words in similar fashion or use Cassette 5B/CD-5.
B. You may wish to give the following *dictée:*
   Non, les cinq copains ne vont pas à la consigne.
   Mon copain annonce le départ du train.
C. Other resources you may wish to use in presenting this section include: Student Tape Manual, Teacher's Edition, *Activités F–H,* pages 96–97, Cassette 5B/CD-5: *Prononciation,* and Pronunciation Transparency P-8.

Activités de communication orale
PRESENTATION *(page 213)*
Activités A and B
   Have students choose the activity they would like to do. Since students are pretending that they are using their French in the real world, don't be concerned if they make some errors.

ANSWERS
Activités A and B
   Answers will vary.

INDEPENDENT PRACTICE

Assign any of the following:
1. Exercises and activities, pages 212–213
2. CD-ROM, Disc 2, pages 212–213

LECTURE ET CULTURE

Bell Ringer Review

Write the following on the board or use BRR Blackline Master 8-5. Write as many words as you can that are connected to the following topics.

le train   l'avion
la gare   l'aéroport

READING STRATEGIES
(*page 214*)

Pre-reading

Have students quickly scan the reading to find as many cognates as they can.

Reading

A. To vary the presentation of the *Lecture,* have students close books and listen as you read to them. Have them tell you what they understood.

B. Read the story again as they follow along in their books. This is beneficial for the acquisition of receptive skills.

C. Call on individuals to read two or three sentences. After every two or three sentences, ask comprehension questions.

Post-Reading

After completing the reading selection, go over the exercises.

Note Students may listen to a recorded version of the *Lecture* on the CD-ROM.

GEOGRAPHY CONNECTION

Locate Marseille on the map of France, page 504 (Part A, page 237). Marseille, a very ancient city founded by the Greeks in the 6th century B.C., is France's second-largest city and its largest port. The *Vieux-Port* area is the bustling heart of the city. Marseille's many seafood restaurants serve the city's specialty, bouillabaisse.

LES TRAINS EN FRANCE

Monique Lutz est une élève américaine qui voyage en France. En ce moment elle est à la Gare de Lyon à Paris. Monique part pour Marseille. Sa tante Hélène habite à Marseille. Tous les trains qui partent pour le sud-est partent de cette gare.

Monique va au guichet où elle achète un billet aller-retour en seconde classe. Elle a de la chance[1]. Il n'y a pas de queue devant le guichet. Monique va passer toute la nuit[2] dans le train. Elle va dormir dans le train. Elle réserve (loue) une couchette.

Monique entend l'annonce du départ du train. Elle va sur le quai et monte dans le train. C'est un vieux train à compartiments. En seconde il y a huit places dans chaque compartiment. Monique trouve sa place et met son sac à dos dans le filet au-dessus de sa tête[3].

Le train part à l'heure précise comme toujours en France. Les trains sont excellents. Le train commence à rouler vite[4]. Le contrôleur arrive. Il vérifie les billets. Il parle un peu à Monique. Il est sympa, le contrôleur. Il explique: Si elle a faim et veut manger quelque chose, il y a une voiture-restaurant et un grill-express dans le train. À la voiture-restaurant on sert un dîner complet à prix fixe. Le grill-express offre de la restauration rapide: un petit sandwich, une pizza ou une boisson, par exemple.

[1] a de la chance   *is lucky*
[2] toute la nuit   *the whole night*
[3] filet au-dessus de sa tête   *overhead rack*
[4] commence à rouler vite   *begins to speed up*

LEARNING FROM PHOTOS

1. Have the students say as much as they can about the two photographs on this page.

2. The top photo is of the Gare de Lyon in Paris. It was built between 1895 and 1900, around the same time as the Eiffel Tower, in the Belle Époque style of architecture popular in the late 19th—early 20th century. Why is it called the Gare de Lyon if it is in Paris? Because trains going southeast to Lyon and beyond leave from this station.

3. The bottom photo is of the Perrache station in Lyon. Not far from it is the La Part-Dieu station. The most recent of that city's three train stations, Satolas, with its futuristic architecture, is pictured on page 199.

Étude de mots

A **Quel est le nom?** Trouvez le nom qui correspond au verbe. (*Find the noun that corresponds to the verb.*)

1. réserver   a. la location
2. louer   b. l'explication
3. annoncer   c. la réservation
4. partir   d. l'annonce
5. commencer   e. le service
6. arriver   f. le départ
7. expliquer   g. le commencement
8. servir   h. l'arrivée

Compréhension

B **Le voyage de Monique.** Choisissez la bonne réponse. (*Choose the correct answer.*)

1. Monique est (française, américaine).
2. Elle voyage (avec ses copains, seule).
3. Elle est à (la Gare de Lyon, la Gare Montparnasse).
4. Elle va à (Nice, Marseille).
5. Elle achète un (aller simple en première, aller-retour en seconde).
6. Monique (est debout, a une place dans un compartiment).
7. C'est un (vieux, nouveau) train.
8. Monique met son sac à dos (sous son siège, dans le filet au-dessus de sa tête).
9. Le train part (en retard, à l'heure précise).
10. On sert un dîner à prix fixe (au grill-express, à la voiture-restaurant).

DÉCOUVERTE CULTURELLE

À Paris il y a cinq grandes gares. Les trains qui partent de chaque gare vont dans des directions différentes. Il y a combien de gares dans votre ville (ou une ville près de chez vous)? Qu'est-ce que vous pensez[1]: Le train est un moyen de transport important en France ou aux États-Unis?

Les trains en France partent presque toujours à l'heure. Ils ne partent pas en retard. Les retards ne sont pas du tout fréquents. Et les trains en France sont très propres[2]. Il y a des différences entre les trains en France et les trains en Amérique?

[1] pensez   *think*   [2] propres   *clean*

CRITICAL THINKING ACTIVITY

(*Thinking skill: evaluating information*)
Put the following on the board or on an overhead transparency.

M. Benoît est français. Il fait un voyage aux États-Unis. Il veut aller de New York à Washington, D.C. Il peut aller de New York à Washington en train ou en avion. Le voyage en train dure trois heures et demie; en avion, une heure. En général, les gares sont en ville et les aéroports ne sont pas en ville. Mais à Washington, ce n'est pas le cas. L'aéroport National n'est pas très loin de la ville. Il est tout près. Mais les aéroports de New York sont assez loin du centre-ville. Qu'est-ce que M. Benoît doit (*should*) faire? Pourquoi?

PRESENTATION *(pages 216–217)*

The purpose of this section is to have the students enjoy the photographs. If they want to say something about the photographs in French, let them. If they have any questions, let them ask them.

Note Students will learn more about the TGV in future chapters, particularly in Chapter 4 of *À bord*. For now, they can see another photo of it at the beginning of this chapter, on pages 198–199.

Note In the CD-ROM version, students can listen to the recorded captions and discover a hidden video behind one of the photos.

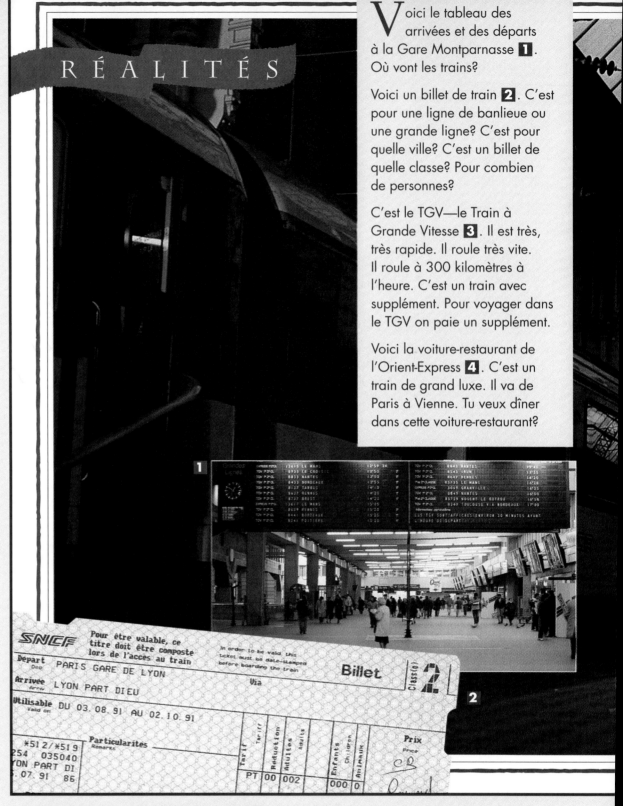

RÉALITÉS

Voici le tableau des arrivées et des départs à la Gare Montparnasse **1**. Où vont les trains?

Voici un billet de train **2**. C'est pour une ligne de banlieue ou une grande ligne? C'est pour quelle ville? C'est un billet de quelle classe? Pour combien de personnes?

C'est le TGV—le Train à Grande Vitesse **3**. Il est très, très rapide. Il roule très vite. Il roule à 300 kilomètres à l'heure. C'est un train avec supplément. Pour voyager dans le TGV on paie un supplément.

Voici la voiture-restaurant de l'Orient-Express **4**. C'est un train de grand luxe. Il va de Paris à Vienne. Tu veux dîner dans cette voiture-restaurant?

DID YOU KNOW?

One of Paris's oldest train stations, the Gare Montparnasse was originally built in 1840. The present station dates from the 1960's. In 1895 a terrible accident occurred: the locomotive and first car of a train went through the glass and crashed onto the sidewalk below.

Trains leaving from the Gare Montparnasse go to Brittany and the west of France.

The streets around the station bear the names of Breton locales. The rue de Rennes, for example, evokes one of Brittany's main cities. (Students can locate the station, the rue de Rennes, and the Tour Montparnasse on the map of Paris, page 505 [Part A, page 238] or use the Map Transparency.)

The enormous Tour Montparnasse, a skyscraper built as part of the new station, outraged many Parisians, who considered it an eyesore that ruined the Paris skyline.

ADDITIONAL PRACTICE

Assign any of the following:
1. Workbook, *Un Peu Plus*, pages 76–77
2. Student Tape Manual, Teacher's Edition, *Deuxième Partie*, pages 97–98
3. Situation Cards, Chapter 8

COOPERATIVE LEARNING

Have students work in small groups and make up as many questions as possible about Communication Transparency C-8. Groups take turns asking and answering their questions.

CULMINATION

RECYCLING

The *Activités de communication orale* and *écrite* combine vocabulary and structures from this and from previous chapters into situations having to do with train travel. Success in these activities should be based on fluency and intelligent use of language for communication rather than on strict grammatical accuracy.

INFORMAL ASSESSMENT

Oral Activities A and B may be used to evaluate speaking. For one-on-one work, you may wish to assume the role of *Élève 1* in Activity B. Use the evaluation criteria given on page 34 of this Teacher's Wraparound Edition.

Activités de communication orale

PRESENTATION (page 218)

Activité A

In the CD-ROM version of this activity, students can interact with an on-screen native speaker.

ANSWERS

Activité A

Answers will vary but may include the following.

1. Oui, je vais à Nice ce soir.
2. Oui, je voudrais réserver une couchette.
3. En première (seconde).
4. Je voudrais un aller simple (un aller-retour).

Activité B

Answers will vary.

Activités de communication écrite

ANSWERS

Activités A and B

Answers will vary.

Activités de communication orale

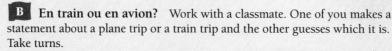

A **Le train de nuit.** You're at the French rail information office in Paris and you'd like to take the night train to Nice. Answer the reservations agent's questions.

1. Alors, vous allez à Nice ce soir?
2. Vous voulez réserver une couchette?
3. En première ou en seconde?
4. Vous voulez un aller simple ou un aller-retour?

B **En train ou en avion?** Work with a classmate. One of you makes a statement about a plane trip or a train trip and the other guesses which it is. Take turns.

> Élève 1: On part de la gare.
> Élève 2: C'est un voyage en train.

Activités de communication écrite

A **Un horrible voyage.** Imagine you're taking a train trip and everything goes wrong in the station and on the train. Write a paragraph about your experience.

B **Un voyage extraordinaire.** Your parents have given you a train trip to the U.S. city of your choice. Write a paragraph telling what city you'd like to visit and why. Say when you'd like to go, who you'd like to go with, and how much time you plan to spend there. Say what you'll do at the train station the day you leave.

218 CHAPITRE 8

LEARNING FROM PHOTOS

Have students find the following information in the photos:
1. station and city passenger is leaving from
2. station and city passenger is going to
3. how long ticket is valid
4. first day ticket is valid
5. class passenger is traveling in
6. type of ticket

DID YOU KNOW?

Le Train Bleu restaurant, the cover of whose menu is featured on page 219, was listed as a historic monument in 1972. It opened in 1901 in the Gare de Lyon and has been left unchanged since then. Forty-one mural paintings of that period adorn the walls, illustrating the cities served by the train that bears the name.

Réintroduction et recombinaison

A **Elle choisit sa place dans le train.** Complétez. *(Complete.)*

1. Madame Lacoste réserve (loue) une place à l'avance. Elle ___ sa place dans le train. (choisir)
2. Elle ___ une place dans un compartiment de première classe. (choisir)
3. Elle ___ à réserver la place qu'elle veut. (réussir)
4. Elle ___ aux règlements. Elle ne fume pas. (obéir)
5. Le train ___ de la gare à l'heure précise. (sortir)
6. Il ___ pour Marseille. (partir)
7. On ___ le dîner à la voiture-restaurant. (servir)
8. Il y a des serveurs qui ___ le dîner. (servir)
9. Les voyageurs ___ dans les couchettes. (dormir)
10. Le voyageur qui ___ dans une voiture-lit paie un supplément. (dormir)

Vocabulaire

NOMS
la gare
le guichet
le billet
l'aller simple (m.)
le billet aller-retour
l'horaire (m.)
le voyageur
la consigne
la consigne automatique
la salle d'attente
le haut-parleur
l'annonce (f.)
le kiosque
le journal
le porteur
le quai
la voie

le train
la voiture
la couchette
la voiture-lit
le lit
le couloir
le contrôleur

VERBES
changer (de)
composter

laisser
monter
attendre
descendre
entendre
mettre
perdre
répondre
vendre

ADJECTIFS
assis(e)
prochain(e)
quelques

AUTRES MOTS
ET EXPRESSIONS
être à l'heure
être en avance
être en retard
faire la queue
mettre le couvert
perdre patience
debout
depuis
en première
en seconde
la plupart

Le Train Bleu
GARE DE PARIS-LYON 1er ÉTAGE

Par Judith GLANCY. Tiré de son livre "Not a Station, but a Place" (Synergistic Press)
Spécialités Lyonnaises et Foréziennes.
Dans le Décor classé "Belle Epoque" le plus étonnant de Paris.
RESERVATIONS : 43.43.09.06

Réintroduction et recombinaison

RECYCLING

Exercise A recycles the forms of regular and irregular *-ir* verbs, allowing students to re-use them and to differentiate between their conjugations.

ANSWERS

Exercice A
1. choisit
2. choisit
3. réussit
4. obéit
5. sort
6. part
7. sert
8. servent
9. dorment
10. dort

ASSESSMENT RESOURCES

1. Chapter Quizzes
2. Testing Program
3. Situation Cards
4. Communication Transparency C-8
5. Performance Assessment
6. Computer Software: Practice/Test Generator

VIDEO PROGRAM

INTRODUCTION (26:15)

EMMANUEL PREND LE TRAIN (26:52)

FOR THE YOUNGER STUDENT

Have students create a play about a train trip and present it to the class. The characters in the play are:
un voyageur
une voyageuse
l'agent de la SNCF
le porteur
le contrôleur

STUDENT PORTFOLIO

Written assignments for your students' portfolios may include the *Activités de communication écrite* on page 218 and the *Mon Autobiographie* section from the Workbook on page 78.

Note Students may create and save both oral and written work using the Electronic Portfolio feature on the CD-ROM.

OPTIONAL MATERIAL

La Cuisine

PRESENTATION *(pages 220–223)*

This cultural material is presented for students to enjoy and to help them gain an appreciation of the francophone world. Since the material is *optional,* you may wish to have students read it on their own as they look at the colorful photographs that accompany it. They can read it at home or you may wish to give them a few minutes in class to read it.

If you prefer to present some of the information in greater depth, you may follow the suggestions given for other reading selections throughout the book. Students can read aloud, answer questions asked by the teacher, ask questions of one another in small groups, and, finally, give a synopsis of the information in their own words using French.

MORE ABOUT THE PHOTOS

Photo 1 Rwanda is a Central African country whose official languages are French and Kinyarwanda. Rwanda and its neighbor Burundi have been in great turmoil in recent years, owing to a longstanding conflict between the two major ethnic groups, the Hutus and the Tutsis. The Hutus are farmers and the Tutsis are herders.

LA CUISINE

Les pays francophones ont en commun la langue qu'on y parle, le français. Mais leurs cuisines respectives sont très différentes parce que la cuisine d'un pays est déterminée par ce que produit la terre (land) de cette région. Ce qu'on cultive dans un pays tropical est totalement différent de ce qu'on cultive dans un pays où il fait presque toujours froid (cold). Pour cette raison, par exemple, les Ivoiriens et les Canadiens ne mangent pas la même nourriture.

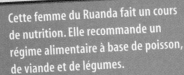

1 Cette femme du Ruanda fait un cours de nutrition. Elle recommande un régime alimentaire à base de poisson, de viande et de légumes.

2 Dans de nombreux pays francophones, on fait ses courses dans un marché en plein air. Sur la place de la Palud à Lausanne, en Suisse, les jours de marché sont le mercredi et le samedi. En France aussi, il y a souvent le marché deux fois par semaine, le mercredi ou le jeudi, et le samedi.

3 Dans ce marché à Conakry, la capitale de la Guinée, on vend des fruits tropicaux, typiques de la région. Les ananas *(pineapples)* et les mangues sont délicieuses. En Guinée, on ne sert pas de gâteaux ou de pâtisseries pour le dessert. On mange des fruits.

4 Les supermarchés existent partout. Dans ce supermarché moderne à Abidjan il y a un très grand choix de produits. Ces deux femmes ivoiriennes bavardent un peu pendant qu'elles font leurs courses.

5 Manger du poisson, c'est bon pour la santé *(health)*. Les gens qui habitent près de la mer mangent beaucoup de poisson. Ces pêcheurs du Bénin ont fait une très belle pêche. Leur filet est plein *(full)*. Ici à Ganvié, la pêche est l'industrie principale. Ganvié est un petit village près de Cotonou, la ville principale du Bénin.

LE MONDE FRANCOPHONE

Photo 3 In many countries the whole family works at the market. Note that in this photo of the outdoor market in Conakry, entire families are there selling the fruit. The stalls at the outdoor markets in Europe are frequently family businesses, too. In Haiti, Martinique, and Guadeloupe, as well as in some other countries, it is the women who work in the markets.

Photo 5 Ganvié is a small town of 12,000 whose inhabitants live in bamboo huts on stilts in Lake Nokué. They earn their living almost exclusively from fishing. The men put branches on the muddy bottom of the lagoon. When the leaves begin to decompose, the fish congregate there to feed. After many days, the men return to catch the fish in a net. Then the women sell them. Since the entire area around Ganvié is swampy because of the many lakes and lagoons, the women navigate the waters in *pirogues*. *Pirogues* are a type of canoe that can be operated by sails or paddles. The *pirogues* (loaded with fish, spices, and fruit) are a very colorful sight.

Photo 6 *Maghreb* is an Arabic word meaning "*là où le soleil se couche.*" The official language of Algeria, Morocco, and Tunisia is Arabic, and the chief religion of the region is Islam.

Le thé à la menthe is a favorite drink in North Africa. People will often stop to have tea at a café. You frequently see a person walking down the street with a tray laden with glasses of tea that he or she is taking to colleagues at the office for a tea break. A glass of tea is almost always offered to customers at a store. Many tourists think of this custom as a way of getting them to linger and therefore to buy more, but it is truly a gesture of hospitality.

Photo 7 A number of West African dishes are made with okra. Rice is eaten with many African dishes, but it is expensive because it must be imported. In the Sahel, millet is a staple. (The Sahel was originally the name given to the area near the coast of Algeria and Tunisia. Today the term is used to encompass the entire geographical area that borders the south of the Sahara.) Sauces are the heart of West African cooking. Each country has its own specialty. Peanuts are used in many sauces. Fresh fruits are eaten for dessert.

It is important for students to understand that the cooking of West Africa is totally different from that of North Africa.

Note that many vegetables used in Southern cooking in the United States are of African origin. They were introduced here by African Americans. Here are several of them and their French translations.

    sweet potato: *la patate douce*
    yam: *l'igname*
    okra: *le gombo, l'okra*
    manioc: *le manioc*

LE MONDE FRANCOPHONE

6 Cet homme est marocain. Il sert du thé à la menthe. Le thé à la menthe est très apprécié dans les pays du Maghreb. Les pays du Maghreb sont les trois pays de l'Afrique du Nord où on parle le français en deuxième langue: le Maroc, l'Algérie et la Tunisie. On sert le thé à la menthe dans un verre, pas dans une tasse. On met beaucoup de sucre dans le thé. Les Maghrébins aiment leur thé bien sucré.

7 La cuisine d'une région dépend des produits ou des aliments disponibles *(available)*. Voici une recette pour un plat africain—les mélongènes *(eggplant)* aux gombos. Le gombo, c'est aussi l'okra. Okra est un mot swahili, une langue parlée par de nombreux Africains. L'okra est à la base de beaucoup de plats africains.

7 *Mélongènes aux gombos*
(Eggplant with okra)

AFRIQUE OCCIDENTALE
Prép: 25 min. - Cuiss: 35 min.
Repos: 1 h. - 4 pers.

500 g de gombos	4 mélongènes
coriandre	2 oignons
cumin	1 gousse d'ail
piment doux	huile d'arachide
gros sel	

Laver les mélongènes et les couper dans le sens de la longueur. Les saupoudrer de gros sel. Les laisser dégorger durant 1 heure.

Les essuyer. Les dorer à l'huile. Les égoutter sur du papier absorbant.

Laver et sécher les gombos. Ôter les queues. Fendre les fruits, les garnir d'épices (coriandre, cumin). Faire cuire le tout pendant 20 minutes dans l'eau bouillante. Égoutter.

Faire fondre dans une cuillerée d'huile une gousse d'ail et les oignons hachés ainsi que le piment doux. Mêler cette préparation aux mélongènes.

Remettre les mélongènes à feu doux en y incorporant les gombos. Servir chaud.

You may wish to give students the following translation of the recipe.

Eggplant and okra

1 lb. okra	4 eggplants
coriander	2 onions
cumin	1 clove garlic
1 sweet pepper	peanut oil
coarse salt	

Wash eggplants and cut in half lengthwise. Sprinkle with coarse salt. Let sweat for an hour.

Dry off and brown in oil. Drain on paper towels.

Wash and dry okra. Remove stem ends. Split okra pods and add spices (coriander and cumin). Cook for 20 minutes in boiling water. Drain.

Heat 1 tbs oil and sauté garlic, chopped onions, and sweet pepper in it. Add to eggplant. Put eggplant back on low heat and stir in okra. Serve warm.

8 La cuisine «cajun» de la Louisiane est renommée (célèbre). Comme la Louisiane est près du Golfe du Mexique, de nombreux plats cajuns sont faits avec des crevettes, des crabes et des écrevisses (*crawfish*). On sert ces plats, qui sont souvent très épicés (*spicy*), avec du riz—le riz qu'on cultive dans les bayous de la Louisiane.

9 Le sirop d'érable (*maple*) est un produit typiquement canadien. Les Canadiens font des desserts et des bonbons délicieux avec le sirop d'érable. Dans l'Ontario et au Québec, on tire la sève (*sap*) des érables au début du printemps (*spring*). Il faut 40 litres de sève pour faire un litre de sirop! Ce petit garçon goûte le sirop. Que c'est bon!

LE MONDE FRANCOPHONE

Photo 8 Cajun cooking is popular in many areas of the United States. It is very spicy and features blackened fish or meat as well as a stew called gumbo, made with okra (*les gombos*). If there is a Cajun restaurant in your area, you may wish to plan a field trip to it.

Photo 9 In many Canadian towns there are fairs and festivities in the Spring, when the maple sap is tapped.

The sap from the maple tree is boiled until it becomes thick enough to make the maple syrup.

During the winter, children like to eat snowballs dipped in maple syrup.

OVERVIEW

This section reviews key grammatical structures and vocabulary from Chapters 5–8. The structure topics were first presented on the following pages: *-ir* verbs, pages 184 and 188; *-re* verbs, page 207; *aller,* page 133; *faire,* page 161; *mettre,* page 211; *pouvoir* and *vouloir,* page 164; the partitive, pages 158 and 159–160; contractions with *à* and *de,* page 135; possessive adjectives, pages 98 and 138.

REVIEW RESOURCES

1. Workbook, Self-Test 2, pages 79–82
2. Videocassette/Videodisc, Unit 2
3. Video Activities Booklet, Unit 2: Chapters 5–8, pages 18–33
4. Computer Software, Chapters 5–8
5. Testing Program, Unit Test: Chapters 5–8, pages 48–51
6. Performance Assessment
7. CD-ROM, Disc 2, *Révision:* Chapters 5–8, pages 224–227
8. CD-ROM, Disc 2, Self-Tests 5–8
9. CD-ROM, Disc 2, Game: *Le Labyrinthe*

Conversation

PRESENTATION *(page 224)*

Present the conversation using the procedures suggested in other chapters.

ANSWERS

Exercice A

1. Mireille est canadienne.
2. Elle va à New York.

224

Conversation   *Mireille va à New York.*

CHRISTIAN: Tu vas à New York, Mireille?
MIREILLE: Oui, j'y vais la semaine prochaine.
CHRISTIAN: Tu y vas en train ou en avion?
MIREILLE: Je préfère l'avion mais je n'ai pas beaucoup d'argent.
CHRISTIAN: Mais il y a un bon train qui fait Montréal–New York.
MIREILLE: Il part de Montréal à quelle heure?
CHRISTIAN: Il part à 10 heures et arrive à New York à 19 heures.

A   **Montréal–New York.**   Répondez d'après la conversation. (*Answer according to the conversation.*)

1. Mireille est canadienne ou américaine?
2. Où est-ce qu'elle va?
3. Quand est-ce qu'elle y va?
4. Comment est-ce qu'elle y va?
5. Elle préfère le train ou l'avion?
6. Il y a un bon train qui fait Montréal–New York?
7. Il part de Montréal à quelle heure?
8. Et il arrive à New York à quelle heure?

Structure

Les verbes en *-ir* et *-re*

1. Review the following forms of regular *-ir* and *-re* verbs.

FINIR	
je fin**is**	nous fin**issons**
tu fin**is**	vous fin**issez**
il/elle/on fin**it**	ils/elles fin**issent**

ATTENDRE	
j'attend**s**	nous attend**ons**
tu attend**s**	vous attend**ez**
il/elle/on attend	ils/elles attend**ent**

Montréal: La Place d'Armes dans la vieille ville

ADDITIONAL PRACTICE

Give the following expressions from the conversation and call on a student to use the expression in a sentence.
la semaine prochaine
beaucoup d'argent
à 10 h.
à 19 h.
à quelle heure

2. Review the following forms of the verbs *sortir*, *partir*, *servir*, and *dormir*.

sortir	je sors, tu sors, il/elle/on sort nous sortons, vous sortez, ils/elles sortent
partir	je pars, tu pars, il/elle/on part nous partons, vous partez, ils/elles partent
servir	je sers, tu sers, il/elle/on sert nous servons, vous servez, ils/elles servent
dormir	je dors, tu dors, il/elle/on dort nous dormons, vous dormez, ils/elles dorment

A **Un voyage en train.** Complétez. (*Complete.*)

Nous ___ (partir) en voyage. Maman ___ (attendre) devant le guichet. Elle
___ (choisir) deux places en seconde. Maman ___ (sortir) de l'argent et achète
les billets. Nous ___ (attendre) le train sur le quai. Le train ___ (partir) à
l'heure. Je ___ (sortir) les billets de mon sac à dos. Je ___ (donner) les billets
au contrôleur. Nous ___ (aller) à la voiture-restaurant. Je ___ (choisir) le
menu à prix fixe et Maman aussi ___ (choisir) le menu à prix fixe. Le serveur
___ (servir) le dîner. Nous ___ (finir) notre dîner. Après le dîner nous ___
(dormir) un peu. Les voyageurs ___ (descendre) du train à Nice, leur
destination.

Les verbes *aller, faire, mettre, pouvoir* et *vouloir*

Review the following forms of the irregular verbs below.

aller	je vais, tu vas, il/elle/on va nous allons, vous allez, ils/elles vont
faire	je fais, tu fais, il/elle/on fait nous faisons, vous faites, ils/elles font
mettre	je mets, tu mets, il/elle/on met nous mettons, vous mettez, ils/elles mettent
pouvoir	je peux, tu peux, il/elle/on peut nous pouvons, vous pouvez, ils/elles peuvent
vouloir	je veux, tu veux, il/elle/on veut nous voulons, vous voulez, ils/elles veulent

3. Elle y va la semaine prochaine.
4. Elle y va en train.
5. Elle préfère l'avion.
6. Oui, il y a un bon train qui fait Montréal–New York.
7. Il part à 10 heures.
8. Il arrive à New York à 19h.

Structure
Les verbes en -ir et -re

PRESENTATION (*pages 224–225*)

A. Have students repeat the forms of *finir* and *attendre,* then give you other -*ir* and -*re* verbs they know.
B. Write all forms of another -*ir* and another -*re* verb on the board. Underline the endings and have the class repeat them.
C. Point out that though the singular forms are pronounced the same, they are spelled differently. The final consonant sound of the *ils/elles* form is pronounced.
D. Review verbs like *sortir.* Have students repeat the plural forms first. Remind them to drop the final sound of the *ils/elles* form to get all the oral singular forms. Though pronounced the same, the singular forms are spelled differently.

ANSWERS
Exercice A

1. partons	9. allons
2. attend	10. choisis
3. choisit	11. choisit
4. sort	12. sert
5. attendons	13. finissons
6. part	14. dormons
7. sors	15. descendent
8. donne	

Les verbes aller, faire, mettre, pouvoir *et* vouloir

PRESENTATION (*page 225*)

Follow the same procedures suggested for the verbs *sortir,* etc.

1. **Elles veulent aller au cinéma?**
   Non, elles ne peuvent pas.
2. **Tu fais les courses?**
   Non, je ne peux pas.
3. **Il fait le dîner?**
   Il veut bien.
4. **Je veux faire un voyage.**
   Mais je ne peux pas.
5. **Vous mettez vos bagages à la consigne?**
   D'accord. Nous voulons bien.
6. **Ils vont prendre l'avion?**
   Non, ils ne veulent pas.

Le partitif

PRESENTATION (*page 226*)

You may wish to dramatize the partitive concept by holding some money in your hand and putting it down as you say: *J'ai de l'argent. Je n'ai pas d'argent* (showing empty hands). *J'ai des magazines. Je n'ai pas de magazines.*

ANSWERS

Exercice C

Answers will vary but may include the following:

Il y a du pain (des baguettes), des œufs, du lait, des bananes, du fromage, des carottes, de la viande.

Exercice D

1. **Non, je ne mange pas de bœuf; je n'aime pas le bœuf.**
2. **Non, je ne mange pas d'œufs aux fines herbes; je n'aime pas les œufs aux fines herbes.**
3. **Non, je ne mange pas de carottes à la crème; je n'aime pas les carottes à la crème.**
4. **Non, je ne mange pas de poulet; je n'aime pas le poulet.**
5. **Non, je ne mange pas de salade; je n'aime pas la salade.**
6. **Non, je ne mange pas de gâteau au chocolat; je n'aime pas le gâteau au chocolat.**

B **On ne peut pas.** Remplacez le mot en italique par le mot indiqué et faites tous les changements nécessaires. (*Replace the italicized word with the cue and make all necessary changes.*)

> *Tu* veux faire du latin? (vous)   Je ne peux pas.
> *Vous voulez faire du latin?*   Nous *ne pouvons pas.*

1. *Vous* voulez aller au cinéma? (elles)   Non, nous ne pouvons pas.
2. *François* fait les courses? (tu)   Non, il ne peut pas.
3. *Vous* faites le dîner? (il)   Nous voulons bien.
4. *Ils* veulent faire un voyage. (je)   Mais ils ne peuvent pas.
5. *Tu* mets tes bagages à la consigne? (vous)   D'accord. Je veux bien.
6. *Il* va prendre l'avion? (ils)   Non, il ne veut pas.

Le partitif

1. Remember that the partitive, "some," "any," is expressed in French by *de* + the definite article. *De* contracts with *le* to form *du* and with *les* to form *des*. In the negative *du*, *de la*, *de l'*, and *des* all become *de* or *d'*.

> Je veux *de l'*argent.   Je *ne* veux *pas d'*argent.
> J'ai *des* croissants.   Je *n'*ai *pas de* croissants.

2. Remember that *un* and *une* also become *de* or *d'* after a negative expression.

> Ils ont *une* maison à Nice.   Ils *n'*ont *pas de* maison à Nice.
> Nous avons *une* orange.   Nous *n'*avons *pas d'*orange.

C **Dans le chariot.** Dites ce qu'il y a dans le chariot. (*Tell what is in the cart.*)

D **J'ai faim.** Répondez d'après le modèle. (*Answer according to the model.*)

> Tu veux du poisson?
> *Non, je ne mange pas de poisson;*
> *je n'aime pas le poisson.*

1. Tu veux du bœuf?
2. Tu veux des œufs aux fines herbes?
3. Tu veux des carottes à la crème?
4. Tu veux du poulet?
5. Tu veux de la salade?
6. Tu veux du gâteau au chocolat?

ADDITIONAL PRACTICE

Ask students if the following items are in the grocery cart pictured on page 226.

1. Il y a des œufs?
2. Il y a de la viande?
3. Il y a du lait?
4. Il y a des carottes?
5. Il y a des verres en plastique?
6. Il y a des bananes?
7. Il y a de l'eau minérale?
8. Il y a du fromage?

Les contractions *au, aux*

Remember that the preposition *à* contracts with *le* to form *au* and with *les* to form *aux*. It remains unchanged with *la* and *l'*.

On va *à la* montagne.　　On va *au* lycée.
On va *à l'*école.　　　　On va *aux* magasins.

E　**Où?**　Répondez d'après les indications. *(Answer according to the cues.)*

1. Où est-ce qu'on achète du saucisson? (charcuterie)
2. Et du pain? (boulangerie)
3. Et de l'eau minérale? (épicerie)
4. Et du poisson? (marché)
5. À qui est-ce qu'on parle au marché? (marchands)

Les adjectifs possessifs

Review the following forms of the possessive adjectives.

notre appartement	notre maison	nos voitures
votre appartement	votre maison	vos voitures
leur appartement	leur maison	leurs voitures

F　**La famille de Pierre et de Louise.**　Complétez avec «notre», «votre» et «leur». *(Complete with* notre, votre, *and* leur.*)*

CAMILLE:　Pierre et Louise, ___ famille est grande ou petite?
PIERRE:　　___ famille est assez grande.
CAMILLE:　Vous avez beaucoup de cousins, n'est-ce pas? Où habitent ___ cousins?
LOUISE:　　Nous avons des cousins à Lyon et des cousins à Strasbourg. ___ cousins à Strasbourg sont étudiants à l'université, mais ils habitent avec ___ parents à Colmar, près de Strasbourg.
CAMILLE:　___ sœur est à l'université de Strasbourg aussi, n'est-ce pas?
PIERRE:　　Non, pas ___ sœur. ___ frère est à Strasbourg.

Activités de communication orale et écrite

A　**Au restaurant.**　With a classmate, make up a conversation between a waiter or waitress and a customer.

B　**Un voyage en train.**　You and your friends are planning a day trip by train. Write a paragraph describing what you're going to do.

ADDITIONAL PRACTICE

Have students give you the name of any place they know other than a country or city. Then have them make up a sentence using the verb *aller* and the particular place.

INDEPENDENT PRACTICE

Assign any of the following:
1. Exercises and activities, pages 224–227
2. Workbook, Self-Test 2, pages 79–82
3. CD-ROM, Disc 2, pages 224–227
4. CD-ROM, Disc 2, Self-Tests 5–8
5. CD-ROM, Disc 2, Game: *Le Labyrinthe*

Les contractions *au, aux*

ANSWERS

Exercice E

1. On achète du saucisson à la charcuterie.
2. On achète du pain à la boulangerie.
3. On achète de l'eau minérale à l'épicerie.
4. On achète du poisson au marché.
5. On parle aux marchands.

Les adjectifs possessifs

PRESENTATION *(page 227)*

Read over the explanation with the students and have them repeat the examples.

Exercice

PRESENTATION *(page 227)*

Exercice F

Call on three students to do Exercise F. One is Camille, another is Pierre, and the third one is Louise.

Extension of *Exercice F*

After completing the conversation in Exercise F, have more able students present it to the class.

ANSWERS

Exercice F

votre
Notre
vos
Nos
leurs
Votre
notre, Notre

Activités de communication orale et écrite

PRESENTATION *(page 227)*

Since students are going to be pretending to use their French in real-life situations, it is recommended that you not correct all their errors.

ANSWERS

Activités A and B

Answers will vary.

Mathématiques: Le Système métrique

OVERVIEW

The three readings in this section are related topically to material in Chapters 5, 6, and 7, which deal with foods, food shopping, and air travel. You may wish to allow students to choose which of the three selections they want to read according to their own interests, or you may prefer to have the entire group read a particular selection.

The readings may be presented at different levels of intensity:

1. The least intensive treatment would be to assign the selection as independent reading and the post-reading exercises as homework. This treatment requires no class time and minimal teacher involvement.

2. For a more intensive treatment, the reading and post-reading exercises can be assigned for homework, which will be gone over orally in class the next day.

3. The most intensive treatment includes a pre-reading presentation of the text by the teacher, an in-class reading and discussion of the passage, the assignment of the exercises for homework, and a discussion of the assignment in class the following day.

Avant la lecture

PRESENTATION *(page 228)*

A. Have students discuss briefly the material in the *Avant la lecture* section.

B. Most students are familiar with the basics of the metric system. Give them the following French equivalents of the metric weights and measures: *un mètre, un gramme, un litre.*

LETTRES ET

MATHÉMATIQUES: LE SYSTÈME MÉTRIQUE

Avant la lecture

1. Make a list of the weights and measures used in the United States.
2. Research what these weights and measures are based on.
3. Find out from your classmates how much they know about the metric system.

Lecture

Les anciennes mesures comme le pied et le pouce (douze pouces dans un pied) sont basées sur des parties du corps[1] humain. Mais les pouces et les pieds varient d'un pays à l'autre. Les pieds des Américains sont certainement plus grands que les pieds des Français! En France, avant la Révolution de 1789, c'est la même chose; les mesures varient d'une région à l'autre. Après 1789, les révolutionnaires décident de créer des mesures communes à toutes les régions de France.

Deux astronomes français, Méchain et Delambre, mesurent la longueur[2] de la partie de méridien qui va de la ville de Dunkerque en France à la ville de Barcelone en Espagne. Ils calculent la longueur totale de ce méridien. La 40.000.000e (quarante millionième) partie de cette longueur est adoptée comme unité de mesure de longueur et reçoit le nom de «mètre». C'est de cette manière que le système métrique

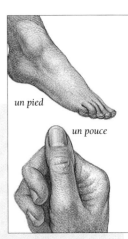

un pied

un pouce

est créé. En 1799, le système métrique est déclaré obligatoire en France. La France est le premier pays à adopter ce système de mesures.

Dunkerque
Barcelone

Aujourd'hui, le système métrique est utilisé par la plupart[3] des pays du monde. Même les pays comme les États-Unis, qui

---

DID YOU KNOW?

Both George Washington and Thomas Jefferson asked Congress to pass a law for a simple system of measurement, but Congress failed to enact their proposals into law. Jefferson wanted to throw out the old system and use the metric system. His proposal was turned over to a committee for study, where it eventually died.

SCIENCES

Médaille commémorative de la Convention du Mètre par Chaplain (1872)

utilisent un autre système normalement, utilisent le système métrique pour les sciences.

C'est un système de poids[4] et mesures. Le mètre mesure la longueur, le gramme le poids, et le litre, dérivé des deux autres unités, est une unité de volume. Les unités plus grandes ou plus petites sont formées avec les préfixes suivants:

kilo	× 1 000	kilogramme = 1 000 grammes
hecto	× 100	hectolitre = 100 litres
déca	× 10	décamètre = 10 mètres
déci	: 10	décigramme = 1/10 gramme
centi	: 100	centilitre = 1/100 litre
milli	: 1 000	millimètre = 1/1 000 mètre

Depuis 1962, le système métrique s'appelle le Système International d'Unités. Il a sept unités de base: le mètre, le kilogramme, la seconde, l'ampère, le kelvin (température), la mole (quantité de matière) et la candela (intensité lumineuse).

[1] corps *body*
[2] longueur *length*
[3] la plupart *majority*
[4] poids *weights*

Après la lecture

A Les poids et les mesures. Vrai ou faux?

1. Il y a des pays qui n'utilisent pas le système métrique.
2. Les États-Unis n'utilisent pas le système métrique pour les sciences.
3. Les anciennes mesures ont des bases scientifiques.
4. À l'origine, le mètre est basé sur la longueur de la partie de méridien qui va de Dunkerque à Barcelone.
5. Le litre est une unité de volume.
6. Le gramme est une unité de longueur.
7. On dérive les unités plus grandes ou plus petites en ajoutant des préfixes.

B Combien font… ? Faites des calculs.

1. 100 cm (centimètres) = ___ mètre(s)
2. 2 kl (kilolitres)   = ___ litre(s)
3. 2 000 g (grammes) = ___ kilogrammes

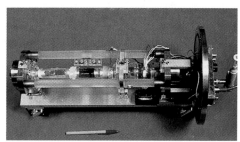

Laser à He-Ne permettant de réaliser le mètre selon la définition adoptée en 1983

C Une nouvelle définition du mètre. Depuis 1983, le mètre est basé sur la longueur du trajet (la distance) parcouru dans le vide par la lumière pendant 1/299.792.458 seconde. Expliquez cette nouvelle définition du mètre en anglais.

C. Explain to students that the reading selection is about the formation and history of the metric system and that the French had a great deal to do with its development.
D. Have students look at the globe. Ask them which lines are longitude and which lines are latitude lines.
E. Have students quickly scan the reading selection for cognates. There are approximately 20.

Lecture

PRESENTATION *(pages 228–229)*

A. Have students read the selection silently.
B. The last paragraph will probably be of interest only to those students who like math and science. It can be omitted for other students. The same applies to Exercise C.

Après la lecture

PRESENTATION *(page 229)*

A. Have students do Exercises A and B.
B. You may wish to ask students: *Combien de décimètres est-ce qu'il y a dans un mètre? (10) Et il y a combien de mètres dans un kilomètre? (1 000) Combien de milligrammes est-ce qu'il y a dans un gramme? (1 000) Combien de centilitres y a-t-il dans un litre? (100)*

ANSWERS

Exercice A

1. vrai	5. vrai
2. faux	6. faux
3. faux	7. vrai
4. vrai	

Exercice B

1. 1   2. 2 000   3. 2

Exercice C

Answers will vary, but may include: **The length of the meter is based on the distance traveled by light in space in 1/299,792,458 of a second.**

INTERDISCIPLINARY CONNECTIONS

Students choose an authentic French recipe and convert the *litres, centilitres, grammes, kilos,* etc. in it into American Standard (quarts, cups, ounces, and so forth) in order to successfully prepare the dish they've chosen.

DIÉTÉTIQUE: UNE ALIMENTATION ÉQUILIBRÉE[1]

OPTIONAL MATERIAL

Diététique: Une Alimentation équilibrée
Avant la lecture

PRESENTATION *(page 230)*

A. Do the pre-reading activities in the textbook or the following variation: Write the names of the six basic nutrients from the reading on the board. Have students work individually or in groups to make a list in French of all the foods they have learned to identify in Chapters 5 and 6. Then have students give you the names of the foods and indicate in which category they belong as you write them on the board.

B. Have students quickly scan the reading selection for cognates. There are approximately 40.

C. Show students pictures of foods from French advertisements that describe the content or benefits of the particular food products. You may also use food packages written in both French and English that have nutritional information.

Lecture

PRESENTATION *(pages 230–231)*

If you are doing the selection intensively, you may wish to ask the following comprehension questions after the reading:

Quelles sont les maladies causées par une mauvaise alimentation? Quel rôle important joue l'alimentation? Il y a combien d'aliments de base? Quels sont les aliments de base? Il y a combien de sortes de vitamines?

Avant la lecture

1. Everybody knows that nutritious foods help you grow. Do you know what the six essential types of nutrients are? If you don't, find out.
2. Make a list of the foods you eat often and of those you rarely eat.
3. Look at the six types of nutrients discussed below and match them with your lists.

Lecture

Le scorbut et le béribéri sont deux maladies[2] causées par une mauvaise alimentation. Elles sont aujourd'hui rares dans les pays industrialisés. Mais il y a encore beaucoup de gens qui ont une mauvaise alimentation. L'alimentation joue un rôle très important dans la préservation de la santé[3].

Quel est le nombre idéal de calories? Tout dépend de la personne, de son métabolisme et de son activité physique. L'âge, le sexe, la taille (grande ou petite) et le climat sont aussi des facteurs. Pour un homme de 25 ans qui fait du sport, c'est 2 900 calories par jour.

Il y a six aliments de base.

1. Les protéines

Les protéines sont particulièrement importantes pour les enfants et les adolescents. Elles aident à fabriquer des cellules. La viande et les œufs contiennent des protéines.

2. Les glucides (les hydrates de carbone en chimie)

Ces aliments sont la source d'énergie la plus efficace pour le corps humain.

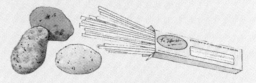

3. Les lipides (les graisses)

Les lipides sont aussi une bonne source d'énergie, mais pour les personnes qui ont un taux de cholestérol élevé, les graisses ne sont pas bonnes. Il faut faire un régime[4] sans graisses, il faut éliminer les graisses.

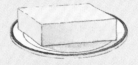 

4. Les minéraux

Beaucoup de minéraux sont essentiels pour le corps humain. Le calcium est absolument nécessaire pour les os et les dents[5].

ADDITIONAL PRACTICE

Have students give the following information: *Les aliments qu'ils mangent souvent. Les aliments qu'ils mangent rarement. Les aliments qu'ils considèrent bons pour la santé. Les aliments qu'ils considèrent mauvais pour la santé.*

5. Les vitamines

Les vitamines sont indispensables. Il y a deux sortes de vitamines: les vitamines solubles dans l'eau (C et B) et les vitamines solubles dans la graisse (A et D).

- La vitamine A (végétaux, graisses animales) est bonne pour les yeux[6].
- La vitamine C (végétaux et fruits) joue un rôle important dans le métabolisme et est bonne pour la résistance aux infections.
- La vitamine D est la vitamine de la croissance, le développement progressif des jeunes. Pour cette raison, elle est bonne pour les enfants et les adolescents.
- La vitamine B (céréales, légumes) joue un rôle important dans le fonctionnement du foie[7] et des cellules nerveuses.

6. L'eau

L'eau est absolument essentielle au corps humain qui est fait de 65% d'eau.

D'une façon générale, une alimentation équilibrée est essentielle pour être en bonne santé.

[1] alimentation équilibrée *balanced diet*
[2] maladies *illnesses*
[3] santé *health*
[4] faire un régime *to go on a diet*
[5] les os et les dents *bones and teeth*
[6] yeux *eyes*
[7] foie *liver*

Après la lecture

A **La santé.** Choisissez.

1. Le scorbut est ___.
   a. une maladie b. une alimentation
   c. une vitamine

2. Une bonne alimentation est essentielle pour ___.
   a. l'obésité b. la santé
   c. la maladie

3. Le nombre de calories idéal dépend de ___.
   a. la personne b. la durée de la vie
   c. la vitamine

4. La vitamine A est bonne pour ___.
   a. les os b. les dents
   c. les yeux

5. Le pourcentage d'eau dans le corps humain est de ___.
   a. 20% b. 65% c. 90%

6. Pour les personnes qui ont un taux de cholestérol élevé, il ne faut pas ___.
   a. de lipides b. de minéraux
   c. de vitamines

7. La vitamine D est bonne surtout pour ___.
   a. les malades b. les enfants
   c. les yeux

B **Les aliments.** Faites une liste des aliments que vous connaissez *(know)* en français et classez-les selon les six catégories.

C **Trois régimes.** Composez trois régimes.

1. un régime pour maigrir *(to lose weight)*
2. un régime pour grossir *(to gain weight)*
3. un régime végétarien

Après la lecture

Vocabulary Expansion

Write the following on the board and have students match the corresponding verbs and nouns.

1. causer
2. alimenter
3. préserver
4. développer
5. fonctionner

a. l'aliment, l'alimentation
b. le développement
c. la cause
d. le fonctionnement
e. la préservation

Exercices

PRESENTATION *(page 231)*

Students can prepare the exercises on their own. You may then wish to go over them in class after completing the reading.

ANSWERS

Exercice A

1. a	5. b
2. b	6. a
3. a	7. b
4. c	

Exercices B and C

Answers will vary.

ADDITIONAL PRACTICE

The following are healthy foods that students have not yet learned to identify in French. Since they are all cognates, students can guess their meaning immediately. Ask them whether or not they eat these foods.

du brocoli	un radis
du céleri	une datte
un concombre	une olive
une endive	une poire

COOPERATIVE LEARNING

1. Have students work in groups to list the foods on the school cafeteria menu. Have them put the foods into each of the six categories. Ask them what, if anything, should or should not be there.
2. Divide the class into six categories—one for each of the six food categories. Each group will prepare a list of its favorite and least favorite foods within the category.

LITTÉRATURE: ANTOINE DE SAINT-EXUPÉRY
(1900–1944)

Avant la lecture

1. Do you know the name of the American who made the first non-stop flight from New York to Paris in 1927? Do you know the name of his plane?
2. Four of Saint-Exupéry's works are mentioned in this reading. Their English titles are *The Little Prince; Night Flight; Wind, Sand, and Stars;* and *Southern Mail.* See if you can match the French and English titles.

Lecture

*Antoine de Saint-Exupéry
(1900–1944)*

Saint-Exupéry est un écrivain[1] célèbre. Mais Saint-Exupéry, c'est aussi un homme d'action. Il est né à Lyon en 1900. Pendant son service militaire il apprend à piloter un avion. Il est pilote de ligne entre Toulouse et Dakar en Afrique; il est chef d'aéroplace à Buenos Aires; il participe aux tout premiers vols France–Amérique.

Ses romans[2] reflètent sa carrière de pilote. *Courrier sud* parle de ses vols Toulouse–Casablanca–Dakar; *Vol de Nuit* parle de trois pilotes qui attendent un autre pilote à l'aéroport de Buenos-Aires. Le pilote qu'ils attendent, Fabien, n'arrive pas. Il est en retard. Il est en difficulté dans le ciel noir d'Amérique. Sa femme, Madame Fabien, est affolée, presqu'hystérique. Un des pilotes parle à Madame Fabien: «Madame, je vous en prie. Calmez-vous. Il est fréquent dans notre métier[3] d'attendre longtemps les nouvelles.»

Toulouse

Dakar

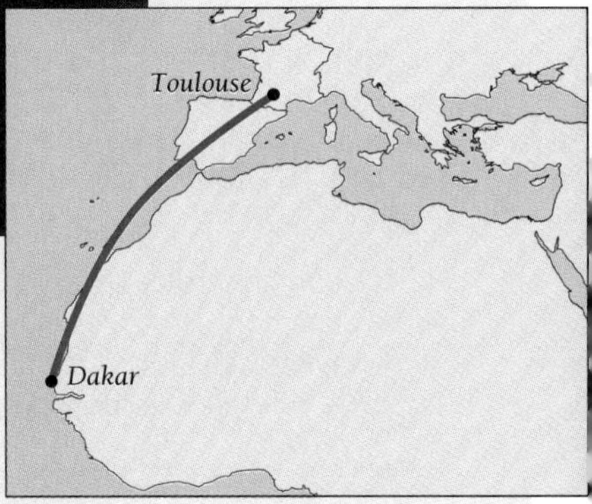

Dans *Terre des Hommes*, Saint-Exupéry parle de sa carrière et de ses camarades qui sont morts[4]. Il parle d'une vie d'action qui unit les hommes pour toujours, même après la mort.

Sidebar (left column)

Littérature: Antoine de Saint-Exupéry (1900–1944)
Avant la lecture

PRESENTATION *(page 232)*

A. Have students look at the photographs on these two pages. Based on the photographs, what do they think Saint-Exupéry's profession was? What famous book did he write? What does the map probably indicate?

B. Have students do the *Avant la lecture* activities in their textbook.

C. You may wish to give the following information in French. *Charles Lindbergh décolle d'un petit aéroport à Long Island, New York, à bord de son avion à un seul moteur—un monomoteur—«L'Espirit de Saint Louis». Trente-trois heures et trente minutes plus tard, il atterrit à l'aéroport Le Bourget à Paris. C'est en 1927.* (Write year on board.)

D. You might have the students skim the reading and look for answers to the following questions: What type of subject matter do Saint-Exupéry's novels reflect? Who was Madame Fabien? At what airport does *Vol de nuit* take place? How did Saint-Exupéry die?

E. Have students look for the cognates that appear in this reading. There are approximately 20.

Lecture

PRESENTATION *(pages 232–233)*

A. Have students read the selection silently.

DID YOU KNOW?

A relatively new French banknote (50 francs) depicts Saint-Exupéry and his plane. Have students turn to page 478 to see it.

Pendant la Deuxième Guerre mondiale, il écrit *Le Petit Prince* (1943) où il évoque sa nostalgie de l'amitié et cherche à définir le sens[5] des actions et des valeurs morales de la société moderne dédiée au progrès technique. Un an plus tard le 13 juillet 1944, il disparaît pour toujours dans une mission aérienne militaire. Il reste pour la légende le courageux, le charmant, l'exceptionnel «Saint-Ex».

[1] écrivain   *writer*
[2] romans   *novels*
[3] métier   *profession*
[4] morts   *dead*
[5] sens   *meaning*

Saint-Exupéry, le 13 juillet 1944

ANTOINE DE SAINT-EXUPÉRY

Le Petit Prince

Avec des aquarelles de l'auteur

nrf

GALLIMARD

Après la lecture

A   «Saint-Ex».   Répondez.

1. Dans quel livre est-il question de l'Argentine?
2. Dans quel livre est-il question du Maroc?
3. Dans *Le Petit Prince*, Saint-Exupéry évoque la nostalgie de l'amitié. Pourquoi, à votre avis?

B   Imaginez.   Imaginez ce qui arrive (*what happens*) dans *Vol de Nuit*.

C   Lindbergh.   Écrivez une courte biographie de Charles Lindbergh en français.

LETTRES ET SCIENCES   **233**

B. Explain to students the following strategies for guessing at meaning.

1. *sa carrière* This word is similar to its English equivalent, but you may not be able to guess at it immediately. The statement says *sa carrière de pilote*, and you know Saint-Exupéry was a pilot. So what does *carrière* mean?

2. *le ciel* You do not know the meaning of this word, but you know the pilot is in trouble. He has not arrived. Where would a pilot be flying a plane and be in trouble? On the ground? No. Where? So what does *ciel* mean?

3. *affolée* Note how this word is used in the sentence: *Sa femme, Madame Fabien, est affolée, presqu'hystérique.* Often a word that follows clarifies the meaning of a previous word—*hystérique* clarifies *affolée*. What do you think *affolée* means?

4. *la Deuxième Guerre mondiale* This phrase is followed by a date (1943). What important historical event was taking place in 1943?

Note Students will read a complete chapter of *Le Petit Prince* in the Level 3 textbook, *En voyage*.

Après la lecture

Exercices

PRESENTATION *(page 233)*

Exercises A and B can be answered in English. Exercise C can be done in French.

ANSWERS

Exercice A

1. *Vol de nuit* (**Night Flight**)
2. *Courrier sud* (**Southern Mail**)
3. Answers will vary.

Exercices B **and** *C*

Answers will vary.

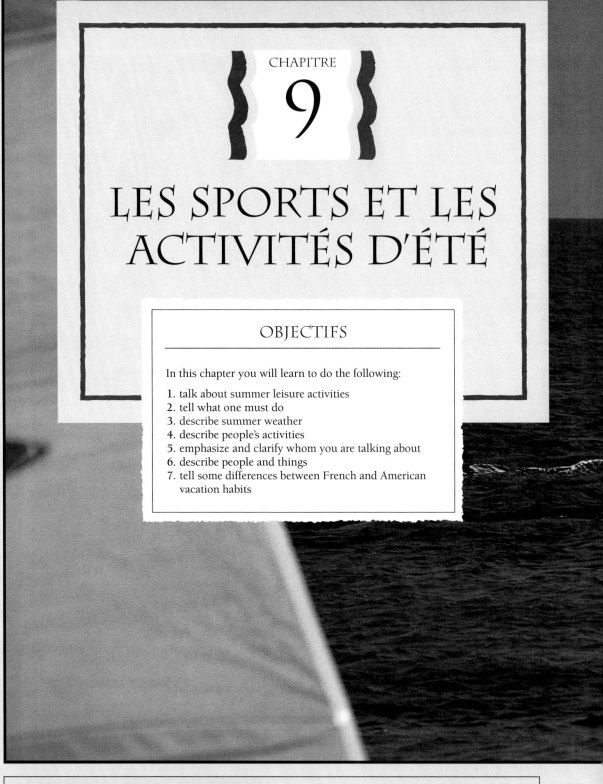

CHAPTER OVERVIEW

Students will learn vocabulary associated with summer weather and sports. They will also learn *prendre* and similar verbs, stress pronouns, and adjectives with a double consonant.

The cultural focus is on French summer vacations and resorts.

CHAPTER OBJECTIVES

By the end of this chapter, students will know:

1. summer weather, beach, and summer sports vocabulary
2. the verbs *prendre, apprendre,* and *comprendre*
3. the *pronoms accentués,* or stress pronouns
4. adjectives with a double consonant in the feminine

CHAPTER 9 RESOURCES

1. Workbook
2. Student Tape Manual
3. Audio Cassette 6A/CD-6
4. Bell Ringer Review Blackline Masters
5. Vocabulary Transparencies
6. Pronunciation Transparency P-9
7. Communication Transparency C-9
8. Communication Activities Masters
9. Map Transparencies
10. Situation Cards
11. Conversation Video
12. Videocassette/Videodisc, Unit 3
13. Video Activities Booklet, Unit 3
14. Lesson Plans
15. Computer Software: Practice/Test Generator
16. Chapter Quizzes
17. Testing Program
18. Internet Activities Booklet
19. CD-ROM Interactive Textbook

CHAPITRE

9

LES SPORTS ET LES ACTIVITÉS D'ÉTÉ

OBJECTIFS

In this chapter you will learn to do the following:

1. talk about summer leisure activities
2. tell what one must do
3. describe summer weather
4. describe people's activities
5. emphasize and clarify whom you are talking about
6. describe people and things
7. tell some differences between French and American vacation habits

CHAPTER PROJECTS

(optional)

1. Have students share their family's vacation habits by bringing in photos and vacation paraphernalia. You may wish to group them according to the kinds of vacations they are accustomed to (mountains, seaside, camping, visiting a city, etc.) and have each group tell as much as they can in French about their type of vacation.

2. Have groups plan the ideal five-week vacation in France. The description should be in French and should include information about transportation, meals, and leisure activities. They may do a poster or collage to illustrate it. Students can then choose the trip they would most like to take.

Pacing

This chapter requires eight to ten class sessions. Pacing will vary according to class length and the age and aptitude of the students.

Note The Lesson Plans offer guidelines for 45- and 55-minute classes and **Block Scheduling.**

NOTE ON DIRECTIONS TO STUDENT EXERCISES AND ACTIVITIES

Beginning in Chapter 9, the English translations of the exercise directions have been dropped in order to increase the amount of authentic communication in French in the classroom and to provide students with further practice in the language. You may wish to avoid even mentioning this change and see if students notice it.

Exercices vs. *Activités*

All exercises (which provide guided practice) are coded in blue. All communicative activities are coded in red.

INTERNET ACTIVITIES

(optional)

You will find these activities, student worksheets, and related teacher information in the *Bienvenue* Internet Activities Booklet and on the Glencoe Foreign Language Home Page at:

http://www.glencoe.com/secondary/fl

DID YOU KNOW?

You may wish to tell students that wind-surfing *(la planche à voile)* is a popular sport along France's Atlantic coast. The 1988 windsurfing championships were held at Les Sables d'Olonne, a town on the Atlantic coast, south of the Loire.

Bell Ringer Review

Write the following on the board or use BRR Blackline Master 9-1: Change the underlined words to a word or phrase with the same meaning.

1. un vol d'un pays à l'autre
2. un vol qui arrive de Paris
3. un vol dans le même pays
4. un vol qui va à Tahiti

PRESENTATION (pages 236–237)

A. Show Vocabulary Transparencies 9.1. Point to individual items and have the class repeat the words after you or Cassette 6A/CD-6.
B. As an alternative, you may wish to bring in some props such as sunglasses, suntan lotion, and so on.
C. During your presentation, ask: *C'est la plage? C'est le sable ou la mer? Il y a du sable sur la plage? C'est une vague? Il y a des vagues dans la mer? C'est une station balnéaire? Il y a des stations balnéaires au bord de la mer? Où est-ce qu'il y a des stations*

236

VOCABULAIRE

MOTS 1

EN ÉTÉ

la mer

une vague

une station balnéaire au bord de la mer

des lunettes de soleil

le sable

la plage

un maillot (de bain)

À la plage il faut faire attention.
Il faut mettre de la crème solaire.
André met de la crème solaire.
Il prend un bain de soleil.
Il bronze.

Christine met des lunettes de soleil.
Mais elle attrape un coup de soleil.
Pourquoi? Parce qu'elle ne fait pas attention.
Elle ne met pas de crème solaire.

Note: The impersonal expression *il faut*, "one must," is used often in French. It is followed by the infinitive.

faire de la planche à voile

faire de la plongée sous-marine

faire du ski nautique

236 CHAPITRE 9

TOTAL PHYSICAL RESPONSE

(following the Vocabulary presentation)

TPR 1
____, levez-vous et venez ici, s'il vous plaît.
Vous allez mimer les activités suivantes.
Nagez.
Faites du ski nautique.
Faites de la planche à voile.
Faites de la plongée sous-marine.

Plongez.
Prenez un bain de soleil.
Merci, ____. Retournez à votre place, s'il vous plaît.

faire du surf faire une promenade aller à la pêche

nager

une piscine plonger

un moniteur

Robert aime nager.
Il nage dans la piscine.
Et Caroline plonge dans la piscine.

Laure prend des leçons de natation.
Elle apprend à nager.
Elle comprend les instructions du moniteur.

CHAPITRE 9 **237**

---

balnéaires? En été, tu vas à une station balnéaire? Tu vas à quelle station balnéaire? Quand tu y vas, tu mets un maillot? Tu mets des lunettes de soleil?

Note *Pourquoi* questions (with *parce que* responses) are the most challenging to answer. The following are some questions you may wish to ask more able students while presenting words in this section: *Pourquoi faut-il mettre de la crème solaire à la plage? Pourquoi faut-il faire attention à la plage? André bronze. Pourquoi? Christine ne bronze pas. Elle attrape un coup de soleil. Pourquoi?*

RECYCLING

In this section, the concept of *-er* verbs is reinforced with the new verbs *bronzer, attraper, nager,* and *plonger.* In addition, the verbs *aller, faire,* and *mettre* are reintroduced. You may wish to explain that *nager* and *plonger* have the same spelling change as *manger: nous mangeons, nous nageons, nous plongeons.*

RETEACHING (*Mots 1*)

Refer students back to the party in Chapter 3, *Mots 2.* You may wish to use Vocabulary Transparencies 3.2 (A & B). Then have students imagine they are having a party at the beach. Have them make up a story combining the party vocabulary with the new beach vocabulary.

DRAMATIZATION

You may wish to have students dramatize the following words or expressions: *mettre de la crème solaire, nager, plonger, faire de la planche à voile, faire du ski nautique, aller à la pêche, mettre des lunettes de soleil.* As one student does the dramatization, another student can describe what he/she is doing.

TPR 2

___, venez ici, s'il vous plaît.
Vous allez à la pêche. Prenez la ligne ou la canne à pêche. Tenez!
Mettez la ligne dans l'eau.
Soyez patient(e). Asseyez-vous. Attendez quelques instants.
Ah, voilà! Vous avez attrapé un poisson.
Sortez votre ligne de l'eau. Allez-y doucement.

Regardez le poisson. Oh, qu'il est grand, votre poisson!
Tout le monde est d'accord?
Retirez le poisson de la ligne.
Mettez le poisson sur la table.
Vous allez préparer le poisson? Oui? Non?
Vous voulez manger le poisson?
Merci, ___. Vous avez très bien fait. Vous êtes un(e) très bon(ne) pêcheur/pêcheuse.
Et maintenant vous pouvez retourner à votre place.

Exercices

PRESENTATION (page 238)

Extension of *Exercice A*: Speaking

After completing Exercise A, focus on the speaking skill by making the exercise a TV interview show. One student is the MC who interviews two other students using the questions in the exercise.

Exercice B

Exercise B can be done with books closed, open, or both ways. You may wish to call on one individual to answer two or three items.

Extension of *Exercice B*

After completing the exercise, call on one student to give the information in his/her own words.

Extension of *Exercice C*

In the *Mots 1* section, students only encountered the *il/elle* forms of the new verbs *prendre*, *apprendre*, and *comprendre*. Since there is no difference in sound in the singular forms of these verbs, you may wish to go over Exercise C again orally. Change *Jeanne* to *tu*. The students respond with *je*.

ANSWERS

Exercices A and B

Answers will vary.

Exercice C

1. Oui, Jeanne apprend à nager.
2. Oui, elle prend des leçons de natation.
3. Elle apprend à nager dans une piscine.
4. Oui, elle comprend bien les instructions de la monitrice.

INFORMAL ASSESSMENT
(Mots 1)

Check for understanding by mixing true and false statements about Vocabulary Transparencies 9.1. Students either agree by saying *Je suis d'accord* or they correct the statement.

238

Exercices

A **Tu aimes les activités d'été?** Donnez des réponses personnelles.

1. Tu aimes nager quand il y a de grandes vagues?
2. Tu aimes plonger dans une piscine?
3. Tu aimes faire de la planche à voile?
4. Tu aimes faire de la plongée sous-marine?
5. Tu aimes faire du ski nautique?
6. Tu aimes faire du surf?
7. Tu aimes aller à la pêche?
8. Tu aimes prendre des bains de soleil sur le sable?
9. Tu aimes faire des promenades sur la plage?

B **Qu'est-ce qu'on fait en été?** Donnez des réponses personnelles.

1. En été, tu aimes aller à la plage?
2. Tu vas à quelle station balnéaire?
3. Tu préfères nager dans la mer ou dans une piscine?
4. Quand tu vas à la plage, tu mets un beau maillot?
5. À ton avis, est-ce qu'il faut mettre de la crème solaire?
6. Est-ce que tu mets de la crème solaire?
7. Tu bronzes facilement ou tu attrapes des coups de soleil?
8. Tu mets des lunettes de soleil quand tu vas à la plage?

C **Qu'est-ce qu'elle apprend à faire?** Répondez d'après la photo.

1. Jeanne apprend à nager?
2. Elle prend des leçons de natation?
3. Elle apprend à nager dans la mer ou dans une piscine?
4. Elle comprend bien les instructions de la monitrice?

NATATION – SKI NAUTIQUE

POUR ÉVITER DE MULTIPLES DANGERS:
courants, trous d'eau, épaves, vents, marées, barres, sables mouvants, tourbillons, etc.

CHOISISSEZ UNE PLAGE SURVEILLÉE.

Baignade interdite   Baignade dangereuse   Baignade autorisée

LA BAIGNADE
La natation est un **sport**; n'allez pas au-delà de vos possibilités.
L'hydrocution est un **accident** qui survient le plus souvent après:
• un repas copieux
• un bain de soleil prolongé

238

INDEPENDENT PRACTICE

Assign any of the following:
1. Exercises, page 238
2. Workbook, *Mots 1: A–D*, pages 83–84
3. Communication Activities Masters, *Mots 1: A*, page 43
4. CD-ROM, Disc 3, pages 236–238

LEARNING FROM REALIA

Ask students to look at the flyer. Ask them what it's about. What do the red, yellow (amber), and green flags indicate? Have them look at the flyer again and find any cognates. Now, have them find the French words for the following:

swimming	quicksand
bathing	don't overdo it
currents	

VOCABULAIRE

MOTS 2

LE TENNIS

une balle

une raquette

un court de tennis

une jupette

un tee-shirt

un filet

un short

des chaussures (f.) de tennis

les limites

hors des limites

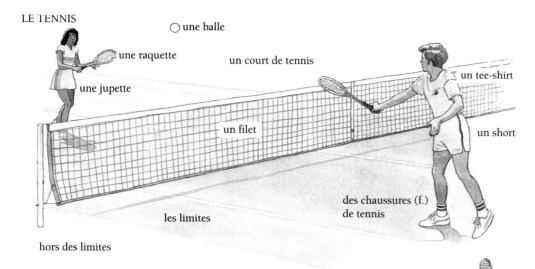

15 | 0

On joue au tennis.
C'est un match de tennis.
Un des joueurs sert.

L'autre joueur renvoie la balle.
Il frappe fort.
Le score est de quinze à zéro.

CHAPITRE 9 **239**

ADDITIONAL PRACTICE

(After completing Exercises A–C, page 238)
1. Have students list things they can do at the beach, using *pouvoir.* Then have them organize their lists in order of preference and report to the class. For example: *À la plage, on peut nager. On peut bronzer. On peut faire du surf.* Students can do the same task with swimming pools and *aimer.* For

example: *À la piscine, j'aime nager. J'aime plonger.*
2. Student Tape Manual, Teacher's Edition, *Activités B–C,* pages 100–101

VOCABULAIRE

MOTS 2

Vocabulary Teaching Resources

1. Vocabulary Transparencies 9.2 (A & B)
2. Audio Cassette 6A/CD-6
3. Student Tape Manual, Teacher's Edition, *Mots 2: D–F,* pages 101–102
4. Workbook, *Mots 2: E–G,* page 85
5. Communication Activities Masters, *Mots 2: B,* page 44
6. Chapter Quizzes, *Mots 2: Quiz 2,* page 49
7. Computer Software, *Vocabulaire*
8. CD-ROM, Disc 3, *Mots 2:* pages 239–242

Bell Ringer Review

Write the following on the board or use BRR Blackline Master 9-2: Sketch people engaged in five sports associated with the beach. Label each sketch.

PRESENTATION *(pages 239–240)*

A. Model the new words using Vocabulary Transparencies 9.2 (A & B). Have students repeat after you or Cassette 6A/CD-6.
B. Using Vocabulary Transparencies 9.2 (A & B), say the words and phrases in random order and have volunteers stand at the screen and point to the appropriate illustrations.
C. Continue this process, expanding your statements about the illustrations in order to recombine the new material and recycle vocabulary from previous chapters. For example: *Elle sert. Il renvoie la balle. Il attend la balle. Elle attend la balle. Il a un microphone. La balle n'est pas hors des limites.*

239

PRESENTATION *(continued)*

On joue une partie en double.
Elle fait une promenade sur la plage quand il fait beau.

D. Ask students for the opposites of the following: *servir (renvoyer la balle)*; (mime this one) *frapper doucement (frapper fort)*; *une partie en simple (une partie en double)*; *perdre le match (gagner le match)*; *une jupette (un short)*; *Il fait beau. (Il fait mauvais.) Il fait chaud. (Il fait froid.) Il fait du soleil. (Il y a des nuages.)*

Vocabulary Expansion

1. The following are some additional weather expressions you may wish to give students. These expressions are often used for the *météo* both on TV and in the newspaper. (See the weather map in *Météorologie: La Prévision du Temps* in *Lettres et sciences,* page 338.)
   Le temps est ensoleillé.
   Il est nuageux.
   Le ciel est couvert.
   Il y a des averses.
   Il y a des éclaircies.

2. If you present the above expressions, you may wish to have the students do the following exercise.
   Exprimez d'une autre façon:
   1. **Il fait du soleil.**
   2. **Il y a des nuages.**
   3. **Il pleut.**
   4. **Le temps est nuageux.**
   5. **Le ciel n'est pas couvert.**

3. With more able groups, do some additional word study. Show students the verb and noun forms of the following words.

jouer	le joueur
	la joueuse
nager	le nageur
	la nageuse
plonger	le plongeur
	la plongeuse
pêcher	le pêcheur
	la pêcheuse
skier	le skieur
	la skieuse

une partie en simple
un match entre deux joueurs

une partie en double
un match entre quatre joueurs

gagner le match

Note: The verb *jouer* takes the preposition *à* when followed by a sport.

On joue au tennis.　　On joue au volley.　　On joue au foot.

LE TEMPS EN ÉTÉ

Quel temps fait-il?　　　Il fait du soleil.　　　Il fait beau.

Il fait chaud.

Il y a des nuages.　　　Il pleut.　　　Il fait mauvais.

Il fait du vent.

Il fait froid.

240　　CHAPITRE 9

ADDITIONAL PRACTICE

1. Have students compile a calendar using photographs, magazine pictures, postcards or other illustrations of summer weather conditions. For each month of summer, they should write a sample date, a weather description, and a sentence about the illustration they have chosen. For example:
   C'est aujourd'hui le 7 juillet. Il fait très chaud. Nous allons à la plage pour faire du surf.

2. Student Tape Manual, Teacher's Edition, *Activités E–F,* page 102

Exercices

A **Un match de tennis.** Donnez des réponses personnelles.

1. Tu aimes le tennis?
2. Tu joues au tennis?
3. Si tu ne joues pas au tennis, tu veux apprendre à jouer au tennis?
4. Tu as une raquette?
5. Il y a un court de tennis près de ta maison ou ton appartement?
6. Ton école a des courts de tennis?

B **Le tennis.** Complétez.

1. Quand un garçon ou un homme joue au tennis, il met un ___, un ___ et des ___.
2. Quand une fille ou une femme joue au tennis, elle met un ___, une ___ et des ___.
3. ___ est un match entre deux joueurs.
4. ___ est un match entre quatre personnes.
5. Quand on joue au tennis, on a une ___ et des ___.
6. 15–"love" est un ___ de quinze à zéro.
7. On ___ ou ___ la balle avec la raquette.
8. Un joueur sert, mais la balle va dans le ___. Quand il sert encore la balle est ___! Il n'a pas de chance!
9. Un des joueurs ___ très fort. Il ___ le match.

C **Le temps.** Répondez.

1. En été, il fait beau ou il fait mauvais dans ta ville?
2. Il fait du soleil?
3. Il pleut souvent?
4. Il fait du vent à la plage?
5. Quel temps fait-il aujourd'hui?

D **Quel temps fait-il?** Répondez d'après les dessins.

1.

2.

3.

4.

5.

6.

INDEPENDENT PRACTICE

Assign any of the following:
1. Exercises, page 241
2. Workbook, *Mots 2: E–G,* page 85
3. Communication Activities Masters, *Mots 2: B,* page 44
4. CD-ROM, Disc 3, pages 239–241

Exercices

PRESENTATION *(page 241)*

Exercice A

This exercise may be done with books open or closed.

Exercice B

Exercise B must be done with books open. Call on an individual to say as much about tennis as he/she can.

Extension of Exercices C and D

After completing the exercises, make statements about the weather and have students say the same thing using a negative form. For example: *Il fait froid. (Il ne fait pas chaud.)*

Extension of Exercice D

After completing Exercise D, have students say all they can about each illustration.

ANSWERS

Exercice A

Answers will vary.

Exercice B

1. tee-shirt, short, chaussures de tennis
2. tee-shirt, jupette, chaussures de tennis
3. Une partie en simple
4. Une partie en double
5. raquette, balles
6. score
7. sert, renvoie
8. filet, hors des limites
9. frappe, gagne

Exercice C

Answers will vary.

Exercice D

1. Il fait chaud. (Il fait du soleil. Il fait beau.)
2. Il y a des nuages. (Il ne fait pas de soleil.)
3. Il pleut. (Il fait mauvais. Il y a des nuages.)
4. Il fait du vent.
5. Il fait beau. (Il fait du soleil. Il fait chaud.)
6. Il fait froid.

INFORMAL ASSESSMENT
(Mots 2)

Show magazine pictures that depict different types of weather. Have students say what the weather is in each one.

PRESENTATION *(page 242)*

Activités A, B, C, and D

It is not necessary to do all the activities. You may select those that are most appropriate for your students, or you may permit the students to select the activity or activities they would like to take part in.

Extension of *Activité D*

All of us have a lot of things we have to do in life. Have students tell some of these things using *Il faut…*

ANSWERS

Activités A, B, C, and D

Answers will vary.

Activités de communication orale
Mots 1 et 2

A **À la plage.** Work with a classmate. One of you describes the weather (sunny, rainy, windy, etc.) on a day at the beach and the other says what he or she likes (or doesn't like) to do there on that kind of day. Take turns.

> Élève 1: Il fait du vent.
> Élève 2: Quand il fait du vent, j'aime faire de la planche à voile.

B **Les vacances parfaites.** Plan a great summer vacation at the beach. Tell your classmate where you want to go and why, and what you like to do there. Then find out your partner's plans.

> Élève 1: Je voudrais aller à Hawaii parce qu'il fait toujours du soleil là-bas. J'aime nager. Et toi?
> Élève 2: Moi aussi, j'aime Hawaii. Je voudrais faire de la plongée sous-marine et du surf.

C **Un match de tennis entre Guy et Nadine.** You're a sports announcer for a local radio station. Describe the tennis match in the illustration below. Don't forget to tell your listeners what the players are wearing.

D **Il faut…** Work with a classmate. One of you chooses a place or a situation from the list below and the other has to name two things that should or shouldn't be done at that place or in that situation. Take turns until all the items on the list have been used.

> à l'école
> Élève 1: À l'école, qu'est-ce qu'il faut faire?
> Élève 2: À l'école, il faut étudier et faire ses devoirs.

à la plage	au cours de français
après un dîner au restaurant	pour gagner un match
avant un examen	pour organiser une fête
avant un voyage	

ADDITIONAL PRACTICE

Have students role-play in pairs: A French student (your partner) asks you for the following information: (**1**) what the weather is like in summer in your town, (**2**) if you go to the beach, (**3**) if you like to swim, (**4**) if you prefer to swim in the ocean or in a pool, (**5**) which sports you engage in at the beach. Answer, then reverse roles.

INDEPENDENT PRACTICE

Assign any of the following:
1. Activities, page 242
2. CD-ROM, Disc 3, page 242

STRUCTURE

STRUCTURE

Les verbes *prendre, apprendre* **et** *comprendre* **au présent**   *Describing People's Activities*

1. The verb *prendre,* "to take," is irregular. Study the following forms.

PRENDRE			
je	prends	nous	prenons
tu	prends	vous	prenez
il elle on	prend	ils elles	prennent

Je prends mes livres quand je quitte la classe.
Vous prenez l'avion pour aller à Boston mais Julie et Marc prennent le train.

Note that the singular forms of *prendre* are the same as those of any regular *-re* verb, but the plural forms are irregular.

2. The verb *prendre* has a number of additional meanings. Here are a few of them.

a. Used with food or beverages, *prendre* means either "to eat" or "to drink."

Au restaurant Marie-Lise prend toujours du poulet.
Au café les enfants prennent toujours de l'eau minérale.

b. *Prendre le petit déjeuner* means "to eat breakfast." Note, however, that you do not use *prendre* with other meals in French. "To eat lunch" is *déjeuner* and "to eat dinner" is *dîner.*

Gérard prend son petit déjeuner à la maison mais il déjeune à la cafétéria.

c. *Prendre les billets* means "to buy tickets."

Je prends mon billet au guichet et j'attends le train.

3. Two other verbs that are conjugated like *prendre* are *apprendre,* "to learn," and *comprendre,* "to understand." You use the preposition *à* after *apprendre* when it is followed by an infinitive.

Ma sœur et mon frère apprennent à jouer au tennis.
Vous comprenez le français, n'est-ce pas?

CHAPITRE 9   **243**

LEARNING FROM PHOTOS

You may wish to ask the following questions about the photo: *C'est un joueur de tennis? Qu'est-ce qu'il a à la main? Il a aussi une raquette? Il a la main sur le filet?*

Structure Teaching Resources

1. Workbook, *Structure: A–E,* pages 86–87
2. Student Tape Manual, Teacher's Edition, *Structure: A–D,* pages 103–104
3. Audio Cassette 6A/CD-6
4. Communication Activities Masters, *Structure: A–C,* pages 45–46
5. Computer Software, *Structure*
6. Chapter Quizzes, *Structure:* Quizzes 3–5, pages 50–52
7. CD-ROM, Disc 3, pages 243–247

Bell Ringer Review

Write the following on the board or use BRR Blackline Master 9-4: Match the word or phrase on the left with the related word or phrase on the right.

1. nager   a. au bord de la mer
2. le sable   b. le match
3. la balle   c. faire du surf
4. la station balnéaire   d. la plage
5. les vagues   e. la piscine
6. le score   f. la raquette
7. bronzer   g. il pleut
8. des nuages   h. prendre un bain de soleil

Les verbes prendre, apprendre et comprendre au présent

PRESENTATION *(page 243)*

Model the pronunciation of the forms of *prendre* and have students repeat them in unison. Pay attention to the pronunciation. The *n* of the root is nasal in the singular forms, but not in the plural forms.

PRESENTATION (*pages 244–245*)

Exercices A–F

All exercises may be done with books open or closed.

Exercices A, B, and C

These exercises use forms that have no pronunciation change. The remainder require sound and spelling changes.

ANSWERS

Exercice A

1. Oui, on prend un bain de soleil à la plage.
2. Non, on ne prend pas de bain de soleil quand il y a des nuages.
3. Oui, on met de la crème solaire quand on prend un bain de soleil.
4. Oui, on bronze quand on prend un bain de soleil.

Exercice B

Answers will vary.

Exercice C

É1 will use **Tu prends…** and É2 will respond with **Oui (Non), je (ne) prends (pas)…**

Exercice D

Answers will employ **Nous (ne) prenons (pas)…**

Exercice E

1. Il prend un crème et ses copains prennent un express.
2. Il prend une salade et ses copains prennent une soupe à l'oignon.
3. Il prend un sandwich au pâté et ses copains prennent un croque-monsieur.
4. Il prend une glace au chocolat et ses copains prennent une glace à la vanille.
5. Il prend de l'eau minérale et ses copains prennent du thé.

Exercices

A **On prend un bain de soleil.** Répondez.

1. On prend un bain de soleil à la plage?
2. On prend un bain de soleil quand il y a des nuages?
3. On met de la crème solaire quand on prend un bain de soleil?
4. On bronze quand on prend un bain de soleil?

B **Moi, en été.** Donnez des réponses personnelles.

1. En été, tu prends des bains de soleil sur le sable?
2. Tu bronzes ou tu attrapes des coups de soleil?
3. Tu préfères nager dans une piscine, dans la mer ou dans un lac?
4. Tu prends des leçons de surf?
5. Tu apprends à faire de la planche à voile?
6. Tu apprends à faire du ski nautique?
7. Tu comprends le moniteur?

C **Qu'est-ce que tu prends?** Posez des questions à un copain ou à une copine d'après le modèle.

> le train
> Élève 1: Tu prends le train?
> Élève 2: Non, je ne prends pas le train. (Oui, je prends le train.)

1. ton billet au guichet à la gare
2. le bus pour aller à l'école
3. l'avion pour aller à New York
4. l'avion pour aller en France

D **Le petit déjeuner.** Répondez en utilisant «nous».

1. Vous prenez votre petit déjeuner à la maison?
2. Vous prenez votre petit déjeuner à quelle heure?
3. Vous prenez votre petit déjeuner quand vous êtes en retard?
4. Vous prenez votre petit déjeuner dans la cuisine ou dans la salle à manger?
5. Vous prenez du lait au petit déjeuner?

E **Qu'est-ce qu'ils prennent au café?** Changez d'après le modèle.

> **Il prend un coca. (un citron pressé)**
> *Il prend un coca et ses copains prennent un citron pressé.*

1. Il prend un crème. (un express)
2. Il prend une salade. (une soupe à l'oignon)
3. Il prend un sandwich au pâté. (un croque-monsieur)
4. Il prend une glace au chocolat. (une glace à la vanille)
5. Il prend de l'eau minérale. (du thé)

ESPACE SOLEIL

BRONZEZ SANS SOLEIL au cœur de Paris AU MULTISTO OPERA

ADDITIONAL PRACTICE

(After completing Exercises A–F)

1. Have students introduce each of the following words or phrases with *j'apprends, je comprends,* or both: *le prof, la leçon, l'article du journal, le français.*
2. Have students make up a list of things they learn and a list of things they learn to do (*J'apprends… , J'apprends à…).*

3. Student Tape Manual, Teacher's Edition, *Activités A–B,* page 103

F **Au cours de français.** Répondez.

1. Au cours de français, les élèves apprennent beaucoup de mots?
2. Ils apprennent le vocabulaire?
3. Ils apprennent des règles de grammaire?
4. Ils apprennent la civilisation française?
5. Et toi, tu apprends à parler français?
6. Tes copains et toi, vous comprenez bien quand le professeur parle français?

Le pronoms accentués

Emphasizing and Clarifying Whom You Are Talking About

1. Compare the subject pronouns below with the corresponding stress pronouns.

STRESS PRONOUNS	SUBJECT PRONOUNS
moi	je
toi	tu
lui	il
elle	elle
nous	nous
vous	vous
eux	ils
elles	elles

La Côte d'Azur: Villefranche-sur-Mer

2. You use stress pronouns in several ways in French.

   a. to reinforce or stress the subject

      Moi, je vais au bord de la mer en été.
      Lui, il reste à la maison.

   b. after a preposition such as *avec, pour, chez,* etc.

      David veut jouer avec nous.
      Les filles rentrent chez elles après la fête.

   c. alone or in a phrase without a verb

      Qui fait du ski nautique? Moi!
      Et eux? Est-ce qu'ils prennent des leçons?

   d. before and after *et* or *ou*

      Marie et moi, nous allons à la plage.
      Qui va faire les courses ce soir? Lui ou elle?

CHAPITRE 9   **245**

Exercice F

1. Oui, au cours de français, les élèves apprennent beaucoup de mots.
2. Oui, ils apprennent le vocabulaire.
3. Oui, ils apprennent des règles de grammaire.
4. Oui, ils apprennent la civilisation française.
5. Oui, j'apprends à parler français.
6. Oui, nous comprenons bien quand le professeur parle français. (Non, nous ne comprenons pas bien quand le professeur parle français.)

Les pronoms accentués

PRESENTATION *(pages 245–246)*

A. Point to yourself and say *moi,* and to a student as you say *toi.* (These are the two stress pronouns the students already know.) Then point to individuals around the room to indicate the meaning of each of the other stress pronouns.
B. Now call on a more able student to do the same thing.
C. Give students additional examples of stress pronoun usage. For example: *Moi, je suis blonde. Mais lui, il est brun. Nous, nous finissons nos devoirs, mais vous, vous n'étudiez pas.* Add each stress pronoun to a list on the board as you use it.
D. Now lead students through steps 1–2 on pages 245–246.

Note In the CD-ROM version, this structure point is presented via an interactive electronic comic strip.

DID YOU KNOW?

Have students locate Nice on the map, page 504, or use the Map Transparency. Tell them that the quiet little town of Ville-franche-sur-Mer is located four miles east of Nice, on the beautiful blue bay shown in the photo. Artists have long been drawn to the town, where they tend to occupy the brightly-colored houses lining the steep narrow streets that wind down to the harbor.

INDEPENDENT PRACTICE

Assign any of the following:
1. Exercises, pages 244–245
2. Workbook, *Structure: A–C,* pages 86–87
3. Communication Activities Masters, *Structure: A,* page 45
4. CD-ROM, Disc 3, pages 243–245

e. after *c'est* or *ce n'est* pas

> C'est toi, Yvonne?
> Oui, c'est moi.
> C'est Jean-Luc?
> Non, ce n'est pas lui.

f. with *-même(s)* to express "myself," "herself," and so forth

> Je vais faire les valises moi-même.
> Ils font la cuisine eux-mêmes.

Exercices

A Moi, toi et les autres.   Complétez.

DAVID: ___, j'adore nager.
CÉLINE: Et ton frère? Il aime nager?
DAVID: ___? Il aime faire du ski nautique.
CÉLINE: Et ta sœur, ___, elle aime faire du ski nautique aussi?
DAVID: Non, mais ___, elle aime faire de la planche à voile.
CÉLINE: Sans blague! Ma copine et ___, nous aimons faire de la planche à voile aussi.
DAVID: ___ aussi, j'aime faire de la planche à voile. Mais mes copains, ___, ils n'aiment pas ça.

B Tu aimes les sports d'été?   Répondez d'après le modèle.

> Tu aimes nager?
> *Moi? Oui, j'adore nager.*

1. Tu aimes aller au bord de la mer?
2. Et ton frère, il aime faire du ski nautique?
3. Et tes sœurs, elles aiment faire de la plongée sous-marine?
4. Et vous, vous aimez nager?
5. Et tes copains, ils aiment faire du surf?

C Une fête.   Complétez.

1. Tu vas donner une fête pour Jean? Oui, je vais donner une fête pour ___.
2. Qui va organiser la fête? Toi? Oui, c'est ___.
3. Et qui va faire les courses? Ta mère? Pas ___! Moi, je vais aller au marché ___-même!
4. Jean va arriver chez toi avec ses copains? Oui, il va arriver chez ___ avec ___.

246   CHAPITRE 9

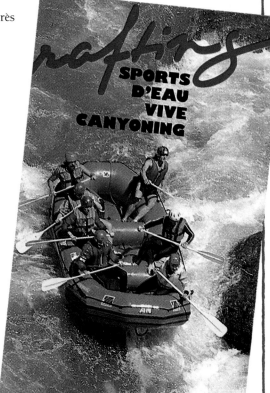

Les adjectifs avec une double consonne

Describing People and Things

1. Note that certain adjectives double their final consonant in the feminine forms. Study the following.

	FÉMININ	MASCULIN
SINGULIER	une compagnie aérienne une voiture européenne	un vol aérien une café européen
PLURIEL	des compagnies aériennes des voitures européennes	des vols aériens des cafés européens

2. Here are some other adjectives that follow the same pattern.

> canadien(ne)   italien(ne)   parisien(ne)

3. The adjective *bon,* which precedes the noun, also doubles its final consonant. Study these forms.

> C'est une très bonne idée.   Robert est un très bon élève.
> Il a de bonnes notes.   Et il a de bons profs.

4. The adjective *gentil,* "nice," also doubles its final consonant.

> une fille gentille   un garçon gentil

Exercices

A **D'après vous.** Répondez.

1. C'est une bonne idée de voyager avec une bonne compagnie aérienne canadienne?
2. C'est une bonne idée de passer une bonne journée sur une belle plage?
3. Est-ce que la compagnie aérienne italienne sert des spécialités italiennes pendant ses vols?
4. Est-ce que les femmes parisiennes font leurs courses dans les beaux magasins parisiens?

B **Une compagnie canadienne.** Complétez.

La compagnie ___ (aérien) ___ (canadien) offre des vols vers des
$\quad\quad\quad\quad\quad\quad\quad\quad 1 \quad\quad\quad\quad 2$
destinations ___ (européen). Le service est très ___ (bon). À bord les
$\quad\quad\quad\quad\quad 3 \quad\quad\quad\quad\quad\quad\quad\quad\quad\quad 4$
stewards sont très ___ (gentil) et les hôtesses de l'air aussi sont très ___
$\quad\quad\quad\quad\quad\quad\quad 5 \quad\quad\quad\quad\quad\quad\quad\quad\quad\quad\quad\quad\quad\quad 6$
(gentil). Il est agréable d'avoir une ___ (bon) place dans un avion ___
$\quad\quad\quad\quad\quad\quad\quad\quad\quad\quad\quad\quad\quad 7 \quad\quad\quad\quad\quad\quad\quad\quad\quad\quad 8$
(canadien) et de faire un ___ (bon) voyage ___ (européen).
$\quad\quad\quad\quad\quad\quad\quad\quad\quad 9 \quad\quad\quad\quad\quad\quad 10$

INDEPENDENT PRACTICE

Assign any of the following:
1. Exercises, pages 246–247
2. Workbook, *Structure: D–E,* page 87
3. Communication Activities Masters, *Structure: B–C,* pages 45–46
4. Computer Software, *Structure*
5. CD-ROM, Disc 3, pages 245–247

LEARNING FROM REALIA

Have students look at the ad on page 246. Ask them how you say "rafting" in French (*le rafting*). This is a fine example of the wholesale takeover of English words into French, an increasingly common phenomenon. Ask them what they think *canyoning* is: does the word exist in English or is it a French invention? How does one say "whitewater" in French? (*eaux vives*).

Les adjectifs avec une double consonne

PRESENTATION (*page 247*)

A. It is recommended that you not spend a great deal of time on this point. Students will need reinforcement of the spelling of these adjectives throughout their study of French.

B. Lead students through steps 1–4 on page 247, modeling the pronunciation of the feminine versus the masculine forms of the adjectives and having students repeat in unison.

Exercices

PRESENTATION (*page 247*)

Exercice B

After going over this exercise, calling on several students to complete the sentences, have one student reread the entire exercise.

ANSWERS

Exercice A

Answers will vary.

Exercice B

1. aérienne
2. canadienne
3. européennes
4. bon
5. gentils
6. gentilles
7. bonne
8. canadien
9. bon
10. européen

DICTATION

You may wish to give the following dictation:
Les filles canadiennes sont gentilles. Elles portent de bonnes lunettes italiennes.

CONVERSATION

PRESENTATION *(page 248)*

A. Have students close their books and watch the Conversation Video or listen as you either read the conversation to them or play Cassette 6A/CD-6.

B. Have them repeat the conversation once or twice in unison.

C. Call on pairs to read the conversation with as much expression as possible.

D. Have pairs act out the conversation, allowing them to make any changes that make sense.

Note In the CD-ROM version, students can play the role of either one of the characters and record the conversation.

ANSWERS

Exercice A

1. **Oui, il fait chaud.**
2. **Oui, Nathalie veut aller à la plage.**
3. **Françoise va chercher son maillot de bain.**
4. **Nathalie n'a pas de crème solaire.**
5. **Nathalie va prendre un bain de soleil.**
6. **Oui, elle aime bronzer.**
7. **Elle va nager et faire du ski nautique.**

Prononciation

PRESENTATION *(page 248)*

A. Model the key phrase *un vieux soleil en maillot* and have students repeat chorally after you. Then model the other words and phrases in similar fashion.

B. For additional practice, use: Cassette 6A/CD-6: *Prononciation*; Pronunciation Transparency P-9; and the Student Tape Manual, Teacher's Edition, *Activités G–I*, page 106.

C. Give the following *dictée:*
La gentille fille a un billet. Le soleil brille. La fille ne travaille pas en maillot.

248

CONVERSATION

Scènes de la vie   *Une belle journée d'été*

NATHALIE: Il fait terriblement chaud!
FRANÇOISE: C'est vrai, c'est horrible!
NATHALIE: Tu veux aller à la plage?
FRANÇOISE: D'accord. Je vais chercher mon maillot de bain.
NATHALIE: Tu as de la crème solaire?
FRANÇOISE: Oui. Pourquoi? Tu vas prendre un bain de soleil?
NATHALIE: Mais bien sûr!
FRANÇOISE: Pas moi.
NATHALIE: Pas toi? Qu'est-ce que tu vas faire alors?
FRANÇOISE: Je vais nager et faire du ski nautique.

A   **La plage.**   Répondez d'après la conversation.

1. Il fait chaud?
2. Nathalie veut aller à la plage?
3. Qu'est-ce que Françoise va chercher?
4. Qui n'a pas de crème solaire?
5. Qui va prendre un bain de soleil?
6. Elle aime bronzer?
7. Et Françoise, qu'est-ce qu'elle va faire?

Prononciation   *Les sons /y/ et /y/ + voyelle*

The sound /y/ occurs in three positions: final, between two vowel sounds, and in combination with another vowel sound. Repeat the following.

fille	soleil	gentille
maillot	travailler	billet
canadien	aérien	vieux

Now repeat the following sentences.

J'ai un vieux maillot.
On ne travaille pas bien au soleil.
C'est un avion canadien.

un vieux soleil en maillot

248   CHAPITRE 9

LEARNING FROM PHOTOS

Ask the following questions about the photo: *La fille à la bicyclette, c'est Nathalie ou Françoise? Quelle fille sur la photo prend un bain de soleil? Elle va prendre un bain de soleil à la plage aussi? À ton avis c'est la maison de Françoise ou de Nathalie?*

Activités de communication orale

A **Qu'est-ce que vous prenez au snack bar?** Divide into small groups and choose a leader. The leader asks the others what they usually buy to eat or drink at the beach snack bar. He or she takes notes, then reports to the class.

> Élève 1: Qu'est-ce que tu prends quand tu as soif (faim)?
> Élève 2: Moi, je prends de l'eau minérale (un sandwich au jambon).
> Élève 1 (*à la classe*): Marc prend de l'eau minérale. Anne et Paul prennent…

B **Moi, je veux apprendre à…** Take turns with a classmate and find out what each of you would like to learn to do and why.

> Élève 1: Qu'est-ce que tu veux apprendre à faire?
> Élève 2: Moi, je voudrais apprendre à bien parler français.
> Élève 1: Pourquoi?
> Élève 2: Parce que je voudrais aller en France.

C **En été.** Tell a classmate some things you do in the summer, then find out what your partner likes to do.

> Élève 1: En été je vais à la plage, je fais du surf et de la planche à voile. Et toi, qu'est-ce que tu aimes faire en été?
> Élève 2: Moi, j'aime aller à la plage aussi, mais je ne fais pas de surf. J'aime nager et j'aime aller à la pêche.

Les chutes de Montmorency à dix kilomètres de Québec

INDEPENDENT PRACTICE

Assign any of the following:
1. Exercises and activities, pages 248–249
2. Workbook, *Un Peu Plus,* pages 88–90
3. CD-ROM, Disc 3, pages 248–249

DID YOU KNOW?

At 83 meters high (270 feet), the Montmorency Falls are one-and-a-half times higher than Niagara Falls. A cable car takes visitors to the top of the falls, where they can view an impressive panorama. In winter, at the foot of the falls, a strange natural phenomenon known as the "sugar loaf" (*pain de sucre*) can be observed.

Activités de communication orale

PRESENTATION *(page 249)*

Activités A, B, and C
Let students choose the activity they want to work with.

ANSWERS

Activités A, B, and C
Answers will vary.

Vocabulary Expansion

Students may ask why the title of the *Conversation* uses the word *journée,* not *jour.* For English speakers the distinction may be a bit subtle. Explain that *la journée* usually refers to a duration of time, a whole day, a more subjective notion of time than that connoted by *le jour,* which is just a simple measure of time. (The same applies to *l'an* and *l'année.*) *La journée* and *l'année* as a result are often qualified with adjectives (as in *Une belle journée d'été*).

READING STRATEGIES
(page 250)

Pre-reading

Have students scan the reading for cognates.

Reading

A. Have the class read the selection once silently.

B. Call on individuals to read one paragraph each.

C. Ask comprehension questions based on each paragraph.

Post-reading

If possible, bring in pictures or slides of popular French beach or mountain vacation areas and share them with your students. Some sources for these materials might be your own collection, that of an acquaintance, your library or a local travel agency.

Note Point out to students that in informal conversational French, people say *Qu'est-ce qu'elles sont belles* or *Ce qu'elles sont belles* for *Qu'elles sont belles.*

Note Students may listen to a recorded version of the *Lecture* on the CD-ROM.

Étude de mots

ANSWERS

Exercice A

1. d	4. e
2. c	5. a
3. b	6. f

Exercice B

1. mois	4. montagnes
2. vacances	5. ouest
3. côtes, belles	

250

LES VACANCES D'ÉTÉ

C'est le premier août. Tout le monde prend la route pour aller au bord de la mer. Les vacances d'été commencent. En France le mois d'août, c'est le mois des vacances. On ne travaille pas. On passe le mois entier au bord de la mer ou à la montagne.

Qu'elles sont belles[1], les plages en France! Il y a des stations balnéaires le long des côtes[2]: sur la Manche au nord, sur l'océan Atlantique à l'ouest, et sur la Côte d'Azur au sud, au bord de la mer Méditerranée.

Qu'est-ce qu'on fait au bord de la mer? On va à la plage, bien sûr. À la plage on prend des bains de soleil. Tout le monde veut rentrer chez soi[3] bien bronzé. Les

Une belle plage bretonne

gens[4] sportifs nagent ou font de la planche à voile. Moi, je fais du ski nautique. Qu'est-ce que tu fais en été?

Vers deux heures on a faim. Après une belle journée à la plage on a une faim de loup. L'air de la mer donne faim. On fait un pique-nique sur la plage ou on va dans un petit restaurant en plein air[5] où on commande des fruits de mer[6].

[1] Qu'elles sont belles *How beautiful they are*
[2] le long des côtes *along the coasts*
[3] chez soi *home*
[4] gens *people*
[5] en plein air *outdoor*
[6] fruits de mer *seafood*

Étude de mots

A **Quel est le mot?** Trouvez une expression équivalente.

1. sportif	a. dans toutes les régions
2. une faim de loup	b. tout le mois
3. le mois entier	c. très faim
4. en plein air	d. qui aime les sports
5. partout	e. à l'extérieur, dehors
6. commencer	f. le contraire de *finir*

CRITICAL THINKING ACTIVITY

(Thinking skills: decision making, evaluating consequences.)

Put the following on the board or on a transparency:

1. Christine va à la plage. Mais elle ne bronze pas. Elle attrape toujours des coups de soleil. Elle doit *(must)* acheter de la crème solaire. Elle a 80 francs. Elle veut acheter des lunettes de soleil qui coûtent 70 francs. Les lunettes sont très jolies. Mais si elle achète les lunettes, elle ne va pas avoir d'argent pour acheter la crème solaire. Qu'est-ce qu'elle doit faire?

2. Christine succombe à la tentation! Elle achète les lunettes de soleil et elle va à la plage. Quelles sont les conséquences de sa décision?

B **Des faits.** Complétez les phrases d'après la lecture.

1. Le ___ d'août a trente et un jours.
2. Le mois d'août est le mois des ___ parce que les gens ne travaillent pas.
3. Le long des ___ de la France, il y a de très ___ plages.
4. Les Pyrénées et les Alpes sont des ___.
5. L'océan Atlantique est à l'___ de la France.

Compréhension

C **Au bord de la mer.** Répondez d'après la lecture.

1. Quelle est la date?
2. Tout le monde prend la route pour aller où?
3. On passe combien de temps au bord de la mer?
4. Il y a des plages partout en France?
5. Qu'est-ce qu'on fait à la plage?
6. Que font les gens sportifs?
7. Tout le monde veut rentrer chez soi comment?
8. Quelle est l'heure du déjeuner?
9. Où est-ce qu'on va manger?
10. Qu'est-ce qu'on commande au bord de la mer?

D **Les vacances.** Trouvez les renseignements suivants dans la lecture.

1. Quel est le mois des vacances, le mois où très peu de gens travaillent?
2. Où est-ce que les Français aiment passer leurs vacances?
3. Les Français passent combien de temps au bord de la mer ou à la montagne?

DÉCOUVERTE CULTURELLE

Les Français sont très travailleurs. Mais les vacances sont très importantes pour eux. Le Français typique a à peu près cinq semaines de vacances par an. Le mois favori pour les vacances d'été, c'est le mois d'août. Le premier août il y a des bouchons et des embouteillages[1] partout. Tout le monde est pressé[2] d'arriver au bord de la mer pour commencer les vacances.

Tes parents ont combien de semaines de vacances? Ta famille et toi, où passez-vous les vacances? Quand est-ce que vous y allez? Vous y passez combien de temps?

[1] des bouchons et des embouteillages   *traffic jams*
[2] est pressé   *is in a hurry*

ANSWERS
Exercice C
1. C'est le premier août.
2. Tout le monde prend la route pour aller au bord de la mer.
3. On passe le mois entier au bord de la mer.
4. Il y a des plages le long des côtes, à l'ouest, au nord et au sud.
5. À la plage, on prend des bains de soleil.
6. Les gens sportifs nagent ou font de la planche à voile ou du ski nautique.
7. Tout le monde veut rentrer chez soi bien bronzé.
8. Deux heures est l'heure du déjeuner.
9. On va manger dans un petit restaurant en plein air ou on va faire un pique-nique sur la plage.
10. On commande des fruits de mer.

Exercice D
1. C'est le mois d'août.
2. Ils aiment passer leurs vacances au bord de la mer ou à la montagne.
3. Ils passent le mois entier au bord de la mer ou à la montagne.

OPTIONAL MATERIAL

Découverte culturelle

PRESENTATION *(page 251)*
Before reading the selection, have students discuss vacations in the U.S.: How long do Americans usually get for their vacation? Do they usually take their vacation in summer? Is one month more popular than the other for summer vacations? How do most American families travel in the summer—by car, plane or train?

Note Students may listen to a recorded version of the *Découverte culturelle* on the CD-ROM.

DID YOU KNOW?

The beautiful beach pictured in the photo on page 250 is just one of many on the coast of Brittany. The clean, sandy beaches are all popular vacation spots in the summer. Dinard, Perros-Guirec, Trégastel-Plage, Douarnenez, Carnac, and La Baule are among the best.

LEARNING FROM PHOTOS

Ask students the following questions about the photo: *Cette fille fait de la planche à voile ou du surf? Est-ce qu'elle fait bien du surf ou est-ce qu'elle apprend à faire du surf? Qui est-ce, derrière elle sur la photo?*

RÉALITÉS

PRESENTATION (*pages 252–253*)

Have the students sit back and enjoy the beautiful photographs as they read the information about them.

Note In the CD-ROM version, students can listen to the recorded captions and discover a hidden video behind one of the photos.

Culture Note In regard to **Photo 2**, you may wish to tell students that in 1991 France won the Davis Cup when Guy Forget defeated the American, Pete Sampras. It was France's first Davis Cup since 1932.

RÉALITÉS

C'est une colonie de vacances en montagne **1**. Les enfants jouent avec les monitrices. Tout le monde adore l'été. C'est la belle saison.

Voici Yannick Noah **2**. C'est un champion de tennis célèbre. Tu voudrais jouer contre lui?

Voici la plage de Nice, une ville sur la Côte d'Azur **3**. Sur la plage à Nice, il y a du sable ou des galets?

Voici Audierne, un joli port de pêche en Bretagne **4**. Il y a beaucoup de bateaux dans le port?

La jeune femme fait une promenade en vélo en montagne **5**. C'est un vélo tout terrain (VTT).

252

DID YOU KNOW?

Nice is a resort city on the French Riviera. It lies at the foot of the Alps near Italy. The Alps protect the city from cold northern winds and give it a mild winter climate. Most tourists visit Nice during the winter or from July to September. Nice has a famous Mardi Gras celebration, *le Carnaval*, which lasts for two weeks.

ADDITIONAL PRACTICE

Assign any of the following:
1. Student Tape Manual, Teacher's Edition, *Deuxième Partie,* pages 107–109
2. Situation Cards, Chapter 9

RECYCLING

The *Activités de communication orale* and *Activités de communica-tion écrite* focus on situations within the contexts of the seaside and sports in order to recycle the words and structures from this chapter and previous ones. Recy-cled language includes food vocabulary, time expressions, the *futur proche*, the verbs *faire, aller, partir, avoir,* and regular verbs.

INFORMAL ASSESSMENT

Oral Activities A and B may serve as a speaking evaluation. They are both guided activities, yet they provide an opportunity for students to come up with their own utterances within the given situations. Use the evaluation cri-teria on page 34 of this Teacher's Wraparound Edition.

Activités de communication orale

PRESENTATION *(page 254)*

Activité A

 In the CD-ROM version of this activity, students can interact with an on-screen native speaker.

ANSWERS

Activités A and B
Answers will vary.

Activité de communication écrite

ANSWERS

Activité A
Answers will vary.

Activités de communication orale

A **Une nouvelle amie.** At the beach in Saint-Tropez you have just met Danielle Lacroix, a French teen around the same age as you. She wants to find out more about you. Answer her questions.

1. Alors tu es comme moi—tu adores la plage! Qu'est-ce que tu aimes faire à la plage?
2. À quelle heure est-ce que tu arrives à la plage?
3. Où est-ce que tu déjeunes?
4. Tu veux déjeuner avec moi demain?

B **Tu veux jouer au tennis avec moi?** The new French exchange student (your partner) wants to know if you play tennis (or would like to learn to play), if you have a racket, and if you'd like to play tennis with him or her tomorrow. Answer, then reverse roles.

Activité de communication écrite

A **Une carte postale.** You're spending two weeks at the beach resort of your choice. Write a postcard to a friend about your vacation. Tell him or her where you are and what the place is like; what the weather's like; what you do every day; a new sport you're learning, and what you think of the instructor; and when you're going to return home.

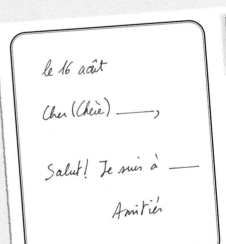

le 16 août

Cher (Chère) ——,

Salut! Je suis à ——

Amitiés

254 CHAPITRE 9

FOR THE YOUNGER STUDENT

1. Have students draw a map of France. On the coastal areas have them write the names of famous beaches. They should highlight these locations by drawing in figures of people engaged in various aquatic sports. When finished, students can write a paragraph explain-ing what is on their map.

2. Have students draw a beach scene on a piece of paper the size of a postcard. They should include figures to represent themselves. On the back of the paper have them write to a friend telling about their vacation at the beach.

3. Using a large map of France or of the U.S., have students play the role of a TV weather forecaster giving the forecast for a typical summer day.

Réintroduction et recombinaison

A **On prend le train.** Complétez.

1. Jacques _____ le train. (prendre)
2. Jacques et ses copains _____ le train. (prendre)
3. Ils _____ au bord de la mer. (aller)
4. Ils _____ le train dans la salle d'attente. (attendre)
5. Jacques _____ au guichet. (aller)
6. Au guichet il _____ les billets pour tous ses copains. (prendre)
7. Les copains _____ l'annonce du départ du train. (entendre)
8. Ils _____ l'annonce. (comprendre)
9. Ils _____ sur le quai. (aller)
10. Ils _____ dans le train. (monter)
11. Ils _____ à la prochaine gare. (descendre)
12. Ils _____ à Deauville à quatorze heures dix-huit. (arriver)

La plage de Deauville en Normandie

Vocabulaire

NOMS
l'été (m.)
la station balnéaire
le bord de la mer
la plage
le sable
la mer
la vague
la crème solaire
les lunettes (f.) de soleil
le maillot (de bain)
la natation
la piscine
la leçon
le moniteur

le tennis
le court de tennis
la balle
le filet
la raquette
le match
le joueur
la partie (en simple, en double)
les limites (f.)
le score
les chaussures (f.) de tennis
le tee-shirt

le short
la jupette

ADJECTIFS
aérien(ne)
bon(ne)
canadien(ne)
européen(ne)
gentil(le)
italien(ne)
parisien(ne)

VERBES
bronzer
frapper
gagner
jouer à
nager
plonger
renvoyer
apprendre (à)
comprendre
prendre

AUTRES MOTS ET EXPRESSIONS
faire de la planche à voile
faire de la plongée sous-marine
faire du ski nautique

faire du surf
faire une promenade
aller à la pêche
attraper un coup de soleil
prendre le petit déjeuner
prendre un bain de soleil
prendre un billet
Il faut + infinitif
entre
fort
hors des limites

pourquoi
parce que

Quel temps fait-il?
Il fait beau.
Il fait chaud.
Il fait du soleil.
Il fait froid.
Il fait mauvais.
Il fait du vent.
Il pleut.
Il y a des nuages.

Réintroduction et recombinaison

PRESENTATION *(page 255)*

Exercise A recycles the forms of the regular *-re* verbs and the three irregular ones introduced in this chapter. It also recycles *aller* and regular *-er* verbs. You may wish to refer students to appropriate pages in their book to look for sample answers to items with which they have trouble.

ANSWERS

Exercice A
1. prend
2. prennent
3. vont
4. attendent
5. va
6. prend
7. entendent
8. comprennent
9. vont
10. montent
11. descendent
12. arrivent

ASSESSMENT RESOURCES

1. Chapter Quizzes
2. Testing Program
3. Situation Cards
4. Communication Transparency C-9
5. Computer Software: Practice/Test Generator

VIDEO PROGRAM

INTRODUCTION (29:08)

C'EST CHOUETTE, (29:51)
SAINT-MALO!

LEARNING FROM PHOTOS

You may wish to give the students the words for "beach (or deck) chair" and "beach umbrella": *un transat, un parasol.*

STUDENT PORTFOLIO

Written assignments that may be included in students' portfolios are *Activité de communication écrite A* on page 254 and the *Mon Autobiographie* section of the Workbook on page 91.

Note Students may create and save both oral and written work using the Electronic Portfolio feature on the CD-ROM.

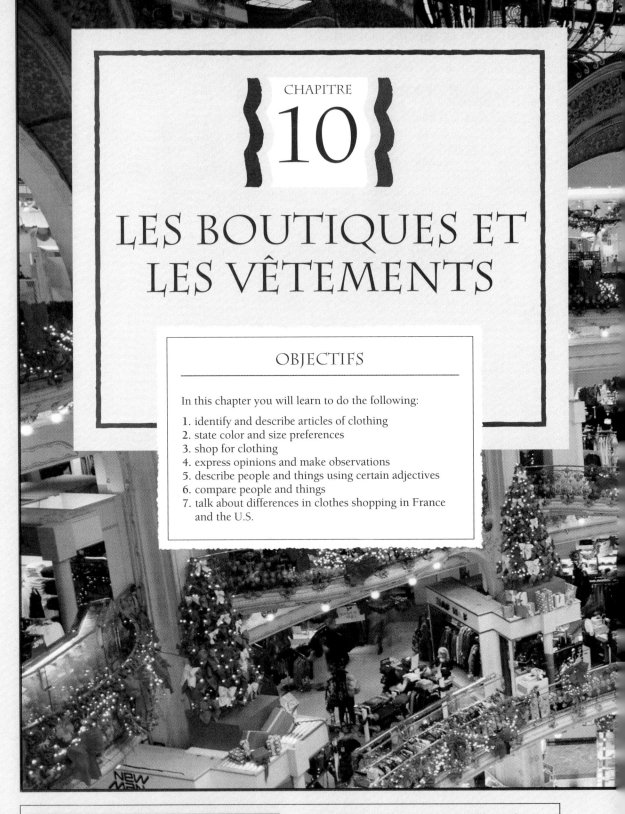

CHAPTER OVERVIEW

In this chapter students will learn to identify and describe articles of clothing, to talk about what they and others wear for different occasions, and to communicate in various situations that arise when shopping for clothes. They will learn to make observations and express opinions using comparative and superlative statements about people and things. In addition, students will learn to use and conjugate verbs such as *croire* and *voir*.

The cultural focus of Chapter 10 is on clothing stores in France, ranging from the boutiques of the *hauts couturiers* to the stalls at the flea markets (*marchés aux puces*).

CHAPTER OBJECTIVES

By the end of this chapter, students will know:

1. vocabulary associated with clothing and its description, including size and color
2. some basic vocabulary and structures necessary for locating items and speaking with salespeople in various types of apparel stores
3. the use of the adverb *trop* in qualitative constructions
4. the present indicative forms of the verbs *croire* and *voir*
5. some adjectives with irregular forms
6. the formation and use of the comparative
7. the formation and use of the superlative

CHAPITRE

10

LES BOUTIQUES ET LES VÊTEMENTS

OBJECTIFS

In this chapter you will learn to do the following:

1. identify and describe articles of clothing
2. state color and size preferences
3. shop for clothing
4. express opinions and make observations
5. describe people and things using certain adjectives
6. compare people and things
7. talk about differences in clothes shopping in France and the U.S.

CHAPTER PROJECTS

(optional)

1. Encourage students to watch any fashion-oriented television shows that are aired in your area. Have them write down any French words they hear to share with the class.
2. Do a group research project on fashion designers, with each group studying a different designer. Students can focus on

such information as when and how the designer started in the industry; if he/she is alive today; what defines the particular label of the designer; what each house is most famous for; if the house designs anything besides clothing. You may want to include American designers as well as European ones.

257

Pacing

This chapter requires eight to ten class sessions. Pacing will depend on class length and the age and aptitude of the students.

Note The Lesson Plans offer guidelines for 45- and 55-minute classes and **Block Scheduling.**

Exercices vs. *Activités*

All exercises (which provide guided practice) are coded in blue. All communicative activities are coded in red.

INTERNET ACTIVITIES

(*optional*)

These activities, student worksheets, and related teacher information are in the *Bienvenue* Internet Activities Booklet and on the Glencoe Foreign Language Home Page at **http://www.glencoe.com/secondary/fl**

LEARNING FROM PHOTOS

Tell students: *C'est un grand magasin à Paris, les Galeries Lafayette. Le magasin est très joli, n'est-ce pas? Il est décoré pour Noël. Ça, c'est un arbre de Noël. C'est quel mois?*

VOCABULAIRE

MOTS 1

LES VÊTEMENTS POUR HOMMES

une veste

une chemise

un pantalon
une cravate
un complet

LES VÊTEMENTS POUR FEMMES

un collant

une jupe
un chemisier

un tailleur
une robe habillée
une robe sport

une chaussette

un blouson
un pull

un jean

une paire de chaussures

Marc porte un sweat-shirt.

LAURENT

la boutique d'un grand couturier

Vocabulary Teaching Resources

1. Vocabulary Transparencies 10.1 (A & B)
2. Audio Cassette 6B/CD-6
3. Student Tape Manual, Teacher's Edition, *Mots 1: A–C,* pages 110–112
4. Workbook, *Mots 1: A–C,* pages 92–93
5. Communication Activities Masters, *Mots 1: A,* page 47
6. Chapter Quizzes, *Mots 1:* Quiz 1, page 53
7. CD-ROM, Disc 3, *Mots 1:* pages 258–261

Bell Ringer Review

Write the following on the board or use BRR Blackline Master 10-1: Make a list in French of as many articles of clothing as you know.

PRESENTATION *(pages 258–259)*

A. Identify items of clothing students are actually wearing. Have the class repeat each item after you once or twice. Ask *Qu'est-ce que c'est?* and have a student respond.

B. Show Vocabulary Transparencies 10.1 (A & B). Have students repeat each item after you or Cassette 6B/CD-6.

C. Ask questions about the material on page 259, referring to the Vocabulary Transparencies. For example: *Lise est une cliente ou une vendeuse? Elle est à quel rayon? Qu'est-ce qu'elle regarde? Qu'est-ce que le garçon regarde? Où est-ce que Madame Laval paie? Combien coûtent les chaussettes? Et le chemisier?*

TOTAL PHYSICAL RESPONSE

(following the Vocabulary presentation)

Getting Ready

Dramatize the meaning of *enlevez.*

TPR 1

(Student 1), **venez ici, s'il vous plaît.**
Vous êtes au rayon des blousons.
Cherchez un joli blouson.
Prenez le blouson.

Essayez le blouson. Mettez le blouson.
Allez au miroir.
Regardez-vous dans le miroir.
Admirez-vous.
Indiquez que vous aimez le blouson.
Et maintenant, enlevez le blouson.
(Student 2), **venez ici, s'il vous plaît.**
Vous êtes le caissier/la caissière.
Allez à la caisse.
Et (Student 1), **allez à la caisse aussi.**

(continued on next page)

AU GRAND MAGASIN

des soldes

un vendeur

une vendeuse

SOLDES
270F

un rayon prêt-à-porter

un client

Lise voit beaucoup de chemisiers.
Elle voit les chemisiers au rayon
prêt-à-porter.
Elle va faire ses achats au rayon
prêt-à-porter.

une cliente

le prix

moins
cher

100F

bon marché

25F

240F

cher

1000F

plus cher

Mme Laval paie à la caisse.
Elle dépense* de l'argent.

* dépenser: employer de l'argent pour
  faire des achats

CHAPITRE 10   259

D. Play a game using the new vocabulary. One student describes what someone in the class is wearing and calls on classmates to guess who is being described.

Extension of D Recycle earlier vocabulary by having students describe the person as well as what that person is wearing.

E. Ask personal questions about yourself and students such as: *Qui porte une jupe aujourd'hui? Je porte une cravate? Marie porte un pantalon ou une robe?* After a few such examples, see if volunteers can come up with similar questions.

CROSS-CULTURAL COMPARISON

Note the invariable adjective *sport (une robe sport)* and the adjective *habillée (une robe habillée).* There is no precise French word that conveys the English "formal" or "dressy" in regard to clothing. *Habillé* is the closest approximation to "formal." Another word students will encounter frequently, if they peruse magazines, is *décontracté,* which conveys the English "casual," "informal."

Vocabulary Expansion

Many other articles of clothing will be presented in later chapters as they are needed. It is therefore recommended that you not give students an extensive list of additional vocabulary now. You may, however, wish to give them the names of a few accessories.
une ceinture
une montre
un bracelet
une bague
une boucle d'oreille

TPR *(continued)*
Sortez de l'argent de votre poche.
Donnez de l'argent au caissier/à la caissière.
(Student 2), mettez le blouson dans un sac.
Donnez le sac au client/à la cliente.
(Student 1), prenez le sac.
Merci (Student 1) et (Student 2). Retournez
   à vos places, s'il vous plaît.

Exercices

Extension of *Exercice A*

After completing Exercise A, have students say what Albert is wearing now. Have students name articles of clothing that Christine is not wearing now.

Exercice B: **Paired Activity**

Have students practice both listening and speaking by doing Exercise B in pairs. One partner reads the questions in random order while the other listens and answers with his/her book closed. Then partners switch roles. Or you can have one student read the questions to the entire class and then call on individuals to respond.

ANSWERS

Exercice A

1. **Albert va mettre une cravate, une veste, un pantalon, une chemise, des chaussettes et des (une paire de) chaussures.**
2. **Christine porte un jean, un pull, un blouson, des chaussettes et des (une paire de) chaussures (de tennis).**

Exercice B

Answers will vary.

Exercices

A   **Albert et Christine.**   Répondez d'après les dessins.

1. Qu'est-ce qu'Albert va mettre?
2. Qu'est-ce que Christine porte?

B   **Qu'est-ce qu'on met?**   Répondez.

1. Ce soir M. Ben-Azar va aller dans un restaurant élégant. Qu'est-ce qu'il va porter?
2. Qu'est-ce que sa femme va mettre?
3. Qu'est-ce que tu portes à l'école?
4. Qu'est-ce que tu portes quand il n'y a pas de cours?
5. Qu'est-ce que tu mets quand il fait froid?
6. Qu'est-ce qu'une femme met quand elle va au travail?
7. Qu'est-ce qu'un homme met quand il va au travail?

NEWS MODE   REVUE DE DÉTAILS   REPÉRÉ AUX QUATRE COINS DE LA MODE, TOUT CE QUI N... PLAÎT. DE LA TÊTE AUX PIEDS.

STRETCH (1) Robe en panne de velours (Capucine Puerari, 1 360 F, 5 tailles, 8 coloris, rens. 45 49 26 90). **SOIR CHIC (2)** Veste croisée, en drap de laine, sur jupe en taffetas de soie (Corinne Sarrut, 1 900 F, 3 tailles, 5 coloris (veste) et 900 F, du 36 au 42, en noir ou bronze (jupe), rens. 42 61 71 60). Gilet en satin (Chacok). **INTÉRIEUR (3)** Robe de chambre en soie (Claudie Pierlot, 800 F, 2 tailles, 3 coloris, rens. 42 36 69 93).

COL HIRONDELLE Très 70, des chemises bicolores (Agnès B., 490 F, 3 tailles, 3 coloris, rens. 45 08 5...

COLORISSIMO (1) Chemise en satin de soie, b...

LEARNING FROM REALIA

1. Ask students to look at the advertisement. Ask them to find as many English words as they can.
2. Have students find as many cognates as they can.

ADDITIONAL PRACTICE

Student Tape Manual, Teacher's Edition, *Activités B–C,* pages 111–112.

C Une boutique ou un grand magasin? Répondez.

1. On vend beaucoup de marchandises différentes dans la boutique d'un grand couturier ou dans un grand magasin?
2. Il y a beaucoup de rayons dans une boutique ou dans un grand magasin?
3. Qui vend des marchandises dans les boutiques et les grands magasins?
4. Et qui fait des achats?
5. Où est-ce qu'on paie dans les boutiques et les grands magasins?
6. Est-ce que les femmes riches achètent leurs vêtements au rayon prêt-à-porter ou chez les grands couturiers?
7. Est-ce qu'on peut acheter des vêtements sport et habillés dans un grand magasin?
8. Est-ce que les gens riches dépensent beaucoup d'argent pour leurs vêtements?

D On va acheter des vêtements. Complétez.

1. Il y a beaucoup de réductions pendant les ___. Les prix sont plus bas, moins élevés.
2. Je préfère faire mes achats quand il y a des ___ parce que je ___ moins d'argent.
3. Le jean est une sorte de ___ sport, pas habillé.
4. Quel est le ___ de ce blouson? 800 francs?
5. Oh là là! Ce blouson n'est pas bon marché! Il est très ___.

LEARNING FROM PHOTOS

You may wish to ask: *C'est une boutique élégante? C'est la boutique de quel grand couturier? On vend quelle sorte de vêtements ici? Est-ce que les vêtements sont en solde? Ils sont chers ou bon marché?*

INDEPENDENT PRACTICE

Assign any of the following:
1. Exercises, pages 260–261
2. Workbook, *Mots 1: A–C,* pages 92–93
3. Communication Activities Masters, *Mots 1: A,* page 47
4. CD-ROM, Disc 3, pages 258–261

PRESENTATION *(page 261)*
Exercice C

You may wish to go over Exercise C orally first and then reinforce it as a reading activity with books open.

ANSWERS
Exercice C

1. **On vend beaucoup de marchandises différentes dans un grand magasin.**
2. **Il y a beaucoup de rayons dans un grand magasin.**
3. **Les vendeurs et les vendeuses vendent des marchandises.**
4. **Les client(e)s font des achats.**
5. **On paie à la caisse.**
6. **Les femmes riches achètent leurs vêtements chez les grands couturiers.**
7. **Oui, on peut acheter des vêtements sport et habillés dans un grand magasin.**
8. **Oui, les gens riches dépensent beaucoup d'argent pour leurs vêtements.**

Exercice D

1. **soldes**
2. **soldes, dépense**
3. **pantalon (vêtement)**
4. **prix**
5. **cher**

INFORMAL ASSESSMENT
(Mots 1)

Check for comprehension by using Vocabulary Transparencies 10.1. Call individuals to the screen. As classmates say words or expressions from *Mots 1,* the student at the screen points to the appropriate image.

RETEACHING *(Mots 1)*

Have students open their books to page 258 and write down what they wear when they do the following things.
dîner en ville
jouer au tennis
aller au café
aller à l'école
travailler à plein temps

262   CHAPITRE 10

Vocabulary Teaching Resources

1. Vocabulary Transparencies 10.2 (A & B)
2. Audio Cassette 6B/CD-6
3. Student Tape Manual, Teacher's Edition, *Mots 2: D–G,* pages 112–114
4. Workbook, *Mots 2: D–H,* pages 93-95
5. Communication Activities Masters, *Mots 2: B,* page 48
6. Chapter Quizzes, *Mots 2: Quiz 2,* page 54
7. Computer Software, *Vocabulaire*
8. CD-ROM, Disc 3, *Mots 2:* pages 262–265

Bell Ringer Review

Write the following on the board or use BRR Blackline Master 10-2: Name five articles of clothing a boy and girl would wear for a job interview.

PRESENTATION *(pages 262–263)*

A. For an activity that's fun, you may wish to bring articles of old clothing to class, or have students bring in articles of old clothing. Have students put on the wrong sizes to convey *large, serré, étroit, long, court, haut, bas.* They should say the appropriate phrase: *Je voudrais la taille au-dessus* or *Je voudrais la taille au-dessous.*

B. Have students open their books and repeat the vocabulary after you or Cassette 6B/CD-6.

C. Model the conversation at the top of page 262. Then have volunteers perform it as a demonstration of *trop* and the various adjectives, substituting different articles of clothing and sizes. Use American sizes here.

TOTAL PHYSICAL RESPONSE

(following the Vocabulary presentation)

Note You may vary the articles of clothing below and their colors according to what the students are wearing.

TPR
Attention, tout le monde. Si vous portez le vêtement que je mentionne, levez-vous.
Je vois un pantalon blanc.
Je vois un sweat-shirt noir.
Je vois une chemise bleue.
Je vois un tee-shirt rouge.
Je vois une jupe verte.
Je vois un short beige.
Je vois un jean noir.
Merci bien, tout le monde! Asseyez-vous.

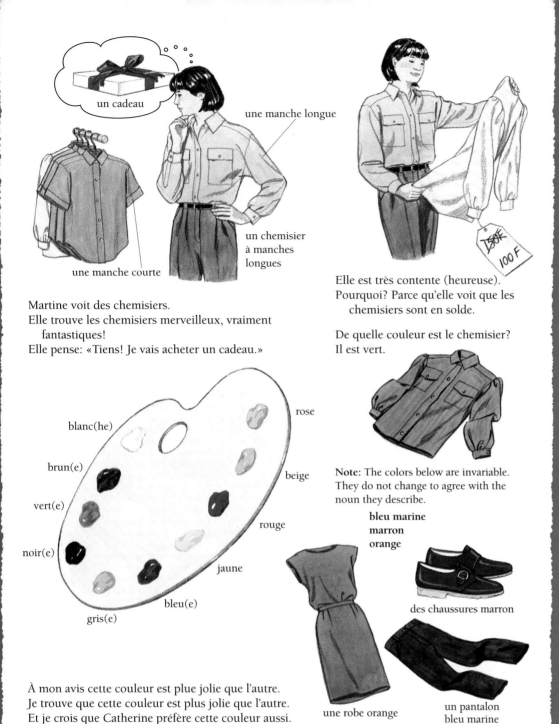

un cadeau

une manche longue

un chemisier
à manches
longues

une manche courte

Martine voit des chemisiers.
Elle trouve les chemisiers merveilleux, vraiment
  fantastiques!
Elle pense: «Tiens! Je vais acheter un cadeau.»

blanc(he)

rose

brun(e)

beige

vert(e)

rouge

noir(e)

jaune

bleu(e)

gris(e)

À mon avis cette couleur est plue jolie que l'autre.
Je trouve que cette couleur est plus jolie que l'autre.
Et je crois que Catherine préfère cette couleur aussi.

150 F
100 F

Elle est très contente (heureuse).
Pourquoi? Parce qu'elle voit que les
  chemisiers sont en solde.

De quelle couleur est le chemisier?
Il est vert.

Note: The colors below are invariable.
They do not change to agree with the
noun they describe.

bleu marine
marron
orange

des chaussures marron

une robe orange

un pantalon
bleu marine

CHAPITRE 10   **263**

Exercices

Exercices

A **Qu'est-ce que Martine voit?** Répondez.

1. Qu'est-ce que Martine veut acheter?
2. Elle aime les chemisiers?
3. Elle trouve que les chemisiers sont merveilleux, vraiment fantastiques?
4. Est-ce que Martine voit que les chemisiers sont en solde?

B **De quelle couleur… ?** Donnez des réponses personnelles.

1. De quelle couleur est ton blouson favori?
2. De quelle couleur est ton jean favori?
3. De quelle couleur est ta chemise favorite ou ton chemisier favori?
4. Qu'est-ce que tu portes aujourd'hui? De quelle couleur sont tes vêtements?

C **De petits problèmes.** Répondez d'après les dessins.

1. Ces chaussures sont trop larges ou trop étroites?

2. Cette jupe est trop longue ou trop courte?

3. Cette chemise a des manches longues ou courtes?

4. Ce pantalon est serré ou large?

Exercices

Extension of *Exercice A*

After completing Exercise A, have students write three questions about the *Mots 2* vocabulary that require negative answers. Then they ask their questions of a partner.

Extension of *Exercice B*

With more able groups you may wish to present the noun forms of colors: *Tu préfères quelle couleur? Moi, je préfère le bleu. Pour une chemise, je préfère le blanc.*

Extension of *Exercice C*

After completing Exercise C, have students quickly write a description of a really ridiculous outfit. Call on volunteers to read their description to the class.

ANSWERS

Exercice A

1. Martine veut acheter un chemisier.
2. Oui, elle aime les chemisiers.
3. Oui, elle trouve que les chemisiers sont merveilleux, vraiment fantastiques.
4. Oui, Martine voit que les chemisiers sont en solde.

Exercice B

Answers will vary.

Exercice C

1. Ces chaussures sont trop étroites.
2. Cette jupe est trop courte.
3. Cette chemise a des manches longues.
4. Ce pantalon est large.

ADDITIONAL PRACTICE

Student Tape Manual, Teacher's Edition, *Activités E–G,* pages 113–114.

PAIRED ACTIVITIES

1. Each student describes the color of his or her clothing. If they are both wearing the same thing, they will say, *Nous deux, nous portons…*
2. Each student makes a list of articles of clothing. Then they combine their lists. Working together, they categorize the clothing under the headings *Sport* and *Habillé.*

D Mes préférences.   Donnez des réponses personnelles.

1. Tu préfères des vêtements sport ou habillés?
2. Tu préfères des chaussures à talons bas ou hauts? Tu fais quelle pointure?
3. Tu préfères une chemise ou un chemisier à manches longues ou courtes?
4. Tu préfères tes vêtements un peu serrés ou larges?
5. Tu préfères un pantalon plus large? Tu voudrais la taille au-dessus?
6. Tu préfères un pantalon plus serré? Tu voudrais la taille au-dessous?
7. Tu préfères faire des achats quand il y a des soldes ou pas?
8. Tu aimes dépenser beaucoup d'argent pour tes vêtements?
9. Tu aimes acheter des cadeaux pour tes copains ou tes copines? Qu'est-ce que tu achètes?

Activités de communication orale
Mots 1 et 2

A Qui est-ce?   Work with a classmate. One of you describes what someone in the class is wearing and the other has to guess who it is. Take turns.

B Une paire de chaussures.   You're in a shoe store in Montreal. Tell the salesperson (your partner) what kind of shoes you want and what size you wear. You try on several pairs before you find the right shoes in the right size at the right price, but the salesperson is very patient.

C Qui porte… ?   Work with a classmate. One of you names an article of clothing and the other has to say who wears it (men, women, or both) and when or where they wear it. Take turns.

> Élève 1: un blouson
> Élève 2: Les hommes et les femmes portent un blouson quand il fait froid.

D Les grands couturiers.   Imagine you and your classmates are costume designers working on a new movie starring two of your favorite actors. You need to create outfits for the stars to wear in various scenes. One person suggests an article of clothing in a certain color, and the next person repeats the item and adds another. Take turns until you have a complete outfit for each star.

Exercice D
Answers will vary.

Activités de communication orale
Mots 1 et 2

PRESENTATION (*page 265*)

Activités A, B, C, and D

The *Activités de communication orale* allow students to use the chapter vocabulary and grammar in open-ended situations. It is not necessary to do all the activities. You may select those that are most appropriate for your students.

ANSWERS

Activité A
Answers will vary.

Activité B
Answers will vary, but may include the following:

É1: Bonjour, Monsieur (Mademoiselle, Madame). Je voudrais des (une paire de) chaussures noires très habillées. Je fais du 38.

É2: Voici une paire de chaussures très élégantes en 38.

É1: Elles sont trop étroites. Je voudrais la pointure au-dessus.

É2: Voilà la pointure au-dessus.

É1: Elles coûtent combien, ces chaussures? Elles sont en solde?

É2: Elles coûtent 500F. Elles ne sont pas en solde.

É1: C'est trop cher. Je voudrais voir une autre paire de chaussures. (Etc.)

Activités C and D
Answers will vary.

Les verbes **croire** *et* voir *au présent*

PRESENTATION *(page 266)*

A. *Croire* and *voir* can be done quickly since there are only three oral forms.
B. Lead students through steps 1–2. Write a few examples on the board and explain: *Une proposition, c'est un groupe de mots qui a un sujet et un verbe.* Ask volunteers for other examples, first of clauses, then of sentences with *croire* or *voir* + a clause.

Exercices

ANSWERS

Exercice A

Answers will vary but will begin with *Elles voient.*

266

Les verbes *croire* et
voir au présent

Expressing Opinions and Making Observations

1. Study the following forms of the irregular verbs *croire*, "to think," "to believe," and *voir*, "to see."

CROIRE	VOIR
je crois	je vois
tu crois	tu vois
il	il
elle croit	elle voit
on	on
nous croyons	nous voyons
vous croyez	vous voyez
ils	ils
elles croient	elles voient

2. The verbs *croire* and *voir* are often followed by a clause. The clause is introduced by *que* which is shortened to *qu'* before a vowel or a silent *h*. In French you must use *que* even though its equivalent, "that," is often omitted in English.

> Je crois que c'est une bonne idée.
> Je vois qu'elle aime cette boutique.

Exercices

A Qu'est-ce qu'elles voient?
Qu'est-ce qu'Annick et Claire voient dans la vitrine de la boutique?

B **La fête.** Répondez d'après le modèle.

> Il va faire beau demain soir?
> *Oui, je crois. Toi, tu ne crois pas?*

1. Il faut porter une robe habillée à la fête?
2. David va inviter Sylvie à la fête?
3. La fête va être amusante?
4. On va servir un gâteau énorme?
5. L'appartement de David est assez grand pour la fête?

C **Tu vois des films?** Donnez des réponses personnelles.

1. Tu vois beaucoup de films?
2. Tu vois des films au cinéma ou à la télé?
3. En général, tu vois des films d'horreur, des films d'aventures ou des films d'amour?
4. Tes parents voient souvent des films?
5. Tu vois tes copains pendant le week-end? Qu'est-ce que tu fais avec eux?

D **Les opinions.** Répondez par «oui».

> Tes copains et toi, vous croyez que le tennis est un sport merveilleux?
> *Oui, nous croyons que le tennis est un sport merveilleux.*

1. Vous croyez que Paris est une belle ville?
2. Vos parents croient que vous êtes intelligents?
3. Votre professeur de français croit que vous travaillez bien?
4. Vos amis croient que vous êtes sympathiques?
5. Vous croyez que les jeans sont chic?
6. Vos grands-parents croient que vous êtes adorables?

E **Des opinions différentes!** Complétez avec «croire».

1. Moi, je ___ que la cousine de Sandra est française mais mes copains ___ qu'elle est italienne.
2. Le professeur ___ que l'examen va être facile mais les élèves ___ que l'examen va être difficile.
3. Tu ___ que les chats sont plus intelligents que les chiens mais ton frère ___ que les chiens sont plus intelligents que les chats.
4. Hélène ___ que Paris est près de Nice mais nous ___ que c'est assez loin de Nice.
5. Tu ___ qu'il va pleuvoir mais je ___ qu'il va faire beau.

GALERIES Lafayette

Le Grand Magasin Capitale de la Mode.

267

ADDITIONAL PRACTICE

Project transparencies from previous chapters out of focus. Ask students what they think they see. Then focus the image so they can see if they were correct. For example: *Prof: Chantal, qu'est-ce que tu crois que tu vois? É1: Je crois que je vois un marchand de fruits et légumes. (Focus image.) Prof: Non, ce n'est pas un marchand de fruits et légumes; c'est une marchande.*

INDEPENDENT PRACTICE

Assign any of the following:
1. Exercises, pages 266–267
2. Workbook, *Structure: A–C*, pages 96–97
3. Communication Activities Masters, *Structure: A*, pages 49–50
4. CD-ROM, Disc 3, pages 266–267

Exercice B

1.– 5. Oui, je crois. Toi, tu ne crois pas?

Exercice C

Answers will vary.

Exercice D

1. Oui, nous croyons que Paris est une belle ville.
2. Oui, nos parents croient que nous sommes intelligents.
3. Oui, notre professeur de français croit que nous travaillons bien.
4. Oui, nos amis croient que nous sommes sympathiques.
5. Oui, nous croyons que les jeans sont chic.
6. Oui, nos grands-parents croient que nous sommes adorables.

Exercice E

1. crois, croient
2. croit, croient
3. crois, croit
4. croit, croyons
5. crois, crois

GAME

Play "I Spy" using items in the room. The student describing a mystery item must use *voir*, and the student(s) guessing must use *croire*. For example: *Je vois un grand objet noir. (Je crois que c'est la chaise.) Non. (Je crois que c'est la porte.) Non. (Je crois que c'est le bureau du professeur.) Oui.*

INFORMAL ASSESSMENT

Distribute magazine pictures of things students can talk about using the French they know. Give some of the pictures to individuals and others to pairs and groups in order to practice the plural forms. Ask students what they see or what another student sees in the pictures. For example: *Martine et Jeannette, qu'est-ce que vous voyez? Thierry, que voit Étienne? Marie, que voient Sophie et Anne?* (You may want to vary the activity by having students ask each other what they see.)

D'autres adjectifs irréguliers

PRESENTATION *(page 268)*

A. Have students close their books and repeat after you only the singular forms of the adjectives in the chart. Start with the feminine forms since students tend to have an easier time dropping the final sound. Then have students open their books and read aloud the four forms of each adjective.

B. Now lead students through step 2 on page 268, modeling the pronunciation and having students repeat chorally.

C. Demonstrate some of the adjectives by using them in questions. For example: *Marie, tu es sérieuse? Henri, tu es sérieux? Isabelle, qui est le premier garçon dans ce rang? Et qui est la première fille?*

Exercices

ANSWERS

Exercice A

Students use the correct pronunciation.

Exercice B

1. Oui, Nathalie est sportive.
2. Oui, son frère est sportif.
3. Oui, Nathalie est active.
4. Oui, il est actif.
5. Oui, le rouge est la couleur favorite de Nathalie.
6. Oui, la planche à voile est son sport favori.
7. Oui, Nathalie est sérieuse.
8. Oui, son frère est un garçon sérieux.
9. Oui, elle est souvent heureuse.
10. Oui, il est souvent heureux.

1. In spoken French the feminine forms of the adjective end in a consonant sound. This consonant sound is dropped in the masculine forms. Here are some irregular adjectives that follow this pattern. Note their spelling changes.

FÉMININ PLURIEL	FÉMININ SINGULIER	MASCULIN PLURIEL	MASCULIN SINGULIER
sérieuses	sérieuse	sérieux	sérieux
délicieuses	délicieuse	délicieux	délicieux
heureuses	heureuse	heureux	heureux
merveilleuses	merveilleuse	merveilleux	merveilleux
basses	basse	bas	bas
favorites	favorite	favoris	favori
longues	longue	longs	long
premières	première	premiers	premier
dernières	dernière	derniers	dernier
entières	entière	entiers	entier
chères*	chère	chers	cher

*All forms of *cher* are pronounced the same way.

2. Here are two adjectives whose endings are pronounced in both the feminine and masculine forms. Note that the feminine ending has a softer sound than the masculine one.

sportives	sportive	sportifs	sportif
actives	active	actifs	actif

Exercices

A **La prononciation.** Prononcez.

1. active / actif
2. favorite / favori
3. longue / long
4. basse / bas
5. merveilleuse / merveilleux
6. délicieuse / délicieux
7. généreuse / généreux
8. première / premier

B **Nathalie et son frère.** Répondez par «oui».

1. Nathalie est sportive?
2. Son frère est sportif?
3. Nathalie est active?
4. Et lui, il est actif?
5. Le rouge est la couleur favorite de Nathalie?
6. La planche à voile est son sport favori?
7. Nathalie est sérieuse?
8. Et son frère est un garçon sérieux?
9. Elle est souvent heureuse?
10. Et lui, il est souvent heureux?

COOPERATIVE LEARNING

Have team members use the adjectives on page 268 to describe at least two persons or things they know. They should make a list of adjectives to describe each person or thing, being careful to use the correct forms. When finished, they read their lists to their teammates, who try to guess what or who is being described. For example: *Bas, long, actif, pas sportif, merveilleux. (mon chien)* Have teammates check each other's spelling.

C **La famille Beauchamp.** Complétez.

La famille Beauchamp est très ___ (sportif). Les parents sont très ___ (actif) et
          1   2
les deux enfants, Véronique et Nicole, sont ___ (actif) aussi. Aujourd'hui, les
                                            3
deux filles sont très ___ (heureux) parce qu'elles partent pour Biarritz, leur
                      4
station balnéaire ___ (favori), où chaque année la famille passe des vacances
                 5
___ (merveilleux). À Biarritz, les filles et les parents vont pratiquer leurs sports
 6
___ (favori), la planche à voile et la natation. Après de ___ (long) journées à la
 7   8
plage, tout le monde est content de manger des fruits de mer ___ (délicieux) à
                                                            9
la terrasse d'un restaurant.

Le comparatif des adjectifs   *Comparing People and Things*

1. You use the comparative to compare two or more people or things. The
   following words are used to express comparisons.

   > (+) *plus… que*
   > (−) *moins… que*
   > (=) *aussi… que*

   Study the following sentences.

   > **Cette vendeuse est plus sympathique que
   >   l'autre vendeuse.**
   > **Ce blouson est moins cher que la veste.**
   > **Les chaussures américaines sont aussi chères
   >   que les chaussures françaises.**

2. Note the liaison after *plus* and *moins* when they are
   followed by a vowel.

   > **plus‿intéressant**
   > **moins‿élégant**

3. If you are comparing people, you use the stress pronouns after *que.*

   > **Il est plus jeune que son ami.**   **Il est plus jeune que *lui.***
   > **Elle est plus âgée que ses amis.**   **Elle est plus âgée qu'*eux.***

4. Note that the adjective *bon* has an irregular form in the comparative, *meilleur.*

   > **Ils trouvent que le pain français est meilleur que le pain américain.**
   > **La robe rose est meilleur marché que la robe blanche.**

Exercices

ANSWERS

Exercice A

1. Oui, le blouson bleu est plus cher que le blouson noir.
2. Non, le blouson bleu est plus joli que le blouson noir. (Oui, le blouson bleu est moins joli que le blouson noir.)
3. Non, la jupe jaune est plus chère que la jupe grise.
4. Oui, la jupe grise est plus courte que la jupe jaune.
5. Non, la robe verte est moins élégante que la robe rouge.
6. Oui, la robe verte est moins habillée que la robe rouge.

Exercice B

1. Non, elle n'est pas plus chère. Mais elle est aussi chère que l'autre.
2. Non, elle n'est pas plus habillée. Mais elle est aussi habillée que l'autre.
3. Non, il n'est pas plus cher. Mais il est aussi cher que l'autre.
4. Non, il n'est pas plus serré. Mais il est aussi serré que l'autre.
5. Non, elles ne sont pas plus larges. Mais elles sont aussi larges que les autres.
6. Non, elles ne sont pas plus courtes. Mais elles sont aussi courtes que les autres.

Exercice C

Answers will vary.

INFORMAL ASSESSMENT

Write the following words on the board:
le chien et le chat
ce garçon et cette fille
le français et l'anglais
la boutique et le grand magasin
Call on students to make up original sentences comparing these items.

270

Exercices

A **Plus ou moins que l'autre.** Répondez d'après les dessins. Suivez le modèle.

Le blouson bleu est aussi long que le blouson noir?
Oui, le blouson bleu est aussi long que le blouson noir.

450F
250F
140F
400F

1. Le blouson bleu est plus cher que le blouson noir?
2. Le blouson bleu est moins joli que le blouson noir?
3. La jupe jaune est moins chère que la jupe grise?
4. La jupe grise est plus courte que la jupe jaune?
5. La robe verte est aussi élégante que la robe rouge?
6. La robe verte est moins habillée que la robe rouge?

B **Non, pas plus.** Répondez d'après le modèle.

Cette chemise est plus chère que l'autre?
Non, elle n'est pas plus chère. Mais elle est aussi chère que l'autre.

1. Cette cravate est plus chère que l'autre?
2. Cette robe est plus habillée que l'autre?
3. Ce pull est plus cher que l'autre?
4. Ce chemisier est plus serré que l'autre?
5. Ces chaussures sont plus larges que les autres?
6. Ces manches sont plus courtes que les autres?

C **À mon avis.** Donnez des réponses personnelles.

1. Le cours de français est plus difficile ou plus facile que le cours de maths?
2. Le professeur de français est plus sévère, moins sévère ou aussi sévère que les autres professeurs?
3. Le football américain est plus intéressant ou moins intéressant que le basket-ball?
4. Une Volkswagen est moins chère ou plus chère qu'une Porsche?
5. Le coca est meilleur que le lait ou le lait est meilleur que le coca?
6. Les fruits et les légumes sont meilleurs pour la santé (*health*) que les pâtisseries?

INDEPENDENT PRACTICE

Assign any of the following:
1. Exercises, pages 270–271
2. Workbook, *Structure: H–K*, pages 99–101
3. Communication Activities Masters, *Structure: C–D*, pages 52–53
4. Computer Software, *Structure*
5. CD-ROM, Disc 3, pages 270–271

Le superlatif

Comparing People and Things

1. You use the superlative to single out one item from the group and compare it to all the others. You form the superlative in French by using *le, la,* or *les* and *plus* or *moins* with the adjective.

> Cette robe est **la plus jolie** de la boutique.
> Cette robe est **la moins chère** de la boutique.

2. Note that the superlative is followed by *de* + a noun.

> Robert est **le plus intelligent** de la classe.
> Carole est **la meilleure** en maths du lycée.
> Les frères Dumas sont **les plus amusants** de tous les élèves.

Exercices

A **La plus chère et la plus grande.** Répondez d'après les indications.

1. Quelle boutique est la plus chère de toute la ville? (cette boutique)
2. Quelle ville est la plus grande de tout le pays? (Paris)
3. Quel magasin est le plus grand du centre commercial? (Monoprix)
4. Quel marché est le moins cher de tous les marchés? (le Village Suisse)
5. Quel couturier est le plus célèbre? (Yves Saint-Laurent)

B **Ma famille.** Donnez des réponses personnelles.

1. Qui est le plus jeune ou la plus jeune de ta famille?
2. Qui est le plus âgé ou la plus âgée de ta famille?
3. Qui est le plus amusant ou la plus amusante de ta famille?
4. Qui est le plus intelligent ou la plus intelligente de ta famille?
5. Qui est le plus beau ou la plus belle de ta famille?
6. Qui est le plus timide ou la plus timide de ta famille?
7. Qui est le plus sportif ou la plus sportive de ta famille?
8. Qui est le plus heureux ou la plus heureuse de ta famille?

MONOPRIX UNIPRIX
On pense à vous tous les jours.

Bell Ringer Review

Write the following on the board or use BRR Blackline Master 10-6: Using the vocabulary you now have for colors and clothes, write four sentences about the people in the Chapter 2, *Mots 1* illustration at the bottom of page 38.

Le superlatif

PRESENTATION *(page 271)*

A. Follow the same suggestions given for presenting the comparative on page 269 of this Teacher's Wraparound Edition.
B. Have students open their books to page 271. Lead them through steps 1 and 2.
C. Tell students that the superlative is followed by *de* in French. Do not explain that it is *de* in French and "in" in English. When this comparison is not made, students tend not to use *dans*.

Note In the CD-ROM version, this structure point is presented via an interactive electronic comic strip.

Exercices

ANSWERS

Exercice A

1. Cette boutique est la plus chère de toute la ville.
2. Paris est la plus grande de tout le pays.
3. Monoprix est le plus grand du centre commercial.
4. Le Village Suisse est le moins cher de tous les marchés.
5. Yves Saint-Laurent est le plus célèbre.

Exercice B

Answers will vary.

ADDITIONAL PRACTICE

1. After completing Exercises A and B on page 271, have students make exaggerated statements using comparative and superlative constructions. They should stick to topics they know in French.
2. Student Tape Manual, Teacher's Edition, *Activités A–C,* page 115

CRITICAL THINKING ACTIVITY

(Thinking skills: making inferences)
Tell students to look at the advertisement on this page and answer these questions in French.

1. What do you think Monoprix is? *(un grand magasin)*
2. What do you think the ducks represent? *(une famille)*

Bell Ringer Review

Write the following on the board or use BRR Blackline Master 10-7: Make at least four comparisons between various members of your family.

PRESENTATION (page 272)

A. Have students open their books to page 272. Have them look at the photo and guess what the conversation is about.

B. Have them watch the Conversation Video or listen to Cassette 6B/CD-6. Then have the class repeat the conversation after you or the cassette/CD. Call on two individuals to read it aloud with as much expression as possible.

C. Now do the comprehension exercise on this page.

Note In the CD-ROM version, students can play the role of either one of the characters and record the conversation.

ANSWERS

Exercice A

1. Oui, Sandrine est dans un grand magasin.
2. Elle est au rayon chemises.
3. Elle est au rayon hommes.
4. Oui, elle veut acheter un cadeau.
5. Le cadeau est pour son père.
6. La vendeuse propose une chemise.
7. Il fait du quarante.
8. Elle préfère le bleu marine ou le blanc.
9. Oui, les chemises sont en solde.
10. La chemise va être moins chère.

Scènes de la vie *Un petit cadeau pour Papa*

LA VENDEUSE: Vous désirez, Mademoiselle?
SANDRINE: Je voudrais un petit cadeau pour mon père.
LA VENDEUSE: Pour la Fête des Pères?
SANDRINE: Non, c'est pour son anniversaire.
LA VENDEUSE: Une chemise, peut-être?
SANDRINE: Oui. Pourquoi pas?
LA VENDEUSE: Il fait quelle taille, votre père?
SANDRINE: Il fait du quarante, je crois. Oui, c'est ça, quarante.
LA VENDEUSE: Vous préférez quelle couleur?
SANDRINE: Bleu marine ou blanc. Il aime le look conservateur.
LA VENDEUSE: Bien, Mademoiselle. Et vous avez de la chance. Toutes les chemises sont en solde aujourd'hui.

A **Un cadeau d'anniversaire.** Répondez d'après la conversation.

1. Sandrine est dans un grand magasin?
2. Elle est au rayon chemises ou complets?
3. Elle est au rayon hommes ou femmes?
4. Elle veut acheter un cadeau?
5. C'est pour qui, le cadeau?
6. Qu'est-ce que la vendeuse propose?
7. Le père de Sandrine fait quelle taille?
8. Sandrine préfère quelle couleur?
9. Les chemises sont en solde?
10. La chemise va être plus chère ou moins chère?

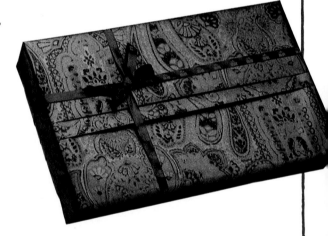

PAIRED ACTIVITY

Have students work in pairs and prepare a skit based on purchasing any item of clothing.

LEARNING FROM PHOTOS

Have students say as much as they can about the *Conversation* photo.

Prononciation   *Les sons /sh/ et /zh/*

It is important to make a distinction between the sound /sh/ as in *chat* and /zh/ as in *joli*. Put your fingers on your throat. When you say the sound /zh/ as in *joli* you should feel a vibration, but not when you say /sh/ as in *chat*. Repeat the following words with the sounds /sh/ and /zh/.

acheter	large
chaussure	jupe
chemise	orange
achats	beige
short	jeune

Now repeat the following sentences that combine both sounds.

J'achète toujours des chaussures bon marché.
Je cherche un joli tee-shirt jaune et un short orange.

chemise orange

Activités de communication orale

A **Une boutique chic.** You're in a boutique on the chic Rue du Faubourg Saint-Honoré in Paris. A salesperson (your partner) greets you. Tell the salesperson what you're looking for. You're a very demanding customer, but after some trial and error the salesperson manages to find something terrific in your size and at a price you can afford.

B **Comparaisons.** Work in small groups. Use the adjectives below on the right to make as many comparisons as you can about each pair on the left. Compare your results with other groups'.

les chiens et les chats

Les chiens sont plus intelligents (moins calmes, aussi beaux, etc.) que les chats.

l'anglais et les maths	meilleur	intelligent
les avions et les trains	difficile	sympathique
une Rolls-Royce et une Toyota	rapide	actif
notre école et une autre école	cher	sportif
les filles et les garçons	heureux	facile

Prononciation
PRESENTATION *(page 273)*

A. Model the key phrase *chemise orange* and have students repeat chorally.
B. Now model the other words and phrases in similar fashion.
C. You may wish to give students the following *dictée*:
   Le short orange est plus large que le short beige. Le jeune joueur a une chemise beige.
D. For additional pronunciation practice you may wish to use the *Prononciation* section on Cassette 6B/CD-6 and the Student Tape Manual, Teacher's Edition, *Activités F–H*, pages 117–118.

Bell Ringer Review
Write the following on the board or use BRR Blackline Master 10-8: Illustrate the following words and label them.
des chaussures
un jean
une robe
une jupe

Activités de communication orale
ANSWERS
Activités A and B
   Answers will vary.

ADDITIONAL PRACTICE

   After they have done Activities A and B, have students work in small groups, share opinions about the topics below, and then summarize their opinions for the class.
la plus belle actrice (le plus bel acteur)
le parfum de glace le plus délicieux
le groupe de rock le plus terrible
le sport le plus amusant
le cours le plus intéressant de l'école

INDEPENDENT PRACTICE

   Assign any of the following:
1. Exercise and activities, pages 272–273
2. CD-ROM, Disc 3, pages 272–273

LECTURE ET CULTURE

Bell Ringer Review

Write the following on the board or use BRR Blackline Master 10-9: Name the colors associated with each topic.

1. les couleurs de l'automne
2. les couleurs du bord de la mer
3. les couleurs de l'école
4. les couleurs que tu aimes porter

READING STRATEGIES
(*page 274*)

Pre-reading

A. Ask students whether they spend their own money for clothes. Do they like malls? Why or why not?
B. Share clothing ads from French magazines (*Elle, Marie-Claire, Homme,* etc.) with students.

Reading

A. Have students read the selection once silently.

Teaching Tip Adherence to pre-set time limits will encourage students to read all the material and not get "bogged down" and stop every time they think they don't know something. Encourage students to read for ideas, rather than word by word.

B. Call on a student to read half a paragraph aloud. Ask some comprehension questions.

Post-reading

Call on volunteers to summarize what they've read.

Note Students may listen to a recorded version of the *Lecture* on the CD-ROM.

Étude de mots

ANSWERS

Exercice A

1. b	4. g	6. f
2. c	5. d	7. e
3. a		

LES ACHATS

*S*i la France est un pays de gastronomie, c'est aussi un pays de haute couture. Les noms des grands couturiers sont célèbres dans le monde entier—Yves Saint-Laurent, Dior, Courrèges, Cardin, Givenchy, Lacroix. Ces couturiers dictent la mode non seulement à Paris, mais à Tokyo, New York et Rio. À Paris on vend les vêtements et accessoires de ces couturiers dans des boutiques Place Vendôme, rue du Faubourg Saint-Honoré ou rue François Ier.

Mais attention[1]! La plupart des Français ne font pas leurs achats chez les grands couturiers. Il y a des grands magasins de toutes les catégories, des plus luxueuses aux plus modestes—les Galeries Lafayette, La Samaritaine, Monoprix, Prisunic, etc. Dans les grands magasins on peut aller d'un rayon à l'autre et acheter toutes sortes de choses dans le même magasin. Beaucoup de gens profitent des soldes quand on vend les marchandises avec d'importantes réductions.

À Paris les jeunes—garçons et filles—achètent leurs vêtements dans les mêmes boutiques unisexe du Quartier Latin. Dans ces boutiques on trouve du prêt-à-porter original et à la mode[2]. Mais si on a très peu d'argent à dépenser on peut aller aux Puces[3] ou au Village Suisse. Nicole adore aller aux Puces ou au Village Suisse où elle trouve presque[4] toujours un chemisier ou un accessoire avec la griffe[5] célèbre d'un grand couturier—et à un prix très bas.

[1] Mais attention! *Careful! Watch out!*
[2] à la mode *in style*
[3] aux Puces *to the flea market*
[4] presque *almost*
[5] griffe *label*

Étude de mots

A Cherchez les mots. Choisissez la définition.

1. le créateur de modèles	a. à la mode
2. fameux	b. le couturier
3. en vogue, populaire	c. célèbre
4. fantastique	d. les marchandises
5. les produits commerciaux	e. fréquemment
6. profiter	f. bénéficier
7. presque toujours	g. merveilleux

ADDITIONAL PRACTICE

After completing the *Étude de mots*, reinforce the lesson with the following:

Have students match each numbered item with its lettered opposite.

1. l'achat	a. haut
2. bon marché	b. la vente
3. bas	c. acheter
4. vendre	d. gagner
5. différent	e. cher
6. dépenser	f. même

Compréhension

B **Boutiques et magasins.** Complétez.

1. La France est un pays de gastronomie et de ___.
2. Deux ___ célèbres sont Pierre Cardin et Yves Saint-Laurent.
3. Les vêtements faits par un couturier portent la ___ du couturier.
4. Les Galeries Lafayette et La Samaritaine sont des ___, pas des boutiques.
5. Les clients dans un grand magasin peuvent aller d'un ___ à l'autre et ils peuvent acheter toutes sortes de choses dans le même magasin.
6. Pendant les ___ il y a d'importantes réductions.
7. Une boutique ___ vend des vêtements pour garçons et filles.

C **Le shopping.** Répondez.

1. On vend les vêtements et les accessoires des grands couturiers au Prisunic?
2. Il y a beaucoup de différents grands magasins en France?
3. Tous les grands magasins sont plus ou moins de la même catégorie?
4. Pendant les soldes, tout est plus cher ou meilleur marché?
5. Qu'est-ce qu'une boutique unisexe?
6. Qui aime les boutiques unisexe?
7. Les marchandises sont chères aux Puces?

D **Les achats.** Trouvez les renseignements suivants dans la lecture.

1. trois couturiers français
2. deux grands magasins français
3. deux marchés parisiens qui ont des prix très bas

DÉCOUVERTE CULTURELLE

En France et en Europe en général les pointures et les tailles ne sont pas les mêmes qu'aux États-Unis. Voici les tailles des vêtements et les pointures des chaussures.

Si vous voulez acheter des chaussures en France, vous demandez quelle pointure? Si vous voulez acheter une chemise ou un chemisier, vous demandez quelle taille?

FEMMES				
CHAUSSURES				
États-Unis	6	7	8	9
France	36	37	38	39
ROBES, TAILLEURS, PULLS, CHEMISIERS				
États-Unis	6	8	10	12	14
France	38	40	42	44	46

HOMMES				
CHEMISES				
États-Unis	14½	15	15½	16	16½
France	37	38	39	40	41
CHAUSSURES				
États-Unis	9	10	11	12
France	40	41	42	43

CRITICAL THINKING ACTIVITY

(Thinking skills: supporting statements with reasons)

Put the following on the board or on a transparency:

1. À votre avis, il faut dépenser beaucoup d'argent pour acheter des vêtements ou pas? Justifiez vos idées.
2. Il faut porter un uniforme à l'école. Discutez des avantages et des inconvénients.

INDEPENDENT PRACTICE

Assign any of the following:
1. *Étude de mots* and *Compréhension* exercises, pages 274–275
2. Workbook, *Un Peu Plus,* pages 102–103
3. CD-ROM, Disc 3, pages 274–275

Compréhension

PRESENTATION *(page 275)*

Exercice B

Have students find the answers to Exercise B in the reading. Call on volunteers to read the completed statements.

ANSWERS

Exercice B

1. haute couture
2. couturiers
3. griffe
4. grands magasins
5. rayon
6. soldes
7. unisexe

Exercice C

1. Non.
2. Oui.
3. Non.
4. Meilleur marché.
5. Une boutique où on vend des vêtements pour garçons et filles.
6. Les jeunes.
7. Non.

Exercice D

Answers will vary but may include the following:

1. Yves Saint-Laurent, Dior, Courrèges, Cardin, Givenchy, Lacroix.
2. Les Galeries Lafayette, la Samaritaine, Monoprix, Prisunic.
3. Le Village Suisse, les Puces.

OPTIONAL MATERIAL

Découverte culturelle

PRESENTATION *(page 275)*

A. Before doing the reading, have students make up a chart of their clothing and shoe sizes using the American system. Tell them to leave space between items—they'll be making additions.

B. Have students read the selection silently and study the chart. Beside each of their American sizes, have them write the equivalent European size. If their size isn't in the chart, they should extrapolate.

Note Students may listen to a recorded version of the *Découverte culturelle* on the CD-ROM.

RÉALITÉS

Bell Ringer Review

Write the following on the board or use BRR Blackline Master 10-10: In what kind of store would the following people shop for clothes?

1. Elizabeth Taylor
2. a university student
3. a lawyer
4. Michael Jordan
5. you and your friends

OPTIONAL MATERIAL

PRESENTATION *(pages 276–277)*

The main objective of this section is to have students enjoy the photographs and absorb some French culture. However, if you would like to do more, do some of the following activities.

A. Ask students to comment on the French shopping scenes depicted in the photos. Are there any differences between these scenes and students' personal shopping experiences?

B. Call on volunteers to read the captions aloud. Then discuss the information as a class.

C. Ask the class these questions: From what you know about young French people, what are some advantages they have when it comes to money and clothes shopping? Some disadvantages? Do any of you shop with credit cards? What are the advantages or disadvantages? Do any of you have your own credit cards? Do you think young French people do? Why or why not?

D. You may wish to explain to students that there is very little difference in the way French and U.S. students dress.

Note In the CD-ROM version, students can listen to the recorded captions and discover a hidden video behind one of the photos.

RÉALITÉS

276

ADDITIONAL PRACTICE

1. Student Tape Manual, Teacher's Edition, *Deuxième Partie,* pages 118–120
2. Situation Cards, Chapter 10

276

Voici une boutique au Quartier Latin. Les deux copines regardent des vêtements ensemble. Tu aimes ces pulls? Tu veux acheter un pull dans cette boutique **1**?

C'est un groupe de jeunes Français **2**. Vous trouvez qu'il y a une grande différence entre les vêtements que vous portez et les vêtements que portent les jeunes Français?

C'est le marché aux puces à Lyon **3**.

On est aux Galeries Lafayette à Noël **4**. À la caisse on paie avec une carte de crédit, un chèque ou en espèces.

Les grands couturiers vendent aussi des articles de luxe comme les foulards en soie dans leurs boutiques **5**.

CHANEL

Christian Dior

FOULARD SOIE DIOR. CARRE 90×90 cm — 650 FF

277

Note Chapter 5 of *À bord* deals with the related topic of hairstyles and beauty products.

HISTORY CONNECTION

In 1852, Aristide Boucicaut opened Au Bon Marché in Paris—the first department store in the world. Boucicaut wanted to offer his customers a large selection of high-quality merchandise as well as inexpensive items, under one roof. Soon other department stores, le Printemps, la Samaritaine, and les Galeries Lafayette, opened in Paris. Today these department stores attract millions of customers every year. More economy-minded French shoppers will be found at Au Bon Marché and la Samaritaine. Monoprix and Prisunic also offer inexpensive merchandise.

Throughout Europe one can see narrow shopping streets in the older sections of cities and towns closed to vehicles and open only to pedestrians. Shopping malls *(centres commerciaux)* are also becoming more popular. They are most frequently found on the outskirts of cities or towns. Today there is a very modern "mall" type of development in the heart of Paris. Called Le Forum, it stands where the famous old food market, Les Halles, once stood before it was moved to the outskirts of Paris.

THE FRANCOPHONE WORLD

Students may be interested in viewing clothing and clothing store windows in other French-speaking countries. See the section entitled *La Mode* in *Le Monde francophone*, pages 326–329.

INDEPENDENT PRACTICE

Have groups create a poster advertising a sale, a special designer collection, or a fashion show. They should use shopping vocabulary, colors, and the comparative and superlative of adjectives.

COOPERATIVE LEARNING

Display Communication Transparency C-10. Have students work in groups of five to create the conversations illustrated on the transparency. Have groups present their conversations to the class.

CULMINATION

INFORMAL ASSESSMENT

Oral Activities A and B may serve as a means to evaluate students' speaking ability.

Teaching Tip For students who are uncomfortable speaking in public, try working with a "cassette mail" system. Students record the assigned task in a private setting and give you the cassette. You can then record your evaluative comments and return the cassette to them. Use the evaluation criteria given on page 34 of this Teacher's Wraparound Edition.

Activités de communication orale

PRESENTATION (*page 278*)

Activité A

In the CD-ROM version of this activity, students can interact with an on-screen native speaker.

ANSWERS

Activités A and B

Answers will vary.

Activités de communication écrite

ANSWERS

Activités A and B

Answers will vary.

OPTIONAL MATERIAL

Réintroduction et recombinaison

PRESENTATION (*page 279*)

Exercices A and B

Exercise A recycles stress pronouns. Exercise B recombines clothing vocabulary with Chapter 9's seaside setting. It also contrasts *mettre* (meaning "to put on") with *porter*.

278

Activités de communication orale

A **Un sondage: Le shopping.** You're in a department store in France and have agreed to answer a few questions for a shopping survey.

1. Quel est votre magasin favori? Pourquoi?
2. Que préférez-vous: les grands magasins ou les boutiques?
3. Quelle sorte de vêtements achetez-vous le plus souvent? Des vêtements habillés ou des vêtements sport?
4. Quand vous achetez des vêtements, préférez-vous aller dans les magasins seul(e) ou avec des amis?

B **Jeu de mémoire.** Study the clothing of all the students in the next row for several minutes. Then turn your back to that row and see if you can answer classmates' questions about what the people in the row are wearing. (*Qui porte un tee-shirt rouge? un jean noir?*, etc.) If you can't answer, the people in the row may help out by giving hints such as *La personne est blonde* or *Elle est assise derrière Suzanne.*

Activités de communication écrite

A **Le catalogue.** Write five descriptions for a clothing catalogue. Using the vocabulary in this chapter, describe the items. Tell the sizes they come in, the colors, the occasions they could be worn for, and the prices.

> **Voici une belle robe longue, très habillée, rouge et noire, parfaite pour les fêtes. Tailles: du 36 au 44. Prix: 1.200F**

B **Le look de ton école.** Write a note to your French friend describing *le look* at your school. Tell him or her what boys and girls usually wear to school and what types of clothing and colors are "in" (*à la mode*).

278 CHAPITRE 10

FOR THE YOUNGER STUDENT	LEARNING FROM PHOTOS
1. Set up a clothing store in the room, using items of clothing or pictures from magazines. Have pairs of students make up skits between a salesperson and a customer buying a gift.	1. Have students identify as many colors as they can find in the photos on this page.
2. Have students make collages using magazine photos, advertisements, fabrics, etc. Then have them describe their collages.	2. Ask students: What dates are on the photo on page 279? What will customers get? What do they have to have?

Réintroduction et recombinaison

A **Des préférences.** Complétez.

1. ___, je préfère un look sportif.
2. Mais ___, il préfère un look conservateur.
3. Les autres, ___, ils font toujours leurs achats dans les boutiques chères.
4. Et ___? Où est-ce que tu fais tes achats?

B **En été.** Donnez des réponses personnelles.

1. Quand est-ce que tu mets un maillot?
2. Qu'est-ce que tu portes quand il fait chaud?
3. Tu vas dans quelle sorte de magasin pour acheter un maillot?
4. Tu voudrais un maillot de quelle couleur?
5. Est-ce qu'on met des lunettes de soleil quand il pleut?
6. Qu'est-ce qu'une femme porte quand elle joue au tennis?

Vocabulaire

NOMS			
les vêtements (m.)	le grand magasin	sérieux, sérieuse	AUTRES MOTS ET EXPRESSIONS
le blouson	la boutique	délicieux, délicieuse	
la chaussette	le rayon (prêt-à-porter)	dernier, dernière	à mon avis
la chaussure	le/la client(e)	entier, entière	beaucoup de
la paire	le vendeur	meilleur(e)	faire des achats
le talon	la vendeuse	beige	trop
le jean	le prix	bleu(e)	vraiment
le pantalon	les soldes (f.)	bleu marine (inv.)	
le pull	le grand couturier	blanc, blanche	
le sweat-shirt		brun(e)	
le chemisier	ADJECTIFS	gris(e)	
la manche	bon marché (inv.)	jaune	
le collant	cher, chère	marron (inv.)	
la jupe	bas(se)	noir(e)	
la robe	haut(e)	orange (inv.)	
le tailleur	long(ue)	rose	
la chemise	court(e)	rouge	
le complet	étroit(e)	vert(e)	
la cravate	serré(e)		
la veste	large	VERBES	
le cadeau	habillé(e)	croire	
la couleur	sport (inv.)	dépenser	
la taille	sportif, sportive	penser	
au-dessus	actif, active	porter	
au-dessous	favori(te)	voir	
la pointure	heureux, heureuse		
	merveilleux, merveilleuse		

ANSWERS

Exercice A

1. Moi   3. eux
2. lui   4. toi

Exercice B

Answers will vary, but may include the following:

1. **Je mets un maillot quand je vais à la plage (piscine). … quand je prends un bain de soleil. … quand je nage.**
2. **Quand il fait chaud, je porte un short (un tee-shirt/une chemise à manches courtes/une jupe…)**
3. **Je vais dans un grand magasin (une boutique).**
4. **Je voudrais un maillot bleu (jaune/rouge…).**
5. **Non, on ne met pas de lunettes de soleil quand il pleut.**
6. **Une femme porte un tee-shirt, une jupette, des chaussettes et des chaussures de tennis quand elle joue au tennis.**

ASSESSMENT RESOURCES

1. Chapter Quizzes
2. Testing Program
3. Situation Cards
4. Communication Transparency C-10
5. Computer Software: Practice/Test Generator

VIDEO PROGRAM

INTRODUCTION	(32:15)
ÇA COUTE CHER!	(35:08)

CHAPTER OVERVIEW

In this chapter students will learn to discuss some aspects of their daily routine, personal hygiene, and keeping in shape. In order to do this they will learn the reflexive verbs, certain verbs with orthographic changes, and the pronoun *qui*.

The cultural focus of Chapter 11 is on French attitudes and activities related to fitness.

CHAPTER OBJECTIVES

By the end of this chapter, students will know:

1. vocabulary associated with morning and evening personal routines and hygiene
2. vocabulary associated with physical fitness and exercise
3. the construction *avoir besoin de*
4. the present indicative forms of reflexive verbs
5. the reflexive pronouns
6. orthographic changes in verbs like *acheter*, verbs that end in *-cer*, verbs like *manger*, and the verb *s'appeler*
7. question formation with *qui* as subject and as object

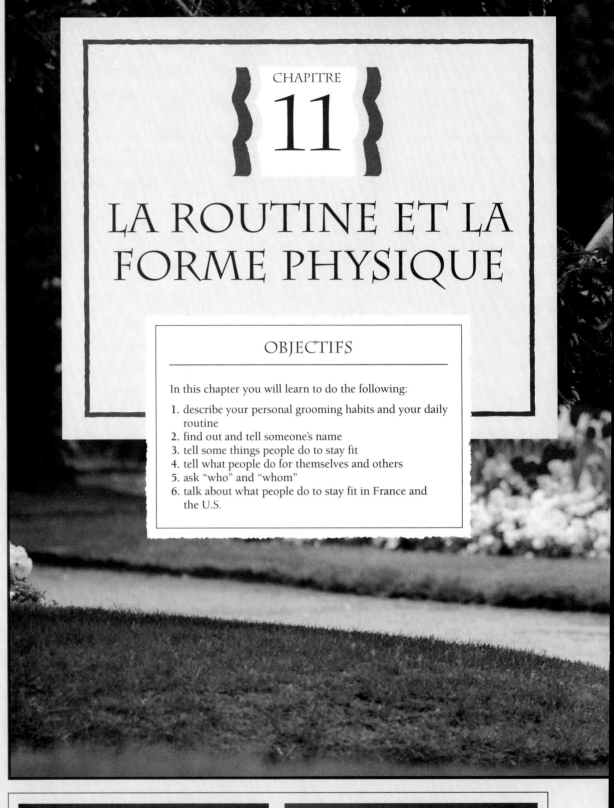

CHAPITRE

{11}

LA ROUTINE ET LA FORME PHYSIQUE

OBJECTIFS

In this chapter you will learn to do the following:

1. describe your personal grooming habits and your daily routine
2. find out and tell someone's name
3. tell some things people do to stay fit
4. tell what people do for themselves and others
5. ask "who" and "whom"
6. talk about what people do to stay fit in France and the U.S.

CHAPTER PROJECTS

(optional)

Do some aerobics with the class in French, using the Total Physical Response approach.

COMMUNITIES

1. Try to obtain video recordings of French sporting events, such as the Tour de France or a Quebec soccer match, and view them with your students.
2. If a French person is available, have students question him/her on the physical education system in French schools, the Tour de France, and so on.

281

CHAPTER 11 RESOURCES

1. Workbook
2. Student Tape Manual
3. Audio Cassette 7A/CD-7
4. Bell Ringer Review
   Blackline Masters
5. Vocabulary Transparencies
6. Pronunciation
   Transparency C-11
7. Grammar Transparency
   G-11
8. Communication
   Transparency C-11
9. Communication Activities
   Masters
10. Map Transparencies
11. Situation Cards
12. Conversation Video
13. Videocassette/Videodisc,
    Unit 3
14. Video Activities Booklet,
    Unit 3
15. Lesson Plans
16. Computer Software:
    Practice/Test Generator
17. Chapter Quizzes
18. Testing Program
19. Internet Activities Booklet
20. CD-ROM Interactive
    Textbook

Pacing

This chapter requires eight to ten class sessions. Pacing will vary according to class length and the age and aptitude of the students.

Note The Lesson Plans offer guidelines for 45- and 55-minute classes and **Block Scheduling.**

Exercices vs. Activités

All exercises (which provide guided practice) are coded in blue. All communicative activities are coded in red.

INTERNET ACTIVITIES

(optional)

These activities, student worksheets, and related teacher information are in the *Bienvenue* Internet Activities Booklet and on the Glencoe Foreign Language Home Page at: http://www.glencoe.com/secondary/fl

LEARNING FROM PHOTOS

After teaching the vocabulary, ask the following questions: *Il y a combien de jeunes gens? Ils sont au parc? Ils font du jogging ou ils se promènent? Qu'est-ce qu'ils portent? Ils veulent rester en forme?*

VOCABULAIRE

MOTS 1

Vocabulary Teaching Resources

1. Vocabulary Transparencies 11.1 (A & B)
2. Audio Cassette 7A/CD-7
3. Student Tape Manual, Teacher's Edition, *Mots 1: A–C*, pages 121–122
4. Workbook, *Mots 1: A–D*, pages 105–107
5. Communication Activities Masters, *Mots 1: A*, page 54
6. Chapter Quizzes, *Mots 1: Quiz 1*, page 59
7. CD-ROM, Disc 3, *Mots 1: pages 282–285*

Bell Ringer Review

Write the following on the board or use BRR Blackline Master 11-1: Use these cues to write a conversation that might take place in a department store.

1. greet the salesperson
2. say what you are looking for
3. give the size, the color, and the style of an article of clothing
4. say how much you would like to spend
5. ask if there is a sale

PRESENTATION *(pages 282–283)*

A. Model the new words using Vocabulary Transparencies 11.1 (A & B). Point to each illustration and have the class repeat after you or Cassette 7A/CD-7.
B. Act out the new words: *se réveiller, se lever, se laver, se laver les cheveux, se brosser les dents, se raser.*

LA ROUTINE

— les cheveux (m.)
la figure
les dents (f.)
la main

se réveiller

se lever

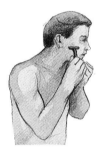

se laver

se laver les cheveux

se brosser les dents

se raser

se peigner

se maquiller

s'habiller

TOTAL PHYSICAL RESPONSE

(following the Vocabulary presentation)

Getting Ready
You may wish to use a chair for a bed and bring in an alarm clock as a prop.

TPR 1
___, venez ici, s'il vous plaît.
Il est sept heures du matin. Vous dormez encore.

Vous entendez le réveil.
Vous vous réveillez.
Vous regardez le réveil.
Vous arrêtez le réveil.
Vous vous levez.
Vous allez dans la salle de bains.
Vous vous lavez.
Vous vous brossez les dents.
Vous vous regardez dans la glace.
Vous vous brossez les cheveux.
(continued on next page)

se coucher

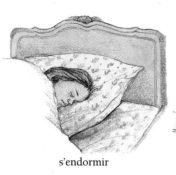

s'endormir

prendre un bain

une glace

prendre une douche

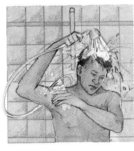

du savon

du déodorant

du dentifrice

1. Elle s'appelle Nathalie.
Nathalie se réveille.
Elle se lève tout de suite.

André va faire sa toilette.
Il a besoin de savon.
Il a besoin de dentifrice.
Il a besoin d'une glace.

Bonjour!
Je m'appelle Christian.
Comment vous appelez-vous?

2. D'abord elle se lave.

3. Ensuite elle se brosse les dents.

4. Enfin elle prend son petit déjeuner.

CHAPITRE 11   283

C. As you present the new vocabulary, ask: *La fille se réveille le matin ou le soir? Ensuite, elle se lève ou elle se couche? Elle se lave la figure et les mains? Elle se brosse les dents? Elle se lave les cheveux tous les matins? Elle s'habille vite? Elle s'habille dans la salle de bains ou dans la chambre à coucher? Elle se lève ou se couche à dix heures et demie du soir? Quand elle se couche, elle s'endort tout de suite?*

D. Call out the following verbs and have students pantomime each one: *se réveiller, se lever, se laver, se laver les cheveux, se brosser les cheveux (les dents), se peigner, s'habiller, se regarder dans la glace, se maquiller.*

E. Demonstrate the meaning of *avoir besoin de* by miming looking for something you need. For example: *Je veux écrire quelque chose. J'ai besoin d'un stylo. Je n'ai pas de stylo. Je ne peux pas écrire.*

F. When you teach *Elle s'appelle Nathalie,* continue down the left column of the page so the meaning of *d'abord* and *ensuite* is clear.

G. **Chain drill** Have one student ask his or her neighbor's name. (*Comment t'appelles-tu?*) The student who responds then asks the next student and so on.

Vocabulary Expansion

You may give students the following additional vocabulary in order to talk about personal grooming:
du shampooing
un peigne
une brosse à dents
des ciseaux
un rasoir
du maquillage

TPR (*continued*)
Vous vous habillez vite.
Vous partez pour l'école. Au revoir et bonne journée!
Merci, ___. Vous pouvez retourner à votre place.

TPR 2
Attention, tout le monde. Levez-vous. Nous allons mimer des actions.
Réveillez-vous.

Lavez-vous la figure et les mains.
Lavez-vous les cheveux.
Rasez-vous.
Brossez-vous les dents.
Brossez-vous les cheveux.
Habillez-vous.
Couchez-vous.
Endormez-vous. Bonne nuit!
Merci, tout le monde. Asseyez-vous.

PRESENTATION *(pages 284–285)*

Exercice A

Exercise A can be done with books either closed or open.

Extension of *Exercice A*: Paired Activity

After completing Exercise A, have pairs of students say or write negative sentences explaining what Nathalie *doesn't* do in her routine.

Extension of *Exercice B*: Paired Activity

Have pairs of students create a morning routine for Gérard. One or both of them can explain the routine to the class.

ANSWERS

Exercice A

1. Oui, le matin Nathalie se réveille à six heures et demie.
2. Oui, elle se lève tout de suite.
3. Oui, d'abord elle va dans la salle de bains pour faire sa toilette.
4. Oui, ensuite elle se lave les mains et la figure avec du savon.
5. Oui, elle se brosse les dents avec du dentifrice et une brosse à dents.
6. À mon avis, elle prend une douche (elle prend un bain).
7. À mon avis, elle se maquille, elle se peigne, elle se regarde dans la glace et elle s'habille.
8. Oui, elle prend son petit déjeuner.

Exercice B

1. Oui, Gérard rentre chez lui vers cinq heures.
2. Oui, il se lave les mains avant le dîner.
3. Il dîne dans la cuisine.
4. Oui, il se brosse les dents après le dîner.
5. Oui, à dix heures il se déshabille.
6. Il prend un bain le soir.
7. Oui, quand il se couche, il s'endort tout de suite.

Exercices

A **La routine de Nathalie.** Répondez.

1. Le matin Nathalie se réveille à six heures et demie?
2. Elle se lève tout de suite?
3. D'abord elle va dans la salle de bains pour faire sa toilette?
4. Ensuite elle se lave les mains et la figure avec du savon?
5. Elle se brosse les dents avec du dentifrice et une brosse à dents?
6. À ton avis, elle prend une douche ou un bain?
7. À ton avis, elle se maquille? Elle se peigne? Elle se regarde dans la glace? Elle s'habille?
8. Elle prend son petit déjeuner?

B **La routine de Gérard.** Répondez d'après les dessins.

1. Gérard rentre chez lui vers cinq heures?
2. Il se lave les mains avant le dîner?
3. Il dîne dans la cuisine ou dans la salle à manger?
4. Il se brosse les dents après le dîner?
5. À dix heures il se déshabille?
6. Il prend un bain le soir ou le matin?
7. Quand il se couche, il s'endort tout de suite?

LEARNING FROM ILLUSTRATIONS

Have students say as much as they can about *le petit chat de Gérard.* You may wish to give them the word for "bathtub," *la baignoire.*

ADDITIONAL PRACTICE

Student Tape Manual, Teacher's Edition, *Activités B–C,* page 122.

C Dans quelle pièce? Complétez.

1. On se brosse les dents dans ___.
2. On s'endort dans ___.
3. On prend une douche dans ___.
4. On se regarde dans la glace dans ___.
5. On se couche dans ___.
6. On prend son petit déjeuner dans ___.
7. La douche est dans ___.
8. Le lit est dans ___.

D Il a besoin de… Choisissez la bonne réponse.

1. Il va se brosser les dents. Il a besoin de ___.
   a. crème   b. dentifrice

2. Il va prendre une douche. Il a besoin de ___.
   a. savon   b. dentifrice

3. Il va se raser. Il a besoin d'un ___.
   a. peigne   b. rasoir

4. Il veut se peigner. Il a besoin d'un ___.
   a. peigne   b. rasoir

5. Il veut se laver les cheveux. Il a besoin de ___.
   a. déodorant   b. shampooing

6 F 90 Bain crème,
parfums au choix, 1 litre

20 F 00 Lot de 3 brosses à dents
GIBBS Intégral

35 F 00 1 brosse + 1 froufrou + 1 peigne
+ 1 miroir, coloris divers

4 F 90 Gel douche, parfums
au choix, 300 ml (le litre : 16,34 F)

CHAPITRE 11   **285**

PRESENTATION *(continued)*

Exercices C and D: Listening

Focus on the listening skill by having students do Exercises C and D in pairs. One partner reads the sentences in random order and the other partner, with his/her book closed, completes the sentences. Then partners switch roles.

Exercice C

1. la salle de bains
2. la chambre à coucher
3. la salle de bains
4. la salle de bains (la chambre à coucher)
5. la chambre à coucher
6. la cuisine (la salle à manger)
7. la salle de bains
8. la chambre à coucher

Exercice D

1. b
2. a
3. b
4. a
5. b

INFORMAL ASSESSMENT
(Mots 1)

Check for comprehension of the expression *avoir besoin de* by making statements about what people want to do and having students say what is needed to do it. For example: *Elle veut se peigner.* (*Elle a besoin d'un peigne.*) *Ils veulent se raser.* (*Ils ont besoin d'un rasoir.*)

RETEACHING *(Mots 1)*

Show Vocabulary Transparencies 11.1 and let students say as much as they can about them in their own words.

LEARNING FROM REALIA

Have students look at the ads and their copy. They will understand almost all the words. The students will probably enjoy the word *froufrou*.

INDEPENDENT PRACTICE

Assign any of the following:

1. Exercises, pages 284–285
2. Workbook, *Mots 1: A–D*, pages 105–107
3. Communication Activities Masters, *Mots 1: A*, page 54
4. CD-ROM, Disc 3, pages 282–285

VOCABULAIRE

MOTS 2

LA FORME PHYSIQUE

grossir

maigrir

un gymnase

un club de forme

faire de la gymnastique

faire de l'exercice

faire de l'aérobic

286   CHAPITRE 11

Vocabulary Teaching Resources

1. Vocabulary Transparencies 11.2 (A & B)
2. Audio Cassette 7A/CD-7
3. Student Tape Manual, Teacher's Edition, *Mots 2: D–F,* pages 123–124
4. Workbook, *Mots 2: E–G,* page 108
5. Communication Activities Masters, *Mots 2: B,* page 55
6. Chapter Quizzes, *Mots 2: Quiz 2,* page 49
7. Computer Software, *Vocabulaire*
8. CD-ROM, Disc 3, *Mots 2:* pages 286–289

Bell Ringer Review

Write the following on the board or use BRR Blackline Master 11-2: Which reflexive verb(s) do you associate with these nouns?

1. **les cheveux**   3. **la figure**
2. **les dents**   4. **les mains**

PRESENTATION (*pages 286–287*)

A. Have students close their books. Show Vocabulary Transparencies 11.2 (A & B). Have students repeat each word after you or Cassette 7A/CD-7 as you point to each item.
B. Ask questions about the statements on page 287: *Robert veut se mettre en forme? Qu'est-ce qu'il fait pour se mettre en forme? Où est-ce qu'il se promène?*

Teaching Tip To demonstrate the meaning of *toujours,* say: *le lundi, le mardi, le mercredi, tous les jours de la semaine—toujours. En décembre, en janvier, en février, etc.—toujours.*

TOTAL PHYSICAL RESPONSE

(*following the Vocabulary presentation*)

TPR

___, levez-vous et venez ici, s'il vous plaît.
Vous êtes au club de forme.
Mettez un short.
Mettez un tee-shirt.
Mettez des tennis.
Montrez que vous avez grossi.
Montrez que vous voulez maigrir.
Faites de l'exercice vigoureux.
Vous êtes très fatigué(e).
Merci. Retournez à votre place et asseyez-vous.

pratiquer un sport

mettre un
survêtement

Robert veut se mettre en forme.
Pour se mettre en forme il se
 promène.
Il se promène dans le parc.

Robert veut rester en forme.
C'est toujours le problème.
Pour rester en forme il fait de
 l'exercice.

Il fait du jogging.

Les copains s'amusent.
Ils s'amusent bien.

C. After presenting the vocabulary with the transparencies, have students open their books and read the words and sentences.

D. After each student reads a sentence, call on a more able student to ask a question about the sentence. He/She calls on another student to answer the question.

Vocabulary Expansion

You may wish to give the students the following expressions when teaching *grossir* and *maigrir: prendre des kilos, perdre des kilos.*

Activity: Have students respond with *On prend des kilos* or *On perd des kilos:*
On mange beaucoup de glace.
On fait du jogging.
On est toujours assis.
On fait beaucoup d'exercice.
On mange beaucoup.
On dort toujours.

ADDITIONAL PRACTICE

As a receptive activity, you may tell students to do the following:
Levez la main si vous aimez aller au gymnase.
Levez la main si vous aimez faire de l'aérobic.
Levez la main si vous faites de l'aérobic.
Levez la main si vous faites de l'exercice.
Levez la main si vous faites de l'exercice au gymnase de l'école.
Levez la main si vous pratiquez un sport.
Levez la main si vous pratiquez un sport après les cours.
Levez la main si vous faites du jogging.
Levez la main si vous habitez près d'un parc.

Exercices

PRESENTATION (*page 288*)

ANSWERS

Exercice A

You may wish to use to recorded version of this exercise.

1. Bien manger, c'est bon pour la santé.
2. Manger beaucoup de pâtisseries, c'est mauvais pour la santé.
3. Prendre du lait, c'est bon pour la santé.
4. Prendre du coca au petit déjeuner, c'est mauvais pour la santé.
5. Ne pas faire d'exercice, c'est mauvais pour la santé.
6. Faire de l'aérobic, c'est bon pour la santé.
7. Prendre des vitamines, c'est bon pour la santé.
8. Fumer, c'est mauvais pour la santé.
9. Pratiquer un sport, c'est bon pour la santé.
10. Grossir, c'est mauvais pour la santé.
11. Se promener tous les jours, c'est bon pour la santé.
12. Se mettre en forme, c'est bon pour la santé.

Exercice B

Answers will vary.

Exercice C

1. a
2. b
3. b
4. b
5. a
6. a

Activités de communication orale

Mots 1 et 2

ANSWERS

Activité A

Answers will vary.

Exercices

A **Pour rester en forme.** C'est bon ou mauvais pour la santé (*health*)?

> manger beaucoup de chocolat
> *Manger beaucoup de chocolat, c'est mauvais pour la santé.*

1. bien manger
2. manger beaucoup de pâtisseries
3. prendre du lait
4. prendre du coca au petit déjeuner
5. ne pas faire d'exercice
6. faire de l'aérobic
7. prendre des vitamines
8. fumer
9. pratiquer un sport
10. grossir
11. se promener tous les jours
12. se mettre en forme

B **En forme.** Donnez des réponses personnelles.

1. Tu aimes être en forme?
2. Tu fais de l'exercice pour rester en forme?
3. Tu fais du jogging? Tu mets un survêtement?
4. Tu pratiques un sport?
5. Tu pratiques quel sport?
6. Tu es membre d'un club de forme?
7. Tu fais de la gymnastique à l'école ou au gymnase?
8. Tu grossis quand tu manges beaucoup?
9. Rester en forme, c'est un problème pour toi?

C **Quel est le mot?** Choisissez.

1. Il prend des kilos. Il ___.
   a. grossit   b. maigrit
2. Il perd des kilos. Il ___.
   a. grossit   b. maigrit
3. Il va faire du jogging. Il met ___.
   a. une chemise   b. un survêtement
4. Il va faire du jogging. Il met ___.
   a. un complet   b. des tennis
5. Il va au parc. Il va ___.
   a. se promener   b. se raser
6. Il va ___ avec ses copains dans le parc.
   a. s'amuser   b. se réveiller

Activités de communication orale

Mots 1 et 2

A **La routine.** Tell your French-Canadian friend (your partner) about a member of your family. Include the information below. Then reverse roles.

1. his or her name
2. what time he or she gets up
3. some of his or her grooming habits
4. what he or she does to stay in shape
5. what sports he or she participates in

COOPERATIVE LEARNING	LEARNING FROM PHOTOS
Have each team of four assemble a composite list of activities from *Mots 1* and 2 that they like or don't like to do when they are on vacation. The list is started by one member and passed on until each member has contributed at least three activities, for a total of twelve. Encourage students to help each other with their contributions.	Have students tell whether you're describing the photo on page 288 or the one on page 289: **La danseuse est en forme.** **C'est un sport nautique.** **Ils font du kayak.** **Elle danse bien.**

B Les sportifs. Work with a classmate. Ask each other the following questions.

1. Qu'est-ce que tu fais pour rester en forme?
2. Où… ?
3. Avec qui… ?
4. Quand… ?

C Tu as besoin de… Play this game in small groups. One person states that he or she wants to do one of the activities listed below. The others have 20 seconds to write down as many things as they can think of that the first student needs in order to do the activity. Players receive one point per correct item.

> Élève 1: Je veux jouer au tennis.
> Élève 2: Tu as besoin d'une raquette, d'une balle, d'un court, d'un short…

aller à la plage	faire un voyage
faire du jogging	jouer au tennis
faire les courses	prendre un bain de soleil
faire les devoirs de…	préparer le dîner

D Madame Nette. Madame Nette is a very organized woman whose daily routine is always the same. With your classmates, take turns describing Madame Nette's day from morning to night. The first student suggests her first activity of the day. The next student repeats that activity and adds another.

> Élève 1: Madame Nette se réveille à six heures.
> Élève 2: Madame Nette se réveille à six heures.
>   Elle se lève tout de suite.

CHAPITRE 11   **289**

Extension of *Activité B*

Call on a student to retell in his/her own words what he/she does to stay in shape, based on the answers to Activity B.

ANSWERS

Activité B

Answers will vary.

Activité C

Answers will vary, but may include the following (all É1 answers begin with *Je veux* and all É2 answers begin with *Tu as besoin de [d']*):

É1: Je veux aller à la plage.
É2: Tu as besoin d'un maillot, (de crème solaire, de lunettes de soleil…).
É1: … faire du jogging.
É2: … d'un survêtement (de tennis, d'un sweat-shirt…).
É1: … faire les courses.
É2: … d'un filet (d'un sac, d'argent, d'une liste…).
É1: … faire les devoirs d'histoire (de biologie…).
É2: … d'un livre d'histoire (d'un livre de biologie…).
É1: … faire un voyage.
É2: … d'une valise (d'un passeport…).
É1: … jouer au tennis.
É2: … de chaussures de tennis (d'une balle…).
É1: … prendre un bain de soleil.
É2: … de crème solaire (de lunettes de soleil, de soleil…).
É1: … préparer le dîner.
É2: … d'un steak (de pommes de terre, d'une cuisine…).

Activité D

Answers will vary.

ADDITIONAL PRACTICE

1. Elicit *Mots 1* and *2* vocabulary by saying days/times of day (especially before and after school and around bedtime) and having individuals say what they're usually doing then. For example: *Le mercredi, 7h30 du matin. Patrick? (Je me brosse les dents.) Georges? (Je prends mon petit déjeuner.)*, etc.
2. Student Tape Manual, Teacher's Edition, *Activité F*, page 124.

INDEPENDENT PRACTICE

Assign any of the following:
1. Exercises and activities, pages 288–289
2. Workbook, *Mots 2: E–G*, page 108
3. Communication Activities Masters, *Mots 2: B*, page 55
4. Computer Software, *Vocabulaire*
5. CD-ROM, Disc 3, pages 286–289

Structure Teaching Resources

1. Workbook, *Structure: A–H*, pages 109–112
2. Student Tape Manual, Teacher's Edition, *Structure: A–D*, pages 124–125
3. Audio Cassette 7A/CD-7
4. Grammar Transparency G-11
5. Communication Activities Masters, *Structure: A–C*, pages 56–58
6. Computer Software, *Structure*
7. Chapter Quizzes, *Structure: Quizzes 3–5*, pages 61–63
8. CD-ROM, Disc 3, pages 290–295

Bell Ringer Review

Write the following on the board or use BRR Blackline Master 11-3: Correct the following sentences.

1. Je prends mon petit déjeuner dans la salle de bains.
2. Je me brosse les dents avec du savon.
3. Je me maquille les mains.
4. Je me regarde dans la figure.

Les verbes réfléchis

PRESENTATION *(pages 290–291)*

A. Have students make a list of the verbs learned in this chapter that describe what they do almost every morning.
B. Write the model verbs shown on page 290 on the board. Underline the reflexive pronouns.
C. Lead students through steps 1–3, calling on volunteers to read the material.

Les verbes réfléchis

Telling What People Do for Themselves

1. Compare the following pairs of sentences.

Chantal lave le bébé.

Chantal se lave.

Chantal regarde le bébé.

Chantal se regarde.

Chantal couche le bébé.

Chantal se couche.

In the sentences on the left Chantal performs the action and the baby receives it. In the sentences on the right Chantal herself is the receiver of the action. In these sentences Chantal both performs and receives the action of the verb. For this reason the pronoun *se* must be used. *Se* refers to Chantal and is called a reflexive pronoun. It indicates that the action of the verb is reflected back to the subject.

290 CHAPITRE 11

2. Each subject pronoun has its corresponding reflexive pronoun. Study the following.

SE LAVER	S'HABILLER
je me lave	je m' habille
tu te laves	tu t' habilles
il se lave	il s' habille
elle se lave	elle s' habille
on se lave	on s' habille
nous nous lavons	nous nous habillons
vous vous lavez	vous vous habillez
ils se lavent	ils s' habillent
elles se lavent	elles s' habillent

Note that *me, te,* and *se* become *m', t',* and *s'* before a vowel or silent *h*.

3. In the negative form of a reflexive verb, *ne* is placed before the reflexive pronoun. *Pas* follows the verb.

> Je me réveille mais je *ne* me lève *pas* tout de suite.
> On *ne* se brosse *pas* les dents avant le dîner.
> Je me couche mais je *ne* m'endors *pas* tout de suite.
> Nous *ne* nous rasons *pas* tous les jours.

Exercices

A **La routine de Charles.** Répétez la conversation.

ROGER: Tu te lèves à quelle heure, Charles?
CHARLES: À quelle heure est-ce que je me lève ou je me réveille?
ROGER: Tu te lèves.
CHARLES: Je me lève à six heures et demie.
ROGER: Et tu quittes la maison à quelle heure?
CHARLES: À sept heures. Je me lave, je me brosse les dents, je me rase et je prends mon petit déjeuner en une demi-heure.
ROGER: Et tu t'habilles aussi?
CHARLES: Bien sûr que je m'habille!

Répondez d'après la conversation.

1. Charles se lève à quelle heure?
2. Il se lave?
3. Il se brosse les dents dans la salle de bains?
4. Il se rase?
5. Il quitte la maison à quelle heure?

Note Point out to students that the forms of these verbs are the same as those of any other *-er* verb. The only addition is the pronoun.

D. Have students refer to the list of verbs they made for Presentation step A. For each verb they wrote down, have them make up sentences with *je*.

E. Ask questions using the reflexive verbs and call on volunteers to respond. For example: *À quelle heure est-ce que tu te réveilles? Tu te couches avant minuit? Tu te brosses les dents trois fois par jour?*, etc.

Note In the CD-ROM version, this structure point is presented via an interactive electronic comic strip.

Exercices

PRESENTATION *(page 291)*

Exercice A

Have a pair of students read the conversation aloud with as much expression as possible. Then call on individuals to answer the questions that follow.

Extension of *Exercice A*

Have students work in pairs. Have them change the information in the conversation in Exercise A to relate to themselves.

ANSWERS

Exercice A

1. Il se lève à six heures et demie.
2. Oui, il se lave.
3. Oui, il se brosse les dents dans la salle de bains.
4. Oui, il se rase.
5. Il quitte la maison à sept heures.

ADDITIONAL PRACTICE

1. Have pairs of students redo the conversation in Exercise A, changing *Charles* to *Charles et Robert* in order to practice the *nous* and *vous* forms.
2. Student Tape Manual, Teacher's Edition, *Activités B–D*, pages 124–125

ANSWERS

Bell Ringer Review

Write the following on the board or use BRR Blackline Master 11-4: Fill in the blanks with a reflexive pronoun or an X if no pronoun is needed.

Ma mère ___ réveille à six heures. Ensuite elle ___ réveille mon frère. Ils ___ préparent le petit déjeuner. Moi, je ___ prépare dans ma chambre et ensuite, je ___ prépare nos sandwichs pour le déjeuner.

Verbes avec changements d'orthographe

PRESENTATION *(pages 292–293)*

A. Go over this topic rather quickly and continue to reinforce good spelling habits throughout your students' study of French.
B. Lead students through steps 1–4 on pages 292–293.

B Jacqueline et Véronique. Changez *Jacqueline* en *Jacqueline et Véronique.*

1. Jacqueline se réveille à sept heures.
2. Jacqueline se lève tout de suite.
3. Jacqueline se brosse les dents.
4. Jacqueline se lave les mains et la figure.
5. Jacqueline se brosse les cheveux.
6. Jacqueline se maquille.

C Je fais ma toilette. Donnez des réponses personnelles.

1. Tu te lèves à quelle heure?
2. Tu vas dans la salle de bains?
3. Tu fais ta toilette?
4. Tu te laves les mains et la figure?
5. Tu prends une douche ou un bain?
6. Tu te laves les cheveux avec du shampooing?
7. Tu te brosses les dents?
8. Tu te peignes?
9. Tu t'habilles vite (rapidement)?

D Marc répond. Complétez.

1. Marc, tu ___? (se raser)
2. Oui, je ___. (se raser)
3. Tu ___ tous les jours? (se raser)
4. Oui, malheureusement il faut ___ tous les jours. (se raser)
5. Tu ___ les cheveux ou tu ___? (se brosser, se peigner)
6. Moi, je ___. Je ne ___ pas les cheveux. (se peigner, se brosser)
7. Tu ___ avant ou après le petit déjeuner? (s'habiller)
8. Je ___ avant le petit déjeuner. (s'habiller)

Verbes avec changements d'orthographe *Verbs with Spelling Changes*

1. The verbs *se promener* and *se lever,* like *acheter,* take an *accent grave* in all forms except the infinitive, *nous,* and *vous.*

SE PROMENER	
je me promène	nous nous promenons
tu te promènes	vous vous promenez
il/elle/on se promène	ils/elles se promènent

SE LEVER	
je me lève	nous nous levons
tu te lèves	vous vous levez
il/elle/on se lève	ils/elles se lèvent

ADDITIONAL PRACTICE

1. Have pairs of students make sentences with the following verbs, taking turns and alternating between reflexive and transitive constructions. For example: É1: *Je me promène.* É2: *Je promène mon chien.*

   amuser   promener   peigner
   laver   brosser   réveiller

2. You might wish to play the following game for more practice with reflexive verbs. Write the reflexive verbs on index cards, one to a card. Put the cards in a deck. A student picks a card and pantomimes the action of the verb. Another student tells him/her what he/she is doing, using the *tu* form.

2. The verb *s'appeler* doubles the *l* in all forms except the infinitive, *nous*, and *vous*.

S'APPELER	
je m'appelle	nous nous appelons
tu t'appelles	vous vous appelez
il/elle/on s'appelle	ils/elles s'appellent

3. Verbs that end in *-ger* such as *manger, nager,* and *voyager* add an *e* in the *nous* form in order to maintain the soft consonant sound.

nous mangeons   **nous nageons**   **nous voyageons**

4. Verbs that end in *-cer,* such as *commencer,* take a cedilla on the *c* in the *nous* form in order to maintain the soft consonant sound.

nous commençons

Exercices

A **Moi et toi.**   Mettez au pluriel.

> **Je me lève à sept heures et tu te lèves à neuf heures.**
> *Nous nous levons à sept heures et vous vous levez à neuf heures.*

1. Je me lève à 8 heures.
2. Je vais au magasin où j'achète un short.
3. Je me promène dans le parc.
4. Ensuite je nage dans la piscine.
5. Je commence à avoir faim.
6. Je rentre chez moi et je mange une pomme.
7. Et toi, tu te lèves à quelle heure?
8. Qu'est-ce que tu achètes au magasin?
9. Tu te promènes dans le parc aussi?
10. Ensuite tu nages dans la piscine?

B **Je m'appelle…**   Complétez avec «s'appeler».

1. Bonjour, je ___ …
2. Mon frère ___ …
3. Et ma sœur ___ …
4. Mon père ___ …
5. Ma mère ___ …
6. Mes meilleurs amis ___ …
7. Et comment ___-vous?
8. Nous ___ Dupont.

CHAPITRE 11   **293**

INDEPENDENT PRACTICE

Assign any of the following:
1. Exercises, pages 292–293
2. Workbook, *Structure: A–F,* pages 109–111
3. Communication Activities Masters, *Structure: A–B,* pages 56–57
4. CD-ROM, Disc 3, pages 290–293

LEARNING FROM PHOTOS

You may wish to ask students: *C'est une piscine? La piscine est grande ou petite? C'est une piscine en plein air ou une piscine couverte? C'est une piscine olympique? La jeune fille aime nager? Et vous, vous aimez nager? Est-ce qu'il y a une piscine couverte dans votre école?*

Exercices
PRESENTATION *(page 293)*

Exercices A and B
   Since the major difficulty with these verbs involves their spelling, you may wish to have students prepare these exercises first in writing. Then call on individuals to read them aloud.

ANSWERS

Exercice A

1. Nous nous levons à 8 heures.
2. Nous allons au magasin où nous achetons un short.
3. Nous nous promenons dans le parc.
4. Ensuite nous nageons dans la piscine.
5. Nous commençons à avoir faim.
6. Nous rentrons chez nous et nous mangeons une pomme.
7. Et vous, vous vous levez à quelle heure?
8. Qu'est-ce que vous achetez au magasin?
9. Vous vous promenez dans le parc aussi?
10. Ensuite vous nagez dans la piscine?

Exercice B
1. m'appelle
2. s'appelle
3. s'appelle
4. s'appelle
5. s'appelle
6. s'appellent
7. vous appelez
8. nous appelons

INFORMAL ASSESSMENT

   Have students study the verbs presented on pages 292–293 for homework. Give a spelling quiz on selected forms the following day.

RETEACHING

   Give the correct form of the verb in parentheses.

1. tu (se promener)
2. vous (manger)
3. ils (acheter)
4. nous (nager)
5. elle (commencer)

Le pronom interrogatif qui

PRESENTATION *(page 294)*

A. Have students close their books. Ask volunteers to supply examples of questions using *qui*. Write them on the board.

B. Now have students open their books. Lead them through steps 1–3 and the accompanying examples on page 294.

Le pronom interrogatif *qui*   *Asking "Who" or "Whom"*

1. You have been using the pronoun *qui* to form a question.

> **Qui est là?**
> **Qui parle?**
> **Qui se lève?**

2. You can also use *qui* as the object of the verb or as the object of a preposition. In this case *qui* means "whom."

> **Tu vois qui?**
> **Vous invitez qui?**
>
> **Vous parlez à qui?**
> **Vous allez au cinéma avec qui?**

3. Note that in the above questions *qui* is at the end of the sentence. In informal French, people put the question word at the end of the sentence and raise the tone of their voice. However, in formal or written French, the pronoun *qui* is placed at the beginning of the question and the subject and verb are inverted. Observe the following differences.

INFORMAL	FORMAL / WRITTEN
Tu vois qui?	**Qui vois-tu?**
Vous invitez qui?	**Qui invitez-vous?**
Vous parlez à qui?	**À qui parlez-vous?**
Vous allez au cinéma avec qui?	**Avec qui allez-vous au cinéma?**

294   CHAPITRE 11

Exercices

A **Pardon? Qui ça?** Posez des questions d'après le modèle.

> **Marie parle.**
> *Pardon? Qui parle?*

1. Son frère arrive.
2. Sa mère va à la porte.
3. Sa mère est très contente.
4. Le frère de Marie s'appelle David.
5. David a un cadeau.

B **Qui?** Posez des questions d'après le modèle.

> **Je regarde Suzanne.**
> *Tu regardes qui?*

1. Je téléphone à Robert.
2. Je parle à Robert.
3. J'invite Alice.
4. Je vois mon ami.
5. Je danse avec Isabelle.

C **Parlons bien.** Récrivez les questions d'après le modèle.

> **Vous ressemblez à qui?**
> *À qui ressemblez-vous?*

1. Vous téléphonez à qui?
2. Vous parlez à qui?
3. Vous invitez qui à la fête?
4. Vous achetez un cadeau pour qui?
5. Vous allez au restaurant avec qui?
6. Vous êtes derrière qui dans la queue?

Bell Ringer Review

Write the following on the board or use BRR Blackline Master 11-6: Use each of the following words or expressions in a logical sentence:

faire de   **s'endormir**
   l'aérobic   **se peigner**
maigrir

PRESENTATION (page 296)

A. Tell students they will hear a conversation between André and Richard.

B. Have them close their books and watch the Conversation Video or listen as you read the conversation or play Cassette 7A/CD-7.

C. Now have them open their books and read the conversation silently.

D. Call on pairs to act out the conversation.

E. Now do Exercise A. Call on individuals to answer each question or call on one student to give all the answers.

Note In the CD-ROM version, students can play the role of either one of the characters and record the conversation.

ANSWERS

Exercice A

1. Il parle à Richard.
2. Il se réveille à six heures et demie.
3. Il reste au lit jusqu'à sept heures.
4. Oui, il aime rester au lit.
5. Il peut vite faire sa toilette et s'habiller.
6. Oui, il va faire du jogging cet après-midi.
7. Richard veut rester en forme.

Scènes de la vie   *Qui est en forme?*

ANDRÉ: Tu te lèves à quelle heure, Richard?
RICHARD: Moi, je me lève à sept heures. Mais je me réveille à six heures et demie.
ANDRÉ: Ah, tu aimes rester un peu au lit.
RICHARD: Oui, mais je peux faire ma toilette, m'habiller et être prêt à quitter la maison en cinq minutes.
ANDRÉ: Tu vas faire du jogging cet après-midi?
RICHARD: Bien sûr. Il faut rester en forme.
ANDRÉ: Rester en forme? Il faut d'abord se mettre en forme!

 A   **La forme.**   Répondez d'après la conversation.

1. André parle à qui?
2. Richard se réveille à quelle heure?
3. Mais il reste au lit jusqu'à quelle heure?
4. Il aime rester au lit?
5. Qu'est-ce qu'il peut vite faire?
6. Richard va faire du jogging cet après-midi?
7. Qui veut rester en forme?

296   CHAPITRE 11

LEARNING FROM PHOTOS

Have students say as much as they can about the photo that accompanies the conversation.

LEARNING FROM REALIA

Ask students: *Ce sont des tennis? De quelle couleur sont-ils? C'est quelle marque (brand) de tennis? Vous avez des tennis? De quelle couleur sont vos tennis?*

Prononciation   *Les sons /s/ et /z/*

It is important to make a distinction between the sounds /s/ and /z/. You would not want to confuse *poisson* with *poison*! Repeat the following words with the sound /s/ as in *assez* and /z/ as in *raser*.

assez	dessert	cassette	boisson	classe
raser	désert	magasin	prise	valise

Now repeat the following sentences. Pay attention to which sounds occur.

Ils s'appellent Dumas.   Ils appellent leur chien.
Elles s'habillent vite.   Elles habillent les bébés.
Ils sont sympathiques.   Ils ont faim.

poisson / poison

Activités de communication orale

A   L'horaire du matin.   Your French friend Sylvie wants to know about your morning routine. Answer her questions.

1. Tu te réveilles à quelle heure?
2. Tu te lèves tout de suite?
3. Tu pars à quelle heure le matin?
4. Tu prends ton petit déjeuner avant de partir?

B   L'horaire du soir.   Find out about a classmate's evening routine: when he or she comes home from school, eats, does homework, goes to bed, etc. Then answer his or her questions about your evening routine.

C   La révolte du samedi et du dimanche.   When the weekend comes, everybody wants a change of pace. In small groups, discuss some of things you do on weekends that are different from the things you do during the week. Then compare results with those of other groups.

Le samedi et le dimanche, on ne se lève pas à sept heures. On se lève
    à neuf heures.
On ne prend pas son petit déjeuner à huit heures mais à 11 heures.
On se promène dans le parc…

Sylvie

PRESENTATION (page 297)

A. Model the key words *poisson/ poison* and have students repeat chorally.
B. Now model the other words and sentences in similar fashion.
C. For additional practice, use Pronunciation Transparency P-11, the *Prononciation* section on Cassette 7A/CD-7, and the Student Tape Manual, Teacher's Edition, *Activités G–I*, pages 126–127.
D. You may wish to give students the following *dictée*:
   Assez de dessert, Cassandre.
   La cassette est dans la valise.

Bell Ringer Review

Write the following on the board or use BRR Blackline Master 11-7: Answer the following questions.

1. **Tu te lèves à quelle heure?**
2. **Où est-ce que tu fais ta toilette?**
3. **Tu te brosses les dents avant le petit déjeuner?**

Activités de communication orale

PRESENTATION (page 297)

Activité A

In the CD-ROM version of this activity, students can interact with an on-screen native speaker.

ANSWERS

Activité A
   Answers will vary.

Activité B
   É2 answers will vary. É1 questions may include:

Tu rentres à quelle heure?
Quand est-ce que tu dînes?
Quand est-ce que tu fais tes devoirs?
Quand est-ce que tu te couches?

Activité C
   Answers will vary.

COOPERATIVE LEARNING

Have pairs of students make up five opinion questions about famous people, using *qui*. Pairs exchange questions and answer them with their own opinions. For example: *Qui est le meilleur joueur américain de tennis? (À mon avis, Michael Chang est le meilleur joueur américain de tennis.)* You can poll the class for answers to any questions that come up frequently.

INDEPENDENT PRACTICE

Assign any of the following:
1. Exercise and activities, pages 296–297
2. Have students rewrite the conversation, changing the information to relate to themselves.
3. CD-ROM, Disc 3, pages 296–297

LECTURE ET CULTURE

READING STRATEGIES
(page 298)

Pre-reading

Have students scan the reading quickly for cognates.

Reading

Have individuals read 2–3 sentences at a time. Ask comprehension questions.

Post-reading

Ask these additional questions: *La forme physique intéresse beaucoup ou peu les Américains? Que font les Américains pour rester en forme? Aux États-Unis, qu'est-ce qui est plus populaire, le jogging ou le cyclisme? Il y a beaucoup de marathons dans les grandes villes américaines? Il y a un marathon près de chez vous? Où? Quand?*

Note Students may listen to a recorded version of the *Lecture* on the CD-ROM.

Étude de mots

ANSWERS

Exercice A

Answers may include any four: forme, physique, moment, obsession, intéressent, sûr, point, degré, exercice, jogging, parcs, clubs, équipement, nécessaire, classes, aérobic, cyclisme, populaire, excellente, internationale, tennis, sport, disciples, marathons, octobre, participent

Exercice B

1. d	3. b	5. e
2. c	4. a	

298

LA FORME PHYSIQUE

Dans beaucoup de pays, la forme physique et la santé sont en ce moment une obsession. La forme physique et la santé intéressent bien sûr les Français mais peut-être pas au même point ou degré qu'aux États-Unis.

Que font les Français pour rester en forme? Les Français estiment qu'il faut faire de l'exercice. On voit des gens qui font du jogging dans les parcs et le long des fleuves[1]. Il y a maintenant de plus en plus de clubs de forme avec tout l'équipement nécessaire pour se mettre en forme. Il y a des classes pour faire de l'aérobic et pour les jeunes il y a des soirées aérobic. Dans les villes, il y a de plus en plus de piscines couvertes[2] pour faire de la natation toute l'année. Le cyclisme est très populaire en France. Le cyclisme est sans aucun doute[3] une excellente forme d'exercice. Le Tour de France est une course[4] cycliste internationale qui a lieu[5] en juillet. Et le tennis? Le tennis est un autre sport qui a de plus en plus de «disciples» en France. On parle toujours des marathons qui ont lieu dans les grandes villes des États-Unis. Il y a aussi un très grand marathon à Paris au mois d'octobre. Beaucoup de coureurs[6] participent au marathon de Paris.

[1] fleuves   *rivers*
[2] piscines couvertes   *indoor pools*
[3] sans aucun doute   *without a doubt*
[4] course   *race*
[5] a lieu   *takes place*
[6] coureurs   *runners*

Étude de mots

A **Le français, c'est facile.** Trouvez quatre mots apparentés dans la lecture.

B **Les noms et les verbes.** Trouvez le verbe qui correspond au nom.

1. l'équipement	a. intéresser
2. une obsession	b. participer
3. la participation	c. obséder
4. l'intérêt	d. équiper
5. le coureur, la course	e. courir

Le marathon de Paris

Compréhension

C **Oui ou non?** Corrigez les phrases fausses.

1. La forme physique intéresse beaucoup plus les Français que les Américains.
2. Les Français ne font pas d'exercice.
3. Il y a des clubs de forme en France.
4. L'aérobic n'est pas du tout populaire en France.
5. Le cyclisme n'est pas populaire chez les Français.
6. Le Tour de France est une course cycliste internationale qui a lieu en France.
7. Le marathon de Paris est une autre course cycliste.
8. Très peu de gens font du tennis en France.

D **La forme physique en France.** Répondez.

1. Qu'est-ce que les Français font pour rester en forme?
2. Qu'est-ce qu'il y a dans les clubs de forme?
3. Où peut-on nager toute l'année?
4. Quel sport a de plus en plus de «disciples»?
5. Il y a un grand marathon dans quelle ville?
6. Le marathon de Paris a lieu quand?
7. Le Tour de France a lieu quand?

E **L'essentiel.** Quelle est l'idée principale de cette lecture?

DÉCOUVERTE CULTURELLE

PETIT DÉJEUNER FRANÇAIS
Croissant au Beurre
Pain, Beurre, Confiture,
Café ou Thé ou Chocolat,
35,00

AMERICAN BREAKFAST
3 Œufs sur le plat, Pain, Beurre,
Jus d'Orange,
Café ou Thé ou Chocolat,
56,00

*A*vant de quitter la maison, André prend son petit déjeuner. Mais qu'est-ce qu'un petit déjeuner typiquement français? C'est du pain, des croissants ou des brioches avec une tasse de café au lait pour les adultes et une tasse de chocolat chaud pour les enfants. Mais des œufs, du bacon, des pommes de terre, absolument pas! Même les céréales ne sont pas très populaires chez les Français.

Comparez les deux petits déjeuners sur la carte d'un café parisien. Qui prend des œufs sur le plat, les Français ou les Américains? Qui prend des croissants?

CHAPITRE 11 **299**

RÉALITÉS

Bell Ringer Review

Write the following on the board or use BRR Blackline Master 11-9: Compare the following people or things, using a form of the adjective in parentheses.

1. (haut) le mont Everest/ le mont McKinley
2. (élégant) une robe/un short
3. (beau) la Côte d'Azur/ la Floride
4. (bon) la glace au chocolat/ la glace à la vanille
5. (âgé) Brad Pitt/ Paul McCartney

PRESENTATION *(pages 300–301)*

Have students look at the photographs for enjoyment. If they would like to talk about them, let them say anything they can.

Note In the CD-ROM version, students can listen to the recorded captions and discover a hidden video behind one of the photos.

Ce nageur français nage la brasse papillon **1**. Il porte des lunettes. À ton avis, c'est un nageur sérieux?

Cette jeune femme fait de l'aérobic **2**. À ton avis, est-elle en bonne forme?

Voici des cyclistes dans le Tour de France **3**. Ils traversent tout le pays—la campagne et les villes.

Ces deux hommes sont au gymnase **4**. Ils jouent au racquetball. Le racquetball ressemble à quels autres sports? Est-ce que le racquetball est un sport populaire aux États-Unis?

Voici des coureurs dans la course de 20 kilomètres à Paris **5**. Est-ce que tu aimes participer à des courses à pied?

300

DID YOU KNOW?

Interest in physical fitness has led to an increased interest in sports among the French. For example, the municipal office for each *arrondissement* of Paris provides a list of addresses where various sports can be practiced. A recent brochure for the 16ème offered the following activities. Share them with the students and have them try to guess the meanings of new words.

Aérobic	Arts martiaux
Gymnastique	Pelote basque
Danse	Pétanque
Tennis	Jogging
Équitation	Boxe
Golf	Plongée sous-marine
Spéléologie	Natation
Judo	Squash
Escrime	Tennis de table
Patinage	Basket-ball

(continued on next page)

4

5

301

(continued)

Ski	Athlétisme
Cyclotourisme	Randonnée
Football	Rugby
Hand-ball	Volley-ball
Hockey sur gazon	Tir

CULMINATION

RECYCLING

These *Activités de communication orale* and *Activités de communication écrite* are designed to aid students in personalizing the vocabulary and structures learned in Chapter 11 and recombine them with material from previous chapters. Recycled language includes food and clothing vocabulary, time expressions, and prepositions of time and location.

INFORMAL ASSESSMENT

Oral Activity B is well suited for evaluating speaking skills. Assign less able students the role of asking the guided questions, perhaps with help from their partner in the construction of the questions. Better students have free rein in creating the answers to the questions. Use the evaluation criteria given on page 34 of this Teacher's Wraparound Edition.

Activités de communication orale

PRESENTATION (page 302)

Activité A

Activity A is designed as a whole-class activity that provides fun and physical action in the classroom as well as practice with *qui* and prepositions of location.

ANSWERS

Activités A and B

Answers will vary.

Activités de communication écrite

ANSWERS

Activité A

Answers will vary but may include the following: **du café au lait, du thé, du chocolat, du pain, des croissants, da la confiture, des brioches**

Activité B

Answers will vary.

302

Activités de communication orale

A **Qui est devant qui?**   Play this game with your classmates. Each student takes a turn going to the front of the class. With his or her back to the class, the student has to answer classmates' questions about where various students are seated. The words below can be used in your questions.

> à côté de   à gauche de   derrière   à droite de   devant
>
> Élève 1: **Isabelle est devant qui?**
> Élève 2: **Elle est devant Paul.**

B **Une interview.**   You're interviewing the new French exchange student (your partner) for the school newspaper. Find out the following: his or her name, where he or she's from, what he or she does after school, his or her friends' names, what he or she does to stay in shape, and what he or she likes to eat.

Activités de communication écrite

A **Qu'est-ce qu'un petit déjeuner typiquement français?**   You're living with a French family for the summer. Write a note to one of your friends describing a typical French breakfast. Tell him or her if you like it or not.

> **En France au petit déjeuner, on mange…**

B **Monsieur Dodu veut se mettre en forme.**
Monsieur Dodu would like to lose some weight and get in shape. As his personal trainer, write out a daily routine for him telling him what time to get up, when to exercise, and what type of exercise to do. Plan his meals for him, too, and suggest what time he should eat them.

> **La routine de M. Dodu**
> 6h   Il se réveille et il se lève tout de suite.
> 6h15 à 7h   Il fait de l'exercice avec moi.
> 7h à 7h05   Il prend une douche froide.

FOR THE YOUNGER STUDENT	PAIRED ACTIVITY
1. Have students in groups of three write a skit about staying in shape and present it to the class. 2. Have students use as much French as possible in preparing a directory of places where sports can be practiced in their area.	Have students work in pairs to draw or make a collage of activities people do to get into shape or to stay in shape. Then have them describe their drawing or collage to the class. They can also mention which of the activities they participate in personally, or which activities they do in physical education class.

302

Réintroduction et recombinaison

A À votre tour. Répondez.

1. Quand tu t'habilles le matin, qu'est-ce que tu mets?
2. Qu'est-ce que tu prends au petit déjeuner?
3. Tu vas à l'école comment? En bus, en voiture ou à pied?
4. Tu fais des achats après les cours?
5. Tu aimes faire des achats dans un grand magasin ou dans une boutique?
6. Tu achètes des cadeaux pour tes copains?
7. De quelle couleur est ton pantalon ou ton tee-shirt favori?
8. Pour les chaussures tu fais quelle pointure?
9. Tu demandes la pointure au-dessus ou au-dessous quand les chaussures sont trop larges?

B L'anniversaire de mon frère. Complétez.

Je ___ (aller) aux Galeries Lafayette. Je ___ (vouloir)
acheter un cadeau pour mon frère. C'est son anniversaire.
Qu'est-ce que je ___ (pouvoir) acheter? Qu'est-ce qu'il
___ (aimer)? Je ___ (aller) au rayon articles de sport. Je
___ (voir) une raquette de tennis. Voilà! C'est une bonne
idée. Mon frère ___ (aimer) bien le tennis. Ses copains
et lui ___ (jouer) souvent au tennis mais mon frère ___
(avoir) une vieille raquette. J'___ (acheter) la raquette
et je ___ (payer) à la caisse.

J'AIME PAS LE SPORT !
J'AIME PAS ME FATIGUER !
J'AIME RIEN !

Vocabulaire

NOMS
les cheveux (m.)
les dents (f.)
la figure
la main

le dentifrice
le savon
le déodorant
la glace

le club de forme
le gymnase
le parc
le survêtement

le lit
le problème

VERBES
s'amuser
s'appeler
se réveiller
se lever
se brosser
se laver
se peigner
s'habiller
se maquiller
se raser
se promener

se coucher
s'endormir

maigrir
grossir

ADVERBES
d'abord
enfin
ensuite
tout de suite

AUTRES MOTS
ET EXPRESSIONS

avoir besoin de
faire de l'exercice

faire de l'aérobic
faire de la gymnastique
faire du jogging
pratiquer un sport
se mettre en forme
rester en forme
faire sa toilette
prendre un bain (une douche)

Réintroduction et recombinaison

ANSWERS

Exercice A
Answers will vary.

Exercice B

1. vais
2. veux
3. peux
4. aime
5. vais
6. vois
7. aime
8. jouent
9. a
10. achète
11. paie

ASSESSMENT RESOURCES

1. Chapter Quizzes
2. Testing Program
3. Situation Cards
4. Communication Transparency C-11
5. Computer Software: Practice/Test Generator

VIDEO PROGRAM

INTRODUCTION   (36:09)

TU ES EN FORME?   (37:20)

LEARNING FROM REALIA

Ask students to give more correct ways of expressing the statements in the realia on page 303.

STUDENT PORTFOLIO

Written assignments that may be included in students' portfolios are the *Activités de communication écrite* on page 302 and the *Mon Autobiographie* section of the Workbook on page 115.

Note Students may create and save both oral and written work using the Electronic Portfolio feature on the CD-ROM.

CHAPTER OVERVIEW

In this chapter students will learn to talk about cars and good driving habits. They will also learn expressions needed to communicate with a gas station attendant. In order to do this, students will learn vocabulary associated with cars, verbs conjugated like *conduire,* and negative constructions.

The cultural focus of Chapter 12 is on French driving customs and the highways of France.

CHAPTER OBJECTIVES

By the end of this chapter, students will know:

1. vocabulary associated with automobile types, features, and basic servicing
2. vocabulary associated with driver's education, driver's licenses, parking, and traffic regulations.
3. the present indicative forms of the verbs *conduire, lire, écrire,* and *dire*
4. the negative constructions *ne… rien, ne… personne,* and *ne… jamais*
5. the formation of questions with inverted word order

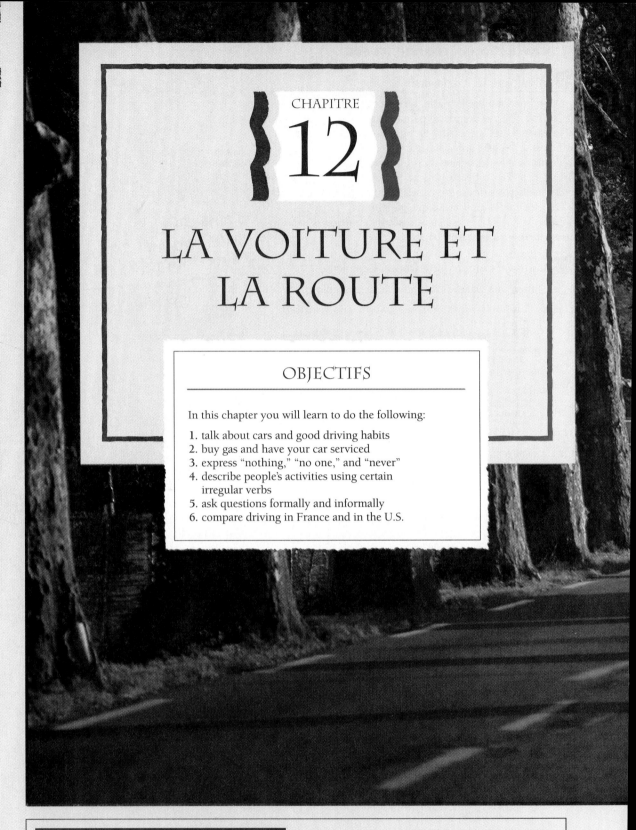

CHAPITRE

12

LA VOITURE ET LA ROUTE

OBJECTIFS

In this chapter you will learn to do the following:

1. talk about cars and good driving habits
2. buy gas and have your car serviced
3. express "nothing," "no one," and "never"
4. describe people's activities using certain irregular verbs
5. ask questions formally and informally
6. compare driving in France and in the U.S.

INTERDISCIPLINARY CONNECTIONS

Borrow a driver's education film from the Driver's Ed. class at your school. Have students view it and have them discuss it in French to the best of their ability. Then have them make a poster illustrating good driving tips and label it in French. Have them share the poster with the Driver's Ed. class.

305

Pacing

This chapter requires eight to ten class sessions. Pacing will vary according to class length and the age and aptitude of the students.

Note The Lesson Plans offer guidelines for 45- and 55-minute classes and **Block Scheduling.**

Exercices vs. *Activités*

All exercises (which provide guided practice) are coded in blue. All communicative activities are coded in red.

INTERNET ACTIVITIES

(optional)
These activities, student worksheets, and related teacher information are in the *Bienvenue* Internet Activities Booklet and on the Glencoe Foreign Language Home Page at: **http://www.glencoe.com/secondary/fl**

DID YOU KNOW?

Tell students that the beautiful tree-lined road in the photo is very typical of France. Plane trees (*les platanes*) often form an arch over the road. The trees grow in temperate regions of the northern hemisphere but are cultivated particularly in Europe because of their rapid growth and attractive bark. The species that grows in North America is called the buttonwood or sycamore tree.

MOTS 1

Bell Ringer Review

Write the following on the board or use BRR Blackline Master 12-1: Write three things you do in the morning and three things you do in the evening.

PRESENTATION *(pages 306–307)*

A. Using Vocabulary Transparencies 12.1 (A & B), point to each illustration and have students repeat the corresponding word two or three times after you or Cassette 7B/CD-7. Dramatize *freiner* and *s'arrêter* to clarify the meaning.

B. Point to items at random and ask *Qu'est-ce que c'est?* Have students identify each item with the appropriate word or expression.

C. Ask students to open their books to pages 306–307 and read the new words and sentences. Call on individuals to read or have students repeat in unison.

VOCABULAIRE

MOTS 1

LA VOITURE

les deux roues

une moto

un vélomoteur

un break

une voiture de sport

PEUGEOT

une marque française

une décapotable

une clé

accélérer

mettre le contact

une conductrice

rouler vite

un conducteur

La voiture s'arrête.

La conductrice freine.

TOTAL PHYSICAL RESPONSE

(following the Vocabulary presentation)

Getting Ready

Set up a chair as a driver's seat. Pieces of paper can serve as accelerator and brake pedals. One student plays the part of a driver, the other a service station attendant. Demonstrate the terms *le capot, le pied,* and *tourner.*

TPR 1

___, levez-vous, s'il vous plaît. Venez ici. C'est votre voiture.
Montez dans votre voiture.
Asseyez-vous.
Fermez la portière.
Mettez votre ceinture de sécurité.
Regardez la carte routière.
Mettez le contact.
Mettez le pied sur l'accélérateur. Accélérez.

(continued on next page)

À LA STATION-SERVICE

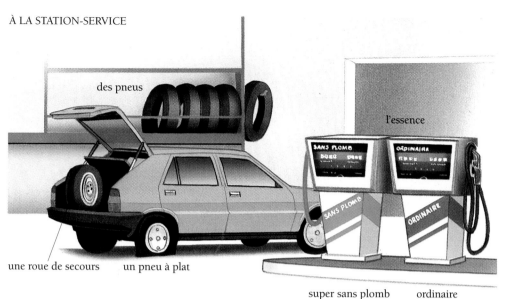

des pneus

l'essence

une roue de secours   un pneu à plat

super sans plomb   ordinaire

les niveaux

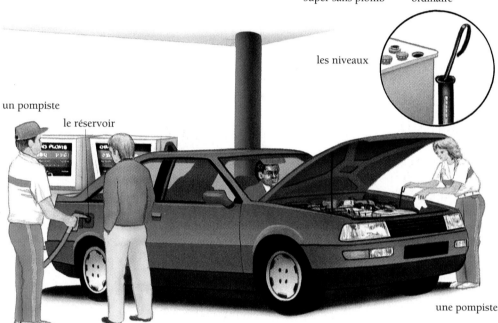

un pompiste

le réservoir

une pompiste

Le pompiste fait le plein.
Il met trente litres de super sans plomb dans le réservoir.
Quelqu'un parle au pompiste.
Une autre pompiste vérifie les niveaux.

RECYCLING

The concept of -er verbs is recycled in *Mots 1* by the introduction of *accélérer, freiner, rouler, s'arrêter,* and *vérifier.* The irregular verbs *mettre* and *faire* are also reintroduced.

Teaching Tip When an individual student cannot respond to a question, try getting the answer from a volunteer by gesturing silently to the class or asking *Qui sait?* Always have the original student repeat the correct model, and come back to him/her soon with the same item.

D. Show a photo of your own car or a magazine picture of one and describe it, using the new vocabulary. For example: *J'ai une voiture de sport rouge. C'est une décapotable américaine. C'est une marque américaine, une Ford. Elle accélère très vite, mais je ne roule pas vite généralement.*

Vocabulary Expansion

You may wish to give the students the word *une bagnole,* which is slang for "car." You may also wish to give students some cognates: *le radiateur, la batterie, l'accélérateur.* Introduce *la ceinture de sécurité, le coffre, le capot, la portière.*

TPR (*continued*)
Voilà un feu rouge. Freinez. Maintenant, accélérez.
Tournez à droite.
Arrêtez-vous.
Merci, ___. C'est très bien. Retournez à votre place, s'il vous plaît.

TPR 2
___, venez ici, s'il vous plaît.
Vous travaillez à mi-temps. Vous êtes pompiste dans une station-service. Une voiture arrive.
Ouvrez le capot.
Vous allez vérifier les niveaux.
Indiquez au conducteur qu'il a besoin d'eau dans son radiateur.
Mettez de l'eau dans le radiateur.
Fermez le capot.
Merci, ___. Retournez à votre place, s'il vous plaît.

Exercices

PRESENTATION *(page 308)*

You may do the exercises while still working with the overhead transparencies since the exercises help the students learn the new words depicted on the transparencies.

Extension of *Exercice B:* Listening

After doing Exercise B as a whole-class activity, have students work in pairs. One partner reads the questions in random order. The second partner listens and answers with book closed. Then they reverse roles.

Extension of *Exercice C*

After completing Exercise C, ask questions such as the following: *Le pompiste ne met pas d'essence dans le radiateur. Qu'est-ce qu'il met dans le radiateur? (de l'eau) Il ne met pas d'eau dans le moteur. Qu'est-ce qu'il met dans le moteur? (de l'huile) Le conducteur ne veut pas rouler plus vite. Il veut rouler moins vite. Qu'est-ce qu'il fait? (Il freine.), etc.*

ANSWERS

Exercice A

1. **C'est une voiture de sport.**
2. **C'est un vélomoteur.**
3. **La moto a deux roues.**
4. **La décapotable, c'est la voiture de sport.**

Exercice B

Answers will vary.

Exercice C

1. b	4. b	7. a
2. b	5. a	8. a
3. b	6. a	9. b

INFORMAL ASSESSMENT
(Mots 1)

Check for comprehension by having one student read vocabulary items at random from pages 306–307 while another points to the appropriate illustrations on the overhead transparencies.

Exercices

A Qu'est-ce que c'est?
Répondez d'après les dessins.

1. C'est une voiture de sport ou un break?
2. C'est un vélomoteur ou une moto?
3. La moto a deux roues ou quatre roues?
4. La décapotable, c'est la voiture de sport ou le break?

B Tu as une voiture? Donnez des réponses personnelles.

1. Tu as une voiture? Tu as quelle marque de voiture?
2. Tu veux une voiture? De quelle marque?
3. Tu préfères les breaks ou les voitures de sport?
4. Tu aimes les décapotables?
5. Tu préfères les voitures ou les motos?
6. Ta mère roule vite? Et ton père?

C Les voitures. Choisissez la bonne réponse.

1. À la station-service le pompiste fait le plein. Il met de l'essence dans ___.
   a. le radiateur   b. le réservoir
2. Il vérifie les niveaux. Il met de l'eau dans ___.
   a. le moteur   b. le radiateur
3. Il met de l'air dans ___.
   a. les roues   b. les pneus
4. Le conducteur veut rouler plus vite. Il ___.
   a. freine   b. accélère
5. La conductrice veut s'arrêter. Elle ___.
   a. freine   b. accélère
6. Quand quelqu'un a un pneu à plat, il ou elle a besoin d'___.
   a. une roue de secours   b. une clé
7. Pour mettre le contact, on a besoin d'___.
   a. une clé   b. un réservoir
8. En général, dans les voitures de sport on met de l'essence ___.
   a. super   b. ordinaire
9. Aux États-Unis les nouvelles voitures consomment de l'essence ___.
   a. avec plomb   b. sans plomb

COOPERATIVE LEARNING

Ask students to write in English four or five characteristics of their car or their family car (model, make, color, year). They then use their prepared lists of features to follow your example and describe their cars to each other in small groups, using the *Mots 1* vocabulary. Do this with books closed. Group members can ask each other for words they can't remember.

INDEPENDENT PRACTICE

Assign any of the following:
1. Exercises, page 308
2. Workbook, *Mots 1: A–C*, pages 116–117
3. Communication Activities Masters, *Mots 1: A*, page 59
4. CD-ROM, Disc 3, pages 306–308

VOCABULAIRE

MOTS 2

LA ROUTE

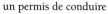

un permis de conduire

l'auto-école (f.)

prendre des leçons de conduite

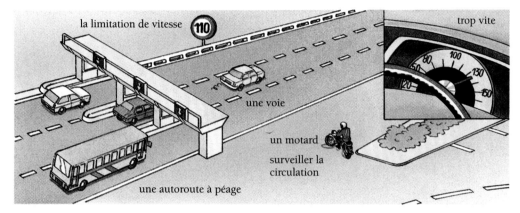

la limitation de vitesse

trop vite

une voie

un motard

surveiller la circulation

une autoroute à péage

un croisement

un carrefour

CHAPITRE 12   **309**

VOCABULAIRE

MOTS 2

Vocabulary Teaching Resources

1. Vocabulary Transparencies 12.2 (A & B)
2. Audio Cassette 7B/CD-7
3. Student Tape Manual, Teacher's Edition, *Mots 2: D–F,* pages 133–134
4. Workbook, *Mots 2: D–G,* pages 117–118
5. Communication Activities Masters, *Mots 2: B,* page 60
6. Computer Software, *Vocabulaire*
7. Chapter Quizzes, *Mots 2:* Quiz 2, page 65
8. CD-ROM, Disc 3, *Mots 2:* pages 309–312

Bell Ringer Review

Write the following on the board or use BRR Blackline Master 12-2: Write the identity of each thing:

1. **Elle a deux roues et un moteur mais pas de pédales.**
2. **On met le contact avec cet objet.**
3. **Elle a quatre roues et un moteur. Elle roule très vite et elle est souvent belle.**

PRESENTATION *(pages 309–310)*

A. Have students close their books. Show Vocabulary Transparencies 12.2 (A & B). Have students repeat each word several times after you or Cassette 7B/CD-7.

B. Ask questions such as: *Il y a combien de personnes dans la voiture? Qui conduit? Qu'est-ce qu'elle prend? Elle a son permis de conduire? Elle va avoir son permis de conduire? L'autoroute a combien de voies? Qui surveille la circulation? Quelle est la limitation de vitesse? Il faut payer un*

309

TOTAL PHYSICAL RESPONSE

(following the Vocabulary presentation)

Getting Ready

Set up a chair as a driver's seat. One student plays the part of a driver and the other a parking enforcement officer.

TPR

(Student 1), **venez ici, s'il vous plaît. Vous êtes le conducteur/la conductrice.**

Montez dans la voiture.
Mettez le contact.
Roulez, mais pas trop vite.
Maintenant, accélérez.
Vous arrivez en ville. Freinez.
Vous voulez garer la voiture.
Cherchez une place.
Voilà une place. Garez la voiture.
Ouvrez la portière.
Descendez.

(continued on page 310)

péage sur l'autoroute? Les conducteurs s'arrêtent pour payer le péage? Où s'arrêtent-ils?

C. Ask the following questions that can be answered with one word. These give excellent practice for oral comprehension: *Qui prend la leçon de conduite, le conducteur ou le moniteur? Qui paie le péage, le motard ou le conducteur? Qui surveille la circulation, le motard ou le conducteur? Qui obéit à la limitation de vitesse, le motard ou le conducteur? Qui traverse la rue, le piéton ou le conducteur? Qui gare la voiture, le piéton ou le conducteur? Qui met la ceinture de sécurité, le piéton ou le conducteur? Qui surveille le stationnement, le motard ou la contractuelle? Qui écrit des contraventions, les piétons ou les contractuelles?*

D. Ask students personalized questions. For example: *Tu as ton permis de conduire, Jessica? Benjamin, tu prends des leçons de conduite? Où? Qui va à l'auto-école? Devant le lycée, quelle est la limitation de vitesse? Sara, tu mets ta ceinture de sécurité? Marc, que fait ta mère à un feu rouge? Qui a des contraventions?*, etc.

Vocabulary Expansion

You may wish to give students the following additional vocabulary in order to talk about driving.

la carte routière
un panneau
une rue à sens unique
la flèche
un agent de police
un parcmètre
doubler

un feu rouge (orange) (vert)   un piéton   les clous (m.)   une piétonne

un trottoir

traverser la rue dans les clous

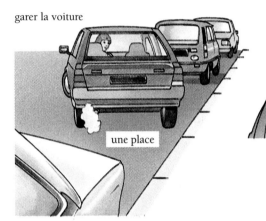

garer la voiture

une place

une ceinture de sécurité

Carole conduit la voiture.
Elle conduit prudemment.
Didier met sa ceinture de sécurité.
Il lit le Guide Michelin.

une contractuelle

Il est interdit de stationner ici.

ZUT!

une contravention

La contractuelle écrit une contravention.
La contractuelle ne dit rien.
Elle ne parle à personne.

Camille lit la contravention.
Elle est fâchée.
Elle dit: «Zut!»

310   CHAPITRE 12

TPR *(continued from page 309)*
Fermez la portière à clé.
Allez faire ce que vous voulez faire.
(Student 2), levez-vous, s'il vous plaît. Vous êtes la contractuelle.
Regardez. Le stationnement est interdit. Écrivez une contravention.
Mettez la contravention ici, sur le pare-brise. (*Indicate.*)

Et (Student 1), venez ici, s'il vous plaît. Vous revenez à votre voiture. Regardez. Prenez la contravention.
Lisez-la.
Qu'est-ce que vous êtes fâché(e)! Dites quelque chose.
Merci, (Student 1) et (Student 2). Vous avez très bien fait. Asseyez-vous, s'il vous plaît.

Exercices

A **En voiture.** Répondez par «oui» ou «non».

1. Il faut payer quand on roule sur une autoroute à péage?
2. Les autoroutes ont souvent quatre ou six voies?
3. À un carrefour il faut faire attention aux piétons?
4. Il faut mettre sa ceinture de sécurité quand on conduit?
5. On peut conduire sans avoir de permis de conduire?
6. Il faut conduire prudemment à un croisement?
7. Il faut respecter la limitation de vitesse?
8. Il faut rouler vite quand le feu est rouge?
9. Il faut s'arrêter quand le feu est vert?
10. Il faut accélérer pour s'arrêter?
11. Il est interdit de garer sa voiture sur le trottoir?

B **À vous de choisir.** Choisissez la bonne réponse.

1. Les ___ traversent la rue dans les clous quand le feu est vert.
   a. motards   b. piétons
2. Les motards surveillent ___.
   a. la circulation   b. le stationnement
3. Les contractuelles surveillent ___.
   a. le stationnement   b. la circulation
4. Les motards donnent des contraventions aux ___ qui roulent trop vite.
   a. conducteurs   b. contractuelles
5. Quand on veut garer sa voiture, on cherche ___.
   a. une place   b. le trottoir

C **Tu conduis ou pas?** Donnez des réponses personnelles.

1. Tu as ton permis de conduire?
2. Tu vas passer ton permis de conduire?
3. Tu as quel âge maintenant?
4. On passe le permis de conduire à quel âge?
5. Tu vas prendre des leçons de conduite?
6. Tu vas prendre des leçons de conduite à l'école ou à une auto-école?

D **Que font-ils?** Complétez en utilisant «conduit», «lit», «dit» ou «écrit».

1. Carole ___ la voiture.
2. Didier ne ___ pas la voiture.
3. Didier ___ le Guide Michelin.
4. Carole ne ___ pas le guide parce qu'elle ___.
5. La contractuelle ___ une contravention.
6. Carole est fâchée. Elle ___: «Zut!»
7. La contractuelle ne ___ rien. Elle ne parle à personne.

CHAPITRE 12   **311**

COOPERATIVE LEARNING

Working in teams of three, students make up a list of activities related to the road. They then separate the activities according to who does them: *le conducteur, le motard, le piéton.* Then each team member plays one of the above roles. For example: *Moi, je suis la conductrice/le conducteur. Je roule dans ma voiture…*

Exercices

PRESENTATION *(page 311)*

Exercices A and C: **Listening**

After doing Exercises A and C as a whole-class activity, focus on the listening skill by having students do them in pairs, one partner reading the questions in random order while the other answers with book closed.

Exercices B and D

After doing Exercises B and D, call on a student to review the information in his or her own words.

ANSWERS

Exercice A

1. Oui.	7. Oui.
2. Oui.	8. Non.
3. Oui.	9. Non.
4. Oui.	10. Non.
5. Non.	11. Oui.
6. Oui.	

Exercice B

1. b
2. a
3. a
4. a
5. a

Exercice C

Answers will vary.

Exercice D

1. conduit
2. conduit
3. lit
4. lit, conduit
5. écrit
6. dit
7. dit

INFORMAL ASSESSMENT *(Mots 2)*

Have students close their books. Check for comprehension by indicating pictures on Vocabulary Transparencies 12.2 and giving students two minutes to note down everything they can remember to say about them. Call on individuals to report orally from their notes.

PRESENTATION *(page 312)*

Activité C

In the CD-ROM version of this activity, students can interact with an on-screen native speaker.

ANSWERS

Activité A

Questions will begin with *C'est une bonne idée de… ?* Answers will begin with *Non, ce n'est pas une bonne idée de…* and end with the cues in the text.

Activités B and C

Answers will vary.

Activités de communication orale
Mots 1 et 2

A **C'est une bonne idée?** Ask a classmate if it's a good idea to do the following things when driving.

> conduire sans avoir de permis de conduire
>
> Élève 1: C'est une bonne idée de conduire sans avoir de permis de conduire?
>
> Élève 2: Non, ce n'est pas une bonne idée de conduire sans avoir de permis de conduire.

1. traverser la rue quand le feu est rouge
2. rouler avec un pneu à plat
3. rouler sans avoir beaucoup d'essence dans le réservoir
4. se maquiller quand on conduit
5. lire le Guide Michelin quand on conduit

B **On conduit bien ou mal?** Ask the Driver's Ed. instructor in your school for a booklet on driving techniques. Work with a classmate and make a list of five good driving habits that you know how to express in French. Make a second list of five things that a good driver shouldn't do. Present your lists to the class in random order and ask your classmates to decide whether the people being described drive well or not.

> Élève 1: On ne respecte pas la limitation de vitesse.
> La classe: On conduit mal.

C **Le permis de conduire.** Your French friend Alain wants to know about driving in the U.S. Answer his questions.

1. Tu as ton permis de conduire?
2. On peut avoir son permis de conduire à quel âge aux États-Unis?
3. Quelle est la limitation de vitesse sur les autoroutes?
4. Les motards donnent beaucoup de contraventions?

Alain

ADDITIONAL PRACTICE	**INDEPENDENT PRACTICE**
1. Have students pretend they had a minor accident. They fill out an accident report including the following details:	Assign any of the following:
	1. Exercises and activities, pages 311–312
	2. Workbook, *Mots 2: D–G,* pages 117–118

<table>
<tr><td>Nom</td><td>Heure</td><td>Marque de</td></tr>
<tr><td>Âge</td><td>Lieu</td><td>   voiture</td></tr>
<tr><td>Sexe</td><td>Vitesse</td><td>Accidentés</td></tr>
<tr><td>Date</td><td></td><td>(victims)</td></tr>
</table>

2. Student Tape Manual, Teacher's Edition, *Activité F,* page 134.

3. Communication Activities Masters, *Mots 2: B,* page 60
4. Computer Software, *Vocabulaire*
5. CD-ROM, Disc 3, pages 309–312

STRUCTURE

STRUCTURE

Les verbes *conduire, lire, écrire* et *dire* au présent

Describing People's Activities

1. Study the following forms of the verbs *conduire*, "to drive," *lire*, "to read," *écrire*, "to write," and *dire*, "to say." Note how similar they are to one another in the present tense.

2. Note the irregular verb form of *dire: vous dites*.

CONDUIRE	LIRE	ÉCRIRE	DIRE
je conduis	je lis	j' écris	je dis
tu conduis	tu lis	tu écris	tu dis
il elle on } conduit	il elle on } lit	il elle on } écrit	il elle on } dit
nous conduisons	nous lisons	nous écrivons	nous disons
vous conduisez	vous lisez	vous écrivez	vous dites
ils elles } conduisent	ils elles } lisent	ils écrivent elles écrivent	ils elles } disent

Exercices

A **Didier va à Bourges.** Répondez par «oui».

1. Didier va à Bourges?
2. Didier conduit prudemment?
3. Avant le voyage Didier lit le Guide Michelin?
4. Didier dit que Bourges est loin?
5. Didier écrit une lettre à son copain Guillaume?
6. Dans sa lettre il dit que Bourges est une jolie ville?
7. Guillaume lit la lettre de Didier?

B **Je lis et j'écris.** Donnez des réponses personnelles.

1. Tu aimes lire?
2. Tu lis beaucoup?
3. Tu lis le journal tous les jours?
4. Tu lis des magazines?
5. Tu lis quels magazines?
6. Tu écris des lettres à tes amis?
7. Tu écris à tes grands-parents ou tu téléphones à tes grands-parents?

ADDITIONAL PRACTICE

Student Tape Manual, Teacher's Edition, *Activités A–B*, page 135.

LEARNING FROM PHOTOS

Have students say as much as they can about the billboard.

Structure Teaching Resources

1. Workbook, *Structure: A–H*, pages 119–122
2. Student Tape Manual, Teacher's Edition, *Structure: A–E*, pages 135–137
3. Audio Cassette 7B/CD-7
4. Communication Activities Masters, *Structure: A–D*, pages 61–63
5. Computer Software, *Structure*
6. Chapter Quizzes, *Structure:* Quizzes 3–5, pages 66–68
7. CD-ROM, Disc 3, pages 313–317

Bell Ringer Review

Write the following on the board or use BRR Blackline Master 12-4: List things you read and things you write.

Les verbes conduire, lire, écrire *et* dire *au présent*

PRESENTATION *(page 313)*

A. Write the third-person plural forms of the verbs on the board. Have students repeat them. Draw a line through the end *-sent* or *-vent* and have students repeat the singular forms.

B. Lead students through steps 1 and 2 on page 313.

Exercices

ANSWERS

Exercice A

1. Oui, il va à Bourges.
2. Oui, il conduit prudemment.
3. Oui, il lit le Guide Michelin.
4. Oui, il dit que Bourges est loin.
5. Oui, il écrit une lettre à son copain Guillaume.
6. Oui, dans sa lettre il dit que Bourges est une jolie ville.
7. Oui, il lit sa lettre.

Exercice B

Answers will vary.

C **Une question, mon ami.** Posez des questions à un copain ou une copine d'après le modèle.

> écrire beaucoup de lettres
> Élève 1: Tu écris beaucoup de lettres?
> Élève 2: Oui, j'écris beaucoup de lettres. (Non, je n'écris pas beaucoup de lettres.)

1. écrire des poèmes
2. écrire des compositions au cours d'anglais
3. lire le journal
4. lire des magazines
5. conduire une nouvelle voiture
6. dire que les voitures de sport sont chouettes

D **Ses amis et lui.** Complétez.

1. Lui, il dit des choses stupides, des bêtises, et ses amis ___ des bêtises aussi.
2. Lui, il conduit une vieille voiture et ses amis ___ de vieilles voitures aussi.
3. Lui, il conduit prudemment et ses amis ___ prudemment aussi.
4. Lui, il écrit une lettre et ses amis ___ une lettre aussi.
5. Lui, il lit un magazine et ses amis ___ un magazine aussi.

E **Oui ou non?** Répondez en utilisant «nous».

1. Vous dites des bêtises?
2. Vous dites des choses sérieuses?
3. Vous dites des choses intéressantes?
4. Vous dites des choses amusantes?
5. Vous conduisez beaucoup?
6. Vous lisez beaucoup?
7. Vous écrivez souvent à vos amis?

F **Qui dit ça?** Complétez avec «dire».

1. On ___ que les autoroutes sont bonnes en France.
2. Je ___ que les autoroutes françaises sont bonnes mais je ___ aussi qu'il y a trop de circulation.
3. Jean ___ que la plupart des autoroutes sont à péage.
4. Tu ___ qu'il faut payer sur les autoroutes à péage?
5. Paul et Monique, qu'est-ce que vous ___? Vous ___ qu'il faut payer sur les autoroutes américaines aussi? Vous ___ que la plupart des autoroutes aux États-Unis sont à péage?
6. Nos amis américains ___ qu'il y a beaucoup d'autoroutes à huit voies, c'est-à-dire quatre voies dans chaque sens (direction).

Les mots négatifs

Expressing "Nothing," "No One," and "Never"

1. You have already learned the negative expression *ne... pas.* Study the following negative expressions that function the same way as *ne... pas.*

AFFIRMATIF	NÉGATIF
Il dit quelque chose.	Il ne dit rien.
Il écrit quelque chose.	Il n'écrit rien.
Il voit quelqu'un.	Il ne voit personne.
Il parle à quelqu'un.	Il ne parle à personne.
Il voyage toujours.	Il ne voyage jamais.
Il lit souvent.	Il ne lit jamais.
Il écrit quelquefois.	Il n'écrit jamais.

2. As with *ne... pas,* when *ne... jamais* is followed by *un, une, des,* or *de la, de l', du,* and *des,* these words change to *de.*

Il fait souvent une promenade.   Il ne fait jamais de promenade.
Elle fait toujours du sport.   Elle ne fait jamais de sport.

Exercices

A **Non, au contraire.** Répondez par «non».

Il voit quelque chose?
Non, il ne voit rien.

1. Il dit quelque chose?
2. Il écrit quelque chose?
3. Il entend quelque chose?
4. Il lit quelque chose?
5. Il vend quelque chose?
6. Il regarde quelque chose?
7. Il voit quelqu'un?
8. Il regarde quelqu'un?
9. Il parle à quelqu'un?
10. Il écrit à quelqu'un?

B **Elle ne voyage jamais.** Répondez d'après le modèle.

Pascale adore nager.
Tu crois? Elle dit ça, mais elle ne nage jamais.

1. Pascale adore conduire.
2. Pascale adore lire.
3. Pascale adore voyager.
4. Pascale adore faire du sport.
5. Pascale adore jouer au tennis.
6. Pascale adore faire du ski nautique.

Les mots négatifs

PRESENTATION *(page 315)*

A. Present negative expressions by holding up an object (pencil, book, etc.) as you say: *J'ai quelque chose (Je vois quelque chose).* Then put the object away and say: *Je n'ai rien (Je ne vois rien).*

B. Have a student stand by you as you say: *Je vois quelqu'un. J'entends quelqu'un.* Then have the person go away as you say: *Je ne vois personne. Je n'entends personne.*

C. Lead students through steps 1 and 2 on page 315.

Note 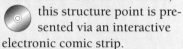 In the CD-ROM version, this structure point is presented via an interactive electronic comic strip.

Note Point out that the words *rien, personne,* and *jamais* can stand alone as short answers. For example: *Qu'est-ce que tu as? Rien! Qui parle? Personne. Tu y vas souvent? Non, jamais.*

Exercices

PRESENTATION *(page 315)*

Exercices A and B

You may wish to use the recorded version of these exercises.

ANSWERS

Exercice A

1. Non, il ne dit rien.
2. Non, il n'écrit rien.
3. Non, il n'entend rien.
4. Non, il ne lit rien.
5. Non, il ne vend rien.
6. Non, il ne regarde rien.
7. Non, il ne voit personne.
8. Non, il ne regarde personne.
9. Non, il ne parle à personne.
10. Non, il n'écrit à personne.

Exercice B

1. Tu crois? Elle dit ça, mais elle ne conduit jamais.
2. ... elle ne lit jamais.
3. ... elle ne voyage jamais.
4. ... elle ne fait jamais de sport.
5. ... elle ne joue jamais au tennis.
6. ... elle ne fait jamais de ski nautique.

ADDITIONAL PRACTICE

After completing Exercises A and B, write the following verbs on the board:

Je vois	J'achète	J'attends
Je veux	J'entends	J'écris
J'ai	Je dis	J'apprends

Have students use either *quelque chose* or *quelqu'un* with each verb. Then have them put the sentences in the negative with *ne... rien* or *ne... personne.*

INDEPENDENT PRACTICE

Assign any of the following:
1. Exercises, pages 313–315
2. Workbook, *Structure: A–E,* pages 119–121
3. Communication Activities Masters, *Structure: A–C,* pages 61–62
4. CD-ROM, Disc 3, pages 313–315

Bell Ringer Review

Write the following on the board or use BRR Blackline Master 12-5: Give the opposite of the following. For example:
jamais — toujours

1. quelqu'un
2. quelquefois
3. rien
4. souvent

Les questions et les mots interrogatifs

You may wish to refer to the **Note on Interrogatives** in Chapter 1, page 13, of this Teacher's Wraparound Edition. Students have now had a great deal of exposure to the various question forms, including both inverted and normal word order. It is suggested that when students actively formulate their own questions, you allow them to use whichever word order they select.

Teaching Tip You can get students started by asking them the three ways they already know to formulate questions in French (rising intonation, *est-ce que,* and inversion). Then ask them what question (interrogative) words they know.

PRESENTATION *(page 316)*

A. Lead students through steps 1–5 on page 316.
B. Have students read aloud all the example questions that appear in the explanation.

---

1. Review the following ways in which questions can be formed in French.

> **Vous parlez français?**
> **Est-ce que vous parlez français?**
> **Parlez-vous français?**

2. Review the following question words you have already learned.

> à quelle heure   comment   où   quand
> combien de   pourquoi   qui

3. Note that you can use these question words in three ways.

   a. In informal, spoken French the question word is often placed at the end of the sentence.

   > **Tu vas où?**
   > **Tu vas au cinéma avec qui?**
   > **Vous allez arriver au cinéma à quelle heure?**

   b. The question word can also be used with *est-ce que*.

   > **Où est-ce que tu vas?**
   > **Avec qui est-ce que tu vas au cinéma?**
   > **Quand est-ce que vous allez arriver au cinéma?**

   c. In more formal conversation and in written French the subject and verb are inverted after a question word.

   > **Où vas-tu?**
   > **Avec qui vas-tu au cinéma?**
   > **Quand allez-vous arriver au cinéma?**

4. With a noun subject, both the noun and *il(s)* or *elle(s)* are used in the inverted question form.

   > **Où** *les copains* **dînent-ils?**
   > **Comment** *Marie* **conduit-elle?**
   > **Combien de roues** *les motos* **ont-elles?**
   > **Pourquoi** *Jean* **vend-il*** la voiture?

   * The final **d** is pronounced as a /t/.

5. In the inverted question form you insert a *t* between *il, elle,* or *on* and any verb that does not end in a *t* or a *d.*

   > **Où Béatrice déjeune-t-elle?**
   > **Comment va-t-on à la rue Racine?**
   > **Pourquoi gare-t-elle la voiture sur le trottoir?**

LEARNING FROM PHOTOS

Have students make up questions about the car. Encourage them to give the same question in various forms. For example: *La voiture est garée sur le trottoir? Est-ce que la voiture est garée sur le trottoir? La voiture est-elle garée sur le trottoir? La voiture est garée où? Où est-ce que la voiture est garée? Où la voiture est-elle garée?*

ADDITIONAL PRACTICE

1. Using question words, students write out dialogues for this situation: Pretend you're the nervous parents of a teenager who's going out with a friend who has recently gotten his/her license. Find out how he/she drives, where they are going, when they are going to return, and so on.
2. Student Tape Manual, Teacher's Edition, *Activité E,* page 137.

Exercices

A **Tu vas où?**   Transposez les questions d'après le modèle.

> **Arlette, où vas-tu?**
> *Tu vas où, Arlette?*

1. Où vas-tu?
2. Comment vas-tu au restaurant?
3. À quelle heure arrives-tu au restaurant?
4. Avec qui dînes-tu?
5. Où es-tu maintenant?

B **Où allez-vous?**   Transposez les questions d'après le modèle.

> **Où est-ce que vous allez?**
> *Où allez-vous?*

1. Où est-ce que vous allez dîner?
2. Comment est-ce que vous allez au restaurant?
3. Est-ce que vous conduisez?
4. Est-ce que vous prenez l'autoroute à péage?
5. Avec qui est-ce que vous allez au restaurant?
6. Est-ce que vous parlez français au serveur?

C **Encore des questions!**   Écrivez des questions d'après le modèle.

> **Marie lit la carte.**
> *Marie lit-elle la carte?*

1. Marie va au restaurant.
2. Marie conduit.
3. Marie gare sa voiture devant le restaurant.
4. Marie regarde la carte.
5. Marie parle au serveur.
6. Marie commande un sandwich au jambon.
7. Le serveur sert le sandwich.
8. Marie mange le sandwich.
9. Le sandwich est bon.
10. Marie paie.
11. Marie laisse un pourboire pour le serveur.

D **À qui Jean parle-t-il?**   Écrivez des questions d'après le modèle.

> **Jean parle à sa copine. (à qui)**
> *À qui Jean parle-t-il?*

1. Jean parle à sa copine au téléphone. (à qui)
2. Il invite sa copine au cinéma. (qui)
3. Ils vont aller au cinéma ce soir. (quand)
4. Jean arrive chez sa copine à sept heures. (à quelle heure)
5. Ils voient le film «Au revoir, les enfants». (quel film)
6. Après le cinéma ils vont au café. (quand)

Exercices

ANSWERS

Exercice A

1. Tu vas où?
2. Tu vas au restaurant comment?
3. Tu arrives au restaurant à quelle heure?
4. Tu dînes avec qui?
5. Tu es où maintenant?

Exercice B

1. Où allez-vous dîner?
2. Comment allez-vous au restaurant?
3. Conduisez-vous?
4. Prenez-vous l'autoroute à péage?
5. Avec qui allez-vous au restaurant?
6. Parlez-vous français au serveur?

Exercice C

1. Marie va-t-elle au restaurant?
2. Marie conduit-elle?
3. Marie gare-t-elle sa voiture devant le restaurant?
4. Marie regarde-t-elle la carte?
5. Marie parle-t-elle au serveur?
6. Marie commande-t-elle un sandwich au jambon?
7. Le serveur sert-il le sandwich?
8. Marie mange-t-elle le sandwich?
9. Le sandwich est-il bon?
10. Marie paie-t-elle?
11. Marie laisse-t-elle un pourboire pour le serveur?

Exercice D

1. À qui Jean parle-t-il au téléphone?
2. Qui invite-t-il au cinéma?
3. Quand vont-ils aller au cinéma?
4. À quelle heure Jean arrive-t-il chez sa copine?
5. Quel film voient-ils?
6. Quand vont-ils au café?

LEARNING FROM REALIA

Have students look at the two toll receipts. Ask them how to say the following in French: "exit," "entrance," "kilometers traveled (covered)." Then ask: Where did the people get on the highway? Where did they get off? How far did they go? How much was the toll?

INDEPENDENT PRACTICE

Assign any of the following:
1. Exercises, page 317
2. Workbook, *Structure: F–H,* pages 121–122
3. Communication Activities Masters, *Structure: D,* page 63
4. Computer Software, *Structure*
5. CD-ROM, Disc 3, pages 316–317

CONVERSATION

CONVERSATION

PRESENTATION *(page 318)*

A. Tell students they will hear a conversation between Francine and Philippe.

B. Have them close their books and watch the Conversation Video or listen as you read the conversation or play Cassette 7B/CD-7.

C. Have pairs of students play the roles of Francine and Philippe. Have the student taking the part of Francine look puzzled.

D. Call on some students to do the conversation orally without looking at their books. They may ad lib. They do not have to know the conversation verbatim.

Note In the CD-ROM version, students can play the role of either one of the characters and record the conversation.

ANSWERS

Exercice A

1. Non, il n'a pas son permis de conduire.

2. Il a seulement quinze ans.

3. Oui, elle croit que Philippe conduit.

4. Oui, il conduit.

5. Elle voit Alain, le frère de Philippe.

6. Elle croit que c'est Philippe.

Prononciation

PRESENTATION *(page 318)*

A. Model the key word *trottoir* and have students repeat chorally.

B. Now model the other words and sentences in similar fashion.

C. For additional practice, use Pronunciation Transparency P-12, the *Prononciation* section on Cassette 7B/CD-7, and the Student Tape Manual, Teacher's Edition, *Activités H–J*, page 139.

318

Scènes de la vie   *Tu as ton permis de conduire?*

FRANCINE: Tu as ton permis de conduire?
PHILIPPE: Non, je n'ai pas mon permis. J'ai seulement quinze ans.
FRANCINE: Mais tu conduis, n'est-ce pas?
PHILIPPE: Tu veux rigoler! Je ne conduis jamais!
FRANCINE: C'est bizarre. Je suis sûre que...
PHILIPPE: Ah... Je comprends! Tu vois mon frère Alain qui conduit et tu crois que c'est moi.

A **Qui conduit?** Répondez d'après la conversation.

1. Philippe a son permis de conduire?
2. Pourquoi pas?
3. Francine croit que Philippe conduit?
4. Le frère de Philippe conduit?
5. Francine voit qui?
6. Elle croit que c'est qui?

Prononciation   *Le son /wa/*

Repeat the following words with the sound /wa/ as in *moi*:

toi	voie	réservoir
croisement	trottoir	pouvoir

Now repeat the following sentences:

Tu ne vois pas le croisement devant toi!
Il va pouvoir partir à trois heures.
Moi, je ne crois pas Antoine!

trottoir

318   CHAPITRE 12

CRITICAL THINKING ACTIVITY

(Thinking skills: drawing conclusions and problem solving)

Read the following to the class or put it on the board or on a transparency:

1. **Faut-il toujours respecter la limitation de vitesse? Si l'on ne respecte pas la limitation de vitesse, quelles peuvent être les conséquences?**

2. **Robert veut aller quelque part. Il veut prendre la voiture. Mais il y a très peu d'essence dans le réservoir. Le réservoir est presque vide. Robert compte son argent. Il n'a pas assez d'argent pour faire le plein. Qu'est-ce qu'il peut faire?**

Activités de communication orale

A **Une enquête.** Divide into small groups and choose a leader. The leader interviews the others to find out how often (*souvent, quelquefois, jamais*) they do the activities listed below. The leader takes notes and reports to the class.

Activité	souvent	quelquefois	jamais
Jouer au tennis			X

jouer au tennis

Élève 1: Tu joues au tennis?
Élève 2: Non, je ne joue jamais au tennis.
Élève 1 (*à la classe*): Patrick ne joue jamais au tennis.

acheter beaucoup de cadeaux   faire des voyages
aller au bord de la mer   faire du jogging
chanter sous la douche   faire les courses
conduire   lire le journal
écouter du jazz   manger des fruits de mer
écrire des lettres   prendre des bains de soleil
faire de la planche à voile   regarder la télé

B **Qu'est-ce que tu dis?** Divide into small groups. Write down several statements that would make your classmates respond with one of the expressions below, then exchange papers with another group member. Take turns reading and responding to the statements.

Absolument pas!   C'est un miracle!   Quelle surprise!
C'est chouette, ça!   Jamais!   Tu veux rigoler!
C'est impossible!   Quelle chance!   Zut!

Élève 1: Suzanne a son permis de conduire!
Élève 2: Je dis: «C'est chouette, ça!» (Je dis: «Quelle surprise!»)

Bonne route.
Soyez prudent !
Le prochain Relais
vous accueillera
à 75 KM (A7)
à 125 KM (A9).

CHAPITRE 12   **319**

LECTURE ET CULTURE

READING STRATEGIES
(page 320)

Reading

A. Read the *Lecture* to the students a paragraph at a time as they follow along in their books.

B. Now call on a student to read. After he/she has read several sentences, ask questions of other students.

Teaching Tip Use other students to answer questions since the student who read aloud was probably concentrating on his/her pronunciation and comprehended little.

C. Allow the class five minutes to reread the *Lecture*.

Post-reading

Have students mention some things they think are different about driving in France.

Note Students may listen to a recorded version of the *Lecture* on the CD-ROM.

Étude de mots

ANSWERS

Exercice A

Answers may include any five: populaires, japonaises, parkings, américaines, direction, routes, nationales, départementales, pittoresques, attention, dangereux, respecter, limitation, surveillent, vigilantes, strictes, circulation, places.

Exercice B

1. a   3. e   5. c
2. d   4. b

ON VA CONDUIRE EN FRANCE?

En France presque[1] tout le monde a une voiture. Les Français conduisent quelles marques de voiture? Il y a deux marques françaises qui sont très populaires, Renault et Peugeot. On voit aussi beaucoup de voitures japonaises sur les autoroutes françaises, mais très peu de voitures américaines.

Les autoroutes en France sont très bonnes. Elles ont trois ou quatre voies dans chaque sens (direction). La plupart des autoroutes sont à péage. Il y a aussi des routes nationales qui sont des routes à grande circulation. Les routes départementales sont plus pittoresques mais il faut faire attention aux croisements, qui peuvent être dangereux.

Si vous conduisez en France, il faut respecter la limitation de vitesse sur les routes et dans les agglomérations[2]. Les motards surveillent la circulation. Si vous roulez trop vite, vous allez avoir une contravention.

Et le stationnement! Il n'y a jamais assez de[3] parkings ou de places pour garer les voitures. Si vous garez votre voiture là où le stationnement est interdit, vous allez trouver une contravention sur le parebrise[4] à votre retour[5]. Les contractuelles sont très vigilantes et très strictes.

[1] presque *almost*
[2] agglomérations *populated areas*
[3] assez de *enough*
[4] parebrise *windshield*
[5] à votre retour *upon your return*

Étude de mots

A **Le français, c'est facile.** Trouvez cinq mots apparentés dans la lecture.

B **Synonymes.** Trouvez les expressions équivalentes.

1. vite
2. la direction
3. surveiller
4. l'agglomération
5. garer

a. rapidement
b. une zone développée
c. stationner
d. le sens
e. contrôler, observer attentivement

Compréhension

C Sur la route en France. Corrigez les phrases.

1. Il y a plus de voitures américaines que de voitures japonaises en France.
2. Il n'y a pas de voitures françaises. L'industrie automobile n'existe pas en France.
3. Beaucoup d'autoroutes en France ne sont pas bonnes.
4. On ne paie jamais sur les autoroutes à péage françaises.
5. La plus grande route c'est la route départementale.
6. Il y a des croisements dangereux sur les autoroutes.
7. Il n'y a pas de limitation de vitesse dans les agglomérations.
8. Les conducteurs aiment avoir des contraventions.

D En route. Répondez.

1. Les grandes autoroutes en France ont combien de voies dans chaque sens?
2. Qu'est-ce qu'il faut payer sur la plupart des autoroutes?
3. Il y a une limitation de vitesse sur les routes en France?
4. Qui surveille les autoroutes?
5. Il y a toujours assez de places pour stationner?
6. Qui a la responsabilité de surveiller le stationnement?
7. Quelles sont deux marques françaises de voiture?

DÉCOUVERTE CULTURELLE

Voici un vélomoteur. Il faut avoir plus de seize ans et un permis spécial pour conduire un vélomoteur.

Le rêve[1] de beaucoup de jeunes, c'est une moto. On peut conduire une moto à partir de seize ans[2] avec un permis spécial moto. Et le casque[3] est obligatoire! Si vous êtes en France, vous pouvez conduire une moto? Pourquoi?

Et pour conduire une voiture il faut avoir dix-huit ans en France. Là où vous habitez, il faut avoir quel âge pour obtenir un permis de conduire?

[1] rêve *dream*
[2] à partir de seize ans *from age 16 on*
[3] casque *helmet*

CHAPITRE 12 **321**

RÉALITÉS

MOUGINS ↑

GRASSE ↑

A 8 ↑

APPEL

PIETONS

322

Bell Ringer Review

Write the following on the board or use BRR Blackline Master 12-8: Write as many words as you can having to do with cars and driving.

OPTIONAL MATERIAL

PRESENTATION *(pages 322–323)*

The main objective of this section is to allow students to enjoy the photographs. However, if you would like to do more with it, you might wish to do the following activities.

Pre-reading

Before reading the captions, have students cover them and look only at the photos. Ask them to describe in as much detail as possible what they see. Use questions as a guide, if necessary.

Reading

Have students read the captions on page 323 silently and be prepared to answer the questions in captions 2 and 5.

Post-reading

Discuss the captions as a class and answer the questions.

Ask students (drivers if possible) what problems they might encounter when driving a car in France for the first time.

Note In the CD-ROM version, students can listen to the recorded captions and discover a hidden video behind one of the photos.

DID YOU KNOW?

Most drivers fear the *motards*, motorcycle police officers. The *motards* mean business when they stop speeding motorists. They ride powerful motorcycles and dress to allow minimum friction while on a speed chase. Some *motards* wear egg-shaped helmets to reduce air drag.

L'agent de police est dans les villes **1**. Les agents de police règlent la circulation.

Voici trois panneaux routiers **2**. Quel panneau indique une autoroute à péage, à ton avis?

Pour traverser la rue, les piétons appuient sur le bouton **3**.

Voici des gendarmes français **4**. Eux, ils sont toujours sur la route, souvent à moto. Ils portent toujours un casque s'ils sont à moto.

Voici quelques signaux importants qu'il faut comprendre pour conduire en France **5**. Quelle est la limitation de vitesse sur cette route?

VOUS N'AVEZ PAS LA PRIORITÉ

45

RAPPEL

323

CROSS-CULTURAL COMPARISON

1. In France, the minimum age requirement for driving various types of vehicles is the same throughout the entire country. (In the United States it varies from state to state.)
   • **le cyclomoteur (une sorte de bicyclette avec un petit moteur): 14 ans**
   • **le vélomoteur: 16 ans**
   • **la moto: 16 ans**
   • **la voiture: 18 ans**
2. You may wish to give students the following information about the French police.

   Le gendarme est toujours sur la route, à bicyclette ou à moto (et dans ce cas il porte un casque).

   L'agent de police règle la circulation dans les villes. Il aide les gens à trouver leur chemin. Et il oblige tout le monde à respecter la loi.

   Un C.R.S. (Compagnie républicaine de sécurité) porte un uniforme de combat. S'il y a une manifestation, le C.R.S. défend l'ordre public.

   La police nationale dépend du ministère de l'Intérieur. La gendarmerie nationale est une force militaire et dépend du ministère de la Défense.

Vocabulary Expansion

Students may be interested to know that the French have an equivalent for the slang word "cop." *C'est un flic. Voilà le flic qui arrive.*

ADDITIONAL PRACTICE

1. Student Tape Manual, Teacher's Edition, *Deuxième Partie*, pages 140–141
2. Situation Cards, Chapter 12
3. Communication Transparency C-12

INDEPENDENT PRACTICE

1. *Étude de mots* and *Compréhension* exercises, pages 320–321
2. Workbook, *Un Peu Plus*, pages 123–124
3. CD-ROM, Disc 3, pages 320–323

RECYCLING

The *Activités de communication orale* and *Activités de communica-tion écrite* provide a forum for students to create and answer their own questions and come up with their own dialogues while working within the automobile context of Chapter 12.

INFORMAL ASSESSMENT

Oral Activities A and B may serve as a means to evaluate students' speaking skills. Activity A is effective as a one-on-one check of spontaneous question-creating abilities. You can designate one part of the illustration at a time to focus attention on the vocabulary and structure necessary to ask a question about that part.

Activity B can be used as a guide for interviews conducted either before the class, or alone with the teacher, or recorded on cassette for an oral grade. Use the evaluation criteria given on page 34 of this Teacher's Wraparound Edition.

Activités de communication orale

ANSWERS

Activités A and B

Answers will vary.

Activités de communication orale

A À la station-service.

1. Make up as many questions about this illustration as you can.
2. Work with a classmate and have him or her answer your questions.

B **La route.**   The new French exchange student (your partner) asks you a lot of questions about driving in the U.S. Answer his or her questions and then reverse roles. You can use the list below for suggestions.

> beaucoup de circulation
> Élève 1: Quand est-ce qu'il y a beaucoup de circulation?
> Élève 2: Il y a beaucoup de circulation le matin de huit heures à neuf heures et le soir de cinq heures à six heures.
>
> des autoroutes à péage
> la limitation de vitesse
> assez de parkings dans la ville
> des vélomoteurs
> beaucoup de stations-service
> de l'essence sans plomb
> des motards
> des contraventions

324   CHAPITRE 12

You may wish to ask the following questions about the photograph: *Quel est le nom du garage? Il y a un mécanicien au garage? Qu'est-ce qu'on vend au garage?*

Ask students to look at the sign and figure out how to say "used."

Have the students draw a bicycle. Even a simple outline will do. Then give them the following words to describe their bicycle: *la selle* (seat); *la pédale; la chaîne; le guidon* (handlebars); *le pneu* (tire); *le frein à main* (hand brake); *les rayons* (spokes); *le garde-boue* (fender).

Activités de communication écrite

A **Mon permis de conduire.** Write a letter to your French friend. Tell him or her how old you are and whether you have your driver's license yet. Tell your friend what you have to do to get a license and find out what people in France have to do to get one.

B **Zut!** A *contractuelle* is writing a parking ticket when the owner of the car comes running up. Write down what they say to each other.

Réintroduction et recombinaison

A **Personnellement.** Donnez des réponses personnelles.

1. Comment t'appelles-tu?
2. Tu es d'où?
3. Tu es de quelle nationalité?
4. Tu vas à quelle école?
5. Qui est ton professeur de français?
6. Qu'est-ce que tu fais au cours de français?
7. Tu aimes être en forme?
8. Qu'est-ce que tu fais pour rester en forme?

B **Jamais!** Répondez en utilisant «ne… jamais».

1. Tes parents se lèvent à midi en semaine?
2. Les élèves se couchent à six heures du soir?
3. Le professeur s'endort en classe?
4. Les garçons se rasent en classe?

Vocabulaire

NOMS
la voiture
la voiture de sport
le break
la décapotable
la marque
les deux roues (f.)
la moto
le vélomoteur
la roue de secours
le pneu (à plat)
le réservoir
la clé
la ceinture de sécurité
le conducteur
la conductrice
l'auto-école (f.)
la leçon de conduite

le permis de conduire
le guide
la route
l'autoroute (f.) à péage
la voie
la limitation de vitesse
le motard
la circulation
le croisement
le carrefour
le trottoir
le piéton
la piétonne
les clous (m.)
le feu
le stationnement
la place
la contractuelle
la contravention

la station-service
le/la pompiste
l'essence (f.)
  super
  ordinaire
  sans plomb
les niveaux (m.)

VERBES
rouler
accélérer
freiner
s'arrêter
traverser
surveiller
conduire
dire
écrire
lire

AUTRES MOTS ET EXPRESSIONS
garer la voiture
faire le plein
vérifier les niveaux
mettre le contact

quelqu'un
ne… jamais
ne… personne
ne… rien

fâché(e)
il est interdit
prudemment
sans
trop
vite
Zut!

CHAPITRE 12   **325**

Activités de communication écrite

ANSWERS

Activités A and B
Answers will vary.

OPTIONAL MATERIAL

Réintroduction et recombinaison

ANSWERS

Exercice A
Answers will vary.

Exercice B
1. **Mes parents ne se lèvent jamais à midi en semaine.**
2. **Les élèves ne se couchent jamais à six heures du soir.**
3. **Le professeur ne s'endort jamais en classe.**
4. **Les garçons ne se rasent jamais en classe.**

ASSESSMENT RESOURCES

1. Chapter Quizzes
2. Testing Program
3. Situation Cards
4. Communication Transparency C-12
5. Performance Assessment
6. Computer Software: Practice/Test Generator

VIDEO PROGRAM

INTRODUCTION   (38:40)

EN ROUTE!   (39:17)

STUDENT PORTFOLIO

Written assignments which may be included in students' portfolios are the *Activités de communication écrite* on page 325 and the *Mon Autobiographie* section of the Workbook on page 125.

Note Students may create and save both oral and written work using the Electronic Portfolio feature on the CD-ROM.

INDEPENDENT PRACTICE

1. Activities and exercises, pages 324–325
2. CD-ROM, Disc 3, pages 324–325
3. Communication Activities Masters, pages 59–63

LE MONDE FRANCOPHONE

La Mode

PRESENTATION *(pages 326–329)*

This cultural material is presented for students to enjoy and to help them gain an appreciation of the francophone world. Since the material is *optional*, you may wish to have students read it on their own as they look at the colorful photographs that accompany it. They can read it at home, or you may wish to give them a few minutes in class to read it.

If you prefer to present some of the information in greater depth, you may follow the suggestions given for other reading selections throughout the book. Students can read aloud, answer questions asked by the teacher, ask questions of one another in small groups and, finally, give a synopsis of the information in their own words in French.

MORE ABOUT THE PHOTOS

Photo 2 Jewelry is important to both men and women in Africa. There is a great variety of jewelry in Africa. In the Sahara and North Africa, the preferred metal for jewelry is silver. Elsewhere in Africa, the preferred metal for jewelry is gold.

Beads and amulets are popular jewelry forms. In areas where there are animistic religious beliefs, these beads and amulets have a religious significance and they can play a role in community rituals. The charms worn around the neck are called *grisgris*.

LA MODE

Aujourd'hui, dans la plupart des grandes villes du monde, on s'habille plus ou moins de la même façon, mais les vêtements traditionnels existent toujours parce qu'ils sont parfaitement adaptés au climat et à la géographie. On s'habille d'une façon en Suisse où il fait froid et d'une autre façon au Sénégal où il fait toujours chaud.

1, 2 La petite fille qui habite à la montagne en Suisse ne porte pas les mêmes vêtements que cette belle Sénégalaise qui habite dans un pays tropical.

326   LE MONDE FRANCOPHONE

3 Montréal est après Paris la deuxième ville francophone du monde. Il y fait très froid *(cold)* pendant de longs mois et, pour cette raison, on a construit une ville souterraine de plusieurs étages. Ce sont des centres commerciaux souterrains—ici celui de la Place du Canada. Tous ces centres sont reliés *(connected)* entre eux par le métro. Il y a en tout plus de 1.700 magasins et 200 restaurants. Alors quand il fait très froid, on n'a plus besoin de sortir dehors pour faire des courses!

4 En Afrique du Nord, on fait ses courses dans un souk, une sorte de marché couvert. Le souk est dans la médina, la partie ancienne d'une ville arabe. Ce souk est à Marrakech au Maroc. Le vendeur est habillé en pull et jean, mais le client porte une djellaba et sa femme porte un caftan. Ce couple regarde des babouches. Les babouches sont des chaussons *(slippers)* qu'on porte souvent dans les pays du Maghreb.

LE MONDE FRANCOPHONE

Photo 4 The *djellaba* is an extremely practical garment. It is very loose and does not hinder movement. It protects the garments worn underneath from rain as well as from the dust that blows in from the Sahara. *Djellabas* are made from all types of fabrics, but particularly gabardine. There are other fancy fabrics woven in the small villages. The *djellaba* has no religious significance.

The veil worn by women in many of the Islamic countries is rarely worn by young urban women of the Maghreb. With the spread of Islamic fundamentalism, however, the veil is becoming more common, particularly in Algeria.

In Tunisia, women wear the *sifsari* everywhere. A *sifsari* resembles a "cloak-shawl." Its loose folds wrap around the woman's head and shoulders. The *sifsari* is extremely practical as it can be pulled across the face to protect the wearer from wind or the sand of the desert. The *sifsari* has no religious significance.

Shoes are taken off when entering a house in the Maghreb. People will often put on a pair of *babouches* to wear in the house or they will go barefoot. Both men and women wear *babouches*. Men's *babouches* are usually white or a light color and they are very simple. There is little or no ornamentation. Women's *babouches* are decorated with gold or silver embroidery.

Photo 5 The beautiful designs on the fabrics of the clothing in this market are distinctively African. Many of the designs are indigo wood-block prints and batiks.

The technique for making a batik fabric is very interesting. It is done by hand, and wax is used as a dye repellent to cover parts of a design while the uncovered fabric is being dyed with one or more colors. After the dyeing process, the wax is dissolved in boiling water.

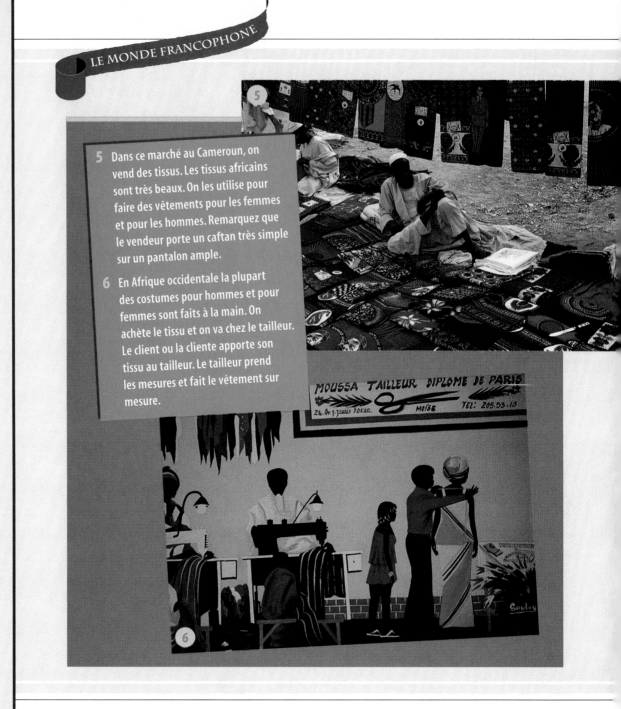

5 Dans ce marché au Cameroun, on vend des tissus. Les tissus africains sont très beaux. On les utilise pour faire des vêtements pour les femmes et pour les hommes. Remarquez que le vendeur porte un caftan très simple sur un pantalon ample.

6 En Afrique occidentale la plupart des costumes pour hommes et pour femmes sont faits à la main. On achète le tissu et on va chez le tailleur. Le client ou la cliente apporte son tissu au tailleur. Le tailleur prend les mesures et fait le vêtement sur mesure.

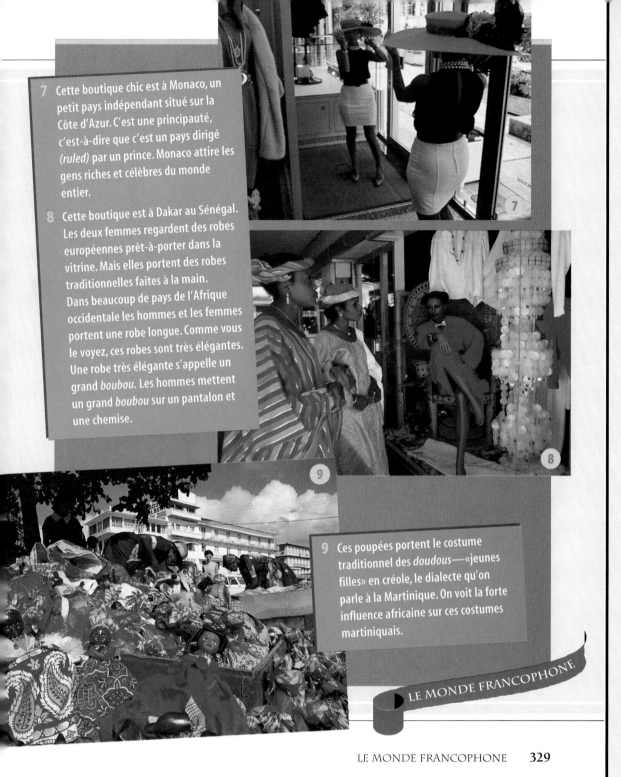

7 Cette boutique chic est à Monaco, un petit pays indépendant situé sur la Côte d'Azur. C'est une principauté, c'est-à-dire que c'est un pays dirigé (ruled) par un prince. Monaco attire les gens riches et célèbres du monde entier.

8 Cette boutique est à Dakar au Sénégal. Les deux femmes regardent des robes européennes prêt-à-porter dans la vitrine. Mais elles portent des robes traditionnelles faites à la main. Dans beaucoup de pays de l'Afrique occidentale les hommes et les femmes portent une robe longue. Comme vous le voyez, ces robes sont très élégantes. Une robe très élégante s'appelle un grand *boubou*. Les hommes mettent un grand *boubou* sur un pantalon et une chemise.

9 Ces poupées portent le costume traditionnel des *doudous*—«jeunes filles» en créole, le dialecte qu'on parle à la Martinique. On voit la forte influence africaine sur ces costumes martiniquais.

LE MONDE FRANCOPHONE

LE MONDE FRANCOPHONE   **329**

Photo 7 The principality of Monaco has been under the rule of the Grimaldi family since 1297. Monaco is a very short distance from the Riviera city of Nice. The citizens of Monaco are called *Monégasques*. Monaco has a population of some 30,000 but only slightly more than 5,000 are *Monégasques*. Citizenship in Monaco is very difficult to obtain. It is not even granted automatically to a person born in Monaco.

The belief that citizens of Monaco pay no personal income taxes because all government expenses are financed by casino revenues is not completely true, but it is true that they are far less burdened with taxes than are outsiders. Non-citizens residing in Monaco pay high taxes.

The official language of the principality is French and the currency is the French franc.

Photo 8 In general, Africans place much importance on dress. They spend a large portion of their after-food budget on clothing.

The *boubou* is worn by both men and women. A woman's elegant *grand boubou* is a long, regal floor-length embroidered dress, while a man's *grand boubou* is an embroidered robe-like garment that reaches the ground.

For everyday dress, women wear a loose top and wrap a length of cloth, called *un pagne*, around themselves as a skirt.

RÉVISION

CHAPITRES 9–12

OPTIONAL MATERIAL

OVERVIEW

This section reviews key grammatical structures and vocabulary from Chapters 9–12. The structure topics were first presented on the following pages: reflexive verbs, page 290; verbs like *prendre*, page 243; *croire* and *voir*, page 266; *lire, dire, écrire,* and *conduire*, page 313; stress pronouns, page 245; the comparative and the superlative, pages 269 and 271.

REVIEW RESOURCES

1. Workbook, Self-Test 3, pages 126–129
2. Videocassette/Videodisc, Unit 3
3. Video Activities Booklet, Unit 3: Chapters 9–12, pages 34–49
4. Computer Software, Chapters 9–12
5. Testing Program, Unit Test: Chapters 9–12, pages 73–78
6. Performance Assessment
7. CD-ROM, Disc 3, *Révision:* Chapters 9–12, pages 330–333
8. CD-ROM, Disc 3, Self-Tests 9–12
9. CD-ROM, Disc 3, Game: *Pour en savoir plus*
10. Lesson Plans

Conversation

PRESENTATION *(page 330)*

Have students repeat each sentence after you. Then call on individuals to dramatize the conversation for the class.

ANSWERS

Exercice A

1. Il voit Stéphanie en robe.
2. La robe est rose.

330

Conversation   *Stéphanie en robe!*

ANTOINE: Qu'est-ce que je vois! Stéphanie en robe!
STÉPHANIE: Euh… tu crois que la robe bleue est plus jolie que la robe rose?
ANTOINE: Mais non. Tu es très jolie en rose.
STÉPHANIE: Je préfère vraiment les pantalons!
ANTOINE: Tu sors avec qui?
STÉPHANIE: Avec Jérôme. On va au restaurant.
ANTOINE: Ah oui, avec lui, c'est toujours les restaurants chic.
STÉPHANIE: Oui, mais on s'amuse bien ensemble.
ANTOINE: Tu pars à quelle heure?
STÉPHANIE: Dans cinq minutes. Je me peigne, je me maquille et je pars.

A   **Stéphanie et son frère.**   Répondez.

1. Qu'est-ce qu'Antoine voit?
2. De quelle couleur est la robe de Stéphanie?
3. D'après Antoine, la robe bleue est plus jolie que la robe rose?
4. Qu'est-ce que Stéphanie préfère, les robes ou les pantalons?
5. À ton avis, est-ce qu'Antoine aime Jérôme?
6. Est-ce que Stéphanie aime sortir avec Jérôme? Pourquoi?
7. Est-ce que Stéphanie va partir dans quelques minutes?
8. Qu'est-ce qu'elle va faire avant de partir?

Structure

Les verbes réfléchis

Review the present tense forms of reflexive verbs.

1. Remember that in reflexive constructions, the subject and the reflexive pronoun refer to the same person.

SE LEVER	
je me **lève**	*nous nous* **levons**
tu te **lèves**	*vous vous* **levez**
il/elle/on se **lève**	*ils/elles se* **lèvent**

LEARNING FROM PHOTOS

Give students the words for the cosmetics pictured: *du rouge à lèvres, un poudrier, de la poudre, de l'ombre à paupières, un flacon de parfum.* Now ask them questions using these words: *Tu mets du rouge à lèvres tous les jours? Quelles couleurs préfères-tu? Tu as un poudrier dans ton sac? Qu'est-ce qu'il y a à à l'intérieur d'un poudrier? Tu mets souvent de l'ombre à paupières? Tu aimes le parfum? Quels parfums?*

2. Review the placement of *ne… pas, ne… plus, ne… jamais.*

> Vous *ne* vous levez *pas*?
> Il *ne* s'endort *jamais* tout de suite.

A **On sort.** Complétez.

Ma sœur et moi, nous ___(1)___ (s'amuser) bien quand nous sortons. Mais elle ___(2)___ (se préparer) pendant des heures, et moi, je ___(3)___ (se laver) et je ___(4)___ (s'habiller) en deux minutes. D'abord, elle, elle ___(5)___ (se brosser) les dents pendant cinq minutes! Puis elle ___(6)___ (s'habiller), mais elle ___(7)___ (se changer) trois fois (*times*) avant de se décider. Puis, elle ___(8)___ (se maquiller) pendant une demi-heure. Enfin, elle ___(9)___ (se peigner). Pendant ce temps, moi, je lis un livre. Quelquefois, je ___(10)___ (s'endormir)!

Les verbes *prendre, croire, voir, lire, dire, écrire* et *conduire*

Review the following forms of some irregular verbs you have learned.

PRENDRE	je prends, tu prends, il/elle/on prend nous prenons, vous prenez, ils/elles prennent
COMPRENDRE	je comprends, tu comprends, il/elle/on comprend nous comprenons, vous comprenez, ils/elles comprennent
CROIRE	je crois, tu crois, il/elle/on croit nous croyons, vous croyez, ils/elles croient
VOIR	je vois, tu vois, il/elle/on voit nous voyons, vous voyez, ils/elles voient
LIRE	je lis, tu lis, il/elle/on lit nous lisons, vous lisez, ils/elles lisent
DIRE	je dis, tu dis, il/elle/on dit nous disons, vous dites, ils/elles disent
ÉCRIRE	j'écris, tu écris, il/elle/on écrit nous écrivons, vous écrivez, ils/elles écrivent
CONDUIRE	je conduis, tu conduis, il/elle/on conduit nous conduisons, vous conduisez, ils/elles conduisent

---

3. Non, il préfère la robe rose.
4. Elle préfère les pantalons.
5. Answers will vary.
6. Elle aime sortir avec lui parce qu'ils s'amusent bien ensemble.
7. Oui, elle va partir dans cinq minutes.
8. Elle va se peigner et se maquiller.

Structure
Les verbes réfléchis
PRESENTATION *(pages 330–331)*

A. Write the forms of *se lever* on the board.
B. Point out the pronunciation change in the stem of *se lever* and the written accent in the forms where it appears.
C. Have students give you other reflexive verbs they know. Select one and write its forms on the board.
D. Have students repeat the forms of *se lever* and the other verb you wrote on the board.

Exercice
PRESENTATION *(page 331)*

Exercice A

Call on several individuals to complete the exercise with books open. Then call on one individual to read the entire exercise.

ANSWERS

Exercice A

1. nous amusons
2. se prépare
3. me lave
4. m'habille
5. se brosse
6. s'habille
7. se change
8. se maquille
9. se peigne
10. m'endors

INFORMAL ASSESSMENT

Have students tell you anything they can about their daily routine.

Les verbes **prendre, croire, voir, lire, dire, écrire** *et* **conduire**
PRESENTATION *(page 331)*

Have students read the verb forms aloud in chorus with you.

COOPERATIVE LEARNING

Have students work in teams of three. One student asks another about his/her routine. The student responds, and the third reports what was said. For example:
É1: Tu te lèves à quelle heure?
É2: Moi, je me lève à sept heures.
É3: ___ se lève à sept heures.

PAIRED ACTIVITY

Students work in pairs and list as many daily activities as they can, classifying them as to the time they occur: *le matin/l'après-midi/le soir.* Then they ask each other how often they do these things at those times.
É1: Tu te maquilles toujours le matin?
É2: Non, je ne me maquille pas toujours le matin. Souvent, quand je vais sortir, je me maquille le soir.

ANSWERS

Exercice B

1. Elles écrivent beaucoup?
2. Elles, elles lisent beaucoup.
3. Tu écris souvent à tes parents?
4. Je vois mes parents toutes les semaines.
5. Tes frères conduisent bien?
6. Non, ils apprennent à conduire.
7. Vous dites déjà au revoir?
8. Oui, nous prenons l'avion dans une heure.
9. Elles conduisent beaucoup?
10. Non, elles voient mal.
11. Ils lisent le journal tous les matins?
12. Nous croyons qu'ils lisent le journal.

Les pronoms accentués

PRESENTATION (page 332)

A. Have students repeat the pronouns while pointing to an appropriate person.
B. When explaining step 2, have students give you additional examples for each use.

Exercice

PRESENTATION (page 332)

Exercice C

Exercise C can be done with books open or closed.

ANSWERS

Exercice C

1. Eux aussi, ils font de la plongée sous-marine.
2. Moi aussi, je fais de la planche à voile.
3. Nous aussi, nous bronzons facilement.
4. Elles aussi, elles plongent bien.
5. Vous aussi, vous sortez ce soir.
6. Toi aussi, tu vas au restaurant.
7. Elle aussi, elle aime les fruits de mer.
8. Nous (Moi) aussi, nous aimons (j'aime) le soleil.

B **Qu'est-ce qu'on fait?** Remplacez les mots en italique et faites les changements nécessaires.

1. *Vous* écrivez beaucoup? (elles)
2. *Moi, je* lis beaucoup. (elles)
3. *Ils* écrivent souvent à leurs parents? (tu)
4. *Ils* voient leurs parents toutes les semaines. (je)
5. *Amélie* conduit bien? (tes frères)
6. Non, *elle* apprend à conduire. (ils)
7. *Tu* dis déjà «au revoir»? (vous)
8. Oui, *je* prends l'avion dans une heure. (nous)
9. *Tu* conduis beaucoup? (elles)
10. Non, *je* vois mal. (elles)
11. *Robert* lit le journal tous les matins? (ils)
12. *Je* crois qu'*il* lit le journal. (nous, ils)

Les pronoms accentués

1. Review the stress pronouns and the corresponding subject pronouns.

STRESS PRONOUNS	SUBJECT PRONOUNS
moi	je
toi	tu
lui	il
elle	elle
nous	nous
vous	vous
eux	ils
elles	elles

2. Remember that you use stress pronouns:

  a. to emphasize the subject   Moi, j'ai faim!
  b. after a preposition   C'est pour moi?
  c. when there is no verb in the sentence   Qui? Moi?
  d. after *c'est* or *ce sont*   C'est lui qui n'écrit jamais.
  e. after *que* in comparisons   Anne est plus grande que toi.

C **En vacances.** Répondez d'après le modèle.

  Sa mère joue au tennis. Et son père?
  Lui aussi, il joue au tennis.

1. Son frère fait de la plongée sous-marine. Et ses cousins?
2. Je fais de la planche à voile. Et toi?
3. Nous bronzons facilement. Et vous deux?
4. Il plonge bien. Et ses sœurs?
5. Vous sortez ce soir. Et nous?
6. Tu vas au restaurant. Et moi?
7. Ils aiment les fruits de mer. Et elle?
8. J'aime le soleil. Et vous?

LEARNING FROM PHOTOS

Have students say as much about the photo as they can. You may wish to give them the French for "wheelchair:" *un fauteuil roulant*.

COOPERATIVE LEARNING

Have students work in groups of four. Group members take turns talking about things that they and others in the group do. Each member makes up several sentences using *moi, vous, eux, elles, toi, lui, elle*. As students make the statements, they should point to the team member(s) they're referring to.

Le comparatif et le superlatif

1. You use the comparative to compare two people or two items.

> **Nathalie est plus (moins, aussi) sportive que son frère.**

2. You use the superlative to single out one person or one item from the group and compare it to all the others.

> **Nathalie est la plus (la moins) sportive de la famille.**
> **Serge est le plus (le moins) sportif de la famille.**
> **Ils sont les plus (les moins) sportifs de la famille.**

3. Remember that the adjective *bon* has an irregular form in the comparative and the superlative: *meilleur(e)*.

> **Mon idée est meilleure que ton idée.**
> **Jean-Claude est le meilleur de la classe.**

D **Bernard et moi.** Répondez d'après le modèle.

> Élève 1: **Bernard est très sérieux.**
> Élève 2: **Il est plus sérieux que moi?**
> Élève 1: **Non, mais il est aussi sérieux que toi.**

1. Bernard est très timide.
2. Bernard est très sportif.
3. Bernard est très généreux.
4. Bernard est très actif.
5. Bernard est très nerveux.
6. Bernard est très grand.
7. Bernard est très patient.
8. Bernard est très intelligent.

E **Nathalie et moi.** Changez *Bernard* en *Nathalie* dans l'Exercice D.

F **Les élèves de Mme Leblond.** Répondez d'après le modèle.

> **Véronique est très amusante.**
> **Véronique est la plus amusante de la classe.**

1. Alain est très timide.
2. Catherine et Émilie sont très intelligentes.
3. Louise est très jolie.
4. Les frères Gautier sont très désagréables.
5. Les sœurs Duhamel sont très gentilles.
6. Olivier est très aimable.
7. Valérie est très réservée.
8. Martine est très bonne.

Activité de communication orale

A **Au Club Med.** Imagine that you're a group leader (*un gentil organisateur* or *un G.O.*) at Club Med. Tell about your daily routine: what time you get up, what you wear, what sports you play, what you eat, and what you do at night.

Le comparatif et le superlatif

PRESENTATION *(page 333)*

Go over steps 1–3. Have students give additional examples.

ANSWERS

Exercice D

1. É2: Il est plus timide que moi?
   É1: Non, mais il est aussi timide que toi.
2. … plus sportif que moi?
   … aussi sportif que toi.
3. … plus généreux que moi?
   … aussi généreux que toi.
4. … plus actif que moi?
   … aussi actif que toi.
5. … plus nerveux que moi?
   … aussi nerveux que toi.
6. … plus grand que moi?
   … aussi grand que toi.
7. …plus patient que moi?
   … aussi patient que toi.
8. … plus intelligent que moi?
   … aussi intelligent que toi.

Exercice E

1. É2: Elle est plus timide que moi?
   É1: Non, mais elle est aussi timide que toi.
2. … plus sportive… aussi…
3. … plus généreuse… aussi …
4. … plus active… aussi…
5. … plus nerveuse… aussi…
6. … plus grande… aussi…
7. … plus patiente… aussi…
8. … plus intelligente… aussi…

Exercice F

1. … le plus timide…
2. … les plus intelligentes…
3. … la plus jolie…
4. … les plus désagréables…
5. … les plus gentilles…
6. … le plus aimable…
7. … la plus réservée…
8. … la meilleure…

Activité de communication orale

PRESENTATION *(page 333)*

Activité A

This activity can be done either orally or in writing.

ANSWERS

Activité A

Answers will vary.

ÉCOLOGIE: LA POLLUTION DE L'EAU

Écologie: La Pollution de l'eau

OVERVIEW

The readings in this *Lettres et sciences* section are related topically to material in Chapters 9 and 10. See page 228 in this Teacher's Wraparound Edition for suggestions on presenting the readings.

Avant la lecture

PRESENTATION *(page 334)*

A. Have students do the *Avant la lecture* activities. Activity 2 asks them to do some scanning.

B. You may wish to give the students the cognates that appear in this reading, or you may prefer to have them scan the passage and look for cognates.

C. Call on a student who has studied biology to describe the water cycle, or draw a diagram of the water cycle on the board and label the following in French: *l'océan, l'eau, les nuages, la pluie, le sol, un arbre, une feuille.* (Water from the oceans evaporates and forms clouds. The clouds produce rain that soaks the ground. Ground water is taken up by the roots of plants. The plants then lose water to the atmosphere through transpiration. This moisture also forms clouds.)

D. Have the students look at the photos to get an idea of what they will be reading about.

Lecture

PRESENTATION *(pages 334–335)*

A. Give students a couple of minutes to read each paragraph. Then ask what the paragraph was about. They can answer in English.

Avant la lecture

1. Is water scarce or abundant where you live? Think about the role that water plays in your town. Are there any regulations concerning the watering of lawns, the washing of cars, the amount of certain substances that can be present in the town water?

2. Here are four titles. Scan the text and see if you can match these titles with the four paragraphs in the text.

> La répartition de l'eau
> Sauvons l'eau!
> La circulation de l'eau
> Les différents genres de pollution

Lecture

L'eau, tu es
la plus grande richesse
qui soit° au monde,   *exists*
et tu es la plus délicate,
toi, si pure
au ventre° de la terre.   *in the depths of*

Antoine de Saint-Exupéry

Nous «sommes» de l'eau. Notre corps est composé de 65% d'eau. On trouve l'eau partout: 96% dans les mers et les océans, 3% dans les glaciers et 1% qui prend part au «cycle de l'eau».

L'eau des lacs et des mers s'évapore. Ensuite elle retombe en pluie et s'infiltre dans le sol. Du sol, elle est absorbée par les arbres où elle arrive dans les feuilles et s'évapore encore, etc.

On ne peut pas vivre (exister) sans eau, mais malheureusement, l'eau est mal distribuée: par exemple, en Afrique certaines régions n'ont pas assez d'eau[1],

Le Gange déborde et cause des inondations.

mais en Inde, quand le Gange, le grand fleuve, déborde, il y a trop d'eau[2]. Aux États-Unis, nous avons quelquefois des périodes de sécheresse quand il n'y a pas de pluie ou, au contraire, des inondations, quand il y a trop de pluie. Mais en général, nous n'avons ni trop, ni trop peu[3] d'eau. Notre problème, c'est la pollution.

La pollution peut prendre plusieurs formes.

1. Les pluies acides

Quand les nuages passent au-dessus des zones industrielles, ils absorbent tous les gaz qui s'échappent (sortent) des cheminées et des voitures. Les nuages transportent ces gaz et les pluies qui tombent un peu plus loin sont des «pluies acides». Ces pluies acides causent la destruction des forêts et contaminent les lacs.

INTERDISCIPLINARY CONNECTIONS

Students form a research partnership with the biology class to find out what people can do to conserve water, prevent pollution, and protect the environment. Then they design posters illustrating these issues and they label them in French.

SCIENCES

Quelques formes de pollution

2. Les engrais[4]

Les agriculteurs utilisent beaucoup d'engrais, en général des phosphates, pour maintenir la fertilité du sol. Ces engrais chimiques sont entraînés[5] par les pluies jusque dans les lacs et les rivières. Ils polluent les rivières et les lacs parce qu'ils font pousser les plantes aquatiques[6]. Ces plantes prennent tout l'oxygène de l'eau. Sans oxygène, les poissons ne peuvent pas vivre et disparaissent. Les engrais polluent aussi les mers. Ils sont entraînés dans les mers par les rivières où ils nourrissent les algues. Ces algues se transforment en véritables «marées[7] rouges» et tuent[8] les poissons.

3. La marée noire

La marée noire est causée par le mazout[9] qui est jeté dans la mer par des pétroliers.

4. Les déchets[10] radioactifs

Il y a à notre époque plus de 100.000 tonnes de déchets radioactifs au fond de l'océan Atlantique et de l'océan Pacifique! Il faut sauver l'eau. Il faut apprendre à conserver les réserves. Et surtout il faut apprendre à ne pas polluer, à ne pas verser les déchets toxiques dans l'eau. C'est le but[11] de beaucoup d'écologistes qui veulent protéger et sauver notre environnement.

[1] assez d'eau *enough water*
[2] trop d'eau *too much water*
[3] nous n'avons ni trop, ni trop peu *we have neither too much nor too little*
[4] engrais *fertilizers*
[5] entraînés *carried*
[6] ils font pousser les plantes aquatiques *they make aquatic vegetation grow*
[7] marées *tides*
[8] tuent *kill*
[9] mazout *fuel oil*
[10] déchets *waste*
[11] but *the goal*

Après la lecture

A **La pollution.**   Vrai ou faux?

1. 65% de l'eau prend part au «cycle de l'eau».
2. On ne peut pas vivre sans eau.
3. Les marées rouges font disparaître les poissons.
4. La marée noire est causée par des algues.
5. Il y a des déchets radioactifs dans l'océan Pacifique.
6. Il faut apprendre à conserver les réserves d'eau.

B **Il faut sauver l'eau.**   Répondez.

1. Quels sont les risques de pollution de l'eau là où vous habitez?
2. Quelles sont les mesures adoptées par votre ville pour ne pas polluer l'eau ou pour la conserver? S'il n'y a pas de mesures adoptées, faites des recommandations.

B. Have students form four groups. Each group reads and discusses a separate topic within the reading and then presents its findings to the rest of the class. The reading can be divided into the following sections: (1) introduction, (2) acid rain, (3) fertilizers, (4) oil spills and radioactive waste.

Après la lecture

PRESENTATION *(page 335)*

A. Go over the *Après la lecture* exercises.

B You may wish to have the students discuss the following question: *Comment utilisez-vous de l'eau?* Prompt them with questions if they cannot come up with the information on their own: *Vous prenez souvent un verre d'eau? Vous vous lavez? Vous vous brossez les dents? Vous vous lavez les cheveux? Vous lavez vos vêtements? Vous lavez la voiture? Vous lavez votre chien? Vous nagez dans une piscine? Vous arrosez vos plantes?*

Exercices

PRESENTATION *(page 335)*

Students can prepare the exercises on their own. You may then go over them in class after completing the reading.

ANSWERS

Exercice A

1. faux
2. vrai
3. vrai
4. faux
5. vrai
6. vrai

Exercice B

Answers will vary.

DID YOU KNOW?

As early as 1971, the French government created the *Ministère de la Protection de la Nature et de l'Environnement*. Laws regulating the disposal of waste products and the protection of water, wild birds, and fish were passed in the 70's. In the 80's additional new laws attempted to control pollution of the oceans and to protect fishing, mountains, and the ozone layer. Even stricter laws—against noise, radioactive waste, certain vehicles, and against water, air, and land pollution in general— followed in the 90's. New agencies, such as the INERIS (Insititut National de l'Environnement), were created to oversee policies and to inform and sensitize the public to environmental issues.

LETTRES ET SCIENCES

Littérature: Apollinaire (1880–1918)

PRESENTATION *(page 336)*

A. Tell students they are going to read about a famous poet who had a very interesting life. His name is Apollinaire.

B. Go over the many cognates that appear in the biographical introduction.

C. Have students read the biography silently.

D. Tell students to look for the following information in the biography: *les villes qu'Apollinaire visite, l'époque où il écrit ses poèmes, des amis d'Apollinaire.*

«La Cravate»
Avant la lecture

PRESENTATION *(page 336)*

A. Have students discuss the *Avant la lecture* activities.

B. Have them look at the poem and figure out the order in which the words should be read.

Lecture

PRESENTATION *(page 336)*

A. Read the poem to the class.

B. Have students read it silently.

C. Have students tell you the message they get from the poem.

D. Ask students if they agree with the sentiments of the poet.

Après la lecture

ANSWERS

Exercices A and B
   Answers will vary.

LITTÉRATURE: APOLLINAIRE (1880–1918)

Guillaume Apollinaire a une vie très fantaisiste et mouvementée. Sa poésie reflète sa vie. Il voyage dans toute l'Europe—à Munich, Berlin, Prague, Vienne. Il s'intéresse à tous les mouvements intellectuels et artistiques de son époque. C'est la période avant la guerre de 1914, une période très riche en idées en tous genres. C'est le début du cubisme, par exemple. Les poètes et les artistes peintres *(painters)* discutent ensemble ces nouvelles idées. Apollinaire est l'ami des peintres Picasso, Vlaminck et Marie Laurencin. Apollinaire est un des premiers grands poètes français modernes. Il annonce les grands mouvements artistiques des années 20 *(1920's)*.

Certains des poèmes d'Apollinaire sont des «caligrammes»: le poème est écrit en forme d'objet. *La cravate* est un exemple de ce genre de poème.

Avant la lecture

1. The poem is written in the shape of a tie. When do men or women wear ties? What impression does a tie convey?

2. What could a tie represent in terms of freedom and society?

Lecture

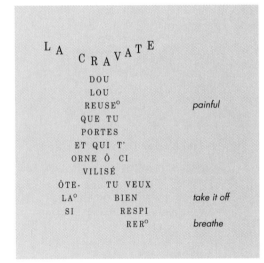

Après la lecture

A **Les vêtements.** Répondez.

1. Qu'est-ce que vous mettez quand vous vous habillez «bien»?

2. Est-ce que vous jugez les gens d'après leurs vêtements?

3. D'après vous, est-ce qu'une école doit *(must)* imposer certaines normes vestimentaires?

B **Êtes-vous poète?**
Avec des amis «poètes», écrivez un calligramme.

INTERDISCIPLINARY CONNECTIONS

1. Bring in a copy of *Calligrammes* and share some of the other poems, such as *"Le jet d'eau,"* with your students.

2. Refer students to page 368 of the Level 4 textbook, *Trésors du temps,* for an example of the cubist art that was popular at the time of Apollinaire. (Picasso's *Les Trois musiciens*). Then, with the help of the art teacher and other samples of cubist art he/she may be able to provide, students do a drawing in the cubist style based on either of the Apollinaire poems on these pages.

En 1901–1902 Apollinaire est en Allemagne et rencontre une jeune Anglaise, Annie Playden. Mais Annie qui est mennonite émigre aux États-Unis.

Avant la lecture

1. Find out about the Mennonites.
2. In French, the word *bouton* means both button (for clothes) and bud (for flowers). In the last stanza of the poem, the poet makes a joke. See if you can explain what the joke is.

Lecture

ANNIE

Sur la côte du Texas
Entre Mobile et Galveston il y a
Un grand jardin tout plein de roses
Il contient aussi une villa
Qui est une grande rose

Une femme se promène souvent
Dans le jardin toute seule
Et quand je passe sur la route
  bordée de tilleuls°   *linden trees*
Nous nous regardons

Comme cette femme est mennonite
Ses rosiers et ses vêtements n'ont
  pas de boutons
Il en manque deux° à mon veston   *two are*
La dame et moi suivons le même rite   *missing*

Apollinaire, blessé à la tête pendant la guerre de 1914

Après la lecture

A **Discutons du poème.** Répondez.

1. The poet imagines his lost love in America. Find examples in the poem that show that this is just a fantasy.
2. In a typically French way, Apollinaire makes light of his emotion and sadness with a "joke." What is the only remaining link between the couple?

LETTRES ET SCIENCES   **337**

Météorologie: La Prévision du temps
Avant la lecture

PRESENTATION *(page 338)*

A. Call on science fans to explain briefly the following:
1. a high-pressure area
2. a low-pressure area
3. an air mass
4. wind
5. clouds

These explanations will assist non-science-oriented students in understanding the reading.

B. Do the *Avant la lecture* activities in the textbook.

C. Have students scan the selection for cognates.

Lecture

PRESENTATION *(pages 338–339)*

A. Have the students read the selection silently.

B. Have them prepare a list of weather terms they remember.

MÉTÉOROLOGIE: LA PRÉVISION DU TEMPS

Avant la lecture

1. Find a weather map in one of your newspapers.

2. Match the French and English terms for weather expressions by comparing the legend of your weather map with that of the French weather map below.

Lecture

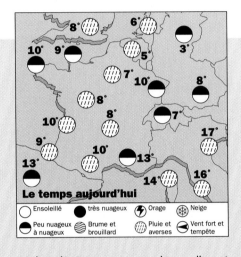

Le matin, beaucoup de gens écoutent le bulletin météorologique à la radio ou le regardent à la télévision pour décider comment s'habiller.

La météo est la science qui étudie l'atmosphère: les vents, les pluies, les dépressions ou zones de basses pressions, et les anticyclones ou zones de hautes pressions. Pour prévoir le temps, les météorologistes doivent savoir[1] le temps qu'il fait sur tout le globe. Il y a trois centres météorologiques dans le monde qui rassemblent toutes les informations météorologiques; ils sont situés à Washington aux États-Unis, à Moscou en Russie et à Melbourne en Australie.

Le soleil chauffe la Terre[2]; la Terre à son tour chauffe l'air et forme l'atmosphère. Mais l'atmosphère n'est pas la même partout: il y a des masses d'air froid au-dessus des pôles, et des masses d'air chaud au-dessus de l'équateur. Quand les masses d'air passent au-dessus des mers ou des océans, elles absorbent de la vapeur d'eau. Il y a donc plusieurs catégories de masses d'air: froides et humides, froides et sèches, chaudes et humides, chaudes et sèches. Ces masses d'air pèsent de manière différente

sur le sol[3]. Dans un anticyclone, elles pèsent lourd[4]: c'est donc une zone de hautes pressions. Dans une dépression, elles ne pèsent pas lourd: c'est donc une zone de basses pressions. Dans un anticyclone, les masses d'air sont trop lourdes et ne peuvent s'affronter[5]. Le temps reste stable. Dans une dépression, les différentes masses d'air s'affrontent si elles sont différentes—une masse d'air froid va contre une masse d'air chaud, par exemple. Le front est la zone où les deux masses s'affrontent; il apporte de la pluie ou du vent.

Les vents sont des mouvements d'air entre les anticyclones (zones de hautes pressions) et les dépressions (zones de

---

INTERDISCIPLINARY CONNECTIONS

With the help of the science class, students consult the weather map in the newspaper for their own state or for the United States as a whole. They then make a weather map for one of these using the map on page 338 as a model. Be sure that they use the French weather symbols and key.

SCIENCES

Les cumulo-nimbus annoncent souvent un orage.

basses pressions). L'air est repoussé[6] par les anticyclones, mais il est aspiré[7] par les dépressions. Ce mouvement d'air est le vent.

Les nuages sont l'ensemble de particules d'eau très fines. Elles sont maintenues en suspension par les mouvements verticaux de l'air.

On peut souvent prévoir le temps d'après les nuages. La forme, la couleur et l'altitude donnent des renseignements relativement précis sur le temps.

[1] doivent savoir   *must know*
[2] chauffe la Terre   *heats the Earth*
[3] pèsent...sol   *exert varying amounts of pressure on the surface of the Earth*
[4] lourd   *heavily*
[5] s'affronter   *collide*
[6] repoussé   *pushed back*
[7] aspiré   *pulled in*

Après la lecture

A **Le bulletin météorologique.**
Donnez une définition en français pour les mots suivants.

1. la météorologie
2. les dépressions
3. les anticyclones
4. le front
5. le vent

B **La météorologie.**   Répondez aux questions.

1. Qu'est-ce que la météorologie étudie?
2. Comment est-ce qu'on obtient les informations nécessaires?
3. Où sont les trois centres météorologiques?
4. Qu'est-ce qui arrive (*happens*) quand les masses d'air passent au-dessus des mers?
5. Quand est-ce que le temps reste stable? Quand est-ce qu'il pleut ou qu'il y a du vent?
6. De quelle autre manière est-ce qu'on peut prévoir le temps?

C **La carte du temps.**   Faites la carte du temps pour les États-Unis pour la journée de demain (en français, bien sûr).

D **Savez-vous que…**   Dans le système Celsius, 0° est la température où l'eau gèle, et 100° est la température où l'eau bout. Si vous voulez passer de degrés Celsius en degrés Fahrenheit ou vice versa, voici deux formules qui peuvent vous aider.

$9/5°C + 32 = °F$	Ex: $(9/5 \times 20°C) + 32 = 68°F$
$(°F - 32) \times 5/9 = °C$	Ex: $(86°F - 32) \times 5/9 = 30°C$

Faites les calculs suivants.

1. 98.6° F = ___ ° C
2. 32° F = ___ ° C
3. 17° C = ___ ° F
4. 25° C = ___ ° F

LETTRES ET SCIENCES   **339**

Après la lecture

PRESENTATION (*page 339*)

Ask students the following questions: *Quel temps fait-il aujourd'hui? Décrivez-le. Donnez tous les détails possibles. Est-ce qu'il y a des nuages? Décrivez-les. Il y a du vent? Il va faire quel temps plus tard aujourd'hui? Et il va faire quel temps demain?*

ANSWERS

Exercice A

Answers will vary but may include the following:

1. **La météorologie est la science qui étudie l'atmosphère.**
2. **Les dépressions sont des zones de basses pressions.**
3. **Les anticyclones sont des zones de hautes pressions.**
4. **Le front est la zone où deux masses d'air s'affrontent.**
5. **Le vent est un mouvement d'air entre les anticyclones et les dépressions.**

Exercice B

1. **La météorologie étudie l'atmosphère.**
2. **On obtient les informations nécessaires des trois centres météorologiques.**
3. **Ils sont à Washington aux États-Unis, à Moscou en Russie et à Melbourne en Australie.**
4. **Les masses d'air absorbent de la vapeur d'eau.**
5. **Le temps reste stable dans un anticyclone. Il pleut ou il y a du vent quand il y a un front.**
6. **On peut prévoir le temps d'après les nuages.**

Exercice C

Answers will vary.

Exercice D

1. 37
2. 0
3. 62,2
4. 77

CHAPTER OVERVIEW

In this chapter students will expand their sports vocabulary. They will learn to talk about cycling as well as team sports such as soccer, basketball, and volleyball. They will also learn to describe the equipment needed for each sport and the seasons in which various sporting events take place. To do this, students will learn the *passé composé* of regular verbs conjugated with *avoir* and the interrogative phrase *qu'est-ce que*.

The cultural focus of Chapter 13 is on soccer and other sports which the French either play or enjoy as fans.

CHAPTER OBJECTIVES

By the end of this chapter, students will know:

1. vocabulary associated with soccer
2. vocabulary associated with basketball, volleyball, cycling, running, football, and baseball
3. the formation and use of the *passé composé* of regular verbs conjugated with *avoir* in the affirmative and negative
4. time expressions frequently used with the *passé composé*
5. interrogative sentence structure using *qu'est-ce que* and *que*

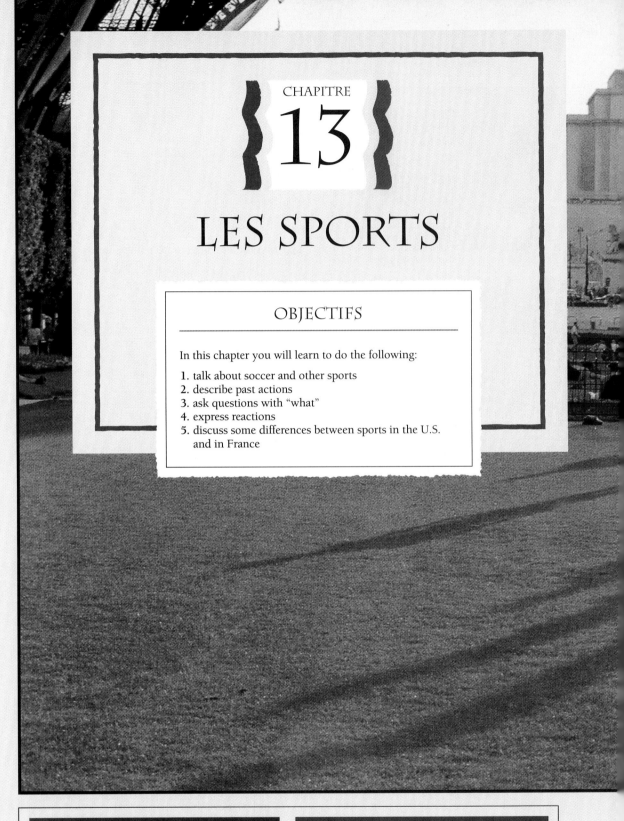

CHAPITRE 13

LES SPORTS

OBJECTIFS

In this chapter you will learn to do the following:

1. talk about soccer and other sports
2. describe past actions
3. ask questions with "what"
4. express reactions
5. discuss some differences between sports in the U.S. and in France

CHAPTER PROJECTS

(optional)

1. Try to get a video of a sporting event in French. Have students watch about five minutes of it. Then have them say as much about it as they can.
2. Have students prepare a TV description of a sporting event.

COMMUNITIES

Attend a soccer match or a basketball game with your students. Comment on the action in French and encourage them to do the same.

341

Pacing

This chapter requires eight to ten class sessions. Pacing will vary according to class length and the age and aptitude of the students.

Note The Lesson Plans offer guidelines for 45- and 55-minute classes and **Block Scheduling.**

Exercices vs. *Activités*

All exercises (which provide guided practice) are coded in blue. All communicative activities are coded in red.

INTERNET ACTIVITIES

(optional)

These activities, student worksheets, and related teacher information are in the *Bienvenue* Internet Activities Booklet and on the Glencoe Foreign Language Home Page at: **http://www.glencoe.com/secondary/fl**

LEARNING FROM PHOTOS

After presenting the *Mots 1* vocabulary, ask students: *C'est quel sport? Dans quelle ville est-ce que ces joueurs jouent? Ils jouent sous quel monument célèbre?*

MOTS 1

Vocabulary Teaching Resources

1. Vocabulary Transparencies 13.1 (A & B)
2. Audio Cassette 8A/CD-8
3. Student Tape Manual, Teacher's Edition, *Mots 1: A–C*, pages 142–144
4. Workbook, *Mots 1: A–C*, page 130
5. Communication Activities Masters, *Mots 1: A*, page 64
6. Chapter Quizzes, *Mots 1: Quiz 1*, page 69
7. CD-ROM, Disc 4, *Mots 1:* pages 342–344

Bell Ringer Review

Write the following on the board or use BRR Blackline Master 13-1: Complete the sentences logically with the correct form of prendre, comprendre, or apprendre.

1. Au lycée nous ___ le français.
2. Mais moi, je ne ___ pas toujours le professeur.
3. Vous ___ un coca?
4. Non, nous ___ du thé.

PRESENTATION *(pages 342–343)*

A. Tell students that the names of many sports are cognates, but don't allow them to anglicize their pronunciation.
B. Using Vocabulary Transparencies 13.1 (A & B), have students repeat the new words after you or Cassette 8A/CD-8.

Note Many masculine nouns ending in *-eur* that formerly did not have a feminine form are now being heard with the *-euse* (or *-rice*) ending. See **CROSS-CULTURAL COMPARISON** on page 359 of this Teacher's Wraparound Edition.

342

LE FOOT(BALL)

le but

siffler

un gardien de but

un arbitre

un terrain de foot(ball)

un ballon

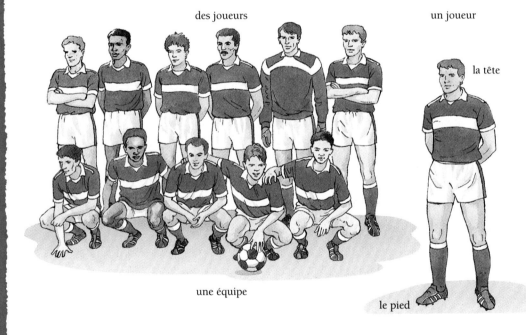

des joueurs

un joueur

la tête

une équipe

le pied

TOTAL PHYSICAL RESPONSE

(following the Vocabulary presentation)

Getting Ready

You will need a lightweight ball, such as one made of foam, or something to serve as an imaginary soccer ball.

TPR 1

(Student 1), **venez ici, s'il vous plaît.**
Montrez-moi la tête. Et le pied.

Voici un ballon. Prenez le ballon.
Donnez un coup de pied dans le ballon.
(Student 2), allez chercher le ballon.
Renvoyez le ballon à (Student 1).
(Student 3), venez ici, s'il vous plaît.
Vous êtes l'arbitre. Prenez le sifflet.
Sifflez.
Merci, ___. Vous avez tous très bien fait.
   Asseyez-vous, s'il vous plaît.

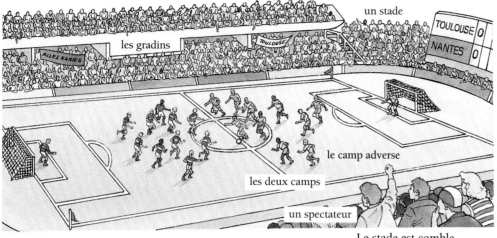

un stade

les gradins

TOULOUSE 0
NANTES 0

le camp adverse

les deux camps

un spectateur

Le stade est comble.
Il y a beaucoup de monde.
Les gradins sont pleins.

21 avril   22 avril

match
Toulouse-
Nantes

hier   aujourd'hui

Hier Nantes a joué contre Toulouse.
Le match a opposé Toulouse à Nantes.

Peyre a donné un coup de pied dans le ballon.

TOULOUSE | 0 | 0
NANTES | 0 | 1

Roland a envoyé le ballon dans le but.
Il a marqué un but.

CHAPITRE 13   **343**

TPR 2

___, venez ici. Vous êtes un spectateur/une
spectatrice. Cherchez une place dans les
gradins.
Prenez votre place. Regardez le match.
    Quelqu'un va marquer un but. Levez-
    vous. Regardez bien.
Ah oui, il a marqué un but. Applaudissez.
Merci, ___. Et maintenant, retournez à votre
    place, s'il vous plaît.

ADDITIONAL PRACTICE

Student Tape Manual, Teacher's Edition,
Activités B–C, pages 143–144

C. Ask questions to elicit the
vocabulary. Begin with yes/no
and either/or questions and
progress to question-word
questions. Encourage full-sen-
tence responses. For example:
*C'est un terrain de football? C'est
un arbitre ou un joueur? Qui
siffle?*

Teaching Tip For the last three
illustrations on page 343, make it
clear that the Nantes/Toulouse
match took place yesterday and
the actions described are in the
past. First demonstrate the meaning of *hier* by constrasting it with
aujourd'hui and the current date
or day. For example: *Aujourd'hui—
mardi. Hier—lundi. Aujourd'hui—
le 28. Hier—le 27.* Point a thumb
back over one shoulder to indicate
hier and the past in general. Use
this gesture from now on to indicate or elicit past structures.

Note The *passé composé* is presented in the *il/elle* form so you
can immediately ask questions
using it while presenting the new
words. Students can respond to
the questions without having to
manipulate forms of the *passé
composé*. They will learn the forms
in this chapter.

D. As you go over the sentences
on page 343, you may wish to
ask questions such as: *Nantes a
joué contre qui? Contre quelle
équipe? Quand ça? Le match a
opposé quelles équipes? Qui a
donné un coup de pied dans le
ballon? Qui a envoyé le ballon
dans le but? Il a marqué un but?
Qu'est-ce qu'il a marqué? Qui a
marqué le but?*

Exercices

Exercices

PRESENTATION *(page 344)*

Exercice C

This exercise allows students to practice using past participles without having to manipulate forms.

ANSWERS

Exercice A

1. Oui, il y a beaucoup de spectateurs dans le stade.
2. Les gradins sont pleins de spectateurs.
3. Oui, le stade est comble.
4. Oui, il y a beaucoup de monde dans le stade.
5. C'est un sport collectif.

Exercice B

1. Dans un match de foot, il y a deux équipes.
2. Chaque équipe a onze joueurs.
3. Il y a vingt-deux joueurs sur le terrain.
4. Dans un match il y a deux camps.
5. Le match est divisé en mi-temps.
6. Il y a deux mi-temps.
7. Chaque mi-temps dure quarante-cinq minutes.
8. Le gardien de but garde le but.
9. Chaque équipe veut marquer un but.
10. Le gardien de but bloque ou arrête le ballon.

Exercice C

1. Oui, Toulouse a joué contre Nantes.
2. Oui, Peyre a donné un coup de pied dans le ballon.
3. Oui, Peyre a passé le ballon à Roland.
4. Oui, Roland a envoyé le ballon dans le but.
5. Oui, Roland a marqué un but.
6. Oui (Non), le gardien n'a pas arrêté le ballon.
7. Oui, il a égalisé le score.
8. Oui, l'arbitre a sifflé.
9. Oui, il a déclaré un penalty contre Toulouse.
10. Oui, Nantes a gagné le match.
11. Oui, Toulouse a perdu le match.

Exercices

A **Le stade est comble.** Répondez.

1. Il y a beaucoup de spectateurs dans le stade?
2. Les gradins sont pleins de spectateurs ou il y a beaucoup de places libres?
3. Le stade est comble?
4. Il y a beaucoup de monde dans le stade?
5. Le foot est un sport d'équipe. C'est un sport individuel ou collectif?

B **Un match de foot.** Répondez d'après les indications.

1. Dans un match de foot, il y a combien d'équipes? (deux)
2. Chaque équipe a combien de joueurs? (onze)
3. Il y a combien de joueurs sur le terrain? (vingt-deux)
4. Dans un match il y a combien de camps? (deux)
5. Le match est divisé en quoi? (mi-temps)
6. Il y a combien de mi-temps? (deux)
7. Chaque mi-temps dure combien de minutes? (quarante-cinq)
8. Qui garde le but? (le gardien de but)
9. Qu'est-ce que chaque équipe veut faire? (marquer un but)
10. Qui bloque ou arrête le ballon? (le gardien de but)

C **Toulouse contre Nantes.** Répondez par «oui».

1. Toulouse a joué contre Nantes?
2. Peyre a donné un coup de pied dans le ballon?
3. Peyre a passé le ballon à Roland?
4. Roland a envoyé le ballon dans le but?
5. Roland a marqué un but?
6. Le gardien n'a pas arrêté le ballon?
7. Roland a égalisé le score?
8. L'arbitre a sifflé?
9. Il a déclaré un penalty contre Toulouse?
10. Nantes a gagné le match?
11. Toulouse a perdu le match?

344   CHAPITRE 13

ADDITIONAL PRACTICE

After completing Exercises A–C, reinforce the lesson with the following: Have students bring in action photographs from sports magazines and describe them in as much detail as possible.

INDEPENDENT PRACTICE

Assign any of the following:
1. Exercises, page 344
2. Workbook, *Mots 1: A–C,* page 130
3. Communication Activities Masters, *Mots 1: A,* page 64
4. CD-ROM, Disc 4, pages 342–344

VOCABULAIRE

MOTS 2

D'AUTRES SPORTS

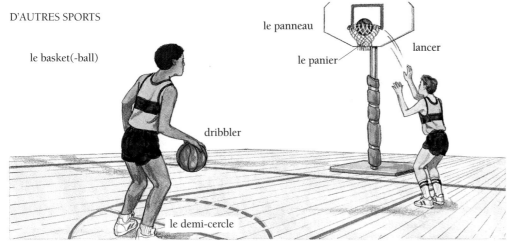

le panneau

le panier

lancer

le basket(-ball)

dribbler

le demi-cercle

Un joueur a dribblé le ballon.
Il a dribblé le ballon jusqu'au demi-cercle.

Un autre joueur a lancé le ballon dans le panier.

le volley(-ball)

le filet

le sol

Un joueur a servi.

par dessus le filet

Un autre joueur a renvoyé le ballon.

COOPERATIVE LEARNING

Each group chooses a leader, who asks the others what their favorite sports are and whether they prefer to play the sports, watch them on TV, or go to games. The leader takes notes and reports to the class. You can follow up with a class survey, grouping names of students on the board according to their preferences and discussing the results.

TOTAL PHYSICAL RESPONSE

(following the Vocabulary presentation)

Getting Ready

If you wish, set up part of a mock basket-ball court on the floor with tape and use a toy basketball and hoop. Demonstrate the meaning of *lentement*.

(continued on page 346)

VOCABULAIRE

MOTS 2

Vocabulary Teaching Resources

1. Vocabulary Transparencies 13.2 (A & B)
2. Audio Cassette 8A/CD-8
3. Student Tape Manual, Teacher's Edition, *Mots 2: D–F,* pages 145–147
4. Workbook, *Mots 2: D–F,* page 131
5. Communication Activities Masters, *Mots 2: B,* page 65
6. Computer Software, *Vocabulaire*
7. Chapter Quizzes, *Mots 2: Quiz 2,* page 70
8. CD-ROM, Disc 4, *Mots 2:* pages 345–349

Bell Ringer Review

Write the following on the board or use BRR Blackline Master 13-2: Make a list of six articles of clothing you would like for your birthday.

PRESENTATION *(pages 345–346)*

A. Using Vocabulary Transparencies 13.2 (A & B) and realia, have students repeat after you or Cassette 8A/CD-8. Stress the correct French pronunciation of words borrowed from English.

B. Have students close their books. Make negative statements about the illustrations on Vocabulary Transparencies 13.2. Ask individuals to respond with an appropriate affirmative statement. For example: *Le joueur n'a pas lancé le ballon. (Le joueur a dribblé le ballon.) Le joueur n'a pas renvoyé le ballon. (Le joueur a servi.)*

C. Have students prepare questions to ask you about which sports you prefer, whether you participate in or watch them, what your favorite teams are, and so on. Hold a question-and-answer session using this material.

le cyclisme

une course cycliste

des coureurs cyclistes

un vélo

un gagnant

un coureur

une piste

Leblanc (27) a gagné la course.
Boulet (28) a perdu la course.

une coupe

Aux États-Unis le football américain est un sport d'automne.

Le base-ball est un sport de printemps.

346 CHAPITRE 13

TPR 1 *(continued from page 345)*
___, venez ici, s'il vous plaît.
Vous allez jouer au basket-ball.
Dribblez le ballon lentement.
Dribblez le ballon vite.
Allez au demi-cercle.
Lancez le ballon dans le panier. Bravo! Vous avez lancé le ballon dans le panier.
Merci, ___. Vous avez très bien joué.
Retournez à votre place, s'il vous plaît.

TPR 2
___, venez ici, s'il vous plaît.
Vous allez jouer au volley-ball.
Prenez le ballon.
Voici le filet. Servez.
Sautez. (Dramatize *sauter.*)
Renvoyez le ballon par-dessus le filet.
Merci, ___. Vous avez très bien joué.
Retournez à votre place, s'il vous plaît.

Exercices

A Un match de basket. Répondez.

1. On joue au basket-ball sur une piste ou sur un terrain?
2. Le basket-ball est un sport individuel ou un sport d'équipe?
3. Il y a cinq ou onze joueurs dans une équipe de basket-ball?
4. Pendant un match de basket les joueurs dribblent le ballon ou donnent un coup de pied dans le ballon?
5. Un joueur a dribblé le ballon jusqu'au panneau ou jusqu'au demi-cercle?
6. Un autre joueur a lancé le ballon dans le panier ou dans le but?

B Le volley-ball. Répondez par «oui» ou «non».

1. Une équipe de volley-ball a six joueurs?
2. Un joueur sert?
3. Un joueur du camp adverse renvoie le ballon?
4. Quand il renvoie le ballon, le ballon peut toucher le filet?
5. On renvoie le ballon par dessus le filet?
6. Le ballon peut toucher le sol?

C C'est quel sport? Identifiez.

le base-ball
le basket-ball
le football
le football américain
le volley-ball

1. Aux États-Unis c'est un sport d'automne.
2. Aux États-Unis c'est un sport de printemps.
3. Le ballon ne peut pas toucher le sol.
4. Il y a cinq joueurs dans l'équipe.
5. Le joueur a donné un coup de pied dans le ballon.
6. Le joueur a renvoyé le ballon par dessus le filet.
7. Le gardien de but a bloqué le ballon.
8. Le joueur a servi.
9. Le joueur a lancé le ballon dans le panier.
10. Le joueur a marqué un but.

Ces jeunes pratiquent des arts martiaux.

Exercices

PRESENTATION (pages 347–348)

Exercices A and B: Listening

After doing Exercises A and B as a whole-class activity, focus on the listening skill by having students do them again in pairs, with one partner reading the questions in random order while the other listens and answers with his/her book closed.

Extension of *Exercice B*

After completing Exercise B, have students use it as a model to make up six new questions about the game of soccer. Then, in pairs, they take turns asking the questions and answering.

Exercice C

You may wish to use the recorded version of this exercise.

ANSWERS

Exercice A

1. On joue au basket-ball sur un terrain.
2. Le basket-ball est un sport d'équipe.
3. Il y a cinq joueurs dans une équipe de basket-ball.
4. Pendant un match de basket les joueurs dribblent le ballon.
5. Un joueur a dribblé le ballon jusqu'au demi-cercle.
6. Un autre joueur a lancé le ballon dans le panier.

Exercice B

1. Oui.	4. Non.
2. Oui.	5. Oui.
3. Oui.	6. Non.

Exercice C

1. le football américain
2. le base-ball
3. le volley-ball
4. le basket-ball
5. le football, le football américain
6. le volley-ball
7. le football
8. le volley-ball
9. le basket-ball
10. le football, le football américain

Exercice D

1. a
2. b
3. c
4. b
5. a
6. a
7. b
8. c

INFORMAL ASSESSMENT
(Mots 2)

Check comprehension by making statements about various sports and having students tell which sport you are talking about. For example: *Il a marqué un but. (C'est le foot.) Les spectateurs applaudissent quand il lance le ballon dans le panier. (C'est le basket.)*

RETEACHING *(Mots 2)*

Tell in which season or seasons people engage in the following sports.

1. **le basket**
2. **le volley**
3. **le cyclisme**
4. **le football américain**
5. **le base-ball**
6. **le foot**

Des coureurs cyclistes aux Jeux Paraolympiques

D **Une course cycliste.** Choisissez.

1. Un vélo est ___.
   a. une bicyclette   b. une voiture   c. un stade

2. ___ roule à vélo.
   a. Une bicyclette   b. Un coureur cycliste   c. Un spectateur

3. Dans une course internationale, chaque équipe ___.
   a. gagne un trophée   b. gagne la coupe   c. représente son pays

4. Le gagnant de la course est ___.
   a. la coupe   b. le champion   c. le coureur

5. ___ gagnent de l'argent.
   a. Les professionnels   b. Les amateurs
             c. Les spectateurs

6. ___ gagne.
   a. Le premier   b. Le dernier
   c. Chaque équipe

7. On donne ___ au gagnant.
   a. la course   b. la coupe
   c. la bicyclette

8. Dans une course cycliste les coureurs roulent sur ___.
   a. des gradins   b. un terrain
   c. une piste

348   CHAPITRE 13

LEARNING FROM PHOTOS

Ask students: *Ces coureurs cyclistes sont dans les Jeux Olympiques ou Paraolympiques? Les Jeux Paraolympiques sont pour les athlètes handicapés?* Tell students: France is behind the U.S. in providing access for the disabled. The newer monuments, hotels, and museums have special facilities. The SNCF has added special cars on some trains for handling wheelchairs.

LEARNING FROM REALIA

Have students look at the ad for the magazine *Vélo.* Ask them if they can guess what *numéros* means in "*11 numéros dont 3 numéros spéciaux.*" Can they guess the meaning of "*27% de remise!*"? Finally, what do they think a "*chronomètre multifonctions*" is? (a stopwatch that measures your performance and is a regular watch as well.)

Activités de communication orale
Mots 1 et 2

A **C'est quel sport?** Give a classmate several details about a sport without mentioning the name of the sport. Your partner has to guess what sport you're describing. Then reverse roles.

> Élève 1: Il y a cinq joueurs dans l'équipe. Les joueurs dribblent le ballon. Les meilleurs joueurs sont souvent très grands. Ils lancent le ballon dans le panier.
> Élève 2: C'est le basket-ball.

B **Ton équipe favorite.** Ask a classmate what his or her favorite team is and why. Then reverse roles and report to the class.

> Élève 1: Quelle est ton équipe favorite? Pourquoi?
> Élève 2: Mon équipe favorite de base-ball, c'est les Expos parce que je suis de Montréal. (Je n'ai pas d'équipe favorite de basket-ball.)

C **Un match de football.** Ask a classmate several questions about the illustration using *qui, quel(le), est-ce que, combien,* and *où.* Then reverse roles.

Bell Ringer Review

Write the following on the board or use BRR Blackline Master 13-3: Sketch the following:
le ballon
la tête
le filet
une coupe

Activités de communication orale
Mots 1 et 2

ANSWERS

Activités A and B
   Answers will vary.

Activité C
Note Stress to students that all their questions need not have a definite answer. The answer to some of their questions may be *Je ne sais pas.* For example: *Quel joueur marque un but?* is a good question, but the answer is not readily apparent from the illustration.
   Questions will vary.

RETEACHING *(Mots 1 and 2)*
   Have students write three sentences describing the sporting activities from *Mots 1* and *2* that can be mimed. Then they choose a partner who must correctly mime the activity in front of the class when it is read only once or twice.

INDEPENDENT PRACTICE

   Assign any of the following:
1. Exercises and activities, pages 347–349
2. Workbook, *Mots 2: D–F,* page 131
3. Communication Activities Masters, *Mots 2: B,* page 65
4. Computer Software, *Vocabulaire*
5. CD-ROM, Disc 4, pages 345–349

Structure Teaching Resources

1. Workbook, *Structure: A–H*, pages 132–135
2. Student Tape Manual, Teacher's Edition, *Structure: A–B*, page 148
3. Audio Cassette 8A/CD-8
4. Communication Activities Masters, *Structure: A–B*, pages 66–67
5. Computer Software, *Structure*
6. Chapter Quizzes, *Structure: Quizzes 3–4*, pages 71–72
7. CD-ROM, Disc 4, pages 350–353

Le passé composé des verbes réguliers

PRESENTATION *(pages 350–351)*

A. Quickly review the forms of *avoir* in the chart on page 350. Students have used this verb many times by now.

B. Show students how the past participle of regular verbs is formed. The more past participles students can repeat the better. Ear training is extremely important. The following are regular verbs they have already learned in the present tense: *chanter, danser, donner, écouter, étudier, gagner, habiter, inviter, parler, regarder, quitter, rigoler, travailler, bavarder, dîner, préparer, chercher, commander, déjeuner, trouver, acheter, payer, laisser, bronzer, jouer, nager, plonger, renvoyer, dépenser, porter, accélérer, freiner, garer, traverser, vérifier, dribbler, envoyer, lancer, opposer, siffler; choisir, finir, remplir, réussir, obéir, punir, atterrir, dormir, servir, maigrir, grossir; attendre, entendre, perdre, vendre, répondre.*

C. Guide students through steps 1–4 on pages 350–351.

Le passé composé des verbes réguliers

Describing Past Actions

1. You use the *passé composé* to express actions completed in the past. The *passé composé* is made up of the present tense of *avoir* and the past participle of the verb. Review the present tense of the verb *avoir.*

AVOIR	
j'ai	nous avons
tu as	vous avez
il/elle/on a	ils/elles ont

2. Study the following forms of the past participle of regular French verbs.

-er ➝ -é		-ir ➝ -i		-re ➝ -u	
regarder	regardé	choisir	choisi	perdre	perdu
parler	parlé	réussir	réussi	vendre	vendu

Almost all past participles of French verbs end in the sound /é/, /i/, or /ü/.

PARLER	FINIR	PERDRE
j'ai parlé	j'ai fini	j'ai perdu
tu as parlé	tu as fini	tu as perdu
il/elle/on a parlé	il/elle/on a fini	il/elle/on a perdu
nous avons parlé	nous avons fini	nous avons perdu
vous avez parlé	vous avez fini	vous avez perdu
ils/elles ont parlé	ils/elles ont fini	ils/elles ont perdu

3. The *passé composé* is often used with time expressions such as:

> avant-hier
> hier
> hier matin
> hier soir
> l'année dernière
> la semaine dernière

ADDITIONAL PRACTICE

After completing Exercises A and B on page 351, you may wish to ask students additional personalized questions to further reinforce the *tu/je* questions and answers in the *passé composé: Jeanne, tu as téléphoné à des amis hier soir? Tu as parlé avec qui? Tu as écouté des cassettes? Tu as étudié? Qu'est-ce que tu as étudié hier soir? Tu as dîné à quelle heure? Qu'est-ce que tu as mangé? Tu as regardé la télé? Qu'est-ce que tu as regardé? À quelle heure?*, etc.

Note When formulating your own questions, be sure to use only regular verbs conjugated with *avoir.*

Study the following examples of the *passé composé.*

> **J'ai regardé le match à la télé hier soir.**
> **Nantes a joué contre Toulouse.**
> **L'année dernière Toulouse a gagné la coupe.**
> **Mais hier soir Toulouse a perdu le match.**
> **L'arbitre a puni Toulouse.**
> **Il a déclaré un penalty contre Toulouse.**
> **Les spectateurs ont applaudi.**

4. Note the placement of *ne… pas* in negative sentences with the *passé composé. Ne… pas* goes around the verb *avoir.*

> **Je *n'ai pas* parlé à Suzanne.**
> **Tu *n'as pas* regardé la télé?**
> **Il *n'a pas* entendu le téléphone.**

Le Niger joue contre l'Argentine pour la Coupe du Monde.

Exercices

A **Quel est le participe passé?** Donnez le participe passé.

1. habiter
2. quitter
3. parler
4. écouter
5. travailler
6. remplir
7. obéir
8. réussir
9. servir
10. dormir
11. perdre
12. vendre
13. attendre
14. répondre

B **Hier ou la semaine dernière.**
Donnez des réponses personnelles.

1. Hier matin tu as quitté la maison à quelle heure?
2. Avant les cours tu as rigolé avec tes copains?
3. Tu as parlé au prof de français?
4. La semaine dernière tu as passé un examen? Tu as réussi à l'examen?
5. Tu as répondu à toutes les questions?
6. Tu as quitté l'école à quelle heure hier?
7. Tu as attendu le bus devant l'école?

Le forcing

Encore une victoire pour l'équipe de Strasbourg. Les Niçois ont perdu leur troisième match.

STRASBOURG ET NICE 6–3

Après l'échec total de Lyon et le demi-échec face à Bourges, Nice a commis une troisième erreur en trois matchs. Les Strasbourgeois ont pratiqué un football collectif de qualité, se montrant patients et prudents pendant la première mi-temps. Le jeu niçois manquait de mouvement et de vitesse. Dortez se pose en rival sérieux de Peyre. Au début du match il a fait le forcing pour égaliser le score.

Ce n'est qu'après la mi-temps qu'il a marqué trois buts de suite. Et quels buts! On n'a jamais vu ça depuis le match légendaire qui a opposé Toulouse et Nantes l'année dernière. Dortez avait du mal à croire ce qu'il venait de faire. Lors d'une interview après le match il a dit: «Je dois être un peu fou. C'est sûrement pour cel que j'intéresse tout le monde». lui est difficile de faire le humb quand son talent saute aux yeu Peyre, par contre, n'essaie mêm pas de cacher son ego. «Je su plus fort que jamais,» a-t-il pr cisé l'autre jour. «Il est vrai q nous avons perdu trois matc mais cela n'a pas d'importanc ou si peu. La semaine procha je vais pouvoir montrer de quo

---

Exercices

PRESENTATION *(pages 351–352)*

Extension of *Exercice B*

After doing Exercise B, call on a student to tell the story in his/her own words.

ANSWERS

Exercice A

1. habité
2. quitté
3. parlé
4. écouté
5. travaillé
6. rempli
7. obéi
8. réussi
9. servi
10. dormi
11. perdu
12. vendu
13. attendu
14. répondu

Exercice B

Answers will vary, but may include the following constructions:

1. J'ai quitté…
2. J'ai rigolé… (Je n'ai pas rigolé…)
3. J'ai parlé… (Je n'ai pas parlé…)
4. J'ai passé… (Je n'ai pas passé…) J'ai réussi… (Je n'ai pas réussi…)
5. J'ai répondu… (Je n'ai pas répondu…)
6. J'ai quitté…
7. J'ai attendu… (Je n'ai pas attendu…)

LEARNING FROM PHOTOS

Have students say as much as they can about the photo using the vocabulary from this chapter.

ADDITIONAL PRACTICE

Student Tape Manual, Teacher's Edition, *Activité A,* page 148

C **La fête d'Élisabeth.** Complétez au passé composé.

1. Vendredi dernier Élisabeth ___ une fête. (donner)
2. Elle ___ à tous ses copains. (téléphoner)
3. Ses copains ___ au téléphone. (répondre)
4. Élisabeth ___ ses copains à la fête. (inviter)
5. Tous ses copains ___ son invitation. (accepter)
6. Yves et moi, nous ___ quelque chose à manger. (préparer)
7. Mais qui ___ les provisions? (acheter)
8. Tu ___ le menu? (choisir)
9. Non, Élisabeth ___ la fête et elle ___ le menu. (donner, choisir)
10. Tout le monde ___. (manger)
11. Vous ___ pendant la fête? (danser)
12. Oui, nous ___ et nous ___. (danser, chanter)

D **Un voyage en avion.** On va imaginer que vous avez voyagé. Répondez.

1. Tu as voyagé l'année dernière?
2. Tu as voyagé avec Air France?
3. Tu as choisi classe économique ou première classe?
4. Tu as choisi une place côté couloir?
5. L'avion a décollé à l'heure?
6. Et il a atterri à l'heure?
7. Tu as voyagé avec un copain ou une copine?
8. Tu as attendu longtemps tes bagages?
9. La compagnie aérienne a perdu tes bagages?

E **Un voyage en train.** Mettez au passé composé.

1. J'attends le train.
2. Je voyage avec ma copine.
3. Nous attendons le train dans la salle d'attente de la gare.
4. J'achète un magazine au kiosque.
5. Je ne choisis pas de journal.
6. Ma copine achète un livre.
7. Nous entendons l'annonce du départ de notre train.
8. On annonce le départ au haut-parleur.
9. Le porteur descend nos bagages sur le quai.
10. Nous réussissons à trouver nos places dans la voiture onze.
11. Le contrôleur vérifie les billets.

Bon Voyage!

ADDITIONAL PRACTICE

1. Give students some regular verbs from the list on page 350 of this Teacher's Wraparound Edition and have them write five sentences telling what they did yesterday and five additional sentences telling what they did last week using these verbs.
2. Have them use the same list of verbs to write several things that they didn't do.

PAIRED ACTIVITY

Have students work in pairs. Give students the list of verbs on page 350 of this Teacher's Wraparound Edition. One student makes up an original question and the other student responds. For example: *Tu as dansé avec ___ à la fête de ___? Oui, j'ai dansé avec ___. (Non, je n'ai pas dansé avec ___.)*

Qu'est-ce que

Asking "What?"; Expressing Reactions

1. *Qu'est-ce que* is another question or interrogative expression. It means "what" and refers to a thing.

> Qu'est-ce que vous voyez?
> Qu'est-ce qu'il regarde?
> Qu'est-ce que vous avez?

Note that *Qu'est-ce que vous avez?* also means "What's the matter?"

2. To ask "what?" in formal or written French you use *que* and invert the subject and verb.

> Que voyez-vous?
> Que regarde-t-il?
> Qu'avez-vous?

3. In informal French *qu'est-ce que* is used in exclamations.

> Qu'est-ce qu'il est beau ce garçon!   *How handsome that boy is!*
> Qu'est-ce qu'elle est belle!   *How beautiful she is!*
> Qu'est-ce que je suis fatigué!   *How tired I am!*

Exercices

A   Comment? Qu'est-ce que tu fais?   Posez des questions d'après le modèle.

> J'écoute la radio.
> *Comment? Qu'est-ce que tu écoutes?*

1. Je lis le journal.
2. Je regarde la télé.
3. Je fais des exercices.
4. Je fais les courses.
5. J'achète un cadeau.
6. Je lave la voiture.
7. Nous écrivons un poème.
8. Nous préparons le petit déjeuner.
9. Nous commandons une boisson.

B   Des mini-conversations.   Posez des questions et répondez d'après le modèle.

> marquer/un but
> Élève 1: Qu'est-ce que les joueurs ont marqué?
> Élève 2: Ils ont marqué un but.

1. lancer/le ballon
2. dribbler/le ballon
3. envoyer/le ballon
4. perdre/le match
5. gagner/la coupe
6. égaliser/le score
7. gagner/de l'argent

VAINQUEUR DE LA COUPE DE FRANCE 1987

CHAPITRE 13   **353**

Qu'est-ce que

PRESENTATION (page 353)

A. Students have been using *qu'est-ce que* for some time. The purpose of this presentation is to have students actively use and produce this interrogative phrase on their own.

B. Lead students through steps 1–3 on page 353.

Note In spoken French one hears both *Qu'est-ce qu'il est beau!* and *Qu'il est beau!* in exclamations.

Exercices

ANSWERS

Exercice A

1. Comment? Qu'est-ce que tu lis?
2. … tu regardes?
3. … tu fais?
4. … tu fais?
5. … tu achètes?
6. … tu laves?
7. … vous écrivez?
8. … vous préparez?
9. … vous commandez?

Exercice B

1. É1: Qu'est-ce que les joueurs ont lancé? É2: Ils ont lancé le ballon.
2. … ont dribblé? … le ballon.
3. … ont envoyé? … le ballon.
4. … ont perdu? … le match.
5. … ont gagné? … la coupe.
6. … ont égalisé? … le score.
7. … ont gagné? … de l'argent.

ADDITIONAL PRACTICE

Show students photographs with prominent features (big mountains, tall buildings or people, fast cars, hot weather, etc.), which will, as much as possible, elicit spontaneous exclamations from students. If necessary, cue an adjective or adverb orally.

Note You can use these same pictures as cues for practicing *que* and/or *qu'est-ce que* as question words.

INDEPENDENT PRACTICE

Assign any of the following:

1. Exercises, pages 351–353
2. Workbook, *Structure: A–H*, pages 132–135
3. Communication Activities Masters, *Structure: A–B*, pages 66–67
4. Computer Software, *Structure*
5. CD-ROM, Disc 4, pages 350–353

CONVERSATION

Scènes de la vie *Une retransmission sportive*

ROMAIN: Tu as regardé la télé hier soir?
CORINNE: Oui, j'ai regardé la retransmission du match France–Brésil.
ROMAIN: Tu parles de la victoire de la France sur le Brésil?
CORINNE: Voilà! La France a gagné un à zéro.
ROMAIN: Le Brésil a fait le forcing pour égaliser le score.
CORINNE: Oui, mais sans succès. À chaque fois Peyre a bloqué le ballon. Ce type est un gardien vachement fort.
ROMAIN: Qui a marqué le but pour la France? J'ai oublié.
CORINNE: Tu as oublié? Tu n'as pas de mémoire! Moi, je ne vais jamais oublier ça! Roland. C'est Roland qui a marqué le but.

A **Quel match alors!** Répondez d'après la conversation.

1. Qui a regardé la télé hier soir?
2. Qu'est-ce qu'elle a regardé à la télé?
3. Qui a joué contre la France?
4. Qui a gagné le match?
5. Quelle équipe a perdu le match?
6. Le Brésil a réussi à égaliser le score? Pourquoi pas?
7. Comment s'appelle le gardien de but français?
8. Qui a marqué le but pour la France?
9. Qui a oublié son nom?

Bell Ringer Review

Write the following on the board or use BRR Blackline Master 13-4: Sketch a soccer stadium including everything and everyone you would see on or around it during a game. Label all items.

PRESENTATION *(page 354)*

A. Tell students they will hear Romain and Corinne discussing a recent international soccer match.

B. Have them watch the Conversation Video or listen as you read the conversation or play Cassette 8A/CD-8.

C. Have pairs of students present the conversation to the class.

D. You may have a more able student retell the conversation in narrative form in his/her own words.

Note In the CD-ROM version, students can play the role of either one of the characters and record the conversation.

ANSWERS

Exercice A

1. Corinne a regardé la télé hier soir.
2. Elle a regardé le match France-Brésil.
3. Le Brésil a joué contre la France.
4. La France a gagné le match.
5. Le Brésil a perdu le match.
6. Non, le Brésil n'a pas réussi à égaliser le score parce que le gardien français a bloqué le ballon.
7. Le gardien de but français s'appelle Peyre.
8. Roland a marqué le but pour la France.
9. Romain a oublié son nom.

20.30

20.25 TF1 22.35

Football

En direct de Rotterdam. Commentaires : Thierry Roland et Jean-Michel Larqué.

Feyenoord/AS Monaco
Demi-finale retour de la **Coupe d'Europe des vainqueurs de Coupes.** «Je crois sincèrement que l'on forme un groupe de joueurs très unis. Quand l'un est en difficulté, l'autre a la volonté de venir l'aider. C'est important comme état d'esprit, car cela veut dire qu'en Coupe d'Europe, où le mental compte énormément, on peut avoir confiance en la solidarité. Je pense qu'on a une équipe capable d'embêter beaucoup de monde.»
Rob Witschge, qui prononce ces paroles pleines de bon sens, sait de quoi il parle. Il connaît aussi bien le football néerlandais que le football français, pour avoir joué pendant deux ans à Saint-Etienne.

Désormais attaquant l'ancien Stéphanois s ment adapté au style son équipe, qui ress mément à celui de l'A tous les défenseurs att les attaquants défen tat : les défenses adve vent confrontées à des lantes bien difficiles Arsène Wenger, l'entr Monégasques, craint ce du rouleau compresse cache pas : «Feyenoo forte impression. Cette est très disciplinée. E d'elle une grande force

En cas d'égalité à la fin du mentaire, il sera procédé a tions et éventuellement aux

20.30 C++ 21.00
Journal du cir
Présentation : Michel De

20.30 M6 20.40
Surprise-parti

DID YOU KNOW?

Soccer is the most important team sport in France. Almost every city has its own team, and competition is fierce for the national title. Soccer fans pack the stadiums on Sundays.

During qualifying games for the Coupe de France, in which the best French teams play for the national title, people are glued to their TV sets or radios. When France plays against other European teams, fans travel in large groups to neighboring countries to watch the games and support their teams. When France wins, French fans throughout the country drive their cars on the main streets of their city, honking their horns, yelling, singing, and waving the French flag. This goes on until the early hours of the morning.

Prononciation *Liaison et élision*

1. You have already seen that in French certain words are pronounced differently depending on whether they are followed by a vowel or a consonant. There is either liaison or elision. Compare the following.

   les copains / les‿amis   je regarde / j'écoute

2. Liaison is the linking of a usually silent consonant to the following word when the word begins with a vowel or silent *h*. Liaison occurs with plural subject pronouns, plural articles, and plural possessive adjectives. Repeat the following.

   ils‿ont gagné   les‿équipes   des‿arbitres   mes‿amis

3. Elision is the linking of a consonant and a vowel sound. It is made by dropping the vowel at the end of a word before a vowel at the beginning of the next. Elision occurs with the articles *le* and *la*, with the pronoun *je*, and with the negative word *ne*. Repeat the following.

   l'arbitre   l'équipe   j'attends   j'ai gagné   Tu n'écoutes pas!

   Now repeat and compare the following pairs of sentences.

   Vous‿avez perdu. / Vous n'avez pas perdu.
   J'ai fini. / Je n'ai pas fini.

l'arbitre

les‿arbitres

Activités de communication orale

A Le week-end dernier.   Your classmate wants to know what you did last weekend. Tell him or her several things you did or didn't do, using the verbs below. Then reverse roles.

acheter	dîner	jouer	regarder	téléphoner
attendre	dormir	manger	rigoler	travailler

B Ton sport d'équipe favori.   Your French friend Nathalie wants to know about your sports interests. Answer her questions.

1. Quelle est ton équipe préférée?
2. C'est une équipe de football?
3. Tu joues à ce sport ou tu préfères regarder les matchs à la télé?
4. Ton équipe préférée a gagné beaucoup de matchs cette année?

Nathalie

355

Prononciation
PRESENTATION *(page 355)*

A. Model the key words *l'arbitre/les arbitres* and have students repeat chorally.
B. Now lead students through steps 1–3 on page 355 and model the other words, phrases, and sentences.
C. You may wish to give the students the following *dictée:* **Mes amis ont joué. Ils ont gagné. L'arbitre a sifflé. L'équipe a marqué un but.**
D. For additional practice, you may wish to use Pronunciation Transparency P-13, the *Prononciation* section on Cassette 8A/CD-8, and the Student Tape Manual, Teacher's Edition, *Activités E–G*, pages 150–151.

Activités de communication orale
PRESENTATION *(page 355)*

Activité B

In the CD-ROM version of this activity, students can interact with an on-screen native speaker.

ANSWERS

Activité A
Answers will vary.

Activité B

1. Answers will vary.
2. Oui (Non), c'est (ce n'est pas) une équipe de football. (C'est une équipe de basket [baseball, etc.])
3. Je joue à ce sport. (Je préfère regarder les matchs à la télé.)
4. Oui (Non), mon équipe préférée (n') a (pas) gagné beaucoup de matchs cette année.

LEARNING FROM REALIA

Have students look at the TV guide on page 354 and scan it for enjoyment.

INDEPENDENT PRACTICE

Assign any of the following:
1. Exercise and activities, pages 354–355
2. Workbook, *Un Peu Plus*, page 136
3. CD-ROM, Disc 4, pages 354–355

LECTURE ET CULTURE

READING STRATEGIES
(page 356)

Note If your students aren't interested in sports, go over the *Lecture* quickly. If they're sports-minded, you can do the reading thoroughly.

Pre-reading
Give students a brief oral synopsis of the reading in French.

Reading
A. Call on individuals to read 2–3 sentences at a time. After each one reads, ask others follow-up questions.

B. Ask 5–6 questions that review the main points. The answers will give a coherent oral review of the *Lecture*.

C. Have a more able student summarize the *Lecture*. Call on slower students to answer questions about the summary. Then have a slower student orally summarize the reading based on the more able student's review.

D. **Writing** Have students in more able classes write their own summary of the *Lecture* in 6–7 minutes.

Post-reading
Have students work in groups to write brief news announcements for three different types of sports and present them to the class.

Note Students may listen to a recorded version of the *Lecture* on the CD-ROM.

Étude de mots

ANSWERS

Exercice A

1. f	4. e
2. d	5. c
3. a	6. b

LES SPORTS EN FRANCE

Est-ce que les Français sont des sportifs sérieux? On peut dire que les sports collectifs intéressent les Français moins que les Américains ou les Russes, par exemple. Mais de nos jours, de plus en plus de Français pratiquent un sport. Le sport d'équipe le plus populaire en France, c'est le football ou, comme on dit souvent, le foot. Chaque grande ville a son équipe de foot. Des championnats nationaux et internationaux attirent[1] des fanas du monde entier. Mais le football en France, et en Europe en général, n'est pas le même que le football américain. D'abord le ballon est rond et les joueurs ne peuvent pas toucher le ballon avec les mains. Ils donnent un coup de tête ou un coup de pied dans le ballon pour envoyer le ballon dans le but de l'équipe adverse.

En France on pratique presque[2] tous les sports—le basket-ball, le volley-ball et le hand-ball. Mais il y a un sport qu'on ne pratique jamais: c'est le base-ball. Le base-ball n'est pas du tout populaire.

La France est le pays du cyclisme. Les courses dans les vélodromes attirent toujours beaucoup de monde. En juillet le célèbre Tour de France a lieu[3]. C'est une course internationale tout autour du[4] pays. Les coureurs cyclistes professionnels de tous les pays du monde participent au Tour de France. On donne au gagnant un trophée. On donne aussi une somme d'argent au nouveau héros international.

[1] attirent *attract*
[2] presque *almost*
[3] a lieu *takes place*
[4] tout autour du *all around*

79ᵉ Tour de France

Étude de mots

A **Quelle est la définition?** Trouvez les mots qui correspondent.

1. un sport collectif	a. le contraire de «différent»
2. un sport individuel	b. faire du sport, jouer
3. le même	c. l'opposition
4. un joueur	d. un sport qu'on pratique seul
5. le camp adverse	e. une personne qui pratique un sport
6. pratiquer un sport	f. un sport d'équipe

CRITICAL THINKING ACTIVITY

(Thinking skills: supporting statements with reasons)

Read the following to the class or put it on the board or on a transparency.

1. **Vous préférez les sports d'équipe ou les sports individuels? Pourquoi?**

2. **Toutes les écoles aux États-Unis ont des équipes de sport. Dans beaucoup d'écoles les sports sont considérés comme très** importants. Il y a beaucoup de compétition. Qu'est-ce que vous en pensez? Cette compétition sportive est bonne ou pas? Justifiez vos arguments.

3. **Le base-ball n'est pas populaire en France. Pourquoi? Qu'en pensez-vous?**

Compréhension

B **Les sports.** Répondez par «oui» ou «non».

1. Les sports collectifs sont plus populaires en France qu'aux États-Unis.
2. Le sport d'équipe le plus populaire en France, c'est le football.
3. Le football est un sport collectif qu'on pratique en compétition.
4. Quand on joue au football américain on peut toucher le ballon avec les mains.
5. Le ballon de football en France est ovale.
6. Le base-ball est assez populaire en France.
7. Le cyclisme est plus populaire aux États-Unis qu'en France.
8. Le Tour de France a lieu au mois de septembre.

C **Les Français aiment les sports.** Répondez.

1. On pratique quels sports d'équipe en France?
2. Quel sport est-ce qu'on ne pratique jamais en France?
3. Quel sport est plus populaire en France qu'aux États-Unis?
4. Qu'est-ce que c'est, le Tour de France?
5. Qui participe au Tour de France?
6. Qu'est-ce qu'on donne au gagnant du Tour de France?

DÉCOUVERTE CULTURELLE

Il y a un sport qu'on pratique en France qui ressemble au football américain? Oui, mais ce n'est pas le foot. C'est le rugby. Le football américain ressemble au rugby.

Aux États-Unis toutes les écoles secondaires ont toujours des équipes de football américain et d'autres sports. En France, ce n'est pas le cas. Les sports ne sont pas très importants dans les lycées français. Il n'y a pas d'équipes organisées. Mais les élèves secondaires en France ont le mercredi après-midi libre et, grâce aux[1] associations sportives scolaires, ils peuvent profiter de leur temps libre pour faire du sport.

Les Françaises et les Français font de la gymnastique, du tennis et du jogging. Mais ce sont surtout les hommes qui jouent au foot.

[1] grâce aux *thanks to*

Compréhension

Exercice B

1. Non.   5. Non.
2. Oui.   6. Non.
3. Oui.   7. Non.
4. Oui.   8. Non.

Exercice C

1. On pratique le football, le basket-ball, et le volley-ball en France.
2. On ne pratique jamais le base-ball en France.
3. Le cyclisme (Le football) est plus populaire en France qu'aux États-Unis.
4. C'est une course cycliste internationale.
5. Les coureurs cyclistes professionnels de tous les pays du monde participent au Tour de France.
6. On donne un trophée et une somme d'argent au gagnant du Tour de France.

OPTIONAL MATERIAL

Découverte culturelle

PRESENTATION *(page 357)*

A. Before reading the selection, focus on the topics by asking students from which game American football is adapted. Also ask them to name in French all of the extracurricular and intramural sports available at your school. Which is the most popular?

B. Have students read the selection silently and restate its three main ideas in their own words.

Note Students may listen to a recorded version of the *Découverte culturelle* on the CD-ROM.

RÉALITÉS

RÉALITÉS

Bell Ringer Review

Write the following on the board or use BRR Blackline Master 13-5: Answer true or false.

1. Beaucoup de Français font du jogging dans les rues des villes.
2. La forme physique intéresse les Français.
3. Le Tour de France a lieu en juin.
4. Il n'y a pas de piscines couvertes en France.
5. Il y a un marathon à Paris en automne.

OPTIONAL MATERIAL

PRESENTATION *(pages 358–359)*

The purpose of this section is to have students enjoy the photographs and gain an appreciation of French culture. If you'd like to do something more, you may do some of the following activities.

A. Have students open their books to pages 358–359 and cover the captions on page 359. For each photo, allow them one minute to come up with one statement about it.

B. Call on volunteers to read the captions on page 359 and answer the questions in numbers 2, 4, and 5.

C. Challenge students to find at least five cognates in the captions and give their English equivalents.

D. Ask personal questions related to the activities shown in the pictures. For example: *Tu fais du cyclisme? Où? Il y a beaucoup de vélodromes aux États-Unis? Qui fait de l'alpinisme? Où? C'est un sport difficile? Dangereux?*

Note In the CD-ROM version, students can listen to the recorded captions and discover a hidden video behind one of the photos.

358

COOPERATIVE LEARNING

Display Communication Transparency C-13. Have students work in groups to make up as many questions as they can about the illustration. Have groups take turns asking and answering the questions.

INDEPENDENT PRACTICE

Assign any of the following:
1. *Étude de mots* and *Compréhension* exercises, pages 356–357
2. Situation Cards, Chapter 13
3. CD-ROM, Disc 4, pages 356–357

L'alpinisme est un sport très pratiqué en France **1**.

Voici des joueuses de volley-ball. Est-ce que l'arbitre regarde attentivement le match **2**?

En France aussi on aime faire du patin à roulettes sur les rampes **3**.

Le Tour de France finit à Paris. Quel monument parisien célèbre est sur la photo **4**?

C'est un match de football à Mulhouse **5**. On joue la nuit. Les gradins sont pleins?

DID YOU KNOW?

The Tour de France, an international bicycle race with competitors from all over the world, lasts for many days and covers several *étapes*, or stages. Each stage lasts one day, at the end of which the winner is awarded the coveted *maillot jaune* (yellow jersey), which he or she can wear the next day. The course is torturous, and many cyclists are eliminated because of heat and fatigue. The final victor carries off the *maillot jaune* and considerable financial winnings as well.

RECYCLING

The *Activités de communication orale* allow students to use the *passé composé* and practice the forms within meaningful contexts. The *Activité de communication écrite* recombines vocabulary and structures from earlier chapters with the sports context of Chapter 13.

INFORMAL ASSESSMENT

The highly guided nature of Oral Activities A and B make them suitable for evaluating the speaking skill. Use either or both as a speaking test. Use the evaluation criteria given on page 34 of this Teacher's Wraparound Edition.

Activités de communication orale

ANSWERS

Activité A

É2 answers will vary. É1 questions will vary, but may include the following:

Qu'est-ce que tu as étudié l'année dernière?

As-tu joué au tennis hier?, etc.

Activité B

Answers will vary.

Activité de communication écrite

ANSWERS

Activité A

Answers will vary, but may include the following:

1. C'est un match de foot (basket, etc.)/une course (à pied, cycliste).
2. Le match (La course) a lieu à ___ (place) le ___ (day of week) à ___ h.
3. L'équipe (de) ___ joue contre l'équipe (de) ___.
4. Nous pouvons y aller à pied (en voiture, en bus, etc.).

CULMINATION

Activités de communication orale

A **Le temps passé.** Ask a classmate several questions about his or her activities, using the verbs and the time expressions below. Then reverse roles and answer your partner's questions.

quitter/hier

Élève 1: À quelle heure as-tu quitté l'école hier?

Élève 2: J'ai quitté l'école à quatre heures hier.

VERBES	EXPRESSIONS DE TEMPS
étudier	ce matin
jouer	hier
perdre	hier soir
quitter	l'année dernière
regarder	la semaine dernière
téléphoner	pendant le week-end
travailler	avant-hier
attendre	

B **Une enquête.** Divide into small groups and choose a leader. Using the list below, the leader asks the others what they did last summer, takes notes, and reports to the class.

Élève 1: Qui a voyagé en Europe l'été dernier?

Élève 2: Moi, j'ai voyagé en Europe l'été dernier.

étudier le français	gagner beaucoup d'argent
jouer au tennis	voyager en avion / train / voiture
travailler	passer quelques semaines à la plage

Activité de communication écrite

A **Une invitation.** Your parents gave you two tickets to a sports event you want to see. Write a short note inviting a friend to go with you. Don't forget to tell your friend:

1. what the event is
2. when and where it's going to be
3. which teams are playing
4. how you plan to get there

FOR THE YOUNGER STUDENT

1. Have groups make posters for a sports day at your school. They should include the date, events, team names, times, etc.
2. Have students pick their favorite athlete and say as much as they can about him or her.

Réintroduction et recombinaison

A   **Mes vêtements.**   Donnez des réponses personnelles.

1. Quelle est la couleur de ta chemise préférée ou de ton chemisier préféré?
2. Quand tu achètes des chaussures, tu fais quelle pointure?
3. Tu achètes des vêtements prêt-à-porter ou sur mesure?
4. Si ton pantalon est trop large, tu as besoin de la taille au-dessus ou de la taille au-dessous?
5. Et s'il est trop serré, tu as besoin de quelle taille?

B   **Raoul.**   Répondez d'après le dessin.

1. Raoul est où?
2. Il parle à qui?
3. Que veut Raoul?
4. Qui met de l'essence dans le réservoir?
5. Qu'est-ce que la pompiste vérifie?

C   **Serge roule en voiture.**   Complétez.

1. Serge ___ bien. (conduire)
2. Il ___ le code de la route. (lire)
3. Il ___ que St.-Brieuc est assez loin d'ici. (dire)
4. Il ___ une carte postale de St.-Brieuc. (écrire)

D   **Et vous aussi!**   Récrivez les phrases de l'Exercice C en utilisant «vous».

Vocabulaire

NOMS
le foot(ball)
le terrain de football
l'équipe (f.)
le camp
le joueur
le gardien de but
le ballon
le but
l'arbitre (m.)
la tête
le pied

le basket(-ball)
le panier
le panneau
le demi-cercle

le base-ball
le volley-ball
le sol

le vélo
le cyclisme
le coureur cycliste
le coureur
la course
la piste
le stade
le gradin
le spectateur
le gagnant
la coupe

l'automne (m.)
le printemps

ADJECTIFS
adverse
comble
plein(e)

VERBES
dribbler
envoyer
lancer
opposer
siffler

AUTRES MOTS ET EXPRESSIONS
donner un coup de pied
marquer un but
contre
par dessus
jusqu'à
beaucoup de monde

hier
hier matin
hier soir
avant-hier
l'année dernière

INDEPENDENT PRACTICE

1. Activities and exercises, pages 360–361
2. Communication Activities Masters, pages 64–67
3. CD-ROM, Disc 4, pages 360–361

Réintroduction et recombinaison

RECYCLING

These exercises review question words and the irregular verbs *conduire, lire, écrire,* and *dire.* They also review vocabulary related to clothing and cars.

ANSWERS

Exercice A
Answers will vary.

Exercice B
1. Raoul est à la station-service.
2. Il parle au pompiste.
3. Il veut faire le plein.
4. Le pompiste met de l'essence dans le réservoir.
5. La pompiste vérifie les niveaux.

Exercice C

1. conduit   3. dit
2. lit   4. écrit

Exercice D
1. Vous conduisez bien.
2. Vous lisez le code de la route.
3. Vous dites que St.-Brieuc est assez loin d'ici.
4. Vous écrivez une carte postale de St.-Brieuc.

ASSESSMENT RESOURCES

1. Chapter Quizzes
2. Testing Program
3. Situation Cards
4. Communication Transparency C-13
5. Computer Software: Practice/Test Generator

VIDEO PROGRAM

INTRODUCTION   (41:44)

QU'EST-CE QU'ON A BIEN JOUÉ!   (42:42)

CHAPTER OVERVIEW

In this chapter students will learn to discuss winter sports: the activities themselves, some of the clothing and equipment needed for them, and some information about winter resorts in the French-speaking world. They will also learn to talk about winter weather. They will increase their ability to talk about actions using the *passé composé* of many irregular verbs that are conjugated with *avoir.* Students will also learn the question words *qui* and *quoi.*

The cultural focus of Chapter 14 is on winter sports facilities and traditions in France, Canada, and Switzerland.

CHAPTER OBJECTIVES

By the end of this chapter, students will know:

1. vocabulary associated with different types of skiing, ski equipment and clothing, and some ski resort personnel and procedures
2. vocabulary associated with ice skating
3. vocabulary associated with winter weather and weather reports in general
4. construction of the *passé composé* of irregular verbs that take *avoir*
5. the interrogative words *qui* and *quoi*

CHAPITRE

14

L'HIVER ET LES SPORTS D'HIVER

OBJECTIFS

In this chapter you will learn to do the following:

1. talk about skiing and ice skating
2. describe winter weather
3. describe past actions
4. ask "whom" or "what"
5. describe French and Canadian ski resorts

CHAPTER PROJECTS

(optional)

1. Have groups or individuals research the winter carnival in Quebec, including such things as dates, events, and the mascot, *Bonhomme Carnaval.*
2. Put on your own class version of a winter carnival, using information gathered from researching the one in Quebec, complete with snow queen, *Bonhomme Carnaval,* etc.
3. Using the formulas on page 339 of *Lettres et sciences (La Météorologie: La Prévision du Temps),* have students convert Fahrenheit temperatures to Celsius and vice versa. Apply Celsius to some familiar temperature ranges in order to give students a "feel" for it. For example, water freezes at 0°C and boils at 100°C. A comfortable Celsius ambient temperature is 24°.

363

CHAPTER 14 RESOURCES

1. Workbook
2. Student Tape Manual
3. Audio Cassette 8B/CD-8
4. Bell Ringer Review
   Blackline Masters
5. Vocabulary Transparencies
6. Pronunciation
   Transparency P-14
7. Communication
   Transparency C-14
8. Communication Activities
   Masters
9. Map Transparencies
10. Situation Cards
11. Conversation Video
12. Videocassette/Videodisc,
    Unit 4
13. Video Activities Booklet,
    Unit 4
14. Lesson Plans
15. Computer Software:
    Practice/Test Generator
16. Chapter Quizzes
17. Testing Program
18. Internet Activities Booklet
19. CD-ROM Interactive
    Textbook

Pacing

This chapter requires eight
to ten class sessions. Pacing
will vary according to class
length and the age and aptitude
of the students.

Note The Lesson Plans offer
guidelines for 45- and 55-
minute classes and **Block
Scheduling.**

Exercices vs. *Activités*

All exercises (which provide
guided practice) are coded in
blue. All communicative activities are coded in red.

INTERNET ACTIVITIES

(optional)
These activities, student worksheets,
and related teacher information are in the
Bienvenue Internet Activities Booklet and on
the Glencoe Foreign Language Home Page at:
http://www.glencoe.com/secondary/fl

LEARNING FROM PHOTOS

After presenting the chapter vocabulary,
you may wish to ask questions about the
photo: *C'est la ville de Québec? C'est quelle
saison? C'est le Palais de Glace? Le Palais est
beau? Est-ce que vous faites des sculptures de
neige ou de glace en hiver? Vous faites des
bonhommes de neige* (snowmen)?

VOCABULAIRE

MOTS 1

UNE STATION DE SPORTS D'HIVER

un sommet

une montagne

une piste très raide

une vallée

des bosses (f.)

un chalet

un télésiège

une skieuse

un skieur — un bonnet

des lunettes (f.)

une écharpe

un anorak

un gant

un bâton

un ski

une chaussure de ski

364 CHAPITRE 14

Bell Ringer Review

Write the following on the board or use BRR Blackline Master 14-1: Make two lists in French: one of summer sports and one of any weather expressions you remember.

PRESENTATION *(pages 364–365)*

A. To vary the procedure for presenting the vocabulary, you may wish to ask students to open their books to pages 364–365. Have them look at the illustrations as you play Cassette 8B/CD-8 once.

B. Show Vocabulary Transparencies 14.1 (A & B). Have students close their books and repeat the new words after you two or three times.

C. Call a student to the front of the room. As you say a new word or phrase, have the student point to the appropriate item on the transparency.

TOTAL PHYSICAL RESPONSE

(following the Vocabulary presentation)

Getting Ready

Demonstrate *faire une chute.*

TPR 1

____, venez ici, s'il vous plaît.
Asseyez-vous ici, s'il vous plaît.
Mettez vos chaussures de ski.
Et maintenant, levez-vous.

Mettez votre anorak.
Mettez votre bonnet.
Mettez vos lunettes et vos gants.
Mettez les skis.
Prenez les bâtons.
Prenez un bâton dans la main droite.
Prenez l'autre bâton dans la main gauche.
Et maintenant, allez faire du ski!
Merci, ____. Vous pouvez retourner à votre place.

le ski de fond

le ski alpin

une piste de slalom

un moniteur   une monitrice

Marie est débutante.
L'hiver dernier elle a pris des leçons de ski.
Elle a appris à faire du ski.
Elle a eu un très bon moniteur.
Le moniteur a appris à faire du ski à Marie.
Elle a compris les instructions du moniteur.

Marie a mis son anorak.
Elle a mis ses gants, son écharpe
   et son bonnet.
Elle a mis ses skis.

Marie a descendu la piste.
Elle a descendu la piste verte.
La piste verte est pour les débutants.

D. Have a student point to items on the transparencies as he or she asks *Qu'est-ce que c'est?* or *Qui est-ce?* and calls on classmates to respond.
E. Now have students read the words and sentences aloud for additional reinforcement.
F. Ask questions about the sentences on page 365. For example: *Marie est débutante ou experte? Quand est-ce qu'elle a pris des leçons de ski? Qu'est-ce qu'elle a fait l'hiver dernier? Qu'est-ce qu'elle a appris? Qui a appris à Marie à faire du ski? Qu'est-ce que Marie a compris?*
G. Bring magazines or catalogs to class that show ski clothing, equipment, and actions. Ask questions about the pictures. For example: *Qui descend la piste? Elle va très vite? Elle a mis quels vêtements? De quelle couleur est son écharpe? De quelle couleur est ce pull? De quelle couleur est l'anorak du skieur? De quelle couleur est son bonnet? Et ses gants?*

TPR 2
____, venez ici, s'il vous plaît.
Vous êtes dans une station de sports d'hiver.
Faites la queue.
Attendez le télésiège.
Le voilà, il arrive. Asseyez-vous sur le
   télésiège.
Prenez les bâtons dans votre main gauche.

À tout à l'heure, ____!
Maintenant vous êtes au sommet de la
   montagne.
Descendez du télésiège.
Prenez un bâton dans chaque main.
Descendez. Faites une chute!
Regardez la jambe.
Non, il n'y a pas de problème. Levez-vous et
   skiez encore.
Merci, ____. Retournez à votre place.

Exercices

PRESENTATION (*page 366*)

Exercice B: Speaking

After doing Exercise B as a whole-class activity, focus on the speaking skill by having students work in pairs. One partner reads the questions to the other in random order. The second partner listens and answers with his/her book closed. Partners then reverse roles.

ANSWERS

Exercice A

1. Oui, Marie a appris à faire du ski.
2. Le moniteur a appris à Marie à faire du ski.
3. Oui, elle a eu un très bon moniteur.
4. Oui, elle a compris les instructions du moniteur.
5. Oui, Marie a mis son anorak.
6. Oui, elle a mis ses gants, son écharpe et son bonnet.
7. Oui, elle a mis ses chaussures de ski et ses skis.
8. Elle a descendu la piste verte.
9. Oui, la piste verte est pour les débutants.

Exercice B

1. Non. (Le ski est un sport d'hiver.)
2. Oui.
3. Non. (Ce n'est pas une piste avec des bosses.)
4. Oui.
5. Non. (Les skieurs prennent le télésiège pour monter.)
6. Oui.
7. Oui.
8. Non. (Les débutants descendent la piste verte.)
9. Oui.
10. Oui.

Exercices

A **Marie a appris à faire du ski.** Répondez.

1. Marie a appris à faire du ski?
2. Qui a appris à Marie à faire du ski?
3. Elle a eu un très bon moniteur?
4. Elle a compris les instructions du moniteur?
5. Marie a mis son anorak?
6. Elle a mis ses gants, son écharpe et son bonnet?
7. Elle a mis ses chaussures de ski et ses skis?
8. Elle a descendu quelle piste?
9. La piste verte est pour les débutants?

B **Un sport fabuleux.** Répondez par «oui» ou «non».

1. Le ski est un sport d'été.
2. Les débutants ne font pas bien de ski.
3. Une piste très raide, c'est une piste avec des bosses.
4. Le moniteur ou la monitrice apprend à faire du ski aux débutants.
5. Les skieurs prennent le télésiège pour descendre la piste.
6. Les skieurs prennent le télésiège pour monter au sommet de la montagne.
7. On n'a pas vraiment besoin de pistes pour faire du ski de fond.
8. Les débutants descendent la piste de slalom.
9. Les skieurs portent souvent des lunettes.
10. Après le ski on va dans le chalet.

Méribel: Des skieurs déjeunent à la terrasse d'un restaurant.

DID YOU KNOW?

You may wish to tell students that the town of Méribel, on the French-Italian border, has been a fashionable ski resort for over 50 years. Because of its location, it is possible to ski in the sun all day long. Like many well-known ski resorts in the Alps, it has a variety of restaurants, shops, and nightclubs.

ADDITIONAL PRACTICE

1. Show a video about skiing to the class. Stop the video when the scene changes and ask students to describe in French what they have seen. Guide them with your own questions and comments.
2. Student Tape Manual, Teacher's Edition, *Activité B,* page 156

C On fait du ski. Répondez d'après les dessins.

1. C'est une station balnéaire ou une station de sports d'hiver?
2. C'est une plage ou une montagne?
3. C'est une piste ou une piscine?
4. C'est un skieur ou un nageur?
5. C'est un ski nautique ou un bâton?
6. C'est un maillot ou un anorak?
7. Elle fait du ski alpin ou du ski nautique?
8. Il fait du ski de fond ou du ski alpin?
9. C'est le sommet de la montagne ou la vallée?
10. C'est un gant ou une écharpe?

PRESENTATION (page 367)

Extension of *Exercice C*: Speaking

After completing Exercise C, have students make at least one additional statement about each of the illustrations. Statements can be affirmative or negative.

ANSWERS

Exercice C

1. C'est une station de sports d'hiver.
2. C'est une montagne.
3. C'est une piste.
4. C'est un skieur.
5. C'est un bâton.
6. C'est un anorak.
7. Elle fait du ski alpin.
8. Il fait du ski de fond.
9. C'est le sommet de la montagne.
10. C'est un gant.

INFORMAL ASSESSMENT
(*Mots 1*)

Check for comprehension by reading sentences, words, or expressions from *Mots 1* in random order and having individuals point to the corresponding illustration on Vocabulary Transparencies 14.1 (A & B).

COOPERATIVE LEARNING

Have teams create composite stories about a ski trip. Start them off with one sentence, such as *Les Dupont arrivent à la station de sports d'hiver.* The team copies the sentence and then passes it around, each member adding a sentence until a story emerges. Teams can pass the story around as many times as they wish. Call on volunteers to read the finished stories.

INDEPENDENT PRACTICE

Assign any of the following:
1. Exercises, pages 366–367
2. Workbook, *Mots 1: A–D*, pages 138–139
3. Communication Activities Masters, *Mots 1: A*, page 68
4. CD-ROM, Disc 4, pages 364–367

VOCABULAIRE

MOTS 2

Vocabulary Teaching Resources

1. Vocabulary Transparencies 14.2 (A & B)
2. Audio Cassette 8B/CD-8
3. Student Tape Manual, Teacher's Edition, *Mots 2: D–F,* pages 157–158
4. Workbook, *Mots 2: E–G,* pages 139–140
5. Communication Activities Masters, *Mots 2: B,* page 68
6. Computer Software, *Vocabulaire*
7. Chapter Quizzes, *Mots 2:* Quiz 2, page 74
8. CD-ROM, Disc 4, *Mots 2:* pages 368–371

Bell Ringer Review

Write the following on the board or use BRR Blackline Master 14-2: Write down these two categories: *faire du ski nautique, faire du ski alpin.* What words do you associate with each?

PRESENTATION (*pages 368–369*)

A. Have students close their books. Briefly review seasons and weather-related vocabulary from Chapter 9.
B. Using an appropriate illustration from a wall calendar, magazine, or ski poster, show a winter scene and introduce as much of the *Mots 2* vocabulary as possible from pages 368–369. Have students repeat the new vocabulary after you.
C. Introduce the temperature vocabulary by drawing a thermometer on the board and asking *Quelle est la température aujourd'hui?* Repeat this a few times, changing the temperature each time.

EN HIVER

Il fait froid.
Le ciel est couvert.
Il neige.
Il gèle.
Le vent est très froid.

Quelle est la température aujourd'hui?
Il fait deux (degrés Celsius).

TOTAL PHYSICAL RESPONSE

TPR

____, venez ici, s'il vous plaît.
Vous allez faire le mime.
Il fait très froid et vous avez froid.
Mettez votre anorak.
Mettez vos patins.
Faites du patin.
Tournez à gauche.
Tournez à droite.
Faites une chute.
Levez-vous.
Vous êtes fatigué(e).
Vous voyez un ami.
Saluez votre ami.
Faites une boule de neige. Lancez la boule de neige à votre ami.
Très bien, ____. Merci. Maintenant, retournez à votre place.

jouer dans la neige

lancer une
boule de neige

une patinoire

une patineuse

un patineur

le patinage

la glace

un patin à glace

Hier Robert a fait du patin.
Il a eu un petit accident.
Il a fait une chute.

D. Referring to Vocabulary Transparencies 14.2 (A & B), have students keep their books closed as they repeat the vocabulary chorally after you or Cassette 8B/CD-8. Then repeat the procedure with books open.

E. Ask yes/no and either/or questions to elicit the vocabulary, referring to the Vocabulary Transparencies. For example: *Il fait chaud ou il gèle? Il fait du soleil ou le ciel est couvert? Est-ce qu'elle lance une boule de neige? C'est de la glace ou de la neige?*

F. Now ask interrogative-word questions. For example: *Quel temps fait-il? Qu'est-ce qu'elle lance? Et lui, qu'est-ce qu'il fait? Qu'est-ce qu'il va faire? Que fait cette patineuse?*, etc.

G. Include some or all of the exercises on page 370 as you present the *Mots 2* vocabulary. These exercises help students learn the new words in *Mots 2*.

COOPERATIVE LEARNING

Have students form teams and write two lists, one of summer and one of winter activities. Each team exchanges lists with another team. Teams then divide the activities on the lists they have received among their members, who write down what gear and clothing are needed for each of the activities as well as short descriptions of locations where each activity can take place. When finished, have teams share all the information. If you wish, have the teams create composite class lists for each activity on the board.

Exercices

PRESENTATION *(page 370)*

Exercices A, B, C, and D

You may want to have students write the answers to all of these exercises or just some of them, after you have gone over them in class.

Extension of Exercices A, B, and C

Call on a student to retell the story in Exercises A, B, and C in his/her own words.

Exercice D

You may wish to use the recorded version of this exercise.

ANSWERS

Exercice A

1. Robert a fait du patin.
2. Il a mis ses patins.
3. Il a fait une chute sur la patinoire.
4. Il a eu un petit accident.

Exercice B

1. En hiver il fait froid.
2. Il neige en hiver.
3. Quand il neige, le ciel est couvert.
4. Quand il neige, il fait froid.
5. Oui, il gèle quelquefois en hiver.
6. Oui, le vent est froid.
7. En général, il fait entre ___ degrés et ___ degrés en hiver dans ma ville.
8. Les températures en hiver sont basses.

Exercice C

Answers will vary.

Exercice D

1. le patinage
2. le ski
3. le ski
4. le ski
5. le patinage
6. le ski
7. le patinage

370

Exercices

 **A** **Le petit accident de Robert.** Répondez.

1. Robert a fait du patin ou du ski?
2. Il a mis ses patins ou ses skis?
3. Il a fait une chute sur la patinoire ou sur la piste de slalom?
4. Il a eu un petit accident ou un accident grave?

B **Le temps en hiver.** Répondez.

1. En hiver il fait froid ou il fait chaud?
2. Il neige en hiver ou en été?
3. Quand il neige, le ciel est couvert ou il fait du soleil?
4. Quand il neige, il fait chaud ou il fait froid?
5. Il gèle quelquefois en hiver?
6. Le vent est froid?
7. En général, quelle est la température dans ta ville en hiver?
8. Les températures en hiver sont basses ou élevées?

C **Les sports d'hiver et d'été.** Donnez des réponses personnelles.

1. Tu préfères l'été ou l'hiver?
2. Quelle est ta saison favorite?
3. Tu préfères les sports d'hiver ou les sports d'été?
4. Qu'est-ce que tu mets quand il fait très froid?
5. Tu as fait du ski? Où?
6. Tu aimes faire du ski?
7. Il y a une station de sports d'hiver près de chez toi?
8. Tu aimes jouer dans la neige?
9. Tu aimes lancer des boules de neige?
10. Tu aimes faire du patin?
11. Tu es bon patineur ou bonne patineuse?
12. Tu as des patins à glace?

D **C'est le ski ou le patinage?** Choisissez.

1. On pratique ce sport sur la glace.
2. On pratique ce sport sur la neige.
3. On descend une piste.
4. Les champions font du slalom.
5. On met des patins à glace.
6. On utilise des bâtons.
7. On pratique ce sport sur une patinoire.

370 CHAPITRE 14

ADDITIONAL PRACTICE

After completing Exercises A–D, tell students they have just won an all-expenses-paid vacation to the ski area of their choice, and they may take along a friend. They should write a note to their friend in which they:

1. explain that they have won a trip to a ski resort
2. say where they want to go and why
3. invite their friend to go along
4. say what the weather is going to be like
5. tell what clothing and equipment their friend will need.

Activités de communication orale
Mots 1 et 2

A **À quels sports joue-t-on?** A French exchange student (your partner) asks you what the weather is like in your town in summer and winter and what people do during these seasons. Give him or her as much information as you can.

B **La météo: Il va faire quel temps demain?** Tomorrow is Saturday, and you'd like to make some plans. Find out if a classmate has heard the weather report (*la météo*) and, if so, what the weather's going to be like. Based on what your partner says about the weather, make some plans with him or her for either an indoor or an outdoor activity.

C **Dans les Alpes.** While skiing at a resort in the French Alps, you meet Jacques Monnier. Answer his questions.

1. Bonjour. Tu es des États-Unis?
2. Tu fais souvent du ski?
3. Tu fais aussi du ski de fond?
4. Tu aimes mieux le ski de fond ou le ski alpin?

Jacques Monnier

Bell Ringer Review

Write the following on the board or use BRR Blackline Master 14-3: You are going on a ski trip for two days. Make a list of the things you will take with you. Include clothing, personal care items, ski equipment, and anything else you may need.

PRESENTATION *(page 371)*

Activité C

In the CD-ROM version of this activity, students can interact with an on-screen native speaker.

ANSWERS

Activité A

Answers will vary.

Activité B

É2 answers will vary. É1 questions will vary but may include the following:

1. **Tu as entendu la météo pour demain?**
2. **Il va faire quel temps demain? (Il va neiger/faire froid?, etc.)**
3. **Tu voudrais faire du ski (jouer dans la neige/faire du patin/te promener, etc.) avec moi demain?**

Activité C

Answers will vary but may include the following:

1. Bonjour. Oui, je suis des États-Unis.
2. Oui (Non), je (ne) fais (pas) souvent du (de) ski.
3. Oui, je fais aussi du ski de fond. (Non, je ne fais pas de ski de fond.)
4. J'aime mieux le ski de fond (alpin).

RETEACHING *(Mots 1 and 2)*

Have students take turns pantomiming words and expressions from *Mots 1* and *2* while the rest of the class guesses.

STRUCTURE

Le passé composé des verbes irréguliers

Describing Past Actions

1. You have already learned the past participles of regular verbs in French which end with an /é/, /i/, or /ü/ sound. Note the past participles of the following irregular verbs which also end with an /i/ or /ü/ sound.

INFINITIF ⟶	PARTICIPE PASSÉ
mettre	mis
permettre	permis
prendre	pris
comprendre	compris
apprendre	appris
dire	dit
écrire	écrit
conduire	conduit
avoir	eu
croire	cru
voir	vu
pouvoir	pu
vouloir	voulu
lire	lu

Un skieur sur les pistes de La Plagne

J'ai pris des leçons de ski.
J'ai appris à faire du ski.
J'ai compris toutes les instructions de la monitrice.
Elle a dit: «Bravo! Vous faites très bien du ski!»
J'ai eu de la chance. J'ai eu une très bonne monitrice.
Elle a écrit un livre sur le ski alpin. J'ai lu son livre.

2. The commonly used verbs *être* and *faire* also have irregular past participles.

être	été
faire	fait

J'ai fait un voyage à Megève l'année dernière.
J'ai été très content de pouvoir faire du ski.

Structure Teaching Resources

1. Workbook, *Structure:* A–E, pages 141–143
2. Student Tape Manual, Teacher's Edition, *Structure:* A–B, pages 159–160
3. Audio Cassette 8B/CD-8
4. Communication Activities Masters, *Structure:* A–C, pages 69–70
5. Computer Software, *Structure*
6. Chapter Quizzes, *Structure:* Quizzes 3–4, pages 75–76
7. CD-ROM, Disc 4, pages 372–375

Bell Ringer Review

Write the following on the board or use BRR Blackline Master 14-4: Write sentences in the passé composé using each of the following verbs.

faire du ski	jouer
faire du patin	lancer

Le passé composé des verbes irréguliers

PRESENTATION *(pages 372–373)*

A. Explain step 1 quickly to students. Have them repeat the /i/ and /ü/ sounds in isolation. Then have them repeat after you the past participles in the list.

B. Write the infinitives from the chart on the board. As you write each one, have the class give you the appropriate past participle. Write it alongside the infinitive.

C. In the case of the /i/ verbs, underline the *-s* and *-t*. Tell students to remember the difference in spelling.

D. Have students read the sentences in unison or call on individuals to read.

DID YOU KNOW?

You may wish to tell students that the skier pictured above is Denis Lechaplain, who began skiing at age 25, seven years after a terrible car accident in which he lost the use of his legs. He now slides down all types of slopes on a "ski-chair" and has been actively involved in ski competitions and in promoting mountain sports among the disabled.

In the photo, he is skiing the slopes of the La Plagne ski resort in the French Alps, site of the 1992 Albertville Olympic luge and bobsled events. La Plagne is actually a conglomeration of 10 villages, each with its own hotels, chalets, restaurants, etc. It is a favorite spot for families because of its child care centers and children's ski schools.

3. Note the position of short adverbs such as *déjà*, *bien*, *trop*, and *vite* with the *passé composé*. They are placed between *avoir* and the past participle.

J'ai *déjà* mangé.	I have already eaten.
Il a *vite* fini son sandwich.	He quickly finished his sandwich.
Il a *bien* choisi son moniteur.	He chose his instructor well.

Adverbs of time such as *hier* and *aujourd'hui* follow the past participle.

Il a fait du ski *hier*.
Mais il n'a pas fait de ski *aujourd'hui*.

Exercices

A **Gilles a fait du ski.** Répondez d'après les dessins.

Bonne chance, Gilles!

1. Gilles a mis son anorak?
2. Il a dit «Bonne chance» à son ami?
3. Son ami a déjà fait du ski aujourd'hui?
4. Gilles a bien fait du ski?
5. Il a eu un accident?
6. Après l'accident Gilles a lu un livre pour les débutants?

E. Lead students through steps 2 and 3.

Note In the CD-ROM version, this structure point is presented via an interactive electronic comic strip.

Exercices

PRESENTATION *(page 373)*

Extension of *Exercice A*
Call on students to give a summary of the story after doing Exercise A.

ANSWERS

Exercice A
1. Oui, Gilles a mis son anorak.
2. Non, son ami a dit «bonne chance» à Gilles.
3. Oui, son ami a déjà fait du ski aujourd'hui.
4. Non, Gilles n'a pas bien fait de ski.
5. Oui, il a eu an accident.
6. Oui, après l'accident Gilles a lu un livre pour les débutants.

ADDITIONAL PRACTICE	**INDEPENDENT PRACTICE**
1. Have students refer to the illustrations that accompany Exercise A and make up their own story based on them. This can be done orally or in writing. 2. Student Tape Manual, Teacher's Edition, *Activités A–B,* pages 159–160	Assign any of the following: 1. Exercises, pages 373–374 2. Workbook, *Structure: A–C,* pages 141–142 3. Communication Activities Masters, *Structure: A,* page 69 4. Computer Software, *Structure* 5. CD-ROM, Disc 4, pages 372–374

ANSWERS

Exercice B

1. a dit, a lu, a écrit
2. avons dit, avons lu, avons écrit
3. as dit, as lu, as écrit
4. avez dit, avez lu, avez écrit
5. ont dit, ont lu, ont écrit

Exercice C

Anwers will vary but may include the following:

1. Oui (Non), j'ai (je n'ai pas) lu… Mes parents (n') ont (pas) lu…
2. Oui (Non), les élèves (n') ont (pas) lu…
3. Oui (Non), j'ai (je n'ai pas) dit…
4. Oui (Non), nous (n') avons (pas) dit…
5. Oui, la femme a dit «Zut!» quand elle a trouvé…
6. Oui (Non), les élèves (n') ont (pas) écrit des (de)…
7. Oui (Non), ils (n') ont (pas) écrit une (de)…
8. Oui (Non), j'ai bien (je n'ai pas bien) écrit…

Exercice D

1. a dit	7. a conduit
2. a lu	8. a fait
3. a vu	9. a été
4. a voulu	10. ont mis
5. ont permis	11. ont pris
6. a pris	12. ont, eu

RETEACHING

Have students write a sentence in the *passé composé* using each of the following verbs. They should change the subject each time. For further practice, have them exchange sentences with a partner and change the partner's sentences to the negative.

dire	lire
prendre	croire
pouvoir	mettre
avoir	faire

B **Tu as dit quoi?** Complétez d'après le modèle avec «dire», «lire» ou «écrire».

> J'___ que j'___ ce que j'___.
> *J'ai dit que j'ai lu ce que j'ai écrit.*

1. Il ___ qu'il ___ ce qu'il ___.
2. Nous ___ que nous ___ ce que nous ___.
3. Tu ___ que tu ___ ce que tu ___.
4. Vous ___ que vous ___ ce que vous ___.
5. Elles ___ qu'elles ___ ce qu'elles ___.

C **Qu'est-ce qu'on a fait?** Répondez.

1. Est-ce que tu as lu le journal ce matin? Et tes parents?
2. Les élèves ont lu leur livre de français avant l'examen?
3. Est-ce que tu as dit «Salut!» à tes copains ce matin?
4. Tes amis et toi, vous avez dit «Au revoir!» à votre professeur de français hier?
5. La femme a dit «Zut!» quand elle a trouvé une contravention sur le parebrise de sa voiture?
6. Les élèves ont écrit des lettres à leurs grands-parents?
7. Ils ont écrit une composition au cours d'anglais?
8. Est-ce que tu as bien écrit cet exercice?

D **En route!** Complétez au passé composé.

Mon ami Laurent ___ (dire) que
Chamonix est une belle station de sports
d'hiver. Il ___ (lire) le Guide Michelin et
il ___ (voir) que Chamonix est loin de
Paris. Mais il ___ (vouloir) y aller. Ses
parents ___ (permettre) à Laurent
de prendre leur voiture. Il ___ (prendre)
leur voiture et il ___ (conduire) jusqu'à
Chamonix. Il ___ (faire) le voyage avec
son copain Alain qui ___ (être) très
content de partir avec lui. Ils ___ (mettre)
leurs skis sur la voiture. Ils ___ (prendre)
l'autoroute. Ils n'___ pas ___ (avoir) de
problème.

La Mer de Glace près de Chamonix

ADDITIONAL PRACTICE

You may wish to ask the following questions to practice the *passé composé* while reviewing previously learned vocabulary: *Gilles a mis son maillot? Il a mis de la crème solaire? Il a pris un bain de soleil à la plage? Il a pris des leçons de natation? Il a eu un bon moniteur? Il a appris à faire du ski nautique? Il a fait du ski nautique? Il a compris tout ce que le moniteur a dit?*

DID YOU KNOW?

At Chamonix (see GEOGRAPHY CONNECTION, page 359) one can board a two-car train that climbs up to La Mer de Glace, a glacier offering dramatic views. Its unusual caves are filled with ice sculptures. Some visitors returning from the top of *l'Aiguille du Midi* (see DID YOU KNOW?, page 376), get off the gondola halfway down and hike along a scenic, winding trail to La Mer de Glace.

Les pronoms *qui* et *quoi*   *Asking "Whom" or "What"*

1. You use the pronouns *qui*, "whom," and *quoi*, "what," with prepositions such as *à*, *de*, *avec*, and *chez* to ask questions in French. *Qui* refers to a person and *quoi* refers to a thing. Study the following examples.

> **Tu parles à qui?**
> **Tu vas chez qui?**
> **Tu parles de quoi?**

2. Note the inversion in formal or written French.

INFORMAL	FORMAL
Vous parlez à qui?	**À qui parlez-vous?**
Vous allez chez qui?	**Chez qui allez-vous?**
Vous avez besoin de quoi?	**De quoi avez-vous besoin?**

Exercices

A   **Comment? Je n'ai pas entendu.**   Répondez d'après le modèle.

> **Elle parle de sa sœur.**
> *Comment? Je n'ai pas entendu. Elle parle de qui?*

1. Elle parle de sa tante.
2. Elle parle de son prof.
3. Elle parle au moniteur.
4. Elle parle à son amie.
5. Elle est chez ses parents.
6. Elle va chez son copain.
7. Elle travaille avec sa cousine.
8. Elle parle de son travail.
9. Elle parle de ses vacances à la montagne.
10. Elle a besoin d'argent.
11. Elle a besoin de skis.

B   **Au téléphone.**   Posez une question d'après le modèle.

> **Vous allez au cinéma avec votre amie.**
> *Avec qui allez-vous au cinéma?*

1. Vous téléphonez à votre amie.
2. Vous parlez à votre amie.
3. Vous parlez de choses sérieuses.
4. Vous laissez un message pour le frère de votre amie.

Un forfait-journée

Les pronoms *qui et quoi*

PRESENTATION (*page 375*)

A. To show that *qui* is for a person and *quoi* is for a thing, draw a stick figure on the board and ask: *Qui?* Then draw a box and ask: *Quoi?*

B. Lead students through steps 1–2 on page 375. Explain the use of *qui* versus *quoi* with prepositions.

Teaching Tip Explain the term "object of a preposition" and provide examples in both French and English.

Exercices

PRESENTATION (*page 375*)

Extension of *Exercice A*

You may want to do this exercise again, this time with formal word order.

ANSWERS

Exercice A

1. Comment? Je n'ai pas entendu. Elle parle de qui?
2. … Elle parle de qui?
3. … Elle parle à qui?
4. … Elle parle à qui?
5. … Elle est chez qui?
6. … Elle va chez qui?
7. … Elle travaille avec qui?
8. … Elle parle de quoi?
9. … Elle parle de quoi?
10. … Elle a besoin de quoi?
11. … Elle a besoin de quoi?

Exercice B

1. À qui téléphonez-vous?
2. À qui parlez-vous?
3. De quoi parlez-vous?
4. Pour qui laissez-vous un message?

ADDITIONAL PRACTICE	**INDEPENDENT PRACTICE**
Have students write questions with *qui* or *quoi* that would elicit the following answers: 1. **Hélène va au cinéma avec son amie.** 2. **Elle a besoin d'argent.** 3. **Max achète un cadeau pour sa mère.** 4. **Les amis vont chez Luc après les cours.** 5. **Madame Martin téléphone à sa fille.**	Assign any of the following: 1. Exercises, page 375 2. Workbook, *Structure: D–E*, page 143 3. Communication Activities Masters, *Structure: B–C*, page 70 4. Computer Software, *Structure* 5. CD-ROM, Disc 4, pages 372–375

PRESENTATION *(page 376)*

A. Tell students they will hear a conversation between Lisette and Michel who are at a ski resort.

B. Have them close their books and watch the Conversation Video, then have them repeat after you or Cassette 8B/CD-8.

C. Call on students to read and dramatize the conversation.

D. Have pairs make up a similar conversation about skiing or skating.

Note In the CD-ROM version, students can play the role of either one of the characters and record the conversation.

ANSWERS

Exercice A

1. Michel a fait du ski hier.
2. Il a descendu la piste noire.
3. La piste noire est difficile.
4. Oui, les pistes noires sont des pistes très raides.
5. Oui, Michel a eu un problème.
6. Oui, il a fait une chute.

Prononciation

PRESENTATION *(page 376)*

A. Model the key word *une radio* and have students repeat chorally.

B. Now model the other words and sentences in similar fashion.

376

Scènes de la vie *Tu as fait du ski?*

LISETTE: Michel, tu as fait du ski hier?
MICHEL: Oui. J'ai descendu la piste noire.
LISETTE: La piste noire? Mais tu es fou! C'est dangereux.
MICHEL: Oui, mais je n'ai pas eu de problème.
LISETTE: Tu n'as pas fait de chute?
MICHEL: Si, une petite chute, rien de grave!

A **La piste noire.** Répondez d'après la conversation.

1. Qui a fait du ski hier?
2. Il a descendu quelle piste?
3. La piste noire est facile ou difficile?
4. Les pistes noires sont des pistes très raides?
5. Michel a eu un problème?
6. Il a fait une chute?

Prononciation *Le son /r/ initial*

You have already practiced saying the /r/ sound in the middle or at the end of a word. You will now practice saying it at the beginning of a word. Repeat the following pairs of words.

opéra / radio	mari / restaurant
favori / rigoler	adoré / rez-de-chaussée

Now repeat the following sentences.

C'est la radio qui réveille Richard.
Pour rester en forme, Raoul ne regarde pas trop la télévison.
Robert roule très vite dans sa Renault rouge.

une radio

DID YOU KNOW?

Although the French like to ski in the Pyrenees, the Vosges, and the Jura, the slopes of the Alps are their favorite places to ski. The Alps offer long ski runs and breathtaking views.

One of the most spectacular and popular ski lifts in all of Europe is *l'Aiguille du Midi*, next to Mont Blanc. To get to the top of the 12,600-foot-high rock needle, skiers take two different gondolas. As they get off the second gondola, they proceed through an ice tunnel to reach the top of the slope. From there, the view is magnificent and the descent exhilarating.

Activités de communication orale

A **Au téléphone.** Find out if a classmate talked on the phone last night. If he or she did, find out who your partner spoke to (*à qui*) and what they talked about (*de quoi*). Then reverse roles.

B **Le week-end dernier.** You want to know if a classmate did one of the activities listed below last weekend. If the answer is "yes," try to get some details. Then reverse roles.

> voir un film
> Élève 1: Tu as vu un film le week-end dernier?
> Élève 2: Oui, j'ai vu *Independence Day*.
> Élève 1: C'est un bon film?

avoir un accident	inviter un copain ou une copine au cinéma
écrire une composition	jouer au football / base-ball / basket-ball, etc.
étudier	lire un journal / un magazine / un livre
faire ses devoirs	parler au téléphone
faire du ski	regarder la télé
faire du patin	

C **J'ai appris à…** Tell a classmate something you learned to do recently (last week, last winter, last summer, etc.). Your partner asks you for the information below. Answer, then reverse roles.

1. when you learned to do the activity
2. where you learned
3. who taught you
4. if you took lessons
5. if you had a good instructor
6. if you understood the instructions

On fait beaucoup de ski au Canada.

INDEPENDENT PRACTICE

Assign any of the following:
1. Exercise and activities, pages 376–377
2. CD-ROM, Disc 4, pages 376–377

LEARNING FROM PHOTOS

Have students say as much about the photo as they can.

C. You may wish to give students the following *dictée*:
Robert regarde la route. Carole écoute une opéra à la radio. Le mari de Carole adore ce restaurant. René roule vite sur la route.

D. For additional practice, use Pronunciation Transparency P-14, the *Prononciation* section on Cassette 8B/CD-8 and *Activité E,* page 161 in the Student Tape Manual, Teacher's Edition.

Bell Ringer Review

Write the following on the board or use BRR Blackline Master 14-6: You are taking your driving exam. The exam asks you for instances when you should slow down instead of accelerating. Write down as many as you can.

Activités de communication orale
ANSWERS

Activité A
É2 answers will vary. É1 questions will vary but may include the following:
Tu as parlé au téléphone hier soir?
Tu as parlé à qui?
Vous avez parlé de quoi?

Activité B
É1 initial questions will follow the model. É2 answers will vary.

Activité C
É2 answers will vary. É1 questions may include:
1. **Quand est-ce que tu as appris à…?**
2. **Où est-ce que tu as appris à…?**
3. **Qui t'a appris à…?**
4. **Tu as pris des leçons de… ?**
5. **Tu as eu un bon moniteur/une bonne monitrice?**
6. **Tu as (bien) compris les instructions?**

LECTURE ET CULTURE

Bell Ringer Review

Write the following on the board or use BRR Blackline Master 14-7: Make a quick sketch of the clothing department of a large store and label the persons, articles of clothing, or signs you might see there.

READING STRATEGIES
(page 378)

Note Find out which members of the class are skiers. If your students do not ski because of geographical or socio-economic reasons, you may wish to go over this *Lecture* very quickly.

Pre-reading

A. Using a wall map, point out the geographic locations mentioned in the reading.

B. There are many instances of the *passé composé* in the *Lecture*. Have students find several.

Reading

Have students open their books to page 378 and follow along as you read the first paragraph. Then have students read the rest of the *Lecture* silently.

Post-reading

Lead students through the exercises that follow the *Lecture*.

Note Students may listen to a recorded version of the *Lecture* on the CD-ROM.

Étude de mots

ANSWERS

Exercice A

1. a
2. b
3. a
4. b
5. b

378

ON VA AUX SPORTS D'HIVER

*E*n février dernier la classe de Madame Carrigan a fait un voyage au Canada. Les élèves ont eu une semaine de vacances. Ils ont pris le train de New York pour aller à Montréal. Ils ont passé trois jours à Montréal où ils ont parlé français. Montréal est la deuxième ville francophone[1] du monde, après Paris.

Après deux jours à Montréal ils ont pris le car[2] jusqu'au Parc du Mont-Sainte-Anne. Le Mont-Sainte-Anne est une station de sports d'hiver tout près de la jolie ville de Québec. Après leur arrivée à Sainte-Anne ils ont tous mis leur anorak et leurs chaussures de ski. Ils ont acheté leur ticket de télésiège. Ils ont pris le télésiège jusqu'au sommet de la montagne. Du sommet ils ont eu une vue splendide sur les montagnes et les vallées couvertes de neige. As-tu jamais[3] vu les montagnes couvertes de neige? C'est vraiment superbe!

Les bâtons à la main et les skis aux pieds, ils ont commencé à descendre une piste. Mais ils ont choisi la mauvaise[4] piste, une piste très raide, trop difficile pour des débutants. Qui a eu un accident? Le casse-cou[5] Michel? Mais oui, c'est lui! Il a fait une chute. Il a glissé jusqu'en bas[6] de la piste. Tous ses copains ont rigolé. Ils ont dit: «Michel, tu es une vraie boule de neige qui roule, roule, roule!»

[1] francophone *French-speaking*
[2] car *bus*
[3] jamais *ever*
[4] mauvaise *wrong*
[5] casse-cou *daredevil*
[6] a glissé jusqu'en bas *slid to the bottom*

Le Mont-Sainte-Anne

Étude de mots

A **Quel est le mot?** Choisissez.

1. Février est _____.
   a. un mois   b. une saison

2. Février est en _____.
   a. été   b. hiver

3. Montréal est une ville _____.
   a. francophone   b. française

4. Les chaussures de ski sont des _____.
   a. tennis   b. bottes

5. On met _____ quand il fait très froid.
   a. un maillot   b. un anorak

378   CHAPITRE 14

DID YOU KNOW?

Quebec City is the oldest city in Canada and the only walled city in North America. Most of the present-day city lies outside the walls. Every February Quebec City holds its *Carnaval de Québec*. Attended by 500,000 tourists, it lasts two weeks and includes costume balls, dog-sled and ice-canoe races, snow and ice sculpture contests, a ski triathlon, and much more, all presided over by a Snow Queen.

Mont-Sainte-Anne, a ski resort of international stature in Quebec province, has the highest drop east of the Rockies, excellent ski jumps, and a gondola lift. The region known as Mont Tremblant in the Laurentians is a well-known ski resort area in Quebec, as is Lac Beauport, which features a beautiful *descente aux flambeaux* (skiers with lighted torches going down the slopes at night) during carnival season.

Compréhension

B **Une excursion.** Corrigez les phrases.

1. Les élèves de Madame Carrigan ont fait un voyage en France.
2. Ils ont pris l'avion.
3. Ils ont passé trois jours à Québec.
4. Québec est la deuxième ville francophone du monde.
5. Le Parc du Mont-Sainte-Anne est une station balnéaire.
6. Les élèves de Madame Carrigan font tous très bien du ski.

C **Un fait important.** Vous avez appris quelque chose d'important au sujet de Montréal. Qu'est-ce que c'est?

DÉCOUVERTE CULTURELLE

Quelques pays francophones ont des stations de sports d'hiver fabuleuses. En France, par exemple, il y a beaucoup de stations de sports d'hiver dans les Alpes et les Pyrénées. La Suisse est un pays célèbre pour le ski. Et n'oubliez pas que le français est une des langues officielles de la Suisse. En Suisse on parle français, allemand et italien. Et au Québec, la province francophone du Canada, il y a des stations de sports d'hiver superbes.

En France les écoles primaires ont des classes de neige. Les élèves vont dans une station de sports d'hiver. Le matin ils ont des cours. Ils étudient les maths, l'anglais, etc. L'après-midi, des moniteurs apprennent à faire du ski aux élèves. Il y a des classes de neige aux États-Unis? Vous croyez que c'est une bonne idée?

Dans les stations de sports d'hiver en France les pistes sont classées selon leur difficulté. Les couleurs indiquent le niveau, ou le degré, de difficulté.

PISTE	NIVEAU	TYPE DE SKIEURS
	facile	débutants
	moyen	bons skieurs
	difficile	très bons skieurs
	très difficile	très, très bons skieurs

CRITICAL THINKING ACTIVITY

(Thinking skill: making inferences)

Read the following to the class or put it on the board or on a transparency.

1. **C'est la première fois que Christophe fait du ski. Qu'est-ce qu'il doit *(must)* faire?**
2. **Mais Christophe est toujours très impatient. La piste verte pour les débutants, ce n'est pas pour lui. Il va descendre la piste rouge. Le panneau indique que c'est** une piste pour les très bons skieurs. **Quelles peuvent être les conséquences de la décision de Christophe?**

Compréhension

PRESENTATION *(page 379)*

Extension of *Exercices B* and *C*

Ask students: *Après leur arrivée à Sainte-Anne, qu'est-ce que les élèves ont mis? Qu'est-ce qu'ils ont acheté? Qu'est-ce qu'ils ont pris pour aller au sommet de la montagne? Qu'est-ce qu'ils ont vu du sommet? Qui a eu un accident? Qu'est-ce qu'il a fait? Il a glissé jusqu'où?*

ANSWERS

Exercice B

1. ... au Canada.
2. ... le train.
3. ... à Montréal.
4. Montréal...
5. ... de sports d'hiver.
6. ... ne font pas tous très bien de ski.

Exercice C

Montréal est la deuxième ville francophone du monde, après Paris.

OPTIONAL MATERIAL

Découverte culturelle

PRESENTATION *(page 379)*

A. Before reading the selection, point out the Vosges, the Alps, the Jura, the Massif Central, and the Pyrenees, as well as Switzerland and Quebec, on the maps on pages 504 and 506.

B. Have students read silently, then ask these *vrai/faux* questions: *La Suisse a trois langues officielles. On ne fait pas de ski dans les Pyrénées. En France, les écoles primaires ont des classes de neige dans les villes. Dans les classes de neige, les élèves apprennent à faire du ski le matin. Une piste rouge est plus facile qu'une piste bleue.*

Note Students may listen to a recorded version of the *Découverte culturelle* on the CD-ROM.

RÉALITÉS

PRESENTATION *(pages 380–381)*

The main objective of this section is to have the students enjoy the photographs and absorb some French culture. However, if you would like to do more, you may wish to do some of the following activities.

A. Before reading the captions, ask students to name as many winter sports as they can think of. How many of these make up part of the Winter Olympics?

B. Have student volunteers read the captions on page 380 and answer any questions they find.

Note In the CD-ROM version, students can listen to the recorded captions and discover a hidden video behind one of the photos.

THE FRANCOPHONE WORLD

Photos of Quebec and additional information about it may be found in *Le Monde francophone*, pages 112, 223, 327, and 436.

RÉALITÉS

La Plagne est une station de sports d'hiver dans les Alpes françaises **1**. Tu veux passer une semaine dans un de ces chalets à La Plagne?

Le hockey sur glace est un sport d'hiver très populaire, surtout au Canada. Et les joueurs canadiens sont parmi les meilleurs joueurs de hockey du monde **2**.

Ces gens sont dans une rue de Méribel pendant les Jeux Olympiques de 1992 **3**. Méribel fait partie des Trois Vallées, un ensemble de stations de sports d'hiver dans les Alpes.

Isabelle et Paul Duchesnay, frère et sœur, sont des patineurs professionnels **4**. Aux Jeux Olympiques de 1992 ils ont gagné la médaille d'argent. Aimes-tu faire du patinage artistique ou le regarder à la télé?

Ces gens font du ski de fond dans la province d'Alberta au Canada **5**.

Ces deux couples prennent le télésiège jusqu'au sommet de la montagne **6**. L'un des couples va faire du ski et l'autre va faire du surf des neiges. As-tu jamais fait du surf des neiges?

380

DID YOU KNOW?

Megève, an elite resort in the Alps, offers skiers a choice of three mountains and 124 slopes, served by 81 lifts. (For information on Méribel and La Plagne, see DID YOU KNOW?, pages 366 and 372, respectively.) In all these resorts *après-ski* activities abound: shopping, restaurants (many serving gourmet-quality food), and night entertainment galore.

COOPERATIVE LEARNING

Display Communication Transparency C-14. Have students work in groups to make up as many questions as they can about the illustration. Have groups take turns asking and answering the questions.

381

ADDITIONAL PRACTICE

1. Student Tape Manual, Teacher's Edition, *Deuxième Partie*, pages 162–163
2. Situation Cards, Chapter 14

INDEPENDENT PRACTICE

Assign any of the following:
1. *Étude de mots* and *Compréhension* exercises, pages 378–379
2. Workbook, *Un peu plus*, pages 144–146
3. CD-ROM, Disc 4, pages 378–381

RECYCLING

The *Activités de communication orale* and *Activités de communication écrite* provide various ways for students to recycle and recombine structures and vocabulary associated with clothing, travel, sports activities, food and restaurants, weather, making plans, and expressing opinions, preferences, and needs.

INFORMAL ASSESSMENT

Oral Activities A and D may serve as a means to evaluate the speaking skill.

Use the evaluation criteria given on page 34 of this Teacher's Wraparound Edition.

Activités de communication orale

ANSWERS

Activité A

Answers will vary but may include the following:

Tu préfères les sports d'été ou les sports d'hiver? Quels sports (en particulier) est-ce que tu aimes? Pourquoi?

Activités B and C

Answers will vary.

Activité D

É1 answers will follow the model. É2 answers will vary but will always begin with *On a besoin de (d')* and may include the following items:

1. une voiture/un permis de conduire/essence/la clé de la voiture
2. un stylo/papier/une bonne idée/instructions
3. skis/bâtons/chaussures de ski/neige
4. billets/une valise/copains/un itinéraire
5. une raquette/un filet/un court/une balle
6. pain/viande/un couteau/beurre

382

Activités de communication orale

A **Sports d'hiver ou sports d'été.** Find out if a classmate prefers winter or summer sports. Then ask which ones he or she likes and why. Reverse roles.

B **Nord et Sud.** Imagine that you're from a city in the South and a classmate is from a city in the North. Contrast the two places you choose in winter. Talk about the weather, clothing, activities, and so on.

> Élève 1: À (*ville du Sud*) il fait chaud en hiver.
> Élève 2: À (*ville du Nord*) il fait froid en hiver.

C **La location de skis.** You need to rent some ski equipment from the attendant (your partner) at a ski resort. Tell your partner what equipment you need and for how long. Your partner will help you find the right equipment for a skier at your level and the right size boots. Discuss price and method of payment.

D **De quoi a-t-on besoin pour… ?** Ask a classmate what people need in order to do one of the activities in the list below. Your partner gets one point for each thing he or she can name. Then reverse roles.

> Élève 1: De quoi a-t-on besoin pour apprendre le français?
> Élève 2: On a besoin d'un bon professeur, d'un livre et de beaucoup de patience!

conduire une voiture	faire un voyage
écrire une composition	jouer au tennis
faire du ski	préparer un sandwich

Le Mont d'Arbois à Megève

Activités de communication écrite

A **Au Canada.** Your Canadian pen pal has invited you to spend a week in Québec during the winter. Write back accepting or declining the invitation. Give several reasons why you can or cannot accept.

FOR THE YOUNGER STUDENT

1. Have students paint a mural of a winter scene including people skiing and skating. Have them label it in French. Hang the mural where it can be seen by other people in your school.
2. Have students draw a winter scene and present it to the class orally.

STUDENT PORTFOLIO

Written assignments that may be included in students' portfolios are *Activité de communication écrite A* on page 382 and the *Mon Autobiographie* section of the Workbook on page 147.

Note Students may create and save both oral and written work using the Electronic Portfolio feature on the CD-ROM.

B Une station de sports d'hiver idéale.

Write a paragraph describing an ideal winter resort (real or imaginary). Be sure to include the following information.

1. where it's located and how to get there
2. what the weather's generally like
3. what facilities there are (lifts, skating rinks, restaurants, etc.)
4. what else you can do there besides ski

Réintroduction et recombinaison

A Un match de foot.   Mettez au passé composé.

1. Je joue au foot.
2. Je passe le ballon à Charles.
3. Il renvoie le ballon.
4. Le gardien bloque le ballon.
5. Nous ne marquons pas de but.
6. L'arbitre déclare un penalty.
7. L'équipe adverse marque un but.
8. Nous faisons le forcing pour égaliser le score.

B La télé.   Complétez au passé composé.

1. Hier soir j'___ la télé. (regarder)
2. J'___ un film intéressant. (voir)
3. J'___ la météo: demain, neige et froid, températures basses. (entendre)
4. À neuf heures mon copain Éric m'___. (téléphoner)
5. Il n'___ pas ___ de bonnes nouvelles. (avoir)
6. Il ___ un examen et il n'___ pas ___ à l'examen. (passer, réussir)

Vocabulaire

NOMS

l'hiver (m.)
le vent
le ski (*skiing*)
le ski alpin
le ski de fond
le skieur
la skieuse
le/la débutant(e)
le moniteur
la monitrice
la piste (raide)
la piste de slalom
la bosse

la station de sports d'hiver
le chalet
le télésiège
la montagne
le sommet
la vallée

le ski (*ski*)
le bâton
la chaussure de ski
l'anorak (m.)
le bonnet
l'écharpe (f.)
le gant
les lunettes (f.)

le patinage
le patin à glace
le patineur
la patineuse
la patinoire
la glace
l'accident (m.)
la chute

VERBES

apprendre à quelqu'un à faire quelque chose
descendre

AUTRES MOTS ET EXPRESSIONS

faire du ski
faire du patin
faire une chute
il fait ___ degrés Celsius
il fait froid
il gèle
il neige

INDEPENDENT PRACTICE

1. Activities and exercises, pages 382–383
2. Communication Activities Masters, pages 68–70
3. CD-ROM, Disc 4, pages 382–383

Activités de communication écrite

ANSWERS

Activités A and B
  Answers will vary.

OPTIONAL MATERIAL

Réintroduction et recombinaison

ANSWERS

Exercice A

1. J'ai joué au foot.
2. J'ai passé le ballon à Charles.
3. Il a renvoyé le ballon.
4. Le gardien a bloqué le ballon.
5. Nous n'avons pas marqué de but.
6. L'arbitre a déclaré un penalty.
7. L'équipe adverse a marqué un but.
8. Nous avons fait le forcing pour égaliser le score.

Exercice B

1. ai regardé
2. ai vu
3. ai entendu
4. a téléphoné
5. a, eu
6. a passé; a, réussi

ASSESSMENT RESOURCES

1. Chapter Quizzes
2. Testing Program
3. Situation Cards
4. Communication Transparency C-14
5. Computer Software: Practice/Test Generator

VIDEO PROGRAM

INTRODUCTION   (43:57)

UNE LETTRE DU CANADA   (44:28)

CHAPITRE 15

CHAPTER OVERVIEW

In this chapter, students will learn to talk about routine illnesses and to describe their symptoms to a doctor. They will learn vocabulary associated with medical exams, prescriptions, and minor ailments such as colds, flu, and headache. Students will learn to talk about themselves and others using the object pronouns *me, te, nous, vous;* the present and *passé composé* of verbs like *ouvrir* and *souffrir;* and the imperative forms of verbs.

The cultural focus of Chapter 15 is on French medical services and facilities, social security, and attitudes towards minor illnesses.

CHAPTER OBJECTIVES

By the end of this chapter, students will know:

1. vocabulary associated with headaches, colds, fevers, and flu
2. body parts associated with various ailments
3. vocabulary associated with a visit to a doctor or a pharmacy
4. informal expressions used to comment on one's own health and that of others
5. the pronouns *me, te, nous,* and *vous* used as direct and indirect objects
6. negative constructions using the object pronouns *me, te, nous,* and *vous*
7. the present tense of verbs like *ouvrir*
8. the past participles of verbs like *ouvrir*
9. the formation of formal and informal imperatives as well as the *nous* imperative
10. negative imperative constructions

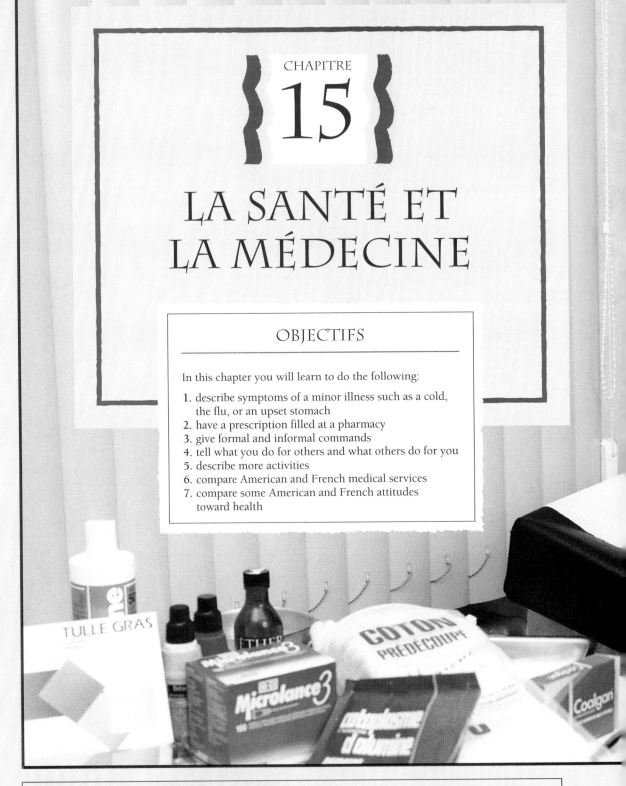

CHAPITRE

15

LA SANTÉ ET LA MÉDECINE

OBJECTIFS

In this chapter you will learn to do the following:

1. describe symptoms of a minor illness such as a cold, the flu, or an upset stomach
2. have a prescription filled at a pharmacy
3. give formal and informal commands
4. tell what you do for others and what others do for you
5. describe more activities
6. compare American and French medical services
7. compare some American and French attitudes toward health

CHAPTER PROJECTS

(optional)

1. Obtain a first-aid film from the health department in your school and use it as a springboard for discussing health and illness using new vocabulary from this chapter. You may also wish to use the film to review the parts of the body in French.
2. Have students create a poster of a man or woman like the kind in doctors' offices, labeling in French as many external and internal body parts as they can. This poster can be displayed in your classroom.

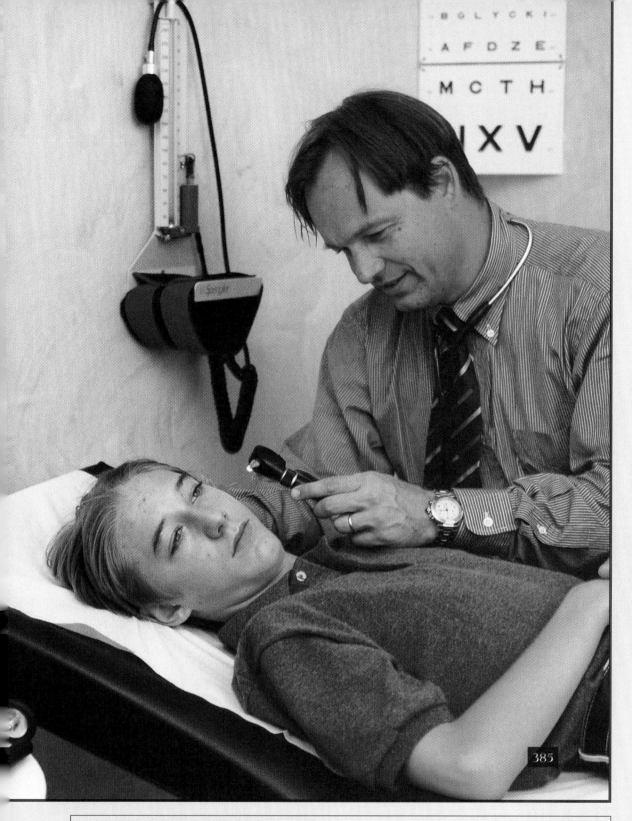

385

CHAPTER 15 RESOURCES

1. Workbook
2. Student Tape Manual
3. Audio Cassette 9A/CD-9
4. Bell Ringer Review Blackline Masters
5. Vocabulary Transparencies
6. Pronunciation Transparency P-15
7. Communication Transparency C-15
8. Communication Activities Masters
9. Map Transparencies
10. Situation Cards
11. Conversation Video
12. Videocassette/Videodisc, Unit 4
13. Video Activities Booklet, Unit 4
14. Lesson Plans
15. Computer Software: Practice/Test Generator
16. Chapter Quizzes
17. Testing Program
18. Internet Activities Booklet
19. CD-ROM Interactive Textbook

Pacing

This chapter requires eight to ten class sessions. Pacing will vary according to class length and the age and aptitude of the students.

Note The Lesson Plans offer guidelines for 45- and 55-minute classes and **Block Scheduling.**

Exercices vs. *Activités*

All exercises (which provide guided practice) are coded in blue. All communicative activities are coded in red.

INTERNET ACTIVITIES

(*optional*)

These activities, student worksheets, and related teacher information are in the *Bienvenue* Internet Activities Booklet and on the Glencoe Foreign Language Home Page at: **http://www.glencoe.com/secondary/fl**

LEARNING FROM PHOTOS

After students have learned the vocabulary of this chapter, you may wish to ask the following questions about the photo: *Le médecin examine le malade? Il l'ausculte? Il regarde sa bouche? Qu'est-ce qu'il examine? Le malade souffre? Il a mal? Il se sent bien? Il a de la fièvre, à ton avis? Qu'est-ce qu'il a, à ton avis? Le médecin va faire un diagnostic au malade? Il va faire une ordonnance au malade?*

385

Vocabulary Teaching Resources

1. Vocabulary Transparencies 15.1 (A & B)
2. Audio Cassette 9A/CD-9
3. Student Tape Manual, Teacher's Edition, *Mots 1: A–C*, pages 164–165
4. Workbook, *Mots 1: A–E*, pages 148–149
5. Communication Activities Masters, *Mots 1: A*, page 71
6. Chapter Quizzes, *Mots 1:* Quiz 1, page 77
7. CD-ROM, Disc 4, *Mots 1:* pages 386–389

Bell Ringer Review

Write the following on the board or use BRR Blackline Master 15-1: Last weekend Maurice had a party at his house. Write sentences explaining at least two things he probably did before the party, two things he or his guests probably did during it, and two things he probably did after it.

PRESENTATION *(pages 386–387)*

Teaching Tip You may wish to bring a handkerchief, tissues, and throat lozenges to class to make the presentation of the *Mots 1* vocabulary more lively for the students.

A. Have students make a list of the parts of the body they already know in French.

B. Point to yourself to model the following parts of the body: *la bouche, le nez, la gorge, l'oreille, les yeux, le ventre.*

C. Use gestures to teach the following expressions: *avoir de la fièvre; avoir des frissons; il est très malade; il n'est pas en bonne*

386

VOCABULAIRE

MOTS 1

ON EST MALADE

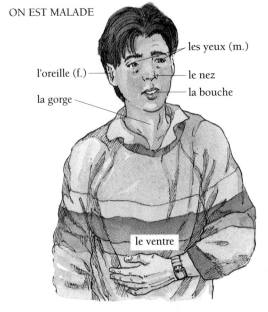

les yeux (m.)

l'oreille (f.)

le nez

la bouche

la gorge

le ventre

avoir de la fièvre

Paul a un rhume.
Il est enrhumé.
Il éternue.

Atchoum!

un kleenex

Il tousse.

Martin n'est pas en bonne santé.
Il est en mauvaise santé.
Il est très malade, le pauvre.
Il ne se sent pas bien.
Qu'est-ce qu'il a, le pauvre garçon?

Note: The expression *Qu'est-ce qu'il a?* means "What's wrong with him?"

un mouchoir

386   CHAPITRE 15

TOTAL PHYSICAL RESPONSE

(following the Vocabulary presentation)

TPR
____, venez ici, s'il vous plaît.
Montrez-moi la bouche.
Montrez-moi la main.
Montrez-moi le nez.
Montrez-moi le pied.
Montrez-moi le ventre.
Montrez-moi la gorge.
Montrez-moi les yeux.
Levez la main.
Ouvrez la bouche.
Fermez les yeux.
Mettez la main sur la tête.
Touchez les pieds avec les mains.
Merci, ____. Retournez à votre place et asseyez-vous, s'il vous plaît.

Miriam a la grippe.
Elle a de la fièvre.
Elle a des frissons.

Elle a mal à la tête.

Elle a mal au ventre.

Elle a mal aux oreilles.

Elle a le nez qui coule.

Elle a les yeux qui piquent.

Elle a la gorge qui gratte.

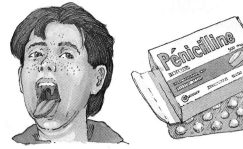

Christophe a très mal à la gorge. Il a une angine.

Note: Study the following cognates related to health and medicine.

allergique	de l'aspirine (f.)
bactérien(ne)	une infection
viral(e)	la pénicilline
une allergie	la température
un antibiotique	

CHAPITRE 15 **387**

---

santé; il a mal au ventre; il tousse; il éternue; il a mal à la tête; il a mal aux oreilles; elle a le nez qui coule; elle a les yeux qui piquent; elle a la gorge qui gratte.

D. Have students repeat the cognates carefully after you or Cassette 9A/CD-9. These are the words they are most likely to anglicize.

Note Remind students that in French the definite article is usually used when talking about parts of the body. Introduce the singular of *les yeux* (*un œil*).

E. Ask several volunteers to come to the front of the room. Have each one mime a different ailment. The rest of the class describes the symptoms and suggests what he/she needs. Use as many props as possible. Guide the class with questions when necessary. For example: *Pauvre Isabelle! Elle a un rhume. (Elle a le nez qui coule.) (Elle a mal à la gorge.) (Elle a un peu de fièvre.) De quoi est-ce qu'elle a besoin? (Elle a besoin de beaucoup de kleenex.) (Elle a besoin d'aspirine.)*, etc.

F. Ask each volunteer to recapitulate his/her illness, symptoms, and needs. Cue key words or ask the class for help as necessary. For example: *J'ai un rhume. J'ai le nez qui coule. J'ai besoin d'aspirine.*, etc.

Vocabulary Expansion

Tourists often experience stomach problems. If you wish, you may give the students the following useful words and expressions.
Vous avez des nausées?
Vous avez de la diarrhée?
Vous êtes constipé(e)?
Vous avez de la constipation?
Vous vomissez?
Vous avez des vomissements?
Vous avez des crampes?

DID YOU KNOW?

You may wish to introduce the colloquial expression *Mon œil!* to students. It means, "Come on, do you think I'm going to believe that?" When people say it they usually put their finger up to their eye.

ADDITIONAL PRACTICE

Play a game of *Jacques a dit*. Call the first round yourself and have students act out the following:
Jacques a dit: Vous avez mal à la tête.
Jacques a dit: Mettez les mains sur la tête.
Jacques a dit: Toussez.
Jacques a dit: Éternuez.
Then call on students to lead the game.

Extension of *Exercice B*

After doing Exercise B, call on a student to give a summary of the story in his or her own words.

ANSWERS

Exercice A

1. la gorge
2. l'oreille
3. la tête
4. les yeux
5. le nez
6. la bouche
7. le ventre
8. la main
9. le pied

Exercice B

1. Oui, Miriam est très malade.
2. Non, elle ne se sent pas bien.
3. Elle a la grippe.
4. Oui, elle a de la fièvre et des frissons.
5. Oui, elle a la gorge qui gratte.
6. Oui, elle a les yeux qui piquent et le nez qui coule.
7. Oui, elle a mal à la tête.
8. Oui, elle a mal au ventre.
9. Oui, elle a mal aux oreilles.

Exercices

A **Qu'est-ce que c'est?**   Identifiez.

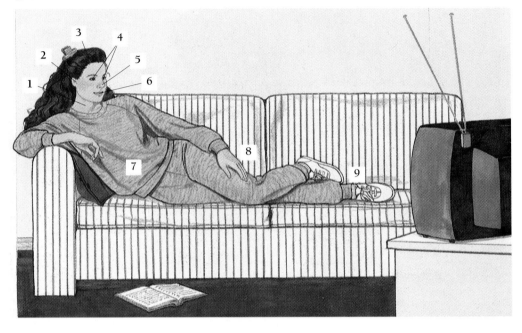

B **Qu'est-ce qu'elle a, la pauvre Miriam?**   Répondez.

1. Miriam est très malade?
2. Elle ne se sent pas bien?
3. Qu'est-ce qu'elle a?
4. Elle a de la fièvre et des frissons?
5. Elle a la gorge qui gratte?
6. Elle a les yeux qui piquent et le nez qui coule?
7. Elle a mal à la tête?
8. Elle a mal au ventre?
9. Elle a mal aux oreilles?

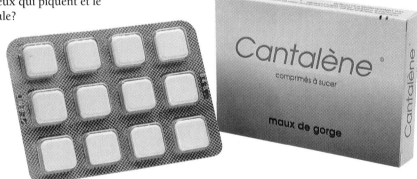

LEARNING FROM ILLUSTRATIONS

Have students say anything they can about the illustration that accompanies Exercise A. Tell students the word "sofa" is the same in French: *un sofa*, but one also frequently hears *un canapé* or *un divan*.

LEARNING FROM REALIA

Have students look at the realia. Tell them to give you the brand name of the medication. Then have them find what it is for. They should be able to guess when they see the word *gorge*. Then ask them for the singular form of the word *maux* (*mal*). Ask in what form the medicine is (*comprimés*). If they are *comprimés*, what does *sucer* mean? What does one do with these *comprimés?*

C 🇫🇷 **La santé.** Donnez des réponses personnelles.

1. Tu es en bonne santé ou en mauvaise santé?
2. Quand tu es enrhumé(e), tu as le nez qui coule?
3. Tu as les yeux qui piquent?
4. Tu as la gorge qui gratte?
5. Tu tousses?
6. Tu éternues?
7. Tu as mal à la tête?
8. Tu ne te sens pas bien?
9. Tu as de la fièvre quand tu as un rhume ou la grippe?
10. Quand tu as de la fièvre, tu as quelquefois des frissons?
11. Quand tu as mal à la tête, tu prends de l'aspirine?

D **On a mal.** Complétez.

1. On prend de l'aspirine. On a mal à la ___.
2. On a très mal à la gorge. On a une ___.
3. La ___ est un antibiotique.
4. On ne peut pas prendre de pénicilline quand on est ___ à la pénicilline.
5. On a une température de 40°C. On a de la ___.
6. Quand on est toujours malade, on est en ___.
7. Les ___ accompagnent souvent la fièvre.
8. On donne des antibiotiques comme la pénicilline pour combattre des infections bactériennes, pas ___.
9. Quand on a le nez qui coule, on a toujours besoin d'un ___ ou d'un ___.
10. Quand on a un rhume, on ___ et on ___.
11. Quand on a de la fièvre, on prend de l'___.
12. Quand on est enrhumé ou quand on écoute trop la musique, on a mal aux ___.

PRESENTATION (*continued*)

Extension of *Exercice C*

After doing Exercise C, call on an individual student to give a summary of the story in his/her own words.

Listening After doing Exercise C, focus on the listening skill by having students do the exercise in pairs. One partner reads the questions in random order. The second partner answers with his/her book closed. Then partners reverse roles.

ANSWERS

Exercice C

Answers will vary but may include the following:

1. Je suis en bonne (mauvaise) santé.
2. Oui, j'ai le nez qui coule quand je suis enrhumé(e).
3. Oui, j'ai les yeux qui piquent.
4. Oui, j'ai la gorge qui gratte.
5. Oui, je tousse.
6. Oui, j'éternue.
7. Oui, j'ai mal à la tête.
8. Non, je ne me sens pas bien.
9. J'ai de la fièvre quand j'ai la grippe. Je n'ai pas de fièvre quand j'ai un rhume.
10. Oui, j'ai quelquefois des frissons.
11. Oui, je prends de l'aspirine. (Non, je ne prends pas d'aspirine.)

Exercice D

1. tête
2. angine
3. pénicilline
4. allergique
5. fièvre
6. mauvaise santé
7. frissons
8. virales
9. mouchoir, kleenex
10. tousse, éternue
11. aspirine
12. oreilles

INFORMAL ASSESSMENT
(*Mots 1*)

Check comprehension by miming the symptoms from *Mots 1* and having students tell you what is wrong with you. For example, mime taking an aspirin, taking your temperature, sneezing, coughing, holding your stomach.

Bell Ringer Review

Write the following on the board or use BRR Blackline Master 15-2: Write down what part(s) of the body you associate with the following.

1. **des lunettes**
2. **un bonnet de ski**
3. **un thermomètre**
4. **un kleenex**

PRESENTATION *(pages 390–391)*

A. Have students close their books. Use TPR, a few props and student volunteers to demonstrate the *Mots 2* vocabulary and act out the sequences as much as possible. Props might include a red cross labeled with a doctor's name for *chez le médecin,* a labeled sign with a green cross for *la pharmacie,* a plastic stethoscope, empty medicine bottles, etc.

B. Have students keep their books closed. Dramatize the following expressions from *Mots 2: ouvrir la bouche; examiner la gorge;*

CHEZ LE MÉDECIN

le médecin — un malade — une malade

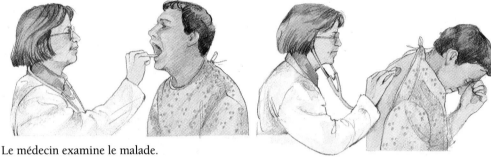

Le médecin examine le malade.
Le malade ouvre la bouche.
Le médecin examine la gorge du malade.

Elle ausculte le malade.
Il souffre, le pauvre.

Où avez-vous mal?

Ouvrez la bouche. — Toussez. — Respirez à fond.

Le médecin parle.

390 CHAPITRE 15

TOTAL PHYSICAL RESPONSE

(following the Vocabulary presentation)

TPR 1

___, levez-vous et venez ici, s'il vous plaît.
Vous allez chez le médecin. Moi, je suis le médecin. Je vais vous examiner.
___, asseyez-vous, s'il vous plaît.
Ouvrez la bouche.
Dites «Ah».

Ouvrez les yeux.
Et maintenant, fermez les yeux.
Levez-vous, s'il vous plaît.
Je vais vous ausculter. Respirez à fond.
Encore une fois.
Merci, ___. Vous avez très bien fait. Vous êtes un(e) bon(ne) patient(e). Retournez à votre place, s'il vous plaît.

une ordonnance

le pharmacien

la pharmacienne

des comprimés (m.)

Le médecin me fait un diagnostic.
Elle me prescrit des antibiotiques.
Elle me fait une ordonnance.

Je suis à la pharmacie.
Qu'est-ce que la pharmacienne te donne?
Elle me donne des médicaments.

Note: You may use the following informal expressions to talk about health.

1. When you are not feeling well, you can say:

Je ne suis pas dans mon assiette aujourd'hui.

2. To tell someone he or she will soon be better, you can say:

Tu vas être vite sur pied.

3. When someone has a high fever, you can say:

Il a une fièvre de cheval.

4. To say "It hurts," you say:

Ça fait mal!

5. When you have a "frog in your throat," you can say:

J'ai un chat dans la gorge.

CHAPITRE 15   **391**

souffrir; tousser; respirer à fond. Ask students to imitate each of your dramatizations and repeat each corresponding word.

C. **Recycling** Bring back previously learned vocabulary by asking *Où avez-vous mal?* and pointing to your hand, foot, eyes, ear, nose, stomach, head, or throat. Have students respond using the correct word.

D. Ask students to open their books to pages 390–391. Have them read along and repeat the new material after you or Cassette 9A/CD-9.

E. Have students close their books. Use Vocabulary Transparencies 15.2 to review all the words and expressions from *Mots 2,* including the informal expressions on page 391. Point to each illustration and have students say as much as they can about it.

F. Practice the expressions on page 391 by miming or supplying the literal statement and having students supply the informal expression. For example: *Je ne peux pas parler. (Vous avez un chat dans la gorge.) Je ne me sens pas bien. (Vous n'êtes pas dans votre assiette.)* Clutch your arm and say «*Aïe!*» (*Ça fait mal.*)

Vocabulary Expansion

You may wish to give students the following expressions related to physical exams.
prendre la tension artérielle
prendre le pouls
faire une piqûre
faire une prise de sang
faire un électrocardiogramme

TPR 2
____, venez ici, s'il vous plaît.
Vous allez faire le mime.
Respirez à fond.
Toussez.
Éternuez.
Vous avez mal à la tête.
Vous avez de la fièvre.
Vous avez des frissons.
Vous avez les yeux qui piquent.
Vous avez le nez qui coule.

Prenez un kleenex.
Prenez un comprimé.
Merci, ____. Très bien. Retournez à votre place, s'il vous plaît.

PRESENTATION *(page 392)*

Exercice A

You may wish to use the recorded version of this exercise.

ANSWERS

Exercice A

1. c	**5.** b	**9.** b
2. b	**6.** a	**10.** c
3. b	**7.** a	**11.** b
4. a	**8.** a	

Exercice B

Answers will vary but may include the following:

1. Oui, je vais chez le médecin quand je suis très malade.
2. Oui, il me demande:«Où avez-vous mal?»
3. Oui, ça fait très mal.
4. Oui, le médecin me dit: «Ouvrez la bouche.»
5. Oui, il m'ausculte.
6. Oui, il me dit: «Respirez à fond.»
7. Oui, le médecin me fait un diagnostic.
8. Oui, il me prescrit des comprimés.
9. Oui, je vais à la pharmacie pour acheter les médicaments.
10. Oui, je prends quelquefois des antibiotiques. (Non, je ne prends jamais d'antibiotiques.)

Exercice C

1. Je ne suis pas dans mon assiette aujourd'hui.
2. Tu vas être vite sur pied.
3. J'ai une fièvre de cheval!
4. J'ai un chat dans la gorge.

INFORMAL ASSESSMENT *(Mots 2)*

Check comprehension by asking students interrogative-word questions about the illustrations on Vocabulary Transparencies 15.2. For example: *Il y a combien de malades chez le médecin? Qui est-ce que le médecin examine? Qui est-ce qu'elle ausculte? Que fait le malade?* Higher skills: *Quel est le diagnostic que le médecin fait? Pourquoi? Qu'est-ce qu'elle prescrit? Pourquoi?*

Exercices

A **Chez le médecin.**   Choisissez.

1. Où est le malade?
   a. À l'hôpital.   b. Chez lui.   c. Chez le médecin.
2. Qui souffre?
   a. Le médecin.   b. Le malade.   c. Le pharmacien.
3. Qu'est-ce que le médecin examine?
   a. La bouche.   b. La gorge.   c. Le ventre.
4. Qu'est-ce que le malade ouvre?
   a. La bouche.   b. La gorge.   c. L'oreille.
5. Le médecin ausculte le malade. Comment respire-t-il?
   a. Il éternue.   b. À fond.   c. Bien.
6. Qui est-ce que le médecin ausculte?
   a. Le malade.   b. Le pharmacien.   c. La pharmacienne.
7. Que fait le médecin?
   a. Un diagnostic.   b. Des comprimés.   c. Des médicaments.
8. Qu'est-ce qu'il a, le pauvre malade?
   a. Une angine.   b. Mal au ventre.   c. Mal aux yeux.
9. Que fait le médecin?
   a. Un pharmacien.   b. Une ordonnance.   c. Un comprimé.
10. Qu'est-ce qu'elle prescrit?
    a. La pharmacie.   b. Des ordonnances.   c. Des antibiotiques.
11. Où va le malade pour acheter des médicaments?
    a. Chez le médecin.   b. À la pharmacie.   c. À l'ordinateur.

B **Le médecin m'examine.**   Donnez des réponses personnelles.

1. Tu vas chez le médecin quand tu es très malade?
2. Le médecin te demande: «Où avez-vous mal?»
3. Quand tu as une angine, ça fait très mal?
4. Le médecin te dit: «Ouvrez la bouche»?
5. Il t'ausculte?
6. Il te dit: «Respirez à fond»?
7. Le médecin te fait un diagnostic?
8. Il te prescrit des comprimés?
9. Tu vas à la pharmacie pour acheter les médicaments?
10. Tu prends quelquefois des antibiotiques?

C **Plus familier, s'il te plaît.**   Dites d'une manière familière.

1. Je ne vais pas très bien aujourd'hui.
2. Tu vas bientôt te sentir mieux.
3. J'ai beaucoup de fièvre!
4. Je ne peux pas parler facilement.

PAIRED ACTIVITY

After completing Exercises A, B, and C, have students work in pairs. Have them take turns telling each other that they think they have the flu or a bad cold. They should explain why they think they are sick by explaining their symptoms. The partner should respond by giving advice.

Activités de communication orale

Mots 1 et 2

Josiane Briand

A **Qu'est-ce que tu as?** You were absent from school today because you had the flu. Josiane Briand, the French exchange student, calls to find out how you are feeling.

1. Alors, tu as la grippe? Qu'est-ce que tu as? Tu as de la fièvre… euh…
2. Tu prends des médicaments?
3. Tu vas aller chez le médecin?
4. Tu vas encore rester à la maison demain?

B **Je ne suis pas dans mon assiette!** Yesterday you did something that made you feel ill today. Using List 1 below, tell a classmate what you did. He or she has to guess what's wrong with you, choosing from List 2.

> trop regarder la télé
> Élève 1: Hier j'ai trop regardé la télé.
> Élève 2: Tu as mal aux yeux.

1	2
lire pendant six heures	être enrhumé(e)
manger trop de chocolat	avoir mal aux yeux
passer beaucoup d'examens	avoir la gorge qui gratte
trop crier au match	avoir mal aux pieds
faire une longue promenade	être fatigué(e)
étudier jusqu'à 3h du matin	avoir mal aux oreilles
trop écouter de la musique	avoir mal à la tête
jouer dans la neige en tee-shirt	avoir mal au ventre

C **Quel médecin?** While on a trip to France, you get sick. Describe your symptoms. A classmate will look at the list of doctors at the Hôpital Saint-Pierre and tell you which one to call and what the phone number is.

> Élève 1: J'ai mal au ventre.
> Élève 2: Appelle le docteur Simonet au 43.89.39.25.

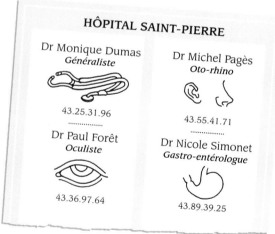

HÔPITAL SAINT-PIERRE

Dr Monique Dumas
Généraliste

43.25.31.96

Dr Paul Forêt
Oculiste

43.36.97.64

Dr Michel Pagès
Oto-rhino

43.55.41.71

Dr Nicole Simonet
Gastro-entérologue

43.89.39.25

CHAPITRE 15 **393**

INDEPENDENT PRACTICE

Assign any of the following:
1. Exercises and activities, pages 392–393
2. Workbook, *Mots 2: F–G,* page 149
3. Communication Activities Masters, *Mots 2: B,* page 72
4. Computer Software, *Vocabulaire*
5. CD-ROM, Disc 4, pages 390–393

Activités de communication orale

Mots 1 et 2

PRESENTATION *(page 393)*

Activité A

 In the CD-ROM version, students can interact with an on-screen native speaker.

ANSWERS

Activité A

Answers will vary, but may include the following:

1. J'ai mal à la tête (à la gorge, etc.). Oui, j'ai de la fièvre.
2. Oui, je prends des médicaments.
3. Oui (Non), je (ne) vais (pas) aller chez le médecin.
4. Oui (Non), je vais encore rester (je ne vais pas rester) à la maison demain.

Activité B

Answers will vary.

Activité C

Answers will vary, but some possible symptoms are listed under the appropriate doctor.

Dr Dumas
J'ai mal au (aux, à la)…
J'ai la grippe.
J'ai une angine, etc.

Dr Forêt
J'ai mal aux yeux.
J'ai les yeux qui piquent.
Je ne vois pas bien.

Dr Pagès
J'ai mal à la gorge (le nez qui coule; une angine; un rhume).
J'éternue.
J'ai mal à l'oreille.
Je n'entends pas bien.

Dr Simonet
J'ai mal au ventre.

STRUCTURE

Les pronoms *me, te, nous, vous*

Telling What You Do for Others and What Others Do for You

1. You have already seen the pronouns *me, te, nous,* and *vous* with reflexive verbs. These same pronouns function as objects of the verb.

Le médecin *te* voit?	Oui, il *me* voit.
Le médecin *t'*examine?	Oui, il *m'*examine.
Le médecin *vous* regarde?	Oui, il *me* regarde.
	Oui, il *nous* regarde.
Le médecin *te* fait une ordonnance?	Oui, il *me* fait une ordonnance.
Il *vous* parle?	Oui, il *me* parle.
	Oui, il *nous* parle.

2. Note that the object pronoun comes right before the verb of which it is the object. This is true even when there is a helping verb, such as *pouvoir, vouloir,* or *aller* in the sentence.

> Il *m'*examine.
> Il va *m'*examiner.
> Il peut *m'*examiner.

3. The object pronoun cannot be separated from the verb by a negative word.

> Il ne *vous fait* pas d'ordonnance.
> Il ne *nous examine* pas.
> Il ne *m'ausculte* jamais.

Exercices

A **Chez le médecin.** Donnez des réponses personnelles.

1. Quand tu vas chez le médecin, il te parle?
2. Il te regarde?
3. Il t'examine?
4. Il t'ausculte?
5. Il te fait un diagnostic?
6. Il te fait une ordonnance?
7. Il te prescrit des médicaments?
8. Il te prescrit des antibiotiques?
9. Le pharmacien te donne des médicaments?

Structure Teaching Resources

1. Workbook, *Structure: A–H,* pages 150–152
2. Student Tape Manual, Teacher's Edition, *Structure: A–C,* pages 168–169
3. Audio Cassette 9A / CD-9
4. Communication Activities Masters, *Structure: A–C,* pages 73–74
5. Computer Software, *Structure*
6. Chapter Quizzes, *Structure:* Quizzes 3–5, pages 79–81
7. CD-ROM, Disc 4, pages 394–399

Bell Ringer Review

Write the following on the board or use BRR Blackline Master 15-4: Make a list of health tips. For example, Il faut faire de l'exercice.

Les pronoms me, te, nous, vous

Note The object pronouns *me, te, nous, vous* are introduced before the third-person pronouns for two reasons. First, they are less complicated than the third-person pronouns since they are both direct and indirect objects. Second, they are the only object pronouns that are truly necessary for communication. For example, if asked a question with *te* or *vous,* one must answer with *me* or *nous.* When speaking in the third person, however, one could respond with a noun instead of a pronoun: *Non, je n'ai pas invité Jean, mais j'ai téléphoné à Marie.* Third-person pronouns will be presented in Chapters 16 and 17.

PRESENTATION (*pages 394–395*)

Lead students through steps 1–3.

ADDITIONAL PRACTICE

1. Have students ask each other for something. For example:
   Robert, tu me donnes ton crayon?
   Oui, je te donne mon crayon.
2. Then have two students ask two others for something.
   Robert et Louise, vous nous donnez ce livre?
   Oui, nous vous donnons ce livre.

PAIRED ACTIVITY

After completing Exercise D on page 395, have students work in pairs and prepare a short skit at a clothing store.

B **Elle nous invite à la fête.**   Répondez d'après le modèle.

> Suzanne vous parle de sa fête?
> *Oui, elle nous parle de sa fête.*

1. Elle vous téléphone?
2. Elle vous parle au téléphone?
3. Elle vous invite à la fête?
4. Elle vous dit l'heure de la fête?
5. Elle vous dit où elle habite?
6. Elle vous donne son adresse?

C **Elle ne nous invite pas à la fête.**   Répondez par «non» aux questions de l'Exercice B.

D **Au rayon prêt-à-porter.**   Complétez avec «vous» ou «me».

Je suis au rayon prêt-à-porter des Galeries Lafayette. La vendeuse ___ parle.
                                                                    1

Elle ___ demande:
      2

—Vous désirez?

—Je voudrais un chemisier, s'il ___ plaît. Je fais du 40.
                                  3

—D'accord. Je peux ___ proposer ces deux types de chemisiers.
                     4

—Ce chemisier bleu marine à manches longues ___ intéresse beaucoup.
                                              5

—Je ___ suggère la taille au-dessous alors. Ces chemisiers sont très grands.
     6

—D'accord. Je peux ___ payer avec une carte de crédit?
                     7

—Mais bien sûr!

E **Pourquoi ça?**   Répondez d'après le modèle.

> Élève 1: Il me regarde.
> Élève 2: Il te regarde? Pourquoi?

1. Il me pose des questions.
2. Il me parle.
3. Il me téléphone.
4. Il me dit son numéro de téléphone.
5. Il me donne son adresse.

F **C'est ton anniversaire.**   Donnez des réponses personnelles.

1. Tes copains vont te téléphoner le jour de ton anniversaire?
2. Ils vont te voir?
3. Ils vont t'inviter au cinéma ou au concert?
4. Ils vont te dire: «Joyeux anniversaire»?
5. Pour ton anniversaire, ils vont te faire un gâteau?

ADDITIONAL PRACTICE	INDEPENDENT PRACTICE
Student Tape Manual, Teacher's Edition, *Activité C,* page 169	Assign any of the following: 1. Exercises, pages 394–395 2. Workbook, *Structure: A–B,* page 150 3. Communication Activities Masters, *Structure: A,* page 73 4. CD-ROM, Disc 4, pages 394–395

Note In the CD-ROM version, this structure point is presented via an interactive electronic comic strip.

Exercices

PRESENTATION *(pages 394–395)*

Exercices B and E

You may use the recorded versions of these exercises.

Extension of *Exercice B*

After finishing Exercise B, have students say as much as they can about a party.

ANSWERS

Exercice A

Answers will vary.

Exercice B

1. Oui, elle nous téléphone.
2. Oui, elle nous parle au téléphone.
3. Oui, elle nous invite à la fête.
4. Oui, elle nous dit l'heure de la fête.
5. Oui, elle nous dit où elle habite.
6. Oui, elle nous donne son adresse.

Exercice C

1. Non, elle ne nous téléphone pas.
2. Non, elle ne nous parle pas au téléphone.
3. Non, elle ne nous invite pas à la fête.
4. Non, elle ne nous dit pas l'heure de la fête.
5. Non, elle ne nous dit pas où elle habite.
6. Non, elle ne nous donne pas son adresse.

Exercice D

1. me	4. vous	6. vous
2. me	5. m'	7. vous
3. vous		

Exercice E

1. Il te pose des questions? Pourquoi?
2. Il te parle? Pourquoi?
3. Il te téléphone? Pourquoi?
4. Il te dit son numéro de téléphone? Pourquoi?
5. Il te donne son adresse? Pourquoi?

Exercice F

Answers will vary.

Les verbes comme ouvrir au présent et au passé composé

PRESENTATION *(page 396)*

Note The material on page 396 should be rather easy since students have already had a great deal of practice with the *-er* verbs. Since all of these verbs, except for *ouvrir,* are of fairly low frequency, it is suggested that you cover them quickly.

Have students open their books to page 396. Lead them through steps 1–2. Have students repeat the forms after you, one verb at a time.

Exercices

ANSWERS

Exercice A
   Answers will vary.

Les verbes comme *ouvrir* au présent et au passé composé

Describing More Activities

1. Although the verbs *ouvrir, souffrir, couvrir,* and *découvrir* have infinitives that end in *-ir,* they follow the same pattern as *-er* verbs in the present tense.

OUVRIR	SOUFFRIR
j' ouvr**e**	je souffr**e**
tu ouvr**es**	tu souffr**es**
il	il
elle } ouvr**e**	elle } souffr**e**
on	on
nous ouvr**ons**	nous souffr**ons**
vous ouvr**ez**	vous souffr**ez**
ils ouvr**ent**	ils
elles ouvr**ent**	elles } souffr**ent**

2. The past participles of these verbs are irregular.

INFINITIF ⟶	PARTICIPE PASSÉ
ouvrir	ouvert
couvrir	couvert
découvrir	découvert
souffrir	souffert
offrir	offert

Pendant la nuit il a ouvert la fenêtre.
Hier le médecin a découvert la cause de la maladie.

Exercices

A **Tu souffres?** Donnez des réponses personnelles.

1. Tu souffres quand tu es enrhumé(e)?
2. Tu souffres plus quand tu as un rhume ou quand tu as la grippe?
3. Tu prends de l'aspirine quand tu souffres d'une allergie?
4. Tu ouvres la bouche quand le médecin t'examine la gorge?
5. Tu ouvres les yeux quand le médecin t'examine les yeux?
6. Tu offres un bouquet de roses à ton amie malade?

B Qu'est-ce qu'on fait? Complétez avec «ouvrir» ou «offrir».

1. Nous ___ les yeux quand nous nous réveillons.
2. Elle ___ un livre à sa mère pour la Fête des Mères.
3. Vous ___ le magazine pour regarder les photos qui vous intéressent?
4. Vous ___ la bouche quand le médecin vous examine la gorge?
5. Ils ___ la bouche pour chanter.
6. J'___ le livre et je commence à lire.
7. J'___ la fenêtre quand il fait chaud.
8. Tu ___ les cadeaux que tes amis t'___ pour ton anniversaire?

C Il a été malade. Répondez par «oui».

1. Charles a été malade?
2. Il a été à l'hôpital?
3. Le médecin a examiné Charles?
4. Charles a ouvert la bouche?
5. Le médecin a découvert la cause de sa maladie?
6. Il a couvert le pauvre Charles?
7. Le médecin a fait un diagnostic?
8. Charles a compris le diagnostic?
9. Le médecin a prescrit des médicaments?
10. Charles a pris les médicaments?
11. Il a pris trois comprimés par jour?
12. Il a beaucoup souffert?
13. Ses amis ont offert un petit cadeau à Charles?

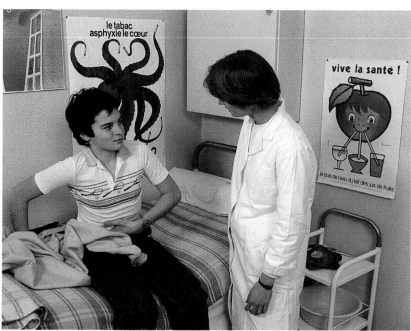

Une femme médecin parle à son jeune patient.

PRESENTATION *(continued)*

Extension of *Exercice C*

After going over Exercise C, have a student retell the story in his/her own words. As a further extension, have other students make up questions based on the student's story.

ANSWERS

Exercice B

1. ouvrons
2. offre
3. ouvrez
4. ouvrez
5. ouvrent
6. ouvre
7. ouvre
8. ouvres, offrent

Exercice C

1. Oui, il a été malade.
2. Oui, il a été à l'hôpital.
3. Oui, il a examiné Charles.
4. Oui, il a ouvert la bouche.
5. Oui, il a découvert la cause de sa maladie.
6. Oui, il a couvert le pauvre Charles.
7. Oui, il a fait un diagnostic.
8. Oui, il a compris le diagnostic.
9. Oui, il a prescrit des médicaments.
10. Oui, il a pris les médicaments.
11. Oui, il a pris trois comprimés par jour.
12. Oui, il a beaucoup souffert.
13. Oui, ils ont offert un petit cadeau à Charles.

LEARNING FROM PHOTOS

1. Ask questions about the photo: *Quel âge a le malade? Où est-il? À qui parle-t-il? Qu'est-ce qu'il a? Quel sport aime-t-il? Pourquoi dites-vous ça?*
2. Have students look at the posters on the walls of the doctor's office. Ask what they are about. What three drinks are being promoted in the poster on the right? Have students give you the French terms from the poster.

L'impératif

PRESENTATION (*page 398*)

Note Students should have little trouble learning the imperative since they are already familiar with the verb forms. The only thing that will be new to them is the dropping of the -s in the spelling of the *tu* form of -er verbs.

A. Have students open their books to page 398. Lead them through steps 1–3.

B. Illustrate the difference between singular and plural imperatives by giving commands to one student and to groups or pairs of students. For example: *Yvonne, prends ton livre de français. Va au tableau. Ouvre le livre à la page 15. Guillaume et Martine, allez à la porte. Ouvrez la porte. Asseyez-vous.*

C. Practice the negative forms by calling out TPR commands and having students change them to the negative. Then reverse the procedure.

L'impératif

Giving Formal and Informal Commands

1. You use the imperative to give commands and make suggestions. The forms are usually the same as the *tu, vous,* or *nous* form of the present tense. Note, however, that you drop the final *s* of the *tu* form of verbs ending in *-er,* including *aller.* The same is true for verbs like *ouvrir* and *souffrir,* which are conjugated like *-er* verbs. In commands the subject is omitted.

INFINITIF	TU	VOUS
regarder	regarde	regardez
aller	va	allez
ouvrir	ouvre	ouvrez
finir	finis	finissez
attendre	attends	attendez
prendre	prends	prenez
faire	fais	faites
dire	dis	dites

Marie, regarde le tableau! Va au tableau!
Madame, prenez des vitamines!
Roger et Vincent, faites attention!

2. To express "Let's…," you use the *nous* form of the verb without the subject.

Dansons!
Choisissons le menu touristique.

3. With commands, negative expressions go around the verb.

Ne parle pas en classe.
N'écoutez jamais ce disque.
Ne disons rien.

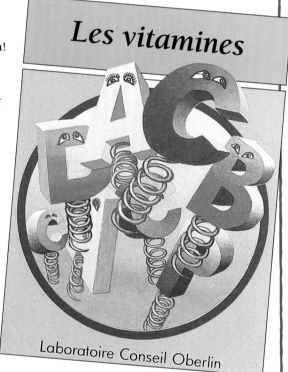

Les vitamines

Laboratoire Conseil Oberlin

ADDITIONAL PRACTICE

(*After completing Exercises A–E*)

1. Students work in groups of three. One student tells another what to do. That student dramatizes the command. The third student describes the scene.

É1: ____, ouvre la bouche.
É2: (*dramatizes*)
É3: ____ ouvre la bouche, mais moi, je n'ouvre pas la bouche.

2. Have students choose three of the following locations. They then make three suggestions of things to do or not to do at each location, using the *nous* imperative.

à la plage	chez le médecin
dans une boutique	au marché
	au café
dans une voiture	à la gare
dans une station de sports d'hiver	

Exercices

A **La loi, c'est moi!** Donnez un ordre à un copain ou à une copine d'après le modèle.

> **chanter**
> *Chante!*

1. danser
2. écouter la musique
3. parler français
4. travailler plus
5. préparer le dîner
6. commander un sandwich
7. ouvrir la porte

B **Et vous aussi!** Refaites l'Exercice A d'après le modèle.

> **chanter**
> *Chantez!*

C **Ne fais pas ça!** Donnez un ordre à un copain ou à une copine d'après le modèle.

> **regarder**
> *Ne regarde pas!*

1. lire le journal
2. écrire une lettre
3. prendre le métro
4. attendre dans la gare
5. descendre
6. aller vite
7. faire attention
8. entrer

D **Ne faites pas ça!** Refaites l'Exercice C d'après le modèle.

> **regarder**
> *Ne regardez pas!*

E **Allons-y!** Répondez d'après le modèle.

> **Vous voulez inviter Marie?**
> *Oui, invitons Marie!*

1. Vous voulez aller à la plage?
2. Vous voulez nager?
3. Vous voulez faire du ski nautique?
4. Vous voulez prendre le petit déjeuner?
5. Vous voulez aller au restaurant?
6. Vous voulez manger des fruits?

le stress

Laboratoire Conseil Oberlin

Stressé? N'oubliez pas de vous relaxer!

CHAPITRE 15　　**399**

Exercices

ANSWERS

Exercice A

1. Danse.
2. Écoute la musique.
3. Parle français.
4. Travaille plus.
5. Prépare le dîner.
6. Commande un sandwich.
7. Ouvre la porte.

Exercice B

1. Dansez.
2. Écoutez la musique.
3. Parlez français.
4. Travaillez plus.
5. Préparez le dîner.
6. Commandez un sandwich.
7. Ouvrez la porte.

Exercice C

1. Ne lis pas le journal.
2. N'écris pas de lettre.
3. Ne prends pas le métro.
4. N'attends pas dans la gare.
5. Ne descends pas.
6. Ne va pas vite.
7. Ne fais pas attention.
8. N'entre pas.

Exercice D

1. Ne lisez pas le journal.
2. N'écrivez pas de lettre.
3. Ne prenez pas le métro.
4. N'attendez pas dans la gare.
5. Ne descendez pas.
6. N'allez pas vite.
7. Ne faites pas attention.
8. N'entrez pas.

Exercice E

1. Oui, allons à la plage.
2. Oui, nageons.
3. Oui, faisons du ski nautique.
4. Oui, prenons le petit déjeuner.
5. Oui, allons au restaurant.
6. Oui, mangeons des fruits.

INFORMAL ASSESSMENT

Have students quickly make up as many commands as they can.

LEARNING FROM REALIA

Have students look at the cards on pages 398 and 399. Explain that these types of cards are provided by drug companies and distributed free at pharmacies. Have students pick out all the cognates and make up sentences using *les vitamines, stressé(e), se relaxer.*

INDEPENDENT PRACTICE

Assign any of the following:
1. Exercises, page 399
2. Workbook, *Structure: F–H,* page 152
3. Communication Activities Masters, *Structure: C,* page 74
4. Computer Software, *Structure*
5. CD-ROM, Disc 4, pages 398–399

PRESENTATION (page 400)

A. Tell students they will hear a conversation between Charlotte and a doctor.

B. Have students close their books and watch the Conversation Video or listen as you read the conversation or play Cassette 9A/CD-9.

C. Now reread the conversation or replay the cassette or CD, stopping after each of the three sections to ask simple comprehension questions.

D. Have students dramatize the conversation.

E. Have a student summarize the conversation in his or her own words.

Note In the CD-ROM version, students can play the role of either one of the characters and record the conversation.

ANSWERS

Exercice A

1. Oui, elle souffre beaucoup.
2. Oui, elle a les yeux qui piquent.
3. Oui, elle a la gorge qui gratte.
4. Oui, elle a mal à la tête.
5. Elle a mal partout.
6. Oui, elle a de la fièvre et des frissons.
7. Il va prendre sa température.
8. Elle ouvre la bouche.
9. Elle a la gorge très rouge.
10. Elle a une angine.
11. Il prescrit des antibiotiques.
12. Oui, elle va être vite sur pied.

CONVERSATION

Scènes de la vie   *Charlotte souffre*

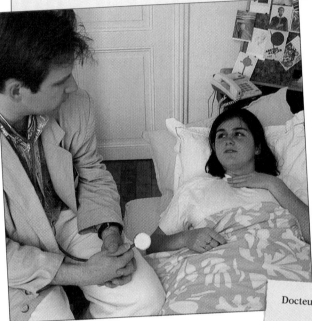

CHARLOTTE: Ah, docteur, qu'est-ce que je peux souffrir!

LE MÉDECIN: Où avez-vous mal? Quels sont vos symptômes?

CHARLOTTE: Qu'est-ce que je suis malade! J'ai les yeux qui piquent et la gorge qui gratte. Ça fait mal!

LE MÉDECIN: Vous avez mal à la tête?

CHARLOTTE: Ah, oui. J'ai mal partout. Et j'ai des frissons. J'ai froid.

LE MÉDECIN: Alors, vous avez de la fièvre. Je vais prendre votre température. Mais d'abord, je vais vous examiner. Ouvrez la bouche, s'il vous plaît... Oui, vous avez la gorge très rouge. Vous avez une angine.

CHARLOTTE: Une angine?!

LE MÉDECIN: Oui, ce n'est pas grave. Je vais vous faire une ordonnance. Je vous prescris des antibiotiques. Vous allez être vite sur pied.

A   Une angine.   Répondez d'après la conversation.

1. Charlotte souffre beaucoup?
2. Elle a les yeux qui piquent?
3. Elle a la gorge qui gratte?
4. Elle a mal à la tête?
5. Où a-t-elle mal?
6. Elle a de la fièvre et des frissons?
7. Qu'est-ce que le médecin va prendre?
8. Qu'est-ce que Charlotte ouvre?
9. Elle a la gorge comment?
10. Qu'est-ce qu'elle a?
11. Qu'est-ce que le médecin prescrit?
12. Charlotte va être vite sur pied?

Docteur Henri ANSART

50, résidence du Bois du Four
78640 NEAUPHLE-LE-CHÂTEAU
(Yvelines)
Tél. 36.89.00.07   36.89.08.95

DUROSEL Charlotte

Hyconcil :

2 Gélules matin et soir pendant 5 jours.

Locabiotal :

3 pulvérisations par jour.

CRITICAL THINKING ACTIVITY	LEARNING FROM REALIA

(Thinking skills: locating causes)

Put the following on the board or on an overhead transparency:

1. Tout le monde parle du stress. Il y a beaucoup de stress dans la société moderne. Quelles sont les causes du stress?

2. Si l'on se sent stressé(e), qu'est-ce qu'on peut faire pour se relaxer?

You may wish to ask questions about the prescription: *Qu'est-ce que c'est? Comment s'appelle le médecin? Quelle est son adresse? Quel est le nom de la malade? L'ordonnance est pour quels médicaments? Quand la malade va-t-elle prendre l'Hyconcil? Combien de fois par jour? Et pendant combien de jours? Comment dit-on en français two capsules et three sprays?*

Prononciation *Le son /ü/*

1. To say the sound /ü/, first say the sound /i/ but round your lips. Repeat the following words.

température	enrhumé	chaussure
voiture	descendu	

2. The sound /ü/ also occurs in combination with other vowels.

éternuer	lui	depuis
aujourd'hui	je suis	

Now repeat the following words and sentences.

Quelle est la température aujourd'hui?
Luc conduit depuis huit ans.
Il a mis ses chaussures dans la voiture.

température

Activités de communication orale

A **Ah docteur, je suis très malade!** Imagine you're sick with a cold, the flu, or a throat infection. Tell the doctor (your partner) what your symptoms are. Your partner makes a diagnosis and tells you what to do to get better.

Élève 1: J'ai mal à la tête et j'éternue tout le temps.
Élève 2: Vous avez un rhume. Prenez de l'aspirine et du bouillon de poulet.

B **Je déteste ce cadeau!** In your worst nightmare, what did the following people give you for your birthday? Your partner will ask you about each person. Answer, then reverse roles.

Élève 1: Qu'est-ce que ta grand-mère t'a offert pour ton anniversaire?
Élève 2: Elle m'a offert des cassettes de Frank Sinatra.

tes parents	tes grands-parents
ton meilleur ami	ton frère
ta meilleure amie	ta sœur

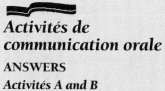

INDEPENDENT PRACTICE

Assign any of the following:
1. Exercise and activities, pages 400–401
2. CD-ROM, Disc 4, pages 400–401

LECTURE ET CULTURE

READING STRATEGIES
(page 402)

Reading

A. Briefly synopsize the *Lecture* in French. Ask a few questions about it.

B. Have a student read 3–4 sentences. Ask a few questions to check comprehension, then have another student read. Continue this way for the rest of the reading.

C. **Finding information:** Have students scan the last paragraph and find the French word for "doctor's fee."

Post-reading

Assign the reading and the exercises that follow as homework. Go over these exercises the next day in class.

Note Students may listen to a recorded version of the *Lecture* on the CD-ROM.

Étude de mots

ANSWERS

Exercice A

1. … une fièvre de cheval.
2. Il n'est pas dans son assiette.
3. … prendre rendez-vous chez le médecin.
4. … ne donne pas de consultations…
5. … à domicile.
6. … ausculte Richard.
7. … grave.
8. … va être vite sur pied.

UNE CONSULTATION OU UNE VISITE

Le pauvre Richard! Qu'est-ce qu'il est malade! Il tousse. Il éternue. Il a mal à la tête. Il a une fièvre de cheval. Il a des frissons. Il n'est pas du tout dans son assiette. Il n'est pas très courageux, notre Richard. Il veut prendre rendez-vous[1] chez le médecin, mais c'est le week-end. Son médecin ne donne pas de consultations.

Alors que faire? Pas de problème! Appelons S.O.S Médecins, un service qui envoie des médecins à domicile. Un médecin arrive chez Richard. Il examine Richard. Il ausculte le malade. Il prend sa température. Le médecin dit que Richard a la grippe. Mais ce n'est pas grave. Il va vite se sentir mieux. Le médecin fait une ordonnance à Richard. Il prescrit des antibiotiques: trois comprimés par jour, un à chaque repas[2].

Richard paie le médecin. Mais en France la Sécurité Sociale rembourse les honoraires des médecins, c'est-à-dire l'argent qu'on donne aux médecins. Les honoraires et tous les frais[3] médicaux sont remboursés de 80 à 100% (pour cent) par la Sécurité Sociale.

[1] prendre rendez-vous *make an appointment*
[2] repas *meal*
[3] frais *expenses*

Étude de mots

A **Autrement dit.** Dites d'une autre manière.

1. Richard a *beaucoup de fièvre*.
2. Il *ne se sent pas bien*.
3. Il veut *aller voir* le médecin.
4. Le médecin *ne voit pas de malades* pendant le week-end.
5. S.O.S Médecins envoie des médecins *chez les malades*.
6. Le médecin *écoute la respiration de* Richard.
7. La grippe n'est pas une maladie *sérieuse*.
8. Richard va vite *se sentir mieux*.

Compréhension

B **Vous avez compris?** Répondez par «oui» ou «non».

1. Richard est très courageux quand il est malade.
2. Il a beaucoup de fièvre.
3. Il a mal au ventre.
4. Richard veut aller chez le médecin.
5. Son médecin donne des consultations tous les jours.
6. Richard prend rendez-vous chez le médecin de S.O.S Médecins.
7. Le médecin prescrit des comprimés d'aspirine.
8. Les frais médicaux ne sont pas remboursés en France.

C **En France.** Qu'est-ce que vous avez appris sur les médecins et les services médicaux en France?

DÉCOUVERTE CULTURELLE

*L*a culture influence la médecine? Certainement. Par exemple, en France tout le monde parle de son foie[1]. Les Français disent souvent, «J'ai mal au foie». En Amérique on n'entend jamais

Les troubles digestifs

«J'ai mal au foie». Pourquoi pas? Parce que, pour les Américains, une maladie du foie est quelque chose de grave. Mais quand un Français dit qu'il a mal au foie, il veut dire tout simplement qu'il a un trouble digestif. Ce n'est rien de grave. Il n'est peut-être pas dans son assiette aujourd'hui mais il va être vite sur pied.

Aux États-Unis on parle d'allergies. Beaucoup d'Américains souffrent d'une petite allergie. Les symptômes d'une allergie ressemblent aux symptômes d'un rhume. On éternue et on a souvent mal à la tête. Une allergie est désagréable, mais pas grave. En France, on parle moins souvent d'allergies. Vive la différence!

[1] foie *liver*

CHAPITRE 15 **403**

OPTIONAL MATERIAL

PRESENTATION (*pages 404–405*)

The main objective of this section is to have students enjoy the photographs and gain an appreciation of France, its people, and its culture. However, if you would like to do more with it, you might want to do some of the following activities.

A. Have students cover the captions. Call on volunteers to say as much as they can about each photo. Encourage them to use *Je pense que* or *Je crois que* to indicate that they are guessing.

B. Call on volunteers to read the captions aloud. What information do students now have that they didn't have before? Which new words can they guess the meaning of from the reading?

C. You may wish to explain to students what the acronym *SIDA* means (*syndrome immuno-déficitaire acquis*).

Note In the CD-ROM version, students can listen to the recorded captions and discover a hidden video behind one of the photos.

RÉALITÉS

DID YOU KNOW?

1. In addition to some 12 medical schools in the Paris region, there are also *Centres hospitaliers universitaires* in the following cities: Brest, Rennes, Caen, Rouen, Angers, Nantes, Lille, Amiens, Reims, Strasbourg, Nancy, Dijon, Besançon, Tours, Poitiers, Lyon, Limoges, Clermont-Ferrand, Grenoble, Bordeaux, Montpellier, Toulouse, Marseille.

2. In France, students who want to pursue a career in medicine can do so at minimum expense. Tuition at French universities is low, but competition in medical school is fierce. Many students drop out before completing the seven years of study required to become a general practitioner.

Voici un médecin généraliste dans son cabinet **1**.

Voici un grand hôpital pédiatrique à Paris **2**. Les grands hôpitaux ont l'équipement le plus moderne et le plus avancé technologiquement. Les services médicaux en France sont généralement excellents.

En France on fait beaucoup de recherches médicales et pharmaceutiques. À l'Insititut Pasteur de Paris, on fait des recherches continues pour trouver des médicaments ou des vaccins pour traiter ou combattre le SIDA, la maladie la plus grave de notre époque **3**.

Voici une ambulance du SAMU (le Service d'Aide Médicale d'Urgence). On appelle le SAMU en cas d'urgence **4**.

Note Information follows in the *Lettres et sciences* section, page 442, about Louis Pasteur and the Pasteur Institute.

405

CRITICAL THINKING ACTIVITY

(*Thinking skill: supporting arguments with reasons*)

Put the following on the board or on a transparency:

Faites une liste des caractéristiques que vous considérez comme importantes pour un médecin. Essayez de dire pourquoi vous les trouvez importantes.

ADDITIONAL PRACTICE

1. Student Tape Manual, Teacher's Edition, *Deuxième Partie*, pages 172–173.
2. Situation Cards, Chapter 15
3. Communication Transparency C-15

RECYCLING

The *Activités de communication orale* and *Activité de communication écrite* provide a forum for students to create and answer their own questions and come up with their own dialogues while working within the health context of Chapter 15.

INFORMAL ASSESSMENT

The guided nature of Oral Activity B, with its numbered cues, makes it more suitable for use as a speaking evaluation, especially for less able students. Use the evaluation criteria given on page 34 of this Teacher's Wraparound Edition.

Activités de communication orale

ANSWERS

Activité A

Answers will be either C'est bon or C'est mauvais pour la santé.

Activité B

É2 answers will vary. É1 questions may include:

1. **Je te téléphone à minuit. Qu'est-ce que tu me dis?**
2. **Je te propose de sortir ensemble. …**
3. **Je te dis que j'ai besoin d'argent. …**
4. **Je te demande de faire mes devoirs pour moi. …**
5. **Je te donne un sandwich au pâté. …**
6. **Je te dis que je ne suis pas dans mon assiette. …**

Activité de communication écrite

ANSWERS

Activité A

Answers will vary.

Activités de communication orale

A **C'est bon ou mauvais pour la santé?** Ask students in the health class for a list of health tips. Then, with a partner, make a list in French of things that people should do to stay healthy. Make a second list of things people should avoid doing. Present your lists to the class in random order and ask your classmates to decide whether the suggestion is good or bad for your health.

B **Qu'est-ce que tu me dis?** Suggest the following situations to your partner and ask what he or she would say in each case.

> **t'offrir un bouquet de roses**
> **Élève 1: Je t'offre un bouquet de roses. Qu'est-ce que tu me dis?**
> **Élève 2: Je te dis: «Merci beaucoup, les roses sont magnifiques!»**

1. te téléphoner à minuit
2. te proposer de sortir ensemble
3. te dire que j'ai besoin d'argent
4. te demander de faire mes devoirs pour moi
5. te donner un sandwich au pâté
6. te dire que je ne suis pas dans mon assiette

Activité de communication écrite

A **Excusez-moi…** You're supposed to take a French test today but you're not feeling well. Write a note to your French teacher explaining why you can't take the test, and mention some symptoms you have. Give the date and time you'd like to take the test.

FOR THE YOUNGER STUDENT

1. Have students make colorful get-well cards using some of the expressions they have learned. If someone they know is ill, they can send him or her the cards.
2. Have students draw their own cartoons to illustrate the following expressions: *Je ne suis pas dans mon assiette aujourd'hui. Il a une fièvre de cheval. Elle a un chat dans la gorge.*
3. Students work in teams to create a composite "monster." They cut out scrap paper and use markers to make body parts, one per member, including facial features as well as limbs, torsos, hair, etc. They should label the back of each cut-out in French. Collect the cut-outs and put them in a bag. Call on students one at a time to draw out one body part and pin it to the bulletin board to create a "monster."

Réintroduction et recombinaison

A **Isabelle se sent très bien aujourd'hui!** Complétez au présent.

Qui ___ (dire) qu'Isabelle n'___ (être) pas dans son assiette aujourd'hui? Ce
___₁_____₂
n'___ (être) pas du tout vrai. Elle ___ (aller) très bien. Elle ___ (se lever) de
___₃_____₄_____₅
bonne heure, ___ (prendre) son petit déjeuner et ___ (quitter) la maison. Elle
_____₆_____₇
___ (vouloir) rester en forme. Elle ___ (aller) au gymnase où elle ___ (faire)
₈_____₉_____₁₀
de l'aérobic. Elle ___ (avoir) beaucoup de copains au gymnase. Ils ___ (mettre)
_____₁₁_____₁₂
un survêtement et ils ___ (faire) de l'exercice ensemble.
_____₁₃

B **Aux sports d'hiver.** Complétez au passé.

1. L'hiver dernier Sylvie et Maryse ___ (passer) une semaine à Val d'Isère dans
   les Alpes françaises.
2. Le premier jour elles ___ (mettre) leur anorak, leurs gants et leurs skis.
3. Elles ___ (prendre) le télésiège jusqu'au sommet de la montagne.
4. Malheureusement elles ___ (choisir) la mauvaise piste—une piste noire, très
   difficile.
5. Sylvie ___ (glisser) et ___ (faire) une chute.
6. Elle ___ (perdre) ses bâtons qui ___ (glisser) jusqu'en bas de la piste.
7. Deux garçons très sympa ___ (trouver) les bâtons et ils ___ (donner) les
   bâtons à Sylvie.
8. Les deux filles ___ (dire) «merci» aux garçons et ils ___ (faire) du ski
   ensemble toute la journée.

Vocabulaire

NOMS
la santé
la médecine
le médicin
le/la malade
le/la pauvre
l'allergie (f.)
l'angine (f.)
la température
la fièvre
les frissons (m.)
la grippe
le rhume
l'infection (f.)

le médicament
l'ordonnance (f.)

l'aspirine (f.)
l'antibiotique (m.)
la pénicilline
le comprimé
la pharmacie
le/la pharmacien(ne)
le kleenex
le mouchoir

les yeux (m.)
le nez
la bouche
l'oreille (f.)
la gorge
le ventre

ADJECTIFS
allergique

bactérien(ne)
enrhumé(e)
malade
viral(e)

VERBES
examiner
ausculter
respirer (à fond)
éternuer
tousser
couvrir
découvrir
offrir
ouvrir
souffrir

se sentir
prescrire

AUTRES MOTS ET EXPRESSIONS
avoir mal à
avoir un chat dans la gorge
avoir de la fièvre
avoir une fièvre de cheval
avoir les yeux qui piquent
avoir le nez qui coule
avoir la gorge qui gratte
ne pas être dans son assiette
être en bonne (mauvaise) santé
être vite sur pied
faire un diagnostic
faire une ordonnance
Ça fait mal.

CHAPITRE 15   **407**

Réintroduction et recombinaison

PRESENTATION *(page 407)*

Exercice A
Exercise A provides practice
with many of the irregular verbs
students have studied.

Exercice B
Exercise B provides practice in
the formation of the *passé composé*
of regular and irregular verbs.

ANSWERS

Exercice A

1. dit	8. veut
2. est	9. va
3. est	10. fait
4. va	11. a
5. se lève	12. mettent
6. prend	13. font
7. quitte	

Exercice B

1. ont passé
2. ont mis
3. ont pris
4. ont choisi
5. a glissé, a fait
6. a perdu, ont glissé
7. ont trouvé, ont donné
8. ont dit, ont fait

ASSESSMENT RESOURCES

1. Chapter Quizzes
2. Testing Program
3. Situation Cards
4. Communication
   Transparency C-15
5. Computer Software:
   Practice/Test Generator

VIDEO PROGRAM

INTRODUCTION   **(46:35)**

QU'EST-CE QU'IL A,   **(47:07)**
ÉTIENNE?

Have students add the *Mon Autobiographie*
section of the Workbook on page 154 to their
portfolios.
Note Students may create and save both oral
and written work using the Electronic
Portfolio feature on the CD-ROM.

INDEPENDENT PRACTICE

1. Activities and exercises, pages 406–407
2. CD-ROM, Disc 4, pages 406–407
3. Communication Activities Masters,
   pages 71–74

CHAPTER OVERVIEW

In this chapter students will learn how to discuss cultural events and express their cultural likes and dislikes. In order to do this, they will learn vocabulary associated with films, museums, the theater, etc. They will also learn to use the verbs *savoir* and *connaître,* the object pronouns *le, la,* and *les,* prepositions with geographical terms, and the verb *venir.*

The cultural focus of this chapter is on French attitudes and preferences with regard to cultural events.

CHAPTER OBJECTIVES

By the end of this chapter, students will know:

1. vocabulary associated with the theater and movies, including some genres of films and plays
2. vocabulary associated with museums and art, including painting and sculpture
3. the present indicative forms of *savoir* and *connaître* and the difference in meaning between the two verbs
4. the use of the direct object pronouns *le, la,* and *les*
5. the names and genders of countries
6. the use of *en, à,* or contractions with *à* to express "to" or "in" a country, city, or continent
7. the use of *de* or contractions with *de* to express "from" a country, city, or continent
8. the present indicative forms of verbs like *venir*

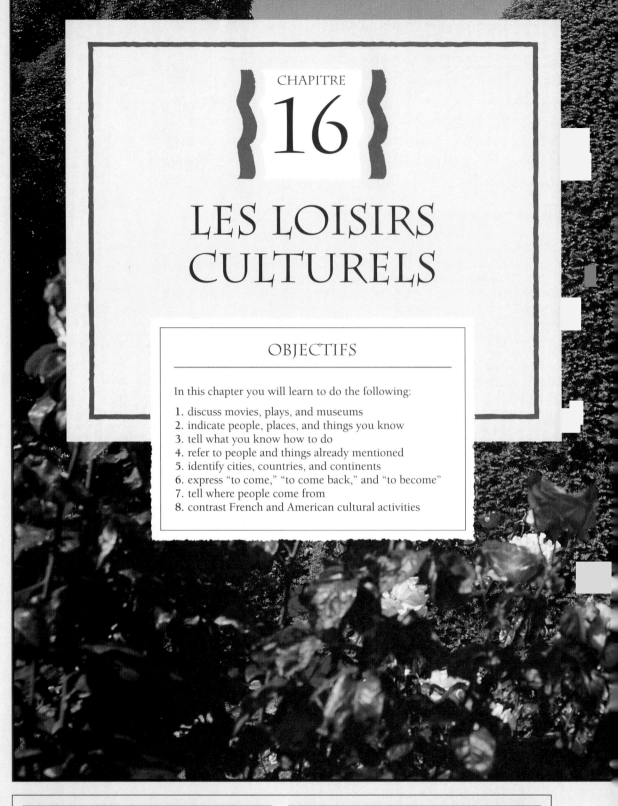

CHAPITRE

16

LES LOISIRS CULTURELS

OBJECTIFS

In this chapter you will learn to do the following:

1. discuss movies, plays, and museums
2. indicate people, places, and things you know
3. tell what you know how to do
4. refer to people and things already mentioned
5. identify cities, countries, and continents
6. express "to come," "to come back," and "to become"
7. tell where people come from
8. contrast French and American cultural activities

CHAPTER PROJECTS

1. Have students visit a local art museum to see different styles of art and the works of French artists.
2. Let students leaf through some French comic books (*Astérix, Tintin,* etc.). Describe a certain character to them or have them focus on French words for noises (*toc toc* = knock, knock; *aïe* = ouch, etc.).

INTERDISCIPLINARY CONNECTIONS

Divide the class into small groups and have them ask students in the art class to help them research several French painters and sculptors. Later, invite the art students to the French class, where each group puts on an "art show" with prints of their artist's most famous works.

409

CHAPTER 16 RESOURCES

1. Workbook
2. Student Tape Manual
3. Audio Cassette 9B/CD-10
4. Bell Ringer Review Blackline Masters
5. Vocabulary Transparencies
6. Pronunciation Transparency P-16
7. Art Transparencies F-5 & F-6
8. Communication Transparency C-16
9. Communication Activities Masters
10. Map Transparencies
11. Situation Cards
12. Conversation Video
13. Videocassette/Videodisc, Unit 4
14. Video Activities Booklet, Unit 4
15. Lesson Plans
16. Computer Software: Practice/Test Generator
17. Chapter Quizzes
18. Testing Program
19. Internet Activities Booklet
20. CD-ROM Interactive Textbook
21. Performance Assessment

Pacing

This chapter requires eight to ten class sessions. Pacing will vary according to class length and the age and aptitude of the students.

Note The Lesson Plans offer guidelines for 45- and 55-minute classes and **Block Scheduling.**

Exercices vs. *Activités*

All exercises (which provide guided practice) are coded in blue. All communicative activities are coded in red.

INTERNET ACTIVITIES

(optional)

These activities, student worksheets, and related teacher information are in the *Bienvenue* Internet Activities Booklet and on the Glencoe Foreign Language Home Page at: http://www.glencoe.com/secondary/fl

DID YOU KNOW?

The Musée Rodin, in the 7th *arrondissement,* is housed in a beautiful 18th-century mansion that was once Rodin's studio. *Le Penseur,* pictured above (with the dome of Napoleon's Tomb behind it), is located in the garden along with many other of Rodin's sculptures, including *Les Bourgeois de Calais,* a photo of which can be found on page 419.

VOCABULAIRE

MOTS 1

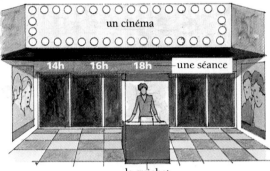

AU CINÉMA

un cinéma

14h 16h 18h — une séance

le guichet

Qui joue dans ce film?

l'écran

une salle de cinéma — un dessin animé

les vedettes (f.)

un acteur

une actrice

un film étranger

Qu'est-ce que tu vas faire?

les sous-titres (m.)

On passe un film étranger à Paris.
On passe le film en V.O., c'est-à-dire en
version originale.
On le voit en version originale avec des
sous-titres français.

Le film est doublé.
La version originale est en anglais.
La version doublée est en français.

410 CHAPITRE 16

Vocabulary Teaching Resources

1. Vocabulary Transparencies 16.1 (A & B)
2. Audio Cassette 9B/CD-10
3. Student Tape Manual, Teacher's Edition, *Mots 1: A–C,* pages 174–176
4. Workbook, *Mots 1: A–E,* pages 155–157
5. Communication Activities Masters, *Mots 1: A,* page 75
6. Chapter Quizzes, *Mots 1:* Quiz 1, page 82
7. CD-ROM, Disc 4, *Mots 1:* pages 410–413

Bell Ringer Review

Write the following on the board or use BRR Blackline Master 16-1: Make a list of activities you like to do in your free time.

PRESENTATION *(pages 410–411)*

A. Using Vocabulary Transparencies 16.1 (A & B), play the *Mots 1* presentation on Cassette 9B/CD-10. Point to the appropriate illustration as you play the cassette or CD.

B. Have students repeat each word or expression after you or the cassette two or three times as you point to the corresponding item on the transparency.

C. Call on individual students to point to the corresponding illustration on the transparency as you say the word or expression.

Teaching Tip Ask questions about students' personal preferences when practicing the vocabulary. For example: *Jacques, tu préfères les drames ou les documentaires? Qui aime les films policiers?*

TOTAL PHYSICAL RESPONSE

(following the Vocabulary presentation)

Getting Ready

Demonstrate the terms *parmi* and *serrer la main à…*

TPR

___, levez-vous et venez-ici, s'il vous plaît.
Vous êtes un acteur/une actrice célèbre.

Habillez-vous.
Mettez votre costume.
Le rideau se lève. Entrez en scène.
Dites «bonjour» aux spectateurs.
Chantez.
Dansez.
Marchez parmi les spectateurs.
Serrez la main à quelqu'un.
Merci, ___. C'est bien. Retournez à votre place, s'il vous plaît.

Tu préfères quels genres de films?

un documentaire   un film policier   un film d'horreur   un film de science-fiction

un film d'aventures   un film d'amour   une comédie   un drame

AU THÉÂTRE

le rideau

une pièce de théâtre

le décor

On monte une pièce.
C'est une comédie.

Comédie-Française

**Molière
Le Tartuffe**

Acte 1   Scène 1
          Scène 2

Entracte
Acte 2   Scène 1
          Scène 2

Acte 3   Scène 1
          Scène 2

la scène

un costume

l'entracte (m.)

La pièce a trois actes.
Chaque acte a deux scènes.
Entre deux actes il y a un entracte.

Voici quelques genres de pièces:

   une tragédie
   un opéra
   une comédie musicale

CHAPITRE 16   **411**

D. Mention some well-known films and ask students to classify them and name their stars. For example: **Frankenstein**— *C'est un très vieux film d'horreur avec Boris Karloff.* **Dead Poets' Society** (*Le Cercle des Poètes disparus*)—*C'est un drame avec Robin Williams.*

E. Ask film trivia questions, incorporating a variety of questioning techniques. For example: **Out of Africa**—*c'est un documentaire? Qui joue dans ce film? C'est un film américain ou un film africain? C'est un film de quelle année?*, etc.

F. Read the following movie summaries to the students and ask them to identify the type of film: *Les acteurs et les actrices habitent sur une autre planète et voyagent dans l'espace. La police recherche un criminel. On étudie la vie des animaux dans leurs milieux. Un couple décide de traverser l'océan Atlantique tout seul dans un canoë.*

Vocabulary Expansion

You may wish to give students the following additional words.
**le fauteuil
l'orchestre
le balcon
le rang
le lever du rideau**

PAIRED ACTIVITY

Have students work in pairs to prepare a skit. One plays the part of a box office clerk at a movie theater; the other plays the part of a ticket buyer.

COOPERATIVE LEARNING

Students work in groups of four. Each person writes down the title of a movie he/she has seen recently. They put their titles together and scramble them. Each person in turn picks one and asks: *Qui a vu ce film?* and then makes up as many present-tense questions about it as possible. The student who saw the film answers. If anyone picks his/her own film, he/she tells the others about it.

PRESENTATION (pages 412–413)

Exercices A, C, E, and F:
Listening

After doing Exercises A, C, E, and F as a whole-class activity, focus on the listening skill by having students do them in pairs. One partner reads the questions to the other in random order. The second partner answers with his/her book closed. Partners then reverse roles.

ANSWERS

Exercice A

Answers will vary.

Exercice B

1. passe
2. sous-titres
3. version
4. séance
5. billet

Exercice C

Answers will vary.

Exercice D

1. monte
2. théâtre
3. actes, actes, scènes
4. entracte
5. acteur
6. actrice
7. décor
8. costumes
9. scène
10. rideau

Exercices

A Fana de cinéma ou pas?   Donnez des réponses personnelles.

1. Tu es fana de cinéma? C'est-à-dire, tu aimes beaucoup voir des films?
2. Tu vas souvent au cinéma?
3. Il y a un cinéma près de chez toi?
4. La première séance est à quelle heure?
5. Il y a toujours un dessin animé avant le film?
6. Tu fais la queue devant le cinéma? Quels soirs spécialement?
7. Où est-ce que tu prends les billets?
8. Dans la salle de cinéma, tu préfères une place près de l'écran ou loin de l'écran?
9. Quel est ton acteur préféré ou ton actrice préférée?
10. Quelle est la vedette de ton film préféré?
11. Si tu vois un film étranger, tu préfères voir la version originale avec des sous-titres ou une version doublée?

B Au cinéma.   Complétez.

1. Ce soir on ___ un très bon film au cinéma Rex.
2. C'est un film étranger. Il n'est pas doublé, il a des ___.
3. On ne passe pas le film en ___ originale.
4. La prochaine ___ commence à quelle heure?
5. Combien coûte le ___?

C Tu aimes quels genres de films?   Donnez des réponses personnelles.

1. Tu préfères les documentaires ou les dessins animés?
2. Tu préfères les films policiers ou les films d'horreur?
3. Tu préfères les films d'aventures ou les films de science-fiction?
4. Tu préfères les comédies ou les drames?
5. Quand tu vas au magasin de vidéos, tu choisis généralement quel genre de films?

D Des pièces et des films.   Complétez.

1. Au théâtre on ___ une pièce.
2. On voit un film au cinéma et on voit une pièce au ___.
3. Une pièce a des ___ et les ___ ont des ___.
4. Entre deux actes il y a un ___.
5. Un ___ joue le rôle de Roméo.
6. Une ___ joue le rôle de Juliette.
7. Le balcon de Juliette est le ___ d'une scène d'amour célèbre.
8. Les acteurs et les actrices portent des ___.
9. Le mot ___ en français signifie (veut dire) *scene* et *stage* en anglais.
10. Le ___ se lève à 20 heures.

E **Au théâtre.** Donnez des réponses personnelles.

1. Tu es fana de théâtre?
2. Tu vas souvent au théâtre?
3. Il y a un théâtre dans ta ville?
4. Ton école a un club d'art dramatique?
5. Tu es membre du club d'art dramatique?
6. Le club monte combien de pièces par an?
7. Cette année le club va monter quelle pièce?

F **Mes préférences.** Donnez des réponses personnelles.

1. Tu préfères les comédies ou les tragédies?
2. Tu aimes l'opéra?
3. Tu aimes les comédies musicales?
4. Tu as déjà joué dans une pièce?
5. Quel rôle as-tu joué?

Collections de la Comédie-Française

CHAPITRE 16   **413**

LEARNING FROM REALIA

Have students look at the theater listings above to see if they have heard of any of the plays. Have them look at the ticket. Ask: *Comment s'appelle le théâtre? (Le Théâtre Français = La Comédie-Française.) Quel est le titre de la pièce? À quelle heure commence la pièce? Combien coûte le billet?*

Note Students will read an excerpt from *Le Malade imaginaire* in *En voyage.*

INDEPENDENT PRACTICE

Assign any of the following:

1. Exercises, pages 412–413
2. Workbook, *Mots 1: A–E,* pages 155–157
3. Communication Activities Masters, *Mots 1: A,* page 75
4. CD-ROM, Disc 4, pages 410–413

MOTS 2

VOCABULAIRE

MOTS 2

AU MUSÉE
une exposition d'art

la peinture

un tableau

une statue

la sculpture

une peintre

un peintre

414 CHAPITRE 16

Vocabulary Teaching Resources

1. Vocabulary Transparencies 16.2 (A & B)
2. Audio Cassette 9B/CD-10
3. Student Tape Manual, Teacher's Edition, *Mots 2: D–E*, page 177
4. Workbook, *Mots 2: F–H*, pages 157–158
5. Communication Activities Masters, *Mots 2: B*, pages 76–77
6. Computer Software, *Vocabulaire*
7. Chapter Quizzes, *Mots 2: Quiz 2*, page 83
8. CD-ROM, Disc 4, *Mots 2:* pages 414–417
9. Art Transparencies F-5 and F-6

Bell Ringer Review

Write the following on the board or use BRR Blackline Master 16-2: Write the titles of the last five movies you saw or the last five books you read. Classify them (comedy, science-fiction, etc.) in French and write one sentence about each one.

PRESENTATION *(pages 414–415)*

A. Introduce the new words by showing French paintings on slides from an art book, or use Art Transparencies F-5 and F-6.
B. Have students close their books. Model the *Mots 2* vocabulary, using Vocabulary Transparencies 16.2 (A & B). Have them repeat the new vocabulary after you or Cassette 9B/CD-10. Repeat the procedure with books open.
C. Ask questions using *Qu'est-ce que c'est?* If a student fails to

TOTAL PHYSICAL RESPONSE

Getting Ready
Demonstrate the term *l'escalier.*

TPR
(Student 1) et (Student 2), **levez-vous et venez ici,** s'il vous plaît.
Vous êtes devant le Musée d'Art Moderne.
Entrez dans le musée.
Allez à l'ascenseur.

(Student 1), **appuyez sur le bouton.**
Montez dans l'ascenseur.
(Student 2), **appuyez sur le bouton du quatrième étage.**
Attendez.
Voilà le quatrième étage.
Sortez de l'ascenseur.
Vous êtes dans une exposition de peinture abstraite.

(continued on the next page)

des sculpteurs (m.)

une œuvre

Je sais le nom du peintre.
C'est Duval.
Je ne connais pas ce peintre personnellement.
Je connais son œuvre, c'est-à-dire ses tableaux.

Musée d'Art Moderne
Ouvert: du mardi au dimanche de 9h à 18h
Fermé: le lundi

Moi, je connais bien le Musée d'Art Moderne.
Je le visite souvent.
Je sais que le musée est fermé le lundi.
Il est ouvert tous les jours sauf le lundi.

CHAPITRE 16 **415**

answer, rephrase the question as an either/or question. For example: *C'est un tableau ou une sculpture?*

COGNATE RECOGNITION

Ask students to identify as many cognates as they can in *Mots* 2. Pay particular attention to their pronunciation of these cognates.

Vocabulary Expansion

You may wish to give students the following additional vocabulary in order to talk about art.
une gravure
une lithographie
de la poterie
une aquarelle
un portrait
une peinture à l'huile

THE FRANCOPHONE WORLD

Information about culture and cultural institutions in other francophone countries can be found in the *Les Arts* section of *Le Monde francophone* on pages 434–437.

TPR (*continued*)
Promenez-vous.
Regardez les tableaux.
(Student 1), **montrez un tableau que vous aimez à** (Student 2).
Promenez-vous encore.
Arrêtez-vous devant un grand tableau.
Regardez le tableau ensemble.
Qu'est-ce qu'il est énorme, ce tableau!
Regardez très haut.

Regardez à gauche. Regardez à droite.
C'est un tableau très bizarre. Vous ne le comprenez pas.
(Student 2), **montrez en gesticulant que vous aimez le tableau.**
(Student 1), **montrez en gesticulant que vous ne l'aimez pas du tout.**
Merci, (Student 1) et (Student 2). Vous avez très bien fait.
Retournez à vos places et asseyez-vous.

PRESENTATION *(page 416)*

Extension of *Exercice B*

Since the *je* and *tu* forms of *savoir* and *connaître* sound the same as the *il/elle* forms, you may wish to personalize Exercise B by asking: *Tu sais le nom d'un artiste? Qu'est-ce que c'est? Tu connais l'artiste? Tu sais sa nationalité? Tu connais l'œuvre de cet artiste? Tu connais un musée? Quel musée connais-tu? Tu sais où est le musée? Tu sais l'adresse du musée?*

ANSWERS

Exercice A

1. C'est un musée.
2. Le musée est ouvert.
3. C'est une exposition de peinture.
4. Elle est sculpteur.
5. C'est une statue.

Exercice B

1. Oui, il sait le nom du peintre.
2. Non, il ne connaît pas le peintre.
3. Oui, il connaît l'œuvre du peintre.
4. Oui, elle sait le nom du musée.
5. Oui, elle connaît le musée.
6. Oui, elle connaît le Musée d'Art Moderne.
7. Oui, elle le visite souvent.
8. Oui, elle sait que le musée est fermé le lundi.
9. Oui, il est ouvert tous les jours sauf le lundi.

Activités de communication orale

Mots 1 et 2

PRESENTATION *(pages 416–417)*

Activité B

In the CD-ROM version of this activity, students can interact with an on-screen native speaker.

Activité F

Your students might enjoy the following movies: *Au revoir les enfants, Les Quatre Cents Coups, L'Enfant sauvage, Le Ballon rouge, Jean de Florette, La Gloire de mon père, Cyrano de Bergerac.*

416

Exercices

A **Un peu de culture.** Répondez d'après les dessins.

1. C'est un musée ou un théâtre?
2. Le musée est ouvert ou fermé?
3. C'est une exposition de peinture ou une exposition de sculpture?

4. Elle est peintre ou sculpteur?
5. C'est un tableau ou une statue?

B **Qui le sait?** Répondez.

1. Robert sait le nom du peintre?
2. Il connaît le peintre?
3. Il connaît l'œuvre du peintre?
4. Annick sait le nom du musée?
5. Elle connaît le musée?
6. Elle connaît le Musée d'Art Moderne?
7. Elle le visite souvent?
8. Elle sait que le musée est fermé le lundi?
9. Le Musée d'Art Moderne est ouvert tous les jours sauf le lundi?

Activités de communication orale

Mots 1 et 2

A **Le théâtre.** A French exchange student at your school (your partner) is interested in theater. He or she wants to know if you like to go to the theater and what kinds of plays you like; if there are theaters in your town; if your school has a drama club and, if so, what play(s) the club is putting on or has put on this year.

ADDITIONAL PRACTICE

1. Have students quickly write down as many words associated with the movies, theater, or museums as they can. (See COOPERATIVE LEARNING on this page for further use of the words they come up with.)
2. Student Tape Manual, Teacher's Edition, *Activité E*, page 177

COOPERATIVE LEARNING

Have students work in groups of four or five. Using the lists they prepared for the ADDITIONAL PRACTICE exercise on this page, one student gives a word and another puts it into a sentence.

B **Tu aimes le cinéma?** You're talking to a French teenager, Hélène Bouvier, in a café in Cannes during the movie festival. You're talking about the movies. Answer her questions.

1. Tu aimes aller au cinéma?
2. Tu y vas souvent?
3. Quels sont tes acteurs préférés?
4. Tu préfères quels genres de films?
5. Ça coûte combien, un billet de cinéma aux États-Unis?

C **Mon film préféré.** Find out what a classmate's favorite movies are and why. Then find out which movies he or she hates and why. Reverse roles.

Hélène Bouvier

D **Au musée.** The French tourist office has sent you the brochure below describing Paris museums. Find out which museum a classmate would like to visit and why. (If he or she doesn't want to visit any of them, find out why.) Then reverse roles.

> Élève 1: Tu veux visiter quel musée?
> Élève 2: Je veux visiter le Centre Pompidou parce que j'aime l'art moderne.

E **Renseignements.** You're in Paris and you'd like to visit one of the museums listed in the brochure on the right. Call the museum and find out from the museum employee (your partner) where it's located, when it opens and closes, what day it's closed, and how much a ticket costs. Your partner can use the information in the brochure to answer your questions.

F **Allons au cinéma!** With your classmates, see a French film that is playing at a local movie theater. Afterwards, go out for a snack together and discuss the movie in French. If there are no French movies playing in your community, ask your teacher to rent a French video that you can watch and discuss in class.

LES MUSÉES

MUSÉE DE L'ARMÉE
Esplanade des Invalides.
45.55.37.70. Tous les jours
de 10h à 18h. Entrée: 27F,
Tarif réduit: 14F. (Musée
accessible aux handicapés
physiques).

**CENTRE POMPIDOU
(BEAUBOURG)**
Rue Rambuteau.
42.77.12.33. Semaine de
12h à 22h. Samedi,
dimanche et fêtes de 10h à
22h. Fermé le mardi. Tarif
musée: 27F. Tarif réduit:
18F. *Le Musée National
d'Art Moderne de l'après-
impressionnisme à nos
jours, plus des expositions
temporaires, concerts,
ballets, cinémathèque.*

MUSÉE DU LOUVRE
Rue de Rivoli. Ouvert
tous les jours sauf le
mardi de 9h à 18h. Entrée:
30F. Tarif réduit: 15F. *Six
musées en un seul: antiqui-
tés gréco-romaines, égypti-
ennes, orientales, beaux-arts
français, italiens et d'autres
encore. En vedette, «la Vénus
de Milo», et «la Joconde».*

MUSÉE DU SPORT
24, rue du Commandant
Guilbaud. 40.45.99.12.
Entrée: 20F. Tarif réduit:
10F. Ouvert tous les jours
de 9h30 à 12h30 et de 14h
à 17h. Fermé mercredi,
samedi et fêtes. *Exposition
permanente: Trésors et
curiosités du sport.*

**MUSÉE DU CINÉMA-
HENRI LANGLOIS**
Palais de Chaillot.
45.53.74.39. Tous les jours
sauf mardi et fêtes. Visites
guidées à 10h, 11h, 14h,
15h, et 16h. Entrée: 22F.
Tarif réduit: 14F. *Documents
sur le cinéma de 1895 à
nos jours.*

ANSWERS

Activité A
É2 answers will vary. É1 questions may include: **Tu aimes aller au théâtre? Tu aimes quels genres de pièces? Il y a des théâtres dans cette ville? Est-ce qu'il y a un club d'art dramatique au lycée? Le club va monter (ou a monté) quelle(s) pièce(s) cette année?**

Activité B
Answers will vary.

Activité C
É2 answers will vary. É1 questions may include: **Quels sont tes films préférés? Pourquoi aimes-tu ces films? Quels films est-ce que tu détestes? Pourquoi?**

Activité D
Answers will vary.

Activité E
É2 answers will vary. É1 questions may include: **Où est (se trouve) le musée? Il ouvre à quelle heure? Et il ferme à quelle heure? Quel jour est-il fermé? C'est combien, l'entrée?**

Structure Teaching Resources

1. Workbook, *Structure: A–J,* pages 159–162
2. Student Tape Manual, Teacher's Edition, *Structure: A–G,* pages 178–181
3. Audio Cassette 9B/CD-10
4. Communication Activities Masters, *Structure: A–F,* pages 78–81
5. Computer Software, *Structure*
6. Chapter Quizzes, *Structure:* Quizzes 3–7, pages 84–88
7. CD-ROM, Disc 4, pages 418–425

Les verbes connaître et savoir *au présent*

PRESENTATION *(page 418)*

A. Write the plural forms of *savoir* and *connaître* on the board and have students repeat after you.

B. Lead students through steps 1–4 and the examples.

C. Make two lists on the board, one of information that follows *connaître* (names of people, cities and other places, artistic and literary works), and the other with facts that follow *savoir* (dates, times, telephone numbers, addresses, infinitives, clauses).

D. Give students the following words or expressions and have them say whether they would use *savoir* or *connaître*: *André, sa famille, son adresse, son numéro de téléphone, le nom de son école, ses professeurs, son quartier.*

Les verbes *connaître* et
savoir au présent

*Indicating People, Places, and Things
You Know and What You Know
How to Do*

1. Study the following forms of the irregular verbs *connaître* and *savoir,* both of which mean "to know."

CONNAÎTRE		SAVOIR	
je	connais	je	sais
tu	connais	tu	sais
il		il	
elle	connaît	elle	sait
on		on	
nous	connaissons	nous	savons
vous	connaissez	vous	savez
ils		ils	
elles	connaissent	elles	savent

2. You use *savoir* to indicate that you know a fact.

> **Je sais le numéro de téléphone et l'adresse du cinéma.**
> **Je sais que le cinéma n'est pas loin d'ici.**
> **Il sait à quelle heure la séance commence.**

3. You use *savoir* + infinitive to indicate that you know how to do something.

> **Elle sait conduire.**
> **Tu sais danser?**

4. *Connaître* means "to know" in the sense of "to be acquainted with." You use it with people, places, and things. Compare the meanings of *savoir* and *connaître* in the sentences below.

> **Je sais son nom. C'est Nathalie. Je connais bien Nathalie.**
> **Je sais où elle habite. Elle habite à Grenoble. Je connais Grenoble.**
> **Je sais le nom de l'auteur. C'est Victor Hugo. Je connais son œuvre.**

418 CHAPITRE 16

LEARNING FROM REALIA

Have students read the information on the sign. Emphasize pronunciation since this is a rather difficult street name for Americans to pronounce. Ask: *L'avenue Victor Hugo est dans quel arrondissement? Victor Hugo est né en quelle année? Il est mort en quelle année? Il a eu beaucoup de professions? Quelles professions?*

COOPERATIVE LEARNING

Have students write five things they know how to do using *savoir* + the infinitive. Team members interview each other, compile a team report, and present it to the class. Students might then practice questioning techniques. For example: *Qui sait nager? Qui ne sait pas nager? Qu'est-ce que Carole sait faire? Tous les membres de l'équipe C savent faire quoi?*

Exercices

A **Qu'est-ce que tu sais?** Donnez des réponses personnelles.

1. Tu sais l'adresse de ton ami(e)? Il/Elle habite quelle ville?
2. Tu connais la ville?
3. Tu sais le nom d'un bon restaurant? Quel est son nom?
4. Tu connais le restaurant?
5. Tu sais le nom de l'auteur de la tragédie de *Macbeth*? Quel est son nom?
6. Tu connais les pièces de Shakespeare?
7. Tu connais *Macbeth*?

B **On sait tout.** Complétez avec «savoir».

1. Moi, je ___ le nom du théâtre.
2. Et Paul ___ le numéro de téléphone du théâtre.
3. Paul et moi, nous ___ l'adresse du théâtre.
4. Mais nous ne ___ pas l'heure du lever de rideau.
5. Voilà Guy et Monique. Ils ___ à quelle heure la pièce commence.
6. Je ___ que le théâtre est fermé le dimanche.
7. Vous ___ quelle pièce on monte maintenant à la Comédie-Française?
8. Et toi, tu ___ qui joue le rôle principal dans cette pièce?

C **Qu'est-ce que tu sais faire?** Donnez des réponses personnelles.

1. Tu sais jouer au tennis?
2. Tu sais faire de l'aérobic?
3. Tu sais faire des costumes?
4. Tu sais organiser une très bonne fête?
5. Tu sais parler français?

D **Qui connaît quoi?** Complétez avec «connaître».

1. Je ___ bien la France.
2. Les élèves de Madame Benoît ___ la peinture française.
3. Mais ils ne ___ pas très bien la littérature française.
4. Tu ___ la culture française?
5. Et Paul, il ___ la culture française contemporaine?
6. Vous ___ l'art français?
7. Nous ___ les Impressionnistes comme Monet, Manet et Renoir.
8. Tu ___ l'œuvre du peintre Degas?
9. Ah, oui. Je ___ son œuvre. J'adore ses danseuses de ballet.

Auguste Rodin: «Les Bourgeois de Calais»

CHAPITRE 16 **419**

INDEPENDENT PRACTICE

Assign any of the following:
1. Exercises, page 419
2. Workbook, *Structure: A–D*, pages 159–160
3. Communication Activities Masters, *Structure: A–B*, page 78
4. CD-ROM, Disc 4, pages 418–419

PRESENTATION *(page 419)*

Extension of *Exercice D*: Higher Skills

After completing the exercise, ask students for some names: *des noms d'artistes, d'auteurs, de poètes, de musiciens, de compositeurs.*

ANSWERS

Exercice A

Answers will vary, but should include **Je (ne) sais (pas)** or **Je (ne) connais (pas)**.

Exercice B

1. sais	5. savent
2. sait	6. sais
3. savons	7. savez
4. savons	8. sais

Exercice C

All answers include **Je (ne) sais (pas)**.

Exercice D

1. connais	6. connaissez
2. connaissent	7. connaissons
3. connaissent	8. connais
4. connais	9. connais
5. connaît	

RETEACHING

Ask students to give a name that they know for each category: *un restaurant que je connais; une famille; un film; un(e) artiste; un médecin; une compagnie aérienne; une école; un professeur.*

HISTORY CONNECTION

Have students find Calais on the map on page 504 or use the Map Transparency. Rodin's sculpture, pictured on this page, commemorates an event of 1347. The English had just captured Calais, and the king of England promised to spare the town if six prominent citizens would give up their lives. The mayor and five other *bourgeois* volunteered. Fortunately, the king relented.

Note You may wish to refer students to the photo of *Le Penseur* on page 409.

Les pronoms le, la, les

PRESENTATION *(page 420)*

Write a few example sentences from step 1 on the board. Put a box around the noun object. Circle the object pronoun. Draw a line from the box to the circle. This helps students grasp the concept.

Exercices

PRESENTATION *(pages 420–421)*

Exercice C

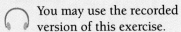 You may use the recorded version of this exercise.

ANSWERS

Exercice A

1. Les gâteaux? Je les aime beaucoup. (Je ne les aime pas. Je les déteste!)
2. … Je l'aime beaucoup. (Je ne l'aime pas. Je la déteste!)
3. … Je l'aime beaucoup. (Je ne l'aime pas. Je la déteste!)
4. … Je l'aime beaucoup. (Je ne l'aime pas. Je le déteste!)
5. … Je l'aime beaucoup. (Je ne l'aime pas. Je le déteste!)
6. … Je l'aime beaucoup. (Je ne l'aime pas. Je la déteste!)
7. … Je les aime beaucoup. (Je ne les aime pas. Je les déteste!)
8. … Je les aime beaucoup. (Je ne les aime pas. Je les déteste!)
9. … Je l'aime beaucoup. (Je ne l'aime pas. Je le déteste!)
10. … Je les aime beaucoup. (Je ne les aime pas. Je les déteste!)

Les pronoms *le, la, les*

Referring to People and Things Already Mentioned

1. You have already learned to use *le, la, l'*, and *les* as definite articles. These same words are also used as direct object pronouns. A direct object pronoun can replace either a person or a thing. Note that the direct object pronoun in French comes right before the verb.

Je sais le nom du film.	Je *le* sais.
Je vois le film.	Je *le* vois.
J'aime le film.	Je *l'aime.*
Je ne connais pas la vedette.	Je ne *la* connais pas.
Je lis les sous-titres.	Je *les* lis.
J'admire les costumes.	Je *les* admire.

2. Note the placement of the direct object pronoun in negative sentences. It cannot be separated from the verb by the negative word.

Tu connais l'auteur?	Non, je ne *le connais* pas.
Tu regardes la télé?	Je ne *la regarde* jamais.
Tu aimes les tragédies?	Je ne *les aime* pas du tout.

3. Remember that in sentences with a verb + infinitive, the pronoun comes right before the infinitive.

Nous pouvons lire les sous-titres.	Nous pouvons *les* lire.
Il ne peut pas comprendre le film.	Il ne peut pas *le* comprendre.

Exercices

A **Tu aimes les pâtisseries?** Donnez des réponses personnelles d'après le modèle.

> les pâtisseries
> *Les pâtisseries? Je les aime beaucoup.*
> *(Je ne les aime pas. Je les déteste!)*

1. les gâteaux
2. l'eau minérale
3. la viande
4. le bœuf
5. le poisson
6. la glace
7. les fruits
8. les crevettes
9. le poulet
10. les haricots verts

Paul Cézanne: «L'Assiette bleue—Abricots et cerises»

DID YOU KNOW?

You may wish to give students some information about Cézanne: *Paul Cézanne est un peintre français du 19ème siècle. Comme les autres Impressionnistes, Cézanne aime pratiquer la peinture en plein air. Il peint des portraits, des natures mortes et des paysages. Ce tableau,* L'Assiette bleue—Abricots et cerises, *est une nature morte. Qu'est-ce qu'une nature morte en anglais?* (still life)

ADDITIONAL PRACTICE

Student Tape Manual, Teacher's Edition, *Activités A–B,* page 178

B On voit le film en version originale. Complétez.

1. — On voit le film doublé ou en version originale?
   — On ___ voit en version originale.
2. — Tu sais le nom de la vedette?
   — Oui, je ___ sais.
3. — Tu connais la vedette?
   — Tu veux rigoler! Mais non, je ne ___ connais pas.
4. — Tu comprends le français?
   — Oui, je ___ comprends.
5. — Tu ___ comprends assez bien pour comprendre le film?
   — Non, mais il n'y a pas de problème. Il y a des sous-titres et je ___ lis quand je ne comprends pas le dialogue.

C Qu'est-ce qu'il est beau! Répondez d'après le modèle.

> Tu vois la statue?
> *Oui, je la vois. Qu'est-ce qu'elle est belle!*

1. Tu vois le théâtre?
2. Tu aimes la pièce?
3. Tu vois le tableau?
4. Tu entends le concert?
5. Tu vois le ballet?
6. Tu vois le film?
7. Tu lis le poème?
8. Tu vois le décor?
9. Tu regardes les costumes?
10. Tu vois la vedette?
11. Tu vois l'actrice?
12. Tu regardes les tableaux?

D Qu'est-ce qu'on va faire?
Répondez en utilisant «le», «la» ou «les».

1. Après les cours tu vas prendre le bus?
2. Tu vas écouter la radio?
3. Tu vas faire les devoirs de français ce soir?
4. Tu vas regarder la télé?
5. Ton père ou ta mère va préparer le dîner?
6. Tes parents vont lire le journal?

Edgar Degas: «Deux Danseuses en scène»

CHAPITRE 16 **421**

Les prépositions avec les noms géographiques

PRESENTATION *(page 422)*

After steps 1–4, have students read the examples aloud. Then ask: *Où est la Tunisie? Où est Shanghaï? Où est la Chine?*

Note In the CD-ROM version, this structure point is presented via an interactive electronic comic strip.

Note You may want to give students the preposition for your state and some nearby states. To say "in" with the name of a state in French, use *dans* before most states preceded by *le* or *l'*, for example: *dans le Connecticut, dans l'Oregon* (exceptions: *au Nouveau-Mexique, au Texas*). Use *en* without the article with states preceded by *la*, for example: *en Virginie.* With Hawaii, use *à: à Hawaii.* Here are the state names: l'Alabama, l'Alaska, l'Arizona, l'Arkansas, la Californie, la Caroline du Nord, la Caroline du Sud, le Colorado, le Connecticut, le Dakota du Nord, le Dakota du Sud, le Delaware, la Floride, la Géorgie, Hawaii, l'Idaho, l'Illinois, l'Indiana, l'Iowa, le Kansas, le Kentucky, la Louisiane, le Maine, le Maryland, le Massachusetts, le Michigan, le Minnesota, le Mississippi, le Missouri, le Montana, le Nebraska, le Nevada, le New Hampshire, le New Jersey, l'état de New York, le Nouveau-Mexique, l'Ohio, l'Oklahoma, l'Oregon, la Pennsylvanie, le Rhode Island, le Tennessee, le Texas, l'Utah, le Vermont, la Virginie, la Virginie Occidentale, l'état de Washington, le Wisconsin, le Wyoming.

Exercices

PRESENTATION *(pages 422–423)*

Exercice A

Use the recorded version of this exercise, if you wish.

ANSWERS

Exercice A

1. **En France.**
2. **En Italie.**
3. **En Espagne.**
4. **En France.**
5. **Au Japon.**
6. **Aux États-Unis.**
7. **Au Canada.**
8. **Au Sénégal.**

422

Les prépositions avec les noms géographiques

Identifying Cities, Countries, and Continents

You use the following prepositions to express "in" or "to" with geographical names.

1. *à* with the name of a city

> **Le Château de Versailles est bien sûr à Versailles.**
> **Le Musée du Louvre est à Paris.**
> **Je vais à New York pour aller au théâtre.**

2. *en* with the name of feminine countries and continents. Most countries and continents whose names end in silent *-e* are feminine. *Le Mexique* and *le Zaïre* are two of the common exceptions.

> **Henri est en France.**
> **La France est en Europe.**
> **Carole va en Tunisie.**
> **La Tunisie est en Afrique.**
> **Shanghaï est en Chine.**
> **La Chine est en Asie.**

3. *au* with the name of masculine countries. Most countries whose names do not end in silent *-e* are masculine.

> **Il va faire du ski au Canada.**
> **Cancún est au Mexique.**
> **Tokyo est au Japon.**
> **Je passe mes vacances au Maroc.**
> **Dakar est au Sénégal.**

4. *aux* with countries whose name is plural

> **Marc fait un voyage aux États-Unis.**
> **Amsterdam est aux Pays-Bas.**

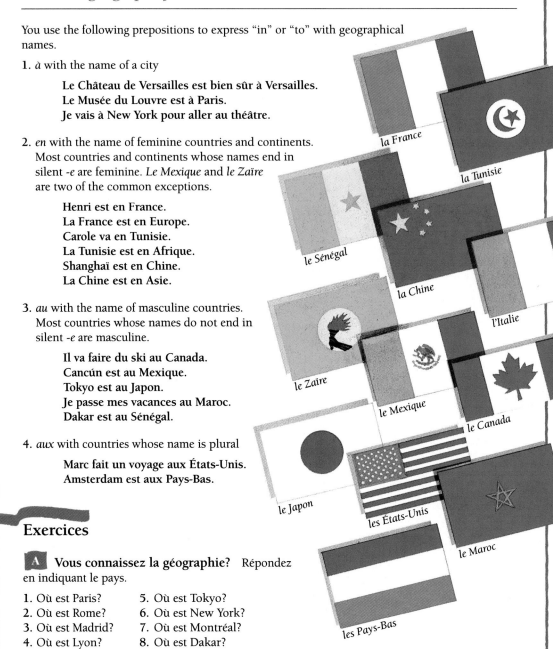

la France
la Tunisie
le Sénégal
la Chine
l'Italie
le Zaïre
le Mexique
le Canada
le Japon
les États-Unis
le Maroc
les Pays-Bas

Exercices

A **Vous connaissez la géographie?** Répondez en indiquant le pays.

1. Où est Paris?
2. Où est Rome?
3. Où est Madrid?
4. Où est Lyon?
5. Où est Tokyo?
6. Où est New York?
7. Où est Montréal?
8. Où est Dakar?

422   CHAPITRE 16

ADDITIONAL PRACTICE

After completing Exercises A–D, designate various areas in the classroom as continents, countries, and cities by taping up signs. Have students move from one place to another while classmates question them on where they or others are going and where they are. For example: É1: *Où vont Luc et Gérard?* É2: *Ils vont en Afrique.* É1: *Où sont-ils maintenant?* É2: *Ils sont à Marseille, en France.*

LEARNING FROM REALIA

1. Have students look at the flags and repeat the names of the countries.
2. Ask students the colors of the flags pictured: *De quelle couleur est le drapeau canadien?*, etc.

B Vous y allez quand? Posez une question d'après le modèle.

Nous allons à Antibes.
Ah oui? Vous allez en France quand?

1. Nous allons à Paris.
2. Nous allons à Cannes.
3. Nous allons à Amsterdam.
4. Nous allons à Barcelone.
5. Nous allons à Québec.
6. Nous allons à Shanghaï.
7. Nous allons à Miami.
8. Nous allons à Casablanca.

C C'est quel continent? Complétez.

1. Le Japon est ___ Asie et la Chine est ___ Asie aussi.
2. L'Italie et l'Espagne sont ___ Europe. Le Portugal est aussi ___ Europe.
3. Le Brésil, le Chili et l'Argentine sont ___ Amérique du Sud.
4. Les États-Unis et le Canada sont ___ Amérique du Nord.
5. Le Sénégal et la Côte-d'Ivoire sont ___ Afrique.

D Les grands musées du monde. Complétez.

1. Le Musée du Prado est ___ Madrid ___ Espagne.
2. Le Musée du Louvre est ___ Paris ___ France.
3. Le Metropolitan Museum est ___ New York ___ États-Unis.
4. Le Musée Britannique est ___ Londres ___ Angleterre, c'est-à-dire ___ Grande-Bretagne.
5. Le Centre Pompidou est ___ Paris ___ France.
6. Le Rijksmuseum est ___ Amsterdam ___ Hollande, c'est-à-dire ___ Pays-Bas.

La Fontaine Stravinski près du Centre Pompidou

PRESENTATION *(continued)*
Extension of *Exercice D*
   After completing Exercise D, have students give other famous sites and say where they are.

ANSWERS
Exercice B

1. Ah oui? Vous allez en France quand?
2. … en France quand?
3. … aux Pays-Bas quand?
4. … en Espagne quand?
5. … au Canada quand?
6. … en Chine quand?
7. … aux États-Unis quand?
8. … au Maroc quand?

Exercice C

1. en, en
2. en, en
3. en
4. en
5. en

Exercice D

1. à, en
2. à, en
3. à, aux
4. à, en, en
5. à, en
6. à, en, aux

INDEPENDENT PRACTICE

Assign any of the following:
1. Exercises, pages 422–423
2. Workbook, *Structure: G–H,* page 161
3. Communication Activities Masters, *Structure: D,* page 80
4. CD-ROM, Disc 4, pages 422–423

DID YOU KNOW?

   Tell students that the whimsical sculptures of the Fontaine Stravinski were created by Nikki de Saint-Phalle in homage to Igor Stravinsky, the Russian-born composer of *The Rite of Spring,* which set off a riot when it was first played in Paris in 1913. The fountain is an integral part of the Beaubourg area around the Pompidou Center, with its perpetual street-fair atmosphere.

Bell Ringer Review

Write the following on the board or use BRR Blackline Master 16-4: Under each column write three sentences. Je connais...   Je sais...

Les verbes irréguliers venir, revenir *et* devenir au présent

PRESENTATION *(page 424)*

A. Write the forms of *venir* on the board. Have students repeat after you, paying particular attention to the changes in pronunciation. Underline the double *n* in the third person plural form: *ils vie**nn**ent.*

B. Have students read the forms once again in their book.

C. Then have them read the example sentences.

Exercices

ANSWERS

Exercice A

1. Oui, Claude vient au cinéma avec nous.
2. Oui, il vient avec Martine.
3. Oui, Liliane vient aussi.
4. Oui, elle vient avec sa copine.
5. Oui, elles viennent à vélomoteur.
6. Oui, je viens au cinéma aussi.
7. Oui, je viens avec un copain.
8. Oui, nous venons à pied.

Exercice B

1. Non, il revient cet après-midi.
2. Non, ils reviennent...
3. Non, elle revient...
4. Non, il revient...

Les verbes irréguliers venir, revenir *et* devenir au présent   *Expressing "to come," "to come back," and "to become"*

1. The verb *venir,* "to come," is irregular in the present tense. Study the following forms.

VENIR			
je	viens	nous	venons
tu	viens	vous	venez
il elle on	} vient	ils elles	} viennent

Tu viens ce soir au théâtre?
Beaucoup de touristes viennent en France en été pour visiter ses musées célèbres.
Venez avec nous!

2. Two other verbs conjugated like *venir* are *revenir,* "to come back," and *devenir,* "to become." *Devenir* is seldom used in the present.

Il revient à trois heures.

Exercices

A   **Qui vient au cinéma?**   Répondez par «oui».

1. Claude vient au cinéma avec nous?
2. Il vient avec Martine?
3. Liliane vient aussi?
4. Elle vient avec sa copine?
5. Elles viennent à vélomoteur?
6. Tu viens au cinéma aussi?
7. Tu viens avec un copain?
8. Ton copain et toi, vous venez à pied?

B   **Ils reviennent cet après-midi.**   Répondez d'après le modèle.

Élève 1: Marie est là?
Élève 2: Non, elle revient cet après-midi.

1. Mon père est là?
2. Mes copains sont là?
3. Sophie est là?
4. Le professeur est là?

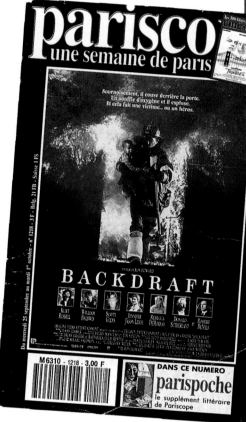

After completing Exercises A and B, ask students the following questions: *Tu viens à l'école à quelle heure le matin? Tu viens à l'école à pied? Comment viens-tu à l'école? Avec qui viens-tu à l'école? Comment est-ce que tes copains viennent à l'école? Vous venez ensemble?*

Have students look at the cover of *Pariscope* and tell you what type of magazine they think it is.

La préposition *de* avec les noms géographiques   *Telling Where People Come From*

You use the following prepositions to express "from" with geographical names.

1. *de* with the name of a city, a feminine country, or a continent

> **Elle est de Bordeaux.**
> **Ses grands-parents viennent d'Italie.**
> **Mes grands-parents viennent d'Amérique du Sud.**

2. *du* with the name of a masculine country

> **Mon amie arrive du Japon ce soir.**
> **Son père revient du Maroc.**

3. *des* with a country whose name is plural

> **Ils arrivent des États-Unis.**

Exercices

A **D'où viennent tous ces touristes?**   Répondez d'après le modèle.

> Italie
> *Ces touristes viennent d'Italie.*

1. Espagne
2. Rome
3. Nice
4. France
5. Tokyo
6. Japon
7. Maroc
8. Mexique
9. New York
10. États-Unis

B **D'où vient ta famille?**   Donnez des réponses personnelles.

1. D'où viens-tu?
2. Ta famille et toi, d'où venez-vous?
3. D'où vient ton père?
4. D'où vient ta mère?
5. D'où viennent tes grands-parents?

Une rue de Fort-de-France à la Martinique

DID YOU KNOW?

Fort-de-France is the capital of the French West Indies island of Martinique. Martinique is a *département* of France and its people are French citizens. See *Le Monde francophone*, pages 113 and 329, for further information on and photos of Martinique. In *À bord*, Chapter 7, students will learn about Martinique in greater detail.

INDEPENDENT PRACTICE

Assign any of the following:
1. Exercises, pages 424–425
2. Workbook, *Structure: I–J*, pages 161–162
3. Communication Activities Masters, *Structure: E–F*, pages 80–81
4. Computer Software, *Structure*
5. CD-ROM, Disc 4, pages 424–425

La préposition *de* avec les noms géographiques

PRESENTATION (page 425)

A. Lead students through the explanation of the usage of *de*. Have them read the example sentences aloud in unison.
B. You may wish to ask questions about the example sentences. For example: *D'où arrive ton ami? D'où revient ton père?*

Exercices

PRESENTATION (page 425)

Exercice A

You may use the recorded version of this exercise.

Extension of *Exercice A*

After going over Exercise A, have students make up original sentences using places they know.

ANSWERS

Exercice A

1. **Ces touristes viennent d'Espagne.**
2. ... **de Rome.**
3. ... **de Nice.**
4. ... **de France.**
5. ... **de Tokyo.**
6. ... **du Japon.**
7. ... **du Maroc.**
8. ... **du Mexique.**
9. ... **de New York.**
10. ... **des États-Unis.**

Exercice B

Answers will vary.

INFORMAL ASSESSMENT

Check comprehension by calling out names of famous people and having students tell where they come from.

THE FRANCOPHONE WORLD

Have students locate the following countries on the map on page 506 or use the Map Transparency: *le Maroc, l'Algérie, la Tunisie*. For views of North African countries, see *Le Monde francophone*, pages 222, 327, 434, and 436.

CONVERSATION

PRESENTATION (page 426)

A. Tell students they will hear a conversation between David and Carole.

B. Have them close books and watch the Conversation Video or listen as you play Cassette 9B/CD-10. Have them listen a second time with books open.

C. Call on pairs to read the conversation to the class.

D. After doing Exercise A, call on a student to retell the story.

Note In the CD-ROM version, students can play the role of either one of the characters and record the conversation.

ANSWERS

Exercice A

1. Carole est fana de cinéma.
2. Elle aime tous les films.
3. On passe un très bon film espagnol.
4. Non, le film n'est pas doublé. (Non, le film est en version originale avec des sous-titres.)
5. Ils peuvent travailler leur espagnol.

Prononciation

PRESENTATION (page 426)

A. Using Pronunciation Transparency P-16, model the key words *une roue* and have students repeat chorally.

B. Now model the other words and sentences in similar fashion.

C. Give students the following *dictée*:
Vous voulez la taille au-dessous? Et tu veux la taille au-dessus? Vous avez vu la roue dans la rue? Tu as vu la statue?

D. For additional practice, use the *Prononciation* section on Cassette 9B/CD-10 and the Student Tape Manual, Teacher's Edition, *Activités J–L,* page 183.

426

Scènes de la vie *On va au cinéma*

DAVID: Carole, tu veux aller au cinéma?
CAROLE: Pourquoi pas? C'est une très bonne idée. On passe quel film?
DAVID: On a le choix. Il y a beaucoup de cinémas, tu sais! Tu préfères quels genres de films?
CAROLE: Moi, j'aime tous les films. Je suis fana de cinéma, une vraie cinéphile.
DAVID: Au Rex on passe un très bon film espagnol—en version originale avec des sous-titres, je crois.
CAROLE: Excellente idée! On peut travailler notre espagnol. La prochaine séance est à quelle heure?

A **Des cinéphiles.** Répondez d'après la conversation.

1. Qui est fana de cinéma?
2. Elle aime quels genres de films?
3. On passe quel film au Rex?
4. Le film est doublé?
5. Qu'est-ce que les deux amis peuvent faire s'ils voient ce film?

Prononciation *Les sons /ü/ et /u/*

It is important to make a clear distinction between /ü/ and /u/ since many words differ only in these two sounds. Repeat the following pairs of words.

vous/vu   dessous/dessus   roue/rue   loue/lu   tout/tu

Now repeat the following sentences.

Vous avez vu ces statues?
Tu vas souvent au musée?
Cette comédie musicale est doublée.

une roue

426   CHAPITRE 16

DID YOU KNOW?

French teenagers can go to the movies very often because movie theaters offer discounts to students. Also, since there are fewer TV channels in France, there is less of a selection of programs to watch on TV. In most French theaters there are ushers who show people to their seats. Movie-goers are expected to tip the ushers. Before the film begins, commercials are shown. Movie theaters don't have concession stands. The ushers come down the aisles with trays of candy and ice cream and one buys items from them after the commercials end. Popcorn is not available in French movie theaters!

Activités de communication orale

A D'où viennent-ils? Play this game in small groups. First make a list of as many foreign celebrities as your group can think of (world leaders, actors, athletes, etc.). Next, take turns asking students from another group where these people are from. Then reverse roles. The group with the most correct answers wins.

> Élève 1: D'où viennent les Beatles?
> Élève 2: Ils viennent d'Angleterre. (Je ne sais pas. Je ne sais pas qui c'est.)

B Où est… ? Ask your classmates in the geography class to give you a list of the top 20 tourist sites in the world. (Ask them to include cities, museums, and monuments.) Bring the list to French class and ask your partner where each site is located.

> Élève 1: Où est Montréal?
> Élève 2: C'est au Canada.

La Place Jacques Cartier à Montréal

C Je connais bien… Think of someone you know well in the class. Using the verbs *savoir* and *connaître*, tell a classmate about this person without saying his or her name. Your partner has to guess who it is you're talking about. Include as much of the following information as you can.

1. son adresse et son numéro de téléphone
2. ses cours
3. ce qu'il y a dans sa chambre
4. les membres de sa famille
5. ses activités

> Élève 1: Je sais qu'il aime le football américain et la musique rock.
>   Je connais son frère Bob. Il habite rue Kennedy…
> Élève 2: C'est Andy.

CRITICAL THINKING ACTIVITY

(Thinking skill: making inferences)
   Put the following on the board or on a transparency:
   Beaucoup de personnes aiment aller voir un film au cinéma. Pourquoi, quand il est possible de rester chez soi regarder une vidéo?

Activités de communication orale

ANSWERS

Activités A and B
   Answers will vary.

Activité C
   Answers will vary but may include the following constructions:

1. Je sais son adresse. Il/Elle habite… Je sais son numéro de téléphone. C'est…
2. Je sais qu'il/elle a le cours de… avec M. (Mme/Mlle…), etc.
3. Je connais bien sa chambre. Il/Elle a… (Je sais qu'il/elle a… dans sa chambre.)
4. Je connais sa sœur… Je sais qu'il y a cinq personnes dans sa famille. Je sais que tu connais sa famille.
5. Je sais qu'il/elle aime (faire)… (jouer au…)

CROSS-CULTURAL COMPARISON

   French cinema and theater customs differ from those in the United States. In a French theater or cinema, one tips the usher after being shown a seat. In U.S. theaters, the house lights are dimmed about three times to announce the beginning of the play or its resumption after the intermission. In France this is done with three "knocks." In France, whistling at the end of a play is a sign of disapproval. And never give an actor or actress carnations in France. It means they are fired!

THE FRANCOPHONE WORLD

   Students can turn to page 437 in the *Les Arts* section of *Le Monde francophone* for information on African film.

LECTURE ET CULTURE

Bell Ringer Review

Write the following on the board or use BRR Blackline Master 16-5: What are the English titles of these movies?

1. La Belle et la Bête
2. Hannah et ses Sœurs
3. L'Arme Fatale
4. S.O.S. Fantômes
5. Les Dents de la Mer

Answers: *Beauty and the Beast, Hannah and her Sisters, Lethal Weapon, Ghostbusters,* and *Jaws.*

READING STRATEGIES
(page 428)

Pre-reading

Ask students: Who goes to the movies? How often? Who attends the theater? Is seeing a play better than watching a movie? Why?

Reading

A. Have students open their books to page 428. Read the *Lecture* to them.
B. Call on a student to read three sentences aloud. Ask others questions about what was read before calling on another student to read.

Post-reading

A. Based on the *Lecture,* what differences and similarities do students see between young French people's reading, TV, movie-going, and theater habits and those of young Americans?
B. After going over the *Lecture* in class, assign it to be read for homework.

Note Students may listen to a recorded version of the *Lecture* on the CD-ROM.

Étude de mots

ANSWERS

Exercice A

Answers will vary.

428

LES LOISIRS CULTURELS EN FRANCE

Il est naturellement difficile de décrire[1] un adolescent américain typique. Et il est difficile aussi de décrire un adolescent français typique. Mais généralisons un peu! Disons que Chantal Brichant est une adolescente française typique. Que fait Chantal quand elle a du temps libre? Est-ce qu'elle lit? Oui, elle lit. Elle lit beaucoup? Pas vraiment. On peut dire que les jeunes Français lisent un peu plus que les jeunes Américains, mais ils ne lisent pas énormément. Quand Chantal lit, qu'est-ce qu'elle choisit? Elle choisit des romans[2] et des bandes dessinées[3].

Chantal va au théâtre? Oui, de temps en temps. Dans toutes les grandes villes de France, et surtout à Paris, il y a des théâtres. Chaque année un certain nombre de pièces sont bien accueillies[4] par le public. Mais Chantal, comme la plupart des adolescents «typiques», va plus souvent au cinéma. Les Français voient beaucoup de films français, bien sûr, mais ils voient aussi pas mal de[5] films étrangers. On passe les grands films étrangers en exclusivité[6] dans les grands cinémas. Ces films sont souvent doublés, mais on peut les voir aussi en version originale avec des sous-titres.

Mais quel est le loisir préféré de Chantal et des Français «typiques»? La télévision? Mais oui! La télévision est de loin[7] le loisir culturel préféré des Français. Et les jeunes gens aiment aussi sortir avec leurs copains. Est-ce qu'il y a beaucoup de différences entre les Américains et les Français? Qu'est-ce que tu en penses?

[1] décrire *to describe*
[2] romans *novels*
[3] bandes dessinées *comic strips*
[4] bien accueillies *well-received*
[5] pas mal de *quite a few*
[6] en exclusivité *first run*
[7] de loin *by far*

Étude de mots

A Le français, c'est facile.
Trouvez cinq mots apparentés dans la lecture.

ADDITIONAL PRACTICE

After completing Exercises B and C, ask students the following questions.
1. Est-ce que les adolescents français et américains aiment faire les mêmes choses quand ils ont du temps libre?
2. Comment est-ce qu'on passe les films étrangers en France?
3. Est-ce que les jeunes Français lisent plus que les jeunes Américains?

CRITICAL THINKING ACTIVITY

(Thinking skill: drawing conclusions)
Put the following on the board or on a transparency:
Où veulent habiter les gens qui aiment beaucoup les activités culturelles? Pourquoi?

Compréhension

B **Vous avez compris?** Répondez.

1. Chantal lit quand elle a du temps libre?
2. Elle lit énormément?
3. Quel est le genre littéraire préféré des Français?
4. Il y a des théâtres en France? Où?
5. Chaque année il y a des pièces que le public aime?
6. Où est-ce qu'on passe les films étrangers en exclusivité?
7. Quel est le loisir préféré des Français?

C **Les adolescents.** Il est extrêmement difficile de décrire un adolescent français ou américain typique. Pourquoi?

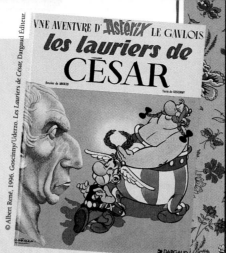

VNE AVENTVRE D'**Astérix** LE GAVLOIS
les lauriers de CÉSAR

© Albert René, 1996. Goscinny/Uderzo. *Les Lauriers de César*, Dargaud Éditeur.

DÉCOUVERTE CULTURELLE

LES MUSÉES

Les musées en France sont très fréquentés par les Français et par les touristes qui viennent du monde entier— d'Europe, d'Asie, d'Australie, d'Amérique et d'Afrique. À Paris il y a beaucoup de musées. Le Musée d'Orsay est une ancienne gare qui est aujourd'hui un musée extraordinaire où il y a une exposition permanente des peintres impressionnistes. Le Centre Pompidou (ou Beaubourg) a toujours des expositions d'art moderne. Il y a un nouveau Musée Picasso. Et la perle des musées français, c'est le Louvre.

Le dimanche, l'entrée dans les musées nationaux est à demi-tarif. Le dimanche, les gens viennent en foule admirer les peintures et les sculptures des artistes de tous les siècles[1] et de tous les pays du monde.

© Hergé/Casterman

LES BANDES DESSINÉES

La lecture préférée des jeunes de 8 à 18 ans est la bande dessinée. La «B.D.» vient en tête[2] des romans policiers, d'espionnage et de science-fiction. Mais les bandes dessinées ne sont pas seulement pour les enfants. Il est certain que la grande majorité des jeunes Français lisent «Tintin», «Astérix» et «Lucky Luke». Quand ils deviennent adultes, ils continuent d'avoir «leurs» bandes dessinées. Beaucoup de bandes dessinées, comme, par exemple, «Les Frustrés» de la dessinatrice humoristique Claire Bretécher, critiquent la vie moderne.

[1] siècles   *centuries*
[2] vient en tête   *rates above*

Tintin et son chien Milou

Compréhension

ANSWERS

Exercice B

1. Oui, Chantal lit quand elle a du temps libre.
2. Pas vraiment. Elle ne lit pas énormément.
3. Les Français préfèrent les romans et les bandes dessinées.
4. Oui, il y a des théâtres dans toutes les grandes villes de France.
5. Oui, chaque année il y a des pièces que le public aime.
6. On passe les films étrangers en exclusivité dans les grands cinémas.
7. La télévision est le loisir préféré des Français.

Exercice C

Answers will vary but may resemble the following:

Il est difficile—même impossible—parce que chaque personne est différente.

OPTIONAL MATERIAL

Découverte culturelle

PRESENTATION *(page 429)*

A. Have students read the information to get whatever they can from it.
B. You may then let them get together in small groups and tell one another in French something they learned from their reading.

Note Point out to students that there are many sophisticated *bandes dessinées* in France. They are popular with adults as well as with children and adolescents.

Note Students may listen to a recorded version of the *Découverte culturelle* on the CD-ROM.

INDEPENDENT PRACTICE

Assign any of the following:
1. *Étude de mots* and *Compréhension* exercises, pages 428–429
2. Workbook, *Un Peu Plus*, pages 163–164
3. CD-ROM, Disc 4, pages 428–429

OPTIONAL MATERIAL

PRESENTATION *(pages 430–431)*

The major objective of this section is to allow students to enjoy the photographs and gain an appreciation of France, its people, and its culture.

A. Ask students to share what they know about the art museums of Paris. Share with them your own knowledge of this topic, showing photographs you may have collected on trips to France or from the library or art department at your school.

B. Call on volunteers to read aloud the captions on page 430.

Note In the CD-ROM version, students can listen to the recorded captions and discover a hidden video behind one of the photos.

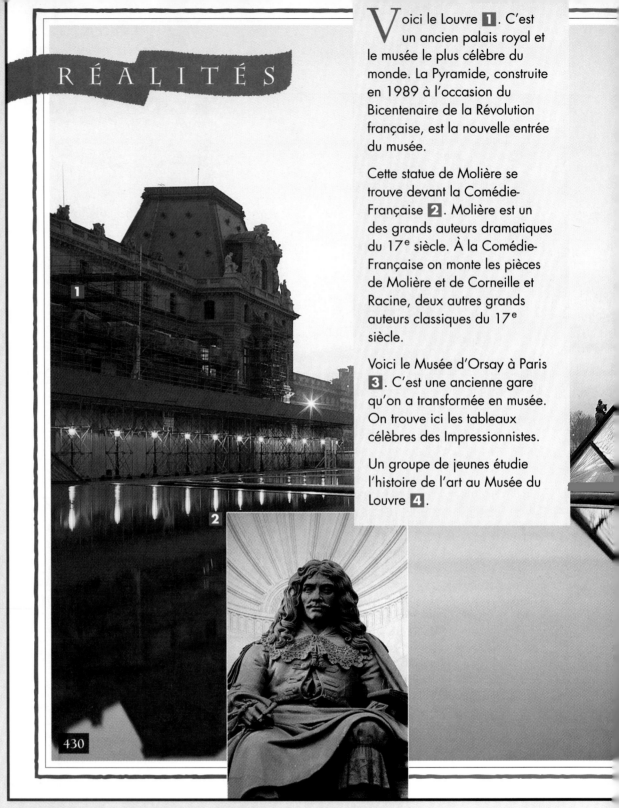

RÉALITÉS

Voici le Louvre **1**. C'est un ancien palais royal et le musée le plus célèbre du monde. La Pyramide, construite en 1989 à l'occasion du Bicentenaire de la Révolution française, est la nouvelle entrée du musée.

Cette statue de Molière se trouve devant la Comédie-Française **2**. Molière est un des grands auteurs dramatiques du 17e siècle. À la Comédie-Française on monte les pièces de Molière et de Corneille et Racine, deux autres grands auteurs classiques du 17e siècle.

Voici le Musée d'Orsay à Paris **3**. C'est une ancienne gare qu'on a transformée en musée. On trouve ici les tableaux célèbres des Impressionnistes.

Un groupe de jeunes étudie l'histoire de l'art au Musée du Louvre **4**.

430

DID YOU KNOW?

In 1680 Louis XIV founded the Comédie-Française by combining several theater companies including that of the actor, director, and dramatist Molière, who had died in 1673.

The present theater has been continuously occupied since 1790. The Comédie-Française has an acting school that prepares future actors for the permanent company. It also offers matinee performances of the classics for school children. The Comédie-Française is subsidized by the French government.

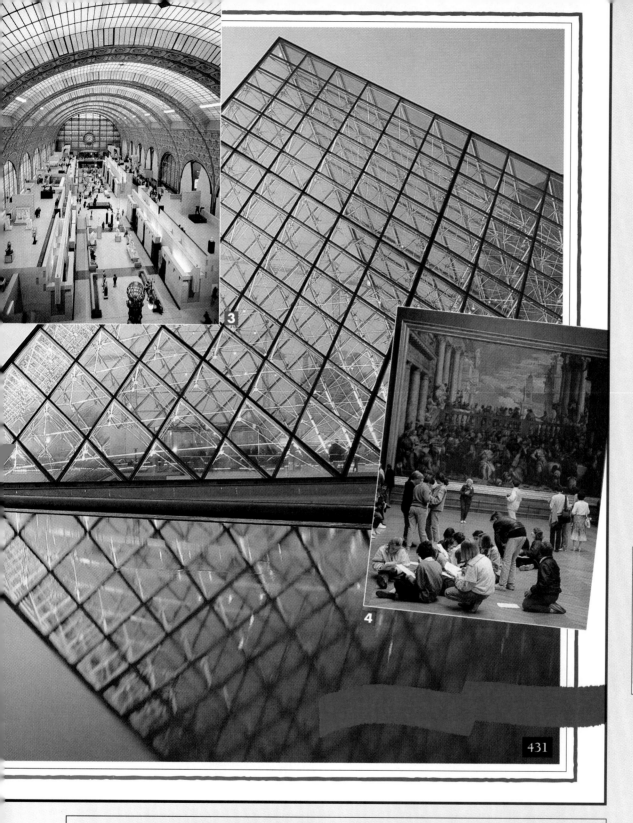

See *Le Monde francophone,* page 436, for photos of a museum and an art institute in other francophone countries.

RECYCLING

The *Activités de communication orale* and *écrite* allow students to use the vocabulary and grammar from this and earlier chapters in open-ended, real-life situations.

INFORMAL ASSESSMENT

Oral Activities A, B, and C provide guided cues and lend themselves to assessing speaking and listening skills. Oral Activity C can also be used to test the skill of reading quickly for information. Use the evaluation criteria given on page 34 of this Teacher's Wraparound Edition.

Activités de communication orale

ANSWERS

Activités A and B

Answers will vary.

Activité C

Answers will vary but may include the following constructions: Tu voudrais voir (*name of movie*) avec moi? On le passe samedi soir au cinéma (*name of movie theater*). Il est doublé/en version originale avec des sous-titres. Je veux voir ce film parce que ma vedette préférée joue dans le film / parce que mon frère m'a dit que c'est très amusant, etc.

Activités de communication orale

A **Dans ta ville.** Since you speak French so well, you've been asked to prepare a radio advertisement to attract French-speaking tourists to a cultural event (real or imaginary) that will take place in your town. Be sure to include:

1. a brief description of the event
2. the date, time, and place
3. how and where to buy tickets
4. the price of the tickets
5. a statement encouraging people to attend

B **La télé.** Divide into small groups and choose a leader. The leader interviews the others to find out how much time they spend watching TV every day and what kinds of shows they like. The leader takes notes and reports to the class.

1. Tu regardes la télé combien d'heures par jour?
2. Quelles sortes de programmes est-ce que tu aimes regarder?

les sports	les clips (vidéos rock)
les comédies	les documentaires
les drames	les dessins animés
le journal télévisé	les séries
les films	

À la classe: Dans mon groupe, tout le monde regarde la télé deux ou trois heures par jour. On préfère les comédies…

C **Tu veux aller au cinéma avec moi?** Look over the ads in this French movie guide. Decide which movie you'd like to see and invite a classmate to see it with you. Tell your partner when and where the movie is playing and whether it's dubbed or in the original language, with subtitles. Discuss whether or not you both want to see the movie or figure out an alternative.

432 CHAPITRE 16

les salles

COMŒDIA
13, avenue Berthelot - Lyon 7e
Tél. 76.58.58.98

ROBIN DES BOIS
(Grand Ecran - Son Dolby Stéréo)
Tlj.: 13h50 - 16h30 - 19h15 - 22h

LA MANIERE FORTE
(Son Dolby Stéréo)
Tlj.: 13h50 - 16h - 18h - 20h15 - 22h15

THELMA ET LOUISE
(Grand Ecran - Son Dolby Stéréo - V.O.)
Tlj.: 14h - 16h45 - 19h30 - 22h

SPARTACUS
(Grand Ecran -Son Dolby Stéréo - V.O.)
Tlj.: 14h30 - 20h15

LES TORTUES NINJA II
Tlj.: 14h - 16h - 18h - 20h - 21h45

UNE EPOQUE FORMIDABLE
+ Court métrage: "Le ridicule tue"
Tlj.: 14h - 16h - 18h - 20h - 22h

FOURMI LAFAYETTE
68, rue P. Corneille angle
cours Lafayette - Tel. 78.60.84.89

BRAZIL
Tlj. (sf. di.): 21h30 - di.: 19h45

SCENES DE MENAGE DANS UN CENTRE COMMERCIAL
Tlj.: 20h

TINTIN ET LE LAC AUX REQUINS
Me., sa., lu.:14h

ASTERIX ET LE COUP DU MENHIR
Me., sa., lu: 14h - di.: 15h 30

MAMAN, J'AI RATÉ L'AVION
Me., sa., di., lu.: 15h30

FANTASIA
Me., sa., di., lu.:15h30

ALICE
Sa.: 18h

JACQUOT DE NANTES
Me., sa., lu.: 15h30 - 21h30 - je., ve., ma.: 21h30 - di.: 17h45

FOR THE YOUNGER STUDENT

1. Have students draw and write dialogue balloons for their own *bandes dessinées*.
2. Have students illustrate a poster that advertises a movie in French (real or imaginary), giving the names of the stars, the film genre, and the time and location of the film.

COMMUNITIES

Have students make a poster in French for the school play. Include the following information: title of the play, date, time, location, actors, price of tickets, and a statement encouraging people to attend.

Activité de communication écrite

A **Des renseignements, s'il vous plaît.** You're going to spend a month in the French city of your choice. Write a letter to the tourist office (*le syndicat d'initiative*) asking for information about cultural events during your stay. Be sure to mention your name and age, what kind of cultural activities you like and the dates of your stay.

Réintroduction et recombinaison

A **Je suis malade.** Donnez des réponses personnelles.

1. Tu te sens bien aujourd'hui?
2. Quand tu es malade, tu te couches?
3. Aux États-Unis le médecin vient chez toi?
4. Le médecin te fait une ordonnance? Tu la donnes au pharmacien?
5. Quand le pharmacien te donne des comprimés, tu les prends avec un verre d'eau?
6. Quand tu es malade, tu ouvres un magazine et tu le lis?
7. Tu souffres beaucoup quand tu as la grippe?
8. Tu éternues et tu tousses quand tu es enrhumé(e)?

Vocabulaire

NOMS
le cinéma
le guichet
la séance
la salle de cinéma
l'écran (m.)
l'acteur (m.)
l'actrice (f.)
la vedette
le film
le film policier
le film d'aventures
le film d'horreur
le film de science-fiction
le film d'amour
le film étranger
le dessin animé
le documentaire
le drame

le film en version
    originale (V.O.)
le film doublé
les sous-titres (m.)

le théâtre
la pièce
la scène (*stage*)
le rideau
le décor
le costume
l'acte (m.)
l'entracte (m.)
la scène (*scene*)
le genre
la comédie
la comédie musicale
la tragédie
l'opéra (m.)

le musée
l'exposition (f.)
la peinture
le/la peintre
le tableau
la sculpture
le sculpteur
la statue
l'œuvre (f.)
le nom

ADJECTIFS
chaque
fermé(e)
ouvert(e)

VERBES
connaître
savoir

venir
revenir
devenir
visiter

**AUTRES MOTS
ET EXPRESSIONS**
monter une pièce
passer un film
c'est-à-dire
entre
personnellement
sauf

CHAPITRE 16   **433**

Activité de communication écrite

ANSWERS

Activité A

Answers will vary but may include the following constructions:

1. Je m'appelle ____. J'ai ____ ans.
2. J'aime le cinéma/le théâtre/la peinture /la sculpture/ aller au musée, etc.
3. Je vais être à ____ du 15 au 25 juillet.

OPTIONAL MATERIAL

Réintroduction et recombinaison

PRESENTATION (*page 433*)

Exercice A

This exercise recycles vocabulary related to medicine presented in Chapter 15. It also reincorporates direct object pronouns.

ANSWERS

Exercice A

Answers will vary.

Les Arts

PRESENTATION *(pages 434–437)*

This cultural material is presented for students to enjoy and to help them gain an appreciation of the francophone world. Since the material is *optional*, you may wish to have students read it on their own as they look at the colorful photographs that accompany it. They can read it at home or you may wish to give them a few minutes in class to read it.

If you prefer to present some of the information in greater depth, you may follow the suggestions given for other reading selections throughout the book. Students can read aloud, answer questions asked by the teacher, ask questions of one another in small groups and, finally, give a synopsis of the information in their own words in French.

MORE ABOUT THE PHOTOS

Photo 1 Tahar Ben Jelloun was born in Fez, Morocco, in 1944. He is a poet, novelist, essayist, and journalist. He writes in both French and Arabic. In his work, he often presents problems faced by North Africans living and working in France.

Photo 2 Senghor was born in 1906 in Western Senegal. Upon receiving his *baccalauréat* in Dakar, he went to Paris and was admitted to the prestigious Lycée Louis-le-Grand where he was a classmate of Georges Pompidou. He then went on to study at the Sorbonne and was the first African to get the difficult *aggrégation* in France.

During the Second World War he was taken prisoner. Upon his release after the war, he taught at

LES ARTS

Chaque société a sa culture. La culture d'une société s'exprime par sa langue, sa religion et ses arts. La culture est l'ensemble des structures sociales, intellectuelles et artistiques qui caractérisent une société.

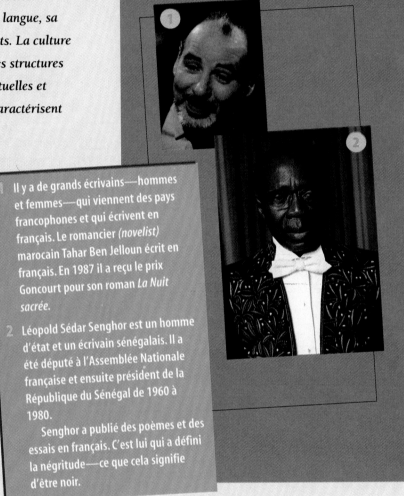

LA LITTÉRATURE

1 Il y a de grands écrivains—hommes et femmes—qui viennent des pays francophones et qui écrivent en français. Le romancier *(novelist)* marocain Tahar Ben Jelloun écrit en français. En 1987 il a reçu le prix Goncourt pour son roman *La Nuit sacrée.*

2 Léopold Sédar Senghor est un homme d'état et un écrivain sénégalais. Il a été député à l'Assemblée Nationale française et ensuite président de la République du Sénégal de 1960 à 1980.
  Senghor a publié des poèmes et des essais en français. C'est lui qui a défini la négritude—ce que cela signifie d'être noir.

the École Nationale de la France d'outre-mer. He was elected a representative of Senegal to the French National Assembly, where he served from 1946 to 1958. He contributed to the French constitution on matters concerning French overseas possessions.

With Aimé Césaire, the writer and politician from Martinique, he launched the important magazine, *Présence Africaine*. Senghor held several government posts and was the president of Senegal from 1960 to 1980.

LA PEINTURE

3 L'art naïf est l'art d'autodidactes—c'est-à-dire, de gens qui n'ont pas fait d'études, qui ont appris seuls. À Haïti il y a un grand nombre de peintres naïfs qui ont beaucoup de talent. Le sujet de leurs tableaux sans perspective est souvent une scène de la vie quotidienne, de la vie de tous les jours. Le petit bus aux couleurs vives s'appelle un *tap tap.* Le *tap tap* est le moyen de transport le plus important à Haïti. Chaque *tap tap* a son nom. Le nom est souvent d'origine religieuse. Et ce *tap tap,* «La Divinité», en est un exemple.

4 Gauguin est un initiateur de la peinture moderne. Pendant toute sa carrière il recherche un paradis exotique. Il va de Paris en Bretagne et en Provence. Finalement il va en Polynésie française où il s'installe à Tahiti.

À Tahiti il commence à peindre des sujets exotiques. Mais il veut donner «carte blanche» à son imagination. «Je ferme les yeux pour voir», dit-il. Les yeux fermés, Gauguin voit des rochers rouges, des arbres dorés et des montagnes violettes. Il les peint comme il les «voit». Il aime utiliser des couleurs vives.

LE MONDE FRANCOPHONE

Photo 4 Gauguin was a successful broker and businessman when he began painting as a hobby. At age 33 he left his well-paying job and turned to painting as a career. His paintings did not sell well, and he and his family were reduced to poverty.

Gauguin traveled from place to place looking for an earthly paradise and finally left his estranged wife and children and took off for the South Seas. He settled in Tahiti, where he lived among the Maori people.

In his art, Gauguin was more interested in creating a decorative pattern than a picture that looked real. His paintings have many flat areas of bright colors. His forms look round and solid.

LE MONDE FRANCOPHONE **435**

Photo 3 Haitian primitive art (*l'art naïf*) can be found in every little Haitian market. Haitians produce art in great abundance, from simple little paintings done by children to works of art exhibited and sold in prestigious galleries around the world. In the 1940's an American, DeWitt Peters, recognized the beauty and value of Haitian art. He opened the Centre d'Art, a gallery that still exists today in Port-au-Prince, the capital of Haiti.

Most Haitian paintings depict scenes from everyday life or the Bible—particularly Adam and Eve. Note the *tap tap* in the painting seen here. *Tap taps* are often decorated with paintings by rather talented artists. Some of these paintings are very unique.

In addition to paintings, Haitians produce wonderful wood carvings and sculptures. Because of the extreme poverty and lack of supplies in Haiti, the sculptures are often made from the metal of old oil drums.

Photo 6 In 1991, the Israeli architect Moshe Safti built a large modern addition to the Musée des Beaux-Arts in Montreal. The addition was constructed with prefabricated modular units that fit into each other like grape clusters. The units served as models for the construction of affordable housing.

Photo 9 *(page 437) Sango Malo* is "The Village Teacher" in English. The film contrasts two views of education, the traditional rigid "Eurocentric" curriculum and a practical-skills approach to building a self-reliant rural community.

This film (94 minutes, in French with English subtitles) and other francophone African and Caribbean videos are available for rent or purchase from the following source:

California Newsreel
149 9th Street
San Francisco, California 94103
(415) 621-6196

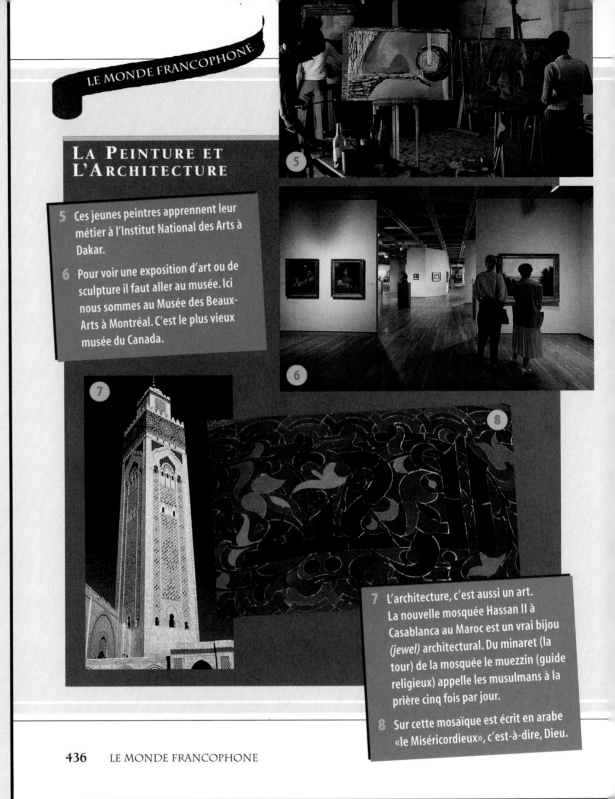

LE MONDE FRANCOPHONE

LA PEINTURE ET L'ARCHITECTURE

5 Ces jeunes peintres apprennent leur métier à l'Institut National des Arts à Dakar.

6 Pour voir une exposition d'art ou de sculpture il faut aller au musée. Ici nous sommes au Musée des Beaux-Arts à Montréal. C'est le plus vieux musée du Canada.

7 L'architecture, c'est aussi un art. La nouvelle mosquée Hassan II à Casablanca au Maroc est un vrai bijou *(jewel)* architectural. Du minaret (la tour) de la mosquée le muezzin (guide religieux) appelle les musulmans à la prière cinq fois par jour.

8 Sur cette mosaïque est écrit en arabe «le Miséricordieux», c'est-à-dire, Dieu.

Le Cinéma et La Musique

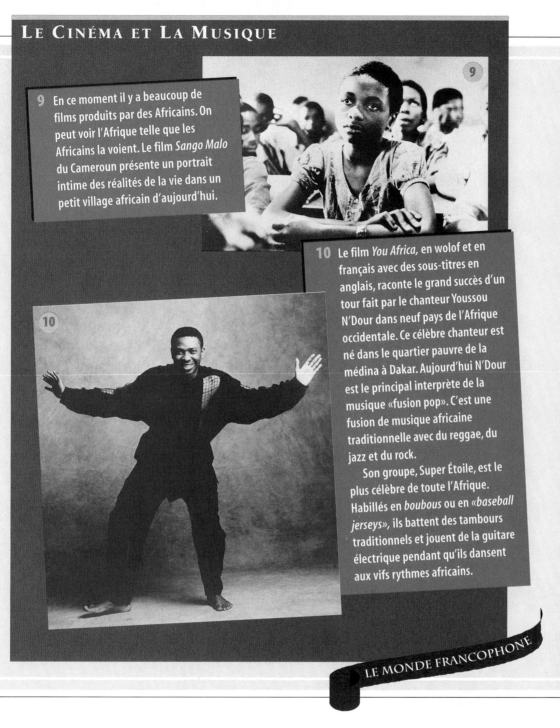

9 En ce moment il y a beaucoup de films produits par des Africains. On peut voir l'Afrique telle que les Africains la voient. Le film *Sango Malo* du Cameroun présente un portrait intime des réalités de la vie dans un petit village africain d'aujourd'hui.

10 Le film *You Africa*, en wolof et en français avec des sous-titres en anglais, raconte le grand succès d'un tour fait par le chanteur Youssou N'Dour dans neuf pays de l'Afrique occidentale. Ce célèbre chanteur est né dans le quartier pauvre de la médina à Dakar. Aujourd'hui N'Dour est le principal interprète de la musique «fusion pop». C'est une fusion de musique africaine traditionnelle avec du reggae, du jazz et du rock.

Son groupe, Super Étoile, est le plus célèbre de toute l'Afrique. Habillés en *boubous* ou en «*baseball jerseys*», ils battent des tambours traditionnels et jouent de la guitare électrique pendant qu'ils dansent aux vifs rythmes africains.

LE MONDE FRANCOPHONE

Photo 10 Music and dance are two extremely interesting aspects of African culture. There are two types of African music: the traditional village music of the bush and modern pop music. Traditional music serves a social purpose. Each social occasion has its own type of music. There are different kinds of music for women, young people, hunters, soldiers, etc. The instruments for traditional music are also different. Pop groups use electric guitars and the like, but instruments for traditional music are hand-made from gourds, animal skins and horns.

One of the most popular forms of African pop music is the modern conga from Zaire and the Congo. African popular music has a Latin sound because it has been heavily influenced by Latin-American music. It is a blend of traditional Latin-American and African-American music with elements of American jazz, rock, and, most recently, reggae.

To purchase recordings of African music, contact:
- African Record Centre Ltd.
  2343 Seventh Ave.
  New York, NY
  (212) 281-2717
  or
- The Kilimanjaro Music Store
  Florida & California Streets, NE
  Washington, D.C.
  (202) 462-8200

OVERVIEW

This section reviews key grammatical structures and vocabulary from Chapters 13–16. The structure topics were first presented on the following pages: object pronouns *me, te, nous, vous,* page 394; direct object pronouns *le, la, les,* page 420; prepositions with geographical names, page 422; *passé composé,* pages 350, 372, and 396; imperative, page 398.

REVIEW RESOURCES

1. Workbook, Self-Test 4, pages 166–169
2. Videocassette/Videodisc, Unit 4
3. Video Activities Booklet, Unit 4: Chapters 13–16, pages 50–63
4. Computer Software, Chapters 13–16
5. Testing Program, Unit Test: Chapters 13–16, pages 102–107
6. Performance Assessment
7. CD-ROM, Disc 4, *Révision:* Chapters 13–16, pages 438–441
8. CD-ROM, Disc 4, Self-Tests 13–16
9. CD-ROM, Disc 4, Game: *Le Labyrinthe*
10. Lesson Plans

Conversation

ANSWERS

Exercice A

1. Non, elles ne le connaissent pas.
2. Oui. Il s'appelle Marc.
3. … dans l'autobus.
4. Il joue au foot(ball).
5. … dans l'équipe du frère…
6. Oui, il est gardien de but.
7. Non, son équipe a perdu parce que l'autre équipe a marqué trois buts.

Conversation   *Le joueur de foot*

CHRISTINE: Tu connais le garçon là-bas?
SABINE: Je sais son nom—c'est Marc. Mais je ne le connais pas.
CHRISTINE: Je le vois tous les jours dans l'autobus.
SABINE: Et tu ne le connais pas!? Tu es trop timide! Je sais qu'il fait du foot tous les mercredis.
CHRISTINE: Comment tu sais ça?
SABINE: Il est dans l'équipe de mon frère. Il est gardien de but.
CHRISTINE: Il joue bien?
SABINE: Pas mal. Mais la semaine dernière, l'autre équipe a marqué trois buts et notre équipe a perdu zero à trois!

A **Trois buts!**   Répondez d'après la conversation.

1. Christine et Sabine connaissent le garçon?
2. Sabine sait son nom? Comment s'appelle-t-il?
3. Où est-ce que Christine le voit tous les jours?
4. Il joue à quoi?
5. Il est dans quelle équipe?
6. Il est gardien de but?
7. Est-ce que son équipe a gagné la semaine dernière? Pourquoi?

Structure

Les pronoms d'objet direct et indirect *me, te, nous* et *vous*

The pronouns *me, te, nous,* and *vous* function as both direct and indirect objects of the verb. Remember that *me* and *te* change to *m'* and *t'* before a vowel or silent *h*. Object pronouns always come right before the verb.

Le professeur *te* regarde?	Oui, il *me* regarde.
Le médecin *t'*examine?	Non, il ne *m'*examine pas.
Il va *vous* faire une ordonnance?	Oui, il va *nous* faire une ordonnance.

A **Qu'est-ce qu'on fait?**   Répondez en utilisant «me» ou «nous».

1. Quand le médecin t'examine, il t'ausculte?
2. Quand tu as une angine, le médecin te prescrit des antibiotiques?
3. Tes copains te téléphonent quand tu es malade?
4. Tes professeurs vous admirent, toi et tes copains?
5. Ils vous donnent beaucoup de devoirs?
6. Ils vont vous voir l'année prochaine?

COOPERATIVE LEARNING

Have students work in teams of three to make up a conversation about a soccer game and present it to the class.

ADDITIONAL PRACTICE

Have students ask each other as many questions as they can that use the object pronouns *me, te, nous, vous.*

Les pronoms d'objet direct *le, la, les*

Review the direct object pronouns *le, la, les*. Remember that *le* and *la* change to *l'* before a vowel or silent *h*. These pronouns can replace either people or things.

Je vois l'acteur.	Je *le* vois.
J'aime beaucoup cet acteur.	Je *l'*aime beaucoup.
Je vais regarder la télé.	Je vais *la* regarder.
Je n'aime pas les romans.	Je ne *les* aime pas.

B **Qu'est-ce qu'on fait?** Répondez en utilisant «le», «la», «l'» ou «les».

1. Vous connaissez les Impressionnistes?
2. Vous savez l'adresse du Musée d'Orsay?
3. Vous aimez les tableaux des Impressionnistes?
4. Qui aime la sculpture?
5. Tes copains et toi, vous aimez voir les films d'horreur?

C **Non.** Mettez à la forme négative d'après les indications.

1. Je les vois souvent. (ne… jamais)
2. Vous les aimez, ces gens? (ne… pas)
3. La télé, nous la regardons de temps en temps. (ne… jamais)
4. Le ballet? Nous voulons le voir. (ne… pas)

Les prépositions avec les noms géographiques

You use the following prepositions to express "in," "to," and "from" with geographical names.

1. *à* and *de* with cities

Il habite à Paris.   Je viens de Rome.

2. *en/au (aux)* and *de/du (des)* with countries, depending on whether the country is masculine or feminine, singular or plural

	FÉMININ	MASCULIN
to	Je vais en France.	Je vais au Brésil. Je vais aux États-Unis.
from	Je viens de France.	Je viens du Brésil. Je viens des États-Unis.

Remember that, except for *le Mexique* and a few others, countries that end in a silent *e* are feminine.

ADDITIONAL PRACTICE

Give students the following nouns and have them make up questions with *tu vois* and a direct object pronoun. For example: *Le livre? Tu le vois?*

les billets	la photo
la lettre	le stylo
le magazine	le disque
les cassettes	la calculatrice
la vidéo	les cartes postales

Exercice E

This exercise can be done as a paired activity.

ANSWERS

Exercice D

1. de, en
2. de, en
3. à, en
4. au
5. à, en
6. à, aux
7. du, de
8. de, aux

Exercice E

Answers will vary but should include one of the following structures: **en Angleterre, en France, en Belgique, aux Pays-Bas, en Allemagne, au Luxembourg, en Suisse, en Autriche, en Espagne, en Italie.**

Le passé composé des verbes réguliers et irréguliers

PRESENTATION (page 440)

A. Have the students repeat the forms of *avoir* after you.
B. Now have students repeat some past participles: *regardé, parlé, joué, gagné, fini, servi, choisi, vendu, attendu, perdu.*
C. Go over steps 1–3 of the explanation in the text.
D. You may wish to ask the following questions: *Hier soir, tu as regardé la télé? Tu as regardé la télé après le dîner? Ta famille a dîné à quelle heure? Tes parents ont regardé la télé aussi? Vous avez regardé la télé dans la salle de séjour?*

D **Quelle ville? Quel pays?** Complétez.

1. Il est ___ Rome. Il habite ___ Italie.
2. Nous venons ___ Londres, mais nous n'habitons pas ___ Angleterre.
3. J'ai un appartement ___ Paris, mais je n'habite pas ___ France.
4. Ils vont tous les ans ___ Mexique.
5. L'Alhambra est ___ Grenade, ___ Espagne.
6. J'ai passé une semaine ___ Amsterdam ___ Pays-Bas.
7. Il vient ___ Maroc. Il est ___ Casablanca.
8. Tu viens ___ New York. Tu habites ___ États-Unis.

E **Dans quel pays?** Regardez la carte à la page 504. Choisissez une ville et répondez d'après le modèle.

> **Bruxelles**
>
> **Élève 1: Dans quel pays est Bruxelles?**
> **Élève 2: Bruxelles est en Belgique.**

Le passé composé des verbes réguliers et irréguliers

1. The *passé composé* is composed of two parts: the present tense of the verb *avoir* and the past participle of the verb. Review the forms of the *passé composé* of regular verbs.

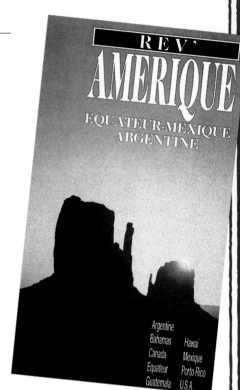

PARLER	FINIR	VENDRE
j'ai parlé	j'ai fini	j'ai vendu
tu as parlé	tu as fini	tu as vendu
il/elle/on a parlé	il/elle/on a fini	il/elle/on a vendu
nous avons parlé	nous avons fini	nous avons vendu
vous avez parlé	vous avez fini	vous avez vendu
ils/elles ont parlé	ils/elles ont fini	ils/elles ont vendu

2. For the past participles of irregular verbs, see p. 372.

3. Remember that *ne… pas, ne… plus, ne… jamais* go around the verb *avoir.*

   Tu n'as pas écouté le prof hier?

ADDITIONAL PRACTICE

Have students complete the following sentence: *Un jour, je voudrais aller (ville) (pays).*

COOPERATIVE LEARNING

Divide the class into four teams. Give students the following topics: *L'hiver; Les sports.* The teams compose as many questions or statements about the topics as they can. One person on each team serves as a secretary and writes down the sentences. After five minutes, collect the sentences from each secretary. The team having the most questions or statements wins.

F Le match de foot. Décrivez un match de foot imaginaire au *passé composé*. Utilisez les verbes et expressions suivants.

regarder	jouer	donner un coup de pied
marquer un but	passer le ballon	arrêter le ballon
égaliser le score	gagner	perdre

G Les achats. Vous avez acheté des vêtements. Décrivez ces vêtements à un copain ou une copine. Utilisez les verbes suivants.

acheter   coûter   prendre   trouver   aimer

L'impératif

Imperative forms are used to give commands or to make suggestions. They are the same as the *tu, nous,* and *vous* forms of the present tense. However, in the case of regular *-er* verbs and *aller,* you drop the final *s* of the *tu* form.

Travaille!	Attends un peu!	Fais ça!
Travaillons!	Attendons un peu!	Faisons ça!
Travaillez!	Attendez un peu!	Faites ça!

H Le jeu de «Jacques a dit» (*Simon says*). Vous donnez des ordres à vos camarades. Si vous dites d'abord «Jacques a dit», ils le font, mais si vous ne dites pas «Jacques a dit», ils ne le font pas.

(*Jacques a dit*): Levez le bras droit! Fermez les yeux!, etc.

Activité de communication orale

A Enquête sur les saisons. You want to know if your partner prefers summer or winter. On a separate sheet of paper, make a chart like the one below. Fill it out for both seasons. Compare your chart with your partner's and try to guess which season he or she prefers by asking questions about his or her choices.

Élève 1: Tu préfères le ski ou le ski nautique?
Élève 2: Je préfère le ski nautique.
Élève 1: Tu préfères l'été.

	L'HIVER	L'ÉTÉ
Vêtements		
Activités	le ski	le ski nautique
Équipement		
Nourriture		

PAIRED ACTIVITY

After reviewing the imperative, have students work in pairs. One suggests an activity, the other agrees or disagrees and says why. For example:

É1: Allons au café.
É2: Bonne idée. J'ai soif.
É1: Regardons la télé.
É2: Non, je ne veux pas. Allons jouer au tennis.

INDEPENDENT PRACTICE

Assign any of the following:
1. Exercises and activities, pages 438–441
2. Workbook, Self-Test 4, pages 166–169
3. CD-ROM, Disc 4, pages 438–441
4. CD-ROM, Disc 4, Self-Tests 13–16
5. CD-ROM, Disc 4, Game: *Le Labyrinthe*

ANSWERS
Exercices F and G
Answers will vary.

L'impératif
PRESENTATION (*page 441*)
Lead students through the explanation on page 441.

Activité de communication orale
PRESENTATION (*page 441*)
Extension of *Activité A*
After students have completed the activity, you may want to have them do a survey of the entire class and present the results of their findings.

ANSWERS
Activité A
Answers will vary.

Microbiologie: Louis Pasteur et L'Institut Pasteur

The three readings in this *Lettres et sciences* section are related topically to material in Chapters 15 and 16. For suggestions on presenting the readings, see page 228 of this Teacher's Wraparound Edition.

Avant la lecture

PRESENTATION *(page 442)*

A. Briefly discuss the *Avant la lecture* topic with the students.

B. Ask students what they know about Louis Pasteur or the Pasteur Institute.

C. Have students scan the reading and pick out the cognates.

Lecture

PRESENTATION *(pages 442–443)*

A. Have students read the selection silently.

B. If you have the students read the selection in class, tell them to look for the following information as they read. Write the questions on the board:
1. Quel nom Pasteur donne-t-il aux microbes?
2. D'après Pasteur, qu'est-ce qui cause les maladies?
3. Pourquoi les savants ne font-ils pas très attention au travail de Pasteur?
4. Qu'est-ce que Pasteur découvre?
5. Qu'est-ce que le BCG?
6. Quel travail le docteur Montagnier a-t-il fait?

C. If you have students read the selection outside of class, give them the following questions in English and have them look for the answers as they read. Have them write the answers (in French or English) and hand them in.

MICROBIOLOGIE:
LOUIS PASTEUR ET L'INSTITUT PASTEUR

Avant la lecture

You have no doubt heard of pasteurized milk. The term comes from the name of the French chemist Louis Pasteur, who invented the method of destroying harmful organisms without altering the milk. Find out how milk and other substances are pasteurized.

Lecture

«La vaccination de Joseph Meister»

Louis Pasteur (1822–1895)

Louis Pasteur est né en 1822 dans le Jura. Au collège, il n'est pas très bon élève. Il n'aime pas beaucoup ses cours, mais il aime le dessin. On l'appelle «l'artiste». Il veut devenir professeur et entre à l'École Normale, un institut qui forme les professeurs. Mais maintenant, il est passionné de sciences et passe son temps à faire de la recherche[1]. Il se spécialise en chimie.

En 1854, il commence à étudier ce que nous connaissons sous le nom de «microbes». Pasteur appelle ces microbes «germes» et il fonde une nouvelle science, la microbiologie. En 1873 Pasteur présente à l'Académie de Médecine un rapport qui révolutionne la médecine. Avant ce rapport de Pasteur, on croit que toutes les maladies terribles comme la typhoïde, le choléra et la fièvre jaune sont créées par le corps humain. C'est la théorie de la «génération spontanée». Mais Pasteur a fait des recherches sur les maladies du vin, de la bière et du ver à soie[2]. Il a compris que ces maladies n'arrivent pas toutes seules. Son

idée, c'est que toutes les maladies sont causées par des micro-organismes. Ce sont des organismes très, très petits. On les baptise «microbes».

Pour Pasteur, les microbes sont partout. Il dit aux chirurgiens[3] de se laver les mains avant d'opérer, de bien laver aussi leurs instruments, c'est-à-dire de pratiquer l'asepsie. Malheureusement peu de[4] gens l'écoutent. Pourquoi? Parce qu'il n'est pas médecin. Il est chimiste et biologiste. Mais Pasteur ne s'arrête pas là. Il continue ses recherches. Il veut lutter[5] contre les microbes. Ses recherches sur les maladies infectieuses des animaux le conduisent à découvrir la vaccination. En 1885, il réalise le vaccin contre la rage[6]. On vaccine alors pour la première fois un être humain, un petit garçon de neuf ans—Joseph Meister—qui a été mordu[7] par un chien

---

DID YOU KNOW?

When Louis Pasteur was nine years old, a neighbor girl was bitten on the leg by a wolf. He knew she was in great danger because other people in his town had died of rabies after similar incidents. At the time, no one knew what caused this disease or how to keep animals from contracting it. Later in life, Pasteur developed a vaccine against rabies.

Louis married Marie Laurent in 1849. She became interested in his work and encouraged him in his research. They had five children. In 1859, Pasteur's nine-year-old daughter Jeanne died of typhoid fever. He lost two other daughters to illness within the next few years. These losses spurred him to work harder to try to prevent other children from dying of disease. He eventually proved that heat could kill the microbes that were responsible for some illnesses.

enragé. C'est la victoire, après 40 ans de recherches.

[1] recherche *research*
[2] vin, de la bière et du ver à soie *wine, beer, and the silkworm*
[3] chirurgiens *surgeons*
[4] peu de *few*
[5] lutter *fight*
[6] rage *rabies*
[7] mordu *bitten*

L'Institut Pasteur (1888)

L'enthousiasme est grand, non seulement en France mais dans le monde entier. L'Académie des Sciences reçoit[1] de l'argent de nombreux pays pour la construction d'un centre de recherches en microbiologie. L'Institut Pasteur est inauguré le 4 novembre 1888. Et qui est son concierge[2]? C'est... Joseph Meister. Les collaborateurs et élèves de Pasteur continuent son travail. En 1891 les docteurs Calmette et Guérin mettent au point[3] le BCG (Bacille de Calmette et Guérin), le vaccin contre la tuberculose. En 1894, le docteur Roux met au point un vaccin contre la diphtérie.

De nos jours, l'Institut Pasteur de Paris est célèbre dans le monde entier. En plus du centre de recherches, il a un hôpital pour les maladies infectieuses et un centre d'enseignement[4]. En 1983, c'est à l'Institut Pasteur que le docteur Montagnier a isolé le virus du SIDA (Syndrome Immuno-Déficitaire Acquis). Aujourd'hui, à l'Institut, on continue à faire des recherches pour trouver une cure ou un vaccin contre cette terrible maladie.

[1] reçoit *receives*
[2] concierge *caretaker, concierge*
[3] mettent au point *come out with*
[4] enseignement *teaching*

Laboratoire de culture cellulaire à l'Institut Pasteur

Pasteur par Robert Thom

Après la lecture

A **Louis Pasteur.** Copiez ce formulaire *(data sheet)* sur Pasteur et remplissez-le.

NOM	
DATES	
ÉCOLE	
SPÉCIALISATION	
SUJET DU RAPPORT EN 1873	
DÉCOUVERTE EN 1885	

B **Enquête.** Que pensent vos camarades? Quelle est pour eux la plus grande découverte de Pasteur? Pourquoi?

C **Savez-vous que... ?** En France les enfants sont en général vaccinés contre les maladies suivantes: la diphtérie, le tétanos et la poliomyélite (un seul vaccin pour les trois); la tuberculose (le BCG); la coqueluche *(whooping cough)*; la rougeole *(measles)*; la rubéole *(German measles)* et les oreillons *(mumps)*. Les deux premiers vaccins sont obligatoires et les autres sont recommandés. Et dans votre pays? Quels sont les vaccins recommandés?

LETTRES ET SCIENCES **443**

---

1. What profession did Pasteur consider before turning to science?
2. What scientific field did Pasteur discover?
3. What vaccine did Pasteur develop?
4. What other vaccines have been developed at the Pasteur Institute?
5. What are the functions of the Pasteur Institute?
6. What virus was first isolated at the Pasteur Institute within the last twenty years?

Après la lecture

PRESENTATION *(page 443)*

You may wish to ask your students the following questions: *Pourquoi est-ce qu'il est important de pasteuriser le lait? D'après vous, est-ce que le SIDA est un grand problème dans votre ville ou dans votre région?*

Vocabulary Expansion

You may wish to have students go through the reading and find words related to the following: **rechercher, une fondation, un malade, une cause, une opération, la lutte, une infection, vacciner, la découverte, construire, collaborer, enseigner.**

ANSWERS

Exercice A

NOM: PASTEUR, Louis
DATES: 1822–1895
ÉCOLE: École Normale
SPÉCIALISATION: Chimie
SUJET DU RAPPORT EN 1873:
Les maladies sont causées par des micro-organismes.
DÉCOUVERTE EN 1885: Vaccin contre la rage

Exercice B

Answers will vary.

Exercice C

Tous les vaccins mentionnés dans le paragraphe sont recommandés.

LETTRES ET SCIENCES

Peinture: Les Impressionnistes
Avant la lecture

PRESENTATION *(page 444)*

A. Go over the *Avant la lecture* activities.

B. Bring in an art book from the library or use Art Transparencies F-7, F-8, and F-9 and show students some paintings by Monet, Renoir, and Degas. Whenever possible, have the students identify the subject of the painting in French.

C. Have students scan the reading for cognates.

Lecture

PRESENTATION *(pages 444–445)*

A. Have students read the selection silently, or break it into parts as suggested in the Cooperative Learning activity below.

B. Tell students to determine what they consider to be the main idea of each section.

PEINTURE: LES IMPRESSIONNISTES

Avant la lecture

1. What does the title of this text refer to?
2. Are any of these paintings familiar to you? Where did you see them?
3. How would you describe them? Realistic? Dreamlike? Colorful?

Lecture

Entre 1870 et 1900, les arts, et en particulier la peinture, commencent à changer. Chaque année, le «Salon» est une grande exposition de peinture. Si les peintres veulent exposer leurs tableaux, ils leur faut être acceptés par un jury.

Nous sommes en 1873. Le jury vient de refuser[1] tout un groupe de jeunes peintres. Ils sont furieux et décident d'avoir leur propre exposition. Elle a lieu[2] en 1874. Le public est scandalisé et crie à la vulgarité: les couleurs sont trop vives, les paysages[3] sont trop «bizarres». Un des tableaux de Monet est intitulé «Impression: soleil levant[4]». De là le terme (péjoratif à l'origine) «les Impressionnistes». Qui sont ces jeunes peintres? En voici trois.

Claude Monet (1840–1926)
Lycéen au Havre, il aime faire les caricatures de ses professeurs sur ses cahiers. Le peintre Eugène Boudin les voit et encourage Monet à faire de la peinture. C'est une révélation pour lui. Il admire les jeux de la lumière[5] sur l'eau, sur tout le paysage. Pour mieux étudier les variations de la forme en fonction de la lumière, il peint le même sujet à différentes heures de la journée. La cathédrale de Rouen est une de ces séries. Il passe la plus grande partie de sa vie dans sa maison de Giverny en Normandie où il reproduit dans son jardin les couleurs de ses tableaux.

Claude Monet: «La Cathédrale de Rouen, le Portail, Harmonie bleue»

COOPERATIVE LEARNING

You may wish to divide the class into four groups. Each group will read one section and report to the other groups about the section they read. The four divisions are: Introduction, Claude Monet, Auguste Renoir, Edgar Degas.

SCIENCES

Auguste Renoir
(1841–1919)

Il commence comme apprenti chez un décorateur de porcelaine à Paris. Il passe ses moments libres au Musée du Louvre où il admire surtout les tableaux du

Auguste Renoir: «Portrait de Margot»

peintre flamand Rubens. Renoir rencontre bientôt Claude Monet, qui l'encourage à peindre avec des couleurs moins sombres. Ils vont ensemble peindre à la campagne. Les Impressionnistes aiment peindre en plein air[6]. Comme tous les Impressionnistes, Renoir reçoit beaucoup de critiques. Il commence à douter, à se demander si les Impressionnistes ont raison[7]. Et pourtant Renoir est le premier Impressionniste reconnu par le public.

Edgar Degas (1834–1917)

Son père est un riche banquier qui est amateur d'art. Degas va régulièrement au Louvre où il copie les grands maîtres[8]. Degas aime le théâtre, l'opéra, la vie facile. Il devient ami avec les autres peintres impressionnistes, mais il n'a pas grand-chose en commun avec eux. Il n'aime pas peindre en plein air et il aime peindre des personnages et pas des paysages. On l'appelle souvent «le peintre des danseuses» parce qu'il a peint beaucoup de scènes où on voit des danseuses s'exercer avant le spectacle.

[1] vient de refuser *has just turned down*
[2] a lieu *takes place*
[3] paysages *landscapes*
[4] soleil levant *sunrise*
[5] jeux de la lumière *play of light*
[6] en plein air *outdoors*
[7] ont raison *are right*
[8] maîtres *masters*

Après la lecture

A **Les Impressionnistes.** Dites qui c'est: Monet, Renoir ou Degas.

1. Il aime beaucoup les tableaux de Rubens.
2. Il peint le même sujet à des heures différentes de la journée.
3. Son père est un riche amateur d'art.
4. Il n'aime pas peindre la nature.
5. Il commence par peindre sur de la porcelaine.
6. C'est un de ses tableaux qui leur donne leur nom.
7. Il peint souvent des danseuses.
8. Il aime beaucoup son jardin.

B **Une «Impressionniste» américaine.** Faites un rapport sur la vie et l'œuvre de l'artiste peintre américaine Mary Cassatt (1845–1926).

Edgar Degas: «Dans les coulisses (Danseuses en bleu)»

PRESENTATION *(page 445)*

After going over the *Après la lecture* activities in the textbook, have students tell which artist's paintings they prefer and why.

Vocabulary Expansion

Have students scan the reading and find equivalent expressions for the following:

1. des peintures
2. ne pas accepter
3. enragé
4. les changements
5. personne qui apprend un métier
6. une femme qui danse

ANSWERS

Exercice A

1. Renoir
2. Monet
3. Degas
4. Degas
5. Renoir
6. Monet
7. Degas
8. Monet

Exercice B

Answers will vary.

ADDITIONAL PRACTICE

If any student is particularly interested in art, you may wish to have him/her prepare a short report on one of the other famous French Impressionists or Post-Impressionists: Cézanne, Manet, Pissarro, Renoir, Sisley, Matisse, Toulouse-Lautrec, Gauguin, Seurat, Van Gogh.

LETTRES ET SCIENCES

Histoire: Trois Explorateurs
Avant la lecture

PRESENTATION *(page 446)*

A. Have students read the *Avant la lecture* information.

B. Have students discuss what they already know from American history about French influence in the United States.

GEOGRAPHY CONNECTION

Have students locate on a map the following places mentioned in the *Lecture*.
France: St-Malo, Rouen.
États-Unis: le Mississippi, le Missouri, les Grands Lacs, l'Illinois, le delta du Mississippi, la Louisiane, Savannah, Saint Louis.
Canada: Terre-Neuve, Québec.

C. Have students look at the map on page 447 and note the routes taken by the three explorers.

Lecture

PRESENTATION *(pages 446–447)*

A. Have students read the *Lecture* or the section assigned to them silently.

B. Ask the following questions about each section:
Section 1 (Cartier): *Où Jacques Cartier est-il né? Quel océan traverse-t-il? Où arrive-t-il? Combien de fois revient-il au Canada?*
Section 2 (La Salle): *Quel est le rêve de La Salle? D'où part son expédition? Quels lacs traversent-ils? Où arrivent-ils?*
Section 3 (Frémont): *Où est-il né? Pourquoi son père vient-il en Amérique? Dans quelle ville Frémont rassemble-t-il ses «voyageurs»? Qui est Kit Carson?*

HISTOIRE: TROIS EXPLORATEURS

Avant la lecture

La Nouvelle-France included territories that covered most of the present-day United States. Although the French presence is not as prevalent as it used to be, it is still very much alive.

Lecture

Jacques Cartier

Jacques Cartier est né à Saint-Malo en Bretagne en 1494. C'est une ville de marins[1] qui traversent souvent l'océan Atlantique pour aller pêcher[2]. Jacques Cartier est un marin audacieux, passionné des voyages: de Saint-Malo, il va au Portugal, au Brésil, à Terre-Neuve[3].

En 1534, le roi de France, François 1er, le charge d'une expédition pour découvrir des pays d'Orient où il y a de l'or et des pierres précieuses[4]. Jacques Cartier part avec deux bateaux et 61 marins. Vingt jours après, ils arrivent à Terre-Neuve. C'est un voyage très rapide pour l'époque[5]. Jacques Cartier revient au Canada encore deux fois. La troisième fois, en 1541, il construit un fort qui est devenu une grande ville: Québec.

Robert Cavelier de La Salle

Robert Cavelier de La Salle est le fils d'un riche marchand de Rouen, un grand port de Normandie. La Salle est passionné de l'Amérique et lit tous les rapports des explorateurs qu'il peut trouver. Il rêve[6] de

Jacques Cartier

Robert Cavelier de La Salle

descendre le Mississippi jusqu'au golfe du Mexique. En 1679, il réalise son rêve: il part de Fort Frontenac sur le Saint-Laurent avec six canots qui transportent 23 Français, 18 Indiens, 10 squaws et 3 enfants. Ils traversent les lacs Ontario, Érié, Huron et Michigan. Ils descendent l'Illinois et le Mississippi, et finalement ils arrivent dans le delta du Mississippi en 1682. La Salle prend possession de la région au nom du roi de France et appelle ces nouveaux territoires «La Louisiane» en l'honneur du roi Louis XIV.

John Charles Frémont

John Charles Frémont est né à Savannah en Géorgie en 1813. Son père est un aristocrate français qui est parti en Amérique pendant la Révolution de 1789

COOPERATIVE LEARNING

Divide the class into three groups. Each group reads one part of the *Lecture* and reports to the rest of the class. The three parts of the *Lecture* are: Jacques Cartier, La Salle, and Frémont.

pour échapper à la guillotine. Le jeune Frémont est très intelligent. Il est surtout très bon en mathématiques, mais il aime aussi l'aventure et le danger. Il devient d'abord professeur de maths, mais sur un bateau de guerre[7]. Il devient ensuite l'assistant d'un mathématicien français, Nicolas Nicollet, qui fait le levé topographique[8] des territoires du Nord, entre le Mississippi et le Missouri. Mais à l'époque, c'est la conquête de l'Ouest qui passionne les esprits. Frémont est le candidat idéal pour cette longue route inconnue de plus de 3 500 kilomètres. Frémont rassemble alors à Saint-Louis une équipe de 19 «voyageurs» canadiens qui connaissent bien les fleuves et les forêts. Il est aussi accompagné par un topographe allemand, Preuss, et un guide, Kit Carson. Ils partent en juin 1842. Lorsqu'il revient dans l'Est, il rapporte beaucoup de notes. Sa femme Jessie écrit deux livres d'après ses notes. Les livres sont aussi illustrés de cartes des régions traversées. Ces deux livres font de Frémont et de son guide Kit Carson des héros nationaux et la conquête de l'Ouest est commencée.

[1] marins *sailors*
[2] aller pêcher *to go fishing*
[3] Terre-Neuve *Newfoundland*
[4] de l'or et des pierres précieuses *gold and precious stones*
[5] époque *the times, the age*
[6] rêve *dreams*
[7] guerre *war*
[8] fait le levé topographique *is surveying*

John Charles Frémont

Après la lecture

A Trois explorateurs.   Vrai ou faux?

1. Jacques Cartier a descendu le Mississippi.
2. Cartier a fondé Québec.
3. Cavelier de La Salle est né en France.
4. Le nom «Louisiane» vient du nom du roi Louis XIV.
5. Le père de Frémont a été guillotiné.
6. Frémont a écrit deux livres.

B Les voyages des explorateurs.
Regardez la carte et racontez les voyages des trois explorateurs.

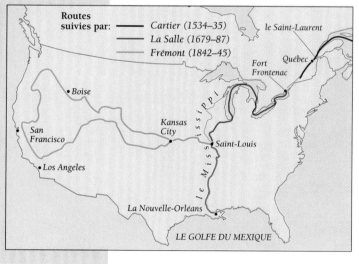

Routes suivies par:	
———	Cartier (1534–35)
———	La Salle (1679–87)
———	Frémont (1842–45)

le Saint-Laurent
Fort Frontenac
Québec
Boise
Kansas City
San Francisco
Saint-Louis
le Mississippi
Los Angeles
La Nouvelle-Orléans
LE GOLFE DU MEXIQUE

Après la lecture

PRESENTATION (page 447)

A. Go over the *Après la lecture* activities.
B. You may wish to give students the following mini-quiz about the reading.
Nommez l'explorateur—Cartier, La Salle ou Frémont— que chaque phrase décrit. 1. Il vit (lived) au 16ème siècle. (JC) 2. Il vit au 17ème siècle. (RLS) 3. Il vit au 19ème siècle. (JF) 4. Il explore le grand fleuve, le Mississippi. (RLS) 5. Il est professeur de mathématiques. (JF) 6. Il prend possession de la Louisiane au nom du roi Louis XIV. (RLS) 7. Il est né à Savannah en Géorgie, mais son père est un aristocrate français. (JF) 8. Il commence son exploration à Fort Frontenac et il traverse les Grands Lacs. (RS) 9. Le roi de France, François Ier, veut qu'il découvre des pays d'Orient. (JC) 10. C'est un des premiers à explorer l'Ouest (des États-Unis). (JF) 11. Il voyage au Portugal, au Brésil et en Terre-Neuve. (JC)

ANSWERS

Exercice A

1. faux
2. vrai
3. vrai
4. vrai
5. faux
6. faux

Exercice B
   Answers will vary.

DID YOU KNOW?

You may wish to give students further information on these French explorers.

Cartier: the first European to sail up the St. Lawrence. Visited the Indian village of Hochelaga in 1535 and climbed up a nearby mountain he called Mont Réal, later to become Montréal. His expeditions formed the basis for the French claim to Canada.

La Salle: immigrated to Canada in 1666 and became a trader, familiar with Indian languages and traditions. Was appointed commander of Fort Frontenac, a trading station at that time.

Frémont: helped with the annexation of California during the Mexican War (1846–48) in which he was an army major. Was elected one of the first two senators from California in 1850. In 1856 ran for President but was defeated by Buchanan. Was nominated again in 1860 but withdrew in favor of Lincoln.

CHAPTER OVERVIEW

In this chapter students will learn to communicate in various types of situations at a hotel. In order to do this they will learn the vocabulary needed to make a reservation, check into a hotel, and check out of a hotel. They will learn the *passé composé* of verbs conjugated with *être* and the indirect object pronouns *lui* and *leur.*

The cultural focus of Chapter 17 is on the many types of hotel accommodations in France.

CHAPTER OBJECTIVES

By the end of this chapter, students will know:

1. vocabulary associated with checking into a hotel, such as requesting different types of accommodations and going through the registration procedure
2. vocabulary associated with checking out of a hotel, as well as hotel features and facilities
3. the formation of the *passé composé* of verbs conjugated with *être*
4. the difference between the use of *avoir* and *être* with certain verbs
5. subject-past participle agreement in the *passé composé* of verbs conjugated with *être*
6. the indirect object pronouns *lui* and *leur*

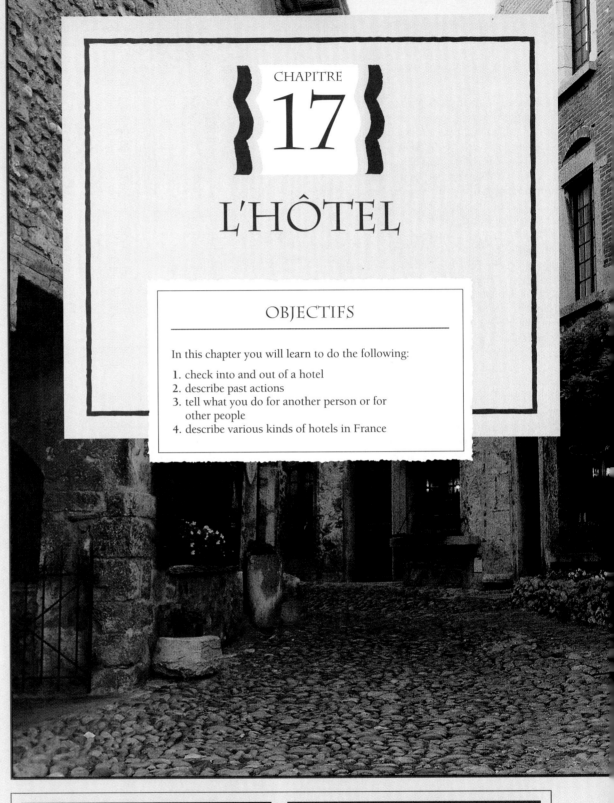

CHAPITRE

17

L'HÔTEL

OBJECTIFS

In this chapter you will learn to do the following:

1. check into and out of a hotel
2. describe past actions
3. tell what you do for another person or for other people
4. describe various kinds of hotels in France

CHAPTER PROJECTS

1. Have students use a Michelin Guide to plan hotel stays in different French cities.
2. Have groups create their own hotel and describe it to the rest of the class, who rate it as to quality, price value, and cuisine.

COMMUNITIES

Go with your class to a local travel agency that specializes in European travel. Have the students listen to a knowledgeable travel agent speak to them about hotels in France and share his/her own (or his/her clients') French hotel experiences. Then have your class compare American and French hotels, based on what they heard.

449

Pacing

This chapter requires eight to ten class sessions. Pacing will vary according to class length and the age and aptitude of the students.

Note The Lesson Plans offer guidelines for 45- and 55-minute classes and **Block Scheduling.**

Exercices vs. *Activités*

All exercises (which provide guided practice) are coded in blue. All communicative activities are coded in red.

INTERNET ACTIVITIES

(optional)

These activities, student worksheets, and related teacher information are in the *Bienvenue* Internet Activities Booklet and on the Glencoe Foreign Language Home Page at: **http://www.glencoe.com/secondary/fl**

DID YOU KNOW?

L'Ostellerie du Vieux-Pérouges, the inn pictured on these pages, is in Pérouges, a perfectly preserved medieval village of crooked, narrow streets, 22 miles from Lyon. The movie of *The Three Musketeers* was filmed in Pérouges. The inn, along with its superb restaurant, is a beautifully restored group of 13th-century timbered buildings.

VOCABULAIRE

MOTS 1

À L'HÔTEL — le hall — un escalier — la réception — une porte — la réceptionniste — le réceptionniste — une fiche d'enregistrement

une chambre avec salle de bains

une chambre à un lit
une chambre pour une personne

une chambre qui donne sur la cour

une chambre à deux lits
une chambre pour deux personnes

Lindsay est arrivée à l'hôtel.
Elle est entrée dans le hall.

Elle est allée à la réception.
Elle a montré son passeport à la réceptionniste.
Elle a rempli la fiche d'enregistrement.
La réceptionniste lui a donné la clé.

450 CHAPITRE 17

Vocabulary Teaching Resources

1. Vocabulary Transparencies 17.1 (A & B)
2. Audio Cassette 10A/CD-11
3. Student Tape Manual, Teacher's Edition, *Mots 1: A-C*, pages 187–189
4. Workbook, *Mots 1: A–C*, pages 170–171
5. Communication Activities Masters, *Mots 1: A*, page 82
6. Chapter Quizzes, *Mots 1: Quiz 1*, page 89
7. CD-ROM, Disc 4, *Mots 1:* pages 450–453

Bell Ringer Review

Write the following on the board or use BRR Blackline Master 17-1: Think of earlier chapters in your textbook and make a list of places in France you know something about.

PRESENTATION *(pages 450–451)*

A. Have students close their books. Use Vocabulary Transparencies 17.1 (A & B) to present *Mots 1*. Lead students through the new vocabulary by asking *Qu'est-ce que c'est?* or *Qui est-ce?* and having students repeat the answer. For example: *C'est un hôtel. C'est un escalier. C'est la réceptionniste.*

Teaching Tip You can also use either/or questions to introduce the new vocabulary by contrasting a new item with one known to students. For example: *C'est un ascenseur ou un escalier? (C'est un escalier.)*

B. Now have students open their books and repeat chorally as you model the entire *Mots 1* vocabulary or play Cassette 10A/CD-11.

TOTAL PHYSICAL RESPONSE

(following the Vocabulary presentation)

Getting Ready

Set up places in a hotel, such as *le hall, l'escalier, le couloir, la réception,* and *la chambre.* As props you might use a briefcase as a suitcase, a room key, and a fake passport.

TPR 1

(Student 1) et (Student 2) **venez ici, s'il vous plaît.**
(Student 1), **vous arrivez à l'hôtel.**
(Student 2), **vous êtes le/la réceptionniste.**
(Student 1), **allez à la réception.**
  Demandez une chambre.
(Student 2), **donnez-lui la fiche d'enregistrement.**
Demandez son passeport.
(Student 1), **mettez votre passeport sur le comptoir.**

(continued on next page)

Elle a monté ses bagages.
Elle est montée au troisième étage.
Elle a pris l'ascenseur, pas l'escalier.

Elle a ouvert la porte de sa chambre avec la clé.

Elle est descendue une heure plus tard.

Elle est sortie.

Elle est rentrée à neuf heures du soir.

C. While showing the Vocabulary Transparencies, call on students to read the vocabulary words and sentences in random order. Have other students take turns going to the screen and identifying the appropriate images.

D. Ask *vrai ou faux* questions such as the following: *Le hall est dans le jardin de l'hôtel? La réception est dans le hall? La réceptionniste remplit une fiche d'enregistrement? On peut monter plus vite dans l'ascenseur? Les clients vont à la réception quand ils quittent l'hôtel?*

Vocabulary Expansion

You may wish to give students the following useful expressions they may need at a hotel.

Le petit déjeuner est compris?
Il y a des messages pour moi?
Il y a des lettres pour moi?

TPR (*continued*)
(Student 2), **ouvrez son passeport et regardez-le.**
(Student 1), **signez la fiche d'enregistrement.**
Donnez la fiche au/à la réceptionniste.
(Student 1), **prenez votre passeport.**
Mettez le passeport dans votre poche.
(Student 2), **donnez la clé au client/à la cliente.**
(Student 1), **prenez vos bagages. Montez l'escalier.**

Merci, (Student 1) **et** (Student 2).

TPR 2
——, venez ici, s'il vous plaît.
Vous êtes touriste. Vous êtes devant la porte de votre chambre dans un hôtel.
Prenez votre clé.
Ouvrez la porte avec la clé.
Prenez vos bagages.
Entrez dans la chambre.

(*continued on next page*)

Exercices

PRESENTATION (*pages 452–453*)

Extension of *Exercice A*:
Speaking

After completing Exercise A, have individual students take turns describing one of the illustrations to the class.

Exercice B

Exercise B prepares students for the *passé composé* with *être*, the grammar topic of this chapter. Note, however, that in the exercise all verb forms are in the third person. Students will learn to manipulate the verbs in the *Structure* section.

Extension of *Exercice B*:
Listening

After doing Exercise B as a whole-class activity, focus on the listening skill by having students work in pairs. One partner reads the questions in random order while the other answers with his/her book closed.

Exercice C

 You may wish to use the recorded version of this exercise.

ANSWERS

Exercice A

1. C'est un hôtel.
2. C'est la réception.
3. C'est une fiche d'enregistrement.
4. C'est un escalier.
5. C'est une chambre.
6. C'est une chambre à deux lits.
7. C'est une chambre qui donne sur la cour.

Exercice B

1. Oui, Lindsay est arrivée à l'hôtel.
2. Oui, elle est entrée dans le hall.
3. Oui, elle a parlé à la réceptionniste.
4. Oui, elle lui a montré son passeport.
5. Oui, elle a rempli la fiche d'enregistrement.
6. Oui, elle lui a donné la clé.
7. Oui, elle a monté ses bagages.
8. Oui, elle est montée par l'ascenseur.

452

Exercices

A Qu'est-ce que c'est?
Répondez d'après les dessins.

1. C'est un hôtel ou une chambre?

Fiche d'enregistrement

Nom: _____   Prénoms: _____
Né le: _____
Département: _____
Profession: _____
Domicile habituel: _____

Nationalité: _____
Signature: _____

2. C'est la réception ou la réceptionniste?

3. C'est une clé ou une fiche d'enregistrement?

4. C'est un ascenseur ou un escalier?

5. C'est une porte ou une chambre?

6. C'est une chambre à un lit ou à deux lits?

7. C'est une chambre qui donne sur la cour ou sur la rue?

TOTAL PHYSICAL RESPONSE

(continued from page 451)
Mettez vos bagages sur le lit.
Fermez la porte.
Regardez bien la chambre.
Allez dans la salle de bains.
Ouvrez la fenêtre. Regardez dehors.
Indiquez que vous êtes content(e) de votre chambre.
Merci, ____. Vous avez bien fait.

COOPERATIVE LEARNING

Have teams write down a series of events that take place when checking into a hotel. They should be able to mime the events. Each team presents its mime one action at a time. The rest of the class asks them questions to determine what they are doing.

9. Oui, la chambre est au troisième étage.
10. Oui, c'est une chambre avec salle de bains.
11. Oui, elle est descendue une heure plus tard.
12. Oui, elle est sortie.
13. Oui, elle est rentrée à neuf heures du soir.

B À l'hôtel. *Répondez.*

1. Lindsay est arrivée à l'hôtel?
2. Elle est entrée dans le hall?
3. Elle a parlé à la réceptionniste?
4. Elle lui a montré son passeport?
5. Lindsay a rempli la fiche d'enregistrement?
6. La réceptionniste lui a donné la clé?
7. Lindsay a monté ses bagages?
8. Elle est montée par l'ascenseur?
9. La chambre est au troisième étage?
10. C'est une chambre avec salle de bains?
11. Lindsay est descendue une heure plus tard?
12. Elle est sortie?
13. Elle est rentrée à neuf heures du soir?

C Le touriste. *Choisissez la bonne réponse.*

1. À l'hôtel le touriste remplit ___.
   a. la fiche   **b.** la chambre   **c.** la clé
2. Pour monter dans sa chambre il prend ___.
   a. le lit   **b.** l'ascenseur   **c.** la porte
3. Il ouvre la porte de sa chambre avec ___.
   a. l'escalier   **b.** le lit   **c.** la clé
4. Il prend une douche dans ___.
   a. la salle de bains   **b.** le hall
   c. le petit déjeuner
5. Il dort dans ___.
   a. le lit   **b.** l'escalier
   c. la salle de bains
6. Le matin il se lève et prend ___.
   a. la fiche   **b.** le petit déjeuner
   c. la cour

Exercice C

1. a	4. a
2. b	5. a
3. c	6. b

INFORMAL ASSESSMENT
(Mots 1)

   Check comprehension by mixing true and false statements about the illustrations on pages 450–451. Students respond with *«C'est vrai.»* or *«C'est faux.»*

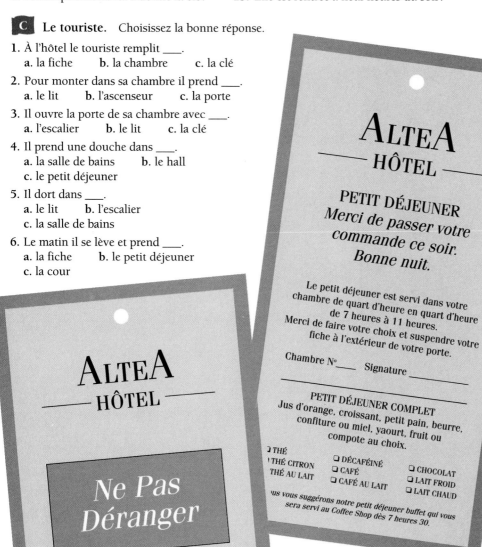

ALTEA
—— HÔTEL ——

Ne Pas Déranger

ALTEA
—— HÔTEL ——

PETIT DÉJEUNER
Merci de passer votre commande ce soir.
Bonne nuit.

Le petit déjeuner est servi dans votre chambre de quart d'heure en quart d'heure de 7 heures à 11 heures.
Merci de faire votre choix et suspendre votre fiche à l'extérieur de votre porte.

Chambre N° ___   Signature ___

PETIT DÉJEUNER COMPLET
Jus d'orange, croissant, petit pain, beurre, confiture ou miel, yaourt, fruit ou compote au choix.

☐ THÉ
☐ THÉ CITRON   ☐ DÉCAFÉINÉ
☐ THÉ AU LAIT   ☐ CAFÉ   ☐ CHOCOLAT
                 ☐ CAFÉ AU LAIT   ☐ LAIT FROID
                                  ☐ LAIT CHAUD

...us vous suggérons notre petit déjeuner buffet qui vous sera servi au Coffee Shop dès 7 heures 30.

CHAPITRE 17   **453**

LEARNING FROM REALIA

   You may wish to ask the following questions about the realia: *Quel est le nom de l'hôtel? À quelle heure est-ce que le petit déjeuner est servi? Il est servi où? Quand est-ce que le client passe sa commande? Où est-ce qu'il y a un petit déjeuner buffet? Le buffet commence à quelle heure? (Le buffet est à partir de quelle heure?) Comment dit-on en français* "Do not disturb"? *et* "Good night"?

INDEPENDENT PRACTICE

Assign any of the following:
1. Exercises, pages 452–453
2. Workbook, *Mots 1: A–C,* pages 170–171
3. Communication Activities Masters, *Mots 1: A,* page 82
4. CD-ROM, Disc 4, pages 450–453

Vocabulary Teaching Resources

1. Vocabulary Transparencies 17.2 (A & B)
2. Audio Cassette 10A/CD-11
3. Student Tape Manual, Teacher's Edition, *Mots 2: D–F,* pages 189–191
4. Workbook, *Mots 2: D–E,* page 171
5. Communication Activities Masters, *Mots 2: B,* page 82
6. Computer Software, *Vocabulaire*
7. Chapter Quizzes, *Mots 2: Quiz 2,* page 90
8. CD-ROM, Disc 4, *Mots 2:* pages 454–457

Bell Ringer Review

Write the following on the board or use BRR Blackline Master 17-2: Briefly sketch each of the following items: **une clé, un escalier, une fiche d'enregistrement, un lit**

PRESENTATION *(pages 454–455)*

A. Have students close their books Model the *Mots 2* vocabulary using Vocabulary Transparencies 17.2 (A & B) and have students repeat each word or expression twice after you or Cassette 10A/CD-11.
B. Point to the appropriate illustration and ask questions such as: *Qu'est-ce que la cliente lit? Qu'est-ce qu'il y a sur la facture? Avec quoi est-ce que la cliente paie?*

VOCABULAIRE

MOTS 2

une facture

les frais (m.)

une carte de crédit

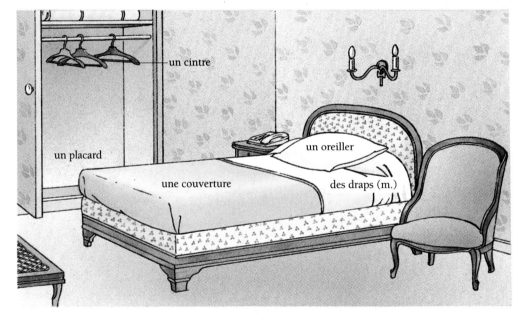

un cintre

un placard

un oreiller

une couverture

des draps (m.)

454   CHAPITRE 17

TOTAL PHYSICAL RESPONSE

TPR 1

Getting Ready

Demonstrate the word *expliquer.*
___, venez ici, s'il vous plaît.
Vous allez quitter l'hôtel.
Allez à la caisse.
Demandez votre facture.
Prenez votre facture.
Regardez et vérifiez les frais.
Il y a quelque chose sur la facture que vous ne comprenez pas. Indiquez le problème au/à la réceptionniste.
Le/La réceptionniste vous l' explique. Vous comprenez. Sortez votre carte de crédit.
Donnez votre carte de crédit au/à la réceptionniste.
Merci, ___. Retournez à votre place.

la salle de bains

un gant de toilette

une serviette

se sécher

du savon

un rouleau de papier hygiénique

Lindsay est restée une semaine à l'hôtel.
Elle a libéré la chambre.

Elle est descendue à la réception.
Elle a demandé la facture.
Elle a vérifié les frais.

Elle a payé avec un chèque de voyage.
Elle n'a pas payé en espèces.

C. Using props as cues (a pillow, a wash mitt, soap, a towel, money, etc.), ask students what one needs in order to do various things. For example: *On va prendre une douche. On va prendre un bain. On va se laver la figure. On va se sécher. On va dormir. On va mettre les vêtements dans le placard.*

D. When presenting the sentences on pages 455, ask questions in order to give students the opportunity to use the words. For example: *Lindsay est allée à l'hôtel. Qui est allé à l'hôtel? Où est-elle allée? Elle est restée une semaine à l'hôtel? Combien de temps est-elle restée à l'hôtel? Elle a libéré la chambre? Elle est descendue où? C'est une facture d'hôtel? Qu'est-ce que c'est? Qui a demandé la facture? Elle a demandé la facture à la réception? Où a-t-elle demandé la facture?*

Teaching Tip When asking the questions above, direct the easier questions to the less able students and the more difficult questions to the more able students.

Vocabulary Expansion

You may wish to give students the following useful expressions.

Je voudrais plus de cintres, s'il vous plaît.

Un autre oreiller, s'il vous plaît.

J'ai besoin d'une autre couverture.

Il n'y a pas de savon.

Il n'y a pas de papier hygiénique.

TPR 2

___, venez ici, s'il vous plaît.
Vous êtes dans un hôtel.
Vous êtes dans votre chambre.
Ouvrez la porte du placard.
Prenez un cintre.
Accrochez votre veste.
Mettez-la dans le placard.
Fermez la porte du placard.
Allez à la salle de bains.
Mettez le gant de toilette.

Prenez le savon. Lavez-vous la figure.
Regardez-vous dans la glace.
Prenez une serviette.
Séchez-vous la figure avec la serviette.
Merci, ___. Vous avez très bien fait.
Maintenant retournez à votre place, s'il vous plaît.

Exercices

PRESENTATION *(page 456)*

Exercice C

You may wish to use the recorded version of this exercise.

ANSWERS

Exercice A

1. Oui, elle a libéré la chambre.
2. Oui, elle est descendue à la réception.
3. Oui, elle a voulu payer.
4. Oui, elle a demandé la facture.
5. Elle a parlé à la réceptionniste.
6. Oui, elle a vérifié les frais.
7. Non, elle n'a pas payé en espèces.
8. Non, elle n'a pas payé avec une carte de crédit.
9. Elle a payé avec un chèque de voyage.

Exercice B

1. savon, gant de toilette
2. serviette
3. cintres
4. couverture
5. oreillers, oreiller
6. papier hygiénique
7. draps

Exercice C

1. dans la chambre
2. dans le placard
3. à la réception
4. dans la salle de bains
5. à la réception
6. dans la salle de bains
7. dans la chambre
8. dans la chambre
9. dans la salle de bains
10. dans la salle de bains

INFORMAL ASSESSMENT
(Mots 2)

Check for comprehension by having students correct the following statements:

On paie dans la chambre.
Les draps sont dans l'ascenseur.
Il y a un oreiller sur le comptoir.
On se sèche avec du savon.
On vérifie le cintre.
On a libéré le placard.

Exercices

A **Elle a libéré la chambre.** Répondez.

1. Lindsay a libéré la chambre?
2. Elle est descendue à la réception?
3. Elle a voulu payer?
4. Elle a demandé la facture?
5. Elle a parlé au réceptionniste ou à la réceptionniste?
6. Elle a vérifié les frais?
7. Elle a payé en espèces?
8. Elle a payé avec une carte de crédit?
9. Elle a payé la facture comment?

B **J'ai besoin de quoi?** Complétez.

1. Je vais me laver. J'ai besoin de ___ et d'un ___.
2. J'ai pris une douche. Maintenant je vais me sécher. J'ai besoin d'une ___.
3. Je vais mettre ma veste et mon pantalon dans le placard. J'ai besoin de ___.
4. Je vais me coucher mais il fait froid dans la chambre. J'ai besoin d'une autre ___.
5. Je préfère dormir avec deux ___. J'ai besoin d'un autre ___.
6. Ah, zut! J'ai besoin d'un rouleau de ___.
7. La chambre n'est pas prête *(ready)*. Il n'y a pas de ___ sur le lit.

C **À l'hôtel?** Où sont les objets suivants—dans la chambre, dans la salle de bains, dans le placard ou à la réception?

1. l'oreiller
2. le cintre
3. la facture
4. le papier hygiénique
5. la carte de crédit
6. le gant de toilette
7. les draps
8. la couverture
9. le savon
10. la serviette

LEARNING FROM PHOTOS

You may wish to ask the following questions about the photo: *Qu'est-ce que c'est? Il y a combien de cintres dans le placard? Il y a combien de lits? Qu'est-ce qu'il y a sur le lit? Qu'est-ce qu'il y a sur la table devant la cheminée? Qu'est-ce qu'il y a sur la petite table à côté du lit? Il y a un téléviseur dans la chambre? Où? Vous trouvez la chambre charmante?*

INDEPENDENT PRACTICE

Assign any of the following:

1. Exercises and activities, pages 456–457
2. Workbook, *Mots 2: D–E*, page 171
3. Communication Activities Masters, *Mots 2: B*, page 82
4. Computer Software, *Vocabulaire*
5. CD-ROM, Disc 4, pages 454–457

Activités de communication orale
Mots 1 et 2

A **On prépare un voyage.** You're in a travel agency in Paris to make reservations for a trip you and a friend are planning to take to the Loire Valley. The travel agent is asking you about your preferences.

1. Vous voulez un petit hôtel confortable ou un grand hôtel de luxe?
2. Vous voulez une chambre pour combien de personnes?
3. Vous voulez une chambre avec ou sans salle de bains?
4. Vous allez rester pendant combien de temps? Quelques jours? Une semaine?

L'agent de voyages

B **Comment réserver une chambre.** A family friend has asked you to phone a Montreal hotel to make reservations for her and her husband. You've already jotted down the information you need (see the card below), and now all you have to do is make sure the hotel employee (your partner) gets it right.

C **Quelle catastrophe!** You have just checked into a French hotel that is under new management. When you walk into the room, you find that some things are missing. Call the desk clerk (your partner), give your name and room number, and then tell what's missing and why you need it. He or she will try to resolve the problem. Then reverse roles.

> Élève 1: Bonjour, monsieur (madame). Je m'appelle M. Scott. Je suis dans la chambre 233. Il n'y a pas de draps sur mon lit et je suis très fatigué(e).
>
> Élève 2: Alors je vous donne des draps tout de suite.

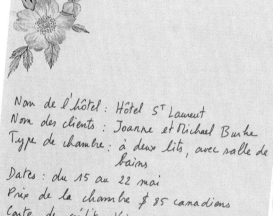

Nom de l'hôtel: Hôtel St Laurent
Nom des clients: Joanne et Michael Burke
Type de chambre: à deux lits, avec salle de bains
Dates: du 15 au 22 mai
Prix de la chambre $ 85 canadiens
Carte de crédit: Visa 550 - 8165 - 98 - 3

CHAPITRE 17   **457**

COOPERATIVE LEARNING

Students work in groups of four. Two members of the team are staying in an expensive hotel and are enjoying everything. The other two are staying in a very cheap hotel and are having a terrible time. Team members meet at a café and take turns describing their experiences. For example:
—Notre chambre est énorme!
—Notre chambre est comme un placard.
—Les lits sont très confortables!
—Nos lits sont horribles!
—La cuisine est excellente!
—Chez nous, la cuisine est infecte!

Bell Ringer Review

Write the following on the board or use BRR Blackline Master 17-3: Write who or what can be seen in each of the following places:
le hall d'un hôtel
une chambre d'hôtel

Activités de communication orale
Mots 1 et 2

PRESENTATION *(page 457)*

Activité A

In the CD-ROM version, students can interact with an on-screen native speaker.

ANSWERS

Activité A

Answers will vary but may include the following:

1. Je voudrais un petit hôtel confortable (un grand hôtel de luxe).
2. Je voudrais une chambre pour deux personnes.
3. Je voudrais une chambre avec (sans) salle de bains.
4. Nous allons rester (pendant) quelques jours (une semaine, etc.).

Activités B and C
Answers will vary.

GEOGRAPHY CONNECTION

The room on page 456 is in the inn shown on page 472. Have students find Angers, Tours, and Orléans on the map of France on page 504 or use the Map Transparency. Ask what river these three cities are on.

Cette auberge s'appelle L'Auberge de Combreux. Elle se trouve dans le Val de Loire. Dans cette région il y a beaucoup de petits hôtels charmants. Le Val de Loire, c'est la région des célèbres châteaux comme par exemple, Chenonceaux, Chambord, Azay-le-Rideau et Blois.

Structure Teaching Resources

1. Workbook, *Structure: A-H,* pages 172–175
2. Student Tape Manual, Teacher's Edition, *Structure: A–B,* pages 191–192
3. Audio Cassette 10A/CD-11
4. Communication Activities Masters, *Structure: A–D,* pages 83-85
5. Computer Software, *Structure*
6. Chapter Quizzes, *Structure:* Quizzes 3–6, pages 91–94
7. CD-ROM, Disc 4, pages 458–465

Bell Ringer Review

Write the following on the board or use BRR Blackline Master 17-4: Complete the following statements.
1. **Sur le lit il y a ___.**
2. **Dans la salle de bains on a besoin ___.**
3. **On peut payer la facture avec ___.**

Le passé composé avec être

PRESENTATION *(page 458)*

A. Have students open their books to page 458. Lead them through steps 1–4. Model the examples and have students repeat chorally.
B. Call on a volunteer to dramatize the verbs listed in step 2.
C. As you write the paradigm in step 4 on the board, you may want to add the endings to the past participle with a different color chalk or underline them for emphasis.

Note: In the CD-ROM version, this structure point is presented via an interactive electronic comic strip.

Le passé composé avec *être*   *Describing Past Actions*

1. You have already learned that you form the *passé composé* of most verbs with the verb *avoir* and the past participle.

> **Elle a parlé au réceptionniste.**
> **Elle a rempli la fiche.**
> **Elle a demandé la facture.**
> **Elle a vérifié les frais.**

2. With certain verbs, however, you use *être* as the helping verb rather than *avoir.* Many verbs that are conjugated with *être* express motion to or from a place.

arriver	Il est arrivé	descendre	Il est descendu.
partir	Il est parti.	aller	Il est allé en ville.
entrer	Il est entré.	venir	Il est venu.
sortir	Il est sorti.	revenir	Il est revenu.
monter	Il est monté.	rentrer	Il est rentré.

3. Remember that with the *passé composé* the *ne... pas* goes around the verb *être.*

> **Paul *n'*est *pas* arrivé à l'heure.**
> **Je *ne* suis *pas* sorti.**

4. The past participle of verbs conjugated with *être* must agree with the subject in number (singular or plural) and gender (masculine or feminine). Study the following forms.

MASCULIN	FÉMININ
Je suis sorti.	**Je suis sortie.**
Tu es sorti.	**Tu es sortie.**
Il est sorti.	**Elle est sortie.**
Nous sommes sortis.	**Nous sommes sorties.**
Vous êtes sorti(s).	**Vous êtes sortie(s).**
Ils sont sortis.	**Elles sont sorties.**

Exercices

A **Un voyage à Avignon.** Répondez par «oui».

1. Monique est allée à Avignon?
2. Elle est arrivée à la Gare de Lyon à 10h?
3. Elle est allée sur le quai?
4. Elle est montée en voiture?
5. Le train pour Avignon est parti à l'heure?
6. Le train est arrivé à Avignon à l'heure?
7. Monique est descendue du train à Avignon?
8. Elle est sortie de la gare?
9. Elle est allée à l'hôtel?
10. Elle est entrée dans le hall de l'hôtel?

B **À l'école.** Donnez des réponses personnelles.

1. Tu es allé(e) à l'école ce matin?
2. Tu es arrivé(e) à l'école à quelle heure?
3. Tu es venu(e) à l'école comment?
4. Tu es entré(e) dans l'école?
5. Tu es allé(e) à ton premier cours?
6. Tu es sorti(e) de l'école à quelle heure hier?
7. Tu es allé(e) manger quelque chose avec tes copains après les cours?
8. Tu es rentré(e) à la maison tout de suite après?

C **Au cinéma.** Mettez au passé composé.

1. Michel et sa sœur vont au cinéma.
2. Ils partent à l'heure.
3. Ils montent dans le bus.
4. Ils arrivent au cinéma.
5. Ils descendent du bus.
6. Ils vont au guichet.
7. Ils entrent dans le cinéma.
8. Ils sortent du cinéma après le film.
9. Ils vont au café.
10. Ils rentrent chez eux à minuit.

D **Qui est sorti?** Donnez des réponses personnelles.

1. Le mois dernier, tes copains et toi, vous êtes allés au cinéma?
2. Vous y êtes allés comment? En voiture? En bus?
3. Vous êtes toujours partis à l'heure?
4. Vous êtes arrivés quelquefois en retard?
5. Après le film vous êtes allés manger quelque chose?
6. Vous êtes souvent rentrés chez vous assez tard?

E **Un séjour.** Complétez au passé composé.

Ce matin Marc ____ (arriver) à Paris avec ses copains. Ils ____ (sortir) de la
gare et ____ (trouver) un taxi. Ils ____ (aller) à l'hôtel. Quand ils ____ (arriver)
à l'hôtel, ils ____ (entrer) dans le hall. Ils ____ (aller) à la réception et tout
le monde ____ (remplir) et ____ (signer) une fiche d'enregistrement. La
réceptionniste ____ (donner) les clés à Marc. Marc et ses copains ____ (monter)
au quatrième étage à pied. Ils ____ (prendre) l'escalier. Ils ____ (mettre) leurs
bagages dans leur chambre et ____ (sortir) tout de suite après.

Exercices

PRESENTATION (*page 459*)

Exercise C

You may wish to use the recorded version of this exercise.

ANSWERS

Exercice A

Answers are the questions transformed into declarative sentences and preceded by «*oui*».

Exercice B

Answers will vary.

Exercice C

1. **Michel et sa sœur sont allés au cinéma.**
2. **Ils sont partis à l'heure.**
3. **Ils sont montés dans le bus.**
4. **Ils sont arrivés au cinéma.**
5. **Ils sont descendus du bus.**
6. **Ils sont allés au guichet.**
7. **Ils sont entrés dans le cinéma.**
8. **Ils sont sortis du cinéma après le film.**
9. **Ils sont allés au café.**
10. **Ils sont rentrés chez eux à minuit.**

Exercice D

Answers will vary.

Exercice E

1. est arrivé
2. sont sortis
3. ont trouvé
4. sont allés
5. sont arrivés
6. sont entrés
7. sont allés
8. a rempli
9. a signé
10. a donné
11. sont montés
12. ont pris
13. ont mis
14. sont sortis

F Une excursion. Mettez au passé composé.

MATHIEU: Tu ___ (aller) en Normandie avec Laure, n'est-ce pas?

THÉRÈSE: Oui, nous y ___ (aller).

MATHIEU: Comment avez-vous trouvé le Mont-Saint-Michel?

THÉRÈSE: C'est vraiment impressionnant. Nous ___ (sortir) de notre petit hôtel à huit heures du matin et nous ___ (arriver) au Mont vers 9h.

MATHIEU: Vous ___ (monter) à la basilique?

THÉRÈSE: Oui, et nous ___ (sortir) sur la terrasse. De là, la vue est superbe.

MATHIEU: Mon frère et moi ___ (aller) au Mont-Saint-Michel l'année dernière et je suis d'accord avec toi—c'est formidable!

Le Mont-Saint-Michel

D'autres verbes avec *être* au passé composé

Describing Past Actions

Although the following verbs do not express motion to or from a place, they are also conjugated with *être*.

rester	**Il est resté huit jours.**	*He stayed a week.*
tomber	**Il est tombé.**	*He fell.*
devenir	**Il est devenu malade.**	*He became sick.*
naître	**Elle est née en France.**	*She was born in France.*
mourir	**Elle est morte en 1991.**	*She died in 1991.*

Exercices

A Être ou ne pas être. Donnez des réponses personnelles.

1. Tu es né(e) quel jour?
2. Tu es né(e) à l'hôpital?
3. Tu es né(e) dans quel hôpital?
4. Ta mère est restée combien de jours à l'hôpital?
5. Où tes parents sont-ils nés?
6. Tu as des grands-parents? Où sont-ils nés?

DID YOU KNOW?

Ask students to look at the photo on page 460 as you share the following information with them. **On a commencé la construction de l'abbaye de Mont-Saint-Michel au début du 8ème siècle. Les pèlerins (*pilgrims*) sont venus au Mont à toutes les époques. Même aujourd'hui le Mont-Saint-Michel est le site le plus visité de France après Paris et Versailles. L'abbaye est entourée de remparts.**

Du haut des remparts il y a une vue splendide sur la baie. Le Mont-Saint-Michel est un îlot, une petite île. Quand la marée (*tide*) est basse, il n'y a pas d'eau dans la baie et on peut aller à pied de la côte jusqu'au Mont. Mais c'est très dangereux à cause des sables mouvants (*quicksand*). Quand la mer monte, elle monte très vite et le Mont-Saint-Michel devient de nouveau un îlot.

B Vous êtes maladroit! Regardez les dessins et dites qui est tombé où.

l'enfant
L'enfant est tombé dans le jardin.

1. tu   2. Michel   3. tes copains

4. nous   5. vous

C Aux Jeux Olympiques. Complétez au passé composé.

1. Sophie ___ (aller) à Albertville en France pour participer aux Jeux Olympiques.
2. Elle ___ (rester) quinze jours dans les Alpes.
3. Elle est patineuse. Pendant la compétition elle n'___ pas ___ (tomber).
4. Mais toutes les autres patineuses ___ (tomber).
5. Alors Sophie ___ (gagner) la médaille d'or.
6. Elle ___ (devenir) championne olympique.
7. Après les Jeux elle ___ (rentrer) au Canada où elle ___ (devenir) très célèbre.

Le passé composé: être ou avoir

A. Read the example sentences to the class. After each sentence that has an object, ask: *Qu'est-ce qu'elle a descendu? Quel est l'objet direct? Qu'est-ce que les filles ont monté? Quel est l'objet direct? Qu'est-ce qu'ils ont sorti? Quel est l'objet direct?*

B. Lead students through the explanation on page 462. Use Grammar Transparency G-17(A) to explain more clearly the illustrations on page 462.

C. Now go on to the exercises on page 463.

Le passé composé: *être* ou *avoir*   *Describing Past Actions*

The verbs *descendre, monter, passer, rentrer,* and *sortir* are conjugated with *être* in the *passé composé* when they are not followed by an object. They are conjugated with *avoir,* however, when they are followed by a direct object. Study the following pairs of sentences. Note the differences in meaning.

WITHOUT OBJECT WITH OBJECT

Elle est descendue.

Elle a descendu *son sac à dos.*

Nous sommes montés au deuxième étage.

Nous avons monté *nos bagages.*

Ils sont sortis hier soir.

Ils ont sorti *leur passeport.*

LEARNING FROM ILLUSTRATIONS

Working in pairs, students write down one question for each illustration on page 462. Then each pair asks their questions of the rest of the class.

Exercices

A **Christine est arrivée.** Répondez par «oui».

1. Christine est arrivée à l'hôtel?
2. Elle est allée à la réception?
3. Elle a sorti son passeport et sa carte de crédit?
4. Elle est montée dans sa chambre?
5. Elle a pris l'escalier?
6. Elle a monté ses bagages?
7. Elle est descendue?
8. Elle est sortie?
9. Elle est allée au musée?
10. Elle est rentrée à l'hôtel à onze heures du soir?

B **En route!** Complétez au passé composé avec «avoir» ou «être».

1. Isabelle et Janine ___ (sortir) de la maison à neuf heures.
2. Elles ___ (sortir) tous leurs bagages sur le trottoir.
3. Elles ___ (attendre) le taxi.
4. Quand le taxi ___(venir), elles ___ (mettre) leurs bagages dans le coffre.
5. Puis les deux filles ___ (monter) dans le taxi.
6. À la gare elles ___ (descendre) du taxi et ___ (descendre) leurs bagages sur le quai.
7. Elles ___ (sortir) leurs billets et ___ (monter) dans le train.

Un hôtel superbe à Tahiti en Polynésie française

INDEPENDENT PRACTICE

Assign any of the following:
1. Exercises, page 463
2. Workbook, *Structure: E,* page 173
3. Communication Activities Masters, *Structure: C,* page 85
4. CD-ROM, Disc 4, pages 462–463

Exercices

PRESENTATION (*page 463*)

Extension of *Exercice A*

After completing Exercise A, have students do it again, changing the subject in both questions and answers first to *Christine et Chantal* and then to *Ma mère et moi.*

Extension of *Exercice B*: Writing

After completing Exercise B, have students retell the story as a writing assignment.

ANSWERS

Exercice A

1. Oui, elle est arrivée à l'hôtel.
2. Oui, elle est allée à la réception.
3. Oui, elle a sorti son passeport et sa carte de crédit.
4. Oui, elle est montée dans sa chambre.
5. Oui, elle a pris l'escalier.
6. Oui, elle a monté ses bagages.
7. Oui, elle est descendue.
8. Oui, elle est sortie.
9. Oui, elle est allée au musée.
10. Oui, elle est rentrée à onze heures du soir.

Exercice B

1. sont sorties
2. ont sorti
3. ont attendu
4. est venu, ont mis
5. sont montées
6. sont descendues, ont descendu
7. ont sorti, sont montées

THE FRANCOPHONE WORLD

The French artist Paul Gauguin lived and painted in Tahiti in the 1890's. For additional information on French Polynesia and Gauguin, refer students to *Le Monde francophone,* pages 113 and 435.

Les pronoms **lui, leur**

PRESENTATION *(pages 464–465)*

A. You may want to write the following sentences on the board. The arrows will help students understand the concept of direct versus indirect objects.

> à Gilles.
> ↑
> Il lance → le ballon
>
> à son copain.
> ↑
> Elle donne → l'argent

As students look at these sentences, say: *Remarquez. Il ne lance pas Gilles. Il lance le ballon. Il lance le ballon à qui? Il lance le ballon à Gilles. «Le ballon», c'est l'objet direct. «Gilles», c'est l'objet indirect.*

B. Lead students through steps 1–5 and the accompanying examples on pages 464–465.

Note Be sure students learn that *lui* and *leur* are both masculine and feminine.

C. You may wish to write the example sentences from step 2 on the board and underline the indirect object once and the direct object twice.

D. You may wish to give some additional sentences and have students indicate if the object is direct or indirect. For example: *J'écris une lettre. Une lettre? J'écris à mon ami. Mon ami? Je lis un livre. Un livre? Je lis à mon petit frère. Mon petit frère? J'achète un cadeau. Un cadeau? J'offre le cadeau à ma mère. Ma mère?*

Les pronoms *lui, leur*

Telling What You Do for Others

---

1. *Lui* and *leur* are indirect object pronouns. Observe the difference between a direct object and an indirect object in the following sentences.

Pierre lance *le ballon à Gilles.*

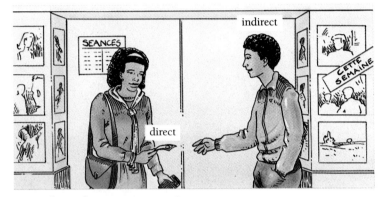

Marie donne *l'argent à son copain.*

In the above sentences, *le ballon* and *l'argent* are direct objects. *Gilles* and *son copain* are indirect objects, introduced by *à*.

2. You use the pronoun *lui* to replace *à* + a person (singular).

Je parle *à Marie.*	Je *lui* parle.
Je parle *à Luc.*	Je *lui* parle.
Il lance le ballon *à l'autre joueur.*	Il *lui* lance le ballon.
Il ne renvoie pas le ballon *à la fille.*	Il ne *lui* renvoie pas le ballon.

464 CHAPITRE 17

3. You use the pronoun *leur* to replace *à* + more than one person.

Je téléphone *à Roger et à Olivier*.	**Je *leur* téléphone.**
Je téléphone *à Catherine et à Jeanne*.	**Je *leur* téléphone.**
L'arbitre parle *aux filles*.	**L'arbitre *leur* parle.**

4. As with other object pronouns, *lui* and *leur* cannot be separated from the verb by a negative word.

> **Je ne *lui* parle pas.**
> **Il ne *leur* téléphone pas.**

5. Remember that in sentences with a verb + infinitive, the pronoun comes right before the infinitive.

> **Je vais *lui* téléphoner.**
> **Je ne veux pas *leur* offrir de cadeaux.**

Exercices

A **Guy offre un cadeau.** Récrivez les phrases d'après le modèle.

> **Guy offre un cadeau *à sa nouvelle amie française*.**
> **Guy *lui offre un cadeau.***

1. Guy offre un cadeau *à Danielle*.
2. Il donne le cadeau *à Danielle* au restaurant.
3. Elle est contente. Elle dit «merci» *à Guy*.
4. Elle téléphone *à sa copine Sandrine* pour décrire le cadeau.
5. Danielle dit *à sa copine*: «Guy est sympa, n'est-ce pas?»
6. Sandrine répond *à Danielle*: «Oh, oui, c'est un garçon vraiment chouette!»

B **Un match de foot.** Complétez avec «lui» ou «leur».

1. Il lance le ballon à Gilles?
   Oui, il ___ lance le ballon.
2. Les joueurs parlent à l'arbitre?
   Oui, ils ___ parlent.
3. Et l'arbitre parle aux joueurs?
   Oui, il ___ parle.
4. L'arbitre explique les règles aux joueurs?
   Oui, il ___ explique les règles.
5. L'employée au guichet parle à un spectateur?
   Oui, elle ___ parle.
6. Le spectateur pose une question à l'employée?
   Oui, il ___ pose une question.
7. L'employée vend des billets aux spectateurs?
   Oui, elle ___ vend des billets.

C **Personnellement.** Répondez en utilisant «lui» ou «leur».

1. Tu parles à tes professeurs?
2. Tu dis «bonjour» à ton professeur de français?
3. Tu vas téléphoner à tes copains ce week-end?
4. Tu aimes parler à tes copains au téléphone?
5. Tu parles souvent à tes copains?
6. Tu vas écrire à ta grand-mère?
7. Tu écris souvent à ta grand-mère?

Exercices
PRESENTATION *(page 465)*
Exercice B
You may wish to use the recorded version of this exercise.

ANSWERS
Exercice A
1. Guy lui offre un cadeau.
2. Il lui donne le cadeau au restaurant.
3. Elle est contente. Elle lui dit «merci».
4. Elle lui téléphone pour décrire le cadeau.
5. Danielle lui dit: «Guy est sympa, n'est-ce pas?»
6. Sandrine lui répond: «Oh, oui, c'est un garçon vraiment chouette!»

Exercice B
1. lui	5. lui
2. lui	6. lui
3. leur	7. leur
4. leur	

Exercice C
Answers will vary but will include the following pronouns:
1. leur	5. leur
2. lui	6. lui
3. leur	7. lui
4. leur	

RETEACHING
You may wish to use Grammar Transparency G-17(B) to reteach the concept of direct versus indirect objects. Have students volunteer additional examples following the pattern of the illustrations.

LEARNING FROM REALIA

Ask students what they think France Télécom is. (The state-owned telecommunications agency) Ask: *France Télécom a combien d'agences? Où sont-elles?*

INDEPENDENT PRACTICE

Assign any of the following:
1. Exercises, page 465
2. Workbook, *Structure F–H*, pages 174–175
3. Communication Activities Masters, *Structure: D*, page 85
4. Computer Software, *Structure*
5. CD-ROM, Disc 4, pages 464–465

CONVERSATION

 ◖ ◗ ▣

Bell Ringer Review

Write the following on the board or use BRR Blackline Master 17-7: Write these sentences in the passé composé.

1. Martine et Corinne vont au lycée.
2. Elles rencontrent leurs amies.
3. Elles parlent au prof de français.
4. Elles n'arrivent pas à l'heure.

PRESENTATION *(page 466)*

A. Tell students they will hear a conversation between Linda and a hotel receptionist.

B. Have students open their books to page 466 and watch the Conversation Video, or you may read them the conversation or play Cassette 10A/CD-11. (Use *Activité D* in the Student Tape Manual to check oral comprehension.)

Note In the CD-ROM version, students can play the role of either one of the characters and record the conversation.

ANSWERS

Exercice A

1. Elle veut une chambre pour deux personnes.
2. Elle parle à la réceptionniste.
3. Oui, elle a réservé une chambre.
4. Elle lui a montré sa confirmation.
5. Elle est au troisième étage.
6. Non, elle donne sur la cour.
7. C'est une chambre à deux lits.
8. Oui, la chambre à une salle de bains privée.
9. C'est 350 francs.
10. Il est compris.

CONVERSATION

Scènes de la vie   *À la réception de l'hôtel*

LINDA: Bonjour, Madame. J'ai réservé une chambre pour deux personnes.
LA RÉCEPTIONNISTE: C'est à quel nom, s'il vous plaît?
LINDA: Au nom de Collins.
LA RÉCEPTIONNISTE: Vous avez votre confirmation?
LINDA: Oui, je l'ai. La voilà. (*Elle lui montre sa confirmation.*)
LA RÉCEPTIONNISTE: Merci. J'ai une très jolie chambre au troisième qui donne sur la cour.
LINDA: C'est une chambre à deux lits?
LA RÉCEPTIONNISTE: Oui, avec salle de bains.
LINDA: C'est combien, la chambre?
LA RÉCEPTIONNISTE: Trois cent cinquante francs. Et le petit déjeuner est compris. Voilà votre clé.

A **Une jolie chambre d'hôtel.** Répondez d'après la conversation.

1. Linda veut une chambre pour combien de personnes?
2. Elle parle à qui?
3. Elle a réservé une chambre?
4. Qu'est-ce qu'elle a montré à la réceptionniste?
5. La chambre est à quel étage?
6. Elle donne sur la rue?
7. C'est une chambre à combien de lits?
8. La chambre a une salle de bains privée?
9. C'est combien la chambre?
10. Le petit déjeuner est compris ou pas?

CRITICAL THINKING ACTIVITY

(Thinking skill: evaluating consequences)
Put the following on the board or on a transparency.

1. Voyager avec beaucoup d'argent en espèces, ce n'est pas une bonne idée. Pourquoi?
2. Est-ce qu'il y a des avantages à payer avec une carte de crédit?

Prononciation   *Les sons /ó/ et /ò/*

It is important to make a clear distinction between the closed sound /ó/ as in *mot* and the open sound /ò/ as in *sort*. Repeat the following pairs of words.

nos / note   mot / mort   dôme / dort   beau / bonne

Now repeat the following sentences.

> Claude ne dort pas beaucoup.
> Paul sort beaucoup trop.
> Il n'y a pas d'eau chaude dans la chambre 14.

Hôtel de Bordeaux

Activités de communication orale

A   **Au voleur!**   Imagine that one of the rooms in your hotel in Paris was burglarized. The house detective (your partner) asks you and all the other guests what you did from the time you left your room this morning until the time you returned this afternoon. Give a full account of your activities.

> Élève 1: À quelle heure est-ce que vous êtes sorti(e) de votre chambre?
> Élève 2: À dix heures et demie.

B   **Qu'est-ce qu'on fait pour toi?**   Think of a friend or family member you like very much. What does this person do for you? What do you do for him or her? Use the following verbs.

acheter	écrire	préparer
apprendre	faire	répondre
dire	parler	servir
donner	poser des questions	téléphoner

> Mon amie Sylvie me téléphone presque tous les soirs… Moi, je lui écris des lettres pendant les vacances…

C   **Une enquête: Tu es né(e) quand?**   Divide into groups and choose a leader. The leader finds out when group members were born and then tells the class who is the oldest and the youngest in the group.

> Élève 1: Judy, tu es née quand?
> Élève 2: Je suis née le 17 août 1985…
> Élève 1 (*à la classe*): Judy est née le 17 août 1985. Meredith est née le 30 janvier 1986. Judy est la plus âgée et Meredith est la plus jeune de notre groupe.

CHAPITRE 17   **467**

INDEPENDENT PRACTICE

1. Exercise and activities, pages 466–467
2. Workbook, *Un Peu Plus,* pages 176–177
3. CD-ROM, Disc 4, pages 466–467

Prononciation

PRESENTATION (*page 467*)

A. Model the key words *Hôtel de Bordeaux* and have students repeat chorally.

B. Now model the other words and sentences similarly.

C. You may wish to give students the following *dictée:*
Je vais au beau château.
Beaucoup de mots sont beaux.
Nous avons reçu nos notes.
Elle dort sous le dôme.

D. For additional practice, you may use Cassette 10A/CD-11: *Prononciation,* Pronunciation Transparency P-17, and the Student Tape Manual, Teacher's Edition, *Activités E–G,* pages 193–194.

Activités de communication orale

ANSWERS

Activités A, B, and *C*
Answers will vary.

CROSS-CULTURAL COMPARISON

Bed and breakfast places have become very popular in the United States in recent years. They offer homey accommodations and family-style breakfasts, where the guests (and sometimes the hosts) eat together. Guests are usually lodged in a wing of the hosts' own home, which is often a restored example of vintage architecture. American bed and breakfast places are not cheap, and it is usually less expensive (but also less interesting) to stay in a motel.

The French have had the bed and breakfast idea for centuries in the form of the *pension.* In a *pension,* the guests also often stay in the host's own home and take their meals together. One advantage of the *pension* over the American bed and breakfast is that it offers the same character for much less money.

LECTURE ET CULTURE

READING STRATEGIES
(page 468)

Pre-Reading

Have students look at the Michelin Guide hotel rating system on page 471, the hotel ads on page 472, or if possible, bring in any literature you have from French or French-Canadian hotels. Discuss ratings, accommodations, and prices with students.

Reading

Call on individuals to read. After each student has read several sentences, ask comprehension questions before going on.

Post-Reading

Ask students if American towns have any organizations resembling the *syndicat d'initiative*. What are they?

Note Students may listen to a recorded version of the *Lecture* on the CD-ROM.

Étude de mots

ANSWERS

Exercice A

Answers will vary but may include any five of the following: arrivée, descendues, train, syndicat, initiative, tourisme, touristes, chambre, hôtel, réservé, avance, employée, problème, téléphoné, confortable, minutes, bagages, visiter

Exercice B

1. c
2. e
3. d
4. b
5. a

468

L'HÔTEL DE LA GARE

Monique est arrivée avec quelques copines à Nice. Elles sont descendues du train et sont allées tout de suite au syndicat d'initiative. Le syndicat d'initiative est un bureau de tourisme qui se trouve souvent dans les gares ou près des gares. Les touristes vont au syndicat d'initiative pour trouver une chambre d'hôtel dans la ville où ils sont arrivés, s'ils n'ont pas réservé de chambre à l'avance.

Monique a expliqué à l'employée du syndicat d'initiative que ses copines et elle sont étudiantes. Elles ne veulent pas aller dans un hôtel de grand luxe qui coûte très cher. Pas de problème: l'employée a téléphoné à l'Hôtel de la Gare où elle a réservé une chambre pour les filles. L'Hôtel de la Gare est un hôtel confortable mais pas trop cher. Et il est où, l'Hôtel de la Gare? En face de[1] la gare, bien sûr! Il y a un Hôtel de la Gare dans beaucoup de villes en France.

Monique et ses copines sont sorties de la gare, elles ont traversé la rue et sont arrivées à l'hôtel en deux minutes. Elles ont rempli les fiches d'enregistrement et ont monté leurs bagages à la chambre. Elles sont redescendues tout de suite après et sont allées visiter la ville de Nice.

[1] en face de *across from*

Hôtel - Restaurant de la Gare

Mʳ OBERHAUSSER

Place de la Gare

54120
BACCARAT
☎ 83 75 12 24

Étude de mots

A **Le français, c'est facile.** Trouvez cinq mots apparentés dans la lecture.

B **C'est-à-dire…** Trouvez les mots ou expressions qui correspondent.

1. le syndicat d'initiative
2. réserver
3. expliquer
4. les bagages
5. cher

a. qui coûte beaucoup
b. les sacs à dos, les valises
c. un bureau de tourisme
d. dire
e. louer à l'avance

LEARNING FROM REALIA

You may wish to ask the following questions about the realia: *Quel est le nom de l'hôtel? Il se trouve dans quelle ville? Quel est son numéro de téléphone? C'est un grand hôtel ou un petit hôtel?*

CRITICAL THINKING ACTIVITY

(Thinking skill: making inferences)

Read the following or write it on the board or on an overhead transparency:

«À Rome il faut vivre comme les Romains.» C'est un proverbe célèbre. Expliquez le proverbe. Qu'est-ce qu'il veut dire? Ensuite, imaginez que vous êtes en France. Dites ce que vous allez faire pour suivre la philosophie de ce proverbe.

Compréhension

C Un séjour à Nice.   Répondez.

1. Monique et ses copines sont arrivées où?
2. Elles sont allées à Nice comment?
3. Quand elles sont arrivées à la gare, où sont-elles allées?
4. Pourquoi sont-elles allées au syndicat d'initiative?
5. L'employée du syndicat d'initiative a téléphoné à quel hôtel?
6. Où est l'hôtel?
7. Les filles sont allées à l'hôtel?
8. Elles sont allées à l'hôtel à pied, en autobus ou en taxi?
9. Qu'est-ce qu'elles ont fait quand elles sont arrivées à l'hôtel?
10. Qu'est-ce que le syndicat d'initiative?
11. Qu'est-ce que vous avez appris au sujet des hôtels de la Gare?

DÉCOUVERTE CULTURELLE

En France les hôtels sont classés par le Ministère du Tourisme selon leur confort et leur luxe.

★★★★ L	HÔTEL DE GRAND LUXE
★★★★	HÔTEL DE PREMIÈRE CLASSE, TOUT CONFORT
★★★	HÔTEL TRÈS CONFORTABLE
★★	HÔTEL CONFORTABLE
★	HÔTEL AU CONFORT MOYEN, SIMPLE MAIS CONVENABLE[1]

LIGUE FRANÇAISE POUR LES AUBERGES DE LA JEUNESSE

38, Bd RASPAIL 75007 PARIS
TÉL. (1) 45 48 69 84
FAX 45 44 57 47

Quelle catégorie d'hôtel est la plus chère? Et la moins chère?

En France, il y a beaucoup de pensions qui sont souvent très agréables. Une pension est un hôtel simple à caractère familial.

Les jeunes qui voyagent en France aiment aller dans des auberges de jeunesse[2]. Les auberges de jeunesse ont des dortoirs[3] et ne coûtent pas très cher. Beaucoup de randonneurs[4] et cyclistes louent une chambre (ou un lit) dans une de ces auberges, qui se trouvent souvent près des villes. Les jeunes voyageurs les aiment beaucoup parce que dans les auberges de jeunesse ils peuvent faire la connaissance de[5] jeunes gens qui viennent de beaucoup de pays différents.

[1] moyen… convenable   *moderately priced, no-frills hotel*
[2] auberges de jeunesse   *youth hostels*
[3] dortoirs   *dormitories*
[4] randonneurs   *hikers*
[5] faire la connaissance de   *meet*

CHAPITRE 17   **469**

RÉALITÉS

OPTIONAL MATERIAL

PRESENTATION
(pages 470–471)

A. You may have students guess the meaning of the Michelin Guide's hotel categories (Photo 4). (In the English-language Michelin hotel-restaurant guide, *Luxe* is translated as "Luxury in the traditional style;" *Grand confort* = "Top class comfort;" *Très confortable* = "Very comfortable;" *De bon confort* = "Quite comfortable"; and *Simple mais convenable* = "Simple comfort.") As for the restaurant categories, explain to students that *la table* is the way the Guide refers to the food served by the restaurant. It does not mean "the table." Here are the categories with their English equivalents: *La table vaut le voyage* = "Exceptional cuisine, worth a special journey" *La table mérite un détour* = "Excellent cooking, worth a detour" *Une très bonne table* = "A very good restaurant in its category" *Repas soigné à prix modérés* = "Good food at moderate prices."

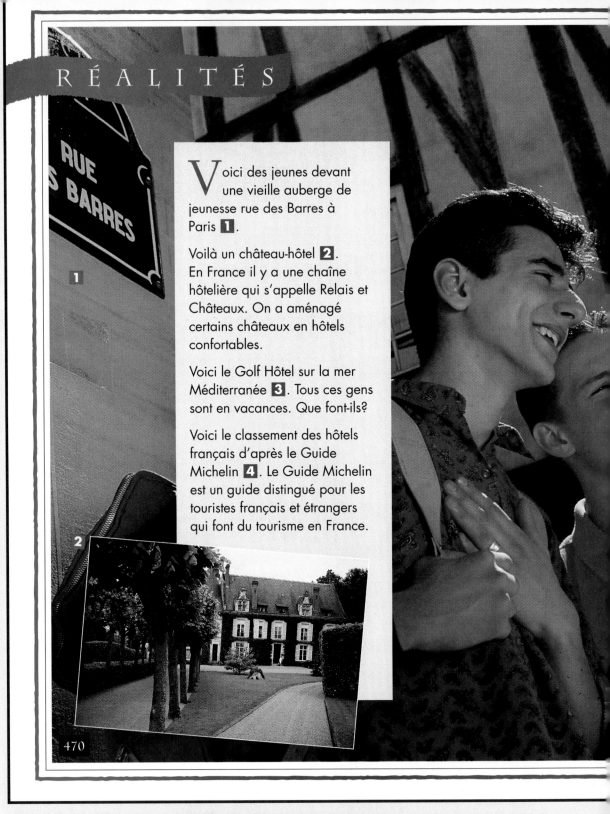

V oici des jeunes devant une vieille auberge de jeunesse rue des Barres à Paris **1**.

Voilà un château-hôtel **2**. En France il y a une chaîne hôtelière qui s'appelle Relais et Châteaux. On a aménagé certains châteaux en hôtels confortables.

Voici le Golf Hôtel sur la mer Méditerranée **3**. Tous ces gens sont en vacances. Que font-ils?

Voici le classement des hôtels français d'après le Guide Michelin **4**. Le Guide Michelin est un guide distingué pour les touristes français et étrangers qui font du tourisme en France.

470

DID YOU KNOW?

Usually, breakfast is included in the price of a room in a French hotel. It is a continental breakfast with a choice of hot tea, coffee, or hot chocolate, a croissant, and bread. The breakfast is served in a small room on the ground floor. If the hotel does not have such a room, breakfast is brought up to the room.

LEARNING FROM PHOTOS

Have students look at the hotels on pages 470–471 and decide what rating they think each would receive.

MICHELIN

	Grand luxe	XXXXX
	Grand confort	XXXX
	Très confortable	XXX
	De bon confort	XX
	Assez confortable	X
	Simple mais convenable	

	La table vaut le voyage
	La table mérite un détour
	Une très bonne table
Repas	Repas soigné à prix modérés 100/130
	Petit déjeuner
enf. 55	Menu enfant

B. Call on volunteers to read the captions on page 470 and answer any questions students may have about the captions or the photographs.

C. Share with students any photos or memorabilia you may have from your own hotel stays in France.

Note In the CD-ROM version, students can listen to the recorded captions and discover a hidden video behind one of the photos.

COOPERATIVE LEARNING

1. Use Communication Transparency C-17 to review winter weather and ski vocabulary as well as the hotel vocabulary of this chapter.
2. Students work in groups of five to make up the conversations of the people in the bottom illustration.

ADDITIONAL PRACTICE

1. Student Tape Manual, Teacher's Edition, *Deuxième Partie,* pages 195–198
2. Situation Cards, Chapter 17

RECYCLING

The *Activités de communication orale* and *Activités de communication écrite* afford students the opportunity to further practice the *passé composé* and chapter vocabulary associated with hotels and traveling. Material presented in earlier chapters is recycled.

INFORMAL ASSESSMENT

Both oral activities can be used as a means to assess speaking skills. Oral Activity A offers guided cues for those students who need practice structuring questions and answers. Oral Activity B asks students to synthesize the material offered in the hotel ads into their own personal presentations. To measure speaking ability you may use the evaluation criteria given on page 34 of this Teacher's Wraparound Edition.

Activités de communication orale

ANSWERS

Activité A

Answers will vary but may include the following constructions:

1. **Tu es allé(e) où pour tes vacances?**
2. **Quand est-ce que tu es allé(e)… ?**
3. **Tu y es allé(e) avec qui?**
4. **Tu as (Vous avez) fait quelles activités pendant tes (vos) vacances?**

Activité B

Answers will vary.

Activités de communication écrite

ANSWERS

Activités A, B, and C

Answers will vary.

Activités de communication orale

A **Des vacances formidables.** Using the following question words and phrases, ask your partner about a great vacation he or she once had. Then reverse roles.

1. où
2. quand
3. avec qui
4. quelles activités

B **Allons en France.** Your partner is planning a trip to France and is going to stay in one of the hotels pictured below. Ask your partner which of the hotels he or she prefers and why. Then reverse roles.

Hôtel de Paris

**34, boulevard d'Alsace
Cannes, Côte d'Azur**

45 chambres de 250 à 580F
Petit déjeuner 35F en salle,
50F en chambre par personne

Hôtel de grand confort en ville. Près de la mer. Chambres avec télé couleurs, radio, salles de bains. Jardin avec piscine.

Hôtel Idéal Mont Blanc

**Combloux
Haute-Savoie**

Ouvert: été et hiver
26 chambres de 310 à 365F
Petit déjeuner 37F
Demi-pension 277F
Pension complète 322 F

Chalet grand confort dans les Alpes. Séjour idéal pour les sports d'été ou d'hiver.

Auberge de Combreux

**Combreux
Val de Loire**

21 chambres de 210 à 350 F
Petit déjeuner 30F
Demi-pension 280F

Auberge pleine de charme en forêt, près des châteaux de la Loire. À l'hôtel vélo, tennis, piscine, practice de golf.

Hostellerie du Châteaux d'Agneaux

**Avenue Sainte-Marie
Agneaux, Normandie**

12 chambres de 300 à 600F
Petit déjeuner 37F
Demi-pension 450F
Pension 550F

Trente km. de la plage. Sur la route du Mont-St.-Michel. Confort, calme avec tennis et sauna.

Activités de communication écrite

A **Une publicité.** Write an ad for a hotel (real or imaginary) using the ads above as a guide.

FOR THE YOUNGER STUDENT

Have students draw a picture of a hotel. Then have them write a brief description of it.

LEARNING FROM REALIA

Have students look at the hotels in the ads and say as much about each of the hotels as they can. You may also wish to have them pick out one hotel and write a description of it.

B **Mon journal intime.** Write a diary entry describing your activities last weekend.

C **Une vie antérieure.** Imagine you lived in another century. Write a short paragraph telling where and when you were born and died.

Réintroduction et recombinaison

A **Les loisirs culturels.** Donnez des réponses personnelles.

1. Tu es sorti(e) le week-end dernier ou tu es resté(e) à la maison?
2. Si tu es sorti(e), avec qui es-tu sorti(e)?
3. Tu es allé(e) au cinéma le mois dernier?
4. Tu as vu quel film? Avec quels acteurs?
5. Tes copains et toi, avez-vous déjà visité un musée? Quel musée?
6. Vous avez admiré quels peintres ou quels sculpteurs?
7. Tu connais la Statue de la Liberté? Tu sais dans quelle ville des États-Unis elle est?
8. Si tu vas en France qu'est-ce que tu veux visiter?
9. Tu veux voir une pièce à la Comédie-Française? Quel genre de pièce, une comédie ou une tragédie?
10. Si tu vas voir un film étranger, tu préfères le voir doublé ou en version originale avec des sous-titres?

Vocabulaire

NOMS
l'hôtel (m.)
le hall
l'escalier (m.)
la réception
le/la réceptionniste
la personne
la fiche d'enregistrement
la facture
les frais (m.)
la carte de crédit
le chèque de voyage

la chambre
  à un lit
  à deux lits
la porte
le placard
le cintre
les draps (m.)

la couverture
l'oreiller (m.)
la salle de bains
le savon
le gant de toilette
la serviette
le rouleau de papier hygiénique

VERBES
monter
montrer
réserver
tomber
se sécher
mourir
naître

AUTRES MOTS ET EXPRESSIONS
donner sur
libérer la chambre
payer en espèces

Hôtel ★★ NN
BELLEVUE
Françoise et Jean Pierre CHODORGE
RESTAURANT - LOGIS DE FRANCE
Restaurant plein air - Salle de réunion
Repas de groupe - Service traiteur
Fermeture hebdomadaire le mercredi (hors saison)
55120 CLERMONT EN-ARGONNE TÉL. 29 87 41 02

CHAPITRE 17   **473**

OPTIONAL MATERIAL

Réintroduction et recombinaison

PRESENTATION *(page 473)*

Exercice A

This exercise recycles much of the cultural vocabulary of Chapter 16, recombining it with such structures as the *passé composé,* the *futur proche,* and verbs such as *vouloir* and *connaître.*

ANSWERS

Exercice A
Answers will vary.

ASSESSMENT RESOURCES

1. Chapter Quizzes
2. Testing Program
3. Situation Cards
4. Communication Transparency C-17
5. Computer Software: Practice / Test Generator

VIDEO PROGRAM

INTRODUCTION   (52:08)

MEREDITH ARRIVE EN FRANCE   (52:44)

STUDENT PORTFOLIO

Written assignments that may be included in students' portfolios are the *Activités de communication écrite* on page 472–473, the ADDITIONAL PRACTICE activities suggested on this page, and the *Mon Autobiographie* section of the Workbook on page 178.

Note Students may create and save both oral and writtwen work using the Electronic Portfolio feature on the CD-

ADDITIONAL PRACTICE

1. Have students get a brochure for a hotel or motel in your area. Have them describe it in French.
2. Have students make up a card for a hotel they have stayed in (or would like to stay in) using the card on page 473 as a model.

CHAPTER OVERVIEW

In this chapter students will learn to exchange money and carry out simple financial and banking transactions. In order to do this, they will learn vocabulary associated with money and money matters such as currency exchange, savings accounts, and checks. Students will learn the pronouns *y* and *en* and the irregular verbs *devoir* and *recevoir*.

The cultural focus of this chapter is on foreign currencies and French spending habits, particularly those of teens.

CHAPTER OBJECTIVES

By the end of this chapter, students will know:

1. vocabulary associated with making change, exchanging money, opening a savings account, and paying by check
2. vocabulary needed to discuss borrowing and lending money as well as personal budgeting
3. formal and familiar expressions used to talk about money
4. the uses and the position of the pronouns *y* and *en*
5. the differences between *y, lui,* and *leur*
6. the present tense and *passé composé* forms of the verbs *recevoir* and *devoir*
7. the meanings of *devoir* and the use of *devoir* with an infinitive

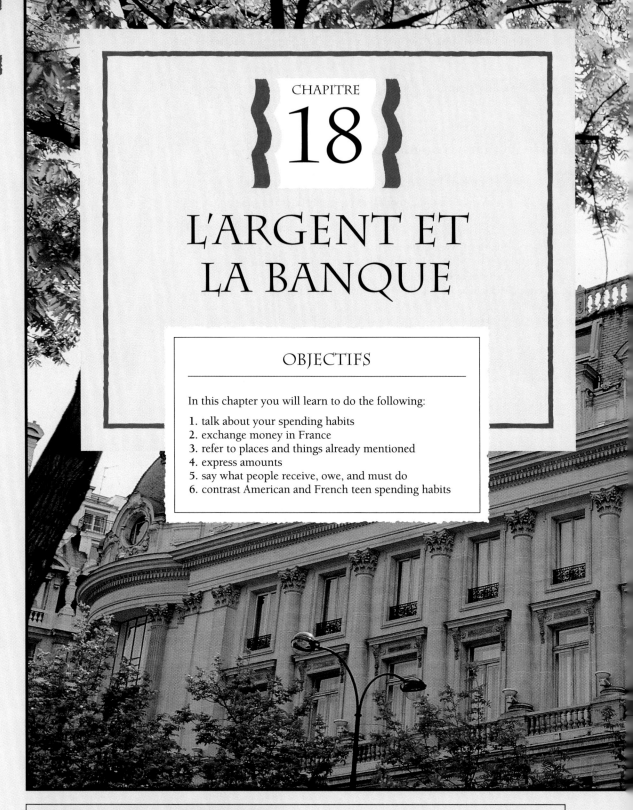

CHAPITRE

18

L'ARGENT ET LA BANQUE

OBJECTIFS

In this chapter you will learn to do the following:

1. talk about your spending habits
2. exchange money in France
3. refer to places and things already mentioned
4. express amounts
5. say what people receive, owe, and must do
6. contrast American and French teen spending habits

CHAPTER PROJECTS

(optional)

1. Check the newspaper for the exchange rates of French francs and those of other countries. Keep track of the rates throughout the chapter.
2. Have students draw up a personal budget and keep a diary of their expenditures while you work on the chapter. At the end of the chapter have them write or tell about what they may have learned about themselves from doing this.
3. If it is available, pass around some real French money or refer students to the photos on pages 9 and 478.

1. Workbook
2. Student Tape Manual
3. Audio Cassette 10B/CD-11
4. Bell Ringer Review Blackline Masters
5. Vocabulary Transparencies
6. Pronunciation Transparency P-18
7. Communication Transparency C-18
8. Communication Activities Masters
9. Map Transparencies
10. Situation Cards
11. Conversation Video
12. Videocassette/Videodisc, Unit 4
13. Video Activities Booklet, Unit 4
14. Lesson Plans
15. Computer Software: Practice/Test Generator
16. Chapter Quizzes
17. Testing Program
18. Internet Activities Booklet
19. CD-ROM Interactive Textbook

Pacing

This chapter requires eight to ten class sessions. Pacing will vary according to class length and the age and aptitude of the students.

Note The Lesson Plans offer guidelines for 45- and 55-minute classes and **Block Scheduling.**

Exercices vs. *Activités*

All exercises (which provide guided practice) are coded in blue. All communicative activities are coded in red.

INTERNET ACTIVITIES

(optional)

These activities, student worksheets, and related teacher information are in the *Bienvenue* Internet Activities Booklet and on the Glencoe Foreign Language Home Page at: **http://www.glencoe.com/secondary/fl**

DID YOU KNOW?

You may wish to give students the following information: *Le Crédit Lyonnais est une grande banque commerciale française. Le Crédit Lyonnais a des succursales* (branch offices) *aux États-Unis.* Ask: *C'est quel drapeau? De quelles couleurs est-il?*

VOCABULAIRE

MOTS 1

Bell Ringer Review

Write the following on the board or use BRR Blackline Master 18-1: Write down a suggestion you might make to a friend who is going with you to each of the following places. Use the *nous* form of the imperative.

1. **au restaurant**   3. **au concert**
2. **en classe**   4. **à la plage**

PRESENTATION *(pages 476–477)*

A. Show Vocabulary Transparencies 18.1 (A & B). Point to each item and have the class repeat after you or Cassette 10B/CD-11.
B. Ask *Qu'est-ce que c'est?* as you point to an item and call on an individual to respond.
C. After going over the vocabulary, have students open their books and read the words and sentences. Ask questions such as: *Sylvie est allée où? Qu'est-ce qu'elle a ouvert? Elle a versé combien d'argent sur son compte?*
D. Show students some French bills and coins or refer them to the photos on pages 9 and 478.

de l'argent liquide

un billet

une pièce

de la monnaie

un sac

un portefeuille

un porte-monnaie

Tu as de la monnaie?

Oui, j'en ai.

Tu peux me faire de la monnaie?

Oui, je peux.

une poche

476   CHAPITRE 18

TOTAL PHYSICAL RESPONSE

(following the Vocabulary presentation)

Getting Ready

Set up a mock bank counter. Dramatize the terms *rendre* and *compter*.

TPR 1

___, venez ici, s'il vous plaît.

J'espère que vous avez un peu d'argent sur vous. Vous en avez ou pas? (If the student has no money, lend him/her some.)

Bon. Montrez-moi votre poche (sac).

Sortez votre portefeuille de votre poche (sac).

Montrez-moi un billet.

Montrez-moi une pièce.

Mettez votre argent dans votre portefeuille.

Mettez le portefeuille dans votre poche (sac).

Très bien, ___, et merci. Retournez à votre place, s'il vous plaît.

À LA BANQUE

un chèque (bancaire)

signer un chèque

toucher un chèque

SOCIÉTÉ NATIONALE

--- 39418 ---   3180

SYLVIE VIDAL
75 BOULEVARD DU TEMPLE
75010 PARIS

RELEVÉ DE COMPTE
11345800PT03941
code banque
3003
code guichet
03182
numéro de compte
0048039532

DATE	NATURE DE L'OPÉRATION	DÉBIT	CRÉDIT	VALEUR
2802	VIREMENT ÉPARGNE DECLIC	2802	608,00	010392

```
* * * * * * * * * * * * * * * *
*  CE RELEVÉ CONCERNE VOTRE   *
*           CODEVI            *
* * * * * * * * * * * * * * * *
```

NOUVEAU SOLDE   2.618.12

un relevé de compte d'épargne

Sylvie est allée à la banque.
Elle a ouvert un compte d'épargne.
Elle a versé de l'argent sur son compte.

AU BUREAU DE CHANGE

le cours du change

ÉTATS UNIS 5,00 F
ITALIE
ALLEMAGNE
JAPON

Steve est allé au bureau de change.
Il y est allé pour changer de l'argent.
Il a changé de l'argent?
Oui, il en a changé.
Il a donné des dollars.
Et il a reçu des francs français.

la monnaie
française

CHAPITRE 18   477

After completing the vocabulary presentation, you may wish to call on students to look at the illustrations and say as much as they can about them.

CROSS-CULTURAL COMPARISON

In France there are coins for denominations as high as 20 francs. Since this results in more coins to be carried, it is common for French people—including men—to carry a coin purse rather than carry coins in their pockets.

Vocabulary Expansion

You may wish to give students the following additional vocabulary about money and banking.
un centime
retirer de l'argent de son compte
endosser un chèque
le taux d'intérêt

TPR 2
_____, venez ici, s'il vous plaît.
Hier, vous êtes allé(e) à la banque.
Entrez dans la banque.
Faites la queue devant la caisse.
Signez deux chèques de voyage.
Donnez les chèques au caissier/à la caissière.
Le caissier/La caissière vous donne des francs français.
Prenez-les.

Comptez-les.
Vous voulez de la monnaie. Donnez un billet au caissier/à la caissière.
Il/Elle vous donne des pièces. Comptez les pièces.
Merci, _____. C'est tout. Retournez à votre place, s'il vous plaît.

PRESENTATION (*pages 478–479*)

Extension of *Exercice B*

    After going over the exercise in class, call on a student to retell the information in his/her own words.

Extension of *Exercice C*

    After going over Exercise C, have a student tell about his/her personal money habits.

ANSWERS

Exercice A

1. C'est un chèque bancaire.
2. C'est de l'argent liquide.
3. C'est une pièce.
4. C'est un porte-monnaie.
5. C'est un sac.
6. C'est un bureau de change.
7. Elle touche le chèque.
8. On fait de la monnaie.

Exercices

A   **Qu'est-ce que c'est?**   Identifiez.

2. C'est de l'argent liquide ou un chèque de voyage?

1. C'est un chèque bancaire ou une carte de crédit?

4. C'est un portefeuille ou un porte-monnaie?

5. C'est un sac ou une poche?

3. C'est un billet ou une pièce?

7. Elle signe le chèque ou elle touche le chèque?

6. C'est une banque ou un bureau de change?

8. On fait de la monnaie ou on change de l'argent?

B Au bureau de change. Répondez.

1. Où est-ce que Steve est allé?
2. Il a de la monnaie américaine ou de la monnaie française?
3. Il a changé de l'argent?
4. Il a changé combien de dollars?
5. Quel est le cours du change?
6. Steve a reçu combien de francs pour ses dollars?

C Mon argent. Donnez des réponses personnelles.

1. Tu as de l'argent sur toi?
2. Tu as combien d'argent sur toi?
3. Tu mets ton argent dans ton portefeuille?
4. Tu mets des pièces ou des billets dans ton portefeuille?
5. Tu mets les pièces dans un portefeuille, dans un porte-monnaie ou dans ta poche?
6. Ton portefeuille est dans ta poche ou dans ton sac?
7. En général, tu paies en espèces, par chèque ou avec une carte de crédit?
8. Tu as un compte d'épargne?
9. Tu regardes ton relevé de compte chaque mois?
10. Tu verses de l'argent sur ton compte?

CHAPITRE 18   **479**

Exercice B

1. Il est allé au bureau de change.
2. Il a de la monnaie américaine.
3. Oui, il a changé de l'argent.
4. Il a changé cinquante dollars.
5. Le dollar est à cinq francs.
6. Il a reçu deux cent cinquante francs.

Exercice C

Answers will vary but may include the following constructions:

1. Oui, j'ai de l'argent. (Non, je n'ai pas d'argent.)
2. J'ai ___ dollars et ___ cents.
3. Oui, je le mets dans mon portefeuille.
4. Je mets des billets dans mon portefeuille.
5. Je les mets dans ma poche (mon portefeuille/porte-monnaie).
6. Il est dans mon sac (ma poche).
7. En général, je paie en espèces (par chèque/avec une carte de crédit).
8. Oui (Non), j'ai un (je n'ai pas de) compte d'épargne.
9. Oui (Non), je (ne) le regarde (pas) chaque mois.
10. Oui (Non), je verse de l'argent (je ne verse pas d'argent) sur mon compte.

RETEACHING (Mots 1)

Place props and pictures from the *Mots 1* presentation in a bag. Have students draw one item from the bag and repeat the corresponding word or phrase.

LEARNING FROM REALIA

1. Ask: *Les comptes Odyssée sont pour les personnes de quel âge?*
2. You may want to tell students that the French Post Office also has banking services. Students 13–16 can have their own accounts with an ATM card. Students will learn more about the postal system and its financial services in Chapter 1 of the Level 2 textbook, *À bord*.

INDEPENDENT PRACTICE

Assign any of the following:
1. Exercises, pages 478–479
2. Workbook, *Mots 1: A–B*, page 179
3. Communication Activities Masters, *Mots 1: A*, pages 86–87
4. CD-ROM, Disc 4, pages 476–479

Vocabulary Teaching Resources

1. Vocabulary Transparencies 18.2 (A & B)
2. Audio Cassette 10B/CD-11
3. Student Tape Manual, Teacher's Edition, *Mots 2: D–E*, pages 201–202
4. Workbook, *Mots 2: C–D*, page 179
5. Communication Activities Masters, *Mots 2: B*, page 88
6. Computer Software, *Vocabulaire*
7. Chapter Quizzes, *Mots 2: Quiz 2*, page 96
8. CD-ROM, Disc 4, *Mots 2:* pages 480–483

Bell Ringer Review

Write the following on the board or use BRR Blackline Master 18-2: In French, write at least three different ways you can earn some extra money.

PRESENTATION *(pages 480–481)*

A. Have students close their books. Present the *Mots 2* vocabulary by showing Vocabulary Transparencies 18.2 (A & B) and having students repeat the items chorally after you or Cassette 10B/CD-11.

B. As you present the sentences on page 480, break them into parts. For example: *de côté* (wave your hand to the side); *de l'argent* (as you point to the money in the illustration); *Elle aime mettre* (no comprehension problems since the words are not new). Now put the sentence together. *Elle aime mettre de l'argent de côté.* You may then want to ask: *Qu'est-ce qu'elle met de côté? Qui aime mettre de l'argent de côté?*

480

VOCABULAIRE

MOTS 2

Voici Lise.
Elle aime mettre de l'argent de côté.
Elle ne dépense pas tout son argent.
Elle fait des économies.

Et voilà Denis.
Denis n'a pas d'argent.
Il est fauché.
Il veut emprunter de
l'argent à Lise.

Tu peux me prêter de l'argent?

Oui, je peux te prêter de l'argent. Tu en veux combien?

480 CHAPITRE 18

TOTAL PHYSICAL RESPONSE

(following the Vocabulary presentation)

Getting Ready

Have students use their own money or give them play money. Demonstrate *vide*.

TPR

(Student 1) et (Student 2), **venez ici, s'il vous plaît.**
(Student 1), **vous êtes fauché(e).**
Montrez vos poches vides à (Student 2).
Demandez-lui de l'argent.
(Student 2), **sortez votre portefeuille.**
Sortez des billets.
Prêtez les billets à (Student 1).
(Student 1), **mettez les billets dans votre poche.**
(Student 1), **vous êtes content(e).**
Dites «merci» à (Student 2).
Serrez-lui la main.

Denis rembourse Lise.
Il lui rend son argent.

Note: Here are some informal words referring to money.

le fric	*l'argent*
50 balles	*50 francs*
Il est fauché.	*Il n'a pas d'argent.*
Il a plein de fric.	*Il a beaucoup d'argent.*

C. When presenting the vocabulary on page 481, you may wish to have two students (who like to act) present the mini-conversation to the class. They can use a great deal of expression.

CROSS-CULTURAL COMPARISON

The words *fric, balle(s),* and *fauché(e)* are used in everyday informal speech by people of all ages. You may wish to point out to students that many languages have lots of slang expressions for money. Have them think of some in English.

Vocabulary Expansion

You may wish to give students the following popular expressions about money.
Il n'a pas un sou.
Il n'a pas un rond.
Il roule sur l'or.
Il jette son argent par les fenêtres.
Il est près de ses sous.

COOPERATIVE LEARNING

Have students work in groups and make up a skit based on the mini-conversation on page 481.

ADDITIONAL PRACTICE

1. Have students look at pages 480–481. Give them sentences such as the following and have them find the opposite. *Lise dépense tout son argent. Denis a beaucoup d'argent. La jeune fille fait des économies. Denis prête de l'argent à Lise. Il a plein de fric.*
2. Student Tape Manual, Teacher's Edition, *Activité E,* page 202.

PRESENTATION (*page 482*)

Exercice A

Exercise A gives students practice in receptive comprehension.

Extension of *Exercice B*

After completing the exercise, call on one student to give all the information in his/her own words.

Extension of *Exercice C*

You may wish to have the students dramatize the mini-conversation in Exercise C using the informal language.

Exercice D: Vocabulary Expansion

Exercise D gives students practice in developing vocabulary building techniques.

ANSWERS

Exercice A

1. Denis
2. Lise
3. Lise
4. Lise
5. Denis
6. Denis
7. Lise
8. Denis

Exercice B

Answers will vary.

Exercice C

… de l'argent
… de l'argent … je n'ai jamais d'argent
… francs
… beaucoup d'argent
… francs

Exercice D

1. d
2. f
3. a
4. h
5. g
6. e
7. c
8. b

INFORMAL ASSESSMENT
(*Mots 2*)

Check for understanding by making true and false statements about the *Mots 2* vocabulary. Students correct false statements and respond with *«C'est vrai»* for true ones. For example: *Denis a toujours beaucoup d'argent. C'est Lise qui est toujours fauchée. Lise n'a pas d'argent. Lise prête de l'argent à Denis. Denis rembourse Lise.*

Exercices

A C'est qui? Décidez si c'est Lise ou Denis.

1. Il/Elle dépense tout son argent.
2. Il/Elle met de l'argent de côté.
3. Il/Elle fait des économies.
4. Il/Elle a un compte d'épargne.
5. Il/Elle est toujours fauché(e).
6. Il/Elle emprunte de l'argent à un(e) ami(e).
7. Il/Elle prête de l'argent.
8. Il/Elle rembourse l'argent qu'il/elle emprunte.

B L'argent et toi! Donnez des réponses personnelles.

1. Tu travailles?
2. Tu gagnes de l'argent? Tu as de l'argent de poche?
3. Qu'est-ce que tu fais pour gagner de l'argent? Tu travailles dans le jardin des voisins? Tu laves des voitures? Tu gardes des enfants? Tu aides ton père ou ta mère?
4. Tu dépenses tout ton argent de poche ou tu en mets de côté?
5. Tu as un compte d'épargne? Dans quelle banque?
6. De temps en temps, tu empruntes de l'argent à tes parents?
7. Quand tu dois de l'argent à tes parents, tu les rembourses toujours?
8. Tu leur rends vite l'argent?

C Un peu d'argot. Dites la même chose d'une autre manière.

DAVID: Tu as *du fric*?
MARIE: Tu me demandes si j'ai *du fric*. Tu sais bien que *je suis toujours fauchée*.
DAVID: Et tu me dois cinquante *balles*.
MARIE: Oui, je sais. Mais tu n'en as pas besoin. Tu as *plein de fric*.
DAVID: C'est pas la question.
MARIE: D'accord. Je te rends les cinquante *balles* demain.

D Quel est le nom? Choisissez le mot qui correspond.

1. épargner
2. économiser
3. verser
4. changer
5. dépenser
6. emprunter
7. prêter
8. rembourser

a. le versement
b. le remboursement
c. le prêt
d. l'épargne
e. l'emprunt
f. des économies
g. la dépense
h. le change

ADDITIONAL PRACTICE

After completing the exercises and activities on pages 482–483, have students role-play exchanging currency. Put some realistic exchange rates on the board for two or three currencies. (These are available in major newspapers.) Have some play money ready in the various currencies or use handwritten cards or pieces of paper. Assign some students to work as "exchange agents" who deal with one kind of currency. Distribute various amounts of play dollars to the rest of the students, who will exchange it for the currency they want. When all the money has been exchanged, have students reverse roles.

Note You may wish to write on the board a few key sentences necessary for the transactions until students become familiar with the language. You can also use slips of paper at the booths as mock exchange receipts, which "customers" must sign.

Activités de communication orale

Mots 1 et 2

Yves Clemenceau

A **Ton argent et toi.** The French exchange student at your school, Yves Clemenceau, wants to know about American teens and money. Answer his questions.

1. Tu travailles pour gagner de l'argent?
2. Tes parents te donnent de l'argent de poche?
3. Qu'est-ce que tu achètes avec ton argent?
4. Tu peux mettre de l'argent de côté?

B **Au bureau de change.** You're at a foreign exchange office in France and want to change 50 dollars into francs. Find out the exchange rate from the teller (your partner). The teller asks you if you have traveler's checks or cash. If you have traveler's checks, you'll have to sign them. You'll also have to show the teller your passport.

C **Quel cours du change!** You and your partner are French tourists visiting the U.S. You'd like to buy a few gifts for your friends and family. Make a list of the items you want and their price in dollars. Your partner will help you figure out how much each of your gifts costs in francs. (The exchange rate is five francs to the dollar.) When you've gone through your list, reverse roles.

> Élève 1: Je voudrais acheter une cassette pour ma sœur. Ça coûte 9 dollars. Ça fait combien en francs?
> Élève 2: Ça fait 45 francs.

D **Un petit problème.** You'd like to buy your mother a birthday present, but you can't afford it at the moment. Your friend (your partner) might be able to help you out. Try to convince your partner to lend you the money.

INDEPENDENT PRACTICE

Assign any of the following:
1. Exercises and activities, pages 482–483
2. Workbook, *Mots 2: C–D,* page 179
3. Communication Activities Masters, *Mots 2: B,* page 88
4. Computer Software, *Vocabulaire*
5. CD-ROM, Disc 4, pages 480–483

Activités de communication orale

Mots 1 et 2

PRESENTATION *(page 483)*

Activité A

In the CD-ROM version, students can interact with an on-screen native speaker.

ANSWERS

Activité A

Answers will vary but may include the following:
1. Oui (Non), je (ne) travaille (pas) pour gagner de l'argent.
2. Oui (Non), mes parents me donnent de l'argent (ne me donnent pas d'argent) de poche.
3. J'achète des compact discs (des vêtements, etc.) avec mon argent.
4. Oui (Non), je peux mettre de l'argent (je ne peux pas mettre d'argent) de côté.

Activité B

Answers will vary but may include the following:
É1: Bonjour, Monsieur/Madame, je voudrais changer cinquante dollars en francs français. Quel est le cours du change aujourd'hui?
É2: Le dollar est à cinq francs. Avez-vous de l'argent liquide ou des chèques de voyage?
É1: J'ai des chèques de voyage.
É2: Signez les chèques, s'il vous plaît, et montrez-moi votre passeport.

Activité C

Answers will vary.

Activité D

Answers will vary.

STRUCTURE

Le pronom **y**   *Referring to Places Already Mentioned*

1. You have already used the pronoun *y* with the verb *aller* to refer to a place just mentioned. *Aller* cannot stand alone.

> **Tu vas au restaurant?**
> **Oui, j'y vais.**
> **On y va ensemble?**

2. You also use the pronoun *y* to replace any location introduced by *à* or another preposition.

> **Tu vas *à* Paris?**   Oui, j'y vais.
> **Henri monte *en haut de la Tour Eiffel*?**   Oui, il y monte.
> **Il est *à l'Hôtel Racine*?**   Oui, il y est.
> **Il veut entrer *dans l'hôtel*?**   Oui, il veut y entrer.
> **Tu veux aller *en France*?**   Oui, je veux y aller.
> **Ils peuvent dîner *chez leurs amis*?**   Oui, ils peuvent y dîner.

3. With the *passé composé, y* comes before the helping verb.

> **Ils sont entrés *dans le musée*?**   Oui, ils y sont entrés.
> **Elle est montée *au troisième étage*?**   Oui, elle y est montée.
> **Elle a vu de beaux tableaux *au musée*?**   Oui, elle y a vu de beaux tableaux.

4. Note the placement of *y* in negative sentences.

> | PRÉSENT | **Je n'y *vais* pas.** |
> | VERBE + INFINITIF | **Je ne vais pas y *aller*.** |
> | PASSÉ COMPOSÉ | **Je n'y *suis* pas allé(e).** |

*La fondation Vasarely à
Aix-en-Provence*

Structure Teaching Resources

1. Workbook, *Structure: A–G,* pages 180–183
2. Student Tape Manual, Teacher's Edition, *Structure: A–F,* pages 203–205
3. Audio Cassette 10B/CD-11
4. Communication Activities Masters, *Structure: A–F,* pages 89–92
5. Computer Software, *Structure*
6. Chapter Quizzes, *Structure:* Quizzes 3–7, pages 97–101
7. CD-ROM, Disc 4, pages 484–491

Bell Ringer Review

Write the following on the board or use BRR Blackline Master 18-4: List at least five ways you spend money in any given week.

Le pronom **y**

PRESENTATION *(page 484)*

A. Briefly review the use of the pronoun *y* with *aller,* taught in Chapter 5, page 133.

B. Lead students through steps 1–3 on page 484. In steps 2 and 3, have one student read the questions on the left and another respond with the sentences on the right.

C. You may wish to write example sentences on the board. Underline the expression replaced by *y.* Circle *y* and draw a line to it from the expression it replaces.

D. Now lead students through step 4.

ADDITIONAL PRACTICE

After completing Exercises A, B, and C, have students tell about a trip they took. They should mention the place by name in the first sentence and use *y* in succeeding ones. For example: *L'année dernière je suis allé à New York. J'y suis allé avec ma famille. Nous y sommes allés en avion. Nous y sommes restés deux semaines,* etc.

DID YOU KNOW?

You may wish to tell students that Vasarely was a 20th-century artist of Hungarian origin who specialized in op art of the type shown in the photo.

Exercices

A **Au gymnase.** Répétez la petite conversation.

BÉATRICE: Tu vas au gymnase?
HÉLÈNE: Oui, j'y vais tous les samedis.
BÉATRICE: Ton copain y va aussi?
HÉLÈNE: Oui, il y va aussi. Il y va souvent.

B **On y va?** Répondez d'après les dessins en utilisant «y».

1. David est allé au bureau de change?
2. Il est allé au bureau de change le matin?
3. Il est allé au bureau de change pour changer de l'argent?
4. Il est arrivé au bureau de change avant l'ouverture?
5. Il a attendu devant le bureau?
6. Il a attendu cinq minutes devant le bureau?
7. Quand le bureau a ouvert, David est entré dans le bureau?
8. Il a fait la queue devant la caisse?

C **À l'école.** Donnez des réponses personnelles en utilisant «y».

1. Tu vas à l'école tous les jours?
2. Tu prépares tes leçons à la maison?
3. Tu parles français au cours de français?
4. Tu parles français au cours d'anglais?
5. Tu attends tes amis dans la cour?
6. Tu mets tes livres dans ton sac à dos?
7. Tu vas à l'école à pied?
8. Tu aimes aller chez tes copains après les cours?
9. Tu veux aller chez eux aujourd'hui?

INDEPENDENT PRACTICE

Assign any of the following:
1. Exercises, page 485
2. Workbook, *Structure: A,* page 180
3. Communication Activities Masters, *Structure: A,* page 89
4. CD-ROM, Disc 4, pages 484–485

Exercices

PRESENTATION

Exercice A

You may wish to use the recorded version of this exercise.

ANSWERS

Exercice A

Students repeat the mini-conversation.

Exercice B

1. Oui, il y est allé.
2. Oui, il y est allé le matin.
3. Oui, il y est allé pour changer de l'argent.
4. Oui, il y est arrivé avant l'ouverture.
5. Oui, il y a attendu.
6. Oui, il y a attendu cinq minutes.
7. Oui, quand le bureau a ouvert, il y est entré.
8. Oui, il y a fait la queue.

Exercice C

Answers will vary but may include the following:

1. Non, je n'y vais pas tous les jours.
2. Oui (Non), j'y (je n'y) prépare (pas) mes leçons.
3. Oui, j'y parle français.
4. Non, je n'y parle pas français.
5. Oui (Non), j'y (je n'y) attends (pas) mes amis.
6. Oui (Non), j'y (je n'y) mets (pas) mes livres.
7. Oui (Non), j'y (je n'y) vais (pas) à pied.
8. Oui (Non), j'aime (je n'aime pas) y aller.
9. Oui (Non), je (ne) veux (pas) y aller aujourd'hui.

INFORMAL ASSESSMENT

Check for understanding by asking questions that would elicit a natural negative response. Ask each question three times: once in the present, once in the *passé composé,* and once in the *futur proche.* For example: *Tu vas à l'école en taxi? (Non, je n'y vais pas en taxi.) Tu y es allé(e) en taxi la semaine dernière? (Non, je n'y suis pas allé[e] en taxi la semaine dernière.) Tu vas y aller en taxi demain? (Non, je ne vais pas y aller en taxi demain.)*

Y, lui *ou* leur

Bell Ringer Review

Write the following on the board or use BRR Blackline Master 18-5: Write down several reasons why it is a good idea to have a savings account.

Y, lui *ou* leur

PRESENTATION *(page 486)*

A. Lead students through steps 1 and 2 on page 486.

B. On the board, write original sentences with prepositional phrases or indirect objects with *à*. Call on students to come to the board and rewrite the sentences using the appropriate pronoun.

Exercices

ANSWERS

Exercice A

1. Oui, il y a téléphoné.
2. Oui, il y a répondu.
3. Oui, il y a obéi.
4. Oui, il y a réussi.
5. Oui, il y a participé.

Exercice B

1. y
2. leur
3. y
4. lui
5. leur
6. lui
7. y

Y, lui *ou* leur

Referring to People and Things Already Mentioned

1. You also use the pronoun *y* to replace *à* + a thing.

Georges répond *à la lettre?*	Oui, il y répond.
Anne a répondu *au téléphone?*	Non, elle n'y a pas répondu.

2. If the preposition *à* is followed by a person, you use *lui* or *leur,* not *y.*

Georges répond *à Marie.*	Il *lui* répond.
Anne a répondu *à ses amis.*	Elle *leur* a répondu.

Exercices

A **Il a téléphoné.** Répondez en utilisant «y».

1. Paul a téléphoné à l'hôtel?
2. Paul a répondu à la question?
3. Il a obéi à la règle?
4. Paul a réussi à l'examen?
5. Paul a participé au match?

B **Y, lui *ou* leur?** Complétez.

1. Tu as répondu à la lettre?
   Oui, j'___ ai répondu.
2. Tu as répondu à tes cousins?
   Oui, je ___ ai répondu.
3. Sa sœur a répondu à une petite annonce? Oui, elle ___ a répondu.
4. Elle a répondu à sa mère?
   Oui, elle ___ a répondu.
5. Elle a téléphoné à ses copains?
   Oui, elle ___ a téléphoné.
6. Tes copains et toi, vous avez obéi au professeur? Oui, nous ___ avons obéi.
7. Vous avez obéi aussi aux règles de l'école?
   Oui, nous ___ avons obéi.

COOPERATIVE LEARNING

Have students form teams of four. Call two members from each team to the board. Dictate a sentence that contains a prepositional phrase with *à* or an indirect object. One team member writes the sentence, and the other rewrites it using *lui, leur,* or *y.* Repeat the procedure with the third and fourth members of each team.

LEARNING FROM PHOTOS

1. The bulletin board has various jobs posted. Ask students to see how many words they can identify.
2. Ask students what Olivier Calendini teaches.

Le pronom *en*

Referring to Things Already Mentioned

1. You use the pronoun *en* to replace a noun that is introduced by *de* or any form of it: *du, de l', de la, des.*

Vous avez *de la monnaie?*	Oui, j'en ai.
	Non, je n'en ai pas.
Richard veut *de l'argent?*	Oui, il en veut.
	Non, il n'en veut pas.
Il va changer *des francs?*	Oui, il va en changer.
	Non, il ne va pas en changer.
Il a besoin *d'argent?*	Oui, il en a besoin.
	Non, il n'en a pas besoin.
Il a parlé *de ses finances?*	Oui, il en a parlé.
	Non, il n'en a pas parlé.
Il est venu *de la banque?*	Oui, il en est venu.
	Non, il n'en est pas venu.
Il y a *des bureaux de change* en ville?	Oui, il y en a.
	Non, il n'y en a pas.

INDEPENDENT PRACTICE

Assign any of the following:
1. Exercises, page 486
2. Workbook, *Structure: B,* pages 180–181
3. Communication Activities Masters, *Structure: B,* page 89
4. CD-ROM, Disc 4, page 486

Bell Ringer Review

Write the following on the board or use BRR Blackline Master 18-6: Rewrite the following sentences, replacing the pronoun *y* with a prepositional phrase that makes sense.

1. Mes amis y vont souvent.
2. Nous y restons tous les ans pendant trois semaines.
3. Mon père y a fait un voyage.
4. Les élèves y apprennent beaucoup.

Le pronom en

PRESENTATION *(page 487)*

Have students open their books to page 487. Lead them through the explanation. Read the questions on the left and have the students read the response with *en* in unison.

Note In the CD-ROM version, 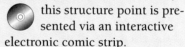 this structure point is presented via an interactive electronic comic strip.

Exercice A

 You may wish to use the recorded version of this exercise.

Extension of *Exercice A*: Speaking

After completing Exercise A, ask additional questions that name foods not included in the picture. Students respond using *en* in negative sentences.

Extension of *Exercice B*: Speaking/Listening

After completing Exercise B, have pairs ask the questions of other pairs in random order, so *nous* and *vous* forms can be practiced.

ANSWERS

Exercice A

1. Oui, elle en sert.
2. Oui, elle en sert.
3. Oui, elle en sert.
4. Non, elle n'en sert pas.
5. Non, elle n'en sert pas.
6. Non, elle n'en sert pas.
7. Oui, elle en sert.
8. Non, elle n'en sert pas.

Exercice B

1. Oui (Non), j'en (je n'en) ai (pas).
2. Oui (Non), j'en (je n'en) ai (pas).
3. Oui (Non), j'en (je n'en) reviens (pas).
4. Oui (Non), j'en (je n'en) ai (pas).
5. Oui (Non), j'en (je n'en) ai (pas).
6. Oui (Non), j'en (je n'en) ai (pas) besoin.
7. Oui (Non), je (ne) veux (pas) en gagner.
8. Oui (Non), j'en (je n'en) parle (pas).
9. Oui (Non), j'en (je n'en) ai (pas) emprunté à mes parents.
10. Oui (Non), j'en (je n'en) ai (pas) prêté à mes amis.

Exercices

A **La fête de Laurence.** Répondez d'après le dessin.

Laurence sert du coca?
Oui, elle en sert.

1. Elle sert de l'eau minérale?
2. Elle sert des sandwichs?
3. Elle sert de la pizza?
4. Elle sert de la salade?
5. Elle sert du fromage?
6. Elle sert des chocolats?
7. Elle sert de la glace?
8. Elle sert de la mousse au chocolat?

B **Oui, j'en ai.** Donnez des réponses personnelles en utilisant «en».

1. Tu as de l'argent dans ton portefeuille?
2. Tu as de la monnaie dans ta poche?
3. Tu reviens de la banque?
4. Tu as des billets?
5. Tu as des pièces?
6. Tu as besoin d'argent?
7. Tu veux gagner de l'argent?
8. Tu parles de l'argent?
9. Tu as emprunté de l'argent à tes parents?
10. Tu as prêté de l'argent à tes amis?

ADDITIONAL PRACTICE

After completing the exercises on pages 488–489, have students interview a partner on his/her life-style, using questions with *combien*. They should vary the verbs in their questions. Partners respond using *en*. Sample questions are: *Tu écris combien de cartes par an? Tu vois combien de films par mois? Tu manges combien de pizzas par mois? Ta famille a combien de voitures?*

C **Dans le réfrigérateur.** Répondez d'après le modèle.

du coca

Élève 1: Il y du coca dans ton réfrigérateur?

Élève 2: Oui, il y en a dans mon réfrigérateur. (Non, il n'y en a pas.)

1. de l'eau minérale
2. de la glace
3. des légumes surgelés
4. du jambon
5. des tartes
6. de la viande

D'autres emplois du pronom *en*

Expressing Amounts

1. Note that you also use the pronoun *en* with numbers and expressions of quantity. They cannot stand alone in French. They must be accompanied by *en*.

Tu as combien de magazines?	J'en ai *deux.*
Et tu as beaucoup de livres?	Oui, j'en ai *beaucoup.*

2. Here are some other expressions of quantity. Note the use of *en* with them.

J'en ai *une paire.*
J'en ai *une douzaine.*

J'en ai *très peu.*	*very few, very little*
J'en ai *assez.*	*enough*
J'en ai *quelques-uns (-unes).*	*a few*
J'en ai *plusieurs.*	*several*
J'en ai *trop.*	*too much, too many*

Exercice

A **J'en ai assez.** Donnez des réponses personnelles en utilisant «en».

1. Tu as combien de paires de chaussures?
2. Tu en as assez?
3. Tu as combien de billets d'un dollar? Tu en as quelques-uns?
4. Tu en as assez pour acheter un coca?
5. Tu as beaucoup d'argent ou peu d'argent dans ton portefeuille?
6. Tu as plusieurs cours aujourd'hui?
7. Tu as trop de devoirs tous les soirs?
8. Tu as beaucoup de cassettes de ton groupe de rock préféré ou tu en as seulement quelques-unes?

CHAPITRE 18   **489**

Exercice C
   Answers will vary but É2 answers will use either **Oui, il y en a** or **Non, il n'y en a pas.**

D'autres emplois du pronom **en**
PRESENTATION *(page 489)*
A. Have students open their books. Lead them through steps 1 and 2 on page 489.
B. Call on volunteers to supply true sentences about themselves or people they know using the expressions of quantity in step 2.

ANSWERS
Exercice A
   Answers will vary.

Les verbes recevoir et devoir

PRESENTATION *(page 490)*

A. Give students the *ils/elles* forms of *recevoir* and *devoir: ils reçoivent, ils doivent.* Have them drop the final consonant sound to get the sound for all the singular forms.

B. Write all the forms on the board and have students repeat them.

C. Call on a student to read the example sentences in step 1.

D. Lead students through steps 2 and 3.

Les verbes *recevoir* et *devoir*   *Saying What People Receive, Owe, and Must Do*

---

1. Study the following forms of the present tense of the irregular verbs *recevoir,* "to receive," and *devoir,* "to owe."

RECEVOIR	DEVOIR
je reçois	je dois
tu reçois	tu dois
il elle on } reçoit	il elle on } doit
nous recevons	nous devons
vous recevez	vous devez
ils elles } reçoivent	ils elles } doivent

Je reçois beaucoup de cadeaux pour mon anniversaire.
Elle reçoit beaucoup de lettres.

Nous devons de l'argent à la banque.
Mon ami me doit de l'argent.

2. When followed by an infinitive, the verb *devoir* also means "must" or "to have to."

Il m'a prêté de l'argent. Je dois lui rendre son argent.
Elle a un examen difficile demain. Elle doit étudier ce soir.

3. Note the past participles of these verbs.

J'ai *reçu* cent dollars.
J'ai *dû* étudier pour réussir à l'examen.

Exercices

A **Je sais que je lui dois de l'argent.**
Mettez au pluriel d'après le modèle.

> **Je lui dois vingt francs.**
> *Nous lui devons vingt francs.*

1. Je lui dois de l'argent.
2. Je lui dois cent dollars.
3. Si je reçois mon chèque aujourd'hui, je vais le rembourser.
4. Je sais que je dois lui rendre l'argent que je lui dois.

Un distributeur automatique de billets

B **Je dois aller au bureau de change.** Répondez.

1. Si tu as besoin de francs, tu dois aller au bureau de change?
2. Si j'ai besoin de francs, je dois y aller aussi?
3. On doit y aller ensemble?
4. Le dollar est à cinq francs. Si je change vingt dollars, je reçois combien de francs?
5. Si un Français change cent francs, il reçoit combien de dollars?
6. Les Français reçoivent leur salaire en dollars ou en francs?
7. Les Américains reçoivent leur salaire en dollars ou en francs?
8. Tu as déjà reçu un salaire?
9. Tu as reçu combien?

C **On doit faire beaucoup de choses.** Complétez au présent avec «devoir» ou «recevoir».

Dans la vie, on ___ faire beaucoup de choses et ce n'est pas toujours agréable!
₁
Moi, tous les matins je ___ me lever à six heures et demie. Je ___ préparer le
₂   ₃
petit déjeuner. Ma sœur Aurélie ___ donner à manger au chien. Nous ___
₄   ₅
quitter la maison à huit heures pour aller à l'école. Le soir nous ___ aider notre
₆
mère à préparer le dîner. Après le dîner nous ___ faire nos devoirs. Nous ___
₇   ₈
beaucoup travailler tous les jours! Mais chaque semaine nous ___ de l'argent
₉
de poche de nos parents. Tes copains et toi, vous ___ de l'argent de poche de
₁₀
vos parents? Qu'est-ce que vous ___ faire tous les jours pour en avoir? Vous
₁₁
travaillez? Une question de plus! Qu'est-ce que vous faites de l'argent que vous
___? Vous le dépensez ou vous en mettez quelques dollars de côté? Vos parents
₁₂
vous disent que vous ___ faire des économies?
₁₃

Exercices

PRESENTATION (*page 491*)

Exercice A

You may wish to use the recorded version of this exercise.

ANSWERS

Exercice A

1. **Nous lui devons de l'argent.**
2. **Nous lui devons cent dollars.**
3. **Si nous recevons nos chèques aujourd'hui, nous allons le rembourser.**
4. **Nous savons que nous devons lui rendre l'argent que nous lui devons.**

Exercice B

Answers will vary but may include the following:

1. **Oui, je dois y aller.**
2. **Oui, tu dois (vous devez) y aller aussi.**
3. **Oui (Non), on (ne) doit (pas) y aller ensemble.**
4. **Tu reçois (Vous recevez) cent francs.**
5. **Il reçoit vingt dollars.**
6. **Ils le reçoivent en francs.**
7. **Ils le reçoivent en dollars.**
8. **Oui (Non), j'en ai déjà (je n'en ai jamais) reçu un.**
9. **J'ai reçu ___ dollars. (Je n'ai rien reçu.)**

Exercice C

1. doit	8. devons
2. dois	9. recevons
3. dois	10. recevez
4. doit	11. devez
5. devons	12. recevez
6. devons	13. devez
7. devons	

RETEACHING

Have students list the things they have to do each day, using the verb *devoir.* Have them question each other about their lists and do a class summary.

LEARNING FROM PHOTOS

You may wish to ask the following questions about the photo: *Qu'est-ce que c'est? Qu'est-ce que le garçon a mis (a introduit) dans le distributeur automatique? Qu'est-ce que le distributeur automatique lui a rendu?*

INDEPENDENT PRACTICE

Assign any of the following:
1. Exercises, page 491
2. Workbook, *Structure: F–G,* page 183
3. Communication Activities Masters, *Structure: E–F,* page 92
4. Computer Software, *Structure*
5. CD-ROM, Disc 4, pages 490–491

Scènes de la vie *Au bureau de change*

ROBERT: Je voudrais changer vingt dollars en francs français, s'il vous plaît.

LE CAISSIER: Vous avez des chèques de voyage ou de l'argent liquide?

ROBERT: Des chèques de voyage. Le dollar est à combien aujourd'hui?

LE CAISSIER: À cinq francs quatre-vingts.

ROBERT: Très bien.

LE CAISSIER: Votre passeport, s'il vous plaît. Et signez votre chèque. Votre adresse à Paris?

ROBERT: Hôtel Molière, rue Molière dans le 1ᵉʳ arrondissement.

A Des francs, s'il vous plaît. Répondez d'après la conversation.

1. Où est-ce que Robert est allé?
2. Il veut changer combien de dollars?
3. Il va changer des chèques de voyage ou de l'argent liquide?
4. Il veut changer des dollars en quelle monnaie?
5. Quel est le cours du change?
6. Le caissier veut voir son passeport?
7. Robert est à quel hôtel à Paris?

B Qu'est-ce qu'il a fait? Corrigez les phrases.

1. Robert est allé à la banque.
2. Il a changé de l'argent liquide.
3. Il a changé cinquante francs.
4. Il a reçu des dollars.
5. Le caissier a voulu voir sa carte de crédit.

492 CHAPITRE 18

LA POSTE
DCV-01

ACHAT ☐ VENTE ☐ DE BILLETS ETRANGE

à M _Johnson, Robert_
(nom, prénom) _Hôtel Molière_
(adresse) _Rue Molière_
Paris   75001

DEVISE	CODE	MONTANT	COURS	CONTRE-VAL
USD	03190	20	5,8000	1,1

A _Paris_, LE _26/06_
SIGNATURE DU CLIENT,
Robert Johnson

COMMISSION

NET   1,0

492

Prononciation *Les sons /p/, /t/, /k/*

1. Repeat the following words with the initial French sounds /p/, /t/, and /k/.

   payer   pour   temps   taxi   quand   calme

2. Repeat the following words with the final French sounds /p/, /t/, and /k/.

   nappe   soupe   carte   contente   banque   fric

3. Now repeat the following sentences.

   Philippe a plein de fric à la banque.
   Tes parents vont payer avec une carte de crédit?

payer avec une
carte de crédit

Activités de communication orale

A **Toujours des excuses!**   Your friend (your partner) wants your help decorating the gym for a dance, a chore you detest. Each time he or she suggests a day and time, say you have to do something else then.

> Élève 1: Tu peux nous aider jeudi à cinq heures?
> Élève 2: Euh, non, je regrette. Jeudi je dois laver la voiture…

B **Où suis-je?**   Think of a place. A classmate has to try and guess which place you're thinking of by asking questions with y. Then reverse roles.

C **Il y en a combien?**   Your partner wants to know if there are a lot of the following at your school. Answer with *beaucoup, assez, quelques-un(e)s, très peu,* or *trop*. Then reverse roles.

> bons professeurs
> élèves sportifs ou sportives
> élèves brillant(e)s
> clubs intéressants
> cours intéressants
> examens difficiles

> Élève 1: À ton avis il y a beaucoup d'élèves amusants à l'école?
> Élève 2: À mon avis il y en a quelques-uns.

Prononciation

PRESENTATION (*page 493*)

A. Model the key expression *payer avec une carte de crédit* and have students repeat chorally.

B. Now model the other words and sentences in similar fashion.

C For additional practice, you may wish to use Pronunciation Transparency P-18, Cassette 10B/CD-11: *Prononciation* and the Student Tape Manual, Teacher's Edition, *Activités I–K*, page 207.

Bell Ringer Review

Write the following on the board or use BRR Blackline Master 18-8: You are being interviewed by the school newspaper. Answer the questions with the pronoun *en* and an expression of quantity if necessary.

Tu as…
> des frères?
> des sœurs?
> une voiture?
> des amis?
> du temps libre?
> des devoirs?

Activités de communication orale

ANSWERS

Activités A, B, and C
   Answers will vary.

DID YOU KNOW?

There are 100 *centimes* in one franc. The denominations of coins are: 5, 10, 20 *centimes,* and 1/2 franc, 1, 5, 10, and 20 francs. Most French coins have the profile or silhouette of Marianne, a young woman who symbolizes the French Republic.

LECTURE ET CULTURE

Bell Ringer Review

Write the following on the board or use BRR Blackline Master 18-9: It is New Year's Eve. Write five resolutions for the coming year using the verb *devoir.*

READING STRATEGIES
(*page 494*)

Pre-Reading

Take a survey to see which students receive a weekly allowance, which ones work part-time for spending money, what they use their money for, etc.

Reading

A. Have students read the *Lecture* silently. Allow five minutes. Encourage them to read for the main ideas and important details only, not to use dictionaries, and not to stop each time they have difficulty. Tell them they will have a chance to reread.

B. Call on volunteers to reread the *Lecture* aloud. Stop occasionally to ask questions.

Post-Reading

Call on a more able student to compare his/her money habits with those of Nathalie.

Note Students may listen to a recorded version of the *Lecture* on the CD-ROM.

Étude de mots

ANSWERS

Exercice A

1. une semaine
2. jeune
3. les parents
4. une façon
5. pas mal de
6. un chanteur
7. faire des économies

494

LA SEMAINE DES JEUNES FRANÇAIS

Qu'est-ce qu'une semaine? Une période de sept jours? Oui, mais une «semaine» peut être aussi quelque chose d'autre. La semaine peut être de l'argent. La semaine est la somme d'argent qu'un jeune Français ou une jeune Française reçoit de ses parents. C'est de l'argent de poche. Les jeunes Français reçoivent combien d'argent pour leur semaine? On ne peut pas répondre d'une façon générale[1] à cette question. Ça dépend d'abord de la générosité des parents et ensuite de la situation économique de la famille. Nathalie Cassis, par exemple, reçoit 50 francs par semaine de ses parents. Qu'est-ce qu'elle fait avec les 50 francs qu'elle reçoit? Nathalie achète de temps en temps un tee-shirt ou une cassette. Elle achète pas mal de[2] cassettes parce qu'elle aime beaucoup la musique. Elle achète aussi des billets pour les concerts de ses chanteurs favoris. De temps en temps elle va au café prendre un pot[3] avec des copains et bien sûr il faut payer.

Ses parents ont ouvert un compte d'épargne pour Nathalie. Elle aime faire des économies et mettre de l'argent de côté. Quand fait-elle des versements sur son compte? Si elle reçoit de l'argent pour son anniversaire, elle en dépense une partie, pas tout, et met le reste de côté. Quand elle reçoit une très bonne note, ses parents lui donnent aussi un peu d'argent. Souvent elle le dépense mais quelquefois elle le verse sur son compte d'épargne. Tu as une semaine? Tu reçois combien d'argent? Qu'est-ce que tu fais avec ta semaine? Tu fais les mêmes choses que Nathalie?

[1] d'une façon générale *in a general way*
[2] pas mal de *a lot of*
[3] prendre un pot *to have a drink (soda, tea, etc.)*

Étude de mots

A **Des définitions.** Trouvez le mot dans la lecture.

1. une période de sept jours
2. pas vieux
3. le père et la mère
4. une manière
5. beaucoup
6. une personne qui chante
7. ne pas dépenser trop d'argent

LEARNING FROM PHOTOS

Have students look at the photograph and say as much as they can about it.

CRITICAL THINKING ACTIVITY

(*Thinking skill: supporting statements with reasons*)

Put the following on the board or on a transparency:

1. À votre avis, il est important de mettre de l'argent de côté, de faire des économies? Pourquoi?
2. Il est important ou pas de devenir riche? Pourquoi?

Compréhension

B **Vrai ou faux?** Répondez par «oui» ou «non».

1. Tous les jeunes Français reçoivent de l'argent de leurs parents.
2. Tous les jeunes Français reçoivent la même somme d'argent.
3. Tous les jeunes Français ont un compte d'épargne.
4. Nathalie Cassis a un compte d'épargne.
5. Elle dépense tout son argent.

C **Vous avez compris?** Répondez d'après la lecture.

1. Combien d'argent de poche est-ce que les jeunes Français reçoivent de leurs parents? Ça dépend de quoi?
2. Qu'est-ce que Nathalie achète avec l'argent qu'elle reçoit?
3. Pourquoi achète-t-elle des cassettes?
4. Avec qui va-t-elle au café?
5. Qu'est-ce qu'elle y prend?
6. Qu'est-ce que les parents de Nathalie ont ouvert pour elle?
7. Quand Nathalie fait-elle des versements sur son compte?
8. Quelles sont les deux définitions du mot «semaine»?

DÉCOUVERTE CULTURELLE

La monnaie change d'un pays à l'autre. C'est le dollar aux États-Unis, mais pas en France. Chaque pays a sa monnaie nationale. Les monnaies étrangères s'appellent des «devises». La monnaie française est le franc français. À propos des francs, il y en a plusieurs: le franc belge, le franc suisse, le franc C.F.A. en Afrique et le franc antillais à la Martinique et à la Guadeloupe. Les devises n'ont pas toujours la même valeur. Il y a des fluctuations. Quelquefois le dollar est à dix francs français et quelquefois il tombe à cinq francs. Quand le dollar est à cinq francs,

tout est très cher pour les Américains en France. Et si le dollar est à dix francs, tout est bon marché pour eux. Pour toi, il vaut mieux aller en France quand le dollar est haut ou quand le dollar est bas? Quand est-ce que tu reçois le plus pour le dollar?

Des francs tahitiens

CHAPITRE 18 **495**

CRITICAL THINKING ACTIVITY

(Thinking skill: problem solving)

Vous êtes en France et vous voyez un tableau, par example, que vous aimez. Vous voulez l'acheter. Il coûte mille francs. Le jour où vous voulez acheter le tableau, le dollar est à 6F. Vous calculez le prix du tableau en dollars. Ça va faire combien? (Answer: $166.) Vous décidez d'acheter le tableau, mais vous ne voulez pas payer en espèces. Vous le payez avec votre carte de crédit. Quelques jours plus tard, le dollar baisse. Le franc monte. Quand vous payez, le dollar est à 5,5 francs. Combien coûte le tableau maintenant? (Answer: $181.) Vous avez gagné ou perdu de l'argent? Combien?

Compréhension (page 495)
ANSWERS

Exercice B
1. Non.   4. Oui.
2. Non.   5. Non.
3. Non.

Exercice C
1. La somme d'argent de poche que les jeunes Français reçoivent de leurs parents dépend de la générosité des parents et de la situation économique de la famille.
2. Nathalie achète des cassettes et des tee-shirts.
3. Elle achète des cassettes parce qu'elle aime beaucoup la musique.
4. Elle y va avec des copains.
5. Elle y prend un pot.
6. Ils ont ouvert un compte d'épargne pour elle.
7. Elle fait des versements sur son compte quand elle reçoit de l'argent.
8. «Une semaine» veut dire (signifie) une période de sept jours et la somme d'argent qu'on reçoit chaque semaine de ses parents.

OPTIONAL MATERIAL

Découverte culturelle
PRESENTATION *(page 495)*

A. Before reading the *Découverte* material, focus on the topic by sharing with students the list of exchange rates taken from a major newspaper. You may wish to photocopy the exchange rates and discuss them with the students.
B. Have students read the information silently.

Note Students may listen to a recorded version of the *Découverte culturelle* on the CD-ROM.

GEOGRAPHY CONNECTION

Have students find *la Belgique, la Suisse, l'Afrique, la Martinique, la Guadeloupe,* and *Tahiti* on the map on page 506.

RÉALITÉS

V oilà la Banque Populaire de Meximieux près de Lyon **1**. Est-ce que tu as un compte de chèque ou un compte d'épargne?

C'est la carte de crédit d'une banque française **2**.

Ce jeune homme veut ouvrir un compte d'épargne ou un compte de chèque dans cette banque **3**. L'employé remplit un formulaire.

Voici des chèques français **4**. Il y a des différences entre les chèques français et les chèques américains?

Bell Ringer Review

Write the following on the board or use BRR Blackline Master 18-10: Rewrite the following sentences, replacing the underlined phrases with the correct pronoun.

1. Nous sommes <u>à la banque</u>.
2. Nous parlons <u>à l'employé</u>.
3. L'employé donne de l'argent <u>à mes parents</u>.
4. Nous retournons <u>à la voiture</u>.

OPTIONAL MATERIAL

PRESENTATION
(*pages 496–497*)

The object of this section is to have students enjoy the photographs. However, if you would like to do more with it, you may wish to do the following activities.

A. Ask students what they know about current economic changes taking place in Europe, such as the formation of the European Community.

B. Ask volunteers to read and discuss the captions and photographs on pages 496–497.

Note In the CD-ROM version, students can listen to the recorded captions and discover a hidden video behind one of the photos.

CRITICAL THINKING ACTIVITY

(*Thinking skill: supporting statements with reasons*)
    Discutez: Comment préférez-vous payer —en espèces, avec une carte de crédit, par chèque ou par chèque de voyage? Expliquez pourquoi.

PAIRED ACTIVITY

Have students work in pairs to make up the conversation between the customer and the *caissier* at the exchange office in the illustration of Communication Transparency C-18.

BANQUE POPULAIRE

497

LEARNING FROM PHOTOS

You may wish to ask the following questions about the photos:

1. Qu'est-ce qu'une Carte bleue?
2. Quel sont les noms de deux banques françaises?
3. Regardez la photo (n° 3) de l'intérieur de la banque. Comment dit-on «*new accounts*»?

ADDITIONAL PRACTICE

1. Student Tape Manual, Teacher's Edition, *Deuxième Partie*, pages 208–210
2. Situation Cards, Chapter 18

RECYCLING

The *Activités de communication orale* and *écrite* allow students to apply the vocabulary and grammar of the chapter to open-ended, real-life situations and to re-use the vocabulary and structures from earlier chapters.

INFORMAL ASSESSMENT

Oral Activity A provides guided cues but still allows for free creation of sentences around the given topics. Oral Activity B allows students even freer rein in that they themselves come up with the topics they will discuss in their group. When using these activities for speaking assessment, you may wish to follow the evaluation criteria given on page 34 of this Teacher's Wraparound Edition.

Activités de communication orale

ANSWERS

Activités A and B
Answers will vary.

Activité de communication écrite

ANSWERS

Activité A
Answers will vary.

Activités de communication orale

A **Devinons.** Play this game in small groups. The items listed below are pieces of mail. Copy the names of the items on separate pieces of paper and fold them. Choose an item, and get your team members to guess what you've just received in the mail by giving them clues. You may not say the words on the paper. The first team to guess three items correctly wins.

une facture	une lettre d'amour
une carte postale	un permis de conduire
un billet d'avion	une carte d'anniversaire

Élève 1: Tiens! Mon oncle m'a donné vingt dollars!
Élève 2: Tu as reçu une carte d'anniversaire.

B **Rêves de voyage, voyages de rêve.** Work in small groups. Write down several places you've visited, then exchange papers with the other group members. Tell whether or not you have visited the places mentioned on the paper you've received. If you have visited a place, tell when and with whom. If you have not visited the place, tell whether or not you would like to.

Montréal

Ah, Montréal. Oui, j'y suis allé l'année dernière avec mes grands-parents. (Ah, non, je ne suis jamais allé à Montréal, mais je voudrais y aller.)

Activité de communication écrite

A **Es-tu comme la cigale ou la fourmi?** Are you careless with your money like the grasshopper or careful with it like the ant? Take the test and see what it reveals about you.

TEST

1. **Pour avoir de l'argent de poche...**
   a. je ne fais rien. Mes parents me donnent de l'argent.
   b. je travaille dans un magasin, dans un restaurant, etc.

2. **Quand je vois quelque chose que j'aime beaucoup...**
   a. je l'achète impulsivement.
   b. je réfléchis avant de l'acheter.

3. **Quand je reçois de l'argent comme cadeau...**
   a. je le dépense tout de suite.
   b. j'en mets de côté.

4. **Quand je veux faire ou acheter quelque chose de spécial...**
   a. j'emprunte de l'argent à mes amis ou à mes parents.
   b. je mets de l'argent de côté à l'avance.

5. **Quand j'emprunte de l'argent à mes copains...**
   a. j'oublie souvent de les rembourser.
   b. je les rembourse tout de suite.

6. **Quand un ami a besoin d'argent...**
   a. je ne peux pas l'aider parce que j'ai déjà dépensé tout mon argent.
   b. je peux lui prêter de l'argent parce que j'en ai mis de côté.

SCORE:

Une majorité de *a*: Tu es une vraie cigale! Tu aimes beaucoup t'amuser dans la vie. Tu dois peut-être essayer de penser un peu plus au futur.

Une majorité de *b*: Tu es une petite fourmi, responsable et toujours bien organisé(e). Tu es sûr(e) de t'amuser assez dans la vie?

498 CHAPITRE 18

FOR THE YOUNGER STUDENT

Set up a "bank" in the classroom, with students playing the role of tellers at a new accounts window and a currency exchange desk. Distribute play money, photocopied checks, travelers' checks, and any other forms you can get from a local bank. Have other students act as customers engaging in various bank transactions covered in this chapter.

INDEPENDENT PRACTICE

1. Activities and exercises, pages 498–499
2. Communication Activities Masters, pages 86–92
3. CD-ROM, Disc 4, pages 498–499

Réintroduction et recombinaison

A **À l'hôtel.** Répondez d'après les indications.

1. Où est-ce que Gilbert est allé? (à l'hôtel)
2. À qui a-t-il parlé? (au réceptionniste)
3. Il a demandé quel type de chambre? (pour une personne)
4. Qu'est-ce qu'il a rempli? (une fiche d'enregistrement)
5. À quel étage est la chambre? (au premier)
6. La chambre donne sur la rue ou sur la cour? (sur la cour)
7. Comment Gilbert est-il monté? (par l'escalier)
8. Qu'est-ce qu'il a monté? (ses bagages)

B **Le séjour de Robert.** Complétez avec «y» ou «en».

1. Robert est allé à l'hôtel?
   Oui, il ___ est allé.
2. Il est entré dans le hall?
   Oui, il ___ est entré.
3. Il est dans le hall maintenant?
   Oui, il ___ est.
4. Il va à la réception?
   Oui, il ___ va.
5. Il a des bagages?
   Oui, il ___ a.
6. Il a combien de valises?
   Il ___ a deux.
7. Robert monte dans sa chambre?
   Oui, il ___ monte.
8. Il reste une semaine à l'hôtel?
   Oui, il ___ reste une semaine.

L'Hôtel Carlton à Cannes

Vocabulaire

NOMS
l'argent de poche (m.)
l'argent liquide (m.)
le billet
la pièce
la monnaie
le chèque (bancaire)
le franc
la balle
le dollar
la banque
le compte d'épargne
le relevé de compte (d'épargne)
le bureau de change
le cours du change
la poche
le sac
le portefeuille
le porte-monnaie

VERBES
changer
emprunter
prêter
rembourser
signer
toucher
verser
devoir
recevoir
rendre

AUTRES MOTS ET EXPRESSIONS
avoir plein de fric
être fauché(e)
faire des économies
faire de la monnaie
mettre de l'argent de côté
assez
peu
plusieurs
quelques-un(e)s
trop

Réintroduction et recombinaison

RECYCLING

Exercise A recycles Chapter 17's hotel vocabulary and asks students to answer questions in the *passé composé* with both *avoir* and *être*. Exercise B has students distinguish between expressions that are replaced by *y* and those that are replaced by *en*.

ANSWERS

Exercice A
1. Gilbert est allé à l'hôtel.
2. Il a parlé au réceptionniste.
3. Il a demandé une chambre pour une personne.
4. Il a rempli une fiche d'enregistrement.
5. Elle est au premier étage.
6. La chambre donne sur la cour.
7. Il est monté par l'escalier.
8. Il a monté ses bagages.

Exercice B

1. y	5. en
2. y	6. en
3. y	7. y
4. y	8. y

ASSESSMENT RESOURCES

1. Chapter Quizzes
2. Testing Program
3. Situation Cards
4. Communication Transparency C-18
5. Computer Software: Practice/Test Generator

VIDEO PROGRAM

INTRODUCTION   (57:38)

BIENVENUE À PARIS,   (58:05)
MEREDITH!

LEARNING FROM PHOTOS

You may wish to ask questions about the photo: *Où est l'hôtel? Cannes est sur la Côte d'Azur? À ton avis, c'est quelle catégorie d'hôtel?*

STUDENT PORTFOLIO

A written assignment which may be included in students' portfolios is the *Mon Autobiographie* section in the Workbook on page 186.

Note Students may create and save both oral and written work using the Electronic Portfolio feature on the CD-ROM.

RÉVISION

CHAPITRES 17–18

OPTIONAL MATERIAL

OVERVIEW

This section reviews key grammatical structures and vocabulary from Chapters 17 and 18. The structure topics were first presented on the following pages: *passé composé* with *être*: page 458; *passé composé* with *être* or *avoir*: page 462; *lui* and *leur*: page 464; the pronoun *y*: page 484; the pronoun *en*: page 487; *recevoir* and *devoir*: page 490.

REVIEW RESOURCES

1. Workbook, Chapters 17–18, pages 170–186
2. Videocassette/Videodisc, Unit 4
3. Video Activities Booklet, Unit 4: Chapters 17–18, pages 64–72
4. Computer Software
5. Testing Program, Chapters 17–18, pages 108–120
6. Performance Assessment
7. CD-ROM, Disc 4, *Révision:* Chapters 17–18, pages 500–503
8. CD-ROM, Disc 4, Self-Tests 17–18
9. CD-ROM, Disc 4, Game: *Le Labyrinthe*
10. Lesson Plans

Conversation

ANSWERS

Exercice A

1. Il s'appelle Michel Boudreau.
2. Il est français.
3. Il doit remplir une fiche.
4. Oui, il est déjà venu plusieurs fois à Montréal.
5. Il est allé au bureau de change où il a changé de l'argent.
6. Il a besoin de monnaie.
7. Il doit prendre un taxi.
8. Non, je ne crois pas qu'il habite au Canada.

500

RÉVISION

CHAPITRES 17–18

Conversation   *L'arrivée à l'hôtel*

M. BOUDREAU: Bonjour, Monsieur. Je m'appelle Michel Boudreau. J'ai réservé une chambre pour ce soir et demain.

L'EMPLOYÉ: Oui, Monsieur. Voilà. Une chambre avec salle de bains pour une personne. Vous êtes de quelle nationalité?

M. BOUDREAU: Je suis français.

L'EMPLOYÉ: Alors, si vous voulez bien remplir cette fiche, s'il vous plaît. *(Il lui donne la fiche.)* C'est votre premier voyage à Montréal?

M. BOUDREAU: Oh non. Je suis déjà venu plusieurs fois.

L'EMPLOYÉ: Si vous voulez changer de l'argent, il y a un bureau de change juste à côté.

M. BOUDREAU: Je sais. J'y suis allé avant de venir ici. Par contre, si vous avez la monnaie de 50 dollars canadiens… Je dois prendre un taxi…

L'EMPLOYÉ: Mais bien sûr, Monsieur.

A **À l'hôtel.** Répondez d'après la conversation.

1. Comment s'appelle le client?
2. Il est de quelle nationalité?
3. Qu'est-ce qu'il doit remplir?
4. M. Boudreau est déjà venu à Montréal?
5. Il est allé où avant d'arriver à l'hôtel? Qu'est-ce qu'il y a fait?
6. De quoi est-ce qu'il a besoin?
7. Pour quoi faire?
8. D'après vous, M. Boudreau habite au Canada?

Structure

Le passé composé avec *être*

1. Review the verbs that use *être* as a helping verb in the *passé composé*. Remember that they are mostly verbs of motion.

arriver	sortir	aller	devenir	tomber
partir	monter	venir	rentrer	naître
entrer	descendre	revenir	rester	mourir

2. Remember that the past participle of verbs conjugated with *être* agrees in gender (masculine or feminine) and in number (singular or plural) with the subject of the verb.

   Elle est arrivée.   **Nous sommes venus.**

A Un groupe de jeunes en visite à Paris.
Répondez d'après le modèle.

> Alain (aller au Louvre)
> *Alain est allé au Louvre.*

1. Caroline et Stéphanie (aller au Musée d'Orsay)
2. Olivier (monter sur la Grande Arche)
3. Bernadette (aller à Versailles)
4. Christian et Marc (descendre à pied du haut de la tour Eiffel)
5. Alain (rester tout l'après-midi au Louvre)

Le passé composé avec *être* ou *avoir*

Les Grandes Eaux du Château de Versailles

In the *passé composé*, the verbs *monter, descendre, sortir,* and *rentrer* take either *être* or *avoir*. They take *avoir* when they are followed by a direct object. Otherwise they take *être*.

> Il a monté ses bagages dans sa chambre.   Il est monté dans sa chambre.

B Le client et l'employée.   M. Delcour est un client de l'hôtel. Mlle Dubois travaille à l'hôtel. Dites qui a fait quoi. (Utilisez le passé composé.)

> monter dans sa chambre   descendre pour changer de l'argent
> descendre les bagages   sortir en ville
> monter le petit déjeuner   sortir sa carte de crédit

Les pronoms d'objet indirect *lui* et *leur*

You use the indirect object pronoun *lui* to replace *à* + a person and the indirect object pronoun *leur* to replace *à* + more than one person. Remember that in negative constructions, the pronoun cannot be separated from the verb by a negative word.

> Je parle *à mon père.*   Je *lui* parle.
> J'écris *à mes parents.*   Je *leur* écris.
> Tu parles souvent *à Marie?*   Je ne *lui* parle jamais!
> Elle va téléphoner *à ses amis?*   Non, elle ne va pas *leur* téléphoner.

C Personnellement.   Répondez en utilisant «lui» ou «leur».

1. Tu téléphones souvent à tes copains?
2. Tu vas téléphoner à un(e) ami(e) ce soir?
3. Tu aimes parler à tes amis?
4. Tes copains obéissent à leurs parents? Et toi?
5. Tu réponds à ton professeur quand il te pose une question?

Structure
Le passé composé avec être
ANSWERS
Exercice A
1. Caroline et Stéphanie sont allées au Musée d'Orsay.
2. Olivier est monté sur la Grande Arche.
3. Bernadette est allée à Versailles.
4. Christian et Marc sont descendus à pied du haut de la tour Eiffel.
5. Alain est resté tout l'après-midi au Louvre.

Le passé composé avec être ou avoir
ANSWERS
Exercice B
M. Delcour est monté dans sa chambre.
Mlle Dubois a descendu les bagages.
Mlle Dubois a monté le petit déjeuner.
M. Delcour est descendu pour changer de l'argent.
M. Delcour est sorti en ville.
M. Delcour a sorti sa carte de crédit.

Les pronoms d'objet indirect lui et leur
PRESENTATION   *(page 501)*
Write the examples on the board. Circle the indirect object nouns and draw a box around the indirect object pronouns. Draw a line from the noun to the corresponding pronoun.

ANSWERS
Exercice C
1. Oui (Non), je (ne) leur téléphone (pas) souvent.
2. Oui (Non), je (ne) vais (pas) lui téléphoner ce soir.
3. Oui (Non), j'aime (je n'aime pas) leur parler.
4. Oui (Non), ils (ne) leur obéissent (pas). Oui (Non), moi, je (ne) leur obéis (pas).
5. Oui (Non), je (ne) lui réponds (pas).

501

PRESENTATION (page 502)

A. Go over steps 1–3 with students.
B. You may wish to have one student read the example sentences in the left-hand column and another student respond with the corresponding sentence with y.

Exercices

PRESENTATION (page 502)

Exercise E is more difficult than Exercise D because it combines the indirect object pronouns with the pronoun y.

ANSWERS

Exercice D

1. Oui (Non), nous (n') y sommes (pas) allés.
2. Oui (Non), nous (n') y allons souvent (pas souvent/jamais).
3. Oui (Non), nous (n') y sommes (pas) montés.
4. Oui (Non), nous (n') y sommes (pas) entrés.
5. Oui (Non), nous (n') y sommes (pas) descendus.
6. Oui (Non), nous (n') y rentron (pas) bientôt.

Exercice E

1. Ils ne leur écrivent jamais.
2. Marie-France y répond.
3. Le professeur leur pose des questions.
4. Les élèves vont y répondre.
5. Vous ne lui obéissez pas toujours.
6. Gilles et Lisa lui disent «Joyeux anniversaire».
7. Michel lui a offert un cadeau.
8. Tu n'y as pas réussi.
9. Carole et Luc n'y ont pas changé 500 francs.
10. Nous y sommes souvent tombés.

Le pronom en

PRESENTATION (page 502)

Go over the explanation with students. Have them repeat the example sentences.

Le pronom y

1. The pronoun y replaces any expression of location introduced by *à* or another preposition (*sur, en, dans, chez, en haut de, en bas de,* etc.).

Tu vas *à Versailles?*	Oui, j'y vais.
Tu es allé *en haut de la tour Eiffel?*	Oui, j'y suis allé.

2. Remember that y can also replace *à* + a thing, not referring to a place.

Je vais répondre *à sa lettre.*	Je vais y répondre.
Elle ne fait pas attention *aux autres voitures.*	Elle n'y fait pas attention.

3. In the *passé composé,* y comes before the helping verb.

Il est entré dans l'hôtel?	Oui, il y est entré.
Elle a vu son nom sur la liste?	Non, elle n'y a pas vu son nom.

D **La visite de Paris continue.** Répondez en utilisant «y».

1. Vous êtes allés à Paris?
2. Vous allez souvent en Europe?
3. Vous êtes montés en haut de la tour Eiffel?
4. Vous êtes entrés dans Notre-Dame?
5. Vous êtes descendus dans les Catacombes?
6. Vous rentrez bientôt aux États-Unis?

E **Y, lui ou leur?** Remplacez les mots en italique par «y», «lui» ou «leur».

1. Ils n'écrivent jamais *à leurs cousins.*
2. Marie-France répond *au téléphone.*
3. Le professeur pose des questions *aux élèves.*
4. Les élèves vont répondre *aux questions du professeur.*
5. Vous n'obéissez pas toujours *à votre mère.*
6. Gilles et Lisa disent «Joyeux anniversaire» *à Olivier.*
7. Michel a offert un cadeau *à Laurence.*
8. Tu n'as pas réussi *à l'examen.*
9. Carole et Luc n'ont pas changé 500 francs *au bureau de change.*
10. Nous sommes souvent tombés *sur la piste noire.*

Le pronom *en*

Review the object pronoun en. It replaces *de (du, de l', de la, des)* + a thing.

Tu as *de l'argent?*	Oui, j'en ai.
Ils ont offert *des boissons?*	Non, ils n'en ont pas offert.

F On fait un pique-nique. Répondez d'après le modèle.

> Je voudrais du pain. (apporter)
> *Qui en apporte?*

1. Je voudrais du coca. (acheter)
2. Je voudrais des sandwichs. (préparer)
3. Je voudrais de la citronnade. (faire)
4. Je voudrais des chips. (apporter)
5. Je voudrais de la limonade. (acheter)
6. Je voudrais de l'orangeade. (faire)

Les verbes *devoir* et *recevoir*

1. Review the forms of these two irregular verbs.

	DEVOIR	RECEVOIR
PRÉSENT	je dois tu dois il/elle/on doit nous devons vous devez ils/elles doivent	je reçois tu reçois il/elle/on reçoit nous recevons vous recevez ils/elles reçoivent
PARTICIPE PASSÉ	dû	reçu

2. Remember that *devoir* means "must" or "ought to" as well as "to owe."

G Questions d'argent. Complétez.

1. Si tu ___ (recevoir) un chèque demain, n'oublie pas de lui rendre l'argent que tu lui ___. (devoir)
2. Vous me ___ (devoir) encore 100 francs.
3. Nous ne voulons pas lui demander de l'argent; nous lui ___ (devoir) déjà 1.000 francs.
4. Vous ___ (recevoir) mon chèque la semaine dernière?
5. Ils ___ (recevoir) de l'argent de leurs parents toutes les semaines.
6. Je ___ (ne… jamais recevoir) votre chèque!

Activité de communication orale

A Au syndicat d'initiative. Working with a partner, make up a conversation between a student looking for an inexpensive hotel in Paris and an agent in the *syndicat d'initiative*.

ANSWERS

Exercice F

1. Qui en achète?
2. Qui en prépare?
3. Qui en fait?
4. Qui en apporte?
5. Qui en achète?
6. Qui en fait?

Les verbes devoir et recevoir

PRESENTATION *(page 503)*

A. Have students read the verbs forms aloud.
B. Go over step 2 with them.

ANSWERS

Exercice G

1. reçois, dois
2. devez
3. devons
4. avez reçu
5. reçoivent
6. n'ai jamais reçu

Activité de communication orale

PRESENTATION *(page 503)*

Extension of *Activité A*

After students have completed the activity, call on volunteers to present their conversations to the class.

ANSWERS

Activité A

Answers will vary.

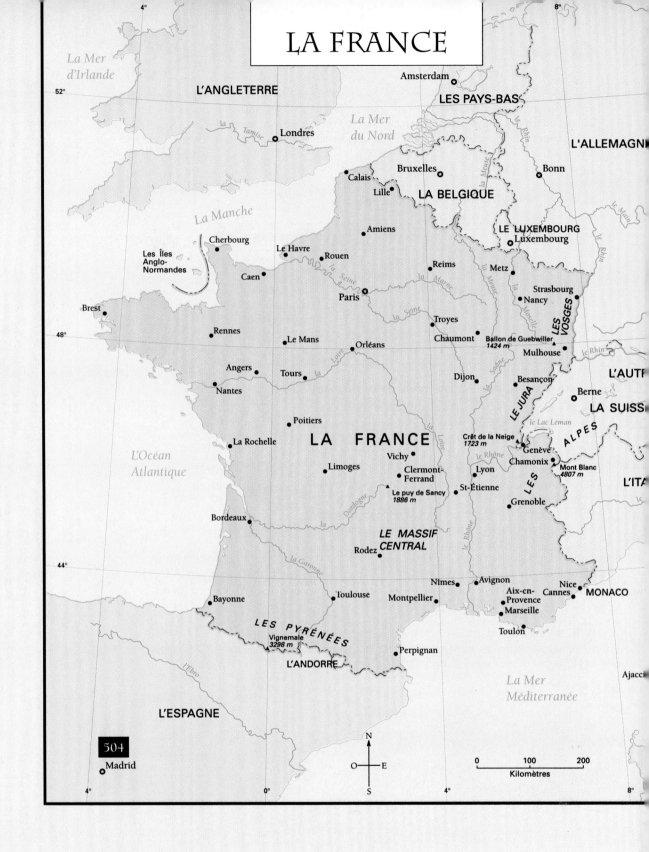

LA FRANCE

La Mer d'Irlande

L'ANGLETERRE

La Mer du Nord

Amsterdam

LES PAYS-BAS

L'ALLEMAGNE

la Tamise

Londres

Bruxelles

Bonn

Calais

Lille

LA BELGIQUE

La Manche

Amiens

LE LUXEMBOURG

Luxembourg

Cherbourg

Le Havre

Rouen

Reims

Metz

Les Îles Anglo-Normandes

Caen

la Seine

la Marne

la Meuse

le Rhin

Strasbourg

Nancy

Brest

Paris

la Seine

Troyes

LES VOSGES

Rennes

Le Mans

Orléans

Chaumont

Ballon de Guebwiller 1424 m

Mulhouse

le Rhin

Angers

Tours

la Loire

Dijon

la Saône

Besançon

L'AUTRICHE

Nantes

LE JURA

Berne

LA SUISSE

Poitiers

LA FRANCE

la Loire

le Lac Léman

LES ALPES

La Rochelle

Crêt de la Neige 1723 m

Genève

L'Océan Atlantique

Vichy

le Rhône

Chamonix

Mont Blanc 4807 m

Limoges

Clermont-Ferrand

Lyon

L'ITALIE

Le puy de Sancy 1886 m

St-Étienne

la Dordogne

Grenoble

Bordeaux

la Garonne

LE MASSIF CENTRAL

Rodez

le Rhône

Nîmes

Avignon

Nice

La Garonne

Toulouse

Montpellier

Aix-en-Provence

Cannes

MONACO

Bayonne

Marseille

LES PYRÉNÉES

Toulon

Vignemale 3298 m

Perpignan

l'Ebro

L'ANDORRE

La Mer Méditerranée

Ajaccio

L'ESPAGNE

504

N

O E

S

Madrid

0   100   200
Kilomètres

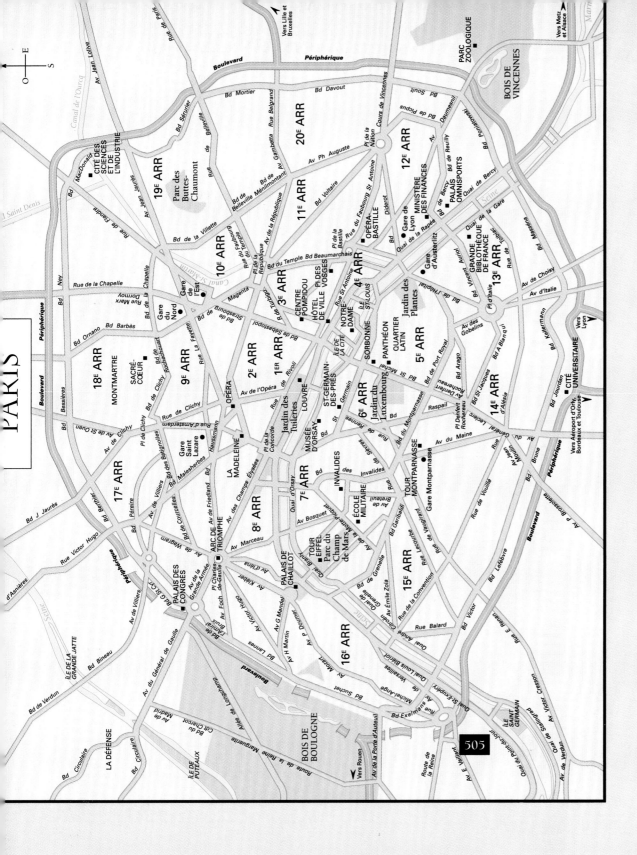

PARIS

17E ARR
18E ARR MONTMARTRE
19E ARR Parc des Buttes-Chaumont
20E ARR
16E ARR
9E ARR
10E ARR
11E ARR
12E ARR
8E ARR
1ER ARR
2E ARR
3E ARR
4E ARR
5E ARR
6E ARR
7E ARR
13E ARR
14E ARR
15E ARR

CITÉ DES SCIENCES ET DE L'INDUSTRIE
SACRÉ-CŒUR
Gare du Nord
Gare de l'Est
Gare Saint Lazare
Gare de Lyon
CENTRE POMPIDOU
HÔTEL DE VILLE
NOTRE-DAME
ÎLE DE LA CITÉ
ÎLE ST-LOUIS
PANTHÉON
QUARTIER LATIN
SORBONNE
ST-GERMAIN-DES-PRÉS
MUSÉE D'ORSAY
Jardin des Tuileries
Jardin du Luxembourg
Jardin des Plantes
LA MADELEINE
OPÉRA
OPÉRA-BASTILLE
PL DES VOSGES
PALAIS DES CONGRÈS
ARC DE TRIOMPHE
INVALIDES
TOUR EIFFEL
Parc du Champ de Mars
ÉCOLE MILITAIRE
PALAIS DE CHAILLOT
TOUR MONTPARNASSE
Gare Montparnasse
CITÉ UNIVERSITAIRE
MINISTÈRE DES FINANCES
PALAIS OMNISPORTS
GRANDE BIBLIOTHÈQUE DE FRANCE
Gare d'Austerlitz

PARC ZOOLOGIQUE
BOIS DE VINCENNES
BOIS DE BOULOGNE
LA DÉFENSE
ÎLE DE LA GRANDE JATTE
ÎLE DE PUTEAUX
ÎLE SAINT GERMAIN

Vers Lille et Bruxelles
Vers Metz et Alsace
Vers Lyon
Vers Aéroport d'Orly
Vers Bordeaux et Toulouse
Vers Rouen

505

LE MONDE FRANCOPHONE

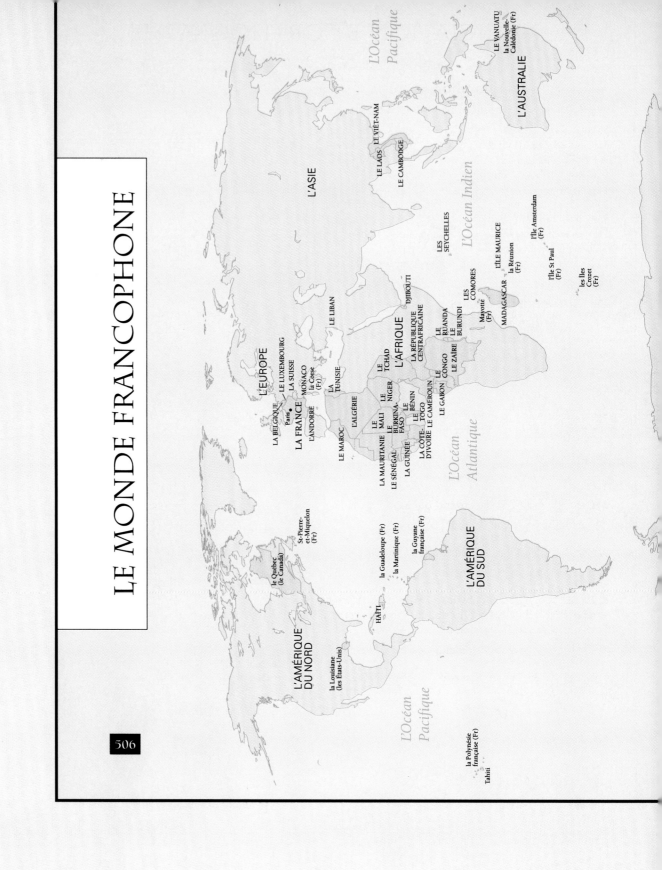

L'AMÉRIQUE
DU NORD

la Louisiane
(les États-Unis)

St-Pierre-
et-Miquelon
(Fr)

le Québec
(le Canada)

la Guadeloupe (Fr)
la Martinique (Fr)

HAÏTI

la Guyane
française (Fr)

L'AMÉRIQUE
DU SUD

L'Océan
Atlantique

L'Océan
Pacifique

la Polynésie
française (Fr)
Tahiti

L'EUROPE

LA BELGIQUE
LE LUXEMBOURG
LA SUISSE

Paris
LA FRANCE
MONACO
la Corse
(Fr)
ANDORRE

LE MAROC

L'ALGÉRIE

LA
TUNISIE

LE LIBAN

L'ASIE

LE LAOS LE VIÊT-NAM
LE CAMBODGE

LA MAURITANIE
LE SÉNÉGAL
LA GUINÉE
LE
MALI
LE
NIGER
LE BURKINA-
FASO
LA CÔTE-
D'IVOIRE
LE
TOGO
LE BÉNIN
LE CAMEROUN
LE GABON

LE
TCHAD

L'AFRIQUE
LA RÉPUBLIQUE
CENTRAFRICAINE

LE
CONGO
LE ZAÏRE

DJIBOUTI

LE
RUANDA
LE
BURUNDI

LES
COMORES

Mayotte
(Fr)

MADAGASCAR

L'ÎLE MAURICE
la Réunion
(Fr)

LES
SEYCHELLES

L'Océan Indien

l'Île Amsterdam
(Fr)

l'Île St Paul
(Fr)

les Îles
Crozet
(Fr)

L'AUSTRALIE

LE VANUATU
la Nouvelle-
Calédonie (Fr)

L'Océan
Pacifique

L'Océan
Atlantique

506

PRONONCIATION ET ORTHOGRAPHE

I. La transcription phonétique

The following are phonetic symbols used in this book.

[a]	la, là, avec	[ã]	dans, encore, temps
[é]	télé, chez, dîner, les	[õ]	non, regardons
[è]	elle, êtes, frère	[ẽ]	fin, demain
[i]	qui, il, lycée, dîne	[œ̃]	un
[ü]	tu, une		
[u]	vous, où, bonjour	[y]	fille, travailler
[ó]	au, beaucoup, allô		
[ò]	homme, alors	[sh]	chez, Michel
[œ]	deux, veut	[zh]	je, âge
[œ]	heure, sœur	[g]	garder, goûter, Guy

II. L'alphabet français

a b c d e f g h i j k l m n o p q r s t u v w x y z
Voyelles: a e i (y) o u
Consonnes: b c d f g h j k l m n p q r s t v w x z

III. Les accents

There are five written accent marks on French letters. These accents are part of the spelling of the word and cannot be omitted.

1. *L'accent aigu* (´) occurs over the letter *e*.

   le téléphone   élémentaire

2. *L'accent grave* (`) occurs over the letters *a, e,* and *u*.

   voilà   frère   où

3. *L'accent circonflexe* (^) occurs over all vowels.

   le château   la fenêtre   le dîner   l'hôtel   août

4. *La cédille* (ç) appears only under the letter *c*. When the letter *c* is followed by an *a, o,* or *u* it has a hard /k/ sound as in *ca*ve, *co*ca, *cu*lmination. The cedilla changes the hard /k/ sound to a soft /s/ sound.

   ça   garçon   commençons   reçu

5. *Le tréma* (¨) indicates that two vowels next to each other are pronounced separately.

   Noël   égoïste

VERBES

A. Verbes réguliers

INFINITIF	**parler** *to speak*	**finir** *to finish*	**répondre** *to answer*
PRÉSENT	je parle tu parles il parle nous parlons vous parlez ils parlent	je finis tu finis il finit nous finissons vous finissez ils finissent	je réponds tu réponds il répond nous répondons vous répondez ils répondent
IMPÉRATIF	parle parlons parlez	finis finissons finissez	réponds répondons répondez
PASSÉ COMPOSÉ	j'ai parlé tu as parlé il a parlé nous avons parlé vous avez parlé ils ont parlé	j'ai fini tu as fini il a fini nous avons fini vous avez fini ils ont fini	j'ai répondu tu as répondu il a répondu nous avons répondu vous avez répondu ils ont répondu

B. Verbes avec changements d'orthographe
(Verbs with spelling changes)

INFINITIF	**acheter**[1] *to buy*	**appeler** *to call*	**commencer** *to begin*
PRÉSENT	j'achète tu achètes il achète nous achetons vous achetez ils achètent	j'appelle tu appelles il appelle nous appelons vous appelez ils appellent	je commence tu commences il commence nous commençons vous commencez ils commencent
INFINITIF	**manger**[2] *to eat*	**payer**[3] *to pay*	**préférer**[4] *to prefer*
PRÉSENT	je mange tu manges il mange nous mangeons vous mangez ils mangent	je paie tu paies il paie nous payons vous payez ils paient	je préfère tu préfères il préfère nous préférons vous préférez ils préfèrent

[1] Verbes similaires: *se lever, se promener*
[2] Verbes similaires: *nager, voyager*
[3] Verbes similaires: *essayer, renvoyer, employer, envoyer*
[4] Verbes similaires: *célébrer, espérer, suggérer*

C. Verbes irréguliers

INFINITIF	**aller** *to go*	**avoir** *to have*	**conduire** *to drive*
PRÉSENT	je vais tu vas il va nous allons vous allez ils vont	j'ai tu as il a nous avons vous avez ils ont	je conduis tu conduis il conduit nous conduisons vous conduisez ils conduisent
PASSÉ COMPOSÉ	je suis allé(e)	j'ai eu	j'ai conduit
INFINITIF	**connaître** *to know*	**croire** *to believe*	**devoir** *to have to, to owe*
PRÉSENT	je connais tu connais il connaît nous connaissons vous connaissez ils connaissent	je crois tu crois il croit nous croyons vous croyez ils croient	je dois tu dois il doit nous devons vous devez ils doivent
PASSÉ COMPOSÉ	j'ai connu	j'ai cru	j'ai dû
INFINITIF	**dire** *to say*	**dormir** *to sleep*	**écrire** *to write*
PRÉSENT	je dis tu dis il dit nous disons vous dites ils disent	je dors tu dors il dort nous dormons vous dormez ils dorment	j'écris tu écris il écrit nous écrivons vous écrivez ils écrivent
PASSÉ COMPOSÉ	j'ai dit	j'ai dormi	j'ai écrit
INFINITIF	**être** *to be*	**faire** *to do, to make*	**lire** *to read*
PRÉSENT	je suis tu es il est nous sommes vous êtes ils sont	je fais tu fais il fait nous faisons vous faites ils font	je lis tu lis il lit nous lisons vous lisez ils lisent
PASSÉ COMPOSÉ	j'ai été	j'ai fait	j'ai lu

INFINITIF	**mettre** *to put*	**ouvrir**[5] *to open*	**partir** *to leave*
PRÉSENT	je mets tu mets il met nous mettons vous mettez ils mettent	j'ouvre tu ouvres il ouvre nous ouvrons vous ouvrez ils ouvrent	je pars tu pars il part nous partons vous partez ils partent
PASSÉ COMPOSÉ	j'ai mis	j'ai ouvert	je suis parti(e)
INFINITIF	**pouvoir** *to be able to*	**prendre**[6] *to take*	**recevoir** *to receive*
PRÉSENT	je peux tu peux il peut nous pouvons vous pouvez ils peuvent	je prends tu prends il prend nous prenons vous prenez ils prennent	je reçois tu reçois il reçoit nous recevons vous recevez ils reçoivent
PASSÉ COMPOSÉ	j'ai pu	j'ai pris	j'ai reçu
INFINITIF	**savoir** *to know*	**servir** *to serve*	**sortir** *to go out*
PRÉSENT	je sais tu sais il sait nous savons vous savez ils savent	je sers tu sers il sert nous servons vous servez ils servent	je sors tu sors il sort nous sortons vous sortez ils sortent
PASSÉ COMPOSÉ	j'ai su	j'ai servi	je suis sorti(e)
INFINITIF	**venir**[7] *to come*	**voir** *to see*	**vouloir** *to want*
PRÉSENT	je viens tu viens il vient nous venons vous venez ils viennent	je vois tu vois il voit nous voyons vous voyez ils voient	je veux tu veux il veut nous voulons vous voulez ils veulent
PASSÉ COMPOSÉ	je suis venu(e)	j'ai vu	j'ai voulu

[5] Verbes similaires: *couvrir, découvrir, offrir, souffrir*
[6] Verbes similaires: *apprendre, comprendre*
[7] Verbes similaires: *devenir, revenir*

D. Verbes avec *être* au passé composé

aller *(to go)*	je suis allé(e)
arriver *(to arrive)*	je suis arrivé(e)
descendre *(to go down, to get off)*	je suis descendu(e)
entrer *(to enter)*	je suis entré(e)
monter *(to go up)*	je suis monté(e)
mourir *(to die)*	je suis mort(e)
naître *(to be born)*	je suis né(e)
partir *(to leave)*	je suis parti(e)
passer *(to go by)*	je suis passé(e)
rentrer *(to go home)*	je suis rentré(e)
rester *(to stay)*	je suis resté(e)
retourner *(to return)*	je suis retourné(e)
revenir *(to come back)*	je suis revenu(e)
sortir *(to go out)*	je suis sorti(e)
tomber *(to fall)*	je suis tombé(e)
venir *(to come)*	je suis venu(e)

VOCABULAIRE FRANÇAIS–ANGLAIS

The *Vocabulaire français–anglais* contains all productive and receptive vocabulary from the text.

The numbers following each productive entry indicate the chapter and vocabulary section in which the word is introduced. For example, 2.2 means that the word first appeared in *Chapitre 2, Mots 2*. BV refers to the introductory *Bienvenue* chapter.

The following abbreviations are used in this glossary.

abbrev.	abbreviation
adj.	adjective
adv.	adverb
conj.	conjunction
dem. adj.	demonstrative adjective
dem. pron.	demonstrative pronoun
dir. obj.	direct object
f.	feminine
fam.	familiar
form.	formal
ind. obj.	indirect object
inf.	infinitive
inform.	informal
inv.	invariable
m.	masculine
n.	noun
pl.	plural
poss. adj.	possessive adjective
prep.	preposition
pron.	pronoun
sing.	singular
subj.	subject

A

à at, in, to, **3.1**
  à l'avance in advance
  à bord de on board, **7.2**
  à côté next door
  à côté de next to, **5**
  À demain. See you tomorrow., **BV**
  à demi-tarif half-price
  à destination de to (plane, train, etc.), **7.1**
  à domicile to the home
  à droite de to, on the right of, **5**
  à l'étranger abroad, in a foreign country
  à gauche de to, on the left of, **5**
  à l'heure on time, **8.1**; an (per) hour (speed)
  à l'intérieur inside
  à mi-temps part-time, **3.2**
  à la mode in style, "in"
  à mon (ton, son, etc.) avis in my (your, his, etc.) opinion, **10.2**
  à l'origine originally
  à partir de from … on; based on
  à peu près about, approximately
  à pied on foot, **5.2**
  à plein temps full-time, **3.2**
  à point medium-rare (meat), **5.2**
  à propos de concerning, as regards
  à quelle heure? at what time?, **2**
  À tout à l'heure. See you later., **BV**
l' abricot (m.) apricot
absolument absolutely
absorber to absorb
accélérer to speed up, go faster, **12.1**
accepter to accept
l' accessoire (m.) accessory
l' accident (m.) accident, **14.2**
accompagné(e) (de) accompanied (by)
accueilli(e): bien accueilli(e) well-received
l' achat (m.) purchase
  faire des achats to shop, **10.1**

acheter to buy, **6.1**
l' acidité (f.) acidity
l' acte (m.) act, **16.1**
l' acteur (m.) actor (m.), **16.1**
actif, active active, **10**
l' action (f.) action
l' activité (f.) activity
l' actrice (f.) actress, **16.1**
l' addition (f.) check, bill (restaurant), **5.2**
admirer to admire
l' adolescent(e) adolescent, teenager
adopter to adopt
adorable adorable
adorer to love, **3.2**
l' adresse (f.) address
l' adulte (m. et f.) adult
adverse opposing, **13.1**
aérien(ne) air, flight (adj.), **9**
  les tarifs aériens airfares
l' aérogare (f.) terminal with bus to airport, **7.2**
l' aéroport (m.) airport, **7.1**
aérospatial(e) aerospace
les affaires (f. pl.) business; belongings
  l'homme (m.) d'affaires businessman
affolé(e) panic-stricken
s' affronter to collide
africain(e) African
l' âge (m.) age, **4.1**
  Tu as quel âge? How old are you? (fam.), **4.1**
âgé(e) old
l' agenda (m.) appointment book, **2.2**
l' agent (m.) agent (m. and f.), **7.1**
  l'agent de police police officer (m. and f.)
l' agglomération (f.) populated area
agité(e) agitated
agréable pleasant
l' agriculteur (m.) farmer (m. and f.)
aider to help
aimable nice (person), **1.2**
aimer to like, love, **3.2**
l' air (m.) air
  en plein air outdoor(s)
ajouter to add
l' algèbre (f.) algebra, **2.2**
l' aliment (m.) food

alimentaire: le régime
    alimentaire diet
l' alimentation (f.) nutrition, diet
l' Algérie (f.) Algeria
l' Allemagne (f.) Germany, 16
l' allemand (m.) German
    (language)
    allemand(e) German
    aller to go, 5.1
      aller à la pêche to go fishing,
      9.1
      aller pêcher to go fishing
l' allergie (f.) allergy, 15.1
    allergique allergic, 15.1
l' aller-retour (m.) round-trip
    ticket, 8.1
l' aller simple (m.) one-way
    ticket, 8.1
    alors so, then, well then
les algues (f. pl.) algae
les Alpes (f. pl.) the Alps
l' alpinisme (m.) mountain
    climbing
l' altitude (f.) altitude
l' amateur (m.): l'amateur d'art
    art lover
    aménager to renovate,
    transform
l' Américain(e) American
    (person)
    américain(e) American, 1.1
l' Amérique (f.) America, 16
    l'Amérique (f.) du Nord
      North America, 16
    l'Amérique (f.) du Sud
      South America, 16
l' ami(e) friend, 1.2
l' amitié (f.) friendship
    ample large, full
    amusant(e) funny, 1.1
s' amuser to have fun, 11.2
l' an (m.): avoir... ans to be ...
    years old, 4.1
l' ananas (m.) pineapple
l' anatomie (f.) anatomy
    ancien(ne) old, ancient; former
l' angine (f.) throat infection,
    tonsillitis, 15.1
l' anglais (m.) English (language),
    2.2
l' Anglais(e) Englishman
    (woman)
l' Angleterre (f.) England, 16
l' animal (m.) animal
    animé(e) lively, animated
l' année (f.) year, 4.1
    l'année dernière last year, 13
l' anniversaire (m.) birthday, 4.1
    Bon (Joyeux) anniversaire!
      Happy birthday!

    C'est quand, ton
      anniversaire? When is
      your birthday? (fam.), 4.1
l' annonce (f.) announcement,
    8.1
    la petite annonce classified
      ad
    annoncer to announce, 8.1
l' anorak (m.) ski jacket, 14.1
    antérieur(e) previous, former
l' anthropologie (f.) anthropology
l' antibiotique (m.) antibiotic,
    15.1
l' anticyclone (m.) high pressure
    area
    antillais(e) West Indian
    antipathique unpleasant
    (person), 1.2
l' Antiquité (f.) ancient times
    anxieux, anxieuse anxious
    août (m.) August, 4.1
    apparenté: le mot apparenté
      cognate
l' appartement (m.) apartment,
    4.2
    appeler to call
s' appeler to be called, be named,
    11.1
    applaudir to applaud
    apporter to bring
    apprendre (à) to learn (to), 9.1
      apprendre à quelqu'un à
        faire quelque chose to
        teach someone to do
        something, 14.1
l' apprenti(e) apprentice
    appuyer sur le bouton to push
      the button
    après after, 3.2
      d'après according to
l' après-midi (m.) afternoon, 2
l' arbitre (m.) referee, 13.1
l' arbre (m.) tree
l' arche (f.) arch
l' archipel (m.) archipelago
l' architecte (m. et f.) architect
l' architecture (f.) architecture
l' argent (m.) money, 3.2
    l'argent liquide cash, 18.1
    l'argent de poche allowance,
      small change
l' Argentine (f.) Argentina, 16
l' argot (m.) slang
l' aristocrate (m. et f.) aristocrat
l' arme (f.) weapon
l' armée (f.) army
s' arrêter to stop, 12.1
l' arrivée (f.) arrival, 7.2
    la ligne d'arrivée finish line
    le tableau des départs et des

      arrivés arrival and
      departure board
    arriver to arrive, 3.1; to happen
l' arrondissement (m.) district
    (in Paris)
l' art (m.) art, 2.2
l' article (m.) article
    les articles de luxe luxury
      items
    les articles de sport sporting
      goods
l' artiste peintre (m. et f.) painter
    artistique artistic
l' ascenseur (m.) elevator, 4.2
l' asepsie (f.): pratiquer l'asepsie
    to sterilize, disinfect
l' Asie (f.) Asia, 16
    aspiré(e) pulled in
l' aspirine (f.) aspirin, 15.1
    assez fairly, quite; enough
      assez de (+ nom) enough
      (+ noun), 18
l' assiette (f.) plate, 5.2
    ne pas être dans son assiette
      to be feeling out of sorts,
      15.1
    assis(e) seated, 8.2
l' assistant(e) assistant
l' association (f.) association
    associer to associate
l' assurance (f.) insurance
l' astronome (m. et f.)
    astronomer
l' atmosphère (f.) atmosphere
    attendre to wait (for), 8.1
l' attente (f.): la salle d'attente
    waiting room, 8.1
l' attention (f.) attention
    Attention! Careful! Watch
      out!
    faire attention to pay
      attention, 6; to be careful,
      9.1
    atterrir to land, 7.1
l' atterrissage (m.) landing
    (plane)
    attirer to attract
    attraper un coup de soleil to
      get a sunburn, 9.1
    au at the, to the, in the, on the
    (sing.), 5
    au bord de la mer by the
      ocean; seaside, 9.1
    au contraire on the contrary
    au debut at the beginning
    au-dessus (de) above
    au-dessous (de) below
    au fond de at the bottom of
    au moins at least
    au revoir good-bye, BV

au sujet de about

l' **auberge (f.) de jeunesse** youth hostel

audacieux, audacieuse audacious, bold

au-dessous: la taille au-dessous the next smaller size, **10.2**

au-dessus: la taille au-dessus the next larger size, **10.2**

augmenter to increase

aujourd'hui today, **2.2**

ausculter to listen with a stethoscope, **15.2**

aussi also, too, **1.1**; as (comparisons), **10**

l' **Australie (f.)** Australia, **16**

l' **auteur (m.)** author (m. and f.)

l' **autocar (m.)** bus, coach, **7.2**

l' **autodidacte (m. et f.)** self-taught person

l' **auto-école (f.)** driving school, **12.2**

automatique: le distributeur automatique de billets automated teller machine (ATM)

l' **automne (m.)** autumn, **13.2**

l' **autoroute (f.)** highway

l'autoroute à péage toll highway, **12.2**

autour de around

autre other, **BV**

Autre chose? Anything else? (shopping), **6.2**

aux at the, to the, in the, on the (pl.), **5**

l' **avance (f.): à l'avance** in advance

en avance early, ahead of time, **8.1**

avancé(e) advanced

avant before, **7.1**

avant de (+ inf.) before (+ verb)

avant-hier the day before yesterday, **13**

avec with, **5.1**

Avec ça? What else? (shopping), **6.2**

l' **aventure (f.)** adventure

l' **avion (m.)** plane, **7.1**

en avion by plane; plane (adj.), **7.1**

l' **avis (m.)** opinion

à mon avis in my opinion, **10.2**

avoir to have, **4.1**

avoir... ans to be ... years old, **4.1**

avoir besoin de to need, **11.1**

avoir de la chance to be lucky

avoir faim to be hungry, **5.1**

avoir une faim de loup to be very hungry

avoir lieu to take place

avoir mal à to have a(n) ... -ache, to hurt, **15.2**

avoir l'occasion de (+ inf.) to have the opportunity (+ inf.)

avoir raison to be right

avoir soif to be thirsty, **5.1**

avoir tendance à (+ inf.) to tend (+ inf.)

avril (m.) April, **4.1**

B

le **baccalauréat** French high school exam

le **bacon** bacon

bactérien(ne) bacterial, **15.1**

les **bagages (m. pl.)** luggage, **7.1**

les bagages à main carry-on luggage, **7.1**

la **baguette** loaf of French bread, **6.1**

le **bain** bath, **11.1**

prendre un bain to take a bath, **11.1**

le bain de soleil: prendre un bain de soleil to sunbathe, **9.1**

le **balcon** balcony, **4.2**

la **balle** ball (tennis, etc.), **9.2**; franc (slang), **18.2**

le **ballon** ball (soccer, etc.), **13.1**

la **banane** banana, **6.2**

la **bande dessinée** comic strip

la **banlieue** suburbs

la **banque** bank, **18.1**

le **banquier, la banquière** banker

baptiser to christen

Barcelone Barcelona, **16**

bas(se) low, **10**

à talons bas low-heeled (shoes), **10**

la **base: de base** basic

le **base-ball** baseball, **13.2**

le **basket(-ball)** basketball, **13.2**

le **bateau** boat

le **bâtiment** building

le **bâton** ski pole, **14.1**

battre des tambours to beat drums

bavarder to chat, **4.2**

beau (bel) beautiful (m.), handsome, **4**

Il fait beau. It's nice weather., **9.2**

beaucoup a lot, **3.1**

beaucoup de a lot of, many, **10.1**

la **beauté** beauty

les **Beaux-Arts (m. pl.)** fine arts

beige (inv.) beige, **10.2**

le/la **Belge** Belgian (person)

belge Belgian

la **Belgique** Belgium

belle beautiful (f.), **4**

le **béribéri** beriberi

le **besoin** need

avoir besoin de to need, **11.1**

la **bêtise** stupid thing, nonsense

le **beurre** butter, **6.2**

le **bicentenaire** bicentennial

bien fine, well, **BV**

bien accueilli(e) well-received

bien cuit(e) well-done (meat), **5.2**

bien élevé(e) well-mannered

bien sûr of course

bientôt soon

Bienvenue! Welcome!

la **bière** beer

le **bijou** jewel

le **billet** bill (currency), **18.1**; ticket, **7.1**

le billet aller-retour round-trip ticket, **8.1**

la **biologie** biology, **2.2**

le/la **biologiste** biologist

bizarre strange, odd

la **blague: Sans blague!** No kidding!

blanc, blanche white, **10.2**

bleu(e) blue, **10.2**

bleu marine (inv.) navy blue, **10.2**

blond(e) blond, **1.1**

bloquer to block

le **blouson** jacket, **10.1**

le **bœuf** beef, **6.1**

la **boisson** beverage, **5.1**

la **boîte de conserve** can of food, **6.2**

bon(ne) correct; good, **9**

bon marché (inv.) inexpensive, **10.1**

Bonjour. Hello., **BV**

le **bonnet** ski cap, hat, **14.1**

le bonnet de bain bathing cap

le **bord: à bord de** aboard (plane, etc.), **7.2**

au bord de la mer by the ocean, seaside, **9.1**

bordé(e) (de) bordered, lined (with)

le **bordereau** receipt
la **bosse** mogul (ski), **14.2**
la **botanique** botany
la **botte** boot
la **bouche** mouth, **15.1**
la **boucherie** butcher shop, **6.1**
le **bouchon** traffic jam
bouger to move
le **bouillon de poulet** chicken
    soup
la **boulangerie-pâtisserie** bakery,
    6.1
la **boule de neige** snowball, **14.2**
le/la **bourgeois(e)** burgher;
    townsperson
bout (inf. bouillir) boils (verb)
la **bouteille** bottle, **6.2**
la **boutique** shop, boutique
le **bouton** button; bud
la **brasse papillon** butterfly (swim
    stroke)
Bravo! Good! Well done!
le **break** station wagon, **12.1**
le **Brésil** Brazil, **16**
la **Bretagne** Brittany
breton(ne) Breton, from
    Brittany
la **brioche** sweet roll
bronzé(e) tan
bronzer to tan, **9.1**
se **brosser** to brush, **11.1**
    se **brosser les dents (f. pl.)**
        to brush one's teeth, **11.1**
le **bruit** noise
brun(e) brunette, **1.1**; brown,
    10.2
le **bulletin** card; report
    le **bulletin de notes** report
        card
    le **bulletin météorologique**
        weather report
le **bureau** desk, **BV**; office, bureau
    le **bureau de change** foreign
        exchange office (for foreign
        currency), **18.1**
le **bus: en bus** by bus, **5**
le **but** goal, **13.1**
    marquer un but to score a
        goal, **13.1**

C
ça that (dem. pron.), **BV**
    Ça coûte cher. It's (That's)
        expensive.
    Ça fait combien? How much
        is it (that)?, **6.2**
    Ça fait... francs. It's (That's)
        ... francs., **6.2**
    Ça fait mal. It (That) hurts.,
        15.2

Ça va. Fine., O.K., **BV**
Ça va? How's it going?, How
    are you? (inform.), **BV**
la **cabine** cabin (plane), **7.1**
le **cabinet** office (doctor's)
le **cadeau** gift, present, **10.2**
le **café** café; coffee, **5.1**
    le **café au lait** coffee with
        milk
le **cahier** notebook, **BV**
la **caisse** cash register, checkout
    counter, **6.2**
le **caissier,** la **caissière** cashier
le **calcium** calcium
le **calcul** calculation
la **calculatrice** calculator, **BV**
calculer to calculate
calme quiet, calm
calmer: Calmez-vous. Calm
    down.
la **calorie** calorie
le/la **camarade** companion, friend
    le/la **camarade de classe**
        classmate
le **camp** side (in a sport or game),
    13.1
    le **camp adverse** opponents,
        other side, **13.1**
la **campagne** country(side)
    la **maison de campagne**
        country house
le **Canada** Canada, **16**
canadien(ne) Canadian, **9**
le/la **candidat(e)** candidate
le **canoë** canoe
la **cantine** school restaurant
la **capitale** capital
le **car** bus (coach)
le **caractère: à caractère familial**
    family-style
la **caractéristique** characteristic
le **carnet** small book
la **carotte** carrot, **6.2**
carré(e) square
les **carreaux (m. pl.)** tiles
le **carrefour** intersection, **12.2**
la **carrière** career
la **carte** menu, **5.1**; map
    la **carte d'anniversaire**
        birthday card
    la **carte de crédit** credit card,
        17.2
    la **carte de débarquement**
        landing card, **7.2**
    la **carte d'embarquement**
        boarding pass, **7.1**
    la **carte postale** postcard
le **cas** case
    en cas d'urgence in an
        emergency

en tout cas in any case
le **casque** helmet
le **casse-cou** daredevil
la **cassette** cassette, **3.2**
la **catégorie** category
la **cathédrale** cathedral
le **cauchemar** nightmare
la **cause** cause
    causer to cause
    ce (cet) (m.) this, that (dem.
        adj.), **8**
        ce que c'est what it is
        Ce n'est rien. You're
            welcome., **BV**
la **ceinture de sécurité** seat belt,
    12.2
célèbre famous, **1.2**
célibataire single, unmarried
cellulaire cellular
la **cellule** cell
    la **cellule nerveuse** nerve cell
cent hundred, **5.2**
les **centaines (f. pl.)** hundreds
le **centre** center, middle
    au centre de in the heart of
    le **centre commercial**
        shopping center
les **céréales (f. pl.)** cereal, grains
la **cérémonie** ceremony
la **cerise** cherry
certainement certainly
certains: pour certains for
    some people
ces (m. et f. pl.) these, those
    (dem. adj.), **8**
c'est it is, it's, **BV**
    c'est-à-dire that is, **1.16**
    C'est ça. That's right.
    C'est combien? How much is
        it?, **BV**
    **C'est quand, ton
        anniversaire?** When is
        your birthday? (fam.), **4.1**
    C'est quel jour? What day is
        it?, **2.2**
    C'est tout? Is that all?, **6.2**
cette (f.) this, that (dem. adj.), **8**
chacun(e) each (one)
la **chaîne** TV channel
    la **chaîne hôtelière** hotel
        chain
la **chaise** chair, **BV**
le **chalet** chalet
la **chambre** room (in a hotel), **17.1**
    la **chambre à coucher**
        bedroom, **4.2**
    la **chambre à deux lits**
        double room, **17.1**
    la **chambre à un lit** single
        room, **17.1**

libérer la chambre to vacate the room, **17.2**

le **champ** field

le champ de manœuvres parade ground

le/la **champion(ne)** champion

le **championnat** championship

la **chance** luck

avoir de la chance to be lucky

changer (de) to change, **8.2**; to exchange, **18.1**

chanter to sing, **3.2**

le **chanteur**, la **chanteuse** singer

chaque each, every, **16.1**

la **charcuterie** deli(catessen), **6.1**

charger to put in charge

le **chariot** shopping cart

charmant(e) charming

le **chat** cat, **4.1**

avoir un chat dans la gorge to have a frog in one's throat, **15.2**

le **château** castle, mansion

chaud(e) warm, hot

Il fait chaud. It's hot. (weather), **9.2**

chauffer to heat

les **chaussettes (f. pl.)** socks, **10.1**

les **chaussons (m. pl.)** slippers

les **chaussures (f. pl.)** shoes, **10.1**

les chaussures de ski ski boots, **14.1**

les chaussures de tennis sneakers, tennis shoes, **9.2**

le **chef** head, boss

la **cheminée** chimney

la **chemise** shirt, **10.1**

le **chemisier** blouse, **10.1**

le **chèque (bancaire)** check, **18.1**

le chèque de voyage traveler's check, **17.2**

le compte de chèque checking account

cher, chère dear; expensive, **10**

Ça coûte cher. It's (That's) expensive.

chercher to look for, seek, **5.1**

le **cheval (pl. les chevaux)** horse

les **cheveux (m. pl.)** hair, **11.1**

chez at the home (business) of, **5**

chez soi home

chic (inv.) chic, stylish

le **chien** dog, **4.1**

le **chiffre** number

le **Chili** Chile, **16**

la **chimie** chemistry, **2.2**

chimique chemical

le/la **chimiste** chemist

la **Chine** China, **16**

chinois(e) Chinese

le **chirurgien** surgeon (m. and f.)

le **chocolat: au chocolat** chocolate (adj.), **5.1**

choisir to choose, **7.1**

le **choix** choice

le **choléra** cholera

le **cholestérol** cholesterol

la **chose** thing

Chouette! Great! (inform.), **2.2**

la **chute: faire une chute** to fall, **14.2**

ciao good-bye (inform.), **BV**

ci-dessus above

le **ciel** sky, **14.2**

la **cigale** grasshopper, cicada

le **cinéma** movie theater, movies, **16.1**

le/la **cinéphile** movie buff

cinq five, **BV**

cinquante fifty, **BV**

le **cintre** hanger, **17.2**

la **circulation** traffic, **12.2**; circulation

la circulation à double sens two-way traffic

citer to cite, mention

le **citron pressé** lemonade, **5.1**

le/la **civilisé(e)** civilized person

la **classe** class (people), **2.1**; class (course)

en classe économique in coach class (plane)

le **classement** classification

classer to classify

la **clé** key, **12.1**

le/la **client(e)** customer, **10.1**

le **climat** climate

les **clous (m. pl.)** pedestrian crossing, **12.2**

le **club** club

le club d'art dramatique drama club

le club de forme health club, **11.2**

le **coca** Coca-Cola, **5.1**

le **cœur** heart

le **coffre** trunk (of car)

le **coin: du coin** neighborhood (adj.)

le **collaborateur**, la **collaboratrice** co-worker, associate

le **collant** pantyhose, **10.1**

le **collège** junior high, middle school

la **colonie de vacances** summer camp

combattre to combat, fight

combien (de) how much, how many, **6.2**

Ça fait combien? How much is it (that)?, **6.2**

C'est combien? How much is it (that)?, **BV**

comble packed (stadium), **13.1**

la **comédie** comedy, **16.1**

la comédie musicale musical comedy, **16.1**

comique funny, **1.2**

commander to order, **5.1**

comme like, as; for

Et comme dessert? What would you like for dessert?

le **commencement** beginning

commencer to begin

comment how, what, **1.2**

Comment est... ? What is ... like? (description), **1.1**

Comment t'appelles-tu? What's your name? (fam.), **11.1**

Comment vas-tu? How are you? (fam.), **BV**

Comment vous appelez-vous? What's your name? (form.), **11.1**

commun(e) common

en commun in common

la **communauté** community

le **compact disc** compact disc, **3.2**

la **compagnie aérienne** airline, **7.1**

le **compartiment** compartment, **7.2**

le **complet** suit (man's), **10.1**

complet, complète full, complete

compléter to complete

le **comportement** behavior

composer to compose

composter to stamp, validate (a ticket), **8.1**

comprendre to understand, **9.1**

le **comprimé** pill, **15.2**

compris(e) included (in the bill)

Le service est compris. The tip is included., **5.2**

le **compte d'épargne** savings account, **18.1**

le **comptoir** counter, **7.1**

le/la **concierge** concierge, caretaker

le **concours** competition, contest

le **conducteur**, la **conductrice** driver, **12.1**

conduire to drive, **12.2**

la **conduite: les leçons (f.) de conduite** driving lessons, **12.2**

confiant(e) confident, **1.1**

le **confort** comfort

confortable comfortable

la **connaissance: faire la connaissance de** to meet
connaître to know, **16.2**
connu(e) known
la **conquête** conquest
conservateur, conservatrice conservative
conserver to conserve
la **conserve: la boîte de conserve** can of food, **6.2**
la **consigne** checkroom, **8.1**
   la **consigne automatique** locker, **8.1**
consommer to consume
construit(e) built
la **consultation** medical visit
le **contact: mettre le contact** to start (a car), **12.1**
contaminer to contaminate
le **conte** tale
contenir to contain
content(e) happy, **1.1**
continu(e) continual, ongoing
continuer to continue
la **contractuelle** meter maid, **12.2**
le **contraire** opposite
   au contraire on the contrary
la **contravention** traffic ticket, **12.2**
   contre against, **13.1**
   par contre on the other hand, however
le **contrôle de sécurité** security (airport), **7.1**
   passer par le contrôle de sécurité to go through security (airport), **7**
le **contrôleur** conductor (train), **8.2**
convenable correct
la **conversation** conversation
la **coopération** cooperation
le **copain** friend, pal (m.), **2.1**
la **copine** friend, pal (f.), **2.1**
la **coqueluche** whooping cough
le **corps** body
correspondre to correspond
corriger to correct
le **costume** costume, **16.1**
la **côte** coast
   la **Côte d'Azur** French Riviera
   la **Côte-d'Ivoire** Ivory Coast, **16**
le **côté** side
   côté couloir aisle (seat) (adj.), **7.1**
   côté fenêtre window (seat) (adj.), **7.1**
coucher to put (someone) to bed, **11**

se **coucher** to go to bed, **11.1**
la **couchette** bunk (on a train), **8.2**
la **couleur** color, **10.2**
   De quelle couleur est... ? What color is … ?, **10.2**
les **coulisses: dans les coulisses** backstage
le **couloir** aisle, corridor, **8.2**
le **coup: donner un coup (de pied, de tête, etc.)** to kick, hit (with one's foot, head, etc.)
la **coupe** winner's cup, **13.2**
la **cour** courtyard, **4.2**; court
courageux, courageuse courageous, brave
le **coureur** runner, **13.2**
   le **coureur cycliste** racing cyclist, **13.2**
couronné(e) crowned
le **courrier** mail service
le **cours** course, class, **2.2**
   le **cours du change** exchange rate, **18.1**
la **course** race, **13.2**
   la **course cycliste** bicycle race
les **courses (f. pl.): faire les courses** to go grocery shopping, **6.1**
court(e) short, **10.2**
le **court de tennis** tennis court, **9.2**
le/la **cousin(e)** cousin, **4.1**
le **couteau** knife, **5.2**
   coûter to cost
   Ça coûte cher. It's (That's) expensive.
la **coutume** custom
le **couturier** designer (of clothes), **10.1**
couvert: Le ciel est couvert. The sky is overcast., **14.2**
le **couvert** table setting, **5.2**
   mettre le couvert to set the table, **8**
la **couverture** blanket, **17.2**
couvrir to cover, **15**
le **crabe** crab, **6.1**
la **craie: le morceau de craie** piece of chalk, **BV**
la **cravate** tie, **10.1**
le **crayon** pencil, **BV**
la **crèche** day-care center
créer to create
la **crème** cream, **6.1**
   la **crème solaire** suntan lotion, **9.1**
le **crème** coffee with cream (in a café), **5.1**
la **crémerie** dairy store, **6.1**

le **créole** creole (language)
la **crêpe** crepe, pancake, **5.1**
la **crêperie** crepe restaurant
crevé(e) exhausted
la **crevette** shrimp, **6.1**
crier to shout
la **crise** crisis
la **critique** criticism
le/la **critique** critic
   critiquer to criticize
   croire to believe, think, **10.2**
le **croisement** intersection, **12.2**
la **croissance** growth
le **croissant** croissant, crescent roll, **6.1**
le **croque-monsieur** grilled ham and cheese sandwich, **5.1**
croustillant(e) crusty
la **croyance** belief
le **cube** cube
le **cubisme** Cubism
la **cuillère** spoon, **5.2**
la **cuisine** kitchen, **4.2**; cuisine (food)
   faire la cuisine to cook, **6**
cuit(e): bien cuit(e) well-done (meat), **5.2**
la **culture** culture
culturel(le) cultural
la **cure** cure
le **cycle: le cycle de l'eau** water cycle
le **cyclisme** cycling, bicycle riding, **13.2**
le/la **cycliste** cyclist, bicycle rider

D

d'abord first, **11.1**
d'accord O.K., **3**
   être d'accord to agree, **2.1**
la **dame** lady
le **danger: en danger** in danger
dangereux, dangereuse dangerous
dans in, **BV**
la **danse** dance
danser to dance, **3.2**
la **danseuse** dancer, ballerina
d'après according to
la **date: Quelle est la date aujourd'hui?** What is today's date?, **4.1**
de from, **1.1**; of, belonging to, **5**
   de bonne heure early
   de côté aside, **17.2**
   de loin by far
   de nos jours today, nowadays
   de plus en plus more and more

De quelle couleur est... ?
What color is … ?, **10.2**
de rêve dream (adj.)
De rien. You're welcome.
(inform.), **BV**
de temps en temps from
time to time, occasionally
le **débarquement** landing,
deplaning
débarquer to get off (plane), **7.2**
déborder to overflow
debout standing, **8.2**
le **début** beginning
au début at the beginning
le/la **débutant(e)** beginner, **14.1**
le **décalage horaire** time
difference
la **décapotable** convertible (car),
12.1
décembre (m.) December, **4.1**
le **déchet** waste
décider (de) to decide (to)
déclarer to declare, call
décoller to take off (plane), **7.1**
le **décor** set (for a play), **16.1**
le **décorateur (de porcelaine)**
painter (of china)
la **découverte** discovery
découvrir to discover, **15**
décrire to describe
dédié(e) dedicated
défense de doubler no passing
(traffic sign)
définir to define
la **définition** definition
le **degré** degree, **14.2**
Il fait... degrés (Celsius).
It's … degrees (Celsius).,
14.2
dehors outside
en dehors de outside (of)
déjà already, **14**
déjeuner to eat lunch, **5.2**
le **déjeuner** lunch
délicieux, délicieuse delicious,
10
le **delta** delta
demain tomorrow, **2.2**
À demain. See you
tomorrow., **BV**
demander to ask (for)
se **demander** to wonder
demi(e) half
et demie half past (time)
le **demi-cercle** semi-circle; top of
the key (on a basketball
court), **13.2**
le **demi-kilo** half a kilo, 500 grams
le **demi-tarif: à demi-tarif** half-
price

la **dent** tooth, **11.1**
avoir mal aux dents to have
a toothache, **15**
se brosser les dents to brush
one's teeth, **11.1**
le **dentifrice** toothpaste, **11.1**
le **déodorant** deodorant, **11.1**
le **départ** departure, **7.1**
le **département d'outre-mer**
French overseas department
dépendre (de) to depend (on)
dépenser to spend (money),
10.1
la **dépression** low-pressure area
(weather)
depuis since, for, **8.2**
dériver to derive
dernier, dernière last, **10**
derrière behind, **BV**
des some, any, **3**; **6**; of the, from
the (pl.), **4**
désagréable unpleasant, **1.2**
descendre to get off (train, bus,
etc.), **8.2**; to take down, **8**;
to go down, **14.1**
la **descente** descent; getting off
(bus, etc.)
le **désert** desert
se **déshabiller** to get undressed
désirer to want
Vous désirez? May I help
you? (store); What would
you like? (café, restaurant)
le **dessert** dessert
desservir to serve, fly to, etc.
(transportation)
le **dessin** illustration, drawing
le **dessin animé** cartoon,
16.1
la **dessinatrice** illustrator (f.)
dessous: au-dessous smaller
(size), **10.2**; below
dessus: au-dessus larger (size),
10.2; above
la **destruction** destruction
le **détergent** detergent
détester to hate, **3.2**
deux two, **BV**
les deux roues (f. pl.) two-
wheeled vehicles
tous (toutes) les deux both
deuxième second, **4.2**
**la Deuxième Guerre
mondiale** World War II
deuxièmement second of all,
secondly
devant in front of, **BV**
le **développement** development
devenir to become, **16**

la **devise** currency
le **devoir** homework (assignment),
BV
faire les devoirs to do
homework, **6**
devoir to owe, **18.2**; must, to
have to (+ verb), **18**
le **diagnostic: faire un diagnostic**
to diagnose, **15.2**
dicter to dictate
la **différence** difference
différent(e) different
difficile difficult, **2.1**
la **difficulté: être en difficulté** to
be in trouble
dimanche (m.) Sunday, **2.2**
le **dîner** dinner, **4.2**
dîner to eat dinner, **4.2**
la **diphtérie** diphtheria
diplômé(e): être diplômé(e)
to graduate
dire to say, tell, **12.2**
la **direction** direction
diriger to direct
discuter to discuss
disparaître to disappear
disponible available
le **disque** record, **3.2**
la **disquette** diskette (computer)
la **distance** distance
distingué(e) distinguished
le **distributeur automatique de
billets** automated teller
machine (ATM)
divisé(e) divided
le **divorce** divorce
dix ten, **BV**
dix-huit eighteen, **BV**
dix-neuf nineteen, **BV**
dix-sept seventeen, **BV**
le **docteur** doctor (title)
le **documentaire** documentary,
16.1
le **dollar** dollar, **3.2**
le **domaine** domain, field
le **domicile: à domicile** to the
home
donner to give, **3.2**
donner à manger à to feed
donner un coup de pied to
kick, **13.1**
donner une fête to throw a
party, **3.2**
donner sur to face, overlook,
17.1
doré(e) golden
dormir to sleep, **7.2**
le **dortoir** dormitory
le **dos** back (body)
la **douane** customs, **7.2**

passer à la douane to go through customs, **7.2**
doublé(e) dubbed (movies), **16.1**
la douche shower
  prendre une douche to take a shower, **11.1**
douloureux, douloureuse painful
le doute: sans aucun doute without a doubt
douter to doubt
la douzaine dozen, **6.2**
douze twelve, **BV**
le drame drama, **16.1**
le drap sheet, **17.2**
le drapeau flag
dribbler to dribble (basketball), **13.2**
droite: à droite de to, on the right of, **5**
du of the, from the (sing.), **5**; some, any, **6**
  du coin neighborhood (adj.)
  pas du tout not at all
la durée length (of time)
durer to last

E

l' eau (f.) water
  l'eau minérale mineral water, **6.2**
l' échange (m.) exchange
s' échapper to escape
l' écharpe (f.) scarf, **14.1**
l' école (f.) school, **1.2**
  l'école primaire elementary school
  l'école secondaire junior high, high school
l' écolier, l'écolière pupil, schoolchild
l' écologiste (m. et f.) ecologist
les économies (f. pl.): faire des économies to save money, **18.2**
économique economical
  en classe économique in coach class (plane), **7**
écouter to listen (to), **3.1**
l' écran (m.) screen, **7.1**
l' écrevisse (f.) crawfish
écrire to write, **12.2**
l' écrivain (m.) writer (m. and f.)
éducatif, éducative educational
l' éducation (f.): l'éducation civique social studies, **2.2**
  l'éducation physique physical education
efficace efficient

égaliser to tie (score)
l' électricité (f.) electricity
électrique electric
l' élément (m.) element
l' élève (m. et f.) student, **1.2**
élevé(e) high, **15**
  bien élevé(e) well brought-up
éliminer to eliminate
elle she, it, **1**; her (stress pron.), **9**
elles they (f.), **2**; them (stress pron.), **9**
l' embarquement (m.) boarding, leaving
embarquer to board (plane, etc.), **7.2**
l' embouteillage (m.) traffic jam
émigrer to emigrate
l' emploi (m.) du temps schedule
l' employé(e) employee (m. and f.)
emprunter to borrow, **18.2**
en of it, of them, etc., **18.2**; in; as
  en avance early, ahead of time, **8.1**
  en avion plane (adj.), by plane, **7.1**
  en baisse coming down (in value)
  en bas to, at the bottom
  en ce moment right now
  en classe in class
  en commun in common
  en dehors de outside (of); besides
  en effet in fact
  en exclusivité first-run (movie)
  en face de across from, opposite
  en fait in fact
  en fonction de in terms of, in accordance with
  en général in general
  en hausse going up (in value)
  en haut de on, to the top of
  en plein(e) (+ nom) right (in, on, etc.) (+ noun)
  en plein air outdoor(s)
  en plus de besides, in addition
  en première (seconde) in first (second) class, **8.1**
  en provenance de arriving from (flight, train), **7.1**
  en retard late, **8.2**
  en solde on sale, **10.2**
  en tout cas in any case

  en version originale original language version, **16.1**
  en ville in town, in the city
encore still (adv.); another; again
encourager to encourage
s' endormir to fall asleep, **11.1**
l' endroit (m.) place
l' énergie (f.) energy
énergique energetic, **1.2**
l' enfant (m. et f.) child (m. and f.), **4.1**
enfin finally
l' engrais (m.) fertilizer
énormément enormously
l' enquête (f.) survey, opinion poll
enragé(e) rabid, enraged
enrhumé(e): être enrhumé(e) to have a cold, **15.1**
l' enseignement (m.) teaching
l' ensemble (m.) body, collection
ensemble together, **5.1**
ensuite then (adv.), **11.1**
entendre to hear, **8.1**
l' enthousiasme (m.) enthusiasm
entier, entière entire, whole, **10**
l' entracte (m.) intermission, **16.1**
entraîner to carry along
entre between, among, **9.2**
l' entrée (f.) entrance, **4.2**; admission
entrer to enter, **3.1**
l' environnement (m.) environment
envoyer to send, **13.1**
l' épargne: le compte d'épargne savings account, **18.1**
épicé(e) spicy
l' épicerie (f.) grocery store, **6.1**
l' époque (f.) period, times
l' équilibre (m.) balance
équilibré(e) balanced
l' équipe (f.) team, **13.1**
l' équipement (m.) equipment
l' érable (m.) maple (tree)
  le sirop d'érable maple syrup
l' escalier (m.) staircase, **17.1**
l' espace (m.) space
l' Espagne (f.) Spain, **16**
l' espagnol (m.) Spanish (language), **2.2**
espagnol(e) Spanish
les espèces (f. pl.): payer en espèces to pay cash, **17.2**
l' espionnage (m.) spying
l' essence (f.) gas(oline), **12.1**
  (l'essence) ordinaire regular gas, **12.1**
  (l'essence) super sans plomb super unleaded gas, **12.1**

essentiel(le) essential
essentiellement essentially
l' est (m.) east
estimer to consider
l' estomac (m.) stomach
et and, 1
   et toi? and you? (fam.), BV
établir to establish
l' étage (m.) floor (of a building), 4.2
l' étal (m.) (market) stall
l' état (m.) state
   l'homme (m.) d'état diplomat, statesman
les États-Unis (m. pl.) United States, 13.2
l' été (m.) summer, 9.1
   en été in summer, 9.1
éternuer to sneeze, 15.1
étranger, étrangère foreign, 16.1
   à l'étranger abroad, in a foreign country
être to be, 2.1
   être à l'heure to be on time, 8.1
   être d'accord to agree, 2.1
   être en avance to be early, 8.1
   être en bonne (mauvaise) santé to be in good (poor) health, 15.1
   être en retard to be late, 8.2
   être enrhumé(e) to have a cold, 15.1
   être vite sur pied to be back on one's feet in no time, 15.2
   ne pas être dans son assiette to be feeling out of sorts, 15.2
l' être (m.) humain human being
étroit(e) tight (shoes), narrow, 10.2
l' étudiant(e) (university) student
étudier to study, 3.1
européen(ne) European, 9
eux them (m. pl. stress pron.), 9
s' évaporer to evaporate
éventuellement possibly
évoquer to evoke
l' examen (m.) test, exam, 3.1
   passer un examen to take a test, 3.1
   réussir à un examen to pass a test, 7
examiner to examine, 15.2
excellent(e) excellent
exceptionnel(le) exceptional
l' exemple (m.) example

par exemple for example
s' exercer to practice
l' expansion (f.) expansion
l' expédition (f.) expedition
expliquer to explain
l' explorateur (m.) explorer
explorer to explore
exposer to exhibit
l' exposition (f.) exhibit, show, 16.2
l' express (m.) espresso, black coffee, 5.1
s' exprimer to express oneself
expulser to expel, banish
exquis(e) exquisite
l' extérieur (m.) exterior, outside
extra terrific (inform.), 2.2
extraordinaire extraordinary
extrêmement extremely

F

fabriqué(e) made
fabriquer to make
fabuleux, fabuleuse fabulous
fâché(e) angry, 12.2
facile easy, 2.1
la façon way, manner
   d'une façon générale in a general way
le facteur factor
la facture bill (hotel, etc.), 17.2
facultatif, facultative elective
faire to do, make, 6.1
   faire du (+ nombre) to take size (+ number), 10.2
   faire des achats to shop, make purchases, 10.1
   faire de l'aérobic to do aerobics, 11.2
   faire l'annonce to announce, 8
   faire attention to pay attention, 6; to be careful, 9.1
   faire une chute to fall, take a fall, 14.2
   faire la connaissance de to meet
   faire les courses to do the grocery shopping, 6.1
   faire la cuisine to cook, 6
   faire les devoirs to do homework, 6
   faire un diagnostic to diagnose, 15.2
   faire des économies to save money, 18.2
   faire enregistrer to check (luggage), 7.1
   faire des études to study, 6

faire de l'exercice to exercise, 11.2
faire du français (des maths, etc.) to study French (math, etc.), 6
faire de la gymnastique to do gymnastics, 11.2
faire du jogging to jog, 11.2
faire le levé topographique to survey (land)
faire de la monnaie to make change, 18.1
faire de la natation to swim, go swimming
faire la navette to go back and forth
faire une ordonnance to write a prescription, 15.2
faire partie de to be a part of
faire du patin to skate, 14.2
faire du patin à glace to ice-skate, 14.2
faire du patin à roulettes to roller-skate
faire peur à to frighten
faire un pique-nique to have a picnic, 6
faire de la planche à voile to go windsurfing, 9.1
faire le plein to fill up (a gas tank), 12.1
faire de la plongée sous-marine to go deep-sea diving, 9.1
faire une promenade to take a walk, 9.1
faire la queue to wait in line, 8.1
faire un régime to go on a diet
faire du ski to ski, 14.1
faire du ski nautique to water-ski, 9.1
faire du sport to play sports
faire du surf to go surfing, 9.1
faire du surf des neiges to go snowboarding
faire sa toilette to wash and groom oneself, 11.1
faire les valises to pack (suitcases), 7.1
faire un voyage to take a trip, 7.1
le fait fact
fait(e) à la main handmade
la famille family, 4.1
   la famille à parent unique single-parent family
le/la fana fan

fantaisiste whimsical
fantastique fantastic, **1.2**
fatigué(e) tired
fauché(e) broke (slang), **18.2**
faut: il faut (+ nom) (noun) is (are) necessary
    il faut (+ inf.) one must, it is necessary to (+ verb), **9.1**
la faute mistake
faux, fausse false
favori(te) favorite, **10**
la femme woman, **2.1**; wife, **4.1**
    la femme médecin (woman) doctor
la fenêtre window
    côté fenêtre window (seat) (adj.), **7.1**
fermé(e) closed, **16.2**
la fertilité fertility
la fête party, **3.2**
    donner une fête to throw a party, **3.2**
    la Fête des Mères (Pères) Mother's (Father's) Day
le feu traffic light, **12.2**
    le feu orange yellow traffic light, **12.2**
    le feu rouge red traffic light, **12.2**
    le feu vert green traffic light, **12.2**
la feuille leaf
    la feuille de papier sheet of paper, **BV**
février (m.) February, **4.1**
la fiche d'enregistrement registration card (hotel), **17.1**
la fièvre fever, **15.1**
    la fièvre jaune yellow fever
    avoir une fièvre de cheval to have a high fever, **15.2**
la figure face, **11.1**
le filet net shopping bag, **6.1**; net (tennis, etc.), **9.2**; rack (train)
la fille girl, **BV**; daughter, **4.1**
le film film, movie, **16.1**
    le film d'amour love story, **16.1**
    le film d'aventures adventure movie, **16.1**
    le film étranger foreign film, **16.1**
    le film d'horreur horror film, **16.1**
    le film policier detective movie, **16.1**
    le film de science-fiction science-fiction movie, **16.1**
le fils son, **4.1**

fin(e) fine
    aux fines herbes with herbs, **5.1**
finalement finally
finir to finish, **7**
fixe: à prix fixe at a fixed price
flamand(e) Flemish
flambé(e) flaming
flâner to stroll
le fleuve river
flotter to float
la fluctuation fluctuation
le foie liver
    avoir mal au foie to have indigestion, **15**
la fois time (in a series)
le fonctionnement functioning
fonctionner to function, work
fond: au fond de at the bottom of
le fondateur, la fondatrice founder
fonder to found
la fontaine fountain
le foot(ball) soccer, **13.1**
    le football américain football
la force force, power
le forcing: faire le forcing to put pressure on
la forêt forest
le forfait-journée lift ticket (skiing)
la forme form, shape
    le club de forme health club, **11.2**
    être en forme to be in shape, **11.2**
    la forme (physique) physical fitness
    rester en forme to stay in shape, **11.2**
    se mettre en forme to get in shape, **11.2**
former to form; to train
le formulaire form, data sheet
la formule formula
le fort fort
fort(e) strong; good
fort (adv.) hard, **9.2**
fou, folle crazy
le foulard scarf
la foule: venir en foule to crowd (into)
la fourchette fork, **5.2**
la fourmi ant
les frais (m. pl.) expenses, charges, **17.2**
la fraise strawberry
le franc franc, **18.1**
le français French (language), **2.2**

le/la Français(e) Frenchman (woman)
    français(e) French, **1.1**
la France France, **16**
franchement frankly
francophone French-speaking
frapper to hit, **9.2**
freiner to brake, put on the brakes, **12.1**
fréquemment frequently
fréquent(e) frequent
fréquenter to frequent, patronize
le frère brother, **1.2**
le fric money, dough (slang), **18.2**
    avoir plein de fric to have lots of money (slang), **18.2**
les frissons (m. pl.) chills, **15.1**
les frites (f. pl.) French fries, **5.1**
froid(e) cold, **14.2**
    avoir froid to be cold
    Il fait froid. It's cold. (weather), **9.2**
le fromage cheese, **5.1**
le front front (weather)
la frontière border
le fruit fruit, **6.2**
    les fruits de mer seafood
fumer to smoke
fumeurs (adj. inv.) smoking (section), **7.1**
    non-fumeurs no smoking (section), **7.1**
furieux, furieuse furious
la fusée rocket
le futur future

G

le/la gagnant(e) winner, **13.2**
    gagner to earn, **3.2**; to win, **9.2**
la galaxie galaxy
le galet pebble
le Gange Ganges River
le gant glove, **14.1**
    le gant de toilette washcloth, **17.2**
le garage garage, **4.2**
le garçon boy, **BV**
garder to guard
le gardien de but goalie, **13.1**
la gare train station, **8.1**
    garer la voiture to park the car, **12.2**
gastronomique gastronomic, gourmet
le gâteau cake, **6.1**
gauche: à gauche de to, on the left of, **5**
le gaz gas
geler to freeze

Il gèle. It's freezing. (weather), **14.2**
le **gendarme** police officer
le **général** general, **7**
   général: en général in general
   généralement generally
   généraliser to generalize
   généraliste: le médecin généraliste general practitioner
   généreux, généreuse generous, **10**
la **générosité** generosity
le **genre** type, kind, **16.1**
les **gens (m. pl.)** people
   gentil(le) nice (person), **9**
la **géographie** geography, **2.2**
la **géométrie** geometry, **2.2**
   géométrique geometric
la **glace** ice cream, **5.1**; mirror, **11.1**; ice, **14.2**
   glisser to slip, slide
le **globe** globe
la **glucide** carbohydrate
le **golfe** gulf
la **gorge** throat, **15.1**
   avoir un chat dans la gorge to have a frog in one's throat, **15.2**
   avoir la gorge qui gratte to have a scratchy throat, **15.1**
   avoir mal à la gorge to have a sore throat, **15.1**
   gourmand(e) fond of eating
   goûter to taste
le **gouvernement** government
   grâce à thanks to
le **gradin** bleacher (stadium), **13.1**
la **graisse** fat
   la graisse animale animal fat
la **grammaire** grammar
le **gramme** gram, **6.2**
   grand(e) tall, big, **1.1**
   pas grand-chose not much
   le grand couturier clothing designer, **10.1**
   le grand magasin department store, **10.1**
   de grand standing luxury (adj.)
   la Grande-Bretagne Great Britain, **16**
   les Grands Lacs (m. pl.) the Great Lakes
   grandir to grow (up) (children)
la **grand-mère** grandmother, **4.1**
le **grand-père** grandfather, **4.1**
les **grands-parents (m. pl.)** grandparents, **4.1**
   grave serious

la **Grèce** Greece
la **griffe** label
le **grill-express** snack bar (train)
la **grippe** flu, **15.1**
   gris(e) gray, **10.2**
   grossir to gain weight, **11.2**
la **Guadeloupe** Guadeloupe
la **guerre: la Deuxième Guerre mondiale** World War II
le **guichet** ticket window, **8.1**; box office, **16.1**
le **guide** guidebook, **12.2**
le **gymnase** gym(nasium), **11.2**
la **gymnastique** gymnastics, **2.2**
   faire de la gymnastique to do gymnastics, **11.2**

H

   habillé(e) dressy, **10.1**
s' **habiller** to get dressed, **11.1**
l' **habitant(e)** resident
   habiter to live (in a city, house, etc.), **3.1**
le **hall** lobby, **17.1**
les **haricots (m. pl.) verts** green beans, **6.2**
   haut(e) high, **10.2**
   avoir... mètres de haut to be … meters high
   du haut de from the top of
   en haut de to, at the top of
   la haute couture high fashion
   à talons hauts high-heeled (shoes)
le **haut-parleur** loudspeaker, **8.1**
le **héros** hero
l' **heure (f.)** time (of day), **2**
   à quelle heure? at what time?, **2**
   À tout à l'heure. See you later., **BV**
   de bonne heure early
   être à l'heure to be on time, **8.1**
   Il est quelle heure? What time is it?, **2**
   heureux, heureuse happy, **10.2**
l' **hexagone (m.)** hexagon
   hier yesterday, **13.1**
   avant-hier the day before yesterday, **13**
   hier matin yesterday morning, **13**
   hier soir last night, **13**
l' **histoire (f.)** history, **2.2**
l' **hiver (m.)** winter, **14.1**
   en hiver in winter, **14.2**
le **H.L.M.** low-income housing
le **hockey** hockey

le **hockey sur glace** ice hockey
la **Hollande** Holland, the Netherlands, **16**
l' **homme (m.)** man, **2.1**
   l'homme d'affaires businessman
   l'homme d'état diplomat, statesman
les **honoraires (m. pl.)** fees (doctor)
l' **hôpital (m.)** hospital
l' **horaire (m.)** schedule, timetable, **8.1**
   hors des limites out of bounds, **9.2**
l' **hôtel (m.)** hotel, **17.1**
l' **hôtesse (f.) de l'air** flight attendant (f.), **7.2**
   huit eight, **BV**
   humain(e) human
   humide wet, humid
   humoristique humorous
l' **hydrate (m.) de carbone** carbohydrate
   hystérique hysterical

I

   idéal(e) ideal
l' **idée (f.)** idea
   identifier to identify
   il he, it, **1**
   Il est... heure(s). It's … o'clock., **2**
   Il est quelle heure? What time is it?, **2**
   il faut (+ nom) (noun) is (are) needed
   il faut (+ inf.) one must, it is necessary to (+ verb), **9.1**
   Il n'y a pas de quoi. You're welcome., **BV**
   il vaut mieux it is better
   il y a there is, there are, **4.2**
l' **île (f.)** island
   illustré(e) illustrated
   ils they (m.), **2**
l' **immeuble (m.)** apartment building, **4.2**
l' **immigration (f.)** immigration, **7.2**
   passer à l'immigration to go through immigration (airport), **7.2**
   impatient(e) impatient, **1.1**
   important(e) important
les **Impressionnistes (m. pl.)** Impressionists (painters)
   inauguré(e) inaugurated
   inclure to include

inconnu(e) unknown
incroyable incredible
l' **Inde (f.)** India
l' **indication (f.)** cue
indiquer to indicate
industrialisé(e) industrialized
l' **industrie (f.)** industry
infectieux, infectieuse
    infectious
l' **infection (f.)** infection, **15.1**
infiltrer to seep (into)
influencer to influence
l' **informatique (f.)** computer
    science, **2.2**
l' **inondation (f.)** flood
s' **installer** to settle (down), move
    in
l' **institut (m.)** institute
l' **institution (f.)** institution
les **instructions (f. pl.)**
    instructions, **9.1**
l' **instrument (m.)** instrument
intelligent(e) intelligent, **1.1**
interdit(e) forbidden,
    prohibited
    Il est interdit de stationner.
        No parking., **12.2**
intéressant(e) interesting, **1.1**
intéresser to interest
s' **intéresser à** to be interested in
l' **intérieur (m.)** interior, inside
intérieur(e) domestic (flight)
    (adj.), **7.1**
international(e) international,
    7.1
intitulé(e) titled
inviter to invite, **3.2**
isoler to isolate
l' **Italie (f.)** Italy, **16**
italien(ne) Italian, **9**

J

jamais ever
    ne... jamais never
le **jambon** ham, **5.1**
janvier (m.) January, **4.1**
le **Japon** Japan, **16**
japonais(e) Japanese
le **jardin** garden, **4.2**
jaune yellow, **10.2**
je I, **1.2**
    Je t'en prie. You're welcome.
        (fam.), **BV**
    je voudrais I would like, **5.1**
    Je vous en prie. You're
        welcome. (form.), **BV**;
        Please, I beg of you.
le **jean** jeans, **10.1**
jeter to throw

le **jeu: les jeux de la lumière** play
    of light
jeudi (m.) Thursday, **2.2**
jeune young, **4.1**
la **jeune fille** girl
les **jeunes (m. pl.)** young people
le **jogging: faire du jogging** to jog,
    11.2
joli(e) pretty, **4.2**
jouer to play, to perform, **16.1**
    jouer à (un sport) to play (a
        sport), **9.2**
le **joueur,** la **joueuse** player, **9.2**
le **jour** day, **2.2**
    C'est quel jour? What day is
        it?, **2.2**
    de nos jours today, nowadays
    par jour a (per) day, **3**
    tous les jours every day
le **journal** newspaper, **8.1**
    le journal intime diary
    le journal télévisé newscast
la **journée** day
juillet (m.) July, **4.1**
juin (m.) June, **4.1**
la **jupe** skirt, **10.1**
la **jupette** tennis skirt, **9.2**
le **Jura** Jura Mountains
le **jury** selection committee
jusqu'à (up) to, until, **13.2**
jusqu'en bas de la piste to the
    bottom of the trail

K

le **kilo(gramme)** kilogram, **6.2**
le **kilomètre** kilometer
le **kiosque** newsstand, **8.1**
le **kleenex** tissue, Kleenex, **15.1**

L

la the (f.), **1**; her, it (dir. obj.),
    16
là there
là-bas over there, **BV**
le **laboratoire** laboratory
le **lac** lake
    les Grands Lacs (m. pl.) the
        Great Lakes
laisser to leave (something
    behind), **5.2**
    laisser un pourboire to leave
        a tip, **5.2**
le **lait** milk, **6.1**
la **laitue** lettuce, **6.2**
lancer to throw, **13.2**
la **langue** language, **2.2**
large loose, wide, **10.2**
le **latin** Latin, **2.2**
la **latitude** latitude
laver to wash, **11.1**

se **laver** to wash oneself, **11.1**
    **se laver les cheveux (la
        figure, etc.)** to wash one's
        hair (face, etc.), **11.1**
le the (m.), **1**; him, it (dir. obj.),
    16.1
la **leçon** lesson, **9.1**
    la leçon de conduite driving
        lesson, **12.2**
la **lecture** reading
légendaire legendary
la **légende** legend
le **légume** vegetable, **6.2**
lent(e) slow
lentement slowly
les the (pl.), **2**; them (dir. obj.),
    16
leur their (sing. poss. adj.), **5**;
    (to) them (ind. obj.), **17**
leurs their (pl. poss. adj.), **5**
levant rising
le **levé: faire le levé
    topographique** to survey
se **lever** to get up, **11.1**
le **lexique** vocabulary
libérer la chambre to vacate the
    room, **17.2**
libre free, **2.2**
le **lieu** place
    avoir lieu to take place
la **ligne** line
    les grandes lignes main lines
        (trains)
    les lignes de banlieue
        commuter trains
la **limitation de vitesse** speed
    limit
les **limites (f. pl.)** boundaries (on
    tennis court), **9.2**
    hors des limites out of
        bounds, **9.2**
la **limonade** lemon-lime drink
la **lipide** fat
    lire to read, **12.2**
    Lisbonne Lisbon
le **lit** bed, **8.2**
le **litre** liter, **6.2**
littéraire literary
la **littérature** literature, **2.2**
la **livre** pound, **6.2**
le **livre** book, **BV**
la **location** rental
loin de far from, **4.2**
les **loisirs (m. pl.)** leisure activities,
    16
    Londres London
le **long: le long de** along
long(ue) long, **10.2**
la **longitude** longitude
longtemps (for) a long time

la **longueur** length
lorsque while
louer to rent
lourd(e) heavy
lui him (m. sing. stress pron.), **9**; (to) him, (to) her (ind. obj.), **17.1**
la **lumière** light
lundi (m.) Monday, **2.2**
les **lunettes (f. pl.)** (ski) goggles, **14.1**
    les lunettes de soleil sunglasses, **9.1**
lutter to fight
le **luxe** luxury
luxueux, luxueuse luxurious
le **lycée** high school, **1.2**
le/la **lycéen(ne)** high school student

M

ma my (f. sing. poss. adj.), **4**
Madame (Mme) Mrs., Ms., **BV**
Mademoiselle (Mlle) Miss, Ms., **BV**
le **magasin** store, **3.2**
le **magazine** magazine, **3.2**
magnifique magnificent
mai (m.) May, **4.1**
maigrir to lose weight, **11.2**
le **maillot de bain** bathing suit, **9.1**
la **main** hand, **11.1**
    fait(e) à la main handmade
maintenant now, **2.1**
mais but, **1**
    Mais oui (non)! Of course (not)!
la **maison** house, **3.1**
le **maître** master
    le maître d'hôtel maitre d', **5.2**
mal badly
    avoir mal à to have a(n) … -ache, to hurt, **15.1**
    Où avez-vous mal? Where does it hurt?, **15.2**
    Pas mal. Not bad., **BV**
le/la **malade** sick person, patient, **15.1**
malade sick, **15.1**
la **maladie** illness
malheureusement unfortunately
la **Manche** English Channel
la **manche** sleeve, **10.1**
    à manches longues (courtes) long- (short-)sleeved, **10.2**
manger to eat
la **mangue** mango
la **manière** manner, way

    avoir de bonnes manières to have good manners
manquer: il en manque deux two are missing
se **maquiller** to put on make-up, **11.1**
le **marathon** marathon
le **marbre** marble
le/la **marchand(e) (de fruits et légumes)** (produce) seller, **6.2**; merchant
la **marchandise** merchandise
le **marché** market, **6.2**
mardi (m.) Tuesday, **2.2**
la **marée** tide
le **mari** husband, **4.1**
le **mariage** marriage
marié(e) married
le **marin** sailor
le **Maroc** Morocco, **16**
la **marque** make (of car), **12.1**
marquer un but to score a goal, **13.1**
marron (inv.) brown, **10.2**
mars (m.) March, **4.1**
martiniquais(e) from Martinique
la **Martinique** Martinique
la **masse** mass
le **match** game, **9.2**
les **mathématiques (f. pl.)** mathematics
les **maths (f. pl.)** math, **2.2**
la **matière** subject (school), **2.2**; matter
le **matin** morning, in the morning, **2**
    du matin A.M. (time), **2**
mauvais(e) bad; wrong
    Il fait mauvais. It's bad weather., **9.2**
le **mazout** fuel oil
me (to) me (dir. and ind. obj.), **15.2**
la **médaille** medal
le **médecin** doctor (m. and f.), **15.2**
    chez le médecin at, to the doctor's, **15.2**
    le femme médecin (woman) doctor
la **médecine** medicine (medical profession), **15**
médical(e) medical
le **médicament** medicine (remedy), **15.2**
la **médina** medina (old Arab section of northwestern African town)
meilleur(e) better (adj.), **10**

le **membre** member
même same (adj.), **2.1**; even (adv.)
le/la **mennonite** Mennonite
mental(e) mental
la **menthe: le thé à la menthe** mint tea
le **menu: le menu touristique** budget (fixed price) meal
la **mer** sea, **9.1**
    la mer des Caraïbes Caribbean Sea
    la mer Méditerranée Mediterranean Sea
merci thank you, **BV**
mercredi (m.) Wednesday, **2.2**
la **mère** mother, **4.1**
le **méridien** meridian
merveilleux, merveilleuse marvelous, **10.2**
mes my (pl. poss. adj.), **4**
la **mesure** measurement
    sur mesure tailored (to one's measurements), tailor-made
mesurer to measure
le **métabolisme** metabolism
la **météo** weather forecast
la **météorologie** meteorology, the study of weather
météorologique meteorological
le **métier** profession
le **mètre** meter
métrique metric
le **métro** subway, **4.2**
    en métro by subway, **5.2**
    la station de métro subway station, **4.2**
mettre to put (on), to place, **8.1**; to put on (clothes), **10**; to turn on (appliance), **8**
    mettre au point to come out with, develop
    mettre de l'argent de côté to put money aside, save, **18.2**
    mettre le contact to start the car, **12.1**
    mettre le couvert to set the table, **8**
se **mettre en forme** to get in shape, **11.1**
le **Mexique** Mexico, **16**
le **microbe** microbe
la **microbiologie** microbiology
le **microscope** microscope
midi (m.) noon, **2.2**
le **militaire** soldier
militaire military
mille (one) thousand, **6.2**
les **milliers (m. pl.)** thousands
le **minéral** mineral

le **ministère** ministry
minuit (m.) midnight, **2.2**
la **mission** mission
la **mi-temps** half (sporting event)
moche terrible, ugly, **2.2**
le **modèle** model
moderne modern
moderniser to modernize
modeste modest, reasonably priced
moi me (sing. stress pron.), **1.2; 9**
moins less
au moins at least
Il est une heure moins dix. It's ten to one. (time), **2**
moins… que less … than
le **mois** month, **4.1**
le **moment: en ce moment** right now
mon my (m. sing. poss. adj.), **4**
le **monde** world
beaucoup de monde a lot of people, **13.1**
tout le monde everyone, everybody, **BV**
le **moniteur**, la **monitrice** instructor, **9.1**; camp counselor
la **monnaie** change; currency, **18.1**
faire de la monnaie to make change, **18.1**
Monsieur (M.) Mr., sir, **BV**
la **montagne** mountain, **14.1**
à la montagne in the mountains
monter to go up, get on, get in, **8.2**; to take upstairs, **17.1**
monter une pièce to put on a play, **16.1**
montrer to show, **17.1**
moral(e) moral
le **morceau de craie** piece of chalk, **BV**
mordu(e) bitten
la **mort** death
mort(e) dead
mortel(le) fatal
Moscou Moscow
la **mosquée** mosque
le **mot** word
le mot apparenté cognate
le **motard** motorcycle cop, **12.2**
le **moteur** engine (car, etc.), **12.1**
la **moto** motorcycle, **12.1**
le **mouchoir** handkerchief, **15.1**
mourir to die, **17**
la **moutarde** mustard, **6.2**
le **mouvement** movement
mouvementé(e) eventful

moyen(ne) average, intermediate
le **moyen de transport** mode of transportation
municipal(e) municipal
musclé(e) muscular
le **musée** museum, **16.2**
la **musique** music, **2.2**
la **mythologie** mythology

N

nager to swim, **9.1**
nager la brasse papillon to do the butterfly (swim stroke)
le **nageur**, la **nageuse** swimmer
naître to be born, **17**
la **nappe** tablecloth, **5.2**
la **natation** swimming, **9.1**
la **nation** nation
national(e) national
la **nature** nature
nature plain (adj.), **5.1**
la **navette: faire la navette** to go back and forth
ne: ne… jamais never, **12**
ne… pas not, **1.2**
ne… personne no one, nobody, **12.2**
ne… rien nothing, **12.2**
né: il est né he was born
nécessaire necessary
négatif, négative negative
la **neige** snow, **14.2**
neige (inf. neiger): Il neige. It's snowing., **14.1**
nerveux, nerveuse nervous
les cellules nerveuses (f. pl.) nerve cells
n'est-ce pas? isn't it?, doesn't it (he, she, etc.)?, **1.2**
neuf nine, **BV**
neutraliser to neutralize
le **neveu** nephew, **4.1**
le **nez** nose, **15.1**
avoir le nez qui coule to have a runny nose, **15.1**
ni… ni neither … nor
la **nièce** niece, **4.1**
le **niveau** level
vérifier les niveaux to check under the hood, **12.1**
noir(e) black, **10.2**
le tableau noir blackboard, **3.1**
le **nom** name, **16.2**; noun
le **nombre** number, **5.2**
nombreux, nombreuse numerous
nommer to name, mention

non no
non-fumeurs no smoking (section), **7.1**
non seulement not only
le **nord** north
normal(e) normal
normalement normally, usually
nos our (pl. poss. adj.), **5**
la **nostalgie** nostalgia
la **note** bill (currency), **17.2**; grade
notre our (sing. poss. adj.), **5**
nourrir to feed
la **nourriture** food, nutrition
nous we, **2**; us (stress pron.), **9**; (to) us (dir. and ind. obj.), **15**
nouveau (nouvel) new (m.), **4**
nouvelle new (f.), **4**
les **nouvelles (f. pl.)** news
novembre (m.) November, **4.1**
le **nuage** cloud, **9.2**
la **nuit** night
le **numéro** number
Quel est le numéro de téléphone de… ? What is the phone number of … ?, **5.2**

O

obéir (à) to obey, **7**
l' **objet (m.)** object
obligatoire mandatory
obliger to oblige
obtenir to obtain
occidental(e) western
occupé(e) busy, **2.2**
occuper to occupy
l' **océan (m.)** ocean
octobre (m.) October, **4.1**
l' **odeur (f.)** scent, smell
l' **œil (m., pl. yeux)** eye
l' **œuf (m.)** egg, **6.2**
l'œuf sur le plat fried egg
l' **œuvre (f.)** work (of art), **16**
officiel(le) official
offrir to offer, give, **15**
l' **oignon (m.)** onion, **6.2**
l' **omelette (f.)** omelette, **5.1**
l'omelette aux fines herbes omelette with herbs, **5.1**
l'omelette nature plain omelette, **5.1**
on we, they, people, **3**
On y va.(?) Let's go.; Shall we go?, **5**
l' **oncle (m.)** uncle, **4.1**
onze eleven, **BV**
l' **opéra (m.)** opera, **16.1**
opérer to operate
opposer to oppose, **13.1**
l' **or (m.)** gold

l' **orage** (m.) storm
l' **orange** (f.) orange, 6.2
    orange (inv.) orange (color), 10
l' **Orangina** (m.) orange soda, 5.1
    ordinaire regular (gasoline), 12.1
l' **ordinateur** (m.) computer, BV
l' **ordonnance** (f.) prescription, 15.2
    faire une ordonnance to write a prescription, 15.2
l' **oreille** (f.) ear, 15.1
    avoir mal aux oreilles to have an earache, 15
l' **oreiller** (m.) pillow, 17.2
les **oreillons** (m. pl.) mumps
    organisé(e) organized
l' **organisme** (m.) organism
    original(e) original
l' **origine** (f.): **à l'origine** originally
    d'origine américaine (française, etc.) from the U.S. (France, etc.)
    orner to decorate
l' **os** (m.) bone
    ôter to take off (clothing)
    ou or, 1.1
    où where, BV
    oublier to forget
l' **ouest** (m.) west
    oui yes, 1
    ouvert(e) open, 16
l' **ouverture** (f.) opening
l' **ouvrier, l'ouvrière** worker
    ouvrir to open, 15
    ovale oval
l' **oxygène** (m.) oxygen

P

le **pain** bread, 6.1
la **paire** pair, 10
le **palais** palace
le **panier** basket, 13.2
le **panneau** backboard (basketball), 13.2
    le panneau routier road sign
    panoramique panoramic
le **pantalon** pants, 10.1
la **papeterie** stationery store
le **papier** paper, 6
    la feuille de papier sheet of paper, BV
    le papier hygiénique toilet paper, 17.2
le **paquet** package, 6.2
    par by, through
    par dessus over (prep.), 13
    par exemple for example
    par jour a (per) day, 3

    par semaine a (per) week, 3.2
le **paradis** paradise, heaven
le **paragraphe** paragraph
le **parallèle** parallel
le **parc** park, 11.2
    parce que because, 9.1
    parcourir to travel, go through
    pardon excuse me, pardon me
le **parebrise** windshield, 12
les **parents** (m. pl.) parents, 4.1
    parfait(e) perfect
    parisien(ne) Parisian, 9
le **parking** parking lot
le **parlement** parliament
    parler to speak, talk, 3.1
    parler au téléphone to talk on the phone, 3.2
    parmi among
    participer (à) to participate (in)
    particulièrement particularly
la **partie** game, match, 9.2; part
    faire partie de to be a part of
    la partie en simple (en double) singles (doubles) match (tennis), 9.2
    partir to leave, 7.1
    partout everywhere
    pas not
    pas de (+ nom) no (+ noun)
    Pas de quoi. You're welcome. (inform.), BV
    pas du tout not at all
    Pas mal. Not bad., BV
    pas mal de quite a few
le **passager, la passagère** passenger, 7.1
le **passé** past
le **passeport** passport, 7.1
    passer to spend (time), 3; to pass, go through, 7.2
    passer à la douane to go through customs, 7.2
    passer à l'immigration to go through immigration
    passer par le contrôle de sécurité to go through security (airport), 7
    passer un examen to take an exam, 3.1
    passer un film to show a movie, 16.1
    passionné(e) de excited by
    passionner to excite
le **pâté** pâté, 5.1
    patient(e) patient, 1.1
le **patin** skate; skating, 14.2
    faire du patin to skate, 14.2
    faire du patin à glace to ice-skate, 14.2

    faire du patin à roulettes to roller-skate
    le patin à glace ice skate, 14.2
le **patinage** skating, 14.2
le **patineur, la patineuse** skater, 14.2
la **patinoire** skating rink, 14.2
le/la **pauvre** poor thing, 15.1
    pauvre poor, 15.1
le **pavillon** small house, bungalow
    payer to pay, 6.1
    payer en espèces to pay cash, 17
le **pays** country, 7.1
le **paysage** landscape
les **Pays-Bas** (m. pl.) the Netherlands, 16
le **péage: l'autoroute** (f.) **à péage** toll road
la **pêche** fishing
    aller à la pêche to go fishing, 9.1
    faire une belle pêche to catch a lot of fish
    le port de pêche fishing port
le **peigne** comb
se **peigner** to comb (one's hair), 11.1
    peindre to paint
le/la **peintre** painter, artist, 16.2
la **peinture** painting, 16.2
    péjoratif, péjorative pejorative, disparaging
le **penalty** penalty (soccer)
    pendant during, for (time), 3.2
    pendant que while
la **pénicilline** penicillin, 15.2
    penser to think, 10.1
le **penseur** thinker
la **pension** small hotel
    perdre to lose, 8.2
    perdre des kilos to lose weight
    perdre patience to lose patience, 8.2
le **père** father, 4.1
la **périphérie** outskirts
la **perle** pearl
    permettre to permit, allow, 14
le **permis** license
    le permis de conduire driver's license, 12.2
le **personnage** character
la **personne** person
    ne... personne no one, nobody, 12.2
    personnel(le) personal
le **personnel de bord** flight attendants, 7.2

personnellement personally, **16.2**

la **perte** loss

peser to weigh

petit(e) short, small, **1.1**
  la **petite annonce** classified ad
  le **petit déjeuner** breakfast, **9**
  prendre le petit déjeuner to eat breakfast, **9**

la **petite-fille** granddaughter, **4.1**

le **petit-fils** grandson, **4.1**

le **pétrolier** oil tanker

peu (de) few, little, **18**
  un **peu (de)** a little

la **pharmacie** pharmacy, **15.2**

le/la **pharmacien(ne)** pharmacist, **15.2**

la **photo** photograph

la **phrase** sentence

la **physique** physics, **2.2**
  physique physical
  la **forme physique** physical fitness, **13**

la **pièce** room, **4.2**; play, **16.1**; coin, **18.1**

le **pied** foot, **13.1**
  à **pied** on foot, **5.2**

la **pierre** stone

le/la **piéton(ne)** pedestrian, **12.2**

le/la **pilote** pilot
  le/la **pilote de ligne** airline pilot
  piloter to pilot

le **pique-nique: faire un pique-nique** to have a picnic, **6**

la **piscine** pool, **9.2**
  la **piscine couverte** indoor pool

la **piste** track, **13.2**; ski trail, **14.1**

pittoresque picturesque

le **placard** closet, **17.2**

la **place** seat (plane, train, etc.), **7.1**; parking space, **12.2**; place

la **plage** beach, **9.1**

la **plaine** plain

le **plan** map

la **planche à voile: faire de la planche à voile** to windsurf, **9.1**

la **plante** plant
  les **plantes aquatiques** aquatic vegetation

le **plastique: en plastique** plastic (adj.)

le **plat** dish (food)

le **plateau** plateau

plein(e) full, **13.1**

avoir plein de fric to have lots of money (slang), **18.2**

en **pleine zone tempérée** right in the temperate zone

faire le plein to fill up (a gas tank), **12.1**

pleut (inf. pleuvoir): Il pleut. It's raining., **9.2**

la **plongée sous-marine: faire de la plongée sous-marine** to go deep-sea diving, **9.1**

plonger to dive, **9.1**

la **pluie** rain
  les **pluies acides** acid rain

la **plupart (des)** most (of), **8.2**

le **pluriel** plural

plus more (comparative), **10**
  en **plus de** in addition to
  plus ou moins more or less
  plus tard later

plusieurs several, **18**

le **pneu** tire, **12.1**
  le **pneu à plat** flat tire, **12.1**

la **poche** pocket, **18.1**

le **poème** poem

la **poésie** poetry

le **poète** poet (m. and f.)

le **poids** weight

le **point** point; period
  à **point** medium-rare (meat), **5.2**

la **pointure** size (shoes), **10.2**
  Vous faites quelle pointure? What (shoe) size do you take?, **10.2**

le **poisson** fish, **6.1**

la **poissonnerie** fish store, **6.1**

le **pôle** pole

la **poliomyélite** polio

polluer to pollute

la **pollution** pollution

la **Polynésie française** French Polynesia

la **pomme** apple, **6.2**

la **pomme de terre** potato, **6.2**

le/la **pompiste** gas station attendant, **12.1**

populaire popular, **1.2**

la **porcelaine** porcelaine, china

le **port** port, harbor
  le **port de pêche** fishing port

le **portail** doorway (church)

la **porte** gate (airport), **7.1**; door, **17.1**

le **portefeuille** wallet, **18.1**

le **porte-monnaie** change purse, **18.1**

porter to wear, **10.1**

le **porteur** porter, **8.1**

le **portrait** portrait

le **Portugal** Portugal, **16**

poser une question to ask a question, **3.1**

la **possibilité** possibility

le **pot** jar, **6.2**

le **pouce** inch, thumb

le **poulet** chicken, **6.1**

la **poupée** doll

pour for; in order to, **2**

le **pourboire** tip (restaurant), **5.2**
  laisser un pourboire to leave a tip, **5.2**

le **pourcentage** percentage

pourquoi why, **9.1**

pourtant yet, still, nevertheless

pouvoir to be able to, **6**

pratiquer un sport to play a sport, **11.2**

précieux, précieuse precious

précis(e) precise, exact
  à **l'heure précise** right on time

préféré(e) favorite

préférer to prefer, **5**

le **préfixe** prefix

premier, première first, **4.1**
  en **première** in first class, **8.1**
  les **tout (inv.) premiers (m. pl.)** very first

premièrement first of all

prendre to take, **9.1**; to buy; to eat (drink) (in café, restaurant, etc.)
  prendre un bain (une douche) to take a bath (shower), **11.1**
  prendre un bain de soleil to sunbathe, **9.1**
  prendre un billet to buy a ticket, **9**
  prendre des kilos to gain weight
  prendre part à to take part in
  prendre le petit déjeuner to eat breakfast, **9**
  prendre possession de to take possession of
  prendre un pot to have a drink
  prendre rendez-vous to make an appointment
  prendre le train (l'avion, etc.) to take the train (plane, etc.), **9**

préparer to prepare, **4.2**

près de near, **4.2**

prescrire to prescribe, **15.2**

présenter to present, introduce

la **préservation** preservation

presque almost

pressé(e) in a hurry
la **pression artérielle** blood
pressure
prêt(e) ready
prêt-à-porter ready-to-wear
(adj.), **10**
le **rayon prêt-à-porter** ready-
to-wear department, **10.1**
prêter to lend, **18.2**
la **preuve** proof
la **prévision** prediction
prévoir to predict
prie: Je vous en prie. Please, I
beg of you., You're welcome.,
BV
la **prière** prayer
appeler à la prière to call to
worship
primaire: l'école (f.) primaire
elementary school
principal(e) main, principal
la **principauté** principality
le **printemps** spring, **13.2**
pris(e) taken, **5.1**
privé(e) private
le **prix** price, cost, **10.1**
à prix fixe at a fixed price
probablement probably
le **problème** problem, **11.2**
prochain(e) next, **8.2**
produire to produce
le **produit** product
le/la **prof** teacher (inform.), **2.1**
le **professeur** teacher (m. and f.),
2.1
professionnel(le) professional
profiter de to take advantage of,
profit from
profond(e) deep
le **programme** TV program
le **progrès** progress
progressif, progressive
progressive
le **projet** project, plan
la **promenade: faire une**
promenade to take a walk,
9.1
se **promener** to walk, **11.2**
proposer to suggest
propre own (adj.); clean
protéger to protect
la **protéine** protein
provenance: en provenance de
arriving from (train, plane,
etc.), **7.1**
provençal(e) from Provence,
the south of France
les **provisions (f. pl.)** groceries
prudemment carefully, **12.2**
le **public** public

la **publicité** advertisement
les **puces (f. pl.): le marché aux**
puces flea market
puissant(e) powerful
le **pull** sweater, **10.1**
punir to punish, **7**
pur(e) pure
la **pureté** purity
la **pyramide** pyramid

Q

le **quai** platform (railroad), **8.1**
la **qualité** quality
quand when, **3.1**
quarante forty, **BV**
le **quart: et quart** a quarter past
(time), **2**
moins le quart a quarter to
(time), **2**
le **quartier** neighborhood, district,
4.2
quatorze fourteen, **BV**
quatre four, **BV**
quatre-vingt-dix ninety, **5.2**
quatre-vingts eighty, **5.2**
quel(le) which, what, **7**
Quel est le numéro de
téléphone de… ? What is
the phone number of … ?,
5
Quelle est la date
aujourd'hui? What is
today's date?, **4.1**
Quel temps fait-il? What's
the weather like?, **9.2**
quelque some (sing.)
quelque chose à manger
something to eat, **5.1**
quelquefois sometimes, **5**
quelques some (pl.), **8.2**
quelqu'un somebody, someone,
12
Qu'est-ce que c'est? What is it?,
BV
Qu'est-ce qu'il a? What's wrong
with him?, **15.1**
la **question: poser une question**
to ask a question, **3.1**
la **queue: faire la queue** to wait in
line, **8.1**
qui who, **BV**; whom, **11**; which,
that
Qui ça? Who (do you
mean)?, **BV**
Qui est-ce? Who is it?, **BV**
quinze fifteen, **BV**
quitter to leave (a room, etc.),
3.1
quoi what (after prep.), **14**
quotidien(ne) daily, everyday

R

raconter to tell (about)
le **racquet(-ball)** racquetball
la **radio** radio, **3.2**
radioactif, radioactive
radioactive
la **rage** rabies
raide steep, **14.2**
la **raison** reason
ralentir to slow down
le **randonneur, la randonneuse**
hiker
rapide quick, fast
le **rapport** relationship; report
rapporter to report
la **raquette** racket, **9.2**
rare rare
se **raser** to shave, **11.1**
le **rasoir** razer, shaver
rassembler to collect, gather
together
le **rayon** department (in a store),
10.1
la **réaction** reaction
réaliser to realize (an ambition),
achieve
la **réalité** reality
la **réception** front desk (hotel),
17.1
le/la **réceptionniste** desk clerk, **17.1**
la **recette** recipe
recevoir to receive, **18.1**
la **recherche: faire de la recherche**
to do research
rechercher to seek
recommandé(e) recommended
reconnu(e) recognized
la **récréation** recess
récrire to rewrite
récupérer to claim (luggage),
7.2
refléter to reflect
regarder to look at, **3.1**
se **regarder** to look at oneself,
look at one another
le **régime: faire un régime** to go
on a diet
le régime alimentaire diet
la **région** region
la **règle** rule
le **règlement** rule
régler to direct (traffic)
regretter to be sorry
régulier, régulière regular
régulièrement regularly
relativement relatively
le **relevé de compte** statement
(bank), **18**
relié(e) connected
remarquer to notice

rembourser to pay back, reimburse, **18.2**
remplir to fill out, **7.2**
la **rencontre** meeting
rencontrer to meet
le **rendez-vous: prendre rendez-vous** to make an appointment
rendre to give back, **18.2**
les **renseignements (m. pl.)** information
rentrer to go home, **3.1**
renvoyer to return (tennis ball), **9.2**
la **répartition** distribution
le **repas** meal
répéter to repeat
répondre to answer, **8**
la **réponse** answer
se **reposer** to rest
repoussé(e) pushed back
représenter to represent
la **reprise** reshowing
reproduire to reproduce
la **république** republic, democracy
la **réserve** resource, supply
réservé(e) reserved
réserver to reserve
le **réservoir** gas tank, **12.1**
résidentiel(le) residential
la **résistance** resistance
respecter to respect
la **respiration** breathing
respirer (à fond) to breathe (deeply), **15.2**
ressembler à to resemble
ressentir to feel
le **restaurant** restaurant, **5.2**
la **restauration** food service
rester to stay, remain, **17**
rester en forme to stay in shape, **11.1**
le **retard** delay
en retard late, **8.2**
retomber to fall back down
le **retour** return
à votre retour when you return
la **retransmission** rebroadcast
réunir to bring together
réussir (à) to succeed; to pass (exam), **7**
le **rêve** dream
se **réveiller** to wake up, **11.1**
la **révélation** revelation
revenir to come back, **16**
rêver to dream
la **révolution** revolution
révolutionner to revolutionize
le **rez-de-chaussée** ground floor, **4.2**

le **rhume** cold (illness), **15.1**
avoir un rhume to have a cold, **15.1**
riche rich
la **richesse** wealth
le **rideau** curtain, **16.1**
le lever du rideau at curtain time (theatre)
Rien d'autre. Nothing else., **6.2**
rigoler to joke around, **3.2**
Tu veux rigoler! Are you kidding?!
le **rite** rite, ritual
la **rivière** river
le **riz** rice
la **robe** dress, **10.1**; robe
le **rocher** rock
le **roi** king
le **rôle** role
le **roman** novel
le roman policier detective novel, mystery
le **romancier,** la **romancière** novelist
rond(e) round
rose pink, **10.2**
le **rosier** rosebush
la **roue** wheel, **12.1**
la roue de secours spare tire, **12.1**
les deux roues two-wheeled vehicles
rouge red, **10.2**
la **rougeole** measles
le **rouleau de papier hygiénique** roll of toilet paper, **17.2**
rouler (vite) to go, drive (fast), **12.1**
la **route** road, **12.1**
En route! Let's go!
prendre la route to take to the road
la **rubéole** German measles
la **rue** street, **3.1**
le **rugby** rugby
rural(e) rural
le/la **Russe** Russian (person)
le **rythme** rhythm

S

sa his, her (f. sing. poss. adj.), **4**
le **sable** sand, **9.1**
le **sac** bag, **6.1**; pocketbook, purse, **18.1**
le sac à dos backpack, **BV**
saignant(e) rare (meat), **5.2**
la **saison** season
la belle saison summer
la **salade** salad, **5.1**
le **salaire** salary

la **salle** room
la salle à manger dining room, **4.2**
la salle d'attente waiting room, **8.1**
la salle de bains bathroom, **4.2**
la salle de cinéma movie theatre, **16.1**
la salle de classe classroom, **2.1**
la salle de séjour living room, **4.2**
le **Salon** official art show
Salut. Hi., **BV**
samedi (m.) Saturday, **2.2**
le **sandwich** sandwich, **5.1**
sans without, **12.1**
sans aucun doute without a doubt
Sans blague! No kidding!
sans plomb unleaded, **12.1**
la **santé** health, **15.1**
être en bonne (mauvaise) santé to be in good (poor) health, **15.1**
la **saucisse de Francfort** hot dog, **5.1**
le **saucisson** salami, **6.1**
sauf except, **16.2**
sauver to save
le **savant** scientist
savoir to know (information), **16.2**
le **savon** soap, **11.1**
scandalisé(e) scandalized, shocked
la **scène** stage; scene, **16.1**
les **sciences (f. pl.)** science, **2.2**
les sciences humaines social sciences
les sciences naturelles natural sciences
le **scorbut** scurvy
le **score** score, **9.2**
le **sculpteur** sculptor (m. and f.), **16.2**
la **sculpture** sculpture, **16.2**
la **séance** show (movie), **16.1**
sec, sèche dry
se **sécher** to dry (off), **17.2**
la **sécheresse** dryness, drought
secondaire: l'école (f.) secondaire junior high, high school
la **seconde** second (time)
seconde: en seconde in second class, **8.1**
seize sixteen, **BV**
le **séjour** stay

selon according to

la **semaine** week, **2.2**; allowance
    par semaine a (per) week,
    3.2
sembler to seem
le **Sénégal** Senegal, **16**
le **sens** direction; meaning
    sens interdit (m.) wrong way
    (traffic sign)
    sens unique (m.) one way
    (traffic sign)
se **sentir** to feel (well, etc.), **15.1**
séparer to separate
sept seven, **BV**
septembre (m.) September, **4.1**
la **série** series
    sérieux, sérieuse serious, **10**
serré(e) tight, **10.2**
le **serveur**, la **serveuse** waiter,
    waitress, **5.1**
le **service** tip; service, **5.2**
    Le service est compris. The
    tip is included., **5.2**
la **serviette** napkin, **5.2**; towel,
    17.2
    servir to serve (food), **7.2**; to
    serve (a ball in tennis, etc.),
    9.2
    ses his, her (pl. poss. adj.), **5**
    seul(e) alone; single; only (adj.)
    tout(e) seul(e) all alone, by
    himself/herself
    seulement only (adv.)
la **sève** sap
    tirer la sève to tap (maple
    sugar)
    sévère strict
le **sexe** sex
le **shampooing** shampoo
le **short** shorts, **9.2**
    si if; yes (after neg. question)
le **SIDA (Syndrome Immuno-
Déficitaire Acquis)** AIDS
le **siècle** century
le **siège** seat, **7.1**
    siffler to (blow a) whistle, **13.1**
le **signal** sign
    signer to sign, **18.1**
    signifier to mean
    s'il te plaît please (fam.), **BV**
    s'il vous plaît please (form.),
    BV
    simplement simply
    sincère sincere, **1.2**
le **sirop: le sirop d'érable** maple
    syrup
    situé(e) located
    six six, **BV**
le **ski** ski, skiing, **14.1**
    faire du ski to ski, **14.1**

faire du ski nautique to
    water-ski, **9.1**
le **ski alpin** downhill skiing,
    14.1
le **ski de fond** cross-country
    skiing, **14.1**
le **skieur**, la **skieuse** skier, **14.1**
social(e) social
la **société** society
la **sociologie** sociology
la **sœur** sister, **1.2**
soi: chez soi home
la **soie: en soie** silk (adj.)
le **soir** evening, in the evening, **2**
    du soir in the evening, P.M.
    (time), **2**
la **soirée** evening
soit is, exists (subjunctive)
soixante sixty, **BV**
soixante-dix seventy, **5.2**
le **sol** ground, **13.2**
les **soldes (f. pl.)** sale (in a store),
    10.2
le **soleil** sun
    Il fait du soleil. It's sunny.,
    9.2
    le soleil levant rising sun
soluble dans l'eau water-
    soluble
soluble dans la graisse fat-
    soluble
sombre dark
la **somme** sum
le **sommeil** sleep
le **sommet** summit, mountaintop,
    14.1
son his, her (m. sing. poss.
    adj.), **4**
le **sondage** survey, opinion poll
la **sorte** sort, kind
la **sortie** exit, **7.1**
    sortir to go out, take out, **7**
    souffrir to suffer, **15.2**
la **soupe à l'oignon** onion soup,
    5.1
la **source** source
    sous under, **BV**
les **sous-titres (m. pl.)** subtitles,
    16.1
    souterrain(e) underground
    souvent often, **5**
se **spécialiser** to specialize
le **spectacle** show
le **spectateur** spectator, **13.1**
la **splendeur** splendor
    splendide splendid
le **sport: faire du sport** to play
    sports
    pratiquer un sport to play a
    sport

le **sport collectif** team sport
le **sport d'équipe** team sport
les **sports d'hiver** winter
    sports, skiing, **14.1**
sport (inv.) casual (clothes),
    10.1
sportif, sportive athletic
le **stade** stadium, **13.1**
la **station** station; resort
    la station balnéaire seaside
    resort, **9.1**
    la station de métro subway
    station, **4.2**
    la station-service gas station,
    12.1
    la station de sports d'hiver
    ski resort, **14.1**
le **stationnement** parking
    Stationnement interdit No
    parking (traffic sign)
    stationner to park, **12.2**
    Il est interdit de stationner.
    No parking. (traffic sign),
    12.2
la **statue** statue
le **steak frites** steak and French
    fries, **5.2**
le **steward** flight attendant (m.),
    7.2
    stop stop (traffic sign)
    strict(e) strict
le **stylo** (ballpoint) pen, **BV**
se **succéder** to follow one another
le **succès** success
le **sucre** sugar
    sucré(e) sweet, with sugar
le **sud** south
le **sud-est** southeast
    suffir to suffice, be enough
la **Suisse** Switzerland
    suisse Swiss
    suivant(e) following
    suivre to follow
le **sujet** subject
    super terrific, super, **2.2**; super
    (gasoline), **12.1**
    superbe superb
la **superficie** area (geography)
le **supermarché** supermarket, **6.1**
    supersonique supersonic
le **supplément** surcharge (train
    fare), **8**
    payer un supplément to pay
    a surcharge
    sur on, **BV**
    sûr(e) sure
le **surf: faire du surf (des neiges)**
    to go surfing (snowboarding)
la **surface** surface
    surgelé(e) frozen, **6.2**

surtout especially, above all
surveiller to watch, **12.2**
le **survêtement** warmup suit, **11.2**
le **swahili** Swahili
le **sweat-shirt** sweatshirt, **10.1**
sympa (inv.) nice (abbrev. for **sympathique**), **1.2**
sympathique nice (person), **1.2**
le **symptôme** symptom
le **syndicat d'initiative** tourist office
le **synonyme** synonym
le **système** system

T

ta your (f. sing. poss. adj.), **4**
la **table** table, **BV**
le **tableau** blackboard, **BV**; painting, **16.2**
   le **tableau des départs et des arrivées** arrival and departure board
la **taille** size (clothes), **10.2**
   la **taille au-dessous** next smaller size, **10.2**
   la **taille au-dessus** next larger size, **10.2**
   Vous faites quelle taille? What size do you take?, **10.2**
le **tailleur** suit (woman's), **10.1**; tailor
le **talon** heel, **10.2**
   à talons hauts (bas) high-(low-)heeled (shoes)
le **tambour** drum
la **tante** aunt, **4.1**
tard late
   plus tard later
le **tarif** fare
   les tarifs aériens airfares
la **tarte** pie, tart, **6.1**
   la tarte aux fruits fruit tart, pie
la **tasse** cup, **5.2**
le **taux** rate, level
le **taxi** taxi, **7.2**
   te (to) you (fam.) (dir. and ind. obj.), **15.2**
technique technical
technologiquement technologically
le **tee-shirt** T-shirt, **9.2**
la **télé** TV, **3.2**
   à la télé on TV
le **téléphone** telephone
le **télésiège** chairlift, **14.1**
la **température** temperature, **14.1**
le **temps** weather, **9.2**; time

de temps en temps from time to time
   Quel temps fait-il? What's the weather like?, **9.2**
la **tendance: avoir tendance à (+ inf.)** to tend (+ inf.)
le **tennis** tennis, **9.2**
   les tennis (f. pl.) sneakers
le **terrain de football** soccer field, **13.1**
la **terrasse** terrace, **4.2**
   la terrasse d'un café sidewalk café, **5.1**
la **terre** earth, land
   la Terre the Earth
la **Terre-Neuve** Newfoundland
terrible terrible; terrific (inform.), **2.2**
le **territoire** territory
le **tétanos** tetanus
la **tête** head, **13.1**
   avoir mal à la tête to have a headache, **15.1**
le **thé citron** tea with lemon, **5.1**
le **théâtre** theater, **16.1**
la **théorie** theory
   Tiens! Hey!, Well!, Look!, **10.1**
le **tilleul** linden tree
timide timid, shy, **1.2**
le **tissu** fabric
toi you (sing. stress pron.), **9**
la **toilette: faire sa toilette** to wash and groom oneself, **11.1**
les **toilettes (f. pl.)** bathroom, **4.2**
la **tomate** tomato, **6.2**
tomber to fall, **17**
ton your (m. sing. poss. adj.), **4**
la **tonne** ton
le **topographe** topographer (m. and f.)
tôt early
total(e) total
toucher to cash (a check), **18.1**; to touch
toujours always, **5**; still
la **tour** tower
   la tour Eiffel Eiffel Tower
le **tour: à son tour** in turn
   À votre tour. (It's) your turn.
le/la **touriste** tourist
tous, toutes all, every, **7**
   tous (toutes) les deux both
tousser to cough
tout(e) the whole, the entire, **7**; all, any
   À tout à l'heure. See you later., **BV**
   C'est tout? Is that all?, **6.2**
   en tout cas in any case

tout autour de all around (prep.)
tout de suite right away, **11.1**
tout le monde everyone, everybody, **BV**
tout(e) seul(e) all alone, all by himself/herself, **5.2**
   les tout (inv.) premiers (m. pl.) the very first
toxique toxic
la **tragédie** tragedy, **16.1**
le **train** train, **8.1**
   le train à grande vitesse (TGV) high-speed train, **8**
traiter to treat (illness)
le **trajet** distance
transporter to transport
le **travail** work
travailler to work, **3.1**
travailleur, travailleuse hardworking
traverser to cross, **12.2**
treize thirteen, **BV**
trente thirty, **BV**
très very, **1.2**
le **tricolore** French flag
la **trigonométrie** trigonometry, **2.2**
trois three, **BV**
troisième third, **4.2**
trop too (excessive), **10.2**
   trop de too many, too much
le **trophée** trophy
tropical(e) tropical, **9**
le **trottoir** sidewalk, **12.2**
le **trouble digestif** indigestion, upset stomach
trouver to find, **5.1**; to think (opinion), **10.2**
se **trouver** to be located, found
tu you (fam., subj. pron.), **1**
la **tuberculose** tuberculosis
tuer to kill
la **Tunisie** Tunisia, **16**
le **type** guy (inform.)
le **typhoïde** typhoid
typique typical

U

un, une a, one, **BV**
unique: l'enfant unique only child
unir to unite
unisexe unisex
l' **unité (f.)** unit
universitaire university
l' **université (f.)** university
l' **urgence (f.): en cas d'urgence** in an emergency
l' **ustensile (m.)** utensil

utiliser to use
en utilisant using

V

les **vacances (f. pl.)** vacation
    en vacances on vacation
le **vaccin** vaccination (shot)
la **vaccination** vaccination
    vacciner to vaccinate
    vachement really (inform.)
la **vague** wave, **9.1**
la **valeur** value
la **valise** suitcase, **7.1**
    faire les valises to pack, **7.1**
la **vallée** valley, **14.1**
la **vanille: à la vanille** vanilla (adj.), **5.1**
la **vapeur d'eau** water vapor
la **variation** variation
    varié(e) varied
    varier to vary
la **variété** variety
    vaste vast, enormous
    vaut: il vaut mieux it's better
la **vedette** star (actor or actress), **16.1**
le **végétal** vegetable, plant
    végétarien(ne) vegetarian
le **vélo** bicycle, bike, **13.2**
    à vélo by bicycle
    le vélo tout terrain (VTT) mountain bike
le **vélodrome** bicycle racing track
le **vélomoteur** moped, **12.1**
le **vendeur**, la **vendeuse** salesperson, **10.1**
    vendre to sell, **8.1**
    vendredi (m.) Friday, **2.2**
    venir to come, **16**
    venir de (+ inf.) to have just (done something)
    venir en tête to rate above
le **vent** wind, **14.2**
    Il fait du vent. It's windy., **9.2**
la **vente** sale
le **ventre** abdomen, stomach, **15.1**
    avoir mal au ventre to have a stomachache, **15.1**
    au ventre de in the depths of
le **ver à soie** silkworm
le **verbe** verb
    vérifier to check, verify, **7.1**
    vérifier les niveaux (m. pl.) to check under the hood, **12.1**
    véritable real
le **verre** glass, **5.2**
    vers around (time); towards
le **versement** deposit

verser to deposit; to pour
la **version originale** original language version (of a movie), **16.1**
    vert(e) green, **10.2**
    vertical(e) vertical
la **veste** (sport) jacket, **10.1**
    vestimentaire: les normes vestimentaires (f.) dress code
le **veston** (suit) jacket
les **vêtements (m. pl.)** clothes, **10.1**
la **viande** meat, **6.1**
la **victoire** victory
le **vide** vacuum, space
    vide empty
la **vidéo(cassette)** videocassette, **3.2**
la **vie** life
    vieille old (f.), **4.1**
    vieux (vieil) old (m.), **4.1**
    vif, vive bright (color)
    vigilant(e) vigilant, watchful
le **vignoble** vineyard
la **villa** house
le **village** village, small town
la **ville** city, town
le **vin (rouge, blanc)** (red, white) wine
    vingt twenty, **BV**
    violent(e) violent
    viral(e) viral, **15.1**
la **virgule** comma
le **virus** virus
la **visite** visit
    visiter to visit (a place), **16.2**
la **vitamine** vitamin
    vite fast (adv.), **12.2**
la **vitrine** (store) window
    Vive... ! Long live … !, Hooray for … !
    vivre to live (exist)
    voici here is, here are, **1.1**
la **voie** track (railroad), **8.1**; lane (of a road), **12.1**
    voilà there is, there are (emphatic)
    voir to see, **10.1**
le/la **voisin(e)** neighbor, **4.2**
la **voiture** car, **4.2**
    en voiture by car, **5.2**; "All aboard!", **8**
    la voiture de sport sports car, **12.1**
    monter en voiture to board the train, **8**
la **voiture-lit** sleeping car, **8.2**
la **voiture-restaurant** dining car
le **vol** flight, **7.1**

le **vol intérieur** domestic flight, **7.1**
    le vol international international flight, **7.1**
le **volley-ball** volleyball, **13.2**
le **volume** volume
    vos your (pl. poss. adj.), **5**
    votre your (sing. poss. adj.), **5**
    voudrais: je voudrais I would like, **5.1**
    vouloir to want, **6.1**
    vous you (sing. form. and pl.), **2**; you (stress pron.), **9**; (to) you (dir. and ind. obj.), **15**
le **voyage** trip
    faire un voyage to take a trip, **7.1**
    voyager to travel, **8.1**
le **voyageur**, la **voyageuse** traveler, passenger, **8.1**
    vrai(e) true, real
    vraiment really, **2.1**
la **vue** view
la **vulgarité** vulgarity

W

le **walkman** Walkman, **3.2**
le **week-end** weekend, **2.2**

Y

    y there, **5.2**
le **yaourt** yogurt, **6.1**
les **yeux (m. pl; sing. œil)** eyes, **15.1**
    avoir les yeux qui piquent to have stinging eyes, **15.1**

Z

    zéro zero, **BV**
la **zone** area, zone, section, **7.1**
    en pleine zone tempérée right in the temperate zone
la **zoologie** zoology
    Zut! Darn!, **12.2**

VOCABULAIRE ANGLAIS–FRANÇAIS

The *Vocabulaire anglais–français* contains all productive vocabulary from the text.

The numbers following each entry indicate the chapter and vocabulary section in which the word is introduced. For example, **2.2** means that the word first appeared in *Chapitre 2, Mots 2.* **BV** refers to the introductory *Bienvenue* chapter.

The following abbreviations are used in this glossary.

adj.	adjective
adv.	adverb
conj.	conjunction
dem. adj.	demonstrative adjective
dem. pron.	demonstrative pronoun
dir. obj.	direct object
f.	feminine
fam.	familiar
form.	formal
ind. obj.	indirect object
inf.	infinitive
inform.	informal
interrog. adj.	interrogative adjective
inv.	invariable
m.	masculine
n.	noun
pl.	plural
poss. adj.	possessive adjective
prep.	preposition
pron.	pronoun
sing.	singular
subj.	subject

A

a un, une, **1.1**
  a day (week) par jour (semaine), **3.2**
  a lot beaucoup, **3.1**
abdomen le ventre, **15.1**
accident l'accident (m.), **14.2**
act l'acte, (m.), **16.1**
active actif, active, **10**
actor l'acteur, (m.), **16.1**
actress l'actrice, (f.), **16.1**
aerobics: to do aerobics faire de l'aérobic, **11.2**
after après, **3.2**
afternoon l'après-midi (m.), **2**
against contre, **13.1**
age l'âge (m.), **4.1**
agent (m. and f.) l'agent (m.), **7.1**
to **agree** être d'accord, **2.1**
air aérien(ne) (adj.), **9**
air terminal l'aérogare (f.), **7.2**
airline la compagnie aérienne, **7.1**
airplane l'avion (m.), **7.1**
airport l'aéroport (m.), **7.1**
aisle le couloir, **8.2**
  aisle seat (une place) côté couloir, **7.1**
algebra l'algèbre (f.), **2.2**
all tous, toutes, **7**
  all alone tout(e) seul(e), **5.2**
  all right (agreement) d'accord, **3**
  Is that all? C'est tout?, **6.2**
allergic allergique, **15.1**
allergy l'allergie (f.), **15.1**
already déjà, **14**
also aussi, **1.1**
always toujours, **5**
American (adj.) américain(e), **1.1**
among entre, **9.2**
and et, **1**
  and you? et toi? (fam.), **BV**
angry fâché(e), **12.2**
announcement l'annonce, (f.), **8.1**
to **answer** répondre, **8**
antibiotic l'antibiotique (m.), **15.1**
Anything else? Autre chose?, Avec ça?, **6.2**
apartment l'appartement (m.), **4.2**
  apartment building l'immeuble (m.), **4.2**
apple la pomme, **6.2**
appointment: appointment book l'agenda (m.), **2.2**
April avril (m.), **4.1**
arrival l'arrivée (f.), **7.2**

to **arrive** arriver, **3.1**
  arriving from (flight) en provenance de, **7.1**
art l'art (m.), **2.2**
to **ask (for)** demander, **5**
  to ask a question poser une question, **3.1**
aspirin l'aspirine (f.), **15.1**
at à, **3.1**
  at the au, à la, à l', aux, **5**
  at the home (business) of chez, **5**
  at what time? à quelle heure?, **2**
athletic sportif, sportive, **10**
August août, (m.), **4.1**
aunt la tante, **4.1**
autumn l'automne (m.), **13.2**

B

backboard (basketball) le panneau, **13.2**
backpack le sac à dos, **BV**
bacterial bactérien(ne), **15.1**
bag le sac, **6.1**
bakery la boulangerie-pâtisserie, **6.1**
balcony le balcon, **4.2**
ball (tennis, etc.) la balle, **9.2**; **(soccer, etc.)** le ballon, **13.1**
banana la banane, **6.2**
bank la banque, **18.1**
baseball le base-ball, **13.2**
basket le panier, **13.2**
basketball le basket(-ball), **13.2**
bathing suit le maillot (de bain), **9.1**
bathroom la salle de bains, les toilettes (f. pl.), **4.2**
to **be** être, **2.1**
  to be able to pouvoir, **6**
  to be better soon être vite sur pied, **15.1**
  to be born naître, **17**
  to be called s'appeler, **11.1**
  to be careful faire attention, **9.1**
  to be early être en avance, **8.1**
  to be hungry avoir faim, **5.1**
  to be in shape être en forme, **11.2**
  to be late être en retard, **8.2**
  to be on time être à l'heure, **8.1**
  to be out of sorts ne pas être dans son assiette, **15.2**
  to be thirsty avoir soif, **5.2**
  to be ... years old avoir... ans, **4.1**
beach la plage, **9.1**
beautiful beau (bel), belle, **4**
because parce que, **9.1**
to **become** devenir, **16**

bed le lit, **8.2**
  to go to bed se coucher, **11.1**
bedroom la chambre à coucher, **4.2**
beef le bœuf, **6.1**
before avant, **7.1**
beginner le/la débutant(e), **14.1**
behind derrière, **BV**
beige beige, **10.2**
to **believe** croire, **10.2**
better meilleur(e) (adj.), **10**
between entre, **9.2**
beverage la boisson, **5.2**
bicycle le vélo, **13.2**
  bicycle racer le coureur cycliste, **13.2**
big grand(e), **1.1**
bill (currency) le billet, **18.1**; **(invoice)** la facture, **17.2**
biology la biologie, **2.2**
birthday l'anniversaire (m.), **4.1**
  When is your birthday? C'est quand, ton anniversaire? (fam.), **4.1**
black noir(e), **10.2**
blackboard le tableau, **BV**
blanket la couverture, **17.2**
bleacher le gradin, **13.1**
blond blond(e), **1.1**
blouse le chemisier, **10.1**
to **blow a whistle** siffler, **13.1**
blue bleu(e), **10.2**
  navy blue bleu marine (inv.), **10.2**
to **board (plane)** embarquer, **7.2**; **(train)** monter, **8.2**
boarding pass la carte d'embarquement, **7.1**
book le livre, **BV**
born: to be born naître, **17**
to **borrow** emprunter, **18.2**
bottle la bouteille, **6.2**
boundaries (on a tennis court) les limites (f. pl.), **9.2**
box office le guichet, **16.1**
boy le garçon, **BV**
to **brake** freiner, **12.2**
bread le pain, **6.1**
  loaf of French bread la baguette, **6.1**
to **breathe (deeply)** respirer (à fond), **15.2**
broke (slang) fauché(e), **18.2**
brother le frère, **1.2**
brown brun(e), marron (inv.), **10.2**
brunette brun(e), **1.1**
to **brush (one's teeth, hair, etc.)** se brosser (les dents, les cheveux, etc.), **11.1**

bunk (on a train) la couchette, **8.2**
bus le bus, **5.2**; l'autocar (m.), **7.2**
  by bus en bus, **5.2**
busy occupé(e), **2.2**
but mais, **1**
butcher shop la boucherie, **6.1**
butter le beurre, **6.2**
to **buy** acheter, **6.1**
  to buy a ticket prendre un billet, **7**

C

cabin (plane) la cabine, **7.1**
café le café, **5.1**
cake le gâteau, **6.1**
calculator la calculatrice, **BV**
can of food la boîte de conserve, **6.2**
Canadian (adj.) canadien(ne), **7**
cap (ski) le bonnet, **14.1**
car la voiture, **4.2**
  by car en voiture, **5.2**
  sports car la voiture de sport, **12.2**
carefully prudemment, **12.2**
carrot la carotte, **6.2**
carry-on luggage les bagages (m. pl.) à main, **7.1**
cartoon le dessin animé, **16.1**
cash l'argent liquide (m.), **18.1**
  cash register la caisse, **6.2**
to **cash (a check)** toucher (un chèque), **18.1**
cassette la cassette, **3.2**
casual (clothes) sport (adj. inv.), **10.1**
cat le chat, **4.1**
chair la chaise, **BV**
chairlift le télésiège, **14.1**
chalk: piece of chalk le morceau de craie, **BV**
change la monnaie, **18.1**
  change purse le porte-monnaie, **18.1**
  to make change faire de la monnaie, **18.1**
to **change** changer (de), **8.2**
to **chat** bavarder, **4.2**
check (in restaurant) l'addition (f.), **5.2**; **(bank)** le chèque (bancaire), **18.1**
  traveler's check le chèque de voyage, **17.2**
to **check** vérifier, **7.1**
  to check (luggage) faire enregistrer, **7.1**
  to check out (of hotel) libérer une chambre, **17.2**

to check under the hood
vérifier les niveaux, **12.2**
checkout counter la caisse, **6.2**
checkroom la consigne, **8.1**
cheese le fromage, **5.1**
chemistry la chimie, **2.2**
chicken le poulet, **6.1**
child l'enfant (m. and f.), **4.1**
chills les frissons (m. pl.), **15.1**
chocolate (adj.) au chocolat, **5.1**
to **choose** choisir, **7.1**
to **claim (luggage)** récupérer, **7.2**
class (people) la classe, **2.1**;
(course) le cours, **2.2**
first (second) class en première
(seconde), **8.1**
classroom la salle de classe, **2.1**
closed fermé(e), **16.2**
closet le placard, **17.2**
clothes les vêtements (m. pl.),
10.1
clothing designer le grand
couturier, **10.1**
cloud le nuage, **9.2**
Coca-Cola le coca, **5.1**
coffee le café, **5.1**
black coffee l'express (m.), **5.1**
coffee with cream (in a café)
le crème, **5.1**
coin la pièce, **18.1**
cold froid(e) (adj.), **14.2**; (illness)
le rhume, **15.1**
It's cold (weather). Il fait froid.,
9.2
to have a cold être enrhumé(e),
15.1
color la couleur, **10.2**
What color is ... ? De quelle
couleur est... ?, **10.2**
to **comb (one's hair)** se peigner, **11.1**
to **come** venir, **16**
to come back revenir, **16**
comedy la comédie, **16.1**
musical comedy la comédie
musicale, **16.1**
comic strip la bande dessinée, **16**
compact disc le compact disc, **3.2**
compartment le compartiment,
7.2
computer l'ordinateur (m.), **BV**
computer science
l'informatique (f.), **2.2**
conductor (train) le contrôleur,
8.2
confident confiant(e), **1.1**
convertible (car) la décapotable,
12.2
to **cook** faire la cuisine, **6**
corridor le couloir, **8.2**
costume le costume, **16.1**

to **cough** tousser, **15.1**
counter le comptoir, **7.1**
country le pays, **7.1**
course le cours, **2.2**
courtyard la cour, **4.2**
cousin le/la cousin(e), **4.1**
to **cover** couvrir, **15**
crab le crabe, **6.1**
cream la crème, **6.1**
credit card la carte de crédit, **17.2**
crepe la crêpe, **5.1**
croissant le croissant, **6.1**
to **cross** traverser, **12.2**
crossroads le carrefour, **12.2**
cup la tasse, **5.2**
winner's cup la coupe, **13.2**
currency la monnaie, **18.1**
curtain le rideau, **16.1**
customer le/la client(e), **10.1**
customs la douane, **7.2**
to go through customs passer à
la douane, **7.2**
cycling le cyclisme, **13.2**
cyclist (in race) le coureur
cycliste, **13.2**

D

dairy store la crémerie, **6.1**
to **dance** danser, **3.2**
dark hair brun(e), **1.1**
Darn! Zut!, **12.2**
date la date, **4.1**
What is the date today? Quelle
est la date aujourd'hui?, **4.1**
daughter la fille, **4.1**
day le jour, **2.2**
a (per) day par jour, **3**
What day is it? C'est quel jour?,
2.2
December décembre (m.), **4.1**
degree: It's ... degrees Celsius. Il
fait... degrés Celsius., **14.2**
delicatessen la charcuterie, **6.1**
delicious délicieux, délicieuse, **10**
deodorant le déodorant, **11.1**
department store le grand
magasin, **10.1**
departure le départ, **7.1**
to **deposit** verser, **18.1**
to **descend** descendre, **14.1**
desk le bureau, **BV**
desk clerk le/la réceptionniste,
17.1
diagnosis: to make a diagnosis
faire un diagnostic, **15.2**
to **die** mourir, **17**
difficult difficile, **2.1**
dining car la voiture-restaurant,
8.2
dining room la salle à manger, **4.2**

dinner le dîner, **4.2**
to eat dinner dîner, **4.2**
to **discover** découvrir, **15**
district le quartier, **4.2**; (Paris)
l'arrondissement (m.)
to **dive** plonger, **9.1**
diving: to go deep-sea diving faire
de la plongée sous-marine, **9.1**
to **do** faire, **6.1**
to do the shopping faire les
courses, **6.1**
doctor le médecin (m. et f.), **15.2**
documentary le documentaire,
16.1
dog le chien, **4.1**
dollar le dollar, **3.2**
domestic (flight) intérieur(e), **7.1**
door la porte, **17.1**
dozen la douzaine, **6.2**
drama le drame, **16.1**
dress la robe, **10.1**
dressed: to get dressed s'habiller,
11.1
dressy habillé(e), **10.1**
to **dribble (a basketball)** dribbler,
13.2
to **drive** conduire, **12.2**
driver le conducteur, la
conductrice, **12.2**
driver's license le permis de
conduire, **12.2**
driving lesson la leçon de
conduite, **12.2**
driving school l'auto-école (f.),
12.2
to **dry (off)** se sécher, **17.2**
dubbed (movie) doublé(e), **16.1**
during pendant, **3.2**

E

each (adj.) chaque, **16.1**
ear l'oreille (f.), **15.1**
earache: to have an earache avoir
mal aux oreilles, **15.1**
early: to be early être en avance,
8.1
to **earn** gagner, **3.2**
easy facile, **2.1**
to **eat** manger, **5**
to eat breakfast prendre le petit
déjeuner, **7**
to eat dinner dîner, **4.2**
to eat lunch déjeuner, **5.2**
egg l'œuf (m.), **6.2**
eight huit, **BV**
eighteen dix-huit, **BV**
eighty quatre-vingts, **5.2**
elevator l'ascenseur (m.), **4.2**
eleven onze, **BV**
energetic énergique, **1.2**

goalie le gardien de but, **13.1**
goggles (ski) les lunettes (f. pl.), **14.1**
good bon(ne), **7**
good-bye au revoir, ciao (inform.), BV
gram le gramme, **6.2**
granddaughter la petite-fille, **4.1**
grandfather le grand-père, **4.1**
grandmother la grand-mère, **4.1**
grandparents les grands-parents (m. pl.), **4.1**
grandson le petit-fils, **4.1**
gray gris(e), **10.2**
Great! Chouette! (inform.), **2.2**
green vert(e), **10.2**
  green beans les haricots (m. pl.) verts, **6.2**
grilled ham and cheese sandwich le croque-monsieur, **5.1**
grocery store l'épicerie (f.), **6.1**
ground le sol, **13.2**
  ground floor le rez-de-chaussée, **4.2**
guide(book) le guide, **12.2**
gym(nasium) le gymnase, **11.2**
gymnastics la gymnastique, **2.2**
  to do gymnastics faire de la gymnastique, **11.2**

H

hair les cheveux (m. pl.), **11.1**
half demi(e)
  half past (time) et demie, **2**
ham le jambon, **5.1**
hand la main, **11.1**
handkerchief le mouchoir, **15.1**
hanger le cintre, **17.2**
happy content(e), **1.1**; heureux, heureuse, **10.2**
hard (adv.) fort, **9.2**
hat (ski) le bonnet, **14.1**
to **hate** détester, **3.2**
to **have** avoir, **4.1**
  to have a(n) … -ache avoir mal à (aux)… , **15.2**
  to have a cold être enrhumé(e), **15.1**
  to have a picnic faire un pique-nique, **6**
  to have to devoir, **18.2**
he il, **1**
head la tête, **13.1**
headache: to have a headache avoir mal à la tête, **15.1**
health la santé, **15.1**
  to be in good (poor) health être en bonne (mauvaise) santé, **15.1**

health club le club de forme, **11.2**
to **hear** entendre, **8.1**
heel le talon, **10.2**
  high (low)-heeled (shoes) à talons hauts (bas), **10.2**
hello bonjour, BV
her elle (stress pron.), **9**; la (dir. obj.), **16**; lui (ind. obj.), **17.1**; sa, son, ses (poss. adj.), **4**
here is, here are voici, **1.1**
hi salut, BV
high élevé(e), **15**; haut(e), **10.2**
  high school le lycée, **1.2**
highway l'autoroute (f.), **12**
him le (dir. obj.), **16.1**; lui (stress pron.), **9**; lui (ind. obj.), **17.1**
his sa, son, ses, **4**
history l'histoire (f.), **2.2**
to **hit** frapper, **9.2**
homework (assignment) le devoir, BV
  to do homework faire les devoirs, **6**
hot: hot dog la saucisse de Francfort, **5.1**
  It's hot (weather). Il fait chaud., **9.2**
hotel l'hôtel (m.), **17.1**
house la maison, **3.1**
how: How are you? Ça va? (inform.); Comment vas-tu? (fam.); Comment allez-vous? (form.), BV
  How beautiful they are! Qu'elles (ils) sont belles (beaux)!
  How much? Combien?, **6.2**
  How much is it? C'est combien?, BV
  How much is that? Ça fait combien?, **5.2**
  How's it going? Ça va?, BV
hundred cent, **5.2**
to **hurt** avoir mal à, **15.1**
  It hurts. Ça fait mal., **15.2**
  Where does it hurt (you)? Où avez-vous mal?, **15.2**
husband le mari, **4.1**

I

I je, **1**
ice la glace, **14.2**
  ice cream la glace, **5.1**
  ice skate le patin à glace, **14.2**
  (ice) skating le patinage, **14.2**
to **(ice-)skate** faire du patin (à glace), **14.2**
immigration l'immigration (f.), **7.2**

impatient impatient(e), **1.1**
in dans, BV; à, **3.1**
  in back of derrière, BV
  in first (second) class en première (seconde), **8.1**
  in front of devant, BV
inexpensive bon marché (inv.), **10.1**
infection l'infection (f.), **15.1**
instructor le moniteur, la monitrice, **9.1**
intelligent intelligent(e), **1.1**
interesting intéressant(e), **1.1**
intermission l'entracte (m.), **16.1**
international international(e), **7.1**
intersection le croisement, **12.2**
to **invite** inviter, **3.2**
it (dir. obj.) le, la, **16.1**
  it is, it's c'est, BV
  It's (That's) expensive. Ça coûte cher., **7.2**
  it is necessary (+ inf.) il faut (+ inf.), **9.1**
Italian (adj.) italien(ne), **7**
Italy l'Italie (f.), **16**

J

jacket le blouson, **10.1**
  (suit) jacket la veste, **10.1**
  ski jacket l'anorak (m.), **14.1**
January janvier (m.), **4.1**
jar le pot, **6.2**
jeans le jean, **10.1**
to **jog** faire du jogging, **11.2**
to **joke around** rigoler, **3.2**
July juillet (m.), **4.1**
June juin (m.), **4.1**

K

key la clé, **12.2**; le demi-cercle (basketball), **13.2**
to **kick** donner un coup de pied, **13.1**
kilogram le kilo, **6.2**
kind le genre, **16.1**
kitchen la cuisine, **4.2**
kleenex le kleenex, **15.1**
knife le couteau, **5.2**
to **know** connaître (be acquainted with), savoir (information), **16.2**

L

to **land** atterrir, **7.1**
landing card la carte de débarquement, **7.2**
lane (of a road) la voie, **12.2**
language la langue, **2.2**
last dernier, dernière, **10**
  last night hier soir, **13**

last year l'année (f.) dernière, 13

late: to be late être en retard, 8.2

Latin le latin, 2.2

to learn (to) apprendre (à), 9.1

to leave partir, 7.1

    to leave (a room, etc.) quitter, 3.1

    to leave (something behind) laisser, 5.2

    to leave a tip laisser un pourboire, 5.2

left: to the left of à gauche de, 5

lemonade le citron pressé, 5.1

to lend prêter, 18.2

lesson la leçon, 9.1

lettuce la laitue, 6.2

level le niveau, 12.2

to like aimer, 3.2

    I would like je voudrais, 5.1

line: to wait in line faire la queue, 8.1

to listen (to) écouter, 3.2

    to listen with a stethoscope ausculter, 15.2

liter le litre, 6.2

literature la littérature, 2.2

to live (in a city, house, etc.) habiter, 3.1

living room la salle de séjour, 4.2

lobby le hall, 17.1

locker la consigne automatique, 8.1

long long(ue), 10.2

to look at regarder, 3.1

to look for chercher, 5.1

to lose perdre, 8.2

    to lose patience perdre patience, 8.2

    to lose weight maigrir, 11.2

lot: a lot of beacoup de, 10.1

    a lot of people beaucoup de monde, 13.1

loudspeaker le haut-parleur, 8.1

to love aimer, adorer, 3.2

love story (movie) le film d'amour, 16.1

low bas(se), 10

luggage les bagages (m. pl.), 7.1

    carry-on luggage les bagages à main, 7.1

M

ma'am madame, BV

magazine le magazine, 3.2

maitre d' le maître d'hôtel, 5.2

make (of car) la marque, 12.2

to make faire, 6.1

man l'homme (m.), 2.1

March mars (m.), 4.1

market le marché, 6.2

marvelous merveilleux, merveilleuse, 10.2

match (singles, doubles) (tennis) la partie (en simple, en double), 9.2

math les maths (f. pl.), 2.2

May mai (m.), 4.1

me me (dir. and ind. obj.), 15.2; moi (stress pron.), 1.2

meat la viande, 6.1

medicine (medical profession) la médecine 15; (remedy) le médicament, 15.2

medium-rare (meat) à point, 5.2

menu la carte, 5.1

merchant le/la marchand(e), 6.2

    produce merchant le/la marchand(e) de fruits et légumes, 6.2

meter maid la contractuelle, 12.2

midnight minuit (m.), 2.2

milk le lait, 6.1

mineral water l'eau (f.) minérale, 6.2

mirror la glace, 11.1

Miss (Ms.) Mademoiselle (Mlle), BV

mogul la bosse, 14.1

Monday lundi (m.), 2.2

money l'argent (m.), 3.2

    to have lots of money avoir plein de fric (slang), 18.2

month le mois, 4.1

moped le vélomoteur, 12.2

morning le matin, 2

    in the morning (A.M.) du matin, 2

Morocco le Maroc, 16

most (of) la plupart (des), 8.2

mother la mère, 4.1

motorcycle la moto, 12.2

    motorcycle cop le motard, 12.2

mountain la montagne, 14.1

mouth la bouche, 15.1

movie le film, 16.1

    movie theater le cinéma, la salle de cinéma, 16.1

Mr. Monsieur (M.), BV

Mrs. (Ms.) Madame (Mme), BV

museum le musée, 16.2

music la musique, 2.2

must devoir, 18.2

mustard la moutarde, 6.2

my ma, mon, mes, 4

N

name le nom, 16.2

What is your name? Tu t'appelles comment? (fam.), 11.1

napkin la serviette, 5.2

narrow étroit(e), 10.2

near près de, 4.2

necessary: it is necessary (+ inf.) il faut (+ inf.), 9.1

to need avoir besoin de, 11.1

neighbor le/la voisin(e), 4.2

neighborhood le quartier, 4.2

nephew le neveu, 4.1

net le filet, 9.2

    net bag le filet, 6.1

never ne... jamais, 12

new nouveau (nouvel), nouvelle, 4

newspaper le journal, 8.1

newsstand le kiosque, 8.1

next prochain(e), 8.2

    next to à côté de, 5

nice (person) aimable, sympathique, 1.2; gentil(le), 9

niece la nièce, 4.1

nine neuf, BV

nineteen dix-neuf, BV

ninety quatre-vingt-dix, 5.2

no non, BV

    no one, nobody ne... personne, 12.2

    No parking. Il est interdit de stationner., 12.2

    no smoking (section) (la zone) non-fumeurs, 7.1

noon midi (m.), 2.2

nose le nez, 15.1

    to have a runny nose avoir le nez qui coule, 15.1

not ne... pas, 1

    not bad pas mal, BV

notebook le cahier, BV

nothing ne... rien, 12.2

    Nothing else. Rien d'autre., 6.2

novel le roman, 16

November novembre (m.), 4.1

now maintenant, 2

number le numéro, 5.2

    What is the phone number of ... ? Quel est le numéro de téléphone de... ?, 5.2

O

to obey obéir (à), 7

o'clock: It's ... o'clock. Il est... heure(s)., 2.2

October octobre (m.), 4.1

of (belonging to) de, 5

    of the du, de la, de l', des, 5

to offer offrir, 15

often souvent, 5

O.K. **(health)** Ça va.; **(agreement)** d'accord, **BV**
old vieux (vieil), vieille, **4.1**
　　How old are you? Tu as quel âge? (fam.), **4.1**
omelette (with herbs/plain) l'omelette (f.) (aux fines herbes/nature), **5.1**
on sur, **BV**
　　on board à bord de, **7.2**
　　on foot à pied, **5.2**
　　on time à l'heure, **8.1**
one un, une, **1**
one-way ticket l'aller simple (m.), **8.1**
onion l'oignon (m.), **6.2**
　　onion soup la soupe à l'oignon, **5.1**
open ouvert(e), **16.2**
to **open** ouvrir, **15.2**
opera l'opéra (m.), **16.1**
opinion: in my opinion à mon avis, **10.2**
to **oppose** opposer, **13.1**
opposing adverse, **13.1**
or ou, **1.1**
orange (fruit) l'orange (f.), **6.2**; **(color)** orange (inv.), **10.2**;
　　orange soda l'Orangina (m.), **5.1**
to **order** commander, **5.1**
original language version (of a film) la version originale, **16.1**
other autre, **BV**
our notre, nos, **5**
out of bounds hors des limites, **9.2**
over (prep.) par dessus, **13.2**
　　over there là-bas, **BV**
overcast (cloudy) couvert(e), **14.2**
to **overlook** donner sur, **17.1**
to **owe** devoir, **18.2**

P

to **pack (suitcases)** faire les valises, **7.1**
package le paquet, **6.2**
packed (stadium) comble, **13.1**
painter le/la peintre, **16.2**
painting la peinture; le tableau, **16.2**
pair la paire, **10.1**
pal le copain, la copine, **2.1**
pancake la crêpe, **5.1**
pants le pantalon, **10.1**
pantyhose le collant, **10.1**
paper: sheet of paper la feuille de papier, **BV**
parents les parents (m. pl.), **4.1**
Parisian (adj.) parisien(ne), **7**

park le parc, **11.2**
to **park the car** garer la voiture, **12.2**
parking: No parking. Il est interdit de stationner., **12.2**
part-time à mi-temps, **3.2**
party la fête, **3.2**
　　to throw a party donner une fête, **3.2**
to **pass** passer, **7.2**
　　to pass an exam réussir à un examen, **7**
passenger le passager, la passagère, **7.1**; **(train)** le voyageur, la voyageuse, **8**
passport le passeport, **7.1**
pâté le pâté, **5.1**
patient patient(e), **1.1**
to **pay** payer, **6.1**
　　to pay attention faire attention, **6**
　　to pay back rembourser, **18.2**
　　to pay cash payer en espèces, **17.2**
pedestrian le/la piéton(ne), **12.2**
　　pedestrian crossing les clous (m. pl.), **12.2**
pen (ballpoint) le stylo, **BV**
pencil le crayon, **BV**
penicillin la pénicilline, **15.1**
to **perform** jouer, **16**
to **permit** permettre, **14**
person la personne, **17.1**
personally personnellement, **16.2**
pharmacist le/la pharmacien(ne), **15.2**
pharmacy la pharmacie, **15.2**
physical education l'éducation (f.) physique, **2.2**
physics la physique, **2.2**
picture le tableau, **16.1**
pie la tarte, **6.1**
pill le comprimé, **15.2**
pillow l'oreiller (m.), **17.2**
pink rose, **10.2**
to **place** mettre, **8.1**
plain (adj.) nature, **5.1**
plane l'avion (m.), **7.1**
plate l'assiette (f.), **5.2**
platform (railroad) le quai, **8.1**
to **play (perform)** jouer, **16**
　　to play (a sport) jouer à, **9.2**; pratiquer un sport, **11.2**
play la pièce, **16.1**
　　to put on a play monter une pièce, **16.1**
player le joueur, **9.2**
please s'il vous plaît (form.), s'il te plaît (fam.), **BV**
pocket la poche, **18.1**
pocketbook le sac, **18.1**

pool la piscine, **9.2**
poor pauvre, **15.1**
　　poor thing le/la pauvre, **15.1**
popular populaire, **1.2**
porter le porteur, **8.1**
potato la pomme de terre, **6.2**
pound la livre, **6.2**
to **prepare** préparer, **4.2**
to **prescribe** prescrire, **15.2**
prescription l'ordonnance (f.), **15.2**
　　to write a prescription faire une ordonnance, **15.2**
pretty joli(e), **4.2**
price le prix, **10.1**
problem le problème, **11.2**
to **punish** punir, **7**
purse le sac, **18.1**
to **put (on)** mettre, **8.1**
　　to put money aside mettre de l'argent de côté, **18.2**
　　to put on makeup se maquiller, **11.1**

Q

quarter: quarter after (time) et quart, **2**
　　quarter to (time) moins le quart, **2**
question: to ask a question poser une question, **3.1**

R

race la course, **13.2**
racket la raquette, **9.2**
radio la radio, **3.2**
raining: It's raining. Il pleut., **9.2**
rare (meat) saignante(e), **5.2**
to **read** lire, **12.2**
ready-to-wear department le rayon prêt-à-porter, **10.1**
really vraiment, **2.1**
to **receive** recevoir, **18.1**
reception desk la réception, **17.1**
record le disque, **3.2**
red rouge, **10.2**
referee l'arbitre (m.), **13.1**
registration card (at hotel desk) la fiche d'enregistrement, **17.1**
regular (gasoline) ordinaire, **12.2**
to **reserve** réserver, **17**
restaurant le restaurant, **5.2**
to **return (tennis ball, etc.)** renvoyer, **9.2**
right: to the right of à droite de, **5**
right away tout de suite, **11.1**
road la route, **12.2**
role le rôle, **16**
room (in house) la pièce, **4.2**; **(in hotel)** la chambre, **17.1**

double room la chambre à deux lits, **17.1**

single room la chambre à un lit, **17.1**

to **vacate the room** libérer la chambre, **17.2**

round-trip ticket le billet aller-retour, **8.1**

runner le coureur, **13.2**

S

salad la salade, **5.1**

sales les soldes (f. pl.), **10.2**

salesperson le vendeur, la vendeuse, **10.1**

same même, **2.1**

sand le sable, **9.1**

sandwich le sandwich, **5.1**

grilled ham and cheese sandwich le croque-monsieur, **5.1**

Saturday samedi (m.), **2.2**

sausage le saucisson, **6.1**

to **save money** faire des économies, **18.2**

savings account le compte d'épargne, **18.1**

to **say** dire, **12.2**

scarf l'écharpe (f.), **14.1**

scene la scène, **16.1**

schedule l'horaire (m.), **8.1**

school l'école (f.), **1.2**

high school le lycée, **1.2**

science les sciences (f. pl.), **2.2**

score le score, **9.2**

to **score a goal** marquer un but, **13.1**

screen l'écran (m.), **7.1**

sculptor le sculpteur (m. et f.), **16.2**

sculpture la sculpture, **16.2**

sea la mer, **9.1**

by the sea au bord de la mer, **9.1**

seashore le bord de la mer, **9.1**

seaside resort la station balnéaire, **9.1**

seat le siège, **7.1**

seat (on plane, at movies, etc.) la place, **7.1**

seat belt la ceinture de sécurité, **12.2**

seated assis(e), **8.2**

second (adj.) deuxième, **4.2**

section la zone, **7.1**

smoking (no smoking) section la zone (non-)fumeurs, **7.1**

security (airport) le contrôle de sécurité, **7.1**

to **see** voir, **10.1**

See you later. À tout à l'heure., **BV**

See you tomorrow. À demain., **BV**

to **sell** vendre, **8.1**

to **send (hit)** envoyer, **13.1**

September septembre (m.), **4.1**

to **serve** servir, **7.2**

service le service, **5.2**

service station la station-service, **12.2**

service station attendant le/la pompiste, **12.2**

set (for a play) le décor, **16.1**

to **set the table** mettre le couvert, **8**

seven sept, **BV**

seventeen dix-sept, **BV**

seventy soixante-dix, **5.2**

several plusieurs, **18.2**

Shall we go? On y va?, **5**

to **shave** se raser, **11.1**

she elle, **1**

sheet le drap, **17.2**

sheet of paper la feuille de papier, **BV**

shirt la chemise, **10.1**

shoes les chaussures (f. pl.), **10.1**

shop la boutique, **10.1**

to **shop** faire des achats, **10.1**

short petit(e), **1.1**; court(e), **10.2**

shorts le short, **9.2**

show (movies) la séance, **16.1**

to **show** montrer, **17.1**

to show a movie passer un film, **16.1**

shrimp la crevette, **6.1**

shy timide, **1.2**

sick malade, **15.1**

sick person le/la malade, **15.2**

side (in a sporting event) le camp, **13.1**

sidewalk le trottoir, **12.2**

sidewalk café la terrasse (d'un café), **5.1**

to **sign** signer, **18.1**

since (time) depuis, **8.2**

sincere sincère, **1.2**

to **sing** chanter, **3.2**

sir monsieur, **BV**

sister la sœur, **1.2**

six six, **BV**

sixteen seize, **BV**

sixty soixante, **BV**

size (clothes) la taille; **(shoes)** la pointure, **10.2**

the next larger size la taille au-dessus, **10.2**

the next smaller size la taille au-dessous, **10.2**

to **take size (number)** faire du (nombre), **10.2**

What size do you take? Vous faites quelle pointure (taille)?, **10.2**

skate (ice) le patin à glace, **14.2**

to **skate (ice)** faire du patin (à glace), **14.2**

skater le patineur, la patineuse, **14.2**

skating le patinage, **14.2**

skating rink la patinoire, **14.2**

ski le ski, **14.1**

ski boot la chaussure de ski, **14.1**

ski jacket l'anorak (m.), **14.1**

ski pole le bâton, **14.1**

ski resort la station de sports d'hiver, **14.1**

to **ski** faire du ski, **14.1**

skier le skieur, la skieuse, **14.1**

skiing le ski, **14.1**

cross-country skiing le ski de fond, **14.1**

downhill skiing le ski alpin, **14.1**

skirt la jupe, **10.1**

sky le ciel, **14.2**

to **sleep** dormir, **7.2**

sleeping car la voiture-lit, **8.2**

sleeve la manche, **10.2**

long- (short-)sleeved à manches longues (courtes), **10.2**

small petit(e), **1.1**

smoking (section) (la zone) fumeurs, **7.1**

snack bar (train) le grill-express, **8**

sneakers les chaussures (f. pl.) de tennis, **9.2**

to **sneeze** éternuer, **15.1**

snowball la boule de neige, **14.2**

snowing: It's snowing. Il neige., **14.2**

soap le savon, **11.1**

soccer le foot(ball), **13.1**

soccer field le terrain de football, **13.1**

socks les chaussettes (f. pl.), **10.1**

some quelques (pl.), **8.2**

somebody, someone quelqu'un, **12.2**

something to eat quelque chose à manger, **5.1**

sometimes quelquefois, **5**

son le fils, **4.1**

sore throat: to have a sore throat avoir mal à la gorge, **15.1**

space (parking) la place, **12.2**

Spanish (language) l'espagnol
(m.), 2.2
to speak parler, 3.1
to speak on the telephone
parler au téléphone, 3.2
spectator le spectateur, 13.1
speed limit la limitation de
vitesse, 12.2
to speed up accélérer, 12.2
to spend (money) dépenser, 10.1
spoon la cuillère, 5.2
sporty (clothes) sport (adj. inv.),
10.1
spring (season) le printemps,
13.2
stadium le stade, 13.1
stage la scène, 16.1
staircase l'escalier (m.), 17.1
to stamp (a ticket) composter, 8.1
standing debout, 8.2
star (actor or actress) la vedette,
16.1
to start the car mettre le contact,
12.2
station wagon le break, 12.2
statue la statue, 16.2
to stay in shape rester en forme,
11.1
steak and French fries le steak
frites, 5.2
steep raide, 14.1
stomach le ventre, 15.1
stomachache: to have a
stomachache avoir mal au
ventre, 15.1
to stop s'arrêter, 12.2
store le magasin, 3.2
street la rue, 3.1
student l'élève (m. et f.), 1.2
to study étudier, 3.1; faire des
études, 6
to study French (math, etc.)
faire du français (des maths,
etc.), 6
subject (in school) la matière, 2.2
subtitles les sous-titres (m. pl.),
16.1
subway le métro, 4.2
by subway en métro, 5.2
subway station la station de
métro, 4.2
to succeed réussir (à), 7
to suffer souffrir, 15.2
suit (men's) le complet;
(women's) le tailleur, 10.1
(suit) jacket la veste, 10.1
suitcase la valise, 7.1
summer l'été (m.), 9.1
summit le sommet, 14.1

to sunbathe prendre un bain de
soleil, 9.1
Sunday dimanche (m.), 2.2
sunglasses les lunettes (f. pl.) de
soleil, 9.1
sunny: It's sunny. Il fait du soleil.,
9.2
suntan lotion la crème solaire, 9.1
super extra, super (inform.), 2.2
super (gasoline) (de l'essence)
super, 12.2
supermarket le supermarché, 6.1
to surf faire du surf, 9.1
sweater le pull, 10.1
sweatshirt le sweat-shirt, 10.1
sweatsuit le survêtement, 11.2
to swim nager, 9.1
swimming la natation, 9.1

T

table la table, BV
table setting le couvert, 5.2
to set the table mettre le
couvert, 8.2
tablecloth la nappe, 5.2
to take prendre, 9.1
to take a bath (a shower)
prendre un bain (une
douche), 11.1
to take an exam passer un
examen, 3.1
to take off (plane) décoller, 7.1
to take size (number) faire du
(+ nombre), 10.2
to take something upstairs
monter, 17.1
to take the train (plane, etc.)
prendre le train (l'avion, etc.),
7
to take a trip faire un voyage,
7.1
to take a walk faire une
promenade, 9.1
taken pris(e), 5.1
to talk parler, 3.1
to talk on the phone parler au
téléphone, 3.2
tan bronzer, 9.1
tart la tarte, 6.1
taxi le taxi, 7.2
tea with lemon le thé citron, 5.1
to teach someone to do something
apprendre à quelqu'un à faire
quelque chose, 14.1
teacher le professeur; le/la prof
(inform.), 2.1
team l'équipe (f.), 13.1
television la télé, 3.2
to tell dire, 12.2
temperature la température, 15.1

ten dix, BV
tennis le tennis, 9.2
tennis court le court de tennis,
9.2
tennis shoes les chaussures
(f. pl.) de tennis, 9.2
tennis skirt la jupette, 9.2
terminal (bus to airport)
l'aérogare (f.), 7.2
terrace la terrasse, 4.2
terrible terrible, 2.2
terrific super, extra, terrible, 2.2
test l'examen (m.), 3.1
to pass a test réussir à un
examen, 7
to take a test passer un
examen, 3.1
thank you merci, BV
that ce (cet), cette (dem. adj.), 8;
ça (pron.), BV
that is to say c'est-à-dire, 16.1
That's (It's) expensive. Ça
coûte cher., 18
the la, le, l', 1; les, 2
theater le théâtre, 16.1
their leur, leurs, 5
them elles, eux (stress pron.), 9;
les (dir. obj.), 16; leur (ind.
obj.), 17
then (adv.) ensuite, 11.1
there y, 5
there is, there are il y a, 4.2;
voilà (emphatic), BV
these (dem. adj.) ces (m. and
f. pl.), 8
they elles, ils, 2
to think penser, 10.2
third troisième, 4.2
thirteen treize, BV
thirty trente, BV
this (dem. adj.) ce (cet), cette, 8
those (dem. adj.) ces (m. and
f. pl.), 8
thousand mille, 6.2
throat la gorge, 15.1
to have a frog in one's throat
avoir un chat dans la gorge,
15.2
to have a scratchy throat avoir
la gorge qui gratte, 15.1
to have a sore throat avoir mal
à la gorge, 15.1
to have a throat infection avoir
une angine, 18.1
to throw lancer, 13.2
Thursday jeudi (m.), 2.2
ticket le billet, 7.1
one-way ticket l'aller simple
(m.), 8.1

VOCABULAIRE ANGLAIS–FRANÇAIS **541**

round-trip ticket le billet aller-retour, **8.1**
ticket window le guichet, **8.1**
traffic ticket la contravention, **12.2**
tie la cravate, **10.1**
tight serré(e); **(shoes)** étroit(e), **10.2**
time (of day) l'heure (f.), **2**
  at what time? à quelle heure?, **2**
  to be on time être à l'heure, **8.1**
  What time is it? Il est quelle heure?, **2**
timid timide, **1.2**
tip (restaurant) le service, le pourboire, **5.2**
  The tip is included. Le service est compris., **5.2**
  to leave a tip laisser un pourboire, **5.2**
tire le pneu, **12.2**
  flat tire le pneu à plat, **12.2**
  spare tire la roue de secours, **12.2**
to à, **3.1**; à destination de (flight, etc.), **7.1**
  to the au, à la, à l', aux, **5**
  to the left of à gauche de, **5**
  to the right of à droite de, **5**
today aujourd'hui, **2.2**
together ensemble, **5.1**
toilet (bathroom) les toilettes (f. pl.), **4.2**
  toilet paper: roll of toilet paper le rouleau de papier hygiénique, **17.2**
toll highway l'autoroute (f.) à péage, **12.2**
tomato la tomate, **6.2**
tomorrow demain, **2.2**
  See you tomorrow. À demain., **BV**
too (also) aussi, **1.1**; **(excessively)** trop, **10.2**
tooth la dent, **11.1**
  se brosser les dents to brush one's teeth, **11.1**
toothpaste le dentifrice, **11.1**
towel la serviette, **17.2**
track (race) la piste, **13.2**; **(train)** la voie, **8.1**
traffic la circulation, **12.2**
  traffic light le feu, **12.2**
  green (traffic) light le feu vert, **12.2**
  red (traffic) light le feu rouge, **12.2**
  yellow (traffic) light le feu orange, **12.2**
tragedy la tragédie, **16.1**

trail la piste, **14.1**
  slalom trail la piste de slalom, **14.1**
train le train, **8.1**
  train station la gare, **8.1**
traveler le voyageur, la voyageuse, **8.1**
trigonometry la trigonométrie, **2.2**
T-shirt le tee-shirt, **9.2**
Tuesday mardi (m.), **2.2**
TV la télé, **3.2**
twelve douze, **BV**
twenty vingt, **BV**
two deux, **BV**
type le genre, **16.1**

U

uncle l'oncle (m.), **4.1**
under sous, **BV**
to **understand** comprendre, **9.1**
United States les États-Unis (m. pl.), **9.1**
unleaded sans plomb, **12.2**
unpleasant désagréable, antipathique (person), **1.2**
up to jusqu'à, **13.2**
us nous, **7**

V

to **vacate the room** libérer la chambre, **17.2**
valley la vallée, **14.1**
vanilla (adj.) à la vanille, **5.1**
vegetable le légume, **6.2**
very très, **1.1**
videocassette la vidéo(cassette), **3.2**
viral viral(e), **15.1**
volleyball le volley-ball, **13.2**

W

to **wait (for)** attendre, **8.1**
  to wait in line faire la queue, **8.1**
waiter le serveur, **5.1**
waiting room la salle d'attente, **8.1**
waitress la serveuse, **5.1**
to **wake up** se réveiller, **11.1**
to **walk** se promener, **11.2**
Walkman le walkman, **3.2**
wallet le portefeuille, **18.1**
to **want** vouloir, **6.1**
warm-up suit le survêtement, **11.2**
to **wash (one's face, hair, etc.)** se laver (la figure, les cheveux, etc.), **11.1**
  to wash and groom oneself faire sa toilette, **11.1**

washcloth le gant de toilette, **17.2**
to **watch** surveiller, **12.2**
water l'eau, **6.2**
to **water-ski** faire du ski nautique, **9.1**
wave la vague, **9.1**
we nous, **2**
to **wear** porter, **10.1**
weather le temps, **9.2**
  It's bad weather. Il fait mauvais., **9.2**
  It's nice weather. Il fait beau., **9.2**
  What's the weather like? Quel temps fait-il?, **9.2**
Wednesday mercredi (m.), **2.2**
week la semaine, **2.2**
  a (per) week par semaine, **3.2**
weekend le week-end, **2.2**
weight: to gain weight grossir, **11.2**
  to lose weight maigrir, **11.2**
well bien, **BV**
well-done (meat) bien cuit(e), **5.2**
what quel(le) (interrog. adj.), **7**; qu'est-ce que, **13**; quoi, **14**
  What else? (shopping) Avec ça?, **6.2**
  What is it? Qu'est-ce que c'est?, **BV**
  What is … like? Comment est… ? (description), **1.1**
wheel la roue, **12.2**
when quand, **3.1**
  When is your birthday? C'est quand, ton anniversaire? (fam.), **4.1**
where où, **BV**
which (interrog. adj.) quel(le), **7**
to **whistle (blow a whistle)** siffler, **13.1**
white blanc, blanche, **10.2**
who qui, **BV**
  Who is it? Qui est-ce?, **BV**
  Who(m) (do you mean)? Qui ça?, **BV**
whole (adj.) tout(e)
whom qui, **14**
why pourquoi, **9.1**
wide large, **10.2**
wife la femme, **4.1**
to **win** gagner, **9.2**
wind le vent, **14.2**
window (seat in plane) (une place) côté fenêtre, **7.1**
to **windsurf** faire de la planche à voile, **9.1**
windy: It's windy. Il fait du vent., **9.2**
winner le/la gagnant(e), **13.2**

winter l'hiver (m.), **14.1**
with avec, **5.1**
without sans, **12.2**
woman la femme, **2.1**
work (art) l'œuvre, **16.2**
to work travailler, **3.2**
to write écrire, **12.2**
wrong: What's wrong with him?
Qu'est-ce qu'il a?, **15.1**

Y

year l'an (m.), l'année (f.), **4.1**
years: to be … years old
avoir... ans, **4.1**
yellow jaune, **10.2**

yes oui, **BV**
yesterday hier, **13.1**
the day before yesterday
avant-hier, **13**
yesterday morning hier matin,
13
yogurt le yaourt, **6.1**
you te (dir. and ind. obj.), **15**; toi
(stress pron.), **9**; tu (fam. sing.),
1; vous (sing. form. and pl.), **2**
You're welcome. De rien., Je
t'en prie., Pas de quoi. (fam.);
Ce n'est rien., Il n'y a pas
de quoi., Je vous en prie.
(form.), **BV**

young jeune, **4.1**
your ta, ton, tes (fam.), **4**; votre,
vos (form.), **5**

Z

zero zéro, **BV**

INDEX GRAMMATICAL

INDEX GRAMMATICAL   **545**